Plant Responses and Tolerance to Salt Stress: Physiological and Molecular Interventions

Plant Responses and Tolerance to Salt Stress: Physiological and Molecular Interventions

Editors

Mirza Hasanuzzaman
Masayuki Fujita

MDPI • Basel • Beijing • Wuhan • Barcelona • Belgrade • Manchester • Tokyo • Cluj • Tianjin

Editors
Mirza Hasanuzzaman
Department of Agronomy, Faculty of Agriculture
Sher-e-Bangla Agricultural University
Dhaka
Bangladesh

Masayuki Fujita
Laboratory of Plant Stress Responses, Faculty of Agriculture
Kagawa University
Kagawa
Japan

Editorial Office
MDPI
St. Alban-Anlage 66
4052 Basel, Switzerland

This is a reprint of articles from the Special Issue published online in the open access journal *International Journal of Molecular Sciences* (ISSN 1422-0067) (available at: www.mdpi.com/journal/ijms/special_issues/Salt_Stress).

For citation purposes, cite each article independently as indicated on the article page online and as indicated below:

LastName, A.A.; LastName, B.B.; LastName, C.C. Article Title. *Journal Name* **Year**, *Volume Number*, Page Range.

ISBN 978-3-0365-4138-9 (Hbk)
ISBN 978-3-0365-4137-2 (PDF)

Contents

About the Editors

Mirza Hasanuzzaman

Dr. Mirza Hasanuzzaman is a Professor of Agronomy at Sher-e-Bangla Agricultural University, Dhaka, Bangladesh. He received his Ph.D. with a dissertation on 'Plant Stress Physiology and Antioxidant Metabolism'from the United Graduate School of Agricultural Sciences, Ehime University, Japan. Later, he completed his postdoctoral research at the University of the Ryukyus, Okinawa, Japan. Subsequently, he became an Adjunct Senior Researcher at the University of Tasmania with an Australian Government's Endeavour Research Fellowship. Prof. Hasanuzzaman has over 200 publications in the Web of Science. He has edited over 20 books and written over 30 book chapters. His publications are cited over 11000 times as per Scopus with an h-index of 54 (as of April 2022). He is an Editor and a reviewer for more than 50 peer-reviewed international journals and the recipient of 'Publons Peer Review Award 2017, 2018, and 2019'. He is acting Associate Editor of {Plant Signaling & Behavior} (Taylor and Francis), {Frontiers in Plant Science} (Switzerland), Academic Editor of {PeerJ} (USA), {Plants} (MDPI), {Vegetos} (Springer); and Member of the Editorial Board of {Plant Physiology and Biochemistry} (Elsevier), {Annals of Agricultural Sciences} (Elsevier), etc. He is an active member of 40 professional societies and is the acting Research and Publication Secretary of the Bangladesh JSPS Alumni Association. He received the World Academy of Science (TWAS) Young Scientist Award 2014, University Grants Commission (UGC) Gold Medal 2018, Global Network of Bangladeshi Biotechnologists (GNOBB) Award 2021, and Distinguished Scientists Award (2020) from AETDS, India. He is a fellow of the Bangladesh Academy of Sciences (BAS), Royal Society of Biology, and a Foreign fellow of The Society for Science of Climate Change and Sustainable Environment. He is named a Highly Cited Researcher by Clarivate AG.

Masayuki Fujita

Dr. Masayuki Fujita is Professor in the Laboratory of Plant Stress Responses, Faculty of Agriculture, Kagawa University, Kagawa, Japan. He received his B.Sc. in Chemistry from Shizuoka University, Shizuoka, and M.Agr. and Ph.D. in plant biochemistry from Nagoya University, Nagoya, Japan. His research interests include physiological, biochemical, and molecular biological responses based on secondary metabolism in plants under various abiotic and biotic stresses; phytoalexin, cytochrome P450, glutathione S-transferase, and phytochelatin; and redox reaction and antioxidants. In the last decade, his works were focused on oxidative stress and antioxidant defense in plants under environmental stress. His group investigates the role of different exogenous protectants in enhancing antioxidant defense and methylglyoxal detoxification systems in plants. He has supervised 4 M.S. students and 13 PhD students as the main supervisor. He has about 150 publications in journals and books, and has edited 10 books.

International Journal of
Molecular Sciences

MDPI

Editorial

Plant Responses and Tolerance to Salt Stress: Physiological and Molecular Interventions

Mirza Hasanuzzaman [1,*] and Masayuki Fujita [2,*]

[1] Department of Agronomy, Faculty of Agriculture, Sher-e-Bangla Agricultural University, Dhaka 1207, Bangladesh
[2] Laboratory of Plant Stress Responses, Department of Plant Science, Faculty of Agriculture, Kagawa University, Takamatsu 761-0795, Kagawa, Japan
* Correspondence: mhzsauag@yahoo.com (M.H.); fujita.masayuki@kagawa-u.ac.jp (M.F.)

Salinity is considered one of the most devastating environmental stresses that drastically curtails the productivity and quality of crops across the world. More than 20% of the world's cultivable lands are dealing with the adversity of salt stress and these salt-prone areas are continuously increasing, due to both natural and anthropogenic activities [1]. However, this adversity has become much more severe in arid and semi-arid regions over the last 20 years due to the increasing demand for irrigation water requirements [2].

Salt stress is a major environmental stress that affects plant growth and development. Salt stress increases the intracellular osmotic pressure and can cause the accumulation of sodium to toxic levels [3]. Like other abiotic stresses, salt stress negatively affects plant growth and reproduction in many ways. It produces nutritional and hormonal imbalances, ion toxicity, oxidative and osmotic stress, and an increase in plant susceptibility to diseases. However, plants can be damaged or even die due to salt stress in three major ways. High salt concentration in the soil alters soil porosity and hydraulic conductivity that leads to low soil water potential; this causes water stress as a result of decreased soil water potential, eventually leading to physiological drought conditions and the destabilization of the cell membrane and protein degradation due to the toxic effects of different ions (mainly Na^+).

Coupled with the abovementioned effects, salt stress produces a variety of physiological and metabolic changes in plants, seed germination behavior, photosynthesis, other biosynthetic process inhibition, and growth reduction. Different crops respond to salinity in different ways; glycophytes mostly show growth and total yield reduction, whereas halophytes can grow and reproduce easily under saline conditions. Therefore, at higher osmotic pressures in the root–soil interface, there is a slower impact caused by the build-up of Na^+ and Cl^- in the foliage. This leads to reduced shoot growth coupled with reduced leaf expansion and the inhibition of lateral bud formation. In response to salt stress, plant cells accumulate compatible solutes and redistribute ions, which enables them to acclimate to a low soil water potential. Additionally, the endogenous abscisic acid (ABA) content increases, followed by changes in principle genetic expression in salt stress conditions. Moreover, increased levels of ions, such as Na^+ and Cl^-, trigger ionic toxicity in plants due to the disruption in ion homeostasis and the unavailability of essential nutrients, which are essential for plant growth and metabolism [4]. The combined effect of osmotic stress and ion toxicity is responsible for the generation of secondary stresses, which could have impaired the germination, growth, and development of the plants [5]. Salt-induced water deficit conditions declined stomatal conductance, thus reducing photosynthetic activities of the plants and accelerating the accumulation of reactive oxygen species (ROS). These are the oxygen radicals and their derivatives, such as hydrogen peroxide, H_2O_2; singlet oxygen, $^1O_2$; superoxide radicals, $O_2^{\bullet-}$; and hydroxyl radicals, $OH^{\bullet}$, which are highly reactive and usually toxic. They are capable of disrupting different cellular components, such as proteins, lipids, and nucleic acids and destructing the structural integrity of the plants [6].

Citation: Hasanuzzaman, M.; Fujita, M. Plant Responses and Tolerance to Salt Stress: Physiological and Molecular Interventions. *Int. J. Mol. Sci.* **2022**, *23*, 4810. https://doi.org/10.3390/ijms23094810

Received: 7 April 2022
Accepted: 17 April 2022
Published: 27 April 2022

On contrary, NaCl may also be used as an elicitor. Hawrylak-Nowak et al. [7] applied an abiotic elicitor, i.e., NaCl, to enhance the biosynthesis and accumulation of phenolic secondary metabolites in *Melissa officinalis* L. Plants were subjected to salt stress treatment by the application of NaCl solutions (50 or 100 mM) to the pots. Generally, the NaCl treatments were found to inhibit the growth of plants, simultaneously enhancing the accumulation of phenolic compounds (total phenolics, soluble flavonols, anthocyanins, phenolic acids), especially at 100 mM NaCl. However, the salt stress did not disturb the accumulation of photosynthetic pigments and the proper functioning of the PS II photosystem [7].

In nature, plants usually produce secondary metabolites as a defense mechanism against environmental stresses. Different stresses determine the chemical diversity of plant-specialized metabolism products. In this study, we applied an abiotic elicitor, i.e., NaCl, to enhance the biosynthesis and accumulation of phenolic secondary metabolites in *Melissa officinalis* L. Plants were subjected to salt stress treatment by application of NaCl solutions (0, 50, or 100 mM) to the pots. Therefore, the proposed method of elicitation represents a convenient alternative to cell suspension or hydroponic techniques as it is easier and cheaper with a simple application in lemon balm pot cultivation. The improvement of lemon balm quality by NaCl elicitation can potentially increase the level of health-promoting phytochemicals and the bioactivity of low-processed herbal products.

Plants are sessile and thus have to develop suitable mechanisms to adapt to high-salt environments. Thus, in response to salt stress signals, plants adapt via various mechanisms, including regulating ion homeostasis, activating the osmotic stress pathway, mediating plant hormone signaling, and regulating cytoskeleton dynamics and cell wall composition. Unraveling the mechanisms underlying these physiological and biochemical responses to salt stress could provide valuable strategies to improve agricultural crop yields. Plants, therefore, alter physiologically to deal with this situation. In species exposed to salt stress, common responses include an increase in osmotic adjustment, changes in cell wall elasticity, and an increase in the percentage of apoplastic water, which minimizes saline effects by maintaining foliar turgidity. Several compounds that work in the osmotic regulation of plants are well known, e.g., carbohydrates (sucrose, sorbitol, mannitol, glycerol, pinitol), nitrogen molecules (proteins, betaine, glutamate, aspartate, glycine, proline, choline, 4-gamma aminobutyric acid), and organic acids (malate and oxalate). Proline (Pro) and glycine betaine (GB) are the most essential and efficient compatible solutes among the organic osmolytes which minimize the salt-stress effects and improve plant growth.

Plant metabolic plasticity is also a vital factor in regulating salt stress response in plants. Nosek et al. [8] analyzed the photosynthetic metabolism of the ice plant (*Mesembryanthemum crystallinum* L.) during a 72 h response period following salinity stress removal and found that the presence of salinity stress is required not only for the induction of stress-dependent crassulacean acid metabolism (CAM) photosynthesis but also for maintaining its functioning. The rapid shutdown of the energy-demanding functional CAM seems to be one small component of the flexibility features. As we showed here, the metabolic flexibility of the common ice plant includes rapid and far-reaching changes [8].

It is essential to carry out the necessary research on salt stress and soil pollution mitigation to increase agricultural productivity and feed the world's growing population. As global hunger and salinity stress intensify, there is a rise in interest in ecologically sustainable solutions for salt tolerance. Apart from using freshwater for irrigation there are many other alternative approaches for developing salt-tolerant crops. Developing halophytes as alternative crops, using interspecific hybridization to improve current crop tolerance, utilizing the wide range of varieties of existing crops, the introduction of variation within existing crops through genetic approaches, or improving salt-tolerant varieties.

Plant growth-promoting bacteria, phytohormones, and organic acids can also be used to promote salt tolerance in plants. Moreover, the improvement and/or alteration of some agricultural management practices will also play a vital role in the mitigation of salt stress in the first place. For example, conventional crop and pasture breeding for salt tolerance, soil reclamation, various management practices based on reducing the salt zone for seed germination and seedling establishment, pre-sowing irrigation with good quality water, appropriate use of ridges or beds for planting, and general management practices (mulching, incorporation of organic matter) will reduce the impact of soil salinity on crop performance.

Surowka et al. [9] reported that salt-exposed α-TC accumulating plants were more flexible in regulating chlorophyll, carotenoid, and polysaccharide levels than TC deficient mutants, while the plants overaccumulating γ-TC had the lowest levels of these compounds. They found that salt-stressed TC-deficient mutants and *tmt* transgenic line exhibited greater proline levels than WT plants, lower chlorogenic acid levels, and lower activity of catalase and peroxidases. Plants that accumulate α-TC produced more methylated proline and glycine betaines and showed greater activity of superoxide dismutase than γ-TC deficient plants [9].

Different plant hormones and genes are also associated with the signaling and antioxidant defense system to protect plants when they are exposed to salt stress. Salt-induced ROS overgeneration is one of the major reasons for hampering the morpho-physiological and biochemical activities of plants which can be largely restored by enhancing the antioxidant defense system that detoxifies ROS [6].

Reactive oxygen species signaling is crucial in modulating stress responses in plants, and NADPH oxidases (NOXs) are an important component of signal transduction under salt stress [10]. Pilarska et al. [10] showed that salt-induced expression patterns of two NOX genes, *RBOHD* and *RBOHF*, varied between the halophyte and the glycophyte. The expression of these genes in *E. salsugineum* leaves was induced by abscisic acid (ABA) and ethephon spraying indicates that in the halophyte *E. salsugineum*, the maintenance of the basal activity of NOXs in leaves plays a role during acclimation responses to salt stress.

Combining genomics and transcriptomics is a new approach that deals with the deep understanding and knowledge of the molecular response of a plant to salinity. Modification of genetic expression in salt stress comprises a variety of ways that plants use to induce or suppress the transcription of genes.

Quertani et al. [11] performed transcriptomic analysis of salt-stress-responsive genes in barley and found that, in leaves, the expression of 3585 genes was upregulated and 5586 were downregulated, while in roots the expression of 13,200 genes was upregulated and 10,575 were downregulated. They also found that response to salt stress is mainly achieved through sensing and signaling pathways, strong transcriptional reprogramming, hormonal regulation, osmoregulation, ion homeostasis and increased ROS scavenging. A number of candidate genes involved in hormone and kinase signaling pathways, as well as several transcription factor families and transporters, were identified. This study provides valuable information on early salt-stress-responsive genes in the roots and leaves of barley and identifies several important players in salt tolerance [11].

Proteomic and metabolomic research provides an excellent tool for examining plant adaptation to salinity. Hence, these are frequently utilized to identify the molecular processes of plant response to salt stress.

In mulberries, Gan et al. [12] performed a comparative proteomic analysis of salt tolerant and sensitive varieties and revealed that the phenylpropanoid biosynthesis may play an important role in the salt tolerance of the mulberry. They also clarified the molecular mechanism of mulberry salt tolerance, which is of great significance for the selection of excellent candidate genes for saline–alkali soil management and mulberry stress resistance genetic engineering [12].

He et al. [13] identified *bolTLP1*, a thaumatin-like protein gene that confers tolerance to salt stress in broccoli (*Brassica oleracea* L. var. *Italica*) because *bolTLP1* may function by regulating phytohormone (ABA, ethylene, and auxin)-mediated signaling pathways, hydrolase and oxidoreductase activity, sulfur compound synthesis, and the differential expression of histone variants [13].

However, more information as well as research on genetic engineering, transcriptomics, proteomics, and metabolomics research, as well as their combined responses, is required to determine the salt-tolerance mechanism of plants. Identification of rice salt-tolerance genes and their molecular mechanisms could help breeders genetically improve salt tolerance [14]. Genome-wide identification and functional characterization of the cation proton antiporter (CPA) family were also found to be effective in providing insight into salt tolerance [15]. Wang et al. [15] identified 60 CPA candidate genes of radish on the whole genome level and concluded that these results would be useful to understand the complexity of the RsCPA gene family and could provide a valuable resource to explore the potential functions of RsCPA genes in radish. Similarly, Chen et al. [16] performed genome-wide analysis of the apple CBL family and revealed that Mdcbl10.1 functions positively in modulating apple salt tolerance. It also provided valuable insights for future research examining the function and mechanism of CBL proteins in regulating apple salt tolerance [16].

The study of Shao et al. [17] provided fundamental information on the involvement of the wheat *14-3-3* family in salt stress. They identified a total of 17 potential *14-3-3* gene family members that were identified from the Chinese Spring whole-genome sequencing database. Importantly, most *14-3-3* members in wheat exhibited significantly downregulated expression in response to alkaline stress [17].

Considering the importance of the physiological functions of melatonin, Tan et al. [18] studied the expression of the melatonin-related gene *HIOMT* in apple plants induced by salinity. They found that compared with the wild type, transgenic lines indicated higher melatonin levels and showed reduced salt damage symptoms, lower relative electrolyte leakage, and less total chlorophyll loss from leaves under salt stress. Further, transgenic lines showed a lower amount of ROS due to the enhanced activity of antioxidant enzymes, downregulated the expression of the abscisic acid synthesis gene (*MdNCED3*), accordingly reducing the accumulation of abscisic acid under salt stress [18].

Zhang et al. [19] characterized the *PsnNAC036* gene and found that the overexpression of *PsnNAC036* stimulated plant growth and enhanced salinity and heat tolerance in *Populus simonii* × *P. nigra*.

Next generation gene sequencing opens the avenue to revealing salt tolerance mechanisms more precisely. Min et al. [20] reported that haplotype analysis of *BADH1* by next-generation sequencing reveals an association with salt tolerance in rice during domestication. Despite the unclear association between *BADH1* and salt stress, these findings can be useful for future research development related to its gene expression [20].

Katja et al. [21] detected jacalin-related lectin HvHorcH protein in root extracellular fluid, suggesting that the revealed expression of HvHorcH is involved in the adaptation of plants to salinity.

Yu et al. [22] found that the C2H2-type zinc-finger protein from *Millettia pinnata*, MpZFP1, is a positive regulator of plant responses to salt stress due to its activation of gene expression and efficient scavenging of ROS and enhances salt tolerance in transgenic *Arabidopsis*. The heterologous expression of *MpZFP1* in Arabidopsis increased the seeds' germination rate, seedling survival rate, and biomass accumulation under salt stress [22].

The plant cytoskeleton is associated with plant salt stress responses. Therefore, the molecular mechanism underlying microtubule functions in plant salt stress response is important. Chun et al. [23] found that microtubule dynamics play crucial roles in plant adaptation and tolerance to salt stress. The modulation of microtubule-related gene expression can be an effective strategy for developing salt-tolerant crops [23].

Overall, the 19 contributions in this Special Issue "Plant Responses and Tolerance to Salt Stress: Physiological and Molecular Interventions" discuss the various aspects of salt stress responses in plants. It also discusses various mechanisms and approaches to conferring salt tolerance on plants. These types of research studies provide further directions in the development of crop plants for the saline environment in the era of climate change.

Author Contributions: Conceptualization, M.H. and M.F.; writing—original draft preparation, M.H. and M.F.; writing—review and editing, M.H. All authors have read and agreed to the published version of the manuscript.

Funding: This work received no external funding.

Institutional Review Board Statement: Not applicable.

Informed Consent Statement: Not applicable.

Data Availability Statement: Not applicable.

Acknowledgments: We would like to thank all contributors for their submission to this Special Issue and the reviewers for their valuable input in improving the articles. We also thank Dani Wu and other members of the editorial office for their helpful support during the compilation of this Special Issue. The authors acknowledge Rakib Hossain Raihan, Farzana Nowroz, Ayesha Siddika, Salma Binte Salam, and Shamima Sultana for their help during the organization of the literature and necessary formatting.

Conflicts of Interest: The authors declare no conflict of interest.

References

1. Arora, N.K. Impact of climate change on agriculture production and its sustainable solutions. *Environ. Sustain.* **2019**, *2*, 95–96. [CrossRef]
2. Acosta-Motos, J.R.; Ortuño, M.F.; Bernal-Vicente, A.; Diaz-Vivancos, P.; Sanchez-Blanco, M.J.; Hernandez, J.A. Plant responses to salt stress: Adaptive mechanisms. *Agronomy* **2017**, *7*, 18. [CrossRef]
3. Zhao, S.; Zhang, Q.; Liu, M.; Zhou, H.; Ma, C.; Wang, P. Regulation of plant responses to salt stress. *Int. J. Mol. Sci.* **2021**, *22*, 4609. [CrossRef]
4. Munns, R. Plant adaptations to salt and water stress: Differences and commonalities. *Adv. Bot. Res.* **2011**, *57*, 1–32. [CrossRef]
5. Munns, R.; Tester, M. Mechanisms of salinity tolerance. *Ann. Rev. Plant Biol.* **2008**, *59*, 651–681. [CrossRef]
6. Hasanuzzaman, M.; Raihan, M.R.H.; Masud, A.A.C.; Rahman, K.; Nowroz, F.; Rahman, M.; Nahar, K.; Fujita, M. Regulation of reactive oxygen species and antioxidant defense in plants under salinity. *Int. J. Mol. Sci.* **2021**, *22*, 9326. [CrossRef]
7. Hawrylak-Nowak, B.; Dresler, S.; Stasińska-Jakubas, M.; Wójciak, M.; Sowa, I.; Matraszek-Gawron, R. NaCl-induced elicitation alters physiology and increases accumulation of phenolic compounds in *Melissa officinalis* L. *Int. J. Mol. Sci.* **2021**, *22*, 6844. [CrossRef]
8. Nosek, M.; Gawrońska, K.; Rozpądek, P.; Sujkowska-Rybkowska, M.; Miszalski, Z.; Kornaś, A. At the edges of photosynthetic metabolic plasticity—On the rapidity and extent of changes accompanying salinity stress-induced CAM photosynthesis withdrawal. *Int. J. Mol. Sci.* **2021**, *22*, 8426. [CrossRef]
9. Surówka, E.; Latowski, D.; Dziurka, M.; Rys, M.; Maksymowicz, A.; Żur, I.; Olchawa-Pajor, M.; Desel, C.; Krzewska, M.; Miszalski, Z. ROS-scavengers, osmoprotectants and violaxanthin de-epoxidation in salt-stressed *Arabidopsis thaliana* with different tocopherol composition. *Int. J. Mol. Sci.* **2021**, *22*, 11370. [CrossRef]
10. Pilarska, M.; Bartels, D.; Niewiadomska, E. Differential regulation of NAPDH oxidases in salt-tolerant *Eutrema salsugineum* and salt-sensitive *Arabidopsis thaliana*. *Int. J. Mol. Sci.* **2021**, *22*, 10341. [CrossRef]
11. Nefissi Ouertani, R.; Arasappan, D.; Abid, G.; Ben Chikha, M.; Jardak, R.; Mahmoudi, H.; Mejri, S.; Ghorbel, A.; Ruhlman, T.A.; et al. Transcriptomic analysis of salt-stress-responsive genes in barley roots and leaves. *Int. J. Mol. Sci.* **2021**, *22*, 8155. [CrossRef] [PubMed]
12. Gan, T.; Lin, Z.; Bao, L.; Hui, T.; Cui, X.; Huang, Y.; Wang, H.; Su, C.; Jiao, F.; Zhang, M.; et al. Comparative Proteomic analysis of tolerant and sensitive varieties reveals that phenylpropanoid biosynthesis contributes to salt tolerance in mulberry. *Int. J. Mol. Sci.* **2021**, *22*, 9402. [CrossRef] [PubMed]
13. He, L.; Li, L.; Zhu, Y.; Pan, Y.; Zhang, X.; Han, X.; Li, M.; Chen, C.; Li, H.; Wang, C. *BolTLP1*, a thaumatin-like protein gene, confers tolerance to salt and drought stresses in broccoli (*Brassica oleracea* L. var. *Italica*). *Int. J. Mol. Sci.* **2021**, *22*, 11132. [CrossRef] [PubMed]
14. Fan, X.; Jiang, H.; Meng, L.; Chen, J. Gene mapping, cloning and association analysis for salt tolerance in rice. *Int. J. Mol. Sci.* **2021**, *22*, 11674. [CrossRef] [PubMed]
15. Wang, Y.; Ying, J.; Zhang, Y.; Xu, L.; Zhang, W.; Ni, M.; Zhu, Y.; Liu, L. Genome-wide identification and functional characterization of the cation proton antiporter (CPA) family related to salt stress response in radish (*Raphanus sativus* L.). *Int. J. Mol. Sci.* **2020**, *21*, 8262. [CrossRef]
16. Chen, P.; Yang, J.; Mei, Q.; Liu, H.; Cheng, Y.; Ma, F.; Mao, K. Genome-wide analysis of the apple CBL family reveals that Mdcbl10.1 functions positively in modulating apple salt tolerance. *Int. J. Mol. Sci.* **2021**, *22*, 12430. [CrossRef]
17. Shao, W.; Chen, W.; Zhu, X.; Zhou, X.; Jin, Y.; Zhan, C.; Liu, G.; Liu, X.; Ma, D.; Qiao, Y. Genome-wide identification and characterization of wheat *14-3-3* genes unravels the role of TaGRF6-A in salt stress tolerance by binding MYB transcription factor. *Int. J. Mol. Sci.* **2021**, *22*, 1904. [CrossRef]
18. Tan, K.; Zheng, J.; Liu, C.; Liu, X.; Liu, X.; Gao, T.; Song, X.; Wei, Z.; Ma, F.; Li, C. Heterologous expression of the melatonin-related gene *HIOMT* improves salt tolerance in *Malus domestica*. *Int. J. Mol. Sci.* **2021**, *22*, 12425. [CrossRef]

19. Zhang, X.; Cheng, Z.; Yao, W.; Zhao, K.; Wang, X.; Jiang, T. Functional characterization of *PsnNAC036* under salinity and high temperature stresses. *Int. J. Mol. Sci.* **2021**, *22*, 2656. [CrossRef]
20. Min, M.-H.; Maung, T.Z.; Cao, Y.; Phitaktansakul, R.; Lee, G.-S.; Chu, S.-H.; Kim, K.-W.; Park, Y.-J. Haplotype analysis of *BADH1* by next-generation sequencing reveals association with salt tolerance in rice during domestication. *Int. J. Mol. Sci.* **2021**, *22*, 7578. [CrossRef]
21. Witzel, K.; Matros, A.; Bertsch, U.; Aftab, T.; Rutten, T.; Ramireddy, E.; Melzer, M.; Kunze, G.; Mock, H.-P. The jacalin-related lectin HvHorcH is involved in the physiological response of barley roots to salt stress. *Int. J. Mol. Sci.* **2021**, *22*, 10248. [CrossRef] [PubMed]
22. Yu, Z.; Yan, H.; Liang, L.; Zhang, Y.; Yang, H.; Li, W.; Choi, J.; Huang, J.; Deng, S. A C_2H_2-type zinc-finger protein from *Millettia pinnata*, MpZFP1, enhances salt tolerance in transgenic Arabidopsis. *Int. J. Mol. Sci.* **2021**, *22*, 10832. [CrossRef] [PubMed]
23. Chun, H.J.; Baek, D.; Jin, B.J.; Cho, H.M.; Park, M.S.; Lee, S.H.; Lim, L.H.; Cha, Y.J.; Bae, D.-W.; Kim, S.T.; et al. Microtubule dynamics plays a vital role in plant adaptation and tolerance to salt stress. *Int. J. Mol. Sci.* **2021**, *22*, 5957. [CrossRef] [PubMed]

International Journal of
Molecular Sciences

Review

Regulation of Plant Responses to Salt Stress

Shuangshuang Zhao [1,*], Qikun Zhang [1], Mingyue Liu [1], Huapeng Zhou [2], Changle Ma [1] and Pingping Wang [1,*]

[1] Shandong Provincial Key Laboratory of Plant Stress, College of Life Sciences, Shandong Normal University, Jinan 250014, China; zhangqikun1016@163.com (Q.Z.); lmy312325@163.com (M.L.); machangle@sdnu.edu.cn (C.M.)
[2] Key Laboratory of Bio-Resource and Eco-Environment of Ministry of Education, College of Life Sciences, Sichuan University, Chengdu 610064, China; zhouhuapeng@scu.edu.cn
* Correspondence: zhaoshuangqw12@163.com (S.Z.); pingping.wang@sdnu.edu.cn (P.W.); Tel.: +86-531-8618-0792 (S.Z.); Fax: +86-531-8618-0792 (P.W.)

Abstract: Salt stress is a major environmental stress that affects plant growth and development. Plants are sessile and thus have to develop suitable mechanisms to adapt to high-salt environments. Salt stress increases the intracellular osmotic pressure and can cause the accumulation of sodium to toxic levels. Thus, in response to salt stress signals, plants adapt via various mechanisms, including regulating ion homeostasis, activating the osmotic stress pathway, mediating plant hormone signaling, and regulating cytoskeleton dynamics and the cell wall composition. Unraveling the mechanisms underlying these physiological and biochemical responses to salt stress could provide valuable strategies to improve agricultural crop yields. In this review, we summarize recent developments in our understanding of the regulation of plant salt stress.

Keywords: salt stress; ion transport; osmotic homeostasis; hormone mediation; cell wall regulation

Citation: Zhao, S.; Zhang, Q.; Liu, M.; Zhou, H.; Ma, C.; Wang, P. Regulation of Plant Responses to Salt Stress. *Int. J. Mol. Sci.* **2021**, *22*, 4609. https://doi.org/10.3390/ijms22094609

Academic Editors: Mirza Hasanuzzaman and Masayuki Fujita

Received: 14 March 2021
Accepted: 23 April 2021
Published: 28 April 2021

1. Introduction

The demands on crop yield have risen sharply worldwide to keep up with the rapidly expanding human population over the past twenty years [1]. Thus, how to improve crop yield and quality has become an urgent global agricultural problem. Soil salinization is a major environmental challenge that is threatening agriculture across the world [2]. Approximately 20% of the world's irrigated agricultural lands are adversely affected by soil salinization [3]. Issues with soil salinization are aggravated by natural environment deterioration, poor irrigation practices, and climate changes [4,5]. Thus, to effectively improve crop yields, it is critical to address the increasingly serious threat of soil salinization.

Two kinds of plants exist: halophytes and glycophytes. Halophytes are salinity-tolerant plants, which have adapted to salinized environments and even benefit from high salt concentrations for optimal growth [6]. In contrast, glycophytes are salinity-sensitive plants, in which growth and development are adversely inhibited by soil salinization [7]. Most crops are glycophytes. High salinity hampers glycophytes' growth and development, seriously limiting crop productivity and challenging food security. The cultivation and development of salt-tolerant crop varieties are key strategies for increasing crop productivity and yield and ensuring global food security.

Salt stress adversely impacts plants by hindering seed germination, growth and development, and flowering and fruiting [8,9]. The high concentrations of sodium in saline soil limits water uptake and the absorption of nutrients in the plant [10]. Water deficiency and nutritional imbalance induce primary stresses, including osmotic stress and ionic stress. These primary stresses result in oxidative stress and can cause a series of secondary stresses [11]. Together, salt stress leads to various physiological and molecular changes and impedes plant growth by inhibiting photosynthesis, thus reducing the available resources and repressing cell division and expansion [12]. Salt stress affects light-harvesting complex formation and regulates the state transition of photosynthesis [13]. Importantly, the enzyme activities or protein stabilities of the key enzymes in photosynthesis, such as ribulose-1,5-bisphophate carboxylase/oxygenase (RuBisCO), are affected through modulating the glycation under salt stress condition. Salt stress also influences sugar signaling and alters the levels of sugars, such as sucrose, fructose, and glycolysis [14].

As sessile organisms, plants have to develop various strategies to adapt to saline environments. These strategies include a series of signaling transduction pathways that are involved in activities ranging from salt stress sensing to the expression of many salt-stress-responsive genes, which regulate processes including ion transport, osmotic homeostasis, and detoxification. These mechanisms rely on multiple regulatory elements, such as phytohormones, lipids, the cell wall, and the cytoskeleton [10–12].

This review briefly describes the recent progress in our understanding of salt stress responses and the underlying regulatory mechanisms in plants, focusing on salt stress signal sensing and transduction. Understanding the molecular mechanisms of plant salt stress regulation will provide insight on how to improve plant salt stress resistance and is a critical step in improving agricultural productivity and food security.

2. Salt Stress Sensing

The sensing of salt stress signals initiates a wide array of complex transduction pathways in plants. Early signals that trigger a salt stress response include excess Na^+, the alteration of intracellular Ca^{2+} levels, and the accumulation of reactive oxygen species (ROS) [4]. Under salt stress, excess Na^+ is perceived rapidly and triggers downstream sodium stress responses [10] (Figure 1). Salt stress induces ion and osmotic stress, which leads to the elevation of Ca^{2+} in the cytosol; thus, salt stress and changes in osmotic pressure are always associated with the activation of Ca^{2+} channels. Ca^{2+} functions as an important secondary messenger by binding to and activating Ca^{2+} sensors, which evoke a specific calcium signal cascade. The plasma membrane calcium-permeable channel OSCA1 was identified as a putative osmosensor that is required for osmotic stress-induced Ca^{2+} signaling [15,16]. Under osmotic stress, the loss-of-function mutant *osca1* displays impaired Ca^{2+} signal enhancement. The plastidial K^+ exchange antiporters KEA1/2 and KEA3 also act as an osmosensory component that participates in osmotic stress-induced Ca^{2+} elevation [17]. The *Arabidopsis* monocation-induced Ca^{2+} increases 1 (MOCA1) was identified as a Na^+-gated calcium-permeable channel and participates in ionic stress-induced Ca^{2+} signaling [18]. The *moca1* mutant is hypersensitive to salt stress and lacks Na^+-evoked Ca^{2+} enhancement. MOCA1 encodes a glucuronosyltransferase and functions in the biosynthesis of glycosyl inositol phosphorylceramide (GIPC). GIPCs are monovalent-cation sensors that sense Na^+ and regulate salt stress responses through activating MOCA1 to increase the Ca^{2+} influx [18]. The plasma membrane receptor-like kinase FERONIA (FER) was reported to be required for salt-induced Ca^{2+} spikes and waves to maintain cell wall integrity during salt stress [19]. FER interacts with the pectin component of the cell wall and can sense salt-stress-induced cell wall damage. Cyclic nucleotide-gated ion channels (CNGCs) are calcium-permeable channels that are inhibited by cellular calcium concentrations and are regulated by calmodulin (CaM). Together with BAK1, FER regulates calcium signaling by phosphorylating CNGCs [20,21]. Lastly, salt stress triggers the excessive accumulation of ROS, which also plays a key role in activating calcium signaling. The leucine-rich-repeat receptor kinase, hydrogen-peroxide-induced Ca^{2+} increases 1 (HPCA1), is a hydrogen peroxide sensor that is located in the plasma membrane that detects the increase of H_2O_2 under stress stimuli [22]. HPCA1 is required for stomatal closure by mediating the H_2O_2-triggered influx of Ca^{2+} [22].

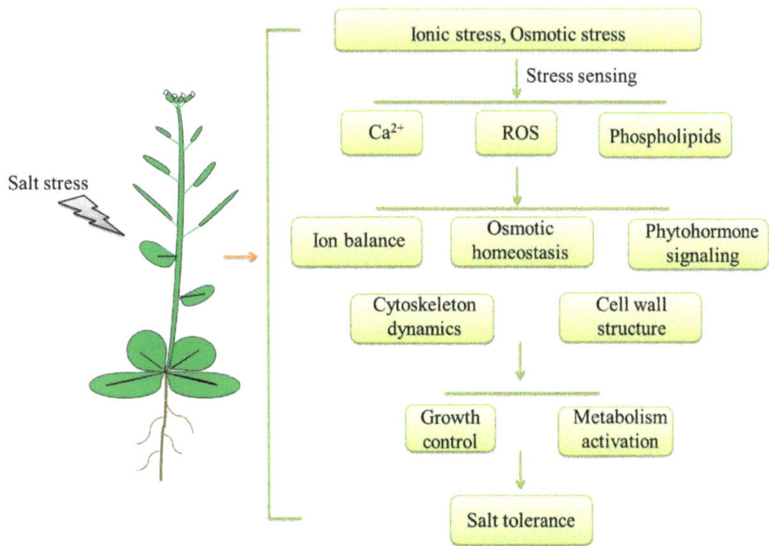

Figure 1. A simplified model of the plant salt stress response. Salt stress primarily causes ionic stress and osmotic stress. After sensing Na^+ and hyperosmolality, plants accumulate Ca^{2+}, activate ROS signaling, and alter their phospholipid composition. These signals activate adaptive processes to alleviate salt stress, including maintaining an ion balance and osmotic homeostasis, inducing phytohormone signaling and regulating cytoskeleton dynamics and the cell wall structure. Subsequently, through an array of signal transduction pathways, plant growth is slowed and metabolism is activated to increase salt tolerance.

3. Regulation of Plant Response to Salt Stress

3.1. Ion Balance

Under salt stress, high concentrations of the sodium ion, Na^+, accumulate in plant cells, ultimately to toxic levels, leading to the disruption of ion homeostasis [4,7]. Plants have developed systems to maintain low levels of Na^+ by removing Na^+ from the cytoplasm. This is mainly achieved using Na^+/H^+ antiporters, which transport Na^+ in exchange for H^+ [11]. The plasma-membrane-localized Na^+/H^+ antiporters transport Na^+ to the apoplast, and the vacuole-localized Na^+/H^+ antiporters are responsible for maintaining Na^+ compartmentation in vacuoles. The salt overly sensitive (SOS) regulatory pathway regulates ion homeostasis through modulating Na^+/H^+ antiporters activity during salt stress [4] (Figure 2).

After being triggered by cytoplasmic Ca^{2+}, the SOS pathway functions to alleviate salt stress by exporting excess Na^+. The SOS pathway is comprised of the Na^+/H^+ antiporter SOS1, the protein kinase SOS2, and two calcium sensors, SOS3 and SCaBP8 (SOS3-like calcium-binding protein 8) [8]. Under salt stress, SOS3/SCaBP8 perceives the increased cytoplasmic calcium signal and transduces it to the downstream serine/threonine protein kinase, SOS2. SOS3/SCaBP8 recruits SOS2 to the plasma membrane and activates it. Subsequently, SOS2 phosphorylates SOS1, which enhances the plant salt tolerance by increasing the Na^+/H^+ exchange activity [12]. SOS1 plays a key role in transporting Na^+ from the cytoplasm to the apoplast. The efflux of Na^+ is driven by the proton gradient that is generated from the plasma membrane H^+-ATPase. Under salt stress, SOS3/SCaBP8-SOS2 also regulates the activities of other transporters involved in ion homeostasis. For instance, the K^+ and Na^+ transporters, vacuolar Na^+/H^+ exchanger (NHX), vacuolar H^+-ATPases, and pyrophosphatases (PPase) were reported to be regulated by the SOS pathway [10].

In summary, the SOS pathway maintains the Na^+ homeostasis and transports excess Na^+ from the cytosol to the apoplast to prevent the accumulation of Na^+ to toxic levels.

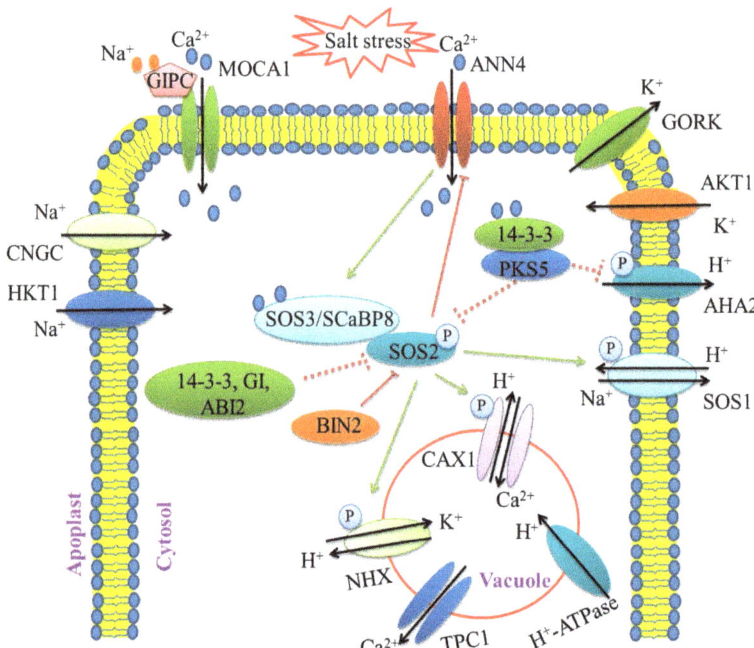

Figure 2. Salt stress triggers ion transport regulation in plant cells. Salt stress induces the accumulation of Na^+ and Ca^{2+} within the cell. The glucuronosyltransferase monocation-induced Ca^{2+} increases 1 (MOCA1) is as a Ca^{2+}-permeable channel in the plasma membrane. Glycosyl inositol phosphorylceramide (GIPC) sphingolipids sense and bind to Na^+ to activate the MOCA1-mediated Ca^{2+} influx. The cyclic nucleotide-gated ion channel (CNGC) and high-affinity potassium (K^+) transporter (HKT1) are required for Na^+ transport into the cell. The inward-rectifier K^+ channel *Arabidopsis* K^+ transporter (AKT1) and the outward-rectifier K^+ channel guard cell outward-rectifier K^+ channel (GORK) help to maintain the Na^+/K^+ balance. The salt overly sensitive (SOS) pathway plays essential roles in Na^+ exclusion. The calcium sensor, SOS3/SCaBP8, recruits SOS2 to the plasma membrane and promotes SOS2-mediated phosphorylation of the Na^+/H^+ antiporter SOS1. Under normal conditions, the kinase SOS2 is repressed by 14-3-3, ABA insensitive 2 (ABI2), and GIGANTEA (GI). Additionally, SOS2-like protein kinase 5 (PKS5) phosphorylates and inhibits SOS2. Under salt stress conditions, 14-3-3 and GI are degraded and release SOS2 to phosphorylate SOS1. Then, 14-3-3 represses PKS5, thereby activating SOS2. Under salt stress conditions, the glycogen synthase kinase 3 (GSK3) kinase, brassinosteroid insensitive 2 (BIN2) fine-tunes the SOS2 activity to prevent overactivation. As a putative Ca^{2+}-permeable transporter, the ANNEXIN protein member, ANN4, interacts with SCaBP8 and SOS2 and regulates calcium signaling under salt stress. During salt stress, SOS2 activates the vacuolar H^+/Ca^{2+} antiporter CAX1 to promote Ca^{2+} enhancement and regulates the vacuolar K^+/H^+ exchanger NHX to maintain the K^+ balance. The arrows and bars indicate positive and negative regulation, respectively. Solid lines and dashed lines indicate direct regulation and indirect regulation, respectively.

How the SOS pathway is regulated has been thoroughly investigated. The kinase activity of SOS2 is specifically activated by salt stress stimuli. Under normal conditions, several protein factors inhibit the SOS2 activity and the SOS pathway, such as SOS2-like protein kinase 5 (PKS5), the phosphatase ABA (abscisic acid) insensitive 2 (ABI2), 14-3-3, and GIGANTEA (GI). PKS5 inhibits SOS2 kinase activity via phosphorylation at Ser294 of SOS2. The 14-3-3 protein functions as a protein-kinase-interacting partner of SOS2 at phosphorylated Ser294 and inhibits the kinase activity of SOS2. Interestingly, 14-3-3 proteins also function as a negative regulator of PKS5. Salt stress promotes the interaction between 14-3-3 proteins and PKS5, causing the inhibition of SOS2 and H^+-ATPase activity [23]. During this process, 14-3-3 proteins bind to Ca^{2+} and are directly modulated by the Ca^{2+} signal. Under non-stress conditions, GI also inhibits SOS2. In response to salt stress, both 14-3-3 and GI are degraded, thereby releasing SOS2 to activate the downstream protein kinase cascade [24–26]. In addition, geminivirus RER-interacting kinase 1 (GRIK1) activates SOS2 by mediating the phosphorylation of SOS2 on Thr168 [27]. The calcium-dependent membrane-binding protein, ANNEXIN4 (ANN4), interacts with the SOS2/SCaBP8 complex to fine-tune calcium signaling under salt stress [28]. Besides these activating mechanisms, the SOS pathway deactivates the regulatory system. Once the salt stress is removed, BIN2, a glycogen synthase kinase 3 (GSK3)-like kinase that is the central component in brassinosteroid (BR) signaling, phosphorylates SOS2 on the Thr172 residue to inhibit SOS2 activity and promotes plant growth via BES1/BZR1-mediated transcriptional networks [29].

Under salt stress, plants have to modulate the Na^+/K^+ homeostasis through maintaining high K^+/Na^+ ratio since excessive Na^+ often leads to K^+ deficiency [4,12]. The *sos* mutants show impaired Na^+/K^+ homeostasis during salt stress. The uptake of K^+ is inhibited in *SCaBP8* mutants under salt stress [30]. The potassium transporters, along with voltage-gated channel proteins and their regulators, play important roles in mediating K^+ absorption, release, and transportation at the cellular and whole-plant levels. For instance, the inward-rectifier K^+ channel *Arabidopsis* K transporter (AKT1) is a major contributor to K^+ uptake and transport. Salt stress inhibits the activity of AKT1 [31]. The calcineurin B-like (CBL) proteins and CBL-interacting protein kinases (CIPKs) interact with and activate AKT1. CIPK23 phosphorylates and activates AKT1 to increase K^+ uptake under K^+-deficient conditions [32,33]. Two CBL proteins, CBL1 and CBL9, activate the phosphorylation of CIPK23 on AKT1 to ensure the activity enhancement of AKT1 [34]. However, CBL10 competes with CIPK23 for binding to AKT1 and negatively modulates AKT1 activity [30]. A protein phosphatase 2C (PP2C) member, AIP1, mediates the dephosphorylation of AKT1 and negatively regulates CIPK23-activated AKT1 [35,36]. Tonoplast-localized K^+ channel (TPK1) is regulated by salt stress and modulates the cytosolic K^+ influx during salinity stress [37]. Salt stress triggers calcium-dependent protein kinase (CDPK) to phosphorylate TPK1 and activate the K^+ influx [38]. High-affinity potassium transporter, HKT1, provides sodium exclusion and the maintenance of high K^+/Na^+ in leaves during salinity stress [39]. Previous studies have indicated that the maintenance of a low Na^+ concentration in leaves is an essential strategy for plants to enhance their salt tolerance [40,41]. HKT1 mediates low-affinity Na^+ transport and plays a role in the distribution of Na^+ from the root-to-shoot xylem sap. ZmHKT1 causes leaf Na^+ exclusion promotion and is identified as a major salt-tolerance quantitative trait locus (QTL) [42]. HKT1 physically interacts with phosphatase PP2C49, which then inhibits the Na^+ permeability of HKT1 and negatively regulates salt tolerance [43]. The salt stress response is regulated by the circadian clock in plants. Several proteins that maintain the circadian clock play key roles in regulating salt stress tolerance [44,45]. Recently, studies have demonstrated a new molecular link between clock components and salt stress tolerance in rice. Oryza sativa pseudo-response regulator (OsPRR73) is induced by salt and specifically confers salt tolerance by recruiting HDAC10 to transcriptionally repress OsHKT2;1 and, therefore, regulates rice salt tolerance [46]. Membrane compartment-localized aquaporins might also participate in ion balance regulation through controlling root water uptake, leaf water transpiration, stomatal closure, and small molecule transport in response to salt stress. For instance, the overexpression of the wheat aquaporin TdPIP2;1 improves salt stress tolerance through retaining a low Na^+/K^+ ratio under high-salt-stress conditions [47].

3.2. Osmotic Homeostasis

Under salt stress, ion imbalance and water deficiency in the plant cell cause osmotic stress. This results in multiple transient biophysical changes, such as the reduction in cell turgor pressure, shrinkage of the plasma membrane, and physical alteration of the cell wall [4]. To alleviate osmotic stress, plants rely on osmotic signaling pathways that regulate processes ranging from gene expression and activation of osmolyte biosynthesis enzymes to water transport systems [41]. Osmolytes, such as proline, polyols, and sugars, accumulate under salt stress. These osmolytes participate in the regulation of osmotic pressure by lowering the osmotic potential in the cytosolic compartment. They also act as signaling molecules to induce ABA accumulation, affect related gene expression, and regulate plant growth under salt stress [48]. Protein kinases act as a convergence point of rapid osmoregulation and salt stress signaling [49]. In response to osmotic treatments, the mRNA levels of histidine kinases, MAPKKK, MAPKK, and MAPK are increased, leading to increased osmolyte synthesis and accumulation [50,51]. Numerous studies have suggested the mitogen-activated protein kinases (MAPKs) are involved in ROS homeostasis [52,53]. For example, ZmMPKs are induced by salt stress and activate oxidative stress regulation to confer salt stress tolerance [51–53]. The receptor-like kinase, salt intolerance 1, is activated by MPK3 and MPK6 and functions in the salt-stress-induced oxidative stress response [54]. Osmotic and salt stresses both induce a rapid increase in cytosolic Ca^{2+}. The copine protein, Ca^{2+}-responsive phospholipid-binding BONZAI1 (BON1), is a critical upstream regulator of osmotic stress signaling since it positively regulates calcium signaling [55]. The disruption of BON1 dampens the cytosolic Ca^{2+} signal in response to osmotic stress.

In plants, salt-stress-triggered ion stress and osmotic stress cause a metabolism imbalance and the toxic accumulation of ROS, inducing plant oxidative damage [40]. Under salt stress, ROS are produced in many plant organelles, such as chloroplast, peroxisomes, mitochondria, and the apoplast. Plant cells sense the accumulated ROS and respond using rapid regulatory mechanisms to scavenge ROS and activate a series of downstream adaptive responses [4,11,12]. ROS function as essential signaling molecules at low levels. Thus, strict control mechanisms are used to balance ROS production and scavenging. Under salt stress, several proteins were found to participate in oxidative stress regulation by activating ROS scavengers or mediating the gene expression of ROS-responsive genes [56]. Several studies have shown that the activities of ROS scavenging enzymes and antioxidants are triggered by salt stress stimuli [57]. For example, the ascorbates peroxidase and catalase are activated by salt stress, improving the tolerance to salinity and oxidative stresses [58]. The overexpression of the ascorbate peroxidases, OsAPXa or OsAPXb, enhances the salt tolerance in rice [59]. Similarly, the constitutive expression of *OsGSTU4* (glutathione *S*–transferase) in *Arabidopsis* also increases the tolerance to salt stress [60]. In rice, the MADS–box transcription factor, OsMADS25, is required for salt tolerance because of its role in maintaining ROS homeostasis [61]. Senescence-associated genes (SAGs) are involved in detoxification in response to salt stress stimuli [62]. The loss of function of SAG29 renders plant seedlings insensitive to salt treatment, while the overexpression of SAG29 results in high sensitivity to salt treatment in *Arabidopsis*. Studies show that NADPH oxidase, respiratory burst oxidase homolog gene, RBOH, mediates ROS synthesis and ROS scavenging to modulate plant development and stress responses [63].

3.3. Phytohormone Signaling Mediation—ABA Signaling and BR Signaling

To withstand constantly changing stress conditions in the environment, plants have developed phytohormone-mediated stress resistance mechanisms. Phytohormones play a crucial role in the plant response to salt stress by regulating plant growth and development adaptation. Phytohormones make great contributions to salt stress signal perception and defense system mediation. Nine plant hormones have been well characterized and are divided into two groups: growth promotion hormones and stress response hormones [64]. The growth promotion hormones are composed of auxin, gibberellin (GA), cytokinins (CKs), brassinosteroids (BRs), and strigolactones (SLs). Some of the growth promotion hormones can also play a role in stress response, such as SLs and BRs [64]. The stress response hormones contain abscisic acid (ABA), ethylene, salicylic acid (SA), and jasmonic acid (JA). The crosstalk between different phytohormones also is important for the salt stress response.

Among the nine plant phytohormones, ABA is the most important hormone regulating stress responses. ABA functions as an important secondary signaling molecule to activate a kinase cascade and mediate gene expression during the salt stress response (Figure 3). Under stress conditions, ABA synthesis is induced quickly leading to rapid increases in ABA levels [65]. A high level of ABA activates kinase cascades and improves stress recognition and stress defense reactions [66]. Salt stress limits water uptake, leading to cell dehydration and changes in cell turgor, generating osmotic stresses. Under high-salinity conditions, the increase in endogenous ABA levels causes stomatal closure to regulate water balance and osmotic homeostasis [67]. Thus, osmoregulation is an important function of the ABA-mediated plant salt stress response.

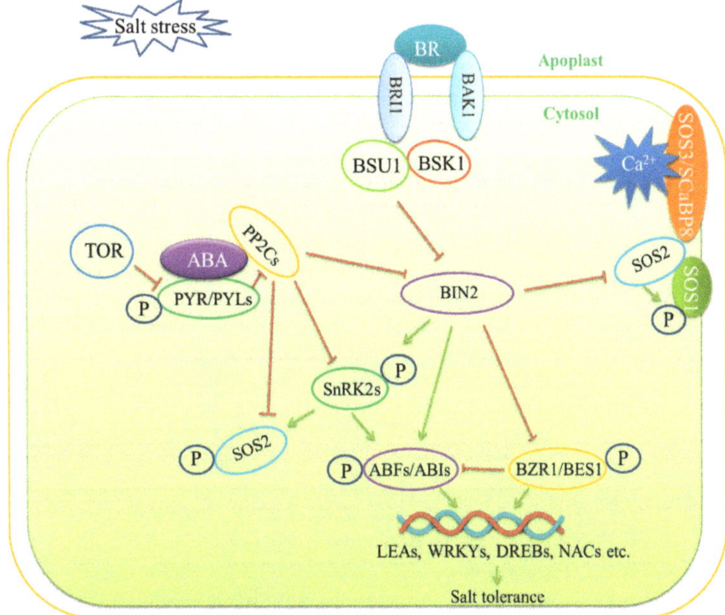

Figure 3. ABA and BR signaling during salt stress. Salt stress promotes abscisic acid (ABA) accumulation. The sucrose nonfermenting-1-related protein kinase2s (SnRK2s) and the clade A type 2C protein phosphatases (PP2Cs) play key roles in mediating the crosstalk between ABA and salt stress signaling. The ABA receptors, PYRABACTIN RESISTANCE/PYR-LIKE (PYR/PYLs) sense ABA and repress PP2Cs, thereby activating the downstream kinase SnRK2. SnRK2s phosphorylate the transcription factors ABSCISIC ACID RESPONSIVE ELEMENT-BINDING FACTORs (ABFs) and ABIs to regulate the expression of stress-responsive genes. The target of rapamycin (TOR) phosphorylates PYL and represses ABA signaling and stress responses. ABI2, a member of the PP2Cs, binds to SOS2 to inhibit its kinase activity, thereby negatively regulating salt tolerance. Additionally, under salt stress, SnRKs phosphorylate SOS2 to activate osmoregulation. Salt stress also upregulates BR biosynthesis. The membrane receptor brassinosteroid insensitive 1 (BRI1) senses BR molecules and acts with its coreceptor BRI1-associated receptor kinase 1 (BAK1) to initiate the downstream phosphorylation cascade. BRI1 and BAK1 transduce the BR signal to BR signaling kinase 1 (BSK1) and activate BRI1 suppressor 1 (BSU1). BSU1 inhibits BIN2 and promotes the transcription factors BZR1/BES1 to induce the expression of BR-responsive genes, which enhances salt tolerance. Under salt stress, BIN2 phosphorylates and inhibits SOS2. This phosphoregulation by BIN2 prevents SOS2 overactivation. Arrows and bars indicate positive and negative regulation, respectively. Solid lines and dashed lines indicate direct regulation and indirect regulation, respectively.

ABA is a 15-carbon isoprenoid that is produced from the methylerythritol 4-phosphate (MEP) pathway via the cleavage of carotenoids. Several enzymes play key roles in the regulation of ABA biosynthesis, such as zeaxanthin oxidase (ZEP), 9-cis-epoxycarotenoid (NCED), and short-chain alcohol dehydrogenase (SCAD) [68]. NCEDs catalyze the rate-limiting carotenoid cleavage reaction. *NCED5* plays an essential role in ABA synthesis and is rapidly induced under salt stress conditions in rice [69]. The small peptide CLAVATA3/ESR-RELATED 25 (CLE25) is secreted from the roots and modulates stomatal control via ABA in root-to-shoot long-distance signaling. The root-derived CLE25 is perceived by the BARELY ANY MERISTEM (BAM) receptors, BAM1 and BAM3, and promotes ABA biosynthesis by upregulating the NCED3 expression in *Arabidopsis* [70]. ABA is primarily synthesized in the root system. ABA is first synthesized in plastids in the root tips and then transported to the shoot and leaves. Changes in ABA levels in roots and leaves have been detected in high-salinity conditions. *Arabidopsis ABA1*, which encodes zeathanxin epoxidase, is the key regulator of ABA synthesis and is also induced by salt stress [71]. Under salt stress, ABA-deficient mutants perform poorly and show salt sensitivity.

Salt-stress-induced osmotic stress also activates ABA signaling transduction pathways. The sucrose-nonfermenting-1-related protein kinase 2s (SnRK2s) are the central components in ABA signaling pathways and play critical roles in osmoregulation [72]. The kinase activities of SnRK2.2/3/6 are inhibited by protein phosphatase 2Cs (PP2Cs) in the absence of ABA. Upon osmotic stress, the ABA receptor proteins, including pyrabactin resistance 1 (PYR1), PYR1-like (PYL), and regulatory component of ABA receptors (RCAR), perceive and bind to the accumulated ABA, which subsequently suppresses the phosphatase activity of PP2Cs [66,72]. As a result, in the absence of inhibition by PP2Cs, SnRK2s are quickly activated. After stress removal, PYL is phosphorylated by the target of rapamycin (TOR) kinase and then disassociates from ABA or PP2C, leading to inactivation of the stress response to promote growth recovery [73]. Recent studies suggest that the upstream kinases, namely, the B2, B3, and B4 Raf-like kinases, are quickly activated and are required for the phosphorylation and activation of SnRK2s in response to early osmotic stress [74]. The Raf-like protein kinases and SnRK2s form the protein kinase cascade that is activated during early osmotic regulation in response to salt stress. The phosphatase ABI1 (abscisic acid insensitive 1) and okadaic acid-sensitive phosphatases of the phosphoprotein phosphatase (PPP) family inhibit the kinase activity of salt-stress-activated SnRK2.4 and regulate primary root growth during the salt stress response [75].

ABA levels increase rapidly under salt stress. Subsequently, salt-stress-induced ABA signaling upregulates the expression of many genes via the targeting of ABA-responsive elements (ABREs) in the regulatory regions of their promoter [76]. ABRE-binding protein/ABRE-binding factor (AREB/ABF) transcription factors are master transcription factors that cooperatively regulate the ABRE-mediated transcription of downstream target genes, enhancing salt stress tolerance. SnRK2s phosphorylate and positively control the AREB/ABF transcription factors [77]. Moreover, the SOS pathway coordinates with ABA signaling. The phosphatase ABI2 (abscisic acid insensitive 2) binds to SOS2 and mediates SOS2 inhibition [78].

Brassinosteroids (BRs) are steroidal hormones that mediate various physiological processes, including cell growth and development, flowering and fruiting [79], and plant stress tolerance. Under salt stress, the biosynthesis of BRs is increased to enhance plant stress tolerance by maintaining ion homeostasis and via osmoregulation. Exogenous BR application reduces ROS production, enhances osmotic regulation and ionic homeostasis, induces the expression of stress-responsive genes, and causes translational changes in stress-responsive proteins. In *Malus hupehensis*, exogenous BR application regulates the activity of Na^+/H^+ antiporters and NHX and alleviates salt stress. Exogenous applications of BRs reduce cytosolic Na^+ levels and increase the absorption of K^+, which is concomitant with higher salt tolerance [80]. Application with 24-epibrassinolide (EBL), a byproduct from the brassinolide biosynthetic pathway, promotes plant growth and development under salt stress [81]. The overexpression of BR-INSENSITIVE 1 (BRI1)-LIKE receptor homolog 3 (BRL3) promotes the accumulation of osmolytes, such as proline and sugars, which play roles in osmoregulation under salt stress [82,83]. The genetic and phenotypic results of BR-related mutants and overexpression transgenic plants indicate that a proper enhancement of BR signaling benefits plants' defense against salt stress [83]. The tomato BZR homolog gene, SlBZR1, positively regulates salt tolerance in tomatoes and upregulates the expression of multiple stress-related genes [84]. The BR receptor SERK2 significantly enhances grain size and salt resistance in rice. Adverse high salinity conditions induce SERK2 accumulation to enhance early BR signaling on the plasma membrane to defend against the stress [85].

Under salt stress, BR exerts anti-stress effects by interacting with other hormones, such as ABA. ABA inhibits the growth-promoting effects of BR during salt stress (Figure 3). ABA and BRs antagonistically fine-tune plant growth under salt stress. The BR receptor BAK1 regulates SnRK2.6 and modulates stomatal closure [86]. BIN2 indirectly activates ABI5, the key transcription factor in the ABA signaling pathway, by phosphorylating SnRK2.6 [87,88]. Brassinazole-resistant 1 (BZR1) and BRI1-EMS-suppressor 1 (BES1) are transcription factors that have been elucidated largely in the BR signaling pathway. BZR1 and BZR2 directly inhibit ABI5 expression [89]. BR shares transcriptional targets with ABA, suggesting that BR antagonistically acts with ABA to regulate the stress response.

The BR pathway can also crosstalk with the SOS pathway. BRs induce the accumulation of calcium in the cytosol, which in turn activates the SOS pathway to regulate ionic and osmotic stresses [81,90]. A recent study has reported that BIN2 inhibits SOS2 kinase activity and negatively regulates salt stress tolerance as a molecular switch in the transition to robust growth after salt stress [29].

3.4. Cytoskeleton Functions

The cytoskeleton plays important roles in a wide variety of cellular processes, including cell shape determination, cell movement, vesicle trafficking, tip growth, and responses to external stress stimuli [91]. The plant cytoskeleton consists of actin filaments (F-actin) and microtubules (MTs), which constantly undergo dynamic changes in architecture. The cytoskeleton has important functions in the plant salt stress response and helps plants to withstand stress conditions through dynamic organizational changes [41]. Cytoskeleton-associated proteins, including MT-associated proteins (MAPs) and actin-binding proteins (ABPs), bind to the cytoskeleton and regulate cytoskeleton organization. Microtubule-associated protein 65-1 (MAP65-1) regulates microtubule stabilization in response to salt stress. Phosphatidic acid (PA) directly binds to MAP65-1 to modulate its microtubule activity [92].

Salt stress triggers changes in the cytoskeleton architecture by modulating dynamic events, such as nucleation and polymerization, severing and depolymerizing, crosslinking/bundling, and growth/shrinkage [93]. During salt stress, the cortical microtubules are first depolymerized and then reorganized. The destabilization of cortical microtubules enhances salt stress tolerances in plants [94,95]. Similarly, actin depolymerization and stabilization are important for plant salt tolerance. The SOS pathway regulates actin dynamics in response to salt stress. Several studies have provided ample pharmacological evidence to link the SOS pathway to cytoskeleton organization [96–98]. The *sos* mutants show abnormal responses to microtubule-associated drugs [99]. The addition of the microtubule-disrupting drug, Oryzalin, causes more death in the *sos1* mutant. Actin reorganization in *sos* mutants is also abnormal in response to salt stress. Disruption of the actin filaments with actin-filament-disrupting drugs, latrunculin A and cytochalasin D (CD), increases death in *sos2* seedlings under salt treatment conditions, while the stabilization of actin filaments with the actin filament stabilizing drug, phalloidin, rescues the lethality phenotype [100].

Calcium is a central secondary messenger that plays an important role in plant salt tolerance. Salt stress induces calcium accumulation in the cytosol and triggers calcium signaling transduction. The cytoskeleton is an important upstream and downstream regulator of calcium signaling [101]. Salt-stress-induced depolymerization of the cortical microtubules leads to the release of Ca^{2+} in the cytosol. The subsequent reorganization of microtubules during salt stress regulates the calcium influx to improve the plant's salt tolerance. Actin dynamics play a role upstream of Ca^{2+} signaling and serve as a signal to induce Ca^{2+} accumulation. Under salt stress, changes in cytoskeleton organization act as a transducer to activate calcium signaling. The actin-related protein2/3 (Arp2/3) complex, as the actin nucleation factor, functions in actin-dynamics-mediated calcium elevation under salt stress [102]. The Arp2/3 complex regulates mitochondrial-dependent Ca^{2+} stimulation via the regulation of the integrity of mitochondria, which is an important organelle for calcium release in response to salt stress. The actin cytoskeleton also plays a role in decoding the downstream calcium signal. The SOS pathway, which is activated by calcium under salt stress, is closely tied to actin dynamics. SOS3, which serves as a calcium sensor in plants, was reported to play a role in regulating actin dynamics under salt stress. The loss of function in SOS3 leads to an abnormal arrangement of actin filaments in response to salt stress, which can be rescued with external calcium application.

Actin dynamics are also important for ROS production under salt stress. Disordered actin organization triggers an accumulation of ROS levels in the *Arabidopsis* root and acts as the initial signal to activate the salt stress response [96].

3.5. Cell Wall Regulation

Accumulating evidence has demonstrated that the cell wall plays an indispensable role in the plant response to salt stress [103]. Salt stress inhibits plant growth and development by repressing cell expansion and division. The cell wall is an important factor for determining cell shape and function and is the first layer of defense against salt stress [12]. Salt stress induces water deficiency in plant cells, causing changes in cell turgor pressure. The cell wall provides mechanical strength to withstand these cell turgor changes [104].

It is believed that the cell wall is one of the early sensors of salt stress. The stress signal is perceived by cell wall sensors localized on the plasma membrane, leading to the induction of downstream responses. One of these cell wall sensors is FER, which is a receptor-like kinase with binding activity to RALF (RAPID ALKALINIZATION FACTOR) peptides [105]. FER senses salt-induced cell wall changes and, in return, sends a downstream signal of cell wall integrity damage. Cell wall leucine-rich repeat extensins (LRX) 3/4/5 function together with RALF peptides and FER to regulate plant growth under salt stress through the modulation of cell wall changes [106].

The cell wall is composed of a complex network of polysaccharides, including cellulose, pectin, and lignin. Secondary cell walls consist of cellulose, hemicelluloses, lignin, and other plant biomass and are distributed in the xylem, fibers, and anther cells. The synthesis of cell wall components is regulated by complicated transcriptional mechanisms under salt stress [107]. The NAC domain and Homeobox HD-ZIP ClassIII (HD-ZIPIII) transcription factors are master regulators of secondary cell wall synthesis [108]. The transcription factors MYB46/MYB83 activates the secondary wall biosynthesis through a transcriptional regulatory program [109]. Cell wall synthesis is tightly regulated by phytohormones, particularly ABA. Both ABA synthesis and signaling are involved in secondary cell wall thickening and lignification [110,111]. SnRK2 kinases, namely, SnRK2.2, 2.3, and 2.6, regulate secondary wall biosynthesis by physically interacting with a NAC family transcription factor, namely NAC secondary wall thickening promoting factor 1 (NST1). SnRK2 phosphorylates NST1 at Ser316, which is a site that is required for the transcriptional activation of downstream secondary wall biosynthesis genes [112].

On the other hand, cell wall components, such as cellulose, lignin, and other polysaccharides, have important biological functions in the plant's response to salt stress [110,113]. Cellulose, the main component of the cell wall, is synthesized by cellulose synthase (CesA) complexes (CSC) at the plasma membrane and tracks along cortical microtubules at a steady pace guided by the protein cellulose synthase interacting (CSI) 1/POM2 [114]. A cellulose synthase–microtubule uncoupling (CMU) protein has been found to affect the function of CSC. CMU is associated with the plasma membrane and interacts with microtubules to regulate cell expansion and development by modulating microtubule displacement [115]. Recent studies have reported that companion of cellulose synthesis (CC) proteins are required for the association of CSCs with microtubules. CC protein1 (CC1) is a MAP that regulates microtubule dynamics to sustain plant growth under salt stress [116]. The mutation of CESA6, CSI1/POM2, or CC1 confers enhanced sensitivity to salt stress in *Arabidopsis* [115–117]. Another cell wall component, namely, lignin, also plays roles in the plant's adaption to saline conditions. Under saline conditions, plant cells adapt to stress by accumulating lignin and thickening the cell wall. As the central component in lignin biosynthesis, caffeoyl-CoA O-methyltransferase 1 (CCoAOMT1) has been reported to play an essential role in the salt stress response [117]. The loss of function of CCoAOMT1 leads to high sensitivity to salt stress. A recent study showed that β-1,4-galactan, a cell wall component, has specific functions in salt hypersensitivity. The synthase GALACTAN SYNTHASE1 (GALS1) catalyzes the biosynthesis of β-1,4-galactan. Salt stress induces the expression of GALS1 and results in the accumulation of β-1,4-galactan levels in plants, which diminishes salt tolerance. This process is transcriptionally regulated. BARLEY B RECOMBINANT/BASIC PENTACYSTEINE transcription factors BPC1/BPC2 repress the expression of GALS1 and positively regulate plant salt tolerance [118].

4. Conclusions

Plants must efficiently adjust their growth to adapt to stress conditions. Salt stress is one of the most serious abiotic stresses experienced worldwide. Identifying the salt stress signaling pathway and characterizing the upstream salt stress sensors could guide approaches to mitigate the negative effects of salt stress on crop yields and ultimately improve agricultural development. Salt stress adversely affects plant growth and development, whereas plants have evolved regulatory mechanisms that allow them to adapt to these adverse conditions. For instance, plant growth is inhibited by salt stress due to decreased photosynthesis. However, the plant also actively slows the growth rate in response to salt stress, leading to increased survival. Plant cells undergo large changes to respond and defend against salt stress. For example, salt stress induces ion stress. In turn, plant cells activate ion transporters and channels to reestablish the ion balance. In the ion transport process, the Na^+ exclusion, K^+ influx, Ca^{2+} pump, and Na^+/H^+ exchange are all important for plant salt tolerance. In addition, strategies for osmotic and oxidative stress alleviations are also utilized in plants under saline conditions. The identification of upstream regulators, the characterization of a high-resolution sensor, transporter, and channel of Na^+ and K^+, and the identification of a novel channel and pool of Ca^{2+} will be areas of active future interest.

High throughput and efficient biotechnologies are important for salt-stress-related gene screening. To date, RNA sequencing has proven to be a fast and effective method for studying the molecular regulation of plant salt tolerance. Transcriptome sequencing techniques have been widely used to identify novel genes that are linked to the regulation of the plant salt stress response [119,120]. The development of next-generation sequencing technology has made it easier to screen for salt-tolerance genes [121,122]. The global survey of transcriptome profiles and microRNA levels of plants in response to salt stress using RNA-Seq has provided useful insights into the mechanisms of salt tolerance [123]. These findings also provide a rich resource for breeding salt-tolerant cultivars through biotechnological approaches using salt-related genes.

Author Contributions: S.Z., H.Z., and P.W. wrote this manuscript. Q.Z. and M.L. participated in the writing and modification of this manuscript. C.M. conceptualized the idea. All authors have read and agreed to the published version of the manuscript.

Funding: This work was supported by the National Natural Science Foundation of China (31700222, 31870241); Shandong Provincial National Science Foundation, China (ZR2017BC026); China Postdoctoral Science Foundation (2017M610444); the Open Project of State Key Laboratory of Plant Physiology and Biochemistry (SKLPPBKF1704).

Institutional Review Board Statement: Not applicable.

Informed Consent Statement: Not applicable.

Data Availability Statement: Not applicable.

Conflicts of Interest: The authors declare no conflict of interest.

References

1. Savary, S.; Akter, S.; Almekinders, C.; Harris, J.; Korsten, L.; Rötter, R.; Waddington, S.; Watson, D. Mapping disruption and resilience mechanisms in food systems. *Food Secur.* **2020**, *12*, 695–717. [CrossRef]
2. Hazell, P.B.R.; Wood, S. Drivers of change in global agriculture. *Philos. Trans. R. Soc. B Biol. Sci.* **2007**, *363*, 495–515. [CrossRef]
3. Qadir, M.; Quillérou, E.; Nangia, V.; Murtaza, G.; Singh, M.; Thomas, R.J.; Drechsel, P.; Noble, A.D. Economics of salt-induced land degradation and restoration. *Nat. Resour. Forum* **2014**, *38*, 282–295. [CrossRef]
4. Park, H.J.; Kim, W.-Y.; Yun, A.D.-J. A New Insight of Salt Stress Signaling in Plant. *Mol. Cells* **2016**, *39*, 447–459. [CrossRef]
5. Ziska, L.H.; Bunce, J.A.; Shimono, H.; Gealy, D.R.; Baker, J.T.; Newton, P.C.D.; Reynolds, M.P.; Jagadish, K.S.V.; Zhu, C.; Howden, M.; et al. Food security and climate change: On the potential to adapt global crop production by active selection to rising atmospheric carbon dioxide. *Proc. R. Soc. B Boil. Sci.* **2012**, *279*, 4097–4105. [CrossRef]
6. Su, T.; Li, X.; Yang, M.; Shao, Q.; Zhao, Y.; Ma, C.; Wang, P. Autophagy: An Intracellular Degradation Pathway Regulating Plant Survival and Stress Response. *Front. Plant Sci.* **2020**, *11*, 164. [CrossRef]
7. Horie, T.; Karahara, I.; Katsuhara, M. Salinity tolerance mechanisms in glycophytes: An overview with the central focus on rice plants. *Rice* **2012**, *5*, 1–18. [CrossRef]
8. Quan, R.; Lin, H.; Mendoza, I.; Zhang, Y.; Cao, W.; Yang, Y.; Shang, M.; Chen, S.; Pardo, J.M.; Guo, Y. SCABP8/CBL10, a Putative Calcium Sensor, Interacts with the Protein Kinase SOS2 to Protect *Arabidopsis* Shoots from Salt Stress. *Plant Cell* **2007**, *19*, 1415–1431. [CrossRef] [PubMed]
9. Park, H.J.; Kim, W.-Y.; Yun, D.-J. A role for GIGANTEA. *Plant Signal. Behav.* **2013**, *8*, e24820. [CrossRef] [PubMed]
10. Gong, Z. Plant abiotic stress: New insights into the factors that activate and modulate plant responses. *J. Integr. Plant Biol.* **2021**, *63*, 429–430. [CrossRef] [PubMed]
11. Zhu, J.K. Salt and drought stress signal transduction in plants. *Annu. Rev. Plant Biol.* **2002**, *53*, 247–273. [CrossRef] [PubMed]
12. Van Zelm, E.; Zhang, Y.; Testerink, C. Salt Tolerance Mechanisms of Plants. *Annu. Rev. Plant Biol.* **2020**, *71*, 403–433. [CrossRef] [PubMed]
13. Chen, Y.; Hoehenwarter, W. Changes in the Phosphoproteome and Metabolome Link Early Signaling Events to Rearrangement of Photosynthesis and Central Metabolism in Salinity and Oxidative Stress Response in *Arabidopsis*. *Plant Physiol.* **2015**, *69*, 3021–3033. [CrossRef] [PubMed]
14. Shumilina, J.; Kusnetsova, A.; Tsarev, A.; Van Rensburg, H.C.J.; Medvedev, S.; Demidchik, V.; Ende, W.V.D.; Frolov, A. Glycation of Plant Proteins: Regulatory Roles and Interplay with Sugar Signalling? *Int. J. Mol. Sci.* **2019**, *20*, 2366. [CrossRef] [PubMed]
15. Yuan, F.; Yang, H.; Xue, Y.; Kong, D.; Ye, R.; Li, C.; Zhang, J.; Theprungsirikul, L.; Shrift, T.; Krichilsky, B.; et al. OSCA1 mediates osmotic-stress-evoked Ca^{2+} increases vital for osmosensing in *Arabidopsis*. *Nat. Cell Biol.* **2014**, *514*, 367–371. [CrossRef] [PubMed]
16. Zhang, S.; Wu, Q.-R.; Liu, L.-L.; Zhang, H.-M.; Gao, J.-W.; Pei, Z.-M. Osmotic stress alters circadian cytosolic Ca^{2+} oscillations and OSCA1 is required in circadian gated stress adaptation. *Plant Signal. Behav.* **2020**, *15*, 1836883. [CrossRef]
17. Stephan, A.B.; Kunz, H.H.; Yang, E.; Schroeder, J.I. Rapid hyperosmotic-induced Ca^{2+} responses in *Arabidopsis thaliana* exhibit sensory potentiation and invovlement of plastidial KEA transporters. *Proc. Natl. Acad. Sci. USA* **2016**, *113*, E5242–E5249. [CrossRef]
18. Jiang, Z.; Zhou, X.; Tao, M.; Yuan, F.; Liu, L.; Wu, F.; Wu, X.; Xiang, Y.; Niu, Y.; Liu, F.; et al. Plant cell-surface GIPC sphingolipids sense salt to trigger Ca^{2+} influx. *Nat. Cell Biol.* **2019**, *572*, 341–346. [CrossRef]
19. Feng, W.; Kita, D.; Peaucelle, A.; Cartwright, H.N.; Doan, V.; Duan, Q.; Liu, M.-C.; Maman, J.; Steinhorst, L.; Schmitz-Thom, I.; et al. The FERONIA Receptor Kinase Maintains Cell-Wall Integrity during Salt Stress through Ca^{2+} Signaling. *Curr. Biol.* **2018**, *28*, 666–675.e5. [CrossRef]
20. Pan, Y.; Chai, X.; Gao, Q.; Zhou, L.; Zhang, S.; Li, L.; Luan, S. Dynamic Interactions of Plant CNGC Subunits and Calmodulins Drive Oscillatory Ca^{2+} Channel Activities. *Dev. Cell* **2019**, *48*, 710–725.e5. [CrossRef] [PubMed]
21. Tian, W.; Hou, C.; Ren, Z.; Wang, C.; Zhao, F.; Dahlbeck, D.; Hu, S.; Zhang, L.; Niu, Q.; Li, L.; et al. A calmodulin-gated calcium channel links pathogen patterns to plant immunity. *Nat. Cell Biol.* **2019**, *572*, 131–135. [CrossRef]
22. Wu, F.; Chi, Y.; Jiang, Z.; Xu, Y.; Xie, L.; Huang, F.; Wan, D.; Ni, J.; Yuan, F.; Wu, X.; et al. Hydrogen peroxide sensor HPCA1 is an LRR receptor kinase in *Arabidopsis*. *Nat. Cell Biol.* **2020**, *578*, 577–581. [CrossRef]
23. Yang, Z.; Wang, C.; Xue, Y.; Liu, X.; Chen, S.; Song, C.; Yang, Y.; Guo, Y. Calcium-activated 14-3-3 proteins as a molecular switch in salt stress tolerance. *Nat. Commun.* **2019**, *10*, 1–12. [CrossRef]

24. Zhou, H.; Lin, H.; Chen, S.; Becker, K.; Yang, Y.; Zhao, J.; Kudla, J.; Schumaker, K.S.; Guo, Y. Inhibition of the *Arabidopsis* Salt Overly Sensitive Pathway by 14-3-3 Proteins. *Plant Cell* **2014**, *26*, 1166–1182. [CrossRef]
25. Kim, W.-Y.; Ali, Z.; Park, H.J.; Park, S.J.; Cha, J.-Y.; Perez-Hormaeche, J.; Quintero, F.J.; Shin, G.; Kim, M.R.; Qiang, Z.; et al. Release of SOS2 kinase from sequestration with GIGANTEA determines salt tolerance in *Arabidopsis*. *Nat. Commun.* **2013**, *4*, 1352. [CrossRef]
26. Tan, T.; Cai, J.; Zhan, E.; Yang, Y.; Zhao, J.; Guo, Y.; Zhou, H. Stability and localization of 14-3-3 proteins are involved in salt tolerance in *Arabidopsis*. *Plant Mol. Biol.* **2016**, *92*, 391–400. [CrossRef] [PubMed]
27. Barajas-Lopez, J.D.D.; Moreno, J.R.; Gamez-Arjona, F.M.; Pardo, J.M.; Punkkinen, M.; Zhu, J.-K.; Quintero, F.J.; Fujii, H. Upstream kinases of plant SnRKs are involved in salt stress tolerance. *Plant J.* **2017**, *93*, 107–118. [CrossRef] [PubMed]
28. Ma, L.; Ye, J.; Yang, Y.; Lin, H.; Yue, L.; Luo, J.; Long, Y.; Fu, H.; Liu, X.; Zhang, Y.; et al. The SOS2-SCaBP8 Complex Generates and Fine-Tunes an AtANN4-Dependent Calcium Signature under Salt Stress. *Dev. Cell* **2019**, *48*, 697–709.e5. [CrossRef] [PubMed]
29. Li, J.; Zhou, H.; Zhang, Y.; Li, Z.; Yang, Y.; Guo, Y. The GSK3-like Kinase BIN2 Is a Molecular Switch between the Salt Stress Response and Growth Recovery in *Arabidopsis thaliana*. *Dev. Cell* **2020**, *55*, 367–380.e6. [CrossRef] [PubMed]
30. Ren, X.-L.; Qi, G.-N.; Feng, H.-Q.; Zhao, S.; Zhao, S.-S.; Wang, Y.; Wu, W.-H. Calcineurin B-like protein CBL10 directly interacts with AKT1 and modulates K+homeostasis in *Arabidopsis*. *Plant J.* **2013**, *74*, 258–266. [CrossRef]
31. Nieves-Cordones, M.; Alemán, F.; Martínez, V.; Rubio, F. K+ uptake in plant roots. The systems involved, their regulation and parallels in other organisms. *J. Plant Physiol.* **2014**, *171*, 688–695. [CrossRef] [PubMed]
32. Xu, J.; Li, H.-D.; Chen, L.-Q.; Wang, Y.; Liu, L.-L.; He, L.; Wu, W.-H. A Protein Kinase, Interacting with Two Calcineurin B-like Proteins, Regulates K+ Transporter AKT1 in *Arabidopsis*. *Cell* **2006**, *125*, 1347–1360. [CrossRef] [PubMed]
33. Sánchez-Barrena, M.J.; Chaves-Sanjuan, A.; Raddatz, N.; Mendoza, I.; Cortés, Á.; Gago, F.; González-Rubio, J.M.; Benavente, J.L.; Quintero, F.J.; Pardo, J.M.; et al. Recognition and Activation of the Plant AKT1 Potassium Channel by the Kinase CIPK23. *Plant Physiol.* **2020**, *182*, 2143–2153. [CrossRef] [PubMed]
34. Cheong, Y.H.; Pandey, G.K.; Grant, J.J.; Batistic, O.; Li, L.; Kim, B.-G.; Lee, S.-C.; Kudla, J.; Luan, S. Two calcineurin B-like calcium sensors, interacting with protein kinase CIPK23, regulate leaf transpiration and root potassium uptake in *Arabidopsis*. *Plant J.* **2007**, *52*, 223–239. [CrossRef] [PubMed]
35. Lee, S.C.; Lan, W.-Z.; Kim, B.-G.; Li, L.; Cheong, Y.H.; Pandey, G.K.; Lu, G.; Buchanan, B.B.; Luan, S. A protein phosphorylation/dephosphorylation network regulates a plant potassium channel. *Proc. Natl. Acad. Sci. USA* **2007**, *104*, 15959–15964. [CrossRef]
36. Lan, W.Z.; Lee, S.C.; Che, Y.; Jiang, Y.; Luan, S. Mechanistic Analysis of AKT1 Regulation by the CBL–CIPK–PP2CA Interactions. *Mol. Plant* **2011**, *4*, 527–536. [CrossRef]
37. Latz, A.; Becker, D.; Hekman, M.; Müller, T.; Beyhl, D.; Marten, I.; Eing, C.; Fischer, A.; Dunkel, M.; Bertl, A.; et al. TPK1, a Ca^{2+}-regulated *Arabidopsis* vacuole two-pore K+ channel is activated by 14-3-3 proteins. *Plant J.* **2007**, *52*, 449–459. [CrossRef]
38. Latz, A.; Mehlmer, N.; Zapf, S.; Mueller, T.D.; Wurzinger, B.; Pfister, B.; Csaszar, E.; Hedrich, R.; Teige, M.; Becker, D. Salt Stress Triggers Phosphorylation of the *Arabidopsis* Vacuolar K+ Channel TPK1 by Calcium-Dependent Protein Kinases (CDPKs). *Mol. Plant* **2013**, *6*, 1274–1289. [CrossRef]
39. Ali, A.; Maggio, A.; Bressan, R.A.; Yun, D.-J. Role and Functional Differences of HKT1-Type Transporters in Plants under Salt Stress. *Int. J. Mol. Sci.* **2019**, *20*, 1059. [CrossRef]
40. Yang, Y.; Guo, Y. Unraveling salt stress signaling in plants. *J. Integr. Plant Biol.* **2018**, *60*, 796–804. [CrossRef]
41. Yang, Y.; Guo, Y. Elucidating the molecular mechanisms mediating plant salt-stress responses. *New Phytol.* **2017**, *217*, 523–539. [CrossRef] [PubMed]
42. Zhang, M.; Cao, Y.; Wang, Z.; Wang, Z.-Q.; Shi, J.; Liang, X.; Song, W.; Chen, Q.; Lai, J.; Jiang, C. A retrotransposon in an HKT1 family sodium transporter causes variation of leaf Na+exclusion and salt tolerance in maize. *New Phytol.* **2017**, *217*, 1161–1176. [CrossRef] [PubMed]
43. Chu, M.; Chen, P.; Meng, S.; Xu, P.; Lan, W. The *Arabidopsis* phosphatase PP2C49 negatively regulates salt tolerance through inhibition of AtHKT1;1. *J. Integr. Plant Biol.* **2021**, *63*, 528–542. [CrossRef] [PubMed]
44. Covington, M.F.; Maloof, J.N.; Straume, M.; Kay, S.A.; Harmer, S.L. Global transcriptome analysis reveals circadian regulation of key pathways in plant growth and development. *Genome Biol.* **2008**, *9*, R130. [CrossRef]
45. Sakuraba, Y.; Bülbül, S.; Piao, W.; Choi, G.; Paek, N. *Arabidopsis* EARLY FLOWERING 3 increases salt tolerance by suppressing salt stress response pathways. *Plant J.* **2017**, *92*, 1106–1120. [CrossRef]
46. Wei, H.; Wang, X.; He, Y.; Xu, H.; Wang, L. Clock component OsPRR73 positively regulates rice salt tolerance by modulating OsHKT2;1 -mediated sodium homeostasis. *EMBO J.* **2021**, *40*, e105086. [CrossRef]
47. Ayadi, M.; Brini, F.; Masmoudi, K. Overexpression of a Wheat Aquaporin Gene, TdPIP2;1, Enhances Salt and Drought Tolerance in Transgenic Durum Wheat cv. Maali. *Int. J. Mol. Sci.* **2019**, *20*, 2389. [CrossRef]
48. Marusig, D.; Tombesi, S. Abscisic Acid Mediates Drought and Salt Stress Responses in *Vitis vinifera*—A Review. *Int. J. Mol. Sci.* **2020**, *21*, 8648. [CrossRef]
49. Chen, X.; Ding, Y.; Yang, Y.; Song, C.; Wang, B.; Yang, S.; Guo, Y.; Gong, Z. Protein kinases in plant responses to drought, salt, and cold stress. *J. Integr. Plant Biol.* **2021**, *63*, 53–78. [CrossRef]
50. Zhou, X.; Naguro, I.; Ichijo, H.; Watanabe, K. Mitogen-activated protein kinases as key players in osmotic stress signaling. *Biochim. Biophys. Acta Gen. Subj.* **2016**, *1860*, 2037–2052. [CrossRef]

51. Moustafa, K.; AbuQamar, S.; Jarrar, M.; Al-Rajab, A.J.; Trémouillaux-Guiller, J. MAPK cascades and major abiotic stresses. *Plant Cell Rep.* **2014**, *33*, 1217–1225. [CrossRef]
52. Pan, J.; Zhang, M.; Kong, X.; Xing, X.; Liu, Y.; Zhou, Y.; Liu, Y.; Sun, L.; Li, D. ZmMPK17, a novel maize group D MAP kinase gene, is involved in multiple stress responses. *Planta* **2011**, *235*, 661–676. [CrossRef]
53. Zhang, D.; Jiang, S.; Pan, J.; Kong, X.; Zhou, Y.; Liu, Y.; Li, D. The overexpression of a maize mitogen-activated protein kinase gene (ZmMPK5) confers salt stress tolerance and induces defence responses in tobacco. *Plant Biol.* **2013**, *16*, 558–570. [CrossRef] [PubMed]
54. Li, C.; Wang, G.; Zhao, J.; Zhang, L.; Ai, L.; Han, Y.; Sun, D.; Zhang, S.; Sun, Y. The Receptor-Like Kinase SIT1 Mediates Salt Sensitivity by Activating MAPK3/6 and Regulating Ethylene Homeostasis in Rice. *Plant Cell* **2014**, *26*, 2538–2553. [CrossRef] [PubMed]
55. Chen, K.; Gao, J.; Sun, S.; Zhang, Z.; Yu, B.; Li, J.; Xie, C.; Li, G.; Wang, P.; Bressan, R.A.; et al. The Calcium-Responsive Phospholipid-Binding BONZAI Proteins Control Global Osmotic Stress Responses in Plants Through Repression of Immune Signaling. *SSRN Electron. J.* **2020**, *30*, 4815. [CrossRef]
56. Nadarajah, K.K. ROS Homeostasis in Abiotic Stress Tolerance in Plants. *Int. J. Mol. Sci.* **2020**, *21*, 5208. [CrossRef]
57. Choudhury, F.K.; Rivero, R.M.; Blumwald, E.; Mittler, R. Reactive oxygen species, abiotic stress and stress combination. *Plant J.* **2016**, *90*, 856–867. [CrossRef]
58. Sofo, A.; Scopa, A.; Nuzzaci, M.; Vitti, A. Ascorbate Peroxidase and Catalase Activities and Their Genetic Regulation in Plants Subjected to Drought and Salinity Stresses. *Int. J. Mol. Sci.* **2015**, *16*, 13561–13578. [CrossRef] [PubMed]
59. Lu, Z.; Liu, D.; Liu, S. Two rice cytosolic ascorbate peroxidases differentially improve salt tolerance in transgenic *Arabidopsis*. *Plant Cell Rep.* **2007**, *26*, 1909–1917. [CrossRef]
60. Sharma, R.; Sahoo, A.; Devendran, R.; Jain, M. Over-Expression of a Rice Tau Class Glutathione S-Transferase Gene Improves Tolerance to Salinity and Oxidative Stresses in *Arabidopsis*. *PLoS ONE* **2014**, *9*, e92900. [CrossRef] [PubMed]
61. Xu, N.; Chu, Y.; Chen, H.; Li, X.; Wu, Q.; Jin, L.; Wang, G.; Huang, J. Rice transcription factor OsMADS25 modulates root growth and confers salinity tolerance via the ABA–mediated regulatory pathway and ROS scavenging. *PLoS Genet.* **2018**, *14*, e1007662. [CrossRef]
62. Seo, P.J.; Park, J.-M.; Kang, S.K.; Kim, S.-G.; Park, C.-M. An *Arabidopsis* senescence-associated protein SAG29 regulates cell viability under high salinity. *Planta* **2010**, *233*, 189–200. [CrossRef]
63. Chapman, J.M.; Muhlemann, J.K.; Gayomba, S.R.; Muday, G.K. RBOH-Dependent ROS Synthesis and ROS Scavenging by Plant Specialized Metabolites To Modulate Plant Development and Stress Responses. *Chem. Res. Toxicol.* **2019**, *32*, 370–396. [CrossRef] [PubMed]
64. Yu, Z.; Duan, X.; Luo, L.; Dai, S.; Ding, Z.; Xia, G. How Plant Hormones Mediate Salt Stress Responses. *Trends Plant Sci.* **2020**, *25*, 1117–1130. [CrossRef] [PubMed]
65. Jia, W.; Wang, Y.; Zhang, S.; Zhang, J. Salt-stress-induced ABA accumulation is more sensitively triggered in roots than in shoots. *J. Exp. Bot.* **2002**, *53*, 2201–2206. [CrossRef] [PubMed]
66. Zhu, J.-K. Abiotic Stress Signaling and Responses in Plants. *Cell* **2016**, *167*, 313–324. [CrossRef] [PubMed]
67. Verma, V.; Ravindran, P.; Kumar, P.P. Plant hormone-mediated regulation of stress responses. *BMC Plant Biol.* **2016**, *16*, 1–10. [CrossRef]
68. Chen, K.; Li, G.; Bressan, R.A.; Song, C.; Zhu, J.; Zhao, Y. Abscisic acid dynamics, signaling, and functions in plants. *J. Integr. Plant Biol.* **2020**, *62*, 25–54. [CrossRef]
69. Huang, Y.; Jiao, Y.; Xie, N.; Guo, Y.; Zhang, F.; Xiang, Z.; Wang, R.; Wang, F.; Gao, Q.; Tian, L.; et al. OsNCED5, a 9-cis-epoxycarotenoid dioxygenase gene, regulates salt and water stress tolerance and leaf senescence in rice. *Plant Sci.* **2019**, *287*, 110188. [CrossRef] [PubMed]
70. Takahashi, F.; Suzuki, T.; Osakabe, Y.; Betsuyaku, S.; Kondo, Y.; Dohmae, N.; Fukuda, H.; Yamaguchi-Shinozaki, K.; Shinozaki, K. A small peptide modulates stomatal control via abscisic acid in long-distance signalling. *Nat. Cell Biol.* **2018**, *556*, 235–238. [CrossRef]
71. Barrero, J.M.; Rodriguez, P.L.; Quesada, V.; Piqueras, P.; Ponce, M.R.; Micol, J.L. Both abscisic acid (ABA)-dependent and ABA-independent pathways govern the induction of NCED3, AAO3 and ABA1 in response to salt stress. *Plant Cell Environ.* **2006**, *29*, 2000–2008. [CrossRef]
72. Zhao, Y.; Zhang, Z.; Gao, J.; Wang, P.; Hu, T.; Wang, Z.; Hou, Y.-J.; Wan, Y.; Liu, W.; Xie, S.; et al. *Arabidopsis* Duodecuple Mutant of PYL ABA Receptors Reveals PYL Repression of ABA-Independent SnRK2 Activity. *Cell Rep.* **2018**, *23*, 3340–3351.e5. [CrossRef] [PubMed]
73. Wang, P.; Zhao, Y.; Li, Z.; Hsu, C.-C.; Liu, X.; Fu, L.; Hou, Y.-J.; Du, Y.; Xie, S.; Zhang, C.; et al. Reciprocal Regulation of the TOR Kinase and ABA Receptor Balances Plant Growth and Stress Response. *Mol. Cell* **2018**, *69*, 100–112.e6. [CrossRef] [PubMed]
74. Lin, Z.; Li, Y.; Zhang, Z.; Liu, X.; Hsu, C.; Du, Y.; Sang, T.; Zhu, C.; Wang, Y.; Satheesh, V.; et al. A RAF-SnRK2 kinase cascade mediates early osmotic stress signaling in higher plants. *Nat. Commun.* **2020**, *11*, 613. [CrossRef] [PubMed]
75. Krzywińska, E.; Bucholc, M.; Kulik, A.; Ciesielski, A.; Lichocka, M.; Dębski, J.; Ludwików, A.; Dadlez, M.; Rodriguez, P.L.; Dobrowolska, G. Phosphatase ABI1 and okadaic acid-sensitive phosphoprotein phosphatases inhibit salt stress-activated SnRK2.4 kinase. *BMC Plant Biol.* **2016**, *16*, 1–12. [CrossRef] [PubMed]

76. Fujita, Y.; Yoshida, T.; Yamaguchi-Shinozaki, K. Pivotal role of the AREB/ABF-SnRK2 pathway in ABRE-mediated transcription in response to osmotic stress in plants. *Physiol. Plant* **2013**, *147*, 15–27. [CrossRef]
77. Yoshida, T.; Fujita, Y.; Sayama, H.; Kidokoro, S.; Maruyama, K.; Mizoi, J.; Shinozaki, K.; Yamaguchi-Shinozaki, K. AREB1, AREB2, and ABF3 are master transcription factors that cooperatively regulate ABRE-dependent ABA signaling involved in drought stress tolerance and require ABA for full activation. *Plant J.* **2010**, *61*, 672–685. [CrossRef]
78. Ohta, M.; Guo, Y.; Halfter, U.; Zhu, J.-K. A novel domain in the protein kinase SOS2 mediates interaction with the protein phosphatase 2C ABI2. *Proc. Natl. Acad. Sci. USA* **2003**, *100*, 11771–11776. [CrossRef]
79. Nolan, T.M.; Vukašinović, N.; Liu, D.; Russinova, E.; Yin, Y. Brassinosteroids: Multidimensional Regulators of Plant Growth, Development, and Stress Responses. *Plant Cell* **2020**, *32*, 295–318. [CrossRef]
80. Su, Q.; Zheng, X.; Tian, Y.; Wang, C. Exogenous Brassinolide Alleviates Salt Stress in Malus hupehensis Rehd. by Regulating the Transcription of NHX-Type Na+(K+)/H+ Antiporters. *Front. Plant Sci.* **2020**, *11*, 38. [CrossRef]
81. Liu, J.; Gao, H.; Zheng, Q.; Wang, C.; Wang, X.; Wang, Q. Effects of 24-epibrassinolide on plant growth, osmotic regulation and ion homeostasis of salt-stressed canola. *Plant Biol.* **2013**, *16*, 440–450. [CrossRef]
82. Krishna, P.; Prasad, B.D.; Rahman, T. Brassinosteroid Action in Plant Abiotic Stress Tolerance. *Methods Mol. Biol.* **2017**, *1564*, 193–202. [CrossRef]
83. Fàbregas, N.; Lozano-Elena, F.; Blasco-Escámez, D.; Tohge, T.; Martínez-Andújar, C.; Albacete, A.; Osorio, S.; Bustamante, M.; Riechmann, J.L.; Nomura, T.; et al. Overexpression of the vascular brassinosteroid receptor BRL3 confers drought resistance without penalizing plant growth. *Nat. Commun.* **2018**, *9*, 1–13. [CrossRef]
84. Jia, C.; Zhao, S.; Bao, T.; Zhao, P.; Peng, K.; Guo, Q.; Gao, X.; Qin, J. Tomato BZR/BES transcription factor SlBZR1 positively regulates BR signaling and salt stress tolerance in tomato and *Arabidopsis*. *Plant Sci.* **2021**, *302*, 110719. [CrossRef]
85. Dong, N.Y.; Liu, D.; Zhang, X.; Yu, Z.; Huang, W.; Liu, J.; Yang, Y.; Meng, W.; Niu, M. Regulation of Brassinosteroid Signaling and Salt Resistance by SERK2 and Potential Utilization for Crop Improvement in Rice. *Front. Plant Sci.* **2020**, *11*, 621859. [CrossRef]
86. Shang, Y.; Yang, D.; Ha, Y.; Shin, H.-Y.; Nam, K.H. RPK1 and BAK1 sequentially form complexes with OST1 to regulate ABA-induced stomatal closure. *J. Exp. Bot.* **2019**, *71*, 1491–1502. [CrossRef]
87. Hu, Y.; Yu, D. BRASSINOSTEROID INSENSITIVE2 Interacts with ABSCISIC ACID INSENSITIVE5 to Mediate the Antagonism of Brassinosteroids to Abscisic Acid during Seed Germination in *Arabidopsis*. *Plant Cell* **2014**, *26*, 4394–4408. [CrossRef]
88. Cai, Z.L.; Liu, J.; Wang, H.; Yang, C.; Chen, Y.; Li, Y.; Pan, S.; Dong, R.; Tang, G.; Barajas-Lopez, J.; et al. GSK3-like kinases positively modulate abscisic acid signaling through phosphorylating subgroup III SnRK2s in *Arabidopsis*. *Proc. Natl. Acad. Sci. USA* **2014**, *111*, 9651–9656. [CrossRef] [PubMed]
89. Yang, X.; Bai, Y.; Shang, J.; Xin, R.; Tang, W. The antagonistic regulation of abscisic acid-inhibited root growth by brassinosteroids is partially mediated via direct suppression of *ABSCISIC ACID INSENSITIVE 5* expression by BRASSINAZOLE RESISTANT 1. *Plant Cell Environ.* **2016**, *39*, 1994–2003. [CrossRef] [PubMed]
90. Yan, J.; Guan, L.; Sun, Y.; Zhu, Y.; Liu, L.; Lu, R.; Jiang, M.; Tan, M.; Zhang, A. Calcium and ZmCCaMK are involved in brassinosteroid-induced antioxidant defense in maize leaves. *Plant Cell Physiol.* **2015**, *56*, 883–896. [CrossRef] [PubMed]
91. Wang, X.; Mao, T. Understanding the functions and mechanisms of plant cytoskeleton in response to environmental signals. *Curr. Opin. Plant Biol.* **2019**, *52*, 86–96. [CrossRef] [PubMed]
92. Zhang, Q.; Lin, F.; Mao, T.; Nie, J.; Yan, M.; Yuan, M.; Zhang, W. Phosphatidic Acid Regulates Microtubule Organization by Interacting with MAP65-1 in Response to Salt Stress in *Arabidopsis*. *Plant Cell* **2012**, *24*, 4555–4576. [CrossRef]
93. Lian, N.; Wang, X.; Jing, Y.; Lin, J. Regulation of cytoskeleton-associated protein activities: Linking cellular signals to plant cytoskeletal function. *J. Integr. Plant Biol.* **2021**, *63*, 241–250. [CrossRef]
94. Wang, C.Z.; Huang, R.D. Cytoskeleton and plant salt stress tolerance. *Plant Signal Behav.* **2014**, *6*, 29–31. [CrossRef] [PubMed]
95. Ma, D.; Han, R. Microtubule organization defects in *Arabidopsis thaliana*. *Plant Biol.* **2020**, *22*, 971–980. [CrossRef] [PubMed]
96. Liu, S.G.; Zhu, D.Z.; Chen, G.H.; Gao, X.-Q.; Zhang, X.S. Disrupted actin dynamics trigger an increment in the reactive oxygen species levels in the *Arabidopsis* root under salt stress. *Plant Cell Rep.* **2012**, *31*, 1219–1226. [CrossRef] [PubMed]
97. Ye, J.; Zhang, W.; Guo, Y. *Arabidopsis* SOS_3 plays an important role in salt tolerance by mediating calcium-dependent microfilament reorganization. *Plant Cell Rep.* **2013**, *32*, 139–148. [CrossRef]
98. Zhou, Y.; Yang, Z.; Guo, G.; Guo, Y. Microfilament Dynamics is Required for Root Growth under Alkaline Stress in *Arabidopsis*. *J. Integr. Plant Biol.* **2010**, *52*, 952–958. [CrossRef]
99. Wang, C.; Li, J.; Yuan, M. Salt Tolerance Requires Cortical Microtubule Reorganization in *Arabidopsis*. *Plant Cell Physiol.* **2007**, *48*, 1534–1547. [CrossRef]
100. Wang, C.; Zhang, L.; Yuan, M.; Ge, Y.; Liu, Y.; Fan, J.; Ruan, Y.; Cui, Z.; Tong, S.; Zhang, S. The microfilament cytoskeleton plays a vital role in salt and osmotic stress tolerance in *Arabidopsis*. *Plant Biol.* **2009**, *12*, 70–78. [CrossRef]
101. Qian, D.; Xiang, Y. Actin Cytoskeleton as Actor in Upstream and Downstream of Calcium Signaling in Plant Cells. *Int. J. Mol. Sci.* **2019**, *20*, 1403. [CrossRef]
102. Zhao, Y.; Pan, Z.; Zhang, Y.; Qu, X.; Yang, Y.; Jiang, X.; Huang, S.; Yuan, M.; Schumaker, K.S.; Guo, Y. The Actin-Related Protein2/3 Complex Regulates Mitochondrial-Associated Calcium Signaling during Salt Stress in *Arabidopsis*. *Plant Cell* **2013**, *25*, 4544–4559. [CrossRef]
103. Endler, A.; Kesten, C.; Schneider, R.; Zhang, Y.; Ivakov, A.; Froehlich, A.; Funke, N.; Persson, S. A Mechanism for Sustained Cellulose Synthesis during Salt Stress. *Cell* **2015**, *162*, 1353–1364. [CrossRef]

104. Monniaux, M.; Hay, A. Cells, walls, and endless forms. *Curr. Opin. Plant Biol.* **2016**, *34*, 114–121. [CrossRef] [PubMed]
105. Liao, H.; Tang, R.; Zhang, X.; Luan, S.; Yu, F. FERONIA Receptor Kinase at the Crossroads of Hormone Signaling and Stress Responses. *Plant Cell Physiol.* **2017**, *58*, 1143–1150. [CrossRef]
106. Zhao, C.; Zayed, O.; Yu, Z.; Jiang, W.; Zhu, P.; Hsu, C.-C.; Zhang, L.; Tao, W.A.; Lozano-Durán, R.; Zhu, J.-K. Leucine-rich repeat extensin proteins regulate plant salt tolerance in *Arabidopsis*. *Proc. Natl. Acad. Sci. USA* **2018**, *115*, 13123–13128. [CrossRef]
107. Zhong, R.; Lee, C.; Ye, Z.-H. Evolutionary conservation of the transcriptional network regulating secondary cell wall biosynthesis. *Trends Plant Sci.* **2010**, *15*, 625–632. [CrossRef]
108. Du, Q.; Wang, H. The role of HD-ZIP III transcription factors and miR165/166 in vascular development and secondary cell wall formation. *Plant Signal. Behav.* **2015**, *10*, e1078955. [CrossRef]
109. Zhong, R.; Ye, Z.-H. MYB46 and MYB83 Bind to the SMRE Sites and Directly Activate a Suite of Transcription Factors and Secondary Wall Biosynthetic Genes. *Plant Cell Physiol.* **2012**, *53*, 368–380. [CrossRef] [PubMed]
110. Endler, A.; Persson, S. Cellulose Synthases and Synthesis in *Arabidopsis*. *Mol. Plant* **2011**, *4*, 199–211. [CrossRef]
111. Wang, L.; Hart, B.E.; Khan, G.A.; Cruz, E.R.; Persson, S.; Wallace, I.S. Associations between phytohormones and cellulose biosynthesis in land plants. *Ann. Bot.* **2020**, *126*, 807–824. [CrossRef]
112. Liu, C.Y.; Rao, X.L.; Li, L.G.; Dixon, R.A. Abscisic acid regulates secondary cell-wall formation and lignin deposition in *Arabidopsis thaliana* through phosphorylation of NST1. *Proc. Natl. Acad. Sci. USA* **2021**, *118*, e2010911118. [CrossRef] [PubMed]
113. Liu, Q.; Luo, L.; Zheng, L. Lignins: Biosynthesis and Biological Functions in Plants. *Int. J. Mol. Sci.* **2018**, *19*, 335. [CrossRef] [PubMed]
114. Endler, A.; Schneider, R.; Kesten, C.; Lampugnani, E.R.; Persson, S. The cellulose synthase companion proteins act non-redundantly with CELLULOSE SYNTHASE INTERACTING1/POM2 and CELLULOSE SYNTHASE 6. *Plant Signal. Behav.* **2016**, *11*, e1135281. [CrossRef]
115. Liu, Z.; Schneider, R.; Kesten, C.; Zhang, Y.; Somssich, M.; Zhang, Y.; Fernie, A.R.; Persson, S. Cellulose-Microtubule Uncoupling Proteins Prevent Lateral Displacement of Microtubules during Cellulose Synthesis in *Arabidopsis*. *Dev. Cell* **2016**, *38*, 305–315. [CrossRef]
116. Kesten, C.; Wallmann, A.; Schneider, R.; McFarlane, H.E.; Diehl, A.; Khan, G.A.; Van Rossum, B.-J.; Lampugnani, E.R.; Szymanski, W.G.; Cremer, N.; et al. The companion of cellulose synthase 1 confers salt tolerance through a Tau-like mechanism in plants. *Nat. Commun.* **2019**, *10*, 1–14. [CrossRef]
117. Chun, H.J.; Baek, D.; Cho, H.M.; Lee, S.H.; Jin, B.J.; Yun, D.-J.; Hong, Y.-S.; Kim, M.C. Lignin biosynthesis genes play critical roles in the adaptation of *Arabidopsis* plants to high-salt stress. *Plant Signal. Behav.* **2019**, *14*, 1625697. [CrossRef]
118. Yan, J.; Liu, Y.; Yang, L.; He, H.; Huang, Y.; Fang, L.; Scheller, H.V.; Jiang, M.; Zhang, A. Cell wall β-1,4-galactan regulated by the BPC1/BPC2-GALS1 module aggravates salt sensitivity in *Arabidopsis thaliana*. *Mol. Plant* **2021**, *14*, 411–425. [CrossRef]
119. Geng, G.; Lv, C.; Stevanato, P.; Li, R.; Liu, H.; Yu, L.; Wang, Y. Transcriptome Analysis of Salt-Sensitive and Tolerant Genotypes Reveals Salt-Tolerance Metabolic Pathways in Sugar Beet. *Int. J. Mol. Sci.* **2019**, *20*, 5910. [CrossRef]
120. Song, Q.; Joshi, M.; Joshi, V. Transcriptomic Analysis of Short-Term Salt Stress Response in Watermelon Seedlings. *Int. J. Mol. Sci.* **2020**, *21*, 6036. [CrossRef]
121. Xiong, Y.; Yan, H.; Liang, H.; Zhang, Y.; Guo, B.; Niu, M.; Jian, S.; Ren, H.; Zhang, X.; Li, Y.; et al. RNA-Seq analysis of *Clerodendrum inerme* (L.) roots in response to salt stress. *BMC Genom.* **2019**, *20*, 1–18. [CrossRef]
122. Tian, X.; Wang, Z.; Zhang, Q.; Ci, H.; Wang, P.; Yu, L.; Jia, G. Genome-wide transcriptome analysis of the salt stress tolerance mechanism in *Rosa chinensis*. *PLoS ONE* **2018**, *13*, e0200938. [CrossRef] [PubMed]
123. Cai, Z.-Q.; Gao, Q. Comparative physiological and biochemical mechanisms of salt tolerance in five contrasting highland quinoa cultivars. *BMC Plant Biol.* **2020**, *20*, 1–15. [CrossRef] [PubMed]

International Journal of
Molecular Sciences

MDPI

Review

Regulation of Reactive Oxygen Species and Antioxidant Defense in Plants under Salinity

Mirza Hasanuzzaman [1,*], Md. Rakib Hossain Raihan [1], Abdul Awal Chowdhury Masud [1], Khussboo Rahman [1], Farzana Nowroz [1], Mira Rahman [1], Kamrun Nahar [2] and Masayuki Fujita [3,*]

1 Department of Agronomy, Sher-e-Bangla Agricultural University, Dhaka 1207, Bangladesh; rakib.raihan1406185@sau.edu.bd (M.R.H.R.); chy.masud3844@sau.edu.bd (A.A.C.M.); khussboorahman1305594@sau.edu.bd (K.R.); farzana.nowroz@sau.edu.bd (F.N.); mirarahman73@gmail.com (M.R.)

2 Department of Agricultural Botany, Sher-e-Bangla Agricultural University, Dhaka 1207, Bangladesh; knahar84@yahoo.com

3 Laboratory of Plant Stress Responses, Faculty of Agriculture, Kagawa University, Miki-cho 761-0795, Japan

* Correspondence: mhzsauag@yahoo.com (M.H.); fujita@ag.kagawa-u.ac.jp (M.F.)

Abstract: The generation of oxygen radicals and their derivatives, known as reactive oxygen species, (ROS) is a part of the signaling process in higher plants at lower concentrations, but at higher concentrations, those ROS cause oxidative stress. Salinity-induced osmotic stress and ionic stress trigger the overproduction of ROS and, ultimately, result in oxidative damage to cell organelles and membrane components, and at severe levels, they cause cell and plant death. The antioxidant defense system protects the plant from salt-induced oxidative damage by detoxifying the ROS and also by maintaining the balance of ROS generation under salt stress. Different plant hormones and genes are also associated with the signaling and antioxidant defense system to protect plants when they are exposed to salt stress. Salt-induced ROS overgeneration is one of the major reasons for hampering the morpho-physiological and biochemical activities of plants which can be largely restored through enhancing the antioxidant defense system that detoxifies ROS. In this review, we discuss the salt-induced generation of ROS, oxidative stress and antioxidant defense of plants under salinity.

Keywords: abiotic stress; antioxidant defense; climate change; hydrogen peroxide; lipid peroxidation; oxidative stress; phytohormones; stress signaling

Citation: Hasanuzzaman, M.; Raihan, M.R.H.; Masud, A.A.C.; Rahman, K.; Nowroz, F.; Rahman, M.; Nahar, K.; Fujita, M. Regulation of Reactive Oxygen Species and Antioxidant Defense in Plants under Salinity. *Int. J. Mol. Sci.* **2021**, *22*, 9326. https://doi.org/10.3390/ijms22179326

Academic Editor: Abir U. Igamberdiev

Received: 5 August 2021
Accepted: 26 August 2021
Published: 28 August 2021

Publisher's Note: MDPI stays neutral with regard to jurisdictional claims in published maps and institutional affiliations.

1. Introduction

Abiotic stresses are closely associated with climate change, and they hamper the growth and development of plants; consequently, the yield and quality of crops are also negatively affected. Therefore, the sustainability of global agricultural production is threatened by abiotic stresses [1,2]. Salt stress is one of the detrimental abiotic stresses which greatly reduces crop growth and productivity [3]. Around the world, about 20–50% of irrigated land areas are affected by salt stress [4]. An increase in the salinity level in plant growing media leads to an increase in endogenous sodium (Na^+) and chloride (Cl^-) contents. Both Na^+ and Cl^- ions can create life-threatening conditions for plants, but between them, Cl^- is more dangerous [5]. Salinity-induced osmotic stress, ionic stress and nutrient imbalance as well as their secondary effects altogether lead to the overgeneration of reactive oxygen species (ROS) [5,6].

Oxygen radicals and their derivatives, called ROS, are produced by different cellular metabolisms in various cellular compartments. The major ROS are hydrogen peroxide (H_2O_2), superoxide ($O_2^{\bullet-}$), singlet oxygen ($^1O_2$) and the hydroxyl radical ($^{\bullet}OH$) [2,7]. Although ROS production is a general phenomenon and a part of cellular metabolism in plants, environmental stresses lead to excess generation of ROS which are not only highly reactive but also toxic in nature and damage lipids, proteins, carbohydrates and DNA [2,8]. Higher concentrations of ROS have an injurious effect on cell organelles and tissues of the shoots and roots [9]. Recent studies revealed that ROS have a dual role in plants [10–12]. However, whether they will act as a signaling molecule or as a stressor depends on the crucial equilibrium between the generation of ROS and their scavenging. Due to disruption of the equilibrium between ROS production and antioxidant defense, oxidative stress occurs under abiotic stresses (including salinity). Non-enzymatic as well as enzymatic components of the antioxidant defense system scavenge or detoxify the excessive ROS which mitigates the negative effect of oxidative stress [2,13]. The most investigated major components of the antioxidant defense system are superoxide dismutase (SOD), peroxidase (POD/POX), catalase (CAT), the ascorbate-glutathione (AsA-GSH) cycle enzymes (ascorbate peroxidase, APX; monodehydroascorbate reductase, MDHAR; dehydroascorbate reductase, DHAR; glutathione reductase, GR), peroxiredoxins (PRX), glutathione peroxidases (GPX) and glutathione *S*-transferases (GST), which act in reducing ROS under abiotic stress including salt stress [2,14]. Along with enzymatic components, non-enzymatic components such as GSH, ascorbic acid (Asc), α-tocopherol, flavonoids, carotenoids, alkaloids, phenolic acids and also non-protein amino acids play a vital role in protecting the plant from ROS-induced oxidative stress and in enhancing the tolerance against stress [2]. The amount of ROS at the cellular level determines the destructive or signaling roles of the ROS [2,15]. Moreover, ROS are associated with several processes in plants such as germination and root, shoot, flower and seed development [16].

Most of the recent investigations focused on physiological, biochemical and molecular approaches to enhance salt tolerance in plants by mitigating ROS-induced oxidative stress [17–19]. The application of many chemical elicitors and biostimulants improves the plant response to salt stress which results in a reduction in the excessive accumulation of ROS [3,19–22]. Moreover, overexpression of several transgenes has been proved to enhance the gas exchange activity, photosynthetic activity, biosynthesis of photosynthetic pigments, antioxidant components and abscisic acid (ABA) biosynthesis, stimulate signaling of hormones and also improve ion homeostasis which altogether improve ROS metabolism and salt tolerance [19]. This review focuses on the generation and consequence of ROS production and the role of the antioxidant defense system under salt stress. We also discuss the hormonal and gene regulation associated with ROS metabolism under salinity.

2. Types of Reactive Oxygen Species

Oxygen exists in the atmosphere as a free molecule (O_2), and the ground state of oxygen (triplet oxygen, $^3O_2$) has two unpaired parallel spin electrons with equal spin numbers which are not reactive in nature [23]. In the aerobic respiration process of plants, the oxygen molecule is the primary acceptor of electrons and is involved in fundamental metabolic and cellular functions such as membrane-linked ATP formation. However, when $^3O_2$ gains extra energy from metabolic processes (biochemical reactions, electron transport chains, ETC, etc.), it overcomes the spin restriction and converts $^3O_2$ into ROS [7,23].

Reactive oxygen species produced in plants are composed of both free radicals and non-radicals [2,13]. The common cellular ROS radicals are $O_2^{\bullet -}$, $^{\bullet}OH$, the perhydroxy radical ($HO_2^{\bullet}$), peroxyl ($RO_2^{\bullet}$), carbonate ($CO_3^{\bullet -}$), semiquinone ($SQ^{\bullet -}$), the alkoxy radical ($RO^{\bullet}$) and the peroxy radical ($ROO^{\bullet}$). The non-radical ROS of plant cells are H_2O_2, $^1O_2$, organic hydroperoxide (ROOH), ozone (O_3), hypoiodous acid (HOI), hypobromous acid (HOBr) and hypochlorous acid (HOCl) (Figure 1; [2,24]).

Figure 1. Types of reactive oxygen species (ROS) in plant cells.

In plant cells, ROS production occurs due to incomplete or partial reduction of oxygen molecules. Therefore, ROS production is a general phenomenon because they are produced as a result of the oxidation–reduction reaction of several metabolic processes in multiple locations and compartments of plant cells. However, abiotic stress including salinity increases ROS generation which exceeds the equilibrium between antioxidant defense and ROS production [2,7]. In plant cells, a prominent source of ROS production is spilling electrons from the ETC of photosynthesis and respiration, transition metal ion decompartmentalization and redox reactions [7]. Several reactions associated with ROS production are depicted in Figure 2.

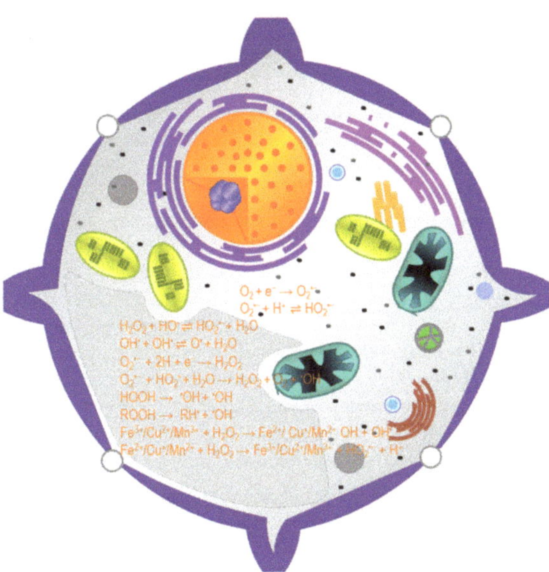

Figure 2. Reactions associated with reactive oxygen species production in plants.

3. ROS Generation in Plant Cells

Reactive oxygen species are the byproduct of aerobic metabolism in different cell organelles such as chloroplasts, mitochondria, peroxisomes, plasma membranes and cell wall (Figure 3 [24,25]). A specific ROS generation in a cell is highly localized, and different pathways are intensively involved in this process [26,27]. For a better understanding of ROS scavenging tactics in different subcellular compartments, at first, it is obligatory to study the subcellular compartment-specific ROS generation in cells.

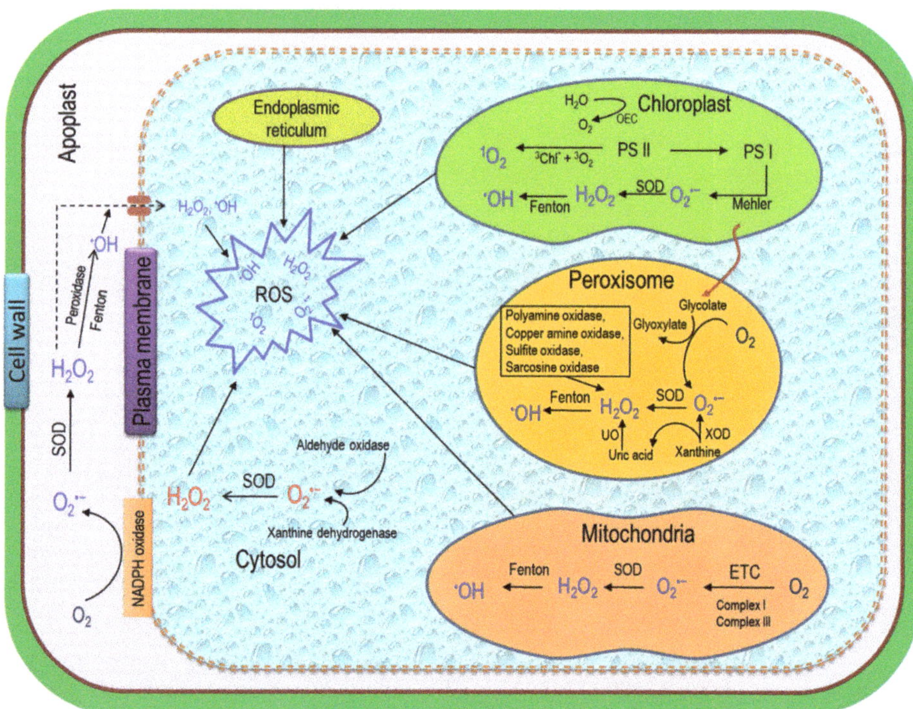

Figure 3. ROS generation process and localization in plant cells. In different cell organelles, ROS are produced through metabolic reactions where different enzymatic and non-enzymatic pathways are involved. ROS—reactive oxygen species; H_2O_2—hydrogen peroxide; $^1O_2$—singlet oxygen; ETC—electron transport chain; $^\bullet OH$—hydroxyl radical; $^3Chl^*$—triplet chlorophyll; PS I—photosystem I; PS II—photosystem II; $O_2^{\bullet-}$—superoxide anion; XOD—xanthine oxidase; SOD—superoxide dismutase; NADPH—nicotinamide adenine dinucleotide phosphate; UO—urate oxidase [24].

Rigid cell wall formation is accelerated under stress conditions when ROS along with POD trigger the polymerization of glycoproteins and phenolic compounds [28]. This cell wall-related POD catalyzes H_2O_2 in the presence of NADH, where the NADH is derived from malate dehydrogenase [28]. In addition, diamine oxidases reduce diamines or polyamines (PAs) to quinine and thus produce ROS in the cell wall [29].

Free radicals are also produced in the plasma membrane under abiotic stress. The higher generation of $O_2^{\bullet-}$ in the plasma membrane is reconciled by NADPH oxidase and quinine reductase where NADPH oxidase transports the electron from the cytoplasmic NADPH and forms $O_2^{\bullet-}$ which, later on, is converted to H_2O_2 [30].

Chloroplasts are considered as the prime site for ROS production that relies on the interactions of chlorophyll (chl) and light [28,31]. Under stress conditions, stomatal conductance (g_s) and the CO_2 assimilation rate are greatly reduced and thus form excited triplet chlorophyll ($^3Chl^*$) that impedes in photosynthetic ETC, and promote ROS overgeneration [32,33]. Photosystems I and II (PS I and PS II) are the major sites in chloroplasts where

$^1O_2$ and $O_2^{\bullet-}$ are largely produced [2]. However, the amount of $O_2^{\bullet-}$ produced in PS I through the Mehler reaction is converted to H_2O_2 with the help of SOD.

In the non-green parts of plants, mitochondria are the major site for ROS generation [2]. ROS produced in mitochondria reduce both mitochondrial energy transportation and other subcellular functions. Respiratory complexes I and III are the main sources of mitochondrial ROS, especially $O_2^{\bullet-}$. However, the produced $O_2^{\bullet-}$ in both complexes due to electron leakage is eventually catalyzed by Mn-SOD and Cu-Zn-SOD and produces H_2O_2 [33,34].

Another vital site for ROS production is the peroxisomes, where a number of oxidases catalyze different reactions and generate H_2O_2 and $O_2^{\bullet-}$ as byproducts. It is considered that glycolate oxidase (GOX) is the main source of ROS production in peroxisomes [35]. This GOX in peroxisomes causes stomatal closure; as a result, the stomatal gas exchange rate is greatly reduced and thus reduces CO_2 for RuBisCO generation, causing photorespiration and H_2O_2 production [36]. In addition, xanthine oxidase (XOD) activity can also generate $O_2^{\bullet-}$ and uric acid in the peroxisomal matrix, which are further dismutased to H_2O_2 by metalloenzymes SOD and urate oxidase, respectively [37,38].

In the endoplasmic reticulum (ER), both $O_2^{\bullet-}$ and H_2O_2 are produced from the fatty acid oxidation by GOX and urate oxidase activities [39]. In addition, a small amount of $O_2^{\bullet-}$ is generated in the ER as a byproduct of oxidation and hydroxylation processes which involve cytochrome P450 and cytochrome P540 reductase in the presence of reduced NADPH [27].

Compared to other cell compartments, the ROS production rate is comparatively lower in the cytosol where the redox balance is highly maintained by cytoplasmic NADPH as a central component. However, besides ROS production, the cytosol conducts a pivotal role in the redox signaling process in plant cells. In general, the ROS signaling from different cell organelles passes through the cytoplasm to modulate gene expression in the cell nucleus [40].

Different enzymes contribute to the generation of ROS in apoplasts, among which quinine reductase, NADPH oxidase, SOD and POD are the most important [10,12].

4. Consequence of ROS in Plant Cells

Overproduction of ROS disrupts the equilibrium between ROS accumulation and scavenging that ultimately damages different cellular biomolecules through protein oxidation, lipid peroxidation, enzyme inactivation, chl degradation and destruction of nucleic acids under stressful conditions (Figure 4; [8,13,41]). However, various factors influence the extent of biomolecular damage which include the concentration of the biomolecule(s), the location of the target biomolecule(s), the site of ROS generation and the reaction rate between the target biomolecule(s) and ROS [13].

The lipid membrane of cells is damaged by oxidative burst, resulting in lipid peroxidation. The extent of peroxidation differs quantitatively between underground and aboveground plant tissues depending on the types of ROS, which is measured by the content of the final product, malondialdehyde (MDA) [29,42]. However, lipid peroxidation not only hampers membrane permeability but also causes electrolyte leakage (EL) and deactivation of enzymes and receptors, as well as accelerating the oxidation process of nucleic acids and proteins [43].

It is well documented that ROS are involved in protein oxidation. However, not all ROS attack all proteins; rather, they cause protein denaturation in a selective manner, among which $^{\bullet}OH$ is the most notorious in nature, causing damage to protein molecules non-selectively [2]. Oxidation of proteins can be both an irreversible and reversible process. Enzymes that contain Fe-S in the center can be damaged irreversibly by the $O_2^{\bullet-}$ radical, and such type of damage causes cellular dysfunction. On the contrary, glutathionylation and *S*-nitrosylation are reversible changes that can mediate the redox regulation in plants [44].

Nucleic acids are the structural components of proteins and DNA which are rapidly oxidized by the action of ROS. Both chloroplastic and mitochondrial DNA are largely oxidized, compared to that of nucleus DNA, due to their proximity to the ROS production site. Among free radicals, $^{\bullet}OH$ is the most pernicious damaging radical for DNA that modifies nucleotide bases (purine and pyrimidine) by oxidizing the sugar residues in the DNA strands [45]. The DNA replication or transcription process is permanently ceased if the damage induced by ROS is completely irreparable [46]. Consequently, many biochemical processes such as irregular protein synthesis and damage of photosynthetic proteins are directly arrested by DNA damage. In addition, signal transduction, transcription, replication errors and, as a whole, genetic instability are the common fate of DNA due to oxidative stress [13].

In plants, the glycolysis pathway and the enzymes of the TCA cycle are the first targets of a free radical attack. For instance, glyceraldehyde 3-phosphate dehydrogenase and fructose-1,6-bisphosphate aldolase are two common enzymes of the pentose phosphate pathway which are inhibited while ROS production is increased due to oxidative stress [2,47].

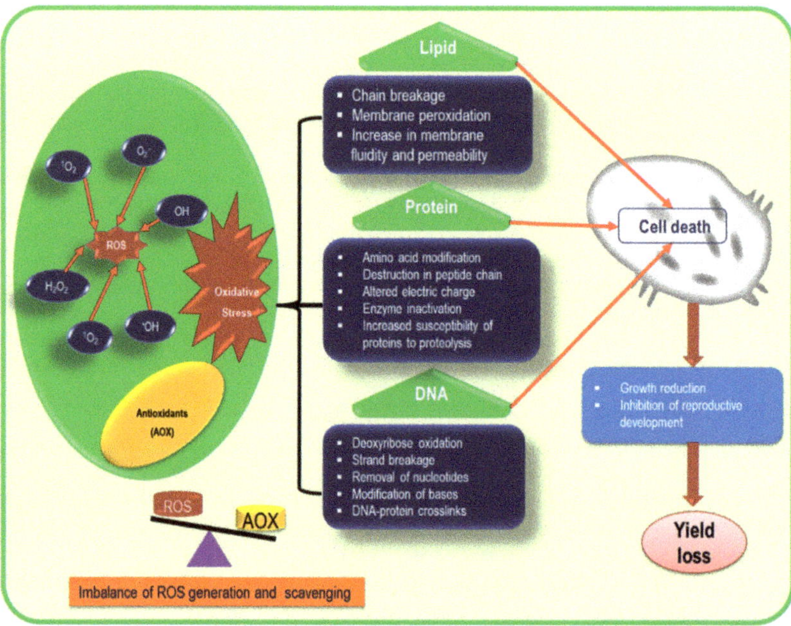

Figure 4. Consequences of oxidative stress on different cellular mechanisms. The imbalance between ROS production and scavenging creates oxidative stress in plants under abiotic stresses. As a consequence, different molecular and cellular damages occur that ultimately cause cell death. ROS—reactive oxygen species; AOX—antioxidants; H_2O_2—hydrogen peroxide; $^{\bullet}OH$—hydroxyl radical; $O_2^{\bullet -}$—superoxide anion; $^1O_2$—singlet oxygen.

5. Plant Responses to Salinity

Plants are posed with salt stress through two mechanisms. One is osmotic stress, which is a rapid mechanism (within minutes to days), hinders the water uptake and is responsible for stomatal closure, ultimately reducing cell expansion and division. Additionally, the other mechanism, which is slower (days to weeks), is ionic toxicity, which creates an ionic imbalance, disrupts ionic homeostasis and cellular functions and also causes premature senescence. Salt stress is detrimental to plants during the germination, growth and development stages and results in significant yield reduction (Figure 5). Excess accumulation of the salt concentration initiates changes in the mineral distribution, vulnerability of the cell membrane, loss of integrity and a reduced turgor pressure as a result of ionic disequilibrium [48], and, in the extreme case, salt stress results in plant death [14].

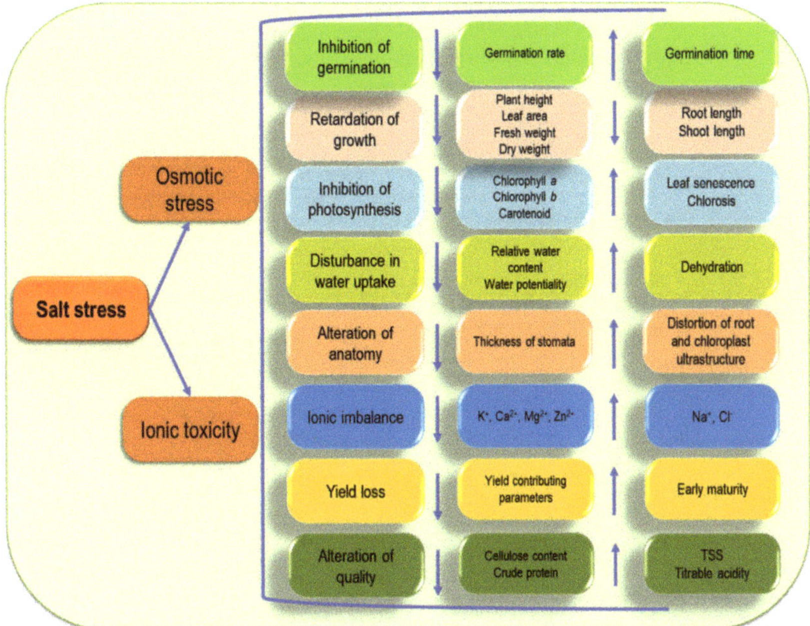

Figure 5. An overview of salt stress-induced changes in plants. TSS—total soluble solids.

Salinity imposes both osmotic stress and ionic toxicity on plants by impairing water uptake, stomatal opening and the ionic balance. As a result, the morphological, physiological and anatomical characteristics of plants start to show changes negatively and finally cause yield loss.

5.1. Germination

Salinity reduces the germination of seeds by its osmotic properties and toxicity mechanism at the time of germination. Moreover, seedling establishment is also greatly hampered in salinity as it impedes imbibition, disturbs active metabolism and hinders embryonic tissue development [49]. In *Oryza sativa*, salt stress can cause an adverse effect on the germination stage by inhibiting gibberellic acid (GA) activity in the seed, which, in turn, can curtail seed germination by up to 71% [50]. A similar trend of decline was reported by Shu et al. [51], where it was explained that salt stress favors ABA synthesis while inhibiting the GA synthesis pathway. This phenomenon creates an imbalance in the GA/ABA ratio and in turn, the GA content is decreased, which is beneficial for the seed germination process [52]. During germination, a higher concentration of salt causes osmotic stress in water-deficient conditions that can reduce the *Triticum aestivum* germination rate [53]. Moreover, high salinity has the potential to increase the mean germination time while decreasing the germination rate and, subsequently, lower the germination percentage of *Helianthus annuus* L. [54].

5.2. Growth

Salt stress hampers cell prolongation in growing tissue, which, in turn, reduces the leaf area and dry matter assimilation in the plant [5]. In the case of long-term salinity, the plant exhibits a lower photosynthesis rate, lower nutrient storage and less growth hormones [55]. Growth reduction upon salt exposure of seedlings can be found in the reduced root and shoot weights of *Solanum lycopersicum*. Additionally, the damage is more prominent in the root fresh weight (FW) and dry weight (DW) by 40 and 35%, respectively, compared to the shoot FW and DW by 19 and 29%, respectively [56]. Accumulation of Na^+ and Cl^- in leaf tissue modified plant growth hormones, enzyme activity, stomatal closure and photosynthetic activity that resulted in lower assimilation of CO_2 and, finally, a reduced plant height and DW under salinity [57].

5.3. Photosynthesis

Salinity generates an unfavorable condition for plant photosynthesis by affecting g_s, sap flow and the transpiration rate (T_r) through accumulating higher concentrations of NaCl within plant cells [58]. Shortage of the net photosynthesis rate (P_n), T_r and g_s resulted in a reduction in the chl *a*, chl *b* and carotenoid contents in *S. lycopersicum*, and, finally, photosynthesis was hampered to a great extent [59]. A decline in the chl *a*, chl *b* and carotenoid contents has also been recorded in *T. aestivum* [60] and *S. lycopersicum* [61] along with a reduced chl *a*+*b* content. In the salt stress condition, the plant develops an increased chl *a*/*b* ratio compared to the stress-free condition, and this imbalanced condition triggers reduced photosynthesis in plants [62]. Importantly, salinity can develop a physiological drought condition in the plant, which acts on stomatal closure and lessens the photosynthetic CO_2 assimilation as well [63]. Plants exhibit chl degradation even at short exposure to salinity and can reach a high severity level with the prolongation of stress [64].

5.4. Ionic Imbalance

Ionic imbalance or toxicity is known as a secondary effect of salt stress that causes disturbance in the plant life cycle. Upon salt exposure, imbalance of nutrition is presumed as a primary phenomenon as it can hamper the supply, uptake and translocation of nutrients in plants [65]. Salinity has the capacity to depolarize the root plasma membrane while inducing the guard cell outward-rectifying K channels, which, in turn, results in a higher Na^+ content with a lower K^+ content in the plant [66]. Excess Na^+ increased the Na^+/K^+ ratio while decreasing calcium (Ca^{2+}) and Mg^{2+} in both the roots and shoots of *S. lycopersicum* seedlings [56]. A similar trend was recorded in *Lens culinaris* [64] and *Vigna radiata* [67]. An excess Na^+ concentration causes disturbance in plant metabolic activities by replacing the K^+ content from the enzymatic components of the cell [68]. Moreover, lower uptake of K^+, Ca^{2+} and Mg^{2+}, caused by excess Na^+ accumulation, subsequently induces an imbalance in mineral homeostasis and ionic stress in the plant [61,69]. Rahman et al. [70] found that under salt stress, ion homeostasis was hampered in *O. sativa* seedlings by the increment in Na^+ and the Na^+/K^+ ratio, compared to the stress-free condition. In this study, the shoots exhibited a higher concentration of Na^+ than the roots. Along with the reduction in K^+, Ca^{2+} and Mg^{2+}, salt stress also negatively affected the zinc (Zn) content, both in the roots and shoots of the plant.

5.5. Water Relation

At a low level of salinity, the plant may not show a difference in water uptake, but at high levels of salt stress, the plant, especially its shoot, is found to be injured greatly with a reduced relative water content (RWC) and a dehydrated condition at the cellular level [71]. *S. lycopersicum* was recorded to be affected by this osmotic stress under salinity, expressed through a reduced RWC, which was also found to be restored after the recovery period [61]. In addition, nutrient imbalance is a common phenomenon in salt stress conditions, caused by reduced water uptake and translocation, as a consequence of accumulated Na^+ and Cl^- ions in the cytoplasm [72]. A reduction in the RWC was also reported in *T. aestivum* [60], *O. sativa* [70,73] and *Sorghum bicolor* [74].

5.6. Anatomical Modifications

Besides morphological and physiological changes, salinity may also alter the anatomical characteristics of a plant through modifying the cell wall and nucleus components and leaf structure, especially the ultrastructure of leaf chloroplasts. Due to the close proximity to the saline-affected soil, salt-induced roots were seen to be hampered by the disrupted nuclei and nuclear membrane of *T. aestivum* [75]. Similar results were recorded in *S. lycopersicum* [76], where salinity changed the root structure through decreasing different layers of the columella, cortex cells and cell sizes and enlarging cell nucleoli, ultimately leading to transforming the cell shape and vacuoles. Interestingly, salt-induced *Brassica napus* roots were seen to produce an apoplastic barrier near the root apex to moderate the over-accumulated toxic ions [77]. Salinity causes an alteration in leaf structure by reducing the thickness of the epidermis and mesophyll and, consequently, causes disturbance in water uptake and turgidity [78]. The stomatal frequency was found to be reduced by 10.5 and 22% in *B. juncea* L. [79] and *T. aestivum* [80], respectively, at the salt-induced condition, with a partially closed stomatal aperture, compared to the stress-free condition. Upon salt exposure, the chloroplast ultrastructure was recorded to change by distorting thylakoids and compressing granular and stroma lamellae with massive plastoglobuli, subsequently resulting in a reduced photosynthetic pigment content of the plant [81].

5.7. Crop Yield

Salt stress causes alterations in the morphological and physiological characteristics of plants and, subsequently, reduces the yield to a great extent. For example, upon exposure to salt, the yield-contributing characteristics of *H. annuus*, e.g., head diameter, 100-seed weight and oil percentage, were found to be reduced 24, 28 and 5%, respectively, with a severe (26%) seed yield reduction [82]. A similar trend was found in *Hordeum vulgare*, where the tiller number, spike length and 100-grain weight were reduced by 53, 40 and 41%, respectively, in saline conditions [83]. Apart from this, the salt-induced condition interrupts the fertilization process at the time of grain filling through reducing pollen viability and stigma receptivity along with providing minimized photoassimilates. This incident ultimately leads to a yield reduction in the grains [84].

5.8. Crop Quality

Salinity causes variation in the qualitative characteristics of plants such as the sugar, citric acid, cellulose and oil contents according to the level of concentration. Upon salt exposure, the cellulose content and sucrose movement are greatly hampered [85]; as cellulose deposition is known as the prime component of fiber quality, this reduction affects the fiber quality in a dose-dependent manner. Salt-induced *Mentha piperita* (a beneficial medicinal and aromatic plant) was seen to lose quality in respect to essential oil and menthol contents compared to the control condition as salinity hampers its growth, photosynthesis and nutrient content to a great extent [86]. Salt-induced *S. lycopersicum* performed positively in respect to electrical conductivity (EC), total soluble solids (TSS), titratable acidity, citric acid content and oxide reduction potential (ORP) [62,87]. As plants produce more soluble sugars and organic acids to compete with the ionic toxicity, generated from Na^+ and Cl^- ions, this can be the reason behind the increment in EC, TSS and titratable acidity in *S. lycopersicum* [88]. Moreover, in saline conditions, plants exhibit early maturation, which promotes sugar accumulation within the fruit and, subsequently, increases TSS [89].

6. Salinity-Induced Oxidative Stress in Plants

When plants are exposed to salinity, the concentration of Na^+ and Cl^- ions increases abundantly in the soil, which are accumulated at a higher rate, thus reducing the essential ions in plants. Hereby, it disrupts the plant–water relation by creating drought-like conditions, resulting in osmotic stress which is liable for the reduction in g_s and photosynthetic enzyme activities, leading to ROS generation in plants [90]. Besides this, it also triggers ionic stress by inhibiting K^+ accumulation and interrupting the nutrient balance in plants; thus, by altering redox homeostasis, salinity hampers the flow of electrons from central transport chains to oxygen reduction pathways in different organelles, which leads to the overgeneration of ROS in plants [14]. Other activities such as carbon metabolism, ion redistribution, ABA accumulation and alkalization are also interfered with upon exposure to salinity and foster ROS generation in plants (Figure 6; [91]).

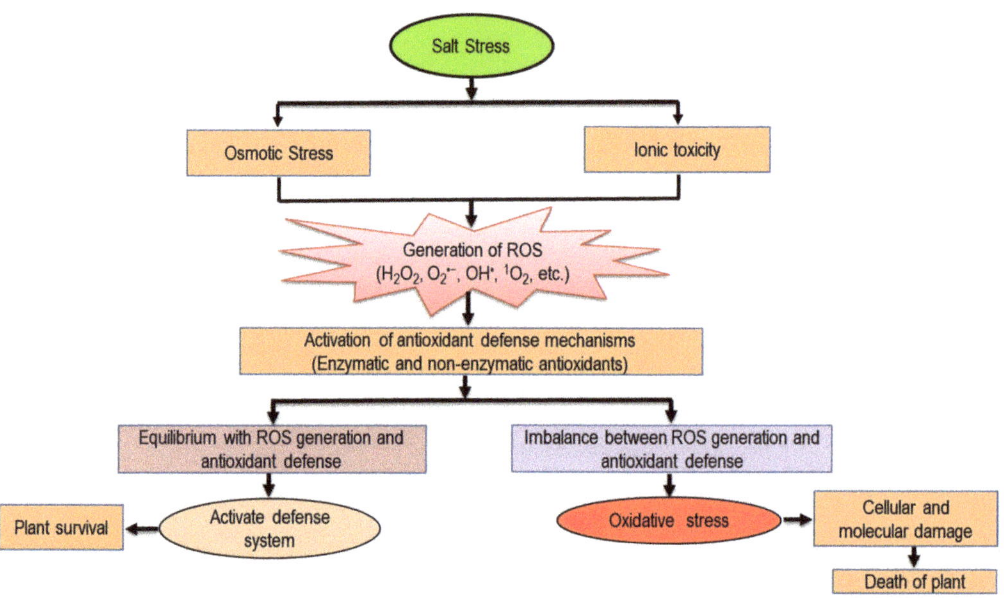

Figure 6. Oxidative stress and antioxidant defense under salinity.

These overproduced ROS are accountable for the damage of nucleic acids and oxidation of different biomolecules, i.e., proteins, lipids, carbohydrates and DNA. Thus, fluctuations occur in their functions and properties, leading to physiological and biochemical alterations and, ultimately, creating oxidative stress in plants, Table 1 [6,10]. Mohsin et al. [60] reported that when *T. aestivum* seedlings are exposed to salt stress at the rate of 150 mM NaCl (mild) and 250 mM NaCl (severe) for 5 d, MDA and H_2O_2 contents were increased by 63 and 116%, and 78 and 108%, respectively, compared to the unstressed seedlings. Similarly, Siddiqui et al. [92] showed increased MDA (by 116%), H_2O_2 (by 198%) and $O_2^{\bullet-}$ (by 263%) under salt-stressed (100 mM NaCl) seedlings of *S. lycopersicum*, while increased H_2O_2 by 58 and 97% and MDA by 74 and 113% were found under 100 mM and 160 mM NaCl stress, respectively, compared to the controls [56]. Recently, a contrasting response was observed by Ali et al. [93], where the MDA content was decreased but H_2O_2 was increased upon exposure to salt stress (150 mM NaCl) in *S. lycopersicum* plants. The extent of damage caused by salinity-induced oxidative stress could be varied among different portions of plants. In most cases, the roots suffer more compared to other tissues as they are directly in contact with the saline conditions, followed by the shoots and leaves [94]. Upon exposure to 100 mM NaCl, a greater increase in the MDA content (by 116%) was observed in the roots than the leaves (by 106%), whereas the H_2O_2 content was greater in the leaves (by 149%) than the roots (by 34%), compared to untreated *S. bicolor* plants [95]. Similarly, Jabeen et al. [96] observed a higher amount of MDA content in the roots (44%) than the leaves (38%) compared to the control when *Glycine max* plants were exposed to 100 mM NaCl. Derbali et al. [97] evaluated four genotypes of *Chenopodium quinoa* (cvs. Tumeko, Red Faro, Kcoito and UDEC-5) under different doses of salt (100, 300 and 500 mM NaCl) and reported that both the MDA and H_2O_2 contents increased in a dose-dependent manner in these four genotypes. However, the lowest accumulation of MDA and H_2O_2 was observed in the salt-resistant genotype (cv. UDEC-5) compared to the salt-sensitive genotypes (cvs. Tumeko, Red Faro, Kcoito) at different NaCl levels. Derbali et al. [98]

further reported that the post-stress recovery capacity was higher in the salt-resistant genotype of *C. quinoa* (cv. UDEC-5) compared to the salt-sensitive genotype (cv. Kcoito). From the above-mentioned examples, it can be stated that plant responses against salinity vary among species, and also different genotypes/varieties of the same species.

Table 1. Oxidative stress in plants under salinity.

Plant Species	Level(s) of Salt Stress	Oxidative Damage	References
T. aestivum cv. Pradip	150 and 300 mM NaCl; 4 d	Lipid peroxidation increased by 60%. H_2O_2 level increased by 73%.	[99]
B. napus cv. BINA Sharisha 3	100 and 200 mM NaCl; 2 d	MDA increased by 69 and 129%. H_2O_2 incremented by 76 and 90%.	[100]
O. sativa cvs. MI-48, IR-28 (salt-sensitive)	100 mM saline solution (mixture of NaCl, $MgCl_2$, $MgSO_4$ and $CaCl_2$ salts); 35 d	H_2O_2 and $O_2^{\bullet-}$ generation increased by 2-fold.	[101]
S. lycopersicum cv. Chibli F1	120 mM NaCl; 8 d	Lipid peroxidation elevated by 35 (leaves) and 37% (roots).	[102]
O. sativa cv. KDML105	60, 120 and 160 mM NaCl; 3 d	H_2O_2 increased in a dose-dependent manner.	[103]
G. max cv. PK9305	100 mM NaCl; 7 d	Increment in MDA found in both leaves and roots. Higher activity of LOX in roots.	[104]
Phaseolus vulgaris cv. Nebraska	2.5 and 5.0 dS m^{-1} ($NaCl/CaCl_2/MgSO_4$ = 2:2:1); 40 d	MDA and H_2O_2 increased with increasing saline concentrations. EL increased, but reduced membrane stability index (MSI).	[105]
B. juncea cv. Pusa Jai Kisan	100 mM NaCl; 30 d	Higher accumulation of H_2O_2 and thiobarbituric acid reactive species (TBARS).	[106]
S. tuberosum cv. Hui 2	50, 75 and 100 mM NaCl; 31 d	Elevation in MDA by 164%.	[107]
O. sativa cv. BRRI dhan47	200 mM NaCl; 3 d	H_2O_2 increased by 82%. Higher production of MDA. The activity of LOX increased by 78%.	[70]
P. vulgaris cvs. Tema (high-yielding) and Djadida (low-yielding)	50, 100 and 200 mM NaCl; 7 d	Increase in MDA by 44 (cv. Tema) and 56% (cv. Djadida).	[108]
O. sativa cv. BRRI dhan47	150 mM NaCl; 3, 6 d	MDA increased by 80 (3 d) and 203% (6 d). Increment in H_2O_2 found, by 74 (3 d) and 92% (6 d). Lipoxygenase (LOX) activity increased by 69 (3 d) and 95% (6 d).	[69]
V. radiata cv. BARI Mung-2	200 mM NaCl; 2 d	Upregulation of MDA and H_2O_2. The activity of LOX also increased.	[67]
Pisum sativum cv. Shubhra IM-9101	100 and 400 mM NaCl; 7 d	Increment in $O_2^{\bullet-}$ by 171–407%. H_2O_2 increased by 191–249%.	[109]
T. aestivum cvs. Kharchia local (salt-tolerant) and HD2329 (salt-sensitive)	10 dS m^{-1} NaCl; 7 d	EL increased more in cv. HD2329 (by 2.6- and 1.5-fold in the roots and shoots) than Kharchia local (by 1.9- and 1.4-fold in the roots and shoots). Higher accumulation of H_2O_2 and MDA in sensitive cultivar.	[110]

Table 1. *Cont.*

Plant Species	Level(s) of Salt Stress	Oxidative Damage	References
B. juncea cvs. CS-52 (salt-tolerant) and RH-8113 (salt-sensitive)	50, 100 and 150 mM NaCl; 21 d	MDA generated more in RH-8113 (138%) than CS-52 (126%). H_2O_2 increased in dose-dependent manner in sensitive cultivar, but at 150 mM NaCl, it increased by 33% in CS-52.	[111]
Cicer arietinum cvs. Flip 97-43c (tolerant), ICC 4958 (tolerant), Flip 97-97c (susceptible) and Flip 97-196c (susceptible)	100 mM NaCl; 3, 7 and 12 d	Accumulation of MDA began to increase after 12 d by 224% in T1 genotype. In S2 genotype, 2.21-, 8.20- and 10.15-fold upregulation of MDA content was found at 3, 7 and 12 d, respectively.	[112]
S. lycopersicum cv. Hezuo 903	150 mM NaCl; 10 d	Elevation in lipid peroxidation by 2.6-fold. H_2O_2 increased by 2.5-fold.	[113]
B. napus cv. BINA Sharisha 3	100 and 200 mM NaCl; 2 d	Gradual increase in MDA by 60–129% in a dose-dependent manner. Production of H_2O_2 elevated by 63–98%.	[114]
S. lycopersicum cv. Pusa Ruby	150 mM NaCl; 5 d	Elevation in H_2O_2 and MDA. Increased LOX activity.	[115]
Zea mays cvs. BARI hybrid Maize-7 and BARI hybrid Maize-9	150 mM NaCl; 15 d	$O_2^{\bullet-}$ and H_2O_2 increased by 130 and 99%. Higher production of MDA by 109%. LOX activity enhanced by 133%.	[116]
Luffa acutangula cv. Pusa Sneha	100 mM NaCl, 10 d	Increase in H_2O_2, $O_2^{\bullet-}$, EL, MDA and LOX activity by 140, 145, 251, 358 and 157%, respectively.	[117]
Gossypium hirsutum	150 mM NaCl; 3, 6, 9 and 12 d	Increased H_2O_2 by 58% (3 d), 34% (6 d), 45% (9 d) and 37% (12 d). $O_2^{\bullet-}$ incremented by 47% (3 d), 25% (6 d), 37% (9 d) and 32% (12 d). MDA augmented by 25% (3 d), 27% (6 d), 36% (9 d) and 41% (12 d).	[118]
T. aestivum	100 mM NaCl; 1 d	1.5- and 1.2-fold upregulation of EL and MDA.	[119]
T. aestivum cv. Norin 61	250 mM NaCl; 5 d	Increase in MDA and H_2O_2 by 62 and 35%. EL increased by 130%.	[120]
G. max cv. Giza 111	75 and 150 mM NaCl; 56 d	Level of MDA increased by 47 and 75%. H_2O_2 augmented by 42 and 50%.	[121]
Lactuca sativa cv. SUSANA	4 dS m^{-1} (low) and 8 dS m^{-1} (high) NaCl; 60 d	MDA increased by 44% under low level of NaCl, while increased under high NaCl (70 and 87%) in both seasons. H_2O_2 augmented by 183 (high) and 166% (low) in both seasons. $O_2^{\bullet-}$ increases more at high dose (208 and 262%) in both seasons.	[122]
L. culinaris cv. BARI Masur-7	110 mM NaCl; 2 d	Content of MDA and H_2O_2 increased by 164 and 229%.	[123]

7. Antioxidant Defense System in Plants under Salinity

Plants have antioxidants that are employed in scavenging ROS by working in coordination with non-enzymatic antioxidants (ascorbate (AsA), GSH, phenolic compounds, flavonoids, alkaloids, α-tocopherol, non-protein amino acids, etc.) and enzymatic antioxidants (SOD, APX, CAT, DHAR, MDHAR, GPX, GR, GST, POD/POX polyphenol oxidase, PPO, PRX, thioredoxin (TRX), etc.), and that protect them from oxidative damage (Figure 7; [13,24]).

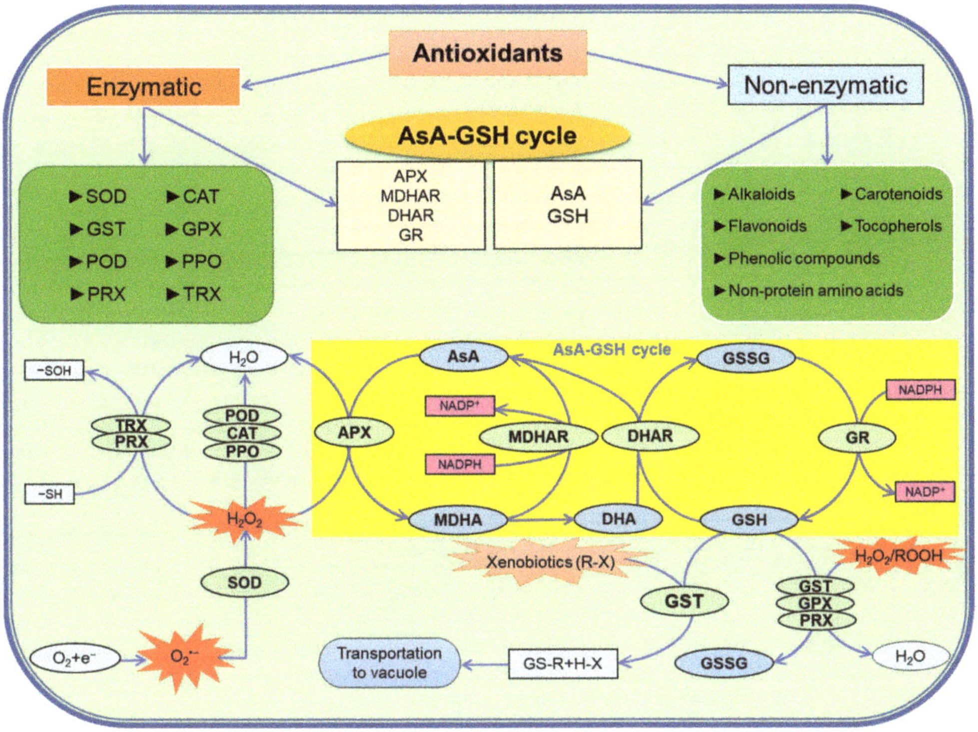

Figure 7. Overview of different types of antioxidants and their combined mechanisms [2]. Detail descriptions are provided in the text. SOD—superoxide dismutase; CAT—catalase; POX—peroxidases; AsA—ascorbate; DHA—dehydroascorbate; GSSG—oxidized glutathione; GSH—reduced glutathione; APX—ascorbate peroxidase; MDHA—monodehydroascorbate; MDHAR—monodehydroascorbate reductase; DHAR—dehydroascorbate reductase; GR—glutathione reductase; GST—glutathione *S*-transferase; GPX—glutathione peroxidase; PPO—polyphenol oxidase; PRX—peroxiredoxins; TRX—thioredoxin; NADPH—nicotinamide adenine dinucleotide phosphate; O_2—oxygen; e^-—electrons; H_2O_2—hydrogen peroxide; $O_2^{\bullet-}$—superoxide anion; R—aliphatic, aromatic or heterocyclic group; X—sulfate, nitrite or halide group; ROOH—hydroperoxides; -SH—thiolate; -SOH—sulfenic acid.

In plants, SOD is the major antioxidant that activates the first line of defense against ROS-induced damage by converting $O_2^{\bullet-}$ into H_2O_2; this is further accumulated by CAT, APX, GPX, PPO, PRX and TRX enzymes or catalyzed in the Asada–Halliwell pathway (AsA-GSH cycle) [124,125]. To detoxify ROS in the AsA-GSH cycle, plants have non-enzymatic antioxidants (AsA and GSH) and four enzymatic antioxidants (APX, DHAR, MDHAR and GR); thus, by minimizing ROS, they can maintain redox homeostasis in plants, Table 2 [2,8,24]. Furthermore, H_2O_2 and other xenobiotics are also detoxified with the help of the GST and GPX enzymes [126,127]. Many researchers have proposed that the activities of these antioxidants depend on the salinity threshold, duration of salinity exposure and growth stages of plants [128]. For instance, Ali et al. [129] found altered antioxidant enzyme activities in scavenging ROS under two concentrations of NaCl (80 mM and 160 mM), with a maximum reduction in SOD, POD, APX and CAT of 49, 43, 39 and 52% in cv. P1574, and 67, 46, 47 and 61% in cv. Hycorn-11, respectively, which are two *Z. mays* cultivars, at 160 mM NaCl stress and concluded that the P1574 cultivar is more salt-tolerant than the Hycorn-11 cultivar. Similarly, upon exposure to 100 mM NaCl, total phenols (by 60%) and AsA (by 55%) were increased in *S. lycopersicum* plants together with enhanced SOD, CAT, POD and PPO activities [130]. Jiang et al. [131] highlighted the *TaSOS1* gene in response to salt stress in two spring *T. aestivum* genotypes, Seri M82 (salt-sensitive) and CIGM90.863 (salt-tolerant), and observed a higher expression of most of the 18 *TaSOS1* genes in the roots of salt-tolerant seedlings than the salt-sensitive seedlings. Recently, several exogenous protectants (i.e., GA, salicylic acid (SA), melatonin (MT), silicon, selenium) have been used to upregulate the antioxidant machinery in plants under salinity [132–135]. For instance, Ahmad et al. [133] found increased activities of SOD (9%), APX (13%), CAT (26%), GR (40%) and POD (98%) when seed priming was conducted with 0.5 mM GA. Foliar spraying of SA (0.5 mM) enhanced AsA, GSH, total phenols and anthocyanin biosynthesis and increased SOD and CAT activities, thus reducing the H_2O_2 and $O_2{}^{\bullet -}$ content in salt-stressed seedlings of *Vicia faba* [134]. Combined application of silicon and selenium has been reported to increase CAT, GR, SOD and APX activities and synthesis of AsA and GSH [135]. Pretreatment of seeds of *Avena nuda* with MT (50 or 100 μM) upregulates CAT, APX, POD and SOD activities, thus reducing H_2O_2 (by 34%), $O_2{}^{\bullet -}$ (by 26%) and MDA (by 51%), compared to salt-treated plants [132]. Chen et al. [136] also reported that seed priming with MT reduces the accumulation of H2O2 and MDA and the percentage of EL and enhances the germination rate under salt stress conditions. Furthermore, Zhang et al. [137] observed that both foliar and root application of MT (100 μM) increased SOD, POD, CAT and APX activities under salt stress and decreased H2O2, O2•− and MDA accumulation in Beta vulgaris.

Table 2. Activities of antioxidant defense system against salinity.

Plant Species	Level(s) of Salt Stress	Antioxidant Defense	References
G. max cv. A3935	50 (low), 100 (medium) and 150 (high) mM NaCl; 30 d	High salinity reduces SOD activity in roots (28%) and leaves (38%). APX activity decreased in leaves by 20% (low) and 57% (high), but slightly increased in roots at low salinity (10%) and decreased by 29% under high salinity. GR and CAT activities decreased in both roots and leaves under high salinity.	[138]
T. aestivum cv. Pradip	150 and 300 mM NaCl; 4 d	AsA content sharply decreased. GSH content and GSH/GSSG ratio increased. Slight increase in APX and GST activities, whereas activities of CAT, DHAR, MDHAR, GR and GPX decreased at 300 mM NaCl stress.	[99]
B. napus cv. BINA Sharisha 3	100 and 200 mM NaCl; 2d	AsA and GSH content decreased, but GSSG content increased at 200 mM NaCl. GPX, GST and GR activities increased at 100 mM NaCl. Increase in APX activity, and decrease in CAT, DHAR, MDHAR, GR, GST and GPX activities at 200 mM NaCl stress.	[100]
O. sativa cv. Pokkali (tolerant)	100 mM saline solution (mixture of NaCl, $MgCl_2$, $MgSO_4$ and $CaCl_2$ salts); 35 d	Increased activities of SOD, CAT, APX and POX.	[101]
O. sativa cv. KDML105	60, 120 and 160 mM NaCl; 3 d	SOD, APX and GR activities increase with increasing NaCl concentrations. CAT activity reduced by 1.6-fold at 160 mM NaCl.	[103]
G. max cv. PK9305	100 mM NaCl; 7 d	Activities of CAT, SOD and PPO observed more in leaves than roots. Higher activity of POX in roots than shoots.	[104]

Table 2. *Cont.*

Plant Species	Level(s) of Salt Stress	Antioxidant Defense	References
S. tuberosum cv. Hui 2	50, 75 and 100 mM NaCl; 31 d	SOD activity increased in a dose-dependent manner. CAT and POD activities decreased at 100 mM NaCl stress, but still higher than non-stressed plants.	[107]
O. sativa cv. BRRI dhan47	200 mM NaCl; 3 d	AsA content reduced by 49%. Reduction in GSH/GSSG ratio by 42%. The activity of SOD increased by 20%. Reduced CAT activity by 33%.	[70]
P. vulgaris cvs. Tema (high-yielding) and Djadida (low-yielding)	50, 100 and 200 mM NaCl; 7 d	GR activity increased by 60% (Tema) and 20% (Djadida). CAT activity increased by 4- and 2-fold in Tema and Djedida. APX activity increased by 9- and 6- fold in Tema and Djedida. Increment in AsA and total flavonoid content in cv. Tema (by 33 and 47%) and cv. Djadida (by 26 and 70%). Total phenolic compounds decreased markedly in cv. Djadida.	[108]
O. sativa cv. BRRI dhan47	150 mM NaCl; 3, 6 d	AsA content decreased, but DHA content increased. Higher level of GSH and GSSG content. Upregulation of MDHAR, DHAR and GR activities. APX activity enhanced at 6 d. SOD and GPX activities increased with increasing duration of stress. Reduction in phenolic and flavonoid contents.	[69]
V. radiata cv. BARI Mung-2	200 mM NaCl; 2 d	Activities of SOD and GST increased by 49 and 88%. CAT activity reduced by 50%. No significant change was observed in GR and GPX activities. Reduction in MDHAR and DHAR activities.	[67]
P. sativum cv. Shubhra IM-9101	100 and 400 mM NaCl; 7 d	Reduction in CAT (94%), POD (57%) and APX (86%) activities. Increase in SOD activity by 174%.	[109]
B. napus cv. BINA Sharisha 3	100 (mild) and 200 (severe) mM NaCl; 2 d	AsA content reduced by 44% under severe stress. GSSG content upgraded by 116% under severe stress. APX activity increased, but reduction in MDHAR, DHAR and GR activities. GPX activity reduced under severe stress. The activity of GST enhanced. CAT activity dropped by 32% (mild) and 41% (severe).	[114]

Table 2. *Cont.*

Plant Species	Level(s) of Salt Stress	Antioxidant Defense	References
S. lycopersicum cv. Pusa Ruby	150 and 250 mM NaCl; 4 d	Upregulation of SOD activity by 30% (150 mM) and 43% (250 mM). CAT and GR activities decreased. Activities of APX, MDHAR, DHAR, GPX and GST upgraded. AsA content reduced, but DHA content increased. Both GSH and GSSG content upgraded.	[61]
G. hirsutum	150 mM NaCl; 3, 6, 9 and 12 d	SOD activity enhanced by 47, 37, 26 and 18% at 3, 6, 9 and 12 d, respectively. Higher activity of POD found, by 103, 63 and 11% at different durations. CAT activity increased by 28, 20, 14 and 16%. Activity of APX enhanced by 126, 104, 67 and 37% at 3, 6, 9 and 12 d, respectively.	[118]
S. lycopersicum cv. Pusa Ruby	150 mM NaCl; 5 d	AsA content decreased, but DHA increased with ratio lowered by 50%. GSH content reduced, but GSSG content increased with decrease in GSH/GSSG ratio by 45%. Activities of APX, MDHAR, GR, GST and SOD increased by 134, 53, 114, 70 and 16%, respectively.	[56]
T. aestivum cv. Norin 61	250 mM NaCl; 5 d	Both AsA and AsA/DHA ratio reduced, but DHA content increased. GSSG content increased, but GSH and GSH/GSSG ratio decreased. Activities of APX, DHAR and GPX increased by 29, 38 and 13%, respectively. Reduction in MDHAR (32%), GR (24%), CAT (37%) and GST (15%) activities.	[120]
G. max cv. Giza 111	75 and 150 mM NaCl; 56 d	Total phenol content notably increased by 24 and 33%. AsA content also enhanced by 32 and 64%.	[121]
L. sativa L. cv. SUSANA	4 dS m^{-1} (low) and 8 dS m^{-1} (high) NaCl; 60 d	CAT activity increased by 87 and 89% at low salinity, whereas at high salinity, it increased by 158 and 162% in both seasons. Elevation in POD, and PPO activities increased significantly.	[122]
L. culinaris Medik cv. BARI Masur-7	110 mM NaCl; 2 d	AsA content reduced by 70%, whereas an increase in GSH (305%) and GSSG (353%) contents was noticed. Reduction in CAT (71%) and APX (41%) activities, while an increase in DHAR (47%), GR (83%) and GPX (162%) activities.	[123]

8. Signaling of ROS in the Regulation of Salinity

Salt stress directly induces primary stresses such as osmotic and ionic stresses as well as imposing secondary stresses such as oxidative stress caused by ROS. These ROS play a signaling function at a certain low concentration which is variable depending upon many factors [139]. In recent decades, research on salt stress signals and adaptation proposed different pathways which still remain hazy because of the complex interaction between and among biomolecules (Figure 8; [139–141]). Under salt stress, ROS production induces mitogen-activated protein kinase (MAP) cascades which mediates ionic, osmotic and ROS homeostasis [139,142].

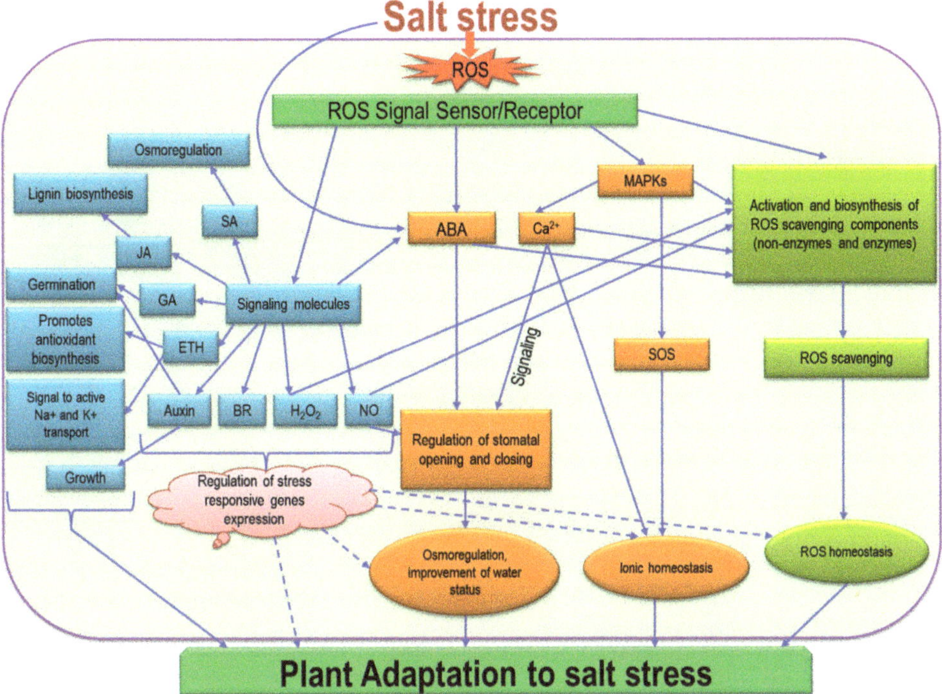

Figure 8. Common proposed ROS signaling pathways during plants' response to salt stress. Under salt stress, ROS production induces mitogen-activated protein kinase (MAP) cascades which mediate different adaptive responses. MAPs, nitric oxide (NO), Ca^{2+} and other signaling molecules have been suggested to be connected in activating antioxidant defense, stomatal movement, membrane properties and ionic homeostasis. The salt overly sensitive signaling (SOS) pathway interacting with other pathways functions in maintaining ionic homeostasis. The ROS-induced ABA production and ROS-induced activation of MAPs and Ca^{2+} signal regulate stomatal opening and closing. ROS signal sensors/receptors can induce activation of biosynthesis/functioning of different hormones and signaling molecules such as salicylic acid (SA), jasmonic acid (JA), gibberellic acid (GA), ethylene (ETH), auxin, brassinosteroid (BR), H_2O_2 and NO. These hormones/signaling molecules can interact with ROS, single or multiple hormones or signaling molecules; can interact with various signal cascades/pathways; and can regulate stress-responsive gene expression which modulates various metabolic and physiological functions, contributing to plant adaptation to salt stress. Dashed lines indicate the mechanisms which are not yet identified.

The salt overly sensitive signaling (SOS) pathway functions in maintaining ionic homeostasis, through extruding Na^+ into the apoplast. Under salinity, excess intracellular or extracellular Na^+ triggers, and ROS-induced MAPs stimulate, a Ca^{2+} signal and production. Ca^{2+} together with the SOS pathway excludes Na^+ to maintain ionic homeostasis [141,143,144]. ROS signals stimulate antioxidant defense and ROS scavenging which confer biomembrane protection and restore membrane function for maintaining ionic and nutrient homeostasis [141,144].

The ROS-induced ABA production and ROS-induced activation of MAPs and the Ca^{2+} signal regulate stomatal opening and closing. The osmoregulation and plant water status are maintained properly as a result [140,142]. Nitric oxide is also considered as one of the regulators of stomatal opening and closing [145]. ROS signal sensors/receptors can induce SA which also has a role in osmoregulation. Different signaling molecules or hormones can be stimulated by ROS which have diversified physiological, growth and developmental functions in the salt adaptation of plants. The ROS signal induces JA to modulate lignin biosynthesis, and GA to affect germination, and it induces auxin to modulate growth [146]. Ethylene–ROS interaction has a differential function. Ethylene enhances ROS generation that activates Na^+ and K^+ transport, where expression of different genes is involved. The modulation of ETH signaling controls AsA biosynthesis under salinity. The synchronizing action of ABA and ETH modulates the AsA biosynthesis during salt exposure. The Ca^{2+} signaling persuaded by ROS signaling participates in AsA biosynthesis [3]. Several research reports demonstrated that exogenous application of hormones and signaling molecules regulated or enhanced the antioxidant defense system, conferred osmoprotection and improved physiology, which defended against oxidative damage under salt stress conditions [147–149]. There might be other hormones/hormone-like molecules or signaling molecules in this complex and hazy salt stress adaptation pathway. Knowledge on the signaling functions of ROS is misty as it is connected to the signaling function and biological function of different biomolecules. The dual role of ROS—creating oxidative stress and having a signaling function—is well known, but the mechanisms, interplay and cross-talk with other molecules and components are complex and yet to be discovered.

9. Hormonal Regulation and ROS under Salinity

Plant hormones or phytohormones are chemicals that are present in plants in small amounts but have a great impact on plant growth and development. Plant hormones play different biological roles in the presence of environmental stress and regulate plant growth positively or negatively. Under salinity, diverse signaling pathways coordinately work to mitigate salt stress [150]. Flexible regulation of phytohormone signaling helps plants to adapt under salinity (Figure 9).

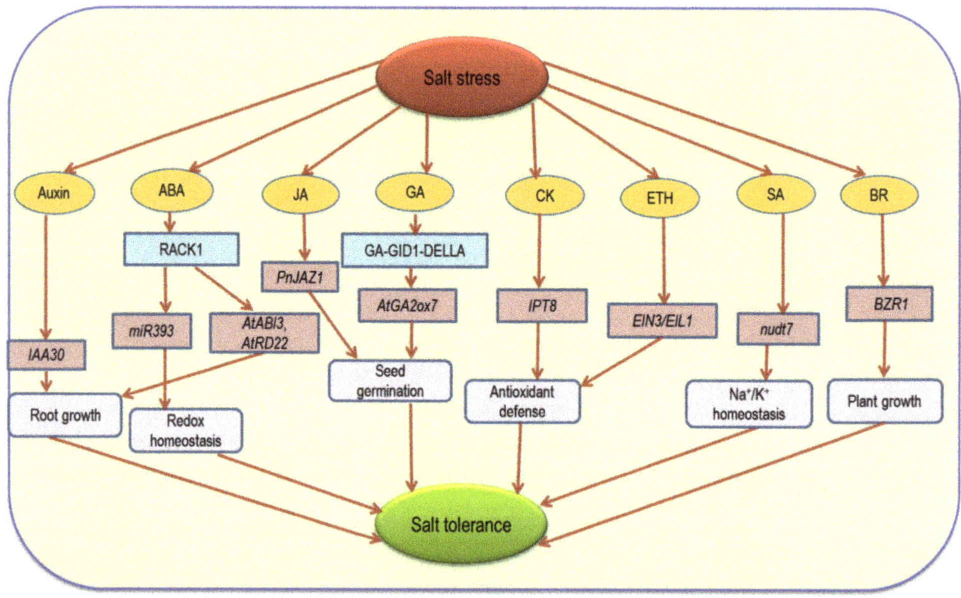

Figure 9. Overview of major plant hormone regulation in *Arabidopsis* under salt stress where different salinity-mitigating traits of phytohormone-regulated genes help to induce salt tolerance in *Arabidopsis*. Detailed descriptions are provided in the text. ABA—abscisic acid; JA—jasmonic acid; GA—gibberellic acid; CK—cytokinin; SA—salicylic acid; ETH—ethylene; BR—brassinosteroid; RACK1—receptor for activated C kinase 1.

In plants, ABA is an important stress-responsive hormone that plays a crucial role in osmotic stress and especially in ROS-mediated signaling pathways. Salt-induced osmotic stress in plants causes a higher accumulation of ABA [151,152]. This higher accumulation of ABA helps to mitigate the negative effect of salinity by improving photosynthesis and the osmolyte content and reducing ROS-mediated toxicity [153]. It also triggers the accumulation of K^+ and Ca^{2+} which inhibit the uptake of Na^+ and Cl^- [154]. ABA also regulates Na^+/K^+ homeostasis and the H_2O_2 content under salt stress [155]. Abscisic acid is known as a vital salinity-mediated signal which regulates salinity-responsive genes. Overexpression of the *MAX2* gene increased the resistance to salt stress, which was associated with redox homeostasis and ABA [156]. Osmotic homeostasis is regulated by ABA-activated SnRK2s which are responsible for the breakdown of starch into sugar and derivate osmolytes [157]. To scavenge ROS, ABA-generated H_2O_2 triggered the accumulation of NO and activated protein kinase (MAPK) [139]. The MAPK cascade signaling upregulates the antioxidant genes and acts against salinity [158]. Salt stress reduced the lateral root length in plants which mediated ABA signaling [159]. Similar to ABA, auxin also alters the root apical meristem under salt stress. In *Arabidopsis thaliana*, receptor for activated C kinase 1 (RACK1) is involved in the biosynthesis of the *miR393* gene which mediates both ABA and auxin regulation to induce redox homeostasis and alter lateral root growth under salinity [160]. During salt exposure, accumulation of excess ROS altered the auxin distribution, induced oxidative damage and caused a reduction in the root apical meristem [161]. Seed germination modulated membrane-bound transcription factor (NTM2), altered auxin signaling in plants which caused overexpression of the *IAA30* gene and induced salt tolerance in the *Arabidopsis* plant [162]. Plant growth and development depend on seed germination which is closely related to GA. DELLA proteins act as a growth suppressor of plants under salt stress. GA binds with the GIBBERELLIN INSENSITIVE DWARF1 (GID1) receptor which recruits DELLA proteins and forms the GA-GID1-DELLA complex [163]. Thus, GA reduces DELLAs and salt tolerance in plants. Some GA-regulated genes, viz., *AtGA2ox7* and *OsGA2ox5*, also help to increase salinity tolerance in *A. thaliana* and *O. sativa* plants, respectively [164,165].

Under the saline condition, plants regulate CK production along with ROS homeostasis to induce salt tolerance in plants. CK is a master phytohormone which acts as a free ROS scavenger and opposite to ABA and induces salt tolerance in plants [166]. CK receptor genes, viz., *AHK2*, *AHK3* and *AHK4*, help to modulate the stress signal in plants which regulates the osmolyte content and membrane integrity [167]. Histidine kinase (HK) of CK receptors plays an important role in CK regulation and stress responses [168]. In the presence of CK, receptor gene *CRE1* changes to its HK form and acts as a negative regulator of stress signaling. However, in the absence of CK, it converts into its phosphatase form and does not show any negative reactions [169]. Due to the higher CK content, receptor genes *AHK2* and *AHK3* antagonistically regulate ABA synthesis in *Arabidopsis* and consequently inhibit the ABA action in germinating seeds and seedling growth through the CK receptor HKs [170]. However, ROS production is reduced by the overexpression of CK biosynthetic gene *IPT8* which helps to upregulate antioxidant enzyme activity under salinity in *Arabidopsis* [171]. Likewise, JA also works against salt stress and helps to mediate ROS. A high concentration of JA was noticed in salt-treated plants, which helps to reduce oxidative damage and shows a positive relationship with salt tolerance [172]. Qiu et al. [173] also observed an increase in the salt tolerance of *T. aestivum* by exogenous application of JA through the reduction in the Na^+ content, ROS and lipid peroxidation. The salt tolerance mechanism of JA is modulated by jasmonate ZIM domain (JAZ) transcription factors which repress MYC2 transcription factors. These MYC2 transcription factors are used by both JA and ABA to regulate salt responses where the functioning of JA activates ABA-regulated genes *RD22* and *AtADH1* under salt stress to induce salt tolerance [173]. In *Arabidopsis*, the *PnJAZ1* gene regulates the JA pathway to induce salt tolerance and cause upregulation of ABA-regulated genes, viz., *AtABI3* and *AtRD22*, to regulate seed germination and seedling growth under salinity [174].

In plants, salinity tolerance is induced by SA via activating the GORK channel which helps to maintain the membrane potential and K^+ loss in plant cells [175]. It also upregulates the activity of the H^+-ATPase enzyme which helps in K^+ retention in the saline condition. However, SA did not retard the accumulation of Na^+ in plant roots but, instead, helped to reduce the Na^+ concentration in the roots [176]. In *Arabidopsis*, expression of the *nudt7* gene reduced the shoot Na^+ concentration during prolonged salt stress [177]. In barley, the application of SA helps to increase P_n, the carotenoid content and membrane integrity and to reduce ROS toxicity which ultimately helps to increase the K^+ and osmolyte concentration in plant roots [178]. It also caused overexpression of the *P5CS* gene which causes Pro accumulation and thus induces salt tolerance in *Dianthus superbus* [179]. ETH homeostasis and signaling have a direct relation with salt stress. Salt stress regulates key components of ETH signaling and also causes overexpression of the *EIN3/EIL1* gene which is known as a defense-related gene. This gene helps to reduce the H_2O_2 content and upregulate antioxidant enzymes to induce salt tolerance in *Arabidopsis* [180]. Similarly, brassinosteroid (BR) also helps to mediate salt stress in *S. lycopersicum* by the overexpression of *BRI1* or *BSK5* [181]. In saline conditions, BRASSINOSTEROID INSENSITIVE1 KINASE INHIBITOR1 (BKI1) translocated BRASSINAZOLE RESISTANT 1 (BZR1) and BRI1-EMSSUPPRESSOR 1 (BES1) and caused overexpression of the *BZR1* gene in the *Arabidopsis* plant which controlled plant growth and development [182]. In *Arabidopsis*, BR-mediated salt tolerance signaling is linked with ABA, ETH and SA pathways which indicates the cross-talk of phytohormones in mitigating salt stress [183].

10. Gene Regulation for Antioxidant Defense under Salinity

Antioxidant enzymes are considered the first line of defense to eliminate accumulated ROS. Antioxidant enzymes are encoded by different gene families which are located in different parts of the cell. The activities of these enzymes are controlled by the expression of genes encoding antioxidant enzymes which are varied in crops. Mishra et al. [184] observed increased SOD activity in a tolerant cultivar of *O. sativa* which occurred due to the expressions of the *Cu/Zn-SOD* genes. In *O. sativa* (cv. BRS AG), expression of *OsCATB* increased with the duration (10, 15 and 20 d) of salt stress, which resulted in increased activity of CAT during the exposure to salinity [185]. *GPX* genes of *T. aestivum* demonstrated different expression patterns at a stressed condition, where *TaGPX* genes showed higher expression at the leaf developmental stage [186]. Glutathione peroxidase is considered as an important antioxidant enzyme involved in the reduction of H_2O_2 with the help of GSH [2]. Salt-tolerant *H. vulgare* genotypes showed higher GPX activity which was regulated by the higher expression of the *GPX* gene [187]. In *B. juncea*, higher expression of ROS detoxification genes, viz., MDHAR and DHAR, was observed in salt-tolerant genotypes compared to the sensitive ones [188]. MDHAR and DHAR help to detoxify plant cells with the help of the AsA-GSH cycle [2]. Under salt stress, antioxidant genes of stressed plants are upregulated or downregulated with the severity of stress. Filiz et al. [189] worked with 36 isoform genes of 10 natural *Arabidopsis* ecotypes under salt stress, where 64% upregulation was demonstrated by *CAT* genes, and 55% downregulation was recorded in *SOD* and *GPX* genes.

Overexpressed genes encoding antioxidant enzymes help to mitigate ROS and induce stress tolerance in plants; some examples are cited in Table 3. Overexpression of *Cu/ZnSOD* in tobacco plants reduced ROS-induced damage in plants and improved the chl content under salinity [190]. Overexpression of *PutAPX* in *A. thaliana* reduced lipid peroxidation and improved plant growth under salt stress [191]. In *PaSOD*- and *RaAPX*-overexpressed potato plants, Shafi et al. [192] observed an improved RWC and chl content, and a reduction in the MDA content, under 150 mM NaCl-induced salt stress.

Table 3. Effects of overexpressed antioxidant genes under salinity in different crops.

Transgenic Plants	Gene Source	Overexpressed Genes	Salt Stress and Duration	Regulatory Roles	References
Nicotiana tabacum	*A. thaliana*	*AtMDAR1*	300 mM NaCl; 6 d	Improved P_n. Decreased H_2O_2 content.	[193]
G. hirsutum	*Agrobacterium tumefaciens*	*GhSOD1, GhCAT1*	200 mM NaCl; 2d	Upregulated SOD, CAT and APX activities. Increased oxidative stress'tolerance.	[194]
A. thaliana	*Tamarix hispida*	*ThGSTZ1*	75, 100 and 125 mM NaCl; 12 d	Upregulated GST, GPX, SOD and POD activities. Reduced EL and MDA content.	[195]
Arachis hypogaea	*A. tumefaciens*	*SbpAPX*	50, 100, 150, 200 and 250 mM NaCl; 21 d	Increased chl content. Improved RWC. Increased plant biomass. Reduced EL.	[196]
N. tabacum	*A. hypogaea*	*AhCuZnSOD*	100 mM NaCl; 15 d	Increased RWC. Reduced MDA by 1.5-fold. Decreased H_2O_2 by 2-fold. Upregulated CAT, GR, APX and SOD activities.	[197]
A. thaliana	*Puccinellia tenuiflora*	*PutAPX*	125, 150 and 175 mM NaCl; 3 d	Improved total chl content. Increased RWC. Reduced MDA content. Upregulated APX activity.	[191]
Ipomoea batatas	*A. tumefaciens*	*CuZnSOD, APX*	100 mM NaCl; 14 d	Pro content increased by 2.7-fold. SOD and APX activities increased by 3.4- and 4.2-fold.	[198]
S. tuberosum	*Potentilla atrosanguinea, Rheum australe*	*PaSOD, RaAPX*	50, 100 and 150 mM NaCl; 15 d	Improved total chl content. Increased RWC. Upregulated SOD and APX activities.	[192]
N. tabacum	*S. lycopersicum*	*SlMDHAR*	200 mM NaCl; 5 d	Reduced MDA and H_2O_2. Reduced Na^+ content.	[199]
B. juncea	*A. tumefaciens*	*AtApx1*	200 mM NaCl; 10 d	Increased chl and carotenoid content. Improved Pro content. Decreased MDA and H_2O_2 content. Increased APX and CAT activities.	[200]

11. Conclusions

The generation of ROS in plants is a natural phenomenon that occurs as a part of different metabolic processes in multiple cellular and subcellular compartments. Salinity-induced osmotic stress and ionic toxicity disrupt redox homeostasis and trigger the over-production of ROS and ROS-induced oxidative stress under salt stress. Several studies revealed that ROS have both negative and positive roles in plants. Along with signaling, ROS aid several basic processes and pathways in plant cells. Lower concentrations of ROS were found to be essential for proliferative pathway activation, cell differentiation and stem cell renewal. On the contrary, excessive accumulation of ROS at the cellular level acts as a stressor and causes oxidative damage of lipids, proteins, carbohydrates and RNA and DNA.

As ROS have a dual role in plants, the metabolism of ROS is crucial for growth, development and adaptation under salinity. The antioxidant defense system plays a vital role in ROS metabolism by scavenging and detoxifying ROS. Plant hormones, amino acids and their derivatives, PAs and vitamin supplementation contributed to ROS metabolism and decreased salt-induced oxidative stress under different levels of salinity by upregulating the antioxidant defense system, biological processes, ion homeostasis, osmolytes and gene expression in several studies. In previous investigations, under different levels of salinity, overexpression of genes helped to mitigate ROS-induced oxidative stress. In recent studies, molecular and genetic tools developed transgenic plants with enhanced activities of antioxidant enzymes and ROS detoxification under salinity.

Although the dual role of ROS is known, the metabolism process of ROS under salinity is complex, including the mechanisms, cross-talk with other molecules and components that are yet to be understood. Therefore, further research is required to understand the metabolism process of ROS more clearly. The understanding of ROS metabolism under salinity will be helpful in mitigating salt-induced oxidative stress. The mechanisms, interplay, cross-talk with other molecules and components are complex and warrant further research.

The present review focuses on ROS production and the antioxidant defense system under salt stress where information about cultivated and common plant species was included mostly from recently published research articles. There are salt-tolerant plant species in nature. Halophytes bear unique morphological, anatomical and physiological behavior, and that is why they are adapted to grow in coastal saline soils, mangrove forests and salt-affected lands of arid and semi-arid regions. Understanding the mechanisms of halophytes, finding out the signaling cascades and identifying salt stress-responsive genes can be useful for managing salt stress-affected cultivated and commonly grown plant species through different agronomic, breeding, biotechnological and other approaches.

Author Contributions: Conceptualization, M.H., M.F.; writing—original draft preparation, M.H., M.R.H.R., A.A.C.M., K.R., F.N., M.R. and K.N.; writing—review and editing, M.H., K.N. and M.F.; visualization, M.H.; supervision, M.H. All authors have read and agreed to the published version of the manuscript.

Funding: This research received no external funding.

Institutional Review Board Statement: Not applicable.

Informed Consent Statement: Not applicable.

Data Availability Statement: Not applicable.

Acknowledgments: We acknowledge Naznin Ahmed for providing some important studies.

Conflicts of Interest: The authors declare no conflict of interest.

References

1. Yadav, S.; Modi, P.; Dave, A.; Vijapura, A.; Patel, D.; Patel, M. Effect of abiotic stress on crops. In *Sustainable Crop Production*; Hasanuzzaman, M., Filho, M.C.M.T., Fujita, M., Nogueira, T.A.R., Eds.; IntechOpen: London, UK, 2020; pp. 3–24.
2. Hasanuzzaman, M.; Bhuyan, M.H.M.B.; Zulfiqar, F.; Raza, A.; Mohsin, S.M.; Mahmud, J.A.; Fujita, M.; Fotopoulos, V. Reactive oxygen species and antioxidant defense in plants under abiotic stress: Revisiting the crucial role of a universal defense regulator. *Antioxidants* **2020**, *9*, 681. [CrossRef] [PubMed]
3. Wang, J.; Huang, R. Modulation of ethylene and ascorbic acid on reactive oxygen species scavenging in plant salt response. *Front. Plant Sci.* **2019**, *10*, 319. [CrossRef]
4. Food and Agriculture Organization of the United Nations. 2021. Available online: http://www.fao.org/global-soil-partnership/resources/highlights/detail/en/c/1412475/ (accessed on 18 July 2021).
5. Hasanuzzaman, M.; Nahar, K.; Fujita, M. Plant response to salt stress and role of exogenous protectants to mitigate salt-induced damages. In *Ecophysiology and Responses of Plants under Salt Stress*; Ahmed, P., Azooz, M.M., Prasad, M.N.V., Eds.; Springer: New'York, NY, USA, 2013; pp. 25–87.

6. Hossain, M.S.; Dietz, K.J. Tuning of redox regulatory mechanisms, reactive oxygen species and redox homeostasis under salinity stress. *Front. Plant Sci.* **2016**, *7*, 548. [CrossRef]
7. Bhattacharjee, S. ROS and oxidative stress: Origin and implication. In *Reactive Oxygen Species in Plant Biology*; Springer: New Delhi, India, 2019; pp. 1–31.
8. Hasanuzzaman, M.; Bhuyan, M.H.M.B.; Anee, T.I.; Parvin, K.; Nahar, K.; Mahmud, J.A.; Fujita, M. Regulation of ascorbate-glutathione pathway in mitigating oxidative damage in plants under abiotic stress. *Antioxidants* **2019**, *8*, 384. [CrossRef] [PubMed]
9. Munns, R.; Gilliham, M. Salinity tolerance of crops—What is the cost? *New Phytol.* **2015**, *208*, 668–673. [CrossRef]
10. Mittler, R. ROS are good. *Trends Plant Sci.* **2017**, *22*, 11–19. [CrossRef]
11. Saini, P.; Gani, M.; Kaur, J.J.; Godara, L.C.; Singh, C.; Chauhan, S.S.; Francies, R.M.; Bhardwaj, A.; Kumar, N.B.; Ghosh, M.K. Reactive oxygen species (ROS): A way to stress survival in plants. In *Abiotic Stress-Mediated Sensing and Signaling in Plants: An Omics Perspective*; Zargar, S., Zargar, M., Eds.; Springer: Singapore, 2018; pp. 127–153.
12. Choudhary, A.; Kumar, A.; Kaur, N. ROS and oxidative burst: Roots in plant development. *Plant Divers.* **2020**, *42*, 33–43. [CrossRef] [PubMed]
13. Sachdev, S.; Ansari, S.A.; Ansari, M.I.; Fujita, M.; Hasanuzzaman, M. Abiotic stress and reactive oxygen species: Generation, signaling, and defense mechanisms. *Antioxidants* **2021**, *10*, 277. [CrossRef]
14. Munns, R.; Tester, M. Mechanisms of salinity tolerance. *Annu. Rev. Plant Biol.* **2008**, *59*, 651–681. [CrossRef]
15. Ahmad, R.; Hussain, S.; Anjum, M.A.; Khalid, M.F.; Saqib, M.; Zakir, I.; Ahmad, S. Oxidative stress and antioxidant defense mechanisms in plants under salt stress. In *Plant Abiotic Stress Tolerance*; Hasanuzzaman, M., Hakeem, K.R., Nahar, K., Alharby, H.F., Eds.; Springer: Cham, Switzerland, 2019; pp. 191–205.
16. Mhamdi, A.; Van Breusegem, F. Reactive oxygen species in plant development. *Development* **2018**, *145*, dev164376. [CrossRef]
17. Guan, L.; Haider, M.S.; Khan, N.; Nasim, M.; Jiu, S.; Fiaz, M.; Zhu, X.; Zhang, K.; Fang, J. Transcriptome sequence analysis elaborates a complex defensive mechanism of grapevine (*Vitis vinifera L.)* in response to salt stress. *Int. J. Mol. Sci.* **2018**, *19*, 4019. [CrossRef] [PubMed]
18. Zhang, X.; Liu, L.; Chen, B.; Qin, Z.; Xiao, Y.; Zhang, Y.; Yao, R.; Liu, H.; Yang, H. Progress in understanding the physiological and molecular responses of Populus to salt stress. *Int. J. Mol. Sci.* **2019**, *20*, 1312. [CrossRef]
19. Hernández, J.A. Salinity tolerance in plants: Trends and perspectives. *Int. J. Mol. Sci.* **2019**, *20*, 2408. [CrossRef]
20. Zhao, G.; Zhao, Y.; Yu, X.; Kiprotich, F.; Han, H.; Guan, R.; Wang, R.; Shen, W. Nitric oxide is required for melatonin-enhanced tolerance against salinity stress in rapeseed (*Brassica napus* L.) seedlings. *Int. J. Mol. Sci.* **2018**, *19*, 1912. [CrossRef]
21. Santos-Sánchez, N.F.; Salas-Coronado, R.; Villanueva-Cañongo, C.; Hernández-Carlos, B. Antioxidant compounds and their antioxidant mechanism. In *Antioxidants*; Shalaby, E., Ed.; IntechOpen: London, UK, 2019; pp. 23–50.
22. Yu, Z.; Duan, X.; Luo, L.; Dai, S.; Ding, Z.; Xia, G. How plant hormones mediate salt stress responses. *Trends Plant Sci.* **2020**, *25*, 1117–1130. [CrossRef]
23. Mailloux, J.R. Application of mitochondria-targeted pharmaceuticals for the treatment of heart disease. *Curr. Pharm. Des.* **2016**, *22*, 4763–4779. [CrossRef]
24. Hasanuzzaman, M.; Bhuyan, M.H.M.; Parvin, K.; Bhuiyan, T.F.; Anee, T.I.; Nahar, K.; Hossen, M.; Zulfiqar, F.; Alam, M.; Fujita, M. Regulation of ROS metabolism in plants under environmental stress: A review of recent experimental evidence. *Int. J. Mol. Sci.* **2020**, *21*, 8695. [CrossRef]
25. Kohli, S.K.; Khanna, K.; Bhardwaj, R.; Abd Allah, E.F.; Ahmad, P.; Corpas, F.J. Assessment of subcellular ROS and NO metabolism in higher plants: Multifunctional signaling molecules. *Antioxidants* **2019**, *8*, 641. [CrossRef] [PubMed]
26. Podgorska, A.; Burian, M.; Szal, B. Extra-cellular but extra-ordinarily important for cells: Apoplastic reactive oxygen species metabolism. *Front. Plant Sci.* **2017**, *8*, 1353. [CrossRef]
27. Janků, M.; Luhová, L.; Petřivalský, M. On the origin and fate of reactive oxygen species in plant cell compartments. *Antioxidants* **2019**, *8*, 105. [CrossRef]
28. Raja, V.; Majeed, U.; Kang, H.; Andrabi, K.I.; John, R. Abiotic stress: Interplay between ROS, hormones and MAPKs. *Environ. Exp. Bot.* **2017**, *137*, 142–157. [CrossRef]
29. Sharma, P.; Jha, A.B.; Dubey, R.S.; Pessarakli, M. Reactive oxygen species, oxidative damage, and antioxidative defense mechanism in plants under stressful conditions. *J. Bot.* **2012**, *2012*, 217037. [CrossRef]
30. Heyno, E.; Mary, V.; Schopfer, P.; Krieger-Liszkay, A. Oxygen activation at the plasma membrane: Relation between superoxide and hydroxyl radical production by isolated membranes. *Planta* **2011**, *234*, 35–45. [CrossRef]
31. Dietz, K.J. Thiol-based peroxidases and ascorbate peroxidases: Why plants rely on multiple peroxidase systems in the photosynthesizing chloroplast? *Mol. Cells* **2016**, *39*, 20.
32. Shakirova, F.M.; Allagulova, C.R.; Maslennikova, D.R.; Klyuchnikova, E.O.; Avalbaev, A.M.; Bezrukova, M.V. Salicylic acid-induced protection against cadmium toxicity in wheat plants. *Environ. Exp. Bot.* **2016**, *122*, 19–28. [CrossRef]
33. Singh, A.; Kumar, A.; Yadav, S.; Singh, I.K. Reactive oxygen species-mediated signaling during abiotic stress. *Plant Gene* **2019**, *18*, 100–173. [CrossRef]
34. Huang, S.; VanAken, O.; Schwarzländer, M.; Belt, K.; Millar, A.H. The roles of mitochondrial reactive oxygen species in cellular signaling and stress responses in plants. *Plant Physiol.* **2016**, *171*, 1551–1559. [CrossRef]

35. Kerchev, P.; Waszczak, C.; Lewandowska, A.; Willems, P.; Shapiguzov, A.; Li, Z. Lack of GLYCOLATE OXIDASE1, but not GLYCOLATE OXIDASE2, attenuates the photorespiratory phenotype of CATALASE2- deficient Arabidopsis. *Plant Physiol.* **2016**, *171*, 1704–1719. [CrossRef] [PubMed]
36. Foyer, C.H.; Noctor, G. Redox regulation in photosynthetic organisms: Signaling, acclimation, and practical implications. *Antioxid. Redox Signal.* **2009**, *11*, 861–905. [CrossRef]
37. Corpas, F.J.; del Río, L.A.; Palma, J.M. Impact of nitric oxide (NO) on the ROS metabolism of peroxisomes. *Plants* **2019**, *8*, 37. [CrossRef]
38. Reumann, S.; Chowdhary, G.; Lingner, T. Characterization, prediction and evolution of plant peroxisomal targeting signals type 1 (PTS1s). *Biochim. Biophys. Acta* **2016**, *1863*, 790–803. [CrossRef]
39. Kumar, S.P.J.; Prasad, R.S.; Banerjee, R.; Thammineni, C. Seed birth to death: Dual functions of reactive oxygen species in seed physiology. *Ann. Bot.* **2015**, *116*, 663–668. [CrossRef]
40. Van Breusegem, F.; Bailey-Serres, J.; Mittler, R. Unraveling the tapestry of networks involving reactive oxygen species in plants. *Plant Physiol.* **2008**, *147*, 978–984. [CrossRef]
41. Shah, K.; Chaturvedi, V.; Gupta, S. Climate change and abiotic stress-induced oxidative burst in rice. In *Advances in Rice Research for Abiotic Stress Tolerance*; Hasanuzzaman, M., Fujita, M., Nahar, K., Biswas, J.K., Eds.; Woodhead Publishing: Cambridge, UK, 2019; pp. 505–535.
42. Alché, J.D. A concise appraisal of lipid oxidation and lipoxidation in higher plants. *Redox Biol.* **2019**, *23*, 101136. [CrossRef]
43. Anjum, N.A.; Sofo, A.; Scopa, A. Lipids and proteins—Major targets of oxidative modifications in abiotic stressed plants. *Environ. Sci. Pollut. Res.* **2015**, *22*, 4099–4121. [CrossRef]
44. Pospíšil, P. Production of reactive oxygen species by photosystem II as a response to light and temperature stress. *Front. Plant Sci.* **2016**, *26*, 1950. [CrossRef]
45. Das, K.; Roychoudhury, A. Reactive oxygen species (ROS) and response of antioxidants as ROS-scavengers during environmental stress in plants. *Front. Environ. Sci.* **2014**, *2*, 53. [CrossRef]
46. Banerjee, A.; Roychoudhury, A. Abiotic stress, generation of reactive oxygen species, and their consequences: An overview. In *Reactive Oxygen Species in Plants: Boon or Bane- Revisiting the Role of ROS*; Singh, V.P., Singh, S., Tripathi, D.K., Prasad, S.M., Chauhan, D.K., Eds.; John Wiley & Sons Ltd.: Chichester, UK, 2018; pp. 23–50.
47. Lehmann, M.; Laxa, M.; Sweetlove, L.J.; Fernie, A.R.; Obata, T. Metabolic recovery of *Arabidopsis thaliana* roots following cessation of oxidative stress. *Metabolomics* **2012**, *8*, 143–153. [CrossRef] [PubMed]
48. Babu, M.A.; Singh, D.; Gothandam, K.M. The effect of salinity on growth, hormones and mineral elements in leaf and fruit of tomato cultivar PKM1. *J. Anim. Plant Sci.* **2012**, *22*, 159–164.
49. Wahid, A.; Farooq, M.; Basra, S.M.A.; Rasul, E.; Siddique, K.H.M. Germination of seeds and propagules under salt stress. In *Handbook of Plant and Crop Stress*; Pessarakli, M., Ed.; CRC Press: Boca Raton, FL, USA, 2010; pp. 321–337.
50. Liu, L.; Xia, W.; Li, H.; Zeng, H.; Wei, B.; Han, S.; Yin, C. Salinity inhibits rice seed germination by reducing α-Amylase activity via decreased bioactive gibberellin content. *Front. Plant Sci.* **2018**, *9*, 275. [CrossRef]
51. Shu, K.; Qi, Y.; Chen, F.; Meng, Y.; Luo, X.; Shuai, H.; Zhou, W.; Ding, J.; Du, J.; Liu, J.; et al. Salt stress represses soybean seed germination by negatively regulating GA biosynthesis while positively mediating ABA biosynthesis. *Front. Plant Sci.* **2017**, *8*, 1372. [CrossRef]
52. Li, Q.-F.; Zhou, Y.; Xiong, M.; Ren, X.-Y.; Han, L.; Wang, J.-D.; Zhang, C.-Q.; Fan, X.-L.; Liu, Q.-Q. Gibberelin recovers seed germination in rice with impaired brassinosteroid signalling. *Plant Sci.* **2020**, *293*, 110435. [CrossRef]
53. Dadshani, S.; Sharma, R.C.; Baum, M.; Ogbonnaya, F.C.; Leon, J.; Ballvora, A. Multidimensional evaluation of response to salt stress in wheat. *PLoS ONE* **2019**, *14*, e0222659. [CrossRef]
54. Wu, G.-Q.; Jiao, Q.; Shui, Q.-Z. Effect of salinity on seed germination, seedling growth, and inorganic and organic solutes accumulation in sunflower (*Helianthus annuus* L.). *Plant Soil Environ.* **2015**, *61*, 220–226.
55. Bistgani, Z.E.; Hashemi, M.; DaCosta, M.; Craker, L.; Maggi, F.; Morshedloo, M.R. Effect of salinity stress on the physiological characteristics, phenolic compounds and antioxidant activity of *Thymus vulgaris* L. and *Thymus daenensis* Celak. *Ind. Crop. Prod.* **2019**, *135*, 311–320. [CrossRef]
56. Parvin, K.; Nahar, K.; Hasanuzzaman, M.; Bhuyan, M.H.M.B.; Mohsin, S.M.; Fujita, M. Exogenous vanillic acid enhances salt tolerance of tomato: Insight into plant antioxidant defense and glyoxalase systems. *Plant Physiol. Biochem.* **2020**, *150*, 109–120. [CrossRef]
57. Mazher, A.M.A.; El-Quesni, E.M.F.; Farahat, M.M. Responses of ornamental and woody trees to salinity. *World J. Agric. Sci.* **2007**, *3*, 386–395.
58. Kwon, O.K.; Mekapogu, M.; Kim, K.S. Effect of salinity stress on photosynthesis and related physiological responses in carnation (*Dianthus caryophyllus*). *Hortic. Environ. Biotechnol.* **2019**, *60*, 831–839. [CrossRef]
59. Taj, Z.; Challabathula, D. Protection of photosynthesis by halotolerant *Staphylococcus sciuri* ET101 in tomato (*Lycoperiscon esculentum*) and rice (*Oryza sativa*) plants during salinity stress: Possible interplay between carboxylation and oxygenation in stress mitigation. *Front. Microbiol.* **2021**, *11*, 547750. [CrossRef] [PubMed]
60. Mohsin, S.M.; Hasanuzzaman, M.; Parvin, K.; Fujita, M. Pretreatment of wheat (*Triticum aestivum* L.) seedlings with 2,4-D improves tolerance to salinity-induced oxidative stress and methylglyoxal toxicity by modulating ion homeostasis, antioxidant defenses, and glyoxalase systems. *Plant Physiol. Biochem.* **2020**, *152*, 221–231. [CrossRef] [PubMed]

61. Parvin, K.; Hasanuzzaman, M.; Bhuyan, M.H.M.B.; Nahar, K.; Mohsin, S.M.; Fujita, M. Comparative physiological and biochemical changes in tomato (*Solanum lycopersicum* L.) under salt stress and recovery: Role of antioxidant defense and glyoxalase systems. *Antioxidants* **2019**, *8*, 350. [CrossRef] [PubMed]
62. Pérez-Labrada, F.; López-Vargas, E.R.; Ortega-Ortiz, H.; Cadenas-Pliego, G.; Benavides-Mendoza, A.; Juárez-Maldonado, A. Responses of tomato plants under saline stress to foliar application of copper nanoparticles. *Plants* **2019**, *8*, 151. [CrossRef]
63. Hasanuzzaman, M.; Inafuku, M.; Nahar, K.; Fujita, M.; Oku, H. Nitric oxide regulates plant growth, physiology, antioxidant defense, and ion homeostasis to confer salt tolerance in the mangrove species, *Kandelia obovata*. *Antioxidants* **2021**, *10*, 611. [CrossRef]
64. Hossain, M.S.; Alam, M.U.; Rahman, A.; Hasanuzzaman, M.; Nahar, K.; Mahmud, J.-A.; Fujita, M. Use of iso-osmotic solution to understand salt stress responses in lentil (*Lens culinaris* Medik.). *S. Afr. J. Bot.* **2017**, *113*, 346–354. [CrossRef]
65. Zeng, Y.; Li, Q.; Wang, H.; Zhang, J.; Du, J.; Feng, H.; Blumwald, E.; Yu, L.; Xu, G. Two NHX-type transporters from *Helianthus tuberosus* improve the tolerance of rice to salinity and nutrient deficiency stress. *Plant Biotechnol. J.* **2018**, *16*, 310–321. [CrossRef]
66. Bose, J.; Rodrigo-Moreno, A.; Shabala, S. ROS homeostasis in halophytes in the context of salinity stress tolerance. *J. Exp. Bot.* **2014**, *65*, 1241–1257. [CrossRef]
67. Nahar, K.; Hasanuzzaman, M.; Rahman, A.; Alam, M.M.; Mahmud, J.A.; Suzuki, T.; Fujita, M. Polyamines confer salt tolerance in mung bean (*Vigna radiata* L.) by reducing sodium uptake, improving nutrient homeostasis, antioxidant defense, and methylglyoxal detoxification systems. *Front. Plant Sci.* **2016**, *7*, 1104. [CrossRef]
68. Mekawya, A.M.M.; Abdelazizc, M.N.; Ueda, A. Apigenin pretreatment enhances growth and salinity tolerance of rice seedlings. *Plant Physiol. Biochem.* **2018**, *130*, 94–104. [CrossRef] [PubMed]
69. Rahman, A.; Nahar, K.; Hasanuzzaman, M.; Fujita, M. Calcium supplementation improves Na^+/K^+ ratio, antioxidant defense and glyoxalase systems in salt-stressed rice seedlings. *Front. Plant Sci.* **2016**, *7*, 609. [CrossRef]
70. Rahman, A.; Hossain, M.S.; Mahmud, J.-A.; Nahar, K.; Hasanuzzaman, M.; Fujita, M. Manganese-induced salt stress tolerance in rice seedlings: Regulation of ion homeostasis, antioxidant defense and glyoxalase systems. *Physiol. Mol. Biol. Plants* **2016**, *22*, 291–306. [CrossRef]
71. Borrelli, G.M.; Fragasso, M.; Nigro, F.; Platani, C.; Papa, R.; Beleggia, R.; Trono, D. Analysis of metabolic and mineral changes in response to salt stress in durum wheat (*Triticum turgidum* ssp. durum) genotypes, which differ in salinity tolerance. *Plant Physiol. Biochem.* **2018**, *133*, 57–70. [CrossRef]
72. Villalta, I.; Reina-Sanchez, A.; Bolarin, M.C.; Cuartero, J.; Belver, A.; Venema, K.; Carbonell, E.A.; Asins, M.J. Genetic analysis of Na^+ and K^+ concentrations in leaf and stem as physiological components of salt tolerance in tomato. *Theor. Appl. Gene* **2008**, *116*, 869–880. [CrossRef]
73. Hasanuzzaman, M.; Alam, M.M.; Rahman, A.; Hasanuzzaman, M.; Nahar, K.; Fujita, M. Exogenous proline and glycine betaine mediated upregulation of antioxidant defense and glyoxalase systems provides better protection against salt-induced oxidative stress in two rice (*Oryza sativa* L.) varieties. *BioMed Res. Int.* **2014**, *2014*, 757219. [CrossRef]
74. Zhu, C.; An, L.; Jiao, X.; Chen, X.; Zhou, G.; McLaughlin, N. Effects of gibberellic acid on water uptake and germination of sweet sorghum seeds under salinity stress. *Chilean J. Agric. Res.* **2019**, *79*, 415–424. [CrossRef]
75. Zeeshan, M.; Lu, M.; Sehar, S.; Holford, P.; Wu, F. Comparison of biochemical, anatomical, morphological, and physiological responses to salinity stress in wheat and barley genotypes deferring in salinity tolerance. *Agronomy* **2020**, *10*, 127. [CrossRef]
76. Bogoutdinova, L.R.; Lazareva, E.M.; Chaban, I.A.; Kononenko, N.V.; Dilovarova, T.; Khaliluev, M.R.; Kurenina, L.V.; Gulevich, A.A.; Smirnova, E.A.; Baranova, E.N. Salt stress-induced structural changes are mitigated in transgenic tomato plants over-expressing superoxide dismutase. *Biology* **2020**, *9*, 297. [CrossRef]
77. Rossi, L.; Zhang, W.; Ma, X. Cerium oxide nanoparticles alter the salt stress tolerance of *Brassica napus* L. by modifying the formation of root apoplastic barriers. *Environ. Pollut.* **2017**, *229*, 132–138. [CrossRef]
78. Boughalleb, F.; Abdellaoui, R.; Nbiba, N.; Mahmoudi, M.; Neffati, M. Effect of NaCl stress on physiological, antioxidant enzymes and anatomical responses of *Astragalus gombiformis*. *Biologia* **2017**, *72*, 1454–1466. [CrossRef]
79. Jahan, B.; AlAjmi, M.F.; Rehman, M.T.; Khan, N.A. Treatment of nitric oxide supplemented with nitrogen and sulfur regulates photosynthetic performance and stomatal behavior in mustard under salt stress. *Physiol. Plant* **2020**, *168*, 490–510. [PubMed]
80. Sehar, Z.; Masood, A.; Khan, N.A. Nitric oxide reverses glucose-mediated photosynthetic repression in wheat (*Triticum aestivum* L.) under salt stress. *Environ. Exp. Bot.* **2019**, *161*, 277–289. [CrossRef]
81. Fatma, M.; Iqbal, N.; Gautam, H.; Sehar, Z.; Sofo, A.; D'Ippolito, I.; Khan, N.A. Ethylene and sulfur coordinately modulate the antioxidant system and ABA accumulation in mustard plants under salt stress. *Plants* **2021**, *10*, 180. [CrossRef]
82. Ramadan, A.A.; Elhamid, E.M.A.; Sadak, M.S. Comparative study for the effect of arginine and sodium nitroprusside on sunflower plants grown under salinity stress conditions. *Bull. Natl. Res. Cent.* **2019**, *43*, 118. [CrossRef]
83. Noreen, S.; Sultan, M.; Akhter, M.S.; Shah, K.H.; Ummara, U.; Manzoor, H.; Ulfat, M.; Alyemeni, M.N.; Ahmad, P. Foliar fertigation of ascorbic acid and zinc improves growth, antioxidant enzyme activity and harvest index in barley (*Hordeum vulgare* L.) grown under salt stress. *Plant Physiol. Biochem.* **2021**, *158*, 244–254. [CrossRef]
84. Farooq, M.; Gogoi, N.; Barthakur, S.; Baroowa, B.; Bharadwaj, N.; Alghamdi, S.S.; Siddique, K.H.M. Draught stress in grain legume during reproduction and grain filling. *J. Agron. Crop. Sci.* **2017**, *203*, 81–102. [CrossRef]
85. Peng, J.; Zhang, L.; Liu, J.; Luo, J.; Zhao, X.; Dong, H.; Ma, Y.; Sui, N.; Zhou, Z.; Meng, Y. Effects of soil salinity on sucrose metabolism in cotton fiber. *PLoS ONE* **2016**, *11*, e0156398. [CrossRef]

86. Khanam, D.; Mohammad, F. Plant growth regulators ameliorate the ill effect of salt stress through improved growth, photosynthesis, antioxidant system, yield and quality attributes in *Mentha piperita* L. *Acta Physiol. Plant* **2018**, *40*, 188. [CrossRef]
87. Costan, A.; Stamatakis, A.; Chrysargyris, A.; Petropoulosc, S.A.; Tzortzakisb, N. Interactive effects of salinity and silicon application on *Solanum lycopersicum* growth, physiology and shelf-life of fruit produced hydroponically. *J. Sci. Food Agric.* **2020**, *100*, 732–743. [CrossRef]
88. Zhang, P.; Senge, M.; Dai, Y. Effects of salinity stress at different growth stages on tomato growth, yield and water use efficiency. *Commun. Soil Sci. Plant Anal.* **2017**, *48*, 624–634. [CrossRef]
89. Islam, M.Z.; Mele, M.A.; Choi, K.Y.; Kang, H.M. Nutrient and salinity concentrations effects on quality and storability of cherry tomato fruits grown by hydroponic system. *Bragantia* **2018**, *77*, 385–393. [CrossRef]
90. Jalil, S.U.; Ansari, M.I. Physiological role of Gamma-aminobutyric acid in salt stress tolerance. In *Salt and Drought Stress Tolerance in Plants*; Hasanuzzaman, M., Tanveer, M., Eds.; Springer: Cham, Switzerland, 2020; pp. 337–350.
91. Geilfus, C.M.; Mithöfer, A.; Ludwig-Müller, J.; Zörb, C.; Muehling, K.H. Chloride-inducible transient apoplastic alkalinazations induce stomata closure by controlling abscisic acid distribution between leaf apoplast and guard cells in salt-stressed *Vicia faba*. *New Phytol.* **2015**, *208*, 803–816. [CrossRef]
92. Siddiqui, M.H.; Alamri, S.; Alsubaie, Q.D.; Ali, H.M. Melatonin and gibberellic acid promote growth and chlorophyll biosynthesis by regulating antioxidant and methylglyoxal detoxification system in tomato seedlings under salinity. *J. Plant Growth Regul.* **2020**, *39*, 1488–1502. [CrossRef]
93. Ali, M.M.; Jeddi, K.; Attia, M.S.; Elsayed, S.M.; Yusuf, M.; Osman, M.S.; Soliman, M.H.; Hessini, K. Wuxal amino (Bio stimulant) improved growth and physiological performance of tomato plants under salinity stress through adaptive mechanisms and antioxidant potential. *Saudi J. Biol. Sci.* **2021**, *28*, 3204–3213. [CrossRef] [PubMed]
94. Ghosh, S.; Mitra, S.; Paul, A. Physiochemical studies of sodium chloride on mungbean (*Vigna radiata* L. Wilczek) and its possible recovery with spermine and gibberellic acid. *Sci. World J.* **2015**, *2015*, 858016. [CrossRef]
95. Nxele, X.; Klein, A.; Ndimba, B.K. Drought and salinity stress alters ROS accumulation, water retention, and osmolyte content in sorghum plants. *S. Afr. J. Bot.* **2017**, *108*, 261–266. [CrossRef]
96. Jabeen, Z.; Fayyaz, H.A.; Irshad, F.; Hussain, N.; Hassan, M.N.; Li, J.; Rehman, S.; Haider, W.; Yasmin, H.; Mumtaz, S.; et al. Sodium nitroprusside application improves morphological and physiological attributes of soybean (*Glycine max* L.) under salinity stress. *PLoS ONE* **2021**, *16*, e0248207. [CrossRef]
97. Derbali, W.; Goussi, R.; Koyro, H.-W.; Abdelly, C.; Manaa, A. Physiological and biochemical markers for screening salt tolerant quinoa genotypes at early seedling stage. *J. Plant Interact.* **2020**, *15*, 27–38. [CrossRef]
98. Derbali, W.; Manaa, A.; Goussi, R.; Derbali, I.; Abdelly, C.; Koyro, H.-W. Post-stress restorative response of two quinoa genotypes differing in their salt resistance after salinity release. *Plant Physiol. Biochem.* **2021**, *164*, 222–236. [CrossRef]
99. Hasanuzzaman, M.; Hossain, M.A.; Fujita, M. Nitric oxide modulates antioxidant defense and the methylglyoxal detoxification system and reduces salinity-induced damage of wheat seedlings. *Plant Biotechnol. Rep.* **2011**, *5*, 353. [CrossRef]
100. Hasanuzzaman, M.; Hossain, M.A.; Fujita, M. Selenium-induced up-regulation of the antioxidant defense and methylglyoxal detoxification system reduces salinity-induced damage in rapeseed seedlings. *Biol. Trace Elem. Res.* **2011**, *143*, 1704–1721. [CrossRef]
101. Chawla, S.; Jain, S.; Jain, V. Salinity induced oxidative stress and antioxidant system in salt-tolerant and salt-sensitive cultivars of rice (*Oryza sativa* L.). *J. Plant Biochem. Biotechnol.* **2013**, *22*, 27–34. [CrossRef]
102. Manai, J.; Kalai, T.; Gouia, H.; Corpas, F.J. Exogenous nitric oxide (NO) ameliorates salinity-induced oxidative stress in tomato (*Solanum lycopersicum*) plants. *J. Soil Sci. Plant Nutr.* **2014**, *14*, 433–446. [CrossRef]
103. Wutipraditkul, N.; Wongwean, P.; Buaboocha, T. Alleviation of salt-induced oxidative stress in rice seedlings by proline and/or glycinebetaine. *Biol. Plant* **2015**, *59*, 547–553. [CrossRef]
104. Kumari, S.; Vaishnav, A.; Jain, S.; Verma, A.; Choudhary, D.K. Bacterial-mediated induction of systemic tolerance to salinity with expression of stress alleviating enzymes in soybean (*Glycine max* L. Merrill). *J. Plant Growth Regul.* **2015**, *34*, 558–573. [CrossRef]
105. Talaat, N.B. Effective microorganisms improve growth performance and modulate the ROS-scavenging system in common bean (*Phaseolus vulgaris* L.) plants exposed to salinity stress. *J. Plant Growth Regul.* **2015**, *34*, 35–46. [CrossRef]
106. Iqbal, N.; Umar, S.; Khan, N.A. Nitrogen availability regulates proline and ethylene production and alleviates salinity stress in mustard (*Brassica juncea*). *J. Plant Physiol.* **2015**, *178*, 84–91. [CrossRef]
107. Hu, Y.; Xia, S.; Su, Y.; Wang, H.; Luo, W.; Su, S.; Xiao, L. Brassinolide increases potato root growth In Vitro in a dose-dependent way and alleviates salinity stress. *BioMed Res. Int.* **2016**, *2016*, 8231873. [CrossRef]
108. Taïbi, K.; Taïbi, F.; Abderrahim, L.A.; Ennajah, A.; Belkhodja, M.; Mulet, J.M. Effect of salt stress on growth, chlorophyll content, lipid peroxidation and antioxidant defence systems in *Phaseolus vulgaris* L. *S. Afr. J. Bot.* **2016**, *105*, 306–312. [CrossRef]
109. Yadu, S.; Dewangan, T.L.; Chandrakar, V.; Keshavkant, S. Imperative roles of salicylic acid and nitric oxide in improving salinity tolerance in *Pisum sativum* L. *Physiol. Mol. Biol. Plants* **2017**, *23*, 43–58. [CrossRef]
110. Kaur, H.; Bhardwaj, R.D.; Grewal, S.K. Mitigation of salinity-induced oxidative damage in wheat (*Triticum aestivum* L.) seedlings by exogenous application of phenolic acids. *Acta Physiol. Plant* **2017**, *39*, 221. [CrossRef]
111. Kumar, M.; Kumar, R.; Jain, V.; Jain, S. Differential behavior of the antioxidant system in response to salinity induced oxidative stress in salt-tolerant and salt-sensitive cultivars of *Brassica juncea* L. *Biocatal. Agric. Biotechnol.* **2018**, *13*, 12–19. [CrossRef]

112. Arefian, M.; Vessal, S.; Shafaroudi, S.M.; Bagheri, A. Comparative analysis of the reaction to salinity of different chickpea (*Cicer arietinum* L.) genotypes: A biochemical, enzymatic and transcriptional study. *J. Plant Growth Regul.* **2018**, *37*, 391–402. [CrossRef]
113. Ahammed, G.J.; Li, Y.; Li, X.; Han, W.-Y.; Chen, S. Epigallocatechin-3-gallate alleviates salinity-retarded seed germination and oxidative stress in tomato. *J. Plant Growth Regul.* **2018**, *37*, 1349–1356. [CrossRef]
114. Hasanuzzaman, M.; Nahar, K.; Rohman, M.M.; Anee, T.I.; Huang, Y.; Fujita, M. Exogenous silicon protects *Brassica napus* plants from salinity-induced oxidative stress through the modulation of AsA-GSH pathway, thiol-dependent antioxidant enzymes and glyoxalase systems. *Gesunde Pflanz.* **2018**, *70*, 185–194. [CrossRef]
115. Parvin, K.; Hasanuzzaman, M.; Bhuyan, M.H.M.B.; Mohsin, S.M.; Fujita, M. Quercetin mediated salt tolerance in tomato through the enhancement of plant antioxidant defense and glyoxalase systems. *Plants* **2019**, *8*, 247. [CrossRef]
116. Rohman, M.M.; Islam, M.R.; Monsur, M.B.; Amiruzzaman, M.; Fujita, M.; Hasanuzzaman, M. Trehalose protects maize plants from salt stress and phosphorus deficiency. *Plants* **2019**, *8*, 568. [CrossRef]
117. Kapoor, R.T.; Hasanuzzaman, M. Exogenous kinetin and putrescine synergistically mitigate salt stress in *Luffa acutangula* by modulating physiology and antioxidant defense. *Physiol. Mol. Biol. Plants* **2020**, *26*, 2125–2137. [CrossRef]
118. Jiang, D.; Lu, B.; Liu, L.; Duan, W.; Chen, L.; Li, J.; Zhang, K.; Sun, H.; Zhang, Y.; Dong, H.; et al. Exogenous melatonin improves salt stress adaptation of cotton seedlings by regulating active oxygen metabolism. *Peer J.* **2020**, *8*, e10486. [CrossRef]
119. Liu, L.; Huang, L.; Lin, X.; Sun, C. Hydrogen peroxide alleviates salinity-induced damage through enhancing proline accumulation in wheat seedlings. *Plant Cell Rep.* **2020**, *39*, 567–575. [CrossRef]
120. Mohsin, S.M.; Hasanuzzaman, M.; Nahar, K.; Hossain, M.S.; Bhuyan, M.H.M.B.; Parvin, K.; Fujita, M. Tebuconazole and trifloxystrobin regulate the physiology, antioxidant defense and methylglyoxal detoxification systems in conferring salt stress tolerance in *Triticum aestivum* L. *Physiol. Mol. Biol. Plants* **2020**, *26*, 1139–1154. [CrossRef]
121. Osman, M.S.; Badawy, A.A.; Osman, A.I.; Latef, A.A.H.A. Ameliorative impact of an extract of the halophyte *Arthrocnemum macrostachyum* on growth and biochemical parameters of soybean under salinity stress. *J. Plant Growth Regul.* **2021**, *40*, 1245–1256. [CrossRef]
122. ALKahtani, M.; Hafez, Y.; Attia, K.; Al-Ateeq, T.; Ali, M.A.M.; Hasanuzzaman, M.; Abdelaal, K. *Bacillus thuringiensis* and silicon modulate antioxidant metabolism and improve the physiological traits to confer salt tolerance in lettuce. *Plants* **2021**, *10*, 1025. [CrossRef]
123. Fardus, J.; Hossain, M.S.; Fujita, M. Modulation of the antioxidant defense system by exogenous L-glutamic acid application enhances salt tolerance in lentil (*Lens culinaris* Medik.). *Biomolecules* **2021**, *11*, 587. [CrossRef] [PubMed]
124. Del Río, L.A.; Corpas, F.J.; López-Huertas, E.; Palma, J.M. Plant superoxide dismutases: Function under abiotic stress conditions. In *Antioxidants and Antioxidant Enzymes in Higher Plants*; Gupta, D., Palma, J., Corpas, F., Eds.; Springer: Cham, Switzerland, 2018; pp. 1–26.
125. Liebthal, M.; Maynard, D.; Dietz, K.-J. Peroxiredoxins and redox signaling in plants. *Antioxid. Redox Signal.* **2018**, *28*, 609–624. [CrossRef]
126. Bela, K.; Horváth, E.; Gallé, Á.; Szabados, L.; Tari, I.; Csiszár, J. Plant glutathione peroxidases: Emerging role of the antioxidant enzymes in plant development and stress responses. *J. Plant Physiol.* **2015**, *176*, 192–201. [CrossRef] [PubMed]
127. Xu, J.; Xing, X.-J.; Tian, Y.-S.; Peng, R.-H.; Xue, Y.; Zhao, W.; Yao, Q.-H. Transgenic *Arabidopsis* plants expressing tomato glutathione *S*-transferase showed enhanced resistance to salt and drought stress. *PLoS ONE* **2015**, *10*, e0136960. [CrossRef]
128. Cunha, J.R.; Neto, M.C.L.; Carvalho, F.E.; Martins, M.O.; Jardim-Messeder, D.; Margis-Pinheiro, M.; Silveira, J.A. Salinity and osmotic stress trigger different antioxidant responses related to cytosolic ascorbate peroxidase knockdown in rice roots. *Environ. Exp. Bot.* **2016**, *131*, 58–67. [CrossRef]
129. Ali, M.; Afzal, S.; Parveen, A.; Kamran, M.; Javed, M.R.; Abbasi, G.H.; Malik, Z.; Riaz, M.; Ahmad, S.; Chattha, M.S.; et al. Silicon mediated improvement in the growth and ion homeostasis by decreasing Na^+ uptake in maize (*Zea mays* L.) cultivars exposed to salinity stress. *Plant Physiol. Biochem.* **2021**, *158*, 208–218. [CrossRef]
130. Attia, M.S.; Osman, M.S.; Mohamed, A.S.; Mahgoub, H.A.; Garada, M.O.; Abdelmouty, E.S.; Latef, A.A.H.A. Impact of foliar application of chitosan dissolved in different organic acids on isozymes, protein patterns and physio-biochemical characteristics of tomato grown under salinity stress. *Plants* **2021**, *10*, 388. [CrossRef]
131. Jiang, W.; Pan, R.; Buitrago, S.; Wu, C.; Abou-Elwafa, S.F.; Xu, Y.; Zhang, W. Conservation and divergence of the *TaSOS1* gene family in salt stress response in wheat (*Triticum aestivum* L.). *Physiol. Mol. Biol. Plants* **2021**, *27*, 1245–1260. [CrossRef] [PubMed]
132. Gao, W.; Feng, Z.; Bai, Q.; He, J.; Wang, Y. Melatonin-mediated regulation of growth and antioxidant capacity in salt-tolerant naked oat under salt stress. *Int. J. Mol. Sci.* **2019**, *20*, 1176. [CrossRef] [PubMed]
133. Ahmad, F.; Kamal, A.; Singh, A.; Ashfaque, F.; Alamri, S.; Siddiqui, M.H.; Khan, M.I.R. Seed priming with gibberellic acid induces high salinity tolerance in *Pisum sativum* through antioxidant system, secondary metabolites and upregulation of antiporter genes. *Plant Biol.* **2020**, *23*, 113–121. [CrossRef]
134. Dawood, M.F.A.; Zaid, A.; Latef, A.A.H.A. Salicylic acid spraying-induced resilience strategies against the damaging impacts of drought and/or salinity stress in two varieties of *Vicia faba* L. seedlings. *J. Plant Growth Regul.* **2021**. [CrossRef]
135. Taha, R.S.; Seleiman, M.F.; Shami, A.; Alhammad, B.A.; Mahdi, A.H.A. Integrated application of selenium and silicon enhances growth and anatomical structure, antioxidant defense system and yield of wheat grown in salt-stressed soil. *Plants* **2021**, *10*, 1040. [CrossRef] [PubMed]

136. Chen, L.; Liu, L.; Lu, B.; Ma, T.; Jiang, D.; Li, J.; Li, J.; Zhang, K.; Sun, H.; Zhang, Y.; et al. Exogenous melatonin promotes seed germination and osmotic regulation under salt stress in cotton (*Gossypium hirsutum* L.). *PLoS ONE* **2020**, *15*, e0228241. [CrossRef] [PubMed]
137. Zhang, P.; Liu, L.; Wang, X.; Wang, Z.; Zhang, H.; Chen, J.; Liu, X.; Wang, Y.; Li, C. Beneficial effects of exogenous melatonin on overcoming salt stress in sugar beets (*Beta vulgaris* L.). *Plants* **2021**, *10*, 886. [CrossRef]
138. Doğan, M. Antioxidative and proline potentials as a protective mechanism in soybean plants under salinity stress. *Afr. J. Biotechnol.* **2011**, *10*, 5972–5978.
139. Yang, Y.; Guo, Y. Elucidating the molecular mechanisms mediating plant salt-stress responses. *New Phytol.* **2018**, *217*, 523–539. [CrossRef]
140. Golldack, D.; Li, C.; Mohan, H.; Probst, N. Tolerance to drought and salt stress in plants: Unraveling the signaling networks. *Front. Plant Sci.* **2014**, *5*, 151. [CrossRef]
141. Kurusu, T.; Kuchitsu, K.; Tada, Y. Plant signaling networks involving Ca^{2+} and Rboh/Nox-mediated ROS production under salinity stress. *Front. Plant Sci.* **2015**, *6*, 427. [CrossRef]
142. Huang, S.; Waadt, R.; Nuhkat, M.; Kollist, H.; Hedrich, R.; Roelfsema, M.R.G. Calcium signals in guard cells enhance the efficiency by which abscisic acid triggers stomatal closure. *New Phytol.* **2019**, *224*, 177–187. [CrossRef]
143. Xiong, L.; Zhu, J.K. Molecular and genetic aspects of plant responses to osmotic stress. *Plant Cell Environ.* **2002**, *25*, 131–139. [CrossRef]
144. Zhang, M.; Smith, J.A.; Harberd, N.P.; Jiang, C. The regulatory roles of ethylene and reactive oxygen species (ROS) in plant salt stress responses. *Plant Mol. Biol.* **2016**, *91*, 651–659. [CrossRef]
145. Hasanuzzaman, M.; Oku, H.; Nahar, K.; Bhuyan, M.B.; Al Mahmud, J.; Baluska, F.; Fujita, M. Nitric oxide-induced salt stress tolerance in plants: ROS metabolism, signaling, and molecular interactions. *Plant Biotechnol. Rep.* **2018**, *12*, 77–92. [CrossRef]
146. Kumar, V.; Khare, T.; Sharma, M.; Wani, S.H. ROS-induced signaling and gene expression in crops under salinity stress. In *Reactive Oxygen Species and Antioxidant Systems in Plants: Role and Regulation under Abiotic Stress*; Khan, M., Khan, N., Eds.; Springer: Singapore, 2017; pp. 159–184.
147. Atia, A.; Barhoumi, Z.; Debez, A.; Hkiri, S.; Abdelly, C.; Smaoui, A.; Haouari, C.C.; Gouia, H. Plant hormones: Potent targets for engineering salinity tolerance in plants. In *Salinity Responses and Tolerance in Plants*; Kumar, V., Wani, S., Suprasanna, P., Tran, L.S., Eds.; Springer: Cham, Switzerland, 2018; pp. 159–184.
148. Singh, V.P.; Prasad, S.M.; Munné-Bosch, S.; Müller, M. Editorial: Phytohormones and the regulation of stress tolerance in plants: Current status and future directions. *Front. Plant Sci.* **2017**, *8*, 1871. [CrossRef]
149. Quamruzzaman, M.; Manik, S.M.N.; Shabala, S.; Zhou, M. Improving performance of salt-grown crops by exogenous application of plant growth regulators. *Biomolecules* **2021**, *11*, 788. [CrossRef]
150. Ryu, H.; Cho, Y.G. Plant hormones in salt stress tolerance. *J. Plant Biol.* **2015**, *58*, 147–155. [CrossRef]
151. Yang, G.; Yu, Z.; Gao, L.; Zheng, C. SnRK2s at the crossroads of growth and stress responses. *Trends Plant Sci.* **2019**, *24*, 672–676. [CrossRef]
152. Yu, Z.; Zhang, D.; Xu, Y.; Jin, S.; Zhang, L.; Zhang, S.; Yang, G.; Huang, J.; Yan, K.; Wu, C.; et al. CEPR2 phosphorylates and accelerates the degradation of PYR/PYLs in Arabidopsis. *J. Exp. Bot.* **2019**, *70*, 5457–5469. [CrossRef]
153. Cabot, C.; Sibole, J.V.; Barceló, J.; Poschenrieder, C. Abscisic acid decreases leaf Na^+ exclusion in salt-treated *Phaseolus vulgaris* L. *J. Plant Growth Regul.* **2009**, *28*, 187–192. [CrossRef]
154. Gurmani, R.; Bano, A.; Khan, S.U.; Din, J.; Zhang, J.L. Alleviation of salt stress by seed treatment with abscisic acid (ABA), 6-benzylaminopurine (BA) and chlormequat chloride (CCC) optimizes ion and organic matter accumulation and increases yield of rice (*Oryza sativa* L.). *Aust. J. Crop. Sci.* **2011**, *5*, 1278–1285.
155. Ma, L.; Zhang, H.; Sun, L.; Jiao, Y.; Zhang, G.; Miao, C.; Hao, F. NADPH oxidase AtrbohD and AtrbohF function in ROS-dependent regulation of Na^+/K^+ homeostasis in Arabidopsis under salt stress. *J. Exp. Bot.* **2012**, *63*, 305–317. [CrossRef] [PubMed]
156. Wang, Y.T.; Chen, Z.Y.; Jiang, Y.; Duan, B.B.; Xi, Z.M. Involvement of ABA and antioxidant system in brassinosteroid-induced water stress tolerance of grapevine (*Vitis vinifera* L.). *Sci. Hortic.* **2019**, *256*, 108596. [CrossRef]
157. Rao, K.P.; Richa, T.; Kumar, K.; Raghuram, B.; Sinha, A.K. In silico analysis reveals 75 members of mitogen-activated protein kinase gene family in rice. *DNA Res.* **2010**, *17*, 139–153. [CrossRef] [PubMed]
158. Lu, S.; Su, W.; Li, H.; Guo, Z. Abscisic acid improves drought tolerance of triploid bermudagrass and involves H_2O_2- and NO induced antioxidant enzyme activities. *Plant Physiol. Biochem.* **2009**, *47*, 132–138. [CrossRef] [PubMed]
159. Duan, L.; Dietrich, D.; Ng, C.H.; Chan, P.M.; Bhalerao, R.; Bennett, M.J.; Dinneny, J.R. Endodermal ABA signaling promotes lateral root quiescence during salt stress in *Arabidopsis* seedlings. *Plant Cell* **2013**, *25*, 324–341. [CrossRef] [PubMed]
160. Denver, J.B.; Ullah, H. miR393s regulate salt stress response pathway in *Arabidopsis thaliana* through scaffold protein RACK1A mediated ABA signaling pathways. *Plant Signal. Behav.* **2019**, *14*, 1600394. [CrossRef] [PubMed]
161. Tognetti, V.B.; Bielach, A.; Hrtyan, M. Redox regulation at the site of primary growth: Auxin, cytokinin and ROS crosstalk. *Plant Cell Environ.* **2017**, *11*, 2586–2605. [CrossRef]
162. Park, J.; Kim, Y.S.; Kim, S.G.; Jung, J.H.; Woo, J.C.; Park, C.M. Integration of auxin and salt signals by the NAC transcription factor NTM2 during seed germination in Arabidopsis. *Plant Physiol.* **2011**, *156*, 537–549. [CrossRef]
163. Bao, S.; Hua, C.; Shen, L.; Yu, H. New insights into gibberellin signaling in regulating flowering in *Arabidopsis*. *J. Integr. Plant Biol.* **2020**, *62*, 118–131. [CrossRef]

164. Magome, H.; Yamaguchi, S.; Hanada, A.; Kamiya, Y.; Oda, K. The DDF1 transcriptional activator upregulates expression of a gibberellin-deactivating gene, *GA2ox7*, under high-salinity stress in *Arabidopsis*. *Plant J.* **2008**, *56*, 613–626. [CrossRef]
165. Shan, C.; Mei, Z.; Duan, J.; Chen, H.; Feng, H.; Cai, W. OsGA2ox5, a gibberellin metabolism enzyme, is involved in plant growth, the root gravity response and salt stress. *PLoS ONE* **2014**, *9*, e87110. [CrossRef] [PubMed]
166. Iqbal, N.; Umar, S.; Khan, N.A.; Khan, M.I.R. A new perspective of phytohormones in salinity tolerance: Regulation of proline metabolism. *Environ. Exp. Bot.* **2014**, *100*, 34–42. [CrossRef]
167. Nishiyama, R.; Le, D.T.; Watanabe, Y.; Matsui, A.; Tanaka, M.; Seki, M.; Yamaguchi-Shinozaki, K.; Shinozaki, K.; Tran, L.S. Transcriptome analyses of a salt-tolerant cytokinin-deficient mutant reveal differential regulation of salt stress response by cytokinin deficiency. *PLoS ONE* **2012**, *7*, e32124. [CrossRef]
168. Joshi, R.; Sahoo, K.K.; Tripathi, A.K.; Kumar, R.; Gupta, B.K.; Pareek, A.; Singla-Pareek, S.L. Knockdown of an inflorescence meristem-specific cytokinin oxidase–*OsCKX2* in rice reduces yield penalty under salinity stress condition. *Plant Cell Environ.* **2018**, *41*, 936–946. [CrossRef]
169. Mähönen, A.P.; Higuchi, M.; Törmäkangas, K.; Miyawaki, K.; Pischke, M.S.; Sussman, M.R.; Helariutta, Y.; Kakimoto, T. Cytokinins regulate a bidirectional phosphorelay network in *Arabidopsis*. *Curr. Biol.* **2006**, *16*, 1116–1122. [CrossRef]
170. Tran, L.S.; Urao, T.; Qin, F.; Maruyama, K.; Kakimoto, T.; Shinozaki, K.; Yamaguchi-Shinozaki, K. Functional analysis of AHK1/ATHK1 and cytokinin receptor histidine kinases in response to abscisic acid, drought, and salt stress in *Arabidopsis*. *Proc. Natl. Acad. Sci. USA* **2007**, *104*, 20623–20628. [CrossRef]
171. Wang, Y.; Shen, W.; Chan, Z.; Wu, Y. Endogenous cytokinin overproduction modulates ROS homeostasis and decreases salt stress resistance in *Arabidopsis thaliana*. *Front. Plant Sci.* **2015**, *6*, 1004. [CrossRef]
172. Javid, M.G.; Sorooshzadeh, A.; Moradi, F.; Sanavy, S.A.M.M.; Allahdadi, I. The role of phytohormones in alleviating salt stress in crop plants. *Aust. J. Crop. Sci.* **2011**, *5*, 726–734.
173. Qiu, Z.; Guo, J.; Zhu, A.; Zhang, L.; Zhang, M. Exogenous jasmonic acid can enhance tolerance of wheat seedlings to salt stress. *Ecotoxicol. Environ. Saf.* **2014**, *104*, 202–208. [CrossRef]
174. Liu, S.; Zhang, P.; Li, C.; Xia, G. The moss jasmonate ZIM-domain protein PnJAZ1 confers salinity tolerance via crosstalk with the abscisic acid signalling pathway. *Plant Sci.* **2019**, *280*, 1–11. [CrossRef] [PubMed]
175. Jayakannan, M.; Bose, J.; Babourina, O.; Rengel, Z.; Shabala, S. Salicylic acid improves salinity tolerance in *Arabidopsis* by restoring membrane potential and preventing salt-induced K^+ loss via a GORK channel. *J. Exp. Bot.* **2013**, *64*, 2255–2268. [CrossRef]
176. Nazar, R.; Iqbal, N.; Syeed, S.; Khan, N.A. Salicylic acid alleviates decreases in photosynthesis under salt stress by enhancing nitrogen and sulfur assimilation and antioxidant metabolism differentially in two mungbean cultivars. *J. Plant Physiol.* **2011**, *168*, 807–815. [CrossRef]
177. Jayakannan, M.; Bose, J.; Babourina, O.; Shabala, S.; Massart, A.; Poschenrieder, C.; Rengel, Z. The NPR1-dependent salicylic acid signalling pathway is pivotal for enhanced salt and oxidative stress tolerance in *Arabidopsis*. *J. Exp. Bot.* **2015**, *66*, 1865–1875. [CrossRef]
178. El-Tayeb, M.A. Response of barley grains to the interactive effect of salinity and salicylic acid. *Plant Growth Regul.* **2005**, *45*, 215–224. [CrossRef]
179. Zheng, J.; Ma, X.; Zhang, X.; Hu, Q.; Qian, R. Salicylic acid promotes plant growth and salt-related gene expression in *Dianthus superbus* L. (Caryophyllaceae) grown under different salt stress conditions. *Physiol. Mol. Biol. Plants* **2018**, *24*, 231–238. [CrossRef]
180. Peng, J.; Li, Z.; Wen, X.; Li, W.; Shi, H.; Yang, L.; Zhu, H.; Guo, H. Salt-induced stabilization of EIN3/EIL1 confers salinity tolerance by deterring ROS accumulation in *Arabidopsis*. *PLoS Genet.* **2014**, *10*, e1004664. [CrossRef]
181. Zhu, T.; Deng, X.; Zhou, X.; Zhu, L.; Zou, L.; Li, P.; Zhang, D.; Lin, H. Ethylene and hydrogen peroxide are involved in brassinosteroid-induced salt tolerance in tomato. *Sci. Rep.* **2016**, *6*, 35392. [CrossRef]
182. Sun, L.; Feraru, E.; Feraru, M.I.; Waidmann, S.; Wang, W.; Passaia, G.; Wang, Z.Y.; Wabnik, K.; Kleine-Vehn, J. PIN-LIKES coordinate brassinosteroid signaling with nuclear auxin input in *Arabidopsis thaliana*. *Curr. Biol.* **2020**, *30*, 1579–1588. [CrossRef] [PubMed]
183. Divi, U.K.; Rahman, T.; Krishna, P. Brassinosteroid-mediated stress tolerance in Arabidopsis shows interaction with absicisic acid, ethylene and salicylic acid pathways. *BMC Plant Biol.* **2010**, *10*, 151. [CrossRef] [PubMed]
184. Mishra, P.; Bhoomika, K.; Dubey, R.S. Differential responses of antioxidative defense system to prolonged salinity stress in salt-tolerant and salt-sensitive Indica rice (*Oryza sativa* L.) seedlings. *Protoplasma* **2013**, *250*, 3–19. [CrossRef] [PubMed]
185. Rossatto, T.; do Amaral, M.N.; Benitez, L.C.; Vighi, I.L.; Braga, E.J.; de Magalhães Júnior, A.M.; Maia, M.A.; da Silva, P.L. Gene expression and activity of antioxidant enzymes in rice plants, cv. BRS AG, under saline stress. *Physiol. Mol. Biol. Plants* **2017**, *23*, 865–875. [CrossRef]
186. Tyagi, S.; Sembi, J.K.; Upadhyay, S.K. Gene architecture and expression analyses provide insights into the role of glutathione peroxidases (GPXs) in bread wheat (*Triticum aestivum* L.). *J. Plant Physiol.* **2018**, *223*, 19–31. [CrossRef]
187. Witzel, K.; Weidner, A.; Surabhi, G.-K.; Börner, A.; Mock, H.-P. Salt stress-induced alterations in the root proteome of barley genotypes with contrasting response towards salinity. *J. Exp. Bot.* **2009**, *60*, 3545–3557. [CrossRef] [PubMed]
188. Sharma, R.; Mishra, M.; Gupta, B.; Parsania, C.; Singla-Pareek, S.L.; Pareek, A. De novo assembly and characterization of stress transcriptome in a salinity-tolerant variety CS52 of *Brassica juncea*. *PLoS ONE* **2015**, *10*, e0126783. [CrossRef] [PubMed]
189. Filiz, E.; Ozyigit, I.I.; Saracoglu, I.A.; Uras, M.E.; Sen, U.; Yalcin, B. Abiotic stress-induced regulation of antioxidant genes in different Arabidopsis ecotypes: Microarray data evaluation. *Biotechnol. Biotechnol. Equip.* **2018**, *33*, 128–143. [CrossRef]

190. Jing, X.; Hou, P.; Lu, Y.; Deng, S.; Li, N.; Zhao, R.; Sun, J.; Wang, Y.; Han, Y.; Lang, T.; et al. Overexpression of copper/zinc superoxide dismutase from mangrove *Kandelia candel* in tobacco enhances salinity tolerance by the reduction of reactive oxygen species in chloroplast. *Front. Plant Sci.* **2015**, *6*, 23. [CrossRef]
191. Guan, Q.; Wang, Z.; Wang, X.; Takano, T.; Liu, S. A peroxisomal APX from *Puccinellia tenuiflora* improves the abiotic stress tolerance of transgenic *Arabidopsis thaliana* through decreasing of H_2O_2 accumulation. *J. Plant Physiol.* **2015**, *175*, 183–191. [CrossRef]
192. Shafi, A.; Pal, A.K.; Sharma, V.; Kalia, S.; Kumar, S.; Ahuja, P.S.; Singh, A.K. Transgenic potato plants overexpressing SOD and APX exhibit enhanced lignification and starch biosynthesis with improved salt stress tolerance. *Plant Mol. Biol. Rep.* **2017**, *35*, 504–518. [CrossRef]
193. Eltayeb, A.E.; Kawano, N.; Badawi, G.H.; Kaminaka, H.; Sanekata, T.; Shibahara, T.; Inanaga, S.; Tanaka, K. Overexpression of monodehydroascorbate reductase in transgenic tobacco confers enhanced tolerance to ozone, salt and polyethylene glycol stresses. *Planta* **2007**, *225*, 1255–1264. [CrossRef]
194. Luo, X.; Wu, J.; Li, Y.; Nan, Z.; Guo, X.; Wang, Y.; Zhang, A.; Wang, Z.; Xia, G.; Tian, Y. Synergistic effects of *GhSOD1* and *GhCAT1* overexpression in cotton chloroplasts on enhancing tolerance to methyl viologen and salt stresses. *PLoS ONE* **2013**, *8*, e54002. [CrossRef]
195. Yang, G.; Wang, Y.; Xia, D.; Gao, C.; Wang, C.; Yang, C. Overexpression of a GST gene (*ThGSTZ1*) from *Tamarix hispida* improves drought and salinity tolerance by enhancing the ability to scavenge reactive oxygen species. *Plant Cell Tissue Organ. Cult.* **2014**, *117*, 99–112. [CrossRef]
196. Singh, N.; Mishra, A.; Jha, B. Ectopic over-expression of peroxisomal ascorbate peroxidase (*SbpAPX*) gene confers salt stress tolerance in transgenic peanut (*Arachis hypogaea*). *Gene* **2014**, *547*, 119–125. [CrossRef] [PubMed]
197. Negi, N.P.; Shrivastava, D.C.; Sharma, V.; Sarin, N.B. Overexpression of *CuZnSOD* from *Arachis hypogaea* alleviates salinity and drought stress in tobacco. *Plant Cell Rep.* **2015**, *34*, 1109–1126. [CrossRef]
198. Yan, H.; Li, Q.; Park, S.C.; Wang, X.; Liu, Y.J.; Zhang, Y.G.; Tang, W.; Kou, M.; Ma, D.F. Overexpression of *CuZnSOD* and *APX* enhance salt stress tolerance in sweet potato. *Plant Physiol. Biochem.* **2016**, *109*, 20–27. [CrossRef] [PubMed]
199. Qi, Q.; Yanyan, D.; Yuanlin, L.; Kunzhi, L.; Huini, X.; Xudong, S. Overexpression of *SlMDHAR* in transgenic tobacco increased salt stress tolerance involving *S*-nitrosylation regulation. *Plant Sci.* **2020**, *299*, 110609. [CrossRef]
200. Saxena, S.C.; Salvi, P.; Kamble, N.U.; Joshi, P.K.; Majee, M.; Arora, S. Ectopic overexpression of cytosolic ascorbate peroxidase gene (*Apx1*) improves salinity stress tolerance in *Brassica juncea* by strengthening antioxidative defense mechanism. *Acta Physiol. Plant.* **2020**, *42*, 45. [CrossRef]

International Journal of
Molecular Sciences

MDPI

Article

NaCl-Induced Elicitation Alters Physiology and Increases Accumulation of Phenolic Compounds in *Melissa officinalis* L.

Barbara Hawrylak-Nowak [1,*], Sławomir Dresler [2], Maria Stasińska-Jakubas [1], Magdalena Wójciak [2], Ireneusz Sowa [2] and Renata Matraszek-Gawron [1]

1 Department of Botany and Plant Physiology, Faculty of Environmental Biology, University of Life Sciences in Lublin, Akademicka 15, 20-950 Lublin, Poland; jakubas.ms@gmail.com (M.S.-J.); renata.matraszek@up.lublin.pl (R.M.-G.)

2 Department of Analytical Chemistry, Medical University of Lublin, Chodźki 4a, 20-093 Lublin, Poland; slawomir.dresler@umlub.pl (S.D.); magdalenawojciak@umlub.pl (M.W.); i.sowa@umlub.pl (I.S.)

* Correspondence: barbara.nowak@up.lublin.pl

Abstract: In nature, plants usually produce secondary metabolites as a defense mechanism against environmental stresses. Different stresses determine the chemical diversity of plant-specialized metabolism products. In this study, we applied an abiotic elicitor, i.e., NaCl, to enhance the biosynthesis and accumulation of phenolic secondary metabolites in *Melissa officinalis* L. Plants were subjected to salt stress treatment by application of NaCl solutions (0, 50, or 100 mM) to the pots. Generally, the NaCl treatments were found to inhibit the growth of plants, simultaneously enhancing the accumulation of phenolic compounds (total phenolics, soluble flavonols, anthocyanins, phenolic acids), especially at 100 mM NaCl. However, the salt stress did not disturb the accumulation of photosynthetic pigments and proper functioning of the PS II photosystem. Therefore, the proposed method of elicitation represents a convenient alternative to cell suspension or hydroponic techniques as it is easier and cheaper with simple application in lemon balm pot cultivation. The improvement of lemon balm quality by NaCl elicitation can potentially increase the level of health-promoting phytochemicals and the bioactivity of low-processed herbal products.

Keywords: phenolic metabolites; lemon balm; chlorophyll fluorescence; medicinal plants; secondary metabolites; abiotic elicitors; salinity

Citation: Hawrylak-Nowak, B.; Dresler, S.; Stasińska-Jakubas, M.; Wójciak, M.; Sowa, I.; Matraszek-Gawron, R. NaCl-Induced Elicitation Alters Physiology and Increases Accumulation of Phenolic Compounds in *Melissa officinalis* L. *Int. J. Mol. Sci.* **2021**, *22*, 6844. https://doi.org/10.3390/ijms22136844

Academic Editor: Francisco Rubio

Received: 1 June 2021
Accepted: 23 June 2021
Published: 25 June 2021

Publisher's Note: MDPI stays neutral with regard to jurisdictional claims in published maps and institutional affiliations.

1. Introduction

Due to their sedentary lifestyle, plants are under constant pressure to adjust their metabolic pathways to changing environmental conditions. Therefore, in addition to primary metabolites, they synthesize a wide range of unique, low-molecular secondary metabolites. Environmental stresses (biotic and abiotic) redirect plant metabolism towards the biosynthesis of these metabolites from primary metabolites and intermediates. The products of specialized metabolism most often have defense and signaling functions. Such compounds are generally toxic or inconsumable for herbivores, possess fungicidal or bactericidal properties, or can detoxify toxic metals and consequently protect plants against stresses [1–3].

Therefore, stress is an important factor determining the chemical composition of plants and thus having a significant impact on their biological activity. The use of plant defense mechanisms to stimulate the biosynthesis of desired secondary metabolites and improve the health-promoting quality of plants is called elicitation [4]. Elicitors are physical factors or chemical substances that can induce responses via modifications of accumulation and/or synthesis of secondary metabolites, mimicking a defensive reaction [5–7]. One of the readily available and cheap abiotic elicitors with proven effectiveness is NaCl [8,9].

Salinity induces both ionic and osmotic stress in plants. The intensity of the salt stress plays a decisive role in the plant salinity response and the possibility of reversible or irreversible changes in plant functioning [10]. The effect of salt stress on plants can be twofold. Excess of salt may have a toxic effect on plants, limiting their growth and development, and may lead to plant death in extreme cases [11]. Nevertheless, a moderate or mild level of salinity may have a stimulating effect on the growth and development of plants (eustress) and/or the accumulation of secondary metabolites, improving the level of pro-health components, antioxidant potential, and nutritional quality of plants [5,8,12]. For example, in safflower (*Carthamus tinctorius* L.) grown at a salinity concentration <100 mM NaCl in hydroponic conditions, the accumulation of flavonoids was enhanced, but the plant growth was not reduced [13]. In turn, Navarro et al. [14] demonstrated improved antioxidant activity in hydrophilic and lipophilic fractions under moderate salinity in red pepper.

Lemon balm (*Melissa officinalis* L.), belonging to the family Lamiaceae, is a valuable herb used as a flavoring agent in food and drinks. Due to the richness of secondary metabolites (mainly monoterpenoids in essential oils and phenolic compounds), it is also used as a medicinal plant, natural insecticide, and an ingredient of cosmetics [15]. The many biologically active components in lemon balm include a number of phenolics, the most important of which is rosmarinic acid [16,17]. Ingredients of extracts from *M. officinalis* revealed a number of pharmacological activities, including clinically proven antiviral, anxiolytic, and antispasmodic effects as well as influence on mood stabilization and memory [18]. Due to the positive effect on cognitive function and agitation, *M. officinalis* extract is valuable in the management of mild to moderate Alzheimer's disease [19]. This species is resistant to relatively low concentrations of salinity (up to 50 mM NaCl) [20]. Irrigation of lemon balm with saline water increased the essential oil yield, free proline, and total soluble sugar levels but decreased plant growth parameters [21]. In turn, a decrease in the yield of essential oils but an increase in the number of their ingredients under NaCl exposure (50–200 mM NaCl) was found by Bonacina et al. [20]. However, the effect of salinity on the level and composition of phenolic compounds in this species is poorly understood. One of the few studies on this topic indicated that NaCl-treated lemon balm accumulated more total phenolics and flavonoids than untreated plants [22].

The hypothesis that NaCl-induced stress affects physiological response and enhances accumulation of (poly)phenolics in lemon balm was tested. This elicitation method may potentially improve the level of health-promoting secondary metabolites and bioactivity of lemon balm-based raw materials. Moreover, this work provides insights into the effect of salt stress on the physiology and phenolic composition of the pharmaceutically significant species *M. officinalis*.

2. Results

2.1. Biomass and Physiological Parameters of Lemon Balm Grown under NaCl Exposure

The use of the NaCl solution for plant irrigation had a significant impact on the plant biomass (Figure 1a). A decrease in the FW of the above-ground organs with the increase in NaCl concentration was observed; however, this reduction (22% in relation to the control) was statistically significant only under the influence of 100 mM NaCl. In comparison to the control plants, an increase (by 6–11%) in the concentration of chlorophyll *b* at 50 and 100 mM NaCl was found. However, the level of chlorophyll *a* and carotenoids did not change significantly under the salt exposure (Figure 1b). The measurements of selected parameters of chlorophyll *a* fluorescence (F_0, F_m, and F_v/F_m) indicated that the applied concentrations of NaCl had no significant effects on the efficiency of photosynthesis (Supplementary Figure S1).

2.2. Total Phenolic Compounds, Flavonoids, Rosmarinic Acid, and Anthocyanin Concentrations under NaCl Elicitation

Salinity influenced the total concentration of phenolic compounds (TPC) in the extracts made from the lemon balm herb. It was found that both of the different NaCl levels enhanced the TPC concentration, but this increase (by 16% compared to the control) was statistically significant only after the application of 100 mM NaCl (Figure 2a). Similarly, the content of total flavonol compounds (TFC) significantly increased by 23% after application of 100 mM NaCl (Figure 2b).

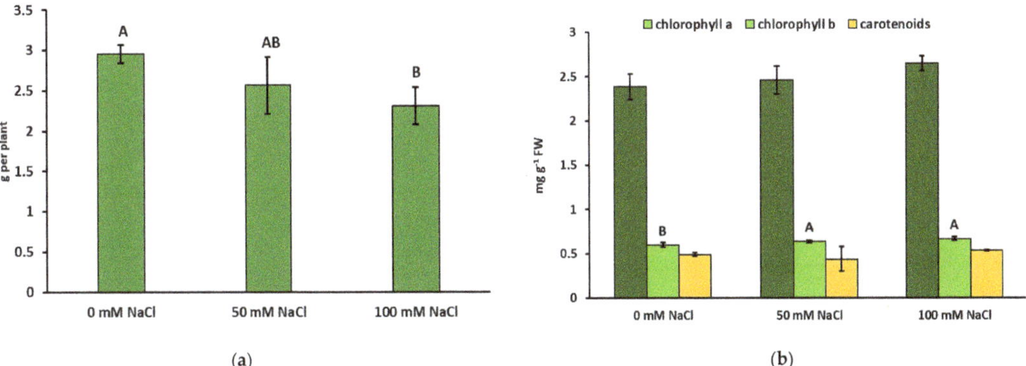

(a) (b)

Figure 1. Effect of the NaCl concentration on (**a**) growth parameters and (**b**) concentrations of photosynthetic pigments in *Melissa officinalis* after 10 days of salt exposure. Data are means ± SD (n = 16 for FW and n = 6 for photosynthetic pigments). Means followed by different letters differ statistically significantly ($p < 0.05$, Tukey's test). The absence of letters indicates that there are no significant differences.

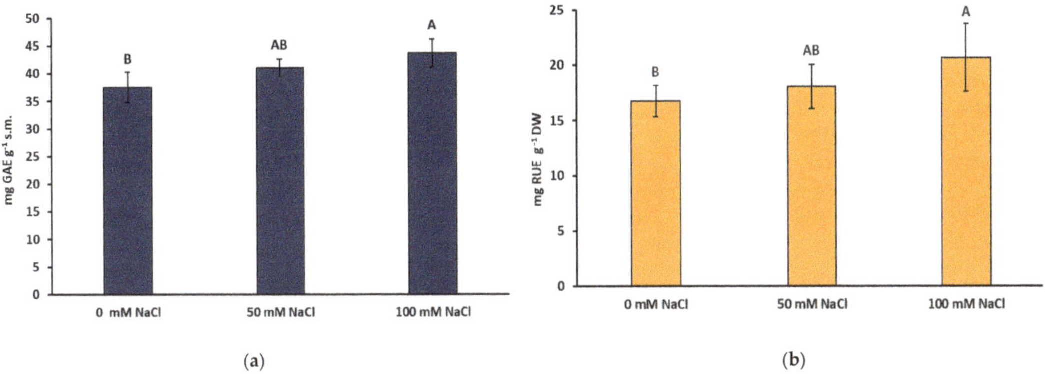

(a) (b)

Figure 2. Effect of the NaCl concentration on the concentration of (**a**) total soluble phenolic compounds and (**b**) total soluble flavonols in *Melissa officinalis* shoots after 10 days of salt exposure. Data are means ± SD (n = 4). Means followed by different letters differ statistically significantly ($p < 0.05$, Tukey's test).

Based on the UPLC-UV-MS analysis, 13 different phenolic acids, mainly hydroxycinnamic acid derivatives including a caffeic acid ester—rosmarinic acid and a caffeate trimer—lithospermic acid, were identified (Figure 3). These acids represented two characteristically high peaks in the chromatograms (Figure 4). All the chromatograms obtained from both the control and the NaCl-treated plants showed a similar phenolic profile, showing differences only in the quantities of each compound. No peaks from the new compounds were observed under the influence of NaCl (Figure 4). It was found that both salinity concentrations induced the accumulation of phenolic acids (Figures 4 and 5). Quantitative determination of rosmarinic acid showed that its level increased by 40% and 67%, respectively, under the influence of 50 and 100 mM NaCl in relation to the control (Figure 5).

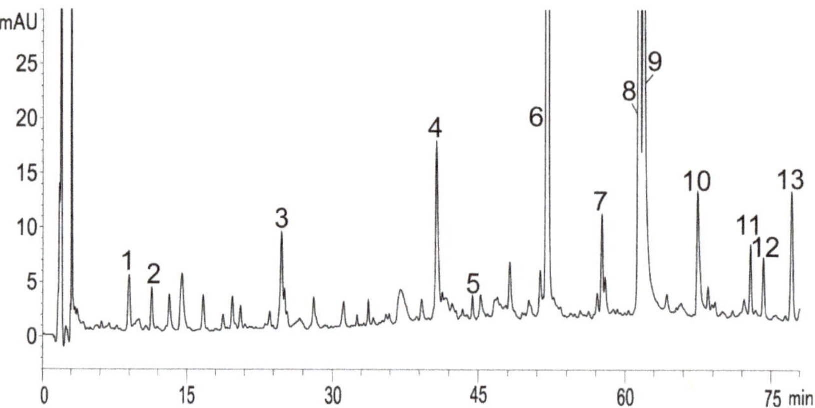

Figure 3. Chromatogram of *Melissa officinalis* leaf extract (control plants) with peaks identified by HPLC-UV-MS. (1) 3-(3,4-dihydroxyphenyl)-lactic acid; (2) caftaric acid; (3) fertaric acid; (4) caftaric acid hexoside; (5) rosmarinic acid hexoside; (6) rosmarinic acid; (7) salvianolic acid H/I (isomer); (8) salvianolic acid C derivative III; (9) lithospheric acid; (10) salvianolic acid C derivative III; (11) salvianolic acid C derivative IV; (12) sagecoumarin 2-hydroxy-3-(3,4-dihydroxyphenyl)-propanoide; and (13) methyl lithospermic acid.

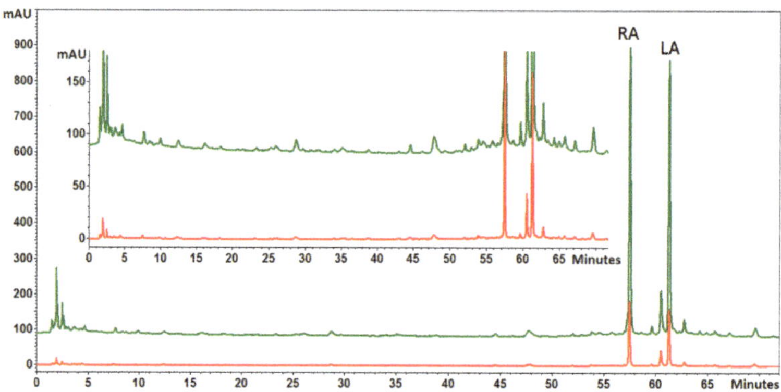

Figure 4. Comparison of chromatograms of *Melissa officinalis* leaf extracts (evaluated by HPLC-UV-MS) from control plants (red line) and from plants treated with 100 mM NaCl (green line); RA, rosmarinic acid; LA, lithospheric acid.

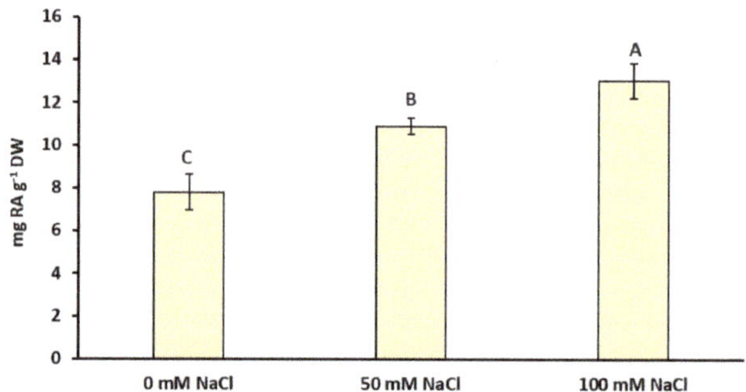

Figure 5. Effect of the NaCl application on the concentration of rosmarinic acid in *Melissa officinalis* shoots after 10 days of exposure. Data are means ± SD ($n = 4$). Means followed by different letters differ statistically significantly ($p < 0.05$, Tukey's test).

The elicitation of the lemon balm with both concentrations of NaCl boosted the accumulation of anthocyanins in lemon balm leaves (Figure 6a), especially in the lower epidermis (Figure 6b). Along with the increase in salinity, over a two-fold or four-fold increase in the concentration of these compounds was found after the exposure to 50 mM NaCl or 100 mM NaCl, respectively.

The heat map (Figure 7) represents graphically the abundance of particular phenolic compounds across the experimental treatments. The standardized parameters are represented by colors (dark blue represents low value while dark red denotes high value). Four individuals are shown for each treatment. Additionally, the heat map also shows the aforementioned enhanced accumulation of anthocyanins, soluble phenols, and soluble flavonols in the NaCl treatments; however, this phenomenon was more notable at the higher concentration of salt (Figure 7). Moreover, the calculated mean contents of rosmarinic acid and anthocyanin per plant in the control plants were 4.6 mg and 7.7 ng, respectively, and increased to 6.6 mg and 33.9 ng per plant after eliciting with 100 mM NaCl. Meanwhile, TPC and TFC were at a similar level per plant in different treatments (Supplementary Table S1). The analysis of the influence of NaCl on the ability of the plant extracts to reduce DPPH radical revealed a significant increase in free radical scavenging activity (FRSA) only in the 100 mM NaCl treatment (Supplementary Figure S2).

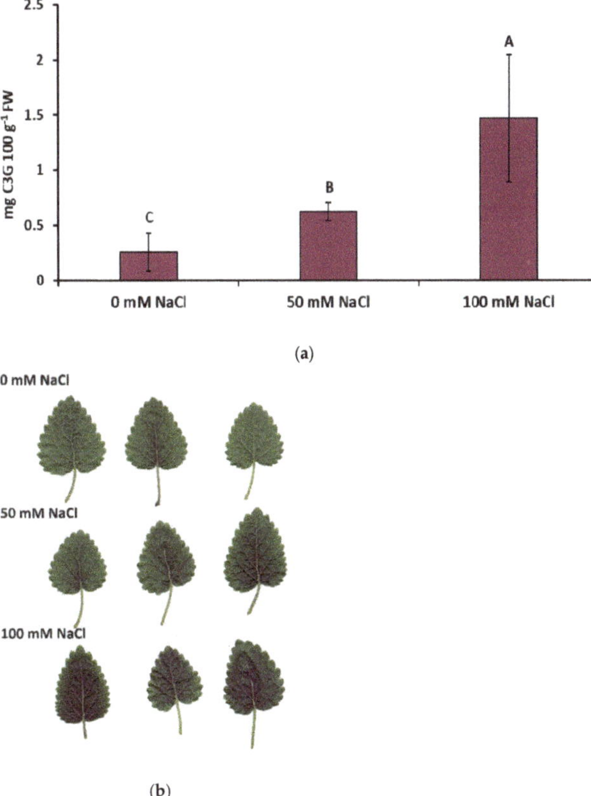

Figure 6. Effect of the NaCl concentration on (**a**) the foliar concentration of anthocyanins (as cyanidin 3-glycoside; C3G) and (**b**) pigmentation of the lower epidermis of *Melissa officinalis* leaves after 10 days of salt exposure. Data are means ± SD (n = 4). Means followed by different letters differ statistically significantly ($p < 0.05$, Tukey's test).

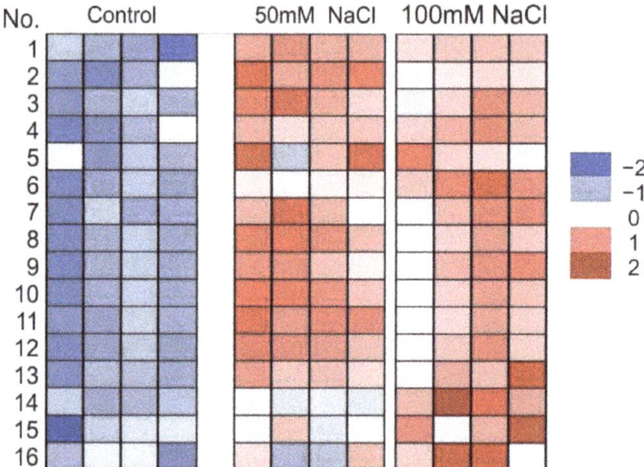

Figure 7. Heat map visualization of changes in the abundance of particular compounds shown in the rows for individual plant samples with different salinity levels (column). The colors range from dark blue (low abundance) to deep red (high abundance); number of compounds: (1) 3-(3,4-dihydroxyphenyl)-lactic acid; (2) caftaric acid; (3) fertaric acid; (4) caftaric acid hexoside; (5) rosmarinic acid hexoside; (6) rosmarinic acid; (7) salvianolic acid H/I (isomer); (8) salvianolic acid C derivative III; (9) lithospheric acid; (10) salvianolic acid C derivative III; (11) salvianolic acid C derivative IV; (12) sagecoumarin 2-hydroxy-3-(3,4-dihydroxyphenyl)-propanoide; (13) methyl lithospermic acid; (14) total anthocyanins; (15) soluble phenols; and (16) soluble flavonols.

3. Discussion

Salt-induced osmotic stress leads to numerous physiological disorders in plants, including water deficit, nutrient imbalance, membrane damage, hormonal imbalance, and oxidative damage [23]. For the above reasons, the biomass of lemon balm can decrease at various NaCl concentrations in the irrigation water used in our experiments (Figure 1a), which was also demonstrated by Khalid and Cai [21]. Similarly, in another medical species, i.e., sage, irrigated with a 100 mM NaCl solution, a significant reduction in growth parameters was demonstrated [24]. Nevertheless, at moderate levels, salt stress can also be an effective method of eliciting plant secondary metabolites [25].

Although the lemon balm biomass was significantly reduced under the influence of 100 mM but not 50 mM NaCl (Figure 1a), the analyzed parameters of chlorophyll *a* fluorescence did not indicate disturbances in photosynthesis in either of the NaCl treatments (Supplementary Figure S1). Also, the concentration of photosynthetic pigments was generally not negatively affected even in the 100 mM NaCl treatment. In the salinity conditions, the level of chlorophyll *b* was even increased (Figure 1a). Therefore, it seems that lemon balm may be more resistant to salinity than previously suggested by Bonacina et al. [20]. In contrast, in a study conducted by Safari et al. [22], a decrease in the content of chlorophyll under salt stress was found in this species. Studies conducted on sunflower have shown that the NaCl-induced reduction of chlorophyll level may be mainly a consequence of inhibited synthesis of 5-aminolaevulinic acid, a precursor of chlorophyll and, to a lesser extent, increased activity chlorophyll-degrading chlorophyllase [26].

However, in the context of the present research, a more interesting issue was the increase in the accumulation of bioactive substances under salinity. Besides essential oils, (poly)phenolic compounds with a broad spectrum of biological activities are the main group of secondary metabolites in the Lamiaceae family [27]. As shown by Ozarowski et al. [28], caffeic acid esters and glycosides of flavones are the major group of active metabolites in *M. officinalis*. Additionally, they indicated that the rosmarinic acid is a dominant active compound in this species. In our study, 13 caffeic acid derivatives were identified. Rosmarinic acid and lithospheric acid were the main phenolic acids identified in the extracts obtained from the shoots of the lemon balm (Figures 3 and 4). It was found that the accumulation of different phenolic compounds, including caffeic acid derivatives, was clearly enhanced under salinity (Figures 4 and 7). The concentration of rosmarinic acid was improved by 40–67% in comparison to the control (Figure 5). In turn, rosmarinic acid concentrations in five genotypes of *M. officinalis* were reduced by osmotic stress induced by drought, while the level of essential oils increased in some genotypes [29]. The phenomenon of an increasing level of (poly)phenolic compounds in response to various abiotic stress factors was observed before [30–32]. Although the mechanisms of (poly)phenol biosynthesis were not investigated here, ample evidence indicated that the phenylpropanoid pathway which generates a majority of compounds is activated by stress factors [33]. Previously, a positive correlation between salinity and phenolic compound concentrations was noted in *Thymus* species [34] or *Fagopyrum esculentum* [33]. In contrast, the content of phenolic compounds decreased in response to NaCl in broccoli [35] or lettuce [36], which indicates differential responses of plant species to salinity in relation to accumulation of (poly)phenolics.

Our recent studies on the application of a biotic elicitor chitosan lactate in basil and lemon balm showed that the foliar application of this compound effectively induced accumulation of phenolic compounds [37]. Here, the NaCl-induced elicitation generally had a positive effect on the level of all (poly)phenolics tested. However, the accumulation of TPC and TFC as well as FRSA was significantly improved only at the higher level of salinity (Figure 2a,b, Supplementary Figure S2). Much more effective elicitation of TPC was achieved in a study with *Mentha pulegium* [38], where a 3.5-fold increase in its level in leaf extracts was found in a 100 mM NaCl treatment. Unfortunately, a drastic reduction of plant biomass (approx. 60%) was recorded in these conditions. In turn, in another medicinal species belonging to the Lamiaceae family, i.e., *Salvia macrosiphon*, the content of TPC decreased under salinity, but increased leaf antioxidant capacity was demonstrated [39]. In a mangrove halophyte *Aegiceras corniculatum* exposed to 250 mM NaCl, the levels of (poly)phenols increased more than twofold, which may indicate their protective role under salt stress [40].

In this study, the accumulation of anthocyanins increased several fold in response to the salinity stress (Figure 6a). Changes in the anthocyanin content were also visible as differences in the color of the lower leaf epidermis (Figure 6b). Analogous results were obtained in a study conducted by Jahantigh et al. [41] on hyssop (Lamiaceae), which indicated a significant increase in the level of both TPC and anthocyanins after application of saline water (2–10 mS cm^{-1}). The enhanced accumulation of anthocyanins in plants exposed to salinity has been largely documented (e.g., [42,43]). The protective role of anthocyanins under salt stress includes their antioxidant capacity in response to ROS overproduction triggered by imbalance in Na^+/K^+ ion homeostasis [44]. Moreover, these compounds can be involved in reduction of the leaf osmotic potential, which is directly related to improved plant water status under osmotic stress, suggesting that accumulation of anthocyanins can be a quick and beneficial plant response to salt stress [45]. It should also be emphasized that the increase in the concentration of anthocyanins and rosmarinic acid, both with well-known health-promoting properties [46,47], overcompensates the decrease in plant biomass under salt stress (Supplementary Table S1).

4. Materials and Methods

4.1. Plant Material and Growth Conditions

Twenty seeds of lemon balm (*Melissa officinalis* L.) were sown in 12 pots with a capacity of 0.5 L and allowed to germinate. The soil substrate was the universal organic COMPO BIO substrate with the addition of guano and compost produced on the basis of high peat, with pH = 5.0–7.0. Seed germination as well as further growth and development of plants took place in a phytotron room equipped with air conditioning and fluorescent lamps. The plants were grown at photosynthetic photon flux density (PPFD) at the level of the tops of the plants of 170–200 µmol m^{-2} s^{-1}, 14-h photoperiod, at day/night temperature of 27/23 °C, and 60–65% relative humidity. Due to the uneven germination of the seeds, the plants were thinned after about 21 days, and 12 plants were left in each pot.

The experiment was differentiated 36 days after sowing (DAS), and three treatments were established. Control plants (in four pots) were watered with 50 mL of distilled water, while the other plants were treated with the same volume of NaCl solutions with a concentration of 50 mM (5.38 mS cm^{-1}; four pots) or 100 mM (10.16 mS cm^{-1}; four pots). A similar irrigation scheme was applied after the next 2, 5, and 7 days. In total, 200 mL of solutions with different NaCl concentrations (0, 50, or 100 mM) were introduced into each pot.

After 10 days of lemon balm growth in the different experimental conditions (10 days after the application of the first dose of NaCl), the biometric (biomass of shoots), physiological (photosynthetic pigments, chlorophyll fluorescence), and phytochemical (concentration of total phenolics, soluble flavonoids, and anthocyanins, quantitative analysis of rosmarinic acid, identification of phenolic acids, free radical scavenging activity) parameters were determined.

4.2. *Determination of Biomass and Physiological Parameters*

The chlorophyll *a* fluorescence parameters (the maximal, F_m; and minimal, F_0 possible level of fluorescence; the maximum quantum yield of PS II, F_v/F_m; where $F_v = F_m - F_0$) were measured on the fourth pair (from the top) of dark-adapted (15 min) leaves. Ten different individuals per treatment were randomly selected, and a chlorophyll fluorimeter (Handy PEA, Hansatech Instruments, Pentney, UK) was used for fluorescence determinations.

The concentration of photosynthetic pigments (chlorophylls and carotenoids) was measured using the method proposed by Lichtenthaler and Wellburn [48]. The leaf samples were taken from the fourth pair from the top, homogenized in 80% (*v*/*v*) acetone, and filtered. Absorbance of the extracts was measured at 663 nm, 646 nm, and 470 nm (Cecil CE 9500, Cecil Instruments, Cambridge, UK).

Then, the aboveground parts of four plants from each pot were collected (16 plants from each treatment), and their fresh weight (FW) was determined using a laboratory balance.

4.3. *Preparation of Extracts for Determination of Total Phenolics, Flavonols, Phenolic Acids, and Free Radical Scavenging Activity*

Plant material from each pot dried at 55 °C was used for preparation of extracts. Samples were extracted with 5 mL of 80% (*v*/*v*) methanol at room temperature for 1/2 h in an ultrasonic bath. The extracts were centrifuged (10 min at 4500× *g*), and the clear supernatant was used for further analysis.

4.4. *Analysis of Phenolic Compounds and Free Radical Scavenging Activity*

The total phenolic content (TPC) was determined using the Folin-Ciocalteu phenol reagent, following the method proposed by Wang [49] with slight modifications. The test sample (0.1 mL) was mixed with 1.9 mL of re-distilled water and 1 mL of Folin-Ciocalteu's reagent. After 5 min, 3 mL of a saturated Na_2CO_3 solution were added. The reaction mixture was kept at 40 °C in the dark for 30 min. Absorbance was measured at 756 nm (Cecil CE 9500, Cecil Instruments, Cambridge, UK) against the reagent blank. The concentration of phenolic compounds was calculated as gallic acid equivalents (GAE) per g of dry plant material.

Soluble flavonols were determined with the colorimetric method as a complex with aluminum ions [50]. The absorbance was read at 425 nm (Cecil CE 9500, Cecil Instruments, Cambridge, UK) after 30 min of dark incubation of the test sample (0.3 mL) with 0.75 mL of a 2% $AlCl_3$ methanolic solution (*w*/*v*) and 0.45 mL of 80% methanol (*v*/*v*) against the reagent blank. The concentration of flavonoids was calculated as rutin equivalents (RE) per g of dry plant material.

The accumulation of anthocyanins in the fresh leaves (fourth pair from the top) was determined using the method described previously [51]. Anthocyanins were extracted by maceration of leaf samples in a methanol:HCl solution (99:1, *v*/*v*). The extracts were centrifuged, and their absorbances were read at 527 nm and 652 nm (Cecil CE 9500, Cecil Instruments, Cambridge, UK). The concentration of anthocyanins was calculated using the extinction coefficient (ε = 29,600 M^{-1} cm^{-1}) for cyanidin 3-glycoside (C3G).

The methanol extracts were analyzed using an Agilent Technology 1290 Infinity Series II ultra-high performance liquid chromatograph (UHPLC) equipped with a DAD detector and an Agilent 6224 ESI/TOF mass detector (Agilent Technologies, Santa Clara, CA, USA). The ion source operating parameters were as follows: drying gas temperature 325 °C, drying gas flow 5.0 l min^{-1}, and capillary voltage 3500 V. Ions were acquired in the range from 100 to 1050 m/z in the negative ion mode. Agilent Technologies Mass Hunter software version 10.00 00 (Agilent Technologies, Santa Clara, CA, USA) was used for data acquisition and data analysis. The separation was performed with a method described previously [52]. Briefly, an RP18e LiChrosper 100 column (Merck, Darmstadt, Germany) (25 cm × 4.9 mm i.d., 5 µm particle size) was used to separate phenolic acids. The linear gradient from 5% to 20% of acetonitrile in water within 45 min was applied. The flow rate was 1.0 mL/min. The column temperature was set at 25 °C. The identity and quantification of rosmarinic acid was performed based on comparison with a standard compound.

The identification of 13 different phenolic acids was possible using UPLC-TOF/MS analysis (Figure 3), as in Ozarowski et al. [28] and Barros et al. [53].

The free radical scavenging activity (FRSA) was determined using DPPH (1,1-diphenyl-2-picrylhydrazyl) stable radical. The test sample (50 µL) was added to 2 mL of a DPPH solution (200 µmol L^{-1}). A total of 50 µL of 80% methanol (*v*/*v*) was added to the control sample. The absorbance of the control and test samples was determined at 517 nm (Cecil CE 9500, Cecil Instruments, Cambridge, UK) after 15 min of dark incubation. Results are reported as percentage DPPH reduction by the plant extracts.

4.5. Statistical Analyses

The data were subjected to one-way ANOVA followed by a Tukey's post-hoc test ($p < 0.05$). Statistica ver. 13.3 software (TIBCO Software Inc. 2017, Palo Alto, CA, USA) was used for the statistical analysis. The heat map was constructed based on standardized data with Microsoft Excel (2010).

5. Conclusions

Our study demonstrates that NaCl irrigation functions as an activator of accumulation of (poly)phenolics. These results show, for the first time, enhanced accumulation of hydroxycinnamic acid derivatives in lemon balm under salinity. The increase in anthocyanin concentration was several fold. It is worth emphasizing that the biomass of the aboveground parts did not decrease significantly under the influence of 50 mM NaCl, and its reduction in the 100 mM NaCl treatment was significant but not very large. In the salt treatments, there were no significant disturbances in photosynthesis parameters and the content of photosynthetic pigments. Therefore, NaCl is a cheap and efficient abiotic elicitor that may be potentially used in elicitation of phenolic metabolites in lemon balm under pot cultivation. However, we do not recommend this elicitation method in field conditions due to the limited possibility of later removal of NaCl from the soil.

Supplementary Materials: Supplementary Materials can be found at https://www.mdpi.com/article/10.3390/ijms22136844/s1.

Author Contributions: B.H.-N., M.S.-J. and S.D. designed the project; M.S.-J. and B.H.-N. cultivated the plants; R.M.-G. and M.S.-J. handled photosynthetic pigments concentrations and chlorophyll fluorescence; B.H.-N. and M.S.-J. performed analysis of total phenolics, soluble flavonols, and FRSA; S.D., M.W. and I.S. performed HPLC-UV-MS analysis; B.H.-N. and S.D. performed statistical analysis and data visualization; B.H.-N. and S.D. prepared the original manuscript draft; R.M.-G. and I.S. critically revised the manuscript. All authors have read and agreed to the published version of the manuscript.

Funding: This research received no external funding.

Institutional Review Board Statement: Not applicable.

Informed Consent Statement: Not applicable.

Data Availability Statement: The data presented in this study are available on request from the corresponding author.

Acknowledgments: Special thanks for technical support and help with the experiments are extended to Weronika Woch.

Conflicts of Interest: The authors declare no conflict of interest.

References

1. Nascimento, N.C.; Fett-Neto, A.G. Plant secondary metabolism and challenges in modifying its operation: An overview. In *Plant Secondary Metabolism Engineering. Methods and Protocols*; Fett-Neto, A.G., Ed.; Humana Press: Totowa, NJ, USA, 2010; Volume 643, pp. 1–13. [CrossRef]
2. Yang, L.; Wen, K.-S.; Ruan, X.; Zhao, Y.-X.; Wei, F.; Wang, Q. Response of Plant Secondary Metabolites to Environmental Factors. *Molecules* **2018**, *23*, 762. [CrossRef]
3. Golkar, P.; Taghizadeh, M.; Yousefian, Z. The effects of chitosan and salicylic acid on elicitation of secondary metabolites and antioxidant activity of safflower under in vitro salinity stress. *Plant Cell Tissue Organ Cult. (PCTOC)* **2019**, *137*, 575–585. [CrossRef]
4. Gorelick, J.; Bernstein, N. Elicitation: An underutilized tool in the development of medicinal plants as a source of therapeutic secondary metabolites. *Adv. Agron.* **2014**, *124*, 201–230. [CrossRef]
5. Hassini, I.; Rios, J.J.; Ibáñez, P.G.; Baenas, N.; Carvajal, M.; Moreno, D.A. Comparative effect of elicitors on the physiology and secondary metabolites in broccoli plants. *J. Plant Physiol.* **2019**, *239*, 1–9. [CrossRef] [PubMed]
6. Baenas, N.; García-Viguera, C.; Moreno, D.A. Elicitation: A Tool for Enriching the Bioactive Composition of Foods. *Molecules* **2014**, *19*, 13541–13563. [CrossRef]
7. Thakur, M.; Bhattacharya, S.; Khosla, P.K.; Puri, S. Improving production of plant secondary metabolites through biotic and abiotic elicitation. *J. Appl. Res. Med. Aromat. Plants* **2019**, *12*, 1–12. [CrossRef]
8. Rouphael, Y.; Petropoulos, S.A.; Cardarelli, M.; Colla, G. Salinity as eustressor for enhancing quality of vegetables. *Sci. Hortic.* **2018**, *234*, 361–369. [CrossRef]
9. Kitayama, M.; Tisarum, R.; Theerawitaya, C.; Samphumphung, T.; Takagaki, M.; Kirdmanee, C.; Cha-Um, S. Regulation on anthocyanins, α-tocopherol and calcium in two water spinach (Ipomoea aquatica) cultivars by NaCl salt elicitor. *Sci. Hortic.* **2019**, *249*, 390–400. [CrossRef]
10. Hernández, J.A. Salinity Tolerance in Plants: Trends and Perspectives. *Int. J. Mol. Sci.* **2019**, *20*, 2408. [CrossRef]
11. Zhu, J.-K. Plant salt tolerance. *Trends Plant Sci.* **2001**, *6*, 66–71. [CrossRef]
12. Świeca, M. Elicitation with abiotic stresses improves pro-health constituents, antioxidant potential and nutritional quality of lentil sprouts. *Saudi J. Biol. Sci.* **2015**, *22*, 409–416. [CrossRef]
13. Gengmao, Z.; Yu, H.; Xing, S.; Shihui, L.; Quanmei, S.; Changhai, W. Salinity stress increases secondary metabolites and enzyme activity in safflower. *Ind. Crops Prod.* **2015**, *64*, 175–181. [CrossRef]
14. Navarro, J.M.; Flores, P.; Garrido, C.; Martinez, V. Changes in the contents of antioxidant compounds in pepper fruits at ripening stages, as affected by salinity. *Food Chem.* **2006**, *96*, 66–73. [CrossRef]
15. Turhan, H. Lemon balm. In *Handbook of Herbs and Spices*; Peter, K.V., Ed.; Woodhead Publishing: Cambridge, UK, 2006; Volume 3, pp. 390–399. [CrossRef]
16. Weitzel, C.; Petersen, M. Enzymes of phenylpropanoid metabolism in the important medicinal plant *Melissa officinalis* L. *Planta* **2010**, *232*, 731–742. [CrossRef]
17. Milevskaya, V.V.; Temerdashev, Z.A.; Butyl'skaya, T.S.; Kiseleva, N.V. Determination of phenolic compounds in medicinal plants from the Lamiaceae family. *J. Anal. Chem.* **2017**, *72*, 342–348. [CrossRef]
18. Shakeri, A.; Sahebkar, A.; Javadi, B. *Melissa officinalis* L.—A review of its traditional uses, phytochemistry and pharmacology. *J. Ethnopharmacol.* **2016**, *188*, 204–228. [CrossRef] [PubMed]
19. Akhondzadeh, S.; Noroozian, M.; Mohammadi, M.; Ohadinia, S.; Jamshidi, A.H.; Khani, M. *Melissa officinalis* extract in the treatment of patients with mild to moderate Alzheimer's disease: A double blind, randomised, placebo controlled trial. *J. Neurol. Neurosurg. Psychiatry* **2003**, *74*, 863–866. [CrossRef] [PubMed]
20. Bonacina, C.; Trevizan, C.B.; Stracieri, J.; dos Santos, T.B.; Goncalves, J.E.; Gazim, Z.C.; de Souza, S.G.H. Changes in growth, oxidative metabolism and essential oil composition of lemon balm ('*Melissa officinalis*' L.) subjected to salt stress. *Aust. J. Crop Sci.* **2017**, *11*, 1665–1674. [CrossRef]
21. Khalid, K.A.; Cai, W. The effects of mannitol and salinity stresses on growth and biochemical accumulations in lemon balm. *Acta Ecol. Sin.* **2011**, *31*, 112–120. [CrossRef]
22. Safari, F.; Akramian, M.; Salehi-Arjmand, H. Physiochemical and molecular responses of salt-stressed lemon balm (*Melissa officinalis* L.) to exogenous protectants. *Acta Physiol. Plant.* **2020**, *42*, 1–10. [CrossRef]

23. Hasanuzzaman, M.; Nahar, K.; Fujita, M. Plant response to salt stress and role of exogenous protectants to mitigate salt-induced damages. In *Ecophysiology and Responses of Plants under Salt Stress*; Ahmad, P., Azooz, M.M., Prasad, M.N.V., Eds.; Springer: New York, NY, USA, 2013; pp. 25–87. [CrossRef]
24. Taarit, M.B.; Msaada, K.; Hosni, K.; Hammami, M.; Kchouk, M.E.; Marzouk, B. Plant growth, essential oil yield and composition of sage (*Salvia officinalis* L.) fruits cultivated under salt stress conditions. *Ind. Crop. Prod.* **2009**, *30*, 333–337. [CrossRef]
25. Ampofo, J.O.; Ngadi, M. Stimulation of the phenylpropanoid pathway and antioxidant capacities by biotic and abiotic elicitation strategies in common bean (*Phaseolus vulgaris*) sprouts. *Process Biochem.* **2021**, *100*, 98–106. [CrossRef]
26. Santos, C.V. Regulation of chlorophyll biosynthesis and degradation by salt stress in sunflower leaves. *Sci. Hortic.* **2004**, *103*, 93–99. [CrossRef]
27. Sytar, O.; Hemmerich, I.; Zivcak, M.; Rauh, C.; Brestic, M. Comparative analysis of bioactive phenolic compounds composition from 26 medicinal plants. *Saudi J. Biol. Sci.* **2018**, *25*, 631–641. [CrossRef]
28. Ozarowski, M.; Mikołajczak, P.; Piasecka, A.; Kachlicki, P.; Kujawski, R.; Bogacz, A.; Bartkowiak-Wieczorek, J.; Szulc, M.; Kaminska, E.; Kujawska, M.; et al. Influence of the *Melissa officinialis* leaf extract on long-term memory in scopolamine animal model with assessment of mechanisms of action. *Evid. Based Complement. Altern. Med.* **2016**, *2016*, 1–17. [CrossRef]
29. Szabó, K.; Radácsi, P.; Rajhárt, P.; Ladányi, M.; Németh, É. Stress-induced changes of growth, yield and bioactive compounds in lemon balm cultivars. *Plant Physiol. Biochem.* **2017**, *119*, 170–177. [CrossRef]
30. Ramakrishna, A.; Ravishankar, G.A. Influence of abiotic stress signals on secondary metabolites in plants. *Plant Signal. Behav.* **2011**, *6*, 1720–1731. [CrossRef]
31. Dresler, S.; Wójciak-Kosior, M.; Sowa, I.; Stanisławski, G.; Bany, I.; Wójcik, M. Effect of short-term Zn/Pb or long-term multi-metal stress on physiological and morphological parameters of metallicolous and nonmetallicolous *Echium vulgare* L. populations. *Plant Physiol. Biochem.* **2017**, *115*, 380–389. [CrossRef]
32. Kuzel, S.; Vydra, J.A.N.; Triska, J.A.N.; Vrchotova, N.; Hruby, M.; Cigler, P. Elicitation of pharmacologically active substances in an intact medical plant. *J. Agric. Food Chem.* **2009**, *57*, 7907–7911. [CrossRef] [PubMed]
33. Lim, J.-H.; Park, K.-J.; Kim, B.-K.; Jeong, J.-W.; Kim, H.-J. Effect of salinity stress on phenolic compounds and carotenoids in buckwheat (*Fagopyrum esculentum* M.) sprout. *Food Chem.* **2012**, *135*, 1065–1070. [CrossRef]
34. Bistgani, Z.E.; Hashemi, M.; DaCosta, M.; Craker, L.; Maggi, F.; Morshedloo, M.R. Effect of salinity stress on the physiological characteristics, phenolic compounds and antioxidant activity of *Thymus vulgaris* L. and *Thymus daenesis* Celak. *Ind. Crop. Prod.* **2019**, *135*, 311–320. [CrossRef]
35. López-Berenguer, C.; Martinez-Ballesta, M.C.; Moreno, D.A.; Carvajal, M.; Garcia-Viguera, C. Growing hardier crops for better health: Salinity tolerance and the nutritional value of broccoli. *J. Agric. Food Chem.* **2009**, *57*, 572–578. [CrossRef] [PubMed]
36. Kim, H.-J.; Fonseca, J.M.; Choi, J.-H.; Kubota, C.; Kwon, D.Y. Salt in irrigation water affects the nutritional and visual properties of romaine lettuce (*Lactuca sativa* L.). *J. Agric. Food Chem.* **2008**, *56*, 3772–3776. [CrossRef] [PubMed]
37. Hawrylak-Nowak, B.; Dresler, S.; Rubinowska, K.; Matraszek-Gawron, R. Eliciting effect of foliar application of chitosan lactate on the phytochemical properties of *Ocimum basilicum* L. and *Melissa officinalis* L. *Food Chem.* **2021**, *342*, 128358. [CrossRef]
38. Oueslati, S.; Karray-Bouraoui, N.; Attia, H.; Rabhi, M.; Ksouri, R.; Lachaâl, M. Physiological and antioxidant responses of Mentha pulegium (Pennyroyal) to salt stress. *Acta Physiol. Plant.* **2010**, *32*, 289–296. [CrossRef]
39. Valifard, M.; Mohsenzadeh, S.; Kholdebarin, B. Salinity effects on phenolic content and antioxidant activity of *Salvia macrosiphon*. *Iran. J. Sci. Technol. Trans. A Sci.* **2017**, *41*, 295–300. [CrossRef]
40. Parida, A.K.; Das, A.B.; Sanada, Y.; Mohanty, P. Effects of salinity on biochemical components of the mangrove, *Aegiceras corniculatum*. *Aquat. Bot.* **2004**, *80*, 77–87. [CrossRef]
41. Jahantigh, O.; Najafi, F.; Badi, H.N.; Khavari-Nejad, R.A.; Sanjarian, F. Changes in antioxidant enzymes activities and proline, total phenol and anthocyanine contents in *Hyssopus officinalis* L. plants under salt stress. *Acta Biol. Hung.* **2016**, *67*, 195–204. [CrossRef]
42. Van Oosten, M.J.; Sharkhuu, A.; Batelli, G.; Bressan, R.A.; Maggio, A. The *Arabidopsis thaliana* mutant air1 implicates SOS_3 in the regulation of anthocyanins under salt stress. *Plant Mol. Biol.* **2013**, *83*, 405–415. [CrossRef]
43. Truong, H.A.; Lee, W.J.; Jeong, C.Y.; Trịnh, C.S.; Lee, S.; Kang, C.S.; Cheong, Y.K.; Hong, S.W.; Lee, H. Enhanced anthocyanin accumulation confers increased growth performance in plants under low nitrate and high salt stress conditions owing to active modulation of nitrate metabolism. *J. Plant Physiol.* **2018**, *231*, 41–48. [CrossRef]
44. Landi, M.; Tattini, M.; Gould, K.S. Multiple functional roles of anthocyanins in plant-environment interactions. *Environ. Exp. Bot.* **2015**, *119*, 4–17. [CrossRef]
45. Chalker-Scott, L. Do anthocyanins function as osmoregulators in leaf tissues? In *Advances in Botanical Research*; Academic Press: Cambridge, MA, USA, 2002; Volume 37, pp. 103–127. [CrossRef]
46. Pojer, E.; Mattivi, F.; Johnson, D.; Stockley, C.S. The case for anthocyanin consumption to promote human health: A review. *Compr. Rev. Food Sci. Food Saf.* **2013**, *12*, 483–508. [CrossRef] [PubMed]
47. Nunes, S.; Madureira, A.R.; Campos, D.; Sarmento, B.; Gomes, A.M.; Pintado, M.; Reis, F. Therapeutic and nutraceutical potential of rosmarinic acid—Cytoprotective properties and pharmacokinetic profile. *Crit. Rev. Food Sci. Nutr.* **2015**, *57*, 1799–1806. [CrossRef] [PubMed]
48. Lichtenthaler, H.K.; Wellburn, A.R. Determinations of total carotenoids and chlorophyll a and b of leaf extracts in different solvents. *Biochem. Soc. Trans.* **1983**, *11*, 591–592. [CrossRef]

49. Wang, C.; Lu, J.; Zhang, S.; Wang, P.; Hou, J.; Qian, J. Effects of Pb stress on nutrient uptake and secondary metabolism in submerged macrophyte *Vallisneria natans*. *Ecotox. Environ. Safe* **2011**, *74*, 1297–1303. [CrossRef] [PubMed]
50. Deng, H.; Van Verkel, G.J. Electrospray mass spectrometry and UV/visible spectrophotometry studies of aluminum (III)-flavonoid complex. *J. Mass Spectrom.* **1998**, *33*, 1080–1087. [CrossRef]
51. Hawrylak-Nowak, B. Changes in anthocyanin content as indicator of maize sensitivity to selenium. *J. Plant Nutr.* **2008**, *31*, 1232–1242. [CrossRef]
52. Dresler, S.; Strzemski, M.; Kováčik, J.; Sawicki, J.; Staniak, M.; Wójciak, M.; Sowa, I.; Hawrylak-Nowak, B. Tolerance of facultative metallophyte *Carlina acaulis* to cadmium relies on chelating and antioxidative metabolites. *Int. J. Mol. Sci.* **2020**, *21*, 2828. [CrossRef]
53. Barros, L.; Dueñas, M.; Dias, M.I.; Sousa, M.J.; Santos-Buelga, C.; Ferreira, I.C. Phenolic profiles of cultivated, in vitro cultured and commercial samples of *Melissa officinalis* L. infusions. *Food Chem.* **2013**, *136*, 1–8. [CrossRef]

International Journal of
Molecular Sciences

Article

ROS-Scavengers, Osmoprotectants and Violaxanthin De-Epoxidation in Salt-Stressed *Arabidopsis thaliana* with Different Tocopherol Composition

Ewa Surówka [1,*], Dariusz Latowski [2,*], Michał Dziurka [1], Magdalena Rys [1], Anna Maksymowicz [1], Iwona Żur [1], Monika Olchawa-Pajor [3], Christine Desel [4], Monika Krzewska [1] and Zbigniew Miszalski [5]

1 The Franciszek Górski Institute of Plant Physiology of the Polish Academy of Sciences, ul. Niezapominajek 21, 30-239 Kraków, Poland; m.dziurka@ifr-pan.edu.pl (M.D.); m.rys@ifr-pan.edu.pl (M.R.); amaksymowicz@ifr-pan.edu.pl (A.M.); i.zur@ifr-pan.edu.pl (I.Ż); m.krzewska@ifr-pan.edu.pl (M.K.)

2 Faculty of Biochemistry, Biophysics and Biotechnology of the Jagiellonian University, ul. Gronostajowa 7, 30-387 Kraków, Poland

3 Department of Environmental Protection, Faculty of Mathematics and Natural Sciences, University of Applied Sciences in Tarnow, Mickiewicza 8, 33-100 Tarnów, Poland; m_olchawa@pwsztar.edu.pl

4 Botanical Institute of the Christian-Albrechts-Universität zu Kiel, Am Botanischen Garten 1-9, 24118 Kiel, Germany; cdesel@bot.uni-kiel.de

5 W. Szafer Institute of Botany, Polish Academy of Sciences, ul. Lubicz 46, 31-512 Kraków, Poland; z.miszalski@botany.pl

* Correspondence: e.surowka@ifr-pan.edu.pl (E.S.); dariusz.latowski@uj.edu.pl (D.L.)

Citation: Surówka, E.; Latowski, D.; Dziurka, M.; Rys, M.; Maksymowicz, A.; Żur, I.; Olchawa-Pajor, M.; Desel, C.; Krzewska, M.; Miszalski, Z. ROS-Scavengers, Osmoprotectants and Violaxanthin De-Epoxidation in Salt-Stressed *Arabidopsis thaliana* with Different Tocopherol Composition. *Int. J. Mol. Sci.* **2021**, *22*, 11370. https://doi.org/10.3390/ijms222111370

Academic Editors: Mirza Hasanuzzaman and Masayuki Fujita

Received: 30 August 2021
Accepted: 18 October 2021
Published: 21 October 2021

Abstract: To determine the role of α- and γ-tocopherol (TC), this study compared the response to salt stress (200 mM NaCl) in wild type (WT) *Arabidopsis thaliana* (L.) Heynh. And its two mutants: (1) totally TC-deficient *vte1*; (2) *vte4* accumulating γ-TC instead of α-TC; and (3) *tmt* transgenic line overaccumulating α-TC. Raman spectra revealed that salt-exposed α-TC accumulating plants were more flexible in regulating chlorophyll, carotenoid and polysaccharide levels than TC deficient mutants, while the plants overaccumulating γ-TC had the lowest levels of these biocompounds. Tocopherol composition and NaCl concentration affected xanthophyll cycle by changing the rate of violaxanthin de-epoxidation and zeaxanthin formation. NaCl treated plants with altered TC composition accumulated less oligosaccharides than WT plants. α-TC deficient plants increased their oligosaccharide levels and reduced maltose amount, while excessive accumulation of α-TC corresponded with enhanced amounts of maltose. Salt-stressed TC-deficient mutants and *tmt* transgenic line exhibited greater proline levels than WT plants, lower chlorogenic acid levels, and lower activity of catalase and peroxidases. α-TC accumulating plants produced more methylated proline- and glycine- betaines, and showed greater activity of superoxide dismutase than γ-TC deficient plants. Under salt stress, α-TC demonstrated a stronger regulatory effect on carbon- and nitrogen-related metabolites reorganization and modulation of antioxidant patterns than γ-TC. This suggested different links of α- and γ-TCs with various metabolic pathways via various functions and metabolic loops.

Keywords: antioxidants; carbohydrates; carotenoids; xanthophyll cycle; osmoprotectants; oxidative stress; ROS-scavengers; salt stress; α-/γ-tocopherols

1. Introduction

Due to human activity and global climate changes, the area of heavily salinized (>2000 ppm) lands is on the increase. By 2050, it is expected to reach about 14% of global lands [1,2].

Salt (NaCl) stress induces ionic imbalance, and osmotic, and oxidative disturbances that affect many physiological processes in several subcellular compartments such as mitochondria [3] and chloroplasts [4–6]. Changes taking place in these organelles, and their impact on other subcellular compartments (e.g., peroxisomes [6,7]) initiate defense signaling pathways and regulate key metabolic processes.

Chloroplasts, through the shikimate/phenylpropanoid pathway, take part in the biosynthesis of chorismate, a precursor of aromatic amino acids (e.g., L-phenylalanine and L-tyrosine), and further intermediates, such as homogentisic acid (HGA, polar precursor of tocopherols), as well as a number of phenolic compounds (Scheme 1). Through the 2-C-Methyl-D-erythritol 4-phosphate (MEP) pathway, chloroplasts are the sites of biosynthesis of geranylgeranyl pyrophosphate (GGPP), an intermediate in the biosynthesis of (poly)isoprenoids, such as e.g., carotenoids (CARs), chlorophylls, plastoquinol-9 or lipophilic polyprenyl—a precursor of tocopherols (TC) [8–14]. All of these metabolites are crucial for the molecular and physiological regulation of plant cell functioning, especially under stress conditions.

Tocopherols (methylated phenols, vitamin E) and carotenoids are two the most abundant groups of non-enzymatic lipophilic antioxidants in plastids. They both affect the physical and biochemical properties of lipid membranes [9,10], and their accumulation reflects salt stress tolerance [15–18]. TCs and CARs: (i) perform synergistic protective functions (e.g., remove reactive oxygen species, lipid soluble by-products of oxidative stress); (ii) help to maintain the balance between various metabolites/biochemical pathways; and (iii) share the same precursors/intermediates (e.g., GGPP) [9–11,13].

TCs can occur into four forms (α-, β- γ- or δ-) that differ in the position of methyl groups in the chromanol ring. TCs participate in removing reactive oxygen species (ROS) (e.g., $^1O_2$, $O_2^{\bullet-}$), protecting chloroplast membranes from photooxidation, regulating metabolite biosynthesis and gene expression, and are also components of the information-rich redox buffer system [9,13,19–22].

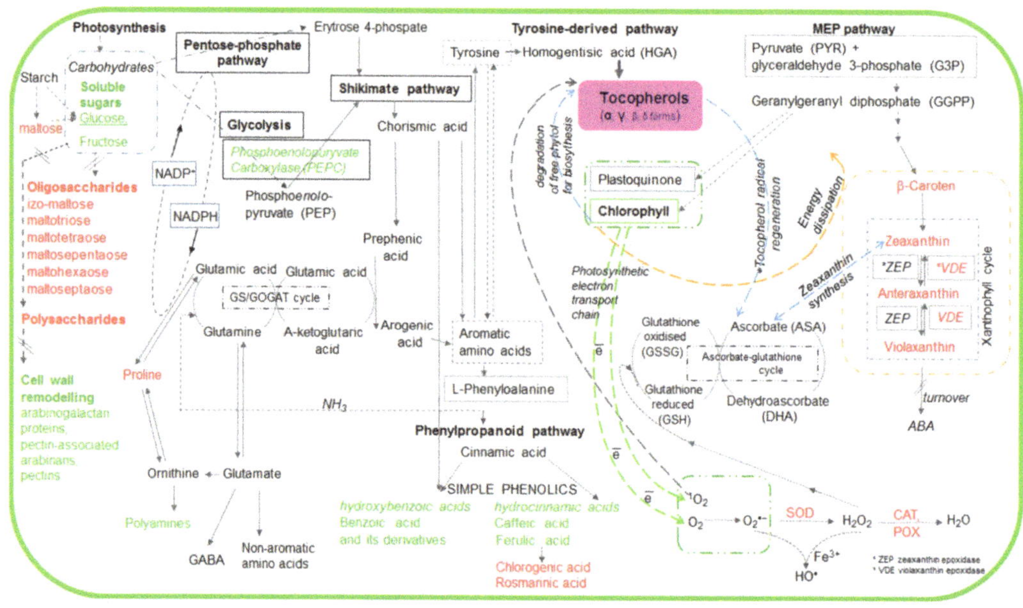

Scheme 1. Simplified scheme of metabolic pathways with marked metabolites the accumulation of which was affected by tocopherol (α– and and γ-) levels. The metabolites marked in red are described in this manuscript, whereas those marked in green were examined in our previous publication [21].

CARs are classified into two main groups: unoxygenated carotenes and oxygenated derivatives – polar xanthophylls [e.g., violaxanthin (Vx), antheraxanthin (Ax), zeaxanthin

(Zx)]. The majority of CARs are located in functional pigment-binding proteins embedded in the thylakoid membranes. Among them, carotenes (mainly β-carotene) are bound to the photosystem reaction centers, while xanthophylls are the most abundant in the light-harvesting complexes. CARs play an indispensable role in energy transfer or dissipation of excess excitation energy, photoprotection by efficient quenching of chlorophyll triplate states, and scavenging ROS or other free radicals [10,11,23]. One of the most common CAR-controlled photoprotective process is xanthophyll cycle, in which Vx is de-epoxidated to Zx, via Ax by Vx de-epoxidase (VDE) under light stress [24,25].

Phenolics belonging to hydroxybenzoic and hydroxycinnamic acids are also involved in ROS regulation. They can function as signaling molecules, precursors of other stress-related, or structural compounds [8,12,14,26].

Additional important compounds determining plant salt stress tolerance include carbohydrates, sugar alcohols [27–29], and hydrophobic compatible solutes (osmoprotectants) including proline (Pro) [30,31], proline-betaine (PB) [8], and glycine betaine (GB) [31–33]. They are involved in the regulation of cellular water potential, (re)allocation of carbon and nitrogen, modulation of ROS and other free radicals, expression of stress response genes, and (de)activation of alternative detoxification pathways.

Primary and secondary metabolites interact with antioxidant enzymes, described as stress tolerance markers, which include superoxide dismutases (SODs, EC 1.14.1.1.) that catalyzes $O_2^{\bullet-}$ disproportionation, and catalase (CAT, EC 1.11.1.6), which modulates levels of H_2O_2. These two antioxidant enzymes cooperate also with peroxidases catalyzing both ROS generation and scavenging [6,7,34].

Although the involvement of TCs in salt stress response was reported in several studies [16,21,35–37], more research is necessary to elucidate the metabolic loops, the links of TCs with different metabolites, and metabolic pathways controlling stress response. Changes in TC content were shown to affect accumulation of small molecular antioxidants [38] and carbon metabolites (e.g., carbohydrates, amino acids) [21,39] that are tightly linked to sulfur and nitrogen metabolism and crucial to plant stress tolerance [40,41]. Our previous study in *Arabidopsis thaliana* with altered TC composition exposed to salt stress showed a slight stimulation of the maximum operating efficiency accompanied by strongly altered cellular osmolarity [21]. Therefore, in this study we investigated the influence of TC composition on selected C-, and N-containing compounds with antioxidant and osmoprotectant properties that can be linked with TC based on their function or biosynthesis pathway (Scheme 1). We also determined the effect of TC composition and salt stress on CAR profile and Vx de-epoxidation in the photoprotective reactions of the xanthophyll cycle. To distinguish between the role of α- and γ-TCs, we compared the response to salt stress under low light in the wild type (WT) *Arabidopsis thaliana* (L.) Heynh., its two mutants: (1) *vte1* totally TCs-deficient, (2) *vte4* accumulating γ-TC instead of α-TC, and (3) *tmt* transgenic line overaccumulating α-TC.

2. Results

2.1. *Leaf Metabolites with Antioxidant and/or Osmotic Properties Detected by FT-Raman Spectroscopy*

FT-Raman spectra obtained for the leaves of different Arabidopsis genotypes growing in control and salt stress conditions revealed several bands (Figure 1) denoting the presence of various chemical compounds in the plant tissues. The spectral pattern is characteristic of CARs with all-trans configuration of the conjugated C=C chain. The assignments of the three prominent Raman bands are well established: the 1005 cm^{-1} band is attributed to the C–CH_3 rocking mode (CH_3 groups attached to the polyene chain and coupled with C–C bonds), the 1156 cm^{-1} band to the C–C stretching mode/vibration (coupled with C–H in-plane bending), and the most intense band at 1525 cm^{-1} to the C=C stretching mode of the conjugated chain in CARs.

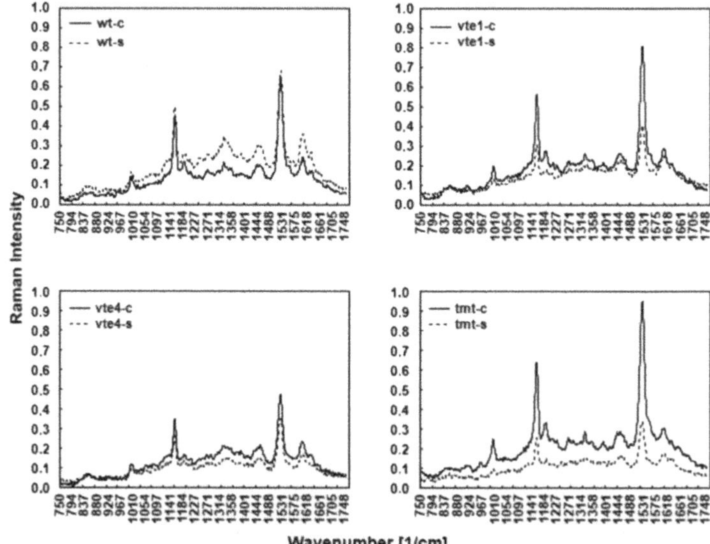

Figure 1. FT-Raman spectra of WT and TC-deficient mutants (*vte1*, *vte4*) and α-TC enriched transgenic (*tmt*) line of *Arabidopsis thaliana* irrigated with water (control) or with NaCl solution after 10 days of the experiment. The plants grown at 100-120 μmol m^{-2} s^{-1} light intensity. The leaves were harvested at the end (8:00 pm.) of the light period.

Changes in the amount of carotenoid compounds visible in the Raman spectra (bands at 1005 cm^{-1}, 1156 cm^{-1}, and 1525 cm^{-1}) varied and depended on the type of mutants (*vte1*, *vte4*) or transgenic line (*tmt*) in both control and salt stress condition. The highest intensity band at 1525 cm^{-1}, medium intensity band at 1156 cm^{-1}, and the band at 1005 cm^{-1} showed in control *vte1* and *tmt* plants were higher, while in *vte4* mutants they were lower than in WT plants. The Raman spectra demonstrated that in control conditions the number of CARs in *vte1* and *tmt* plants was higher, and in *vte4* mutants lower, than in WT plants (Figure 1).

NaCl treatment changed the Raman spectral pattern characteristic of CARs. All plants with altered TC composition showed lowered band intensity, mainly at 1525 and 1156 cm^{-1}, as compared with WT plants (Figure 1). Interestingly, a stronger decline was observed in *vte1* mutants and *tmt* transgenic line than in *vte4* mutants when compared with the controls. The bands at 1525 and 1156 cm^{-1} were almost unchanged in WT plants. Contrary to that, the lowest intensity band at 1005 cm^{-1} slightly declined in *tmt* transgenic line, strongly declined in *vte1* mutants, and remained fairly unchanged in *vte4* mutants and WT plants when compared to the control plants. Under salt stress, the decline in the amount of carotenoid compounds in *tmt* (~60%) and *vte1* (over 40%) plants was stronger than in *vte4* mutants. In contrast, in WT plants, CAR levels were comparable throughout the study under control and salt-stress conditions.

Under control conditions for all the tested plants, a very low intensity of the bands (at 1602, 1328, and 1270 cm^{-1}) representing chlorophylls was observed. This is due to lower scattering of radiation by chlorophylls compared to carotenoids. The intensity of chlorophylls bands was slightly higher in *vte1* and *tmt* plants than in *vte4* and WT plants. NaCl treatment increased the intensity of these bands in WT plants, induced nearly no changes in mutants with α-TC deficiency, and decreased their intensity in the plants overaccumulating α-TC (Figure 1).

The bands visible at 1444 and 1304 cm^{-1} denote the deformation vibrations of CH, CH_2, and CH_3 groups and C-C stretching vibrations of aliphatic hydrocarbons, respectively. Control *vte1* and *tmt* plants showed slightly higher band intensity at 1444 and 1304 cm^{-1}

than *vte4* and WT plants under the same conditions. Salt stress increased the intensity of 1444 and 1304 cm^{-1} in WT plants, and decreased it in the plants with altered TC composition. This decline was the strongest in *tmt* plants (Figure 1).

The bands (at 1188 and 851 cm^{-1}) associated with polysaccharides are also visible in the Raman spectra. In control conditions, the intensity of these bands was higher in *vte1* and *tmt* mutants than in *vte4* and WT plants. NaCl treatment enhanced their intensity in WT plants, and reduced it in the plants with altered TC composition. This reduction was the most prominent in *tmt* plants (Figure 1).

2.2. Chemometrics–Cluster Analysis

Cluster analysis is a method of classifying tested objects into groups (clusters) so that the resulting clusters contain objects that are as similar as possible. We used the cluster analysis to find meaningful and systematic differences among the measured FT-Raman spectra, which show specific groups of the chemical compounds present in the plant tissues. The following dendrograms show the cluster analysis separately for two mutants (*vte1*, *vte4*), transgenic line (*tmt*), and WT plants grouping the control and salt-stressed plants (Figure 2).

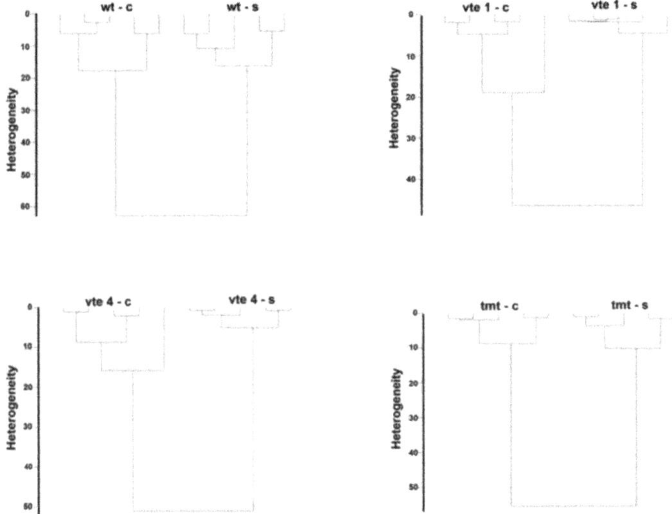

Figure 2. Cluster analysis of the FT-Raman spectra of the WT and TC-deficient mutants (*vte1*, *vte4*) and α-TC accumulated transgenic (*tmt*) line of *Arabidopsis thaliana* irrigated with water (control–c) or with NaCl solution (-s) after 10 days of the experiment.

Distinct discrimination was achieved for the two groups: control and NaCl stressed plants for the entire measuring range for each genotype (with or without α-TC and WT). It was demonstrated that the leaves of *vte1*, *vte4*, *tmt*, and WT plants differed significantly in their content and composition of carotenoids, chlorophylls, or polysaccharides and that these differences depended on salt stress conditions (Figure 2).

2.3. Vx De-Epaoxidation

Under control conditions, only *vte4* plants had higher initial rates of Vx de-epoxidation and Zx formation when compared with WT, while the values for *vte1* and *tmt* plants were comparable with those of WT plants (Figure 3A,B). Initial rates of Ax formation were similar in all tested plants (Figure 3C).

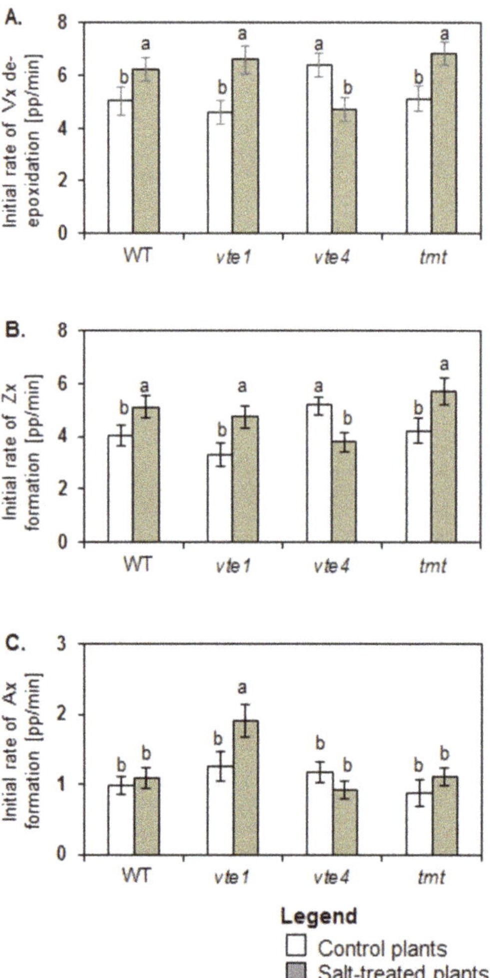

Figure 3. Initial rate of violaxanthin (Vx) de-epoxidation (**A**), zeaxanthin (Zx) (**B**), and antheraxanthin (Ax) formation (**C**). The data are representative of the pooled samples of disc leaves from three independent experiments (mean ± SD; $n = 3$). Different letters above the bars represent the relative significance vs. control according to Tukey's test, at $p \leq 0.05$.

Control *vte4* plants were characterized by a greater drop in Vx (Figure 4A), and greater (8 percentage points (pp)) total Zx production than control WT plants (Figure 4B). No differences between maximal and final Ax levels were found in all tested groups of the control plants (Figure 4C).

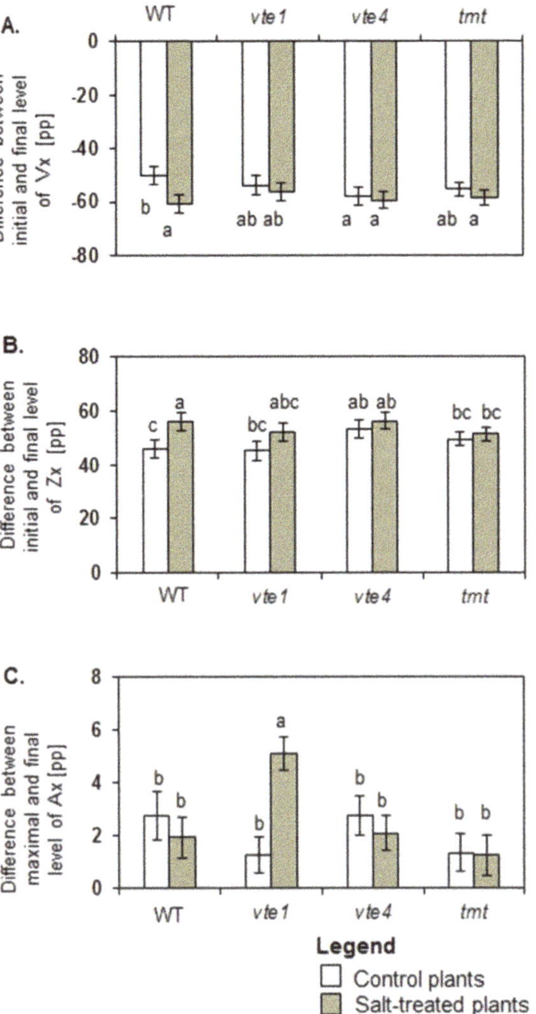

Figure 4. Differences between initial and final levels of violaxanthin (Vx) (**A**), zeaxanthin (Zx) (**B**), and antheraxanthin (Ax) (**C**) during de-epoxidation reaction. The relative significance between samples is indicated as in Figure 3.

Under salt stress, the initial rates of Vx de-epoxidation and Zx formation increased in WT, *vte1* and *tmt* plants but decreased in *vte4* mutants (Figure 3A,B). Significant changes were determined for total Vx loss, calculated as a difference in Vx level at the beginning and end of de-epoxidation, and for the final level of Vx and total Zx production after de-epoxidation between salt treated and control WT plants (Figure 4A,B). In control WT plants after de-epoxidation the total level of Vx dropped by about 50.20 ± 2.36 percentage points (pp) and that of Zx rose by about 45.87 ± 3.44 pp. In the salt-treated WT plants, the total reduction in Vx and rise in Zx were greater and reached 60.67 ± 2.13 pp and 56.12 ± 3.40 pp, respectively. The total decline and final level of Vx after de-epoxidation was similar in all tested experimental genotypes irrespective of the treatment (Figure 4A). No significant changes were also observed in the total decline and final level of Vx after de-epoxidation between the mutants (*vte1*, *vte4*), the transgenic *tmt* line and WT plants following exposure to salt stress. Under salt stress the only significant differences were spotted between WT plants and *tmt* transgenic line regarding a decline in total Zx production after de-epoxidation (Figure 4B). The highest initial rate of Ax formation among all tested plants was detected for salt-stressed *vte1* mutants (Figure 3C). Salt treated *vte1* plants showed also by about 2 pp higher maximal Ax accumulation than control *vte1* plants, whereas the differences between maximal Ax accumulation and the initial rate of Ax formation in the remaining variants were not significant, irrespective of NaCl presence. Salt-stressed *vte1* mutants showed also the highest differences between maximal and final level of Ax during de-epoxidation (Figure 4C).

2.4. Phenolic Acids

In control conditions the levels of chlorogenic acid (CGA) were comparable in WT, *vte1* and *vte4* plants and significantly lower in *tmt* plants (Figure 5A). Rosmarinic acid (RA) was higher in all plants with altered TC composition than in WT plants (Figure 5B).

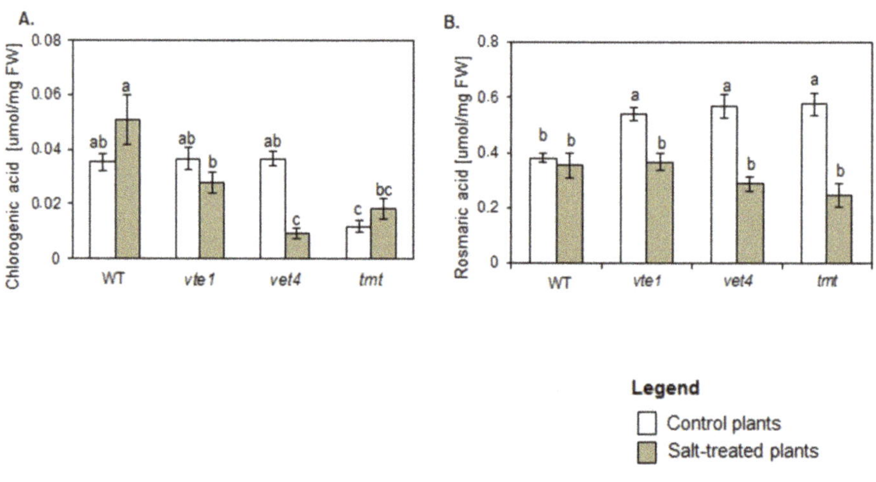

Figure 5. Changes in the content of phenolic acids: (**A**) Chlorogenic acid (CGA), (**B**) Rosmarinic acid (RA) in WT and TC-deficient mutants (*vte1*, *vte4*) and α-TC enriched transgenic *tmt* line of *Arabidopsis thaliana* irrigated with water (control) or with NaCl solution after 10 days of the experiment. The data are representative of the pooled samples of leaves (±SD; $n = 5$) from three independent experiments (mean ± SD; $n = 3$). White bars indicate control plants, and grey plants treated with 200 mM NaCl. Different letters above the bars represent the relative significance vs. control according to Tukey's test, at $p \leq 0.05$.

Under salt stress conditions, the plants with altered TC composition produced less CGA than WT plants (Figure 5A). CGA content in salt-treated WT, *vte1* and *tmt* plants remained unchanged in comparison with their controls, while in *vte4* mutants its drastic drop was detected. NaCl treatment did not affect RA amount in WT plants but lowered its accumulation in plants with altered TC composition to the level comparable to WT plants (Figure 5B).

2.5. Carbohydrates and Polyols

Both under control and salt stress conditions, oligosaccharide content in plants with altered TC composition was lower than in WT plants. When comparing control plants, the lowest oligosaccharide amount was detected in *vte4* mutants (Figure 6A), and maltose content was significantly lower only in *tmt* plants (Figure 6B).

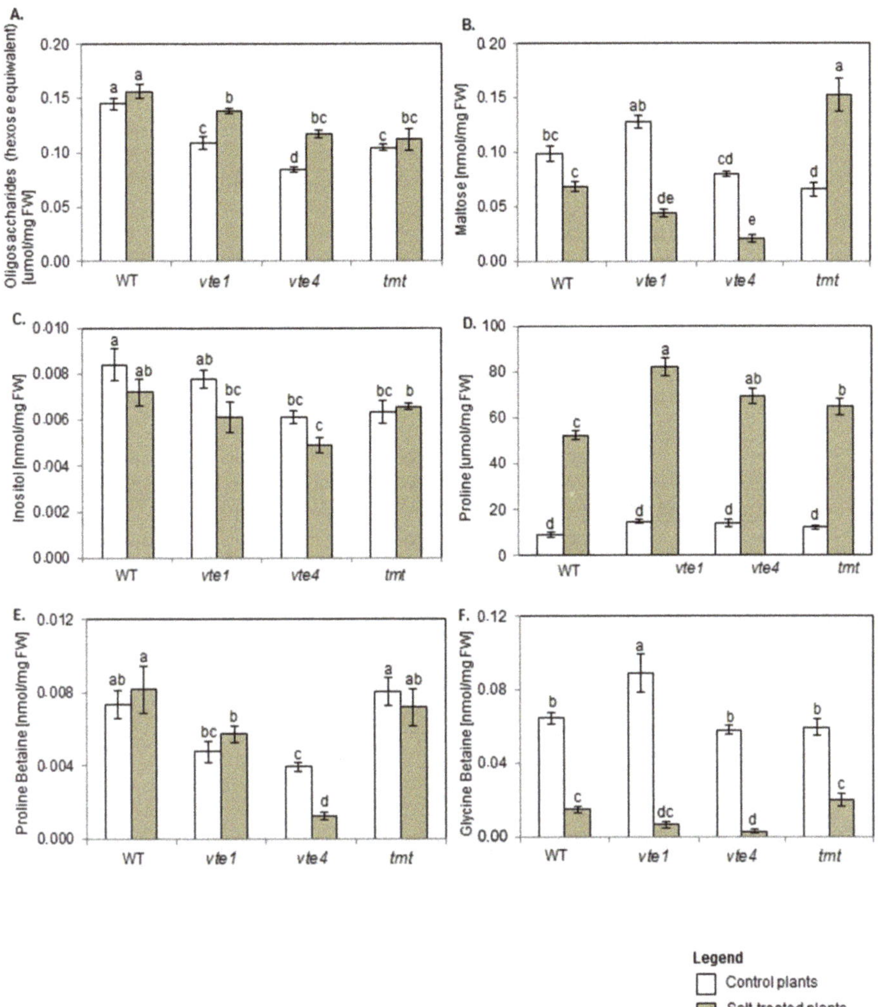

Figure 6. Changes in (**A**) oligosaccharides (total maltose, hexose equivalent), (**B**) maltose, (**C**) inositol, (**D**) proline, (**E**) proline betaine, and (**F**) glycine betaine in WT, α-TC deficient mutants (*vte1*, *vte4*) and α-TC accumulating transgenic *tmt* line of *A. thaliana* irrigated with water (control) or with 200 mM NaCl solution after 10 days of the experiment. The samples are described and marked as in Figure 5. The relative significance between samples is indicated as in Figure 5.

Salt treatment increased the content of oligosaccharides only in α-TC deficient mutants (*vte1*, *vte4*), and their levels remained unaffected in α-TC accumulating *tmt* plants when compared with the controls (Figure 6A). NaCl stress did not affect maltose level in WT plants but reduced its amount in α-TC deficient plants and enhanced it in *tmt* transgenic line (Figure 6B). Among salt-treated plants, *vte4* mutants showed the lowest, while *tmt* plants the highest maltose content.

Inositol content in control *vte4* and *tmt* plants was lower than in WT genotype (Figure 6C). Salt stress had no significant effect on its content in all tested plants, except for *vte4* mutants where the amount of this compound measured after NaCl treatment remained lower in comparison with WT plants.

2.6. Proline and Betaines

Proline (Pro) amount in all control plants was comparable (Figure 6D). Lower accumulation of proline betaine (PB) was detected in α-TC deficient plants when compared with α-TC accumulating plants and WT plants (Figure 6E). Glycine betaine (GB) amount in WT, *vte4* and *tmt* plants was comparable, and significantly higher in *vte1* mutants (Figure 6F).

NaCl treatment strongly boosted Pro content in all tested plants, with a higher amount being recorded in plants with altered TC composition than in WT plants (Figure 6D). Under salt stress, PB content declined in *vte4* plants, while the remaining genotypes were unaffected by NaCl (Figure 6E). α-TC deficient plants (*vte1*, *vte4*) showed a tendency to accumulate more Pro and less PB than *tmt* and WT plants. NaCl presence also strongly declined GB content in all tested plants when compared with their controls (Figure 6F), and this drop was greater in α-TC deficient plants (~10-fold decrease for *vte1*, ~12-fold for *vte4*) than in α-TC accumulating plants (~3-fold for WT, ~2-fold for *tmt*). Among all tested plants, the strongest NaCl-dependent decline and the lowest amount of PB and GB was observed in γ-TC overaccumulating plants (*vte4*).

2.7. Activity of Antioxidant Enzymes

In control conditions, total SOD activity in plants with altered TC composition was higher than in WT plants (Figure 7A).

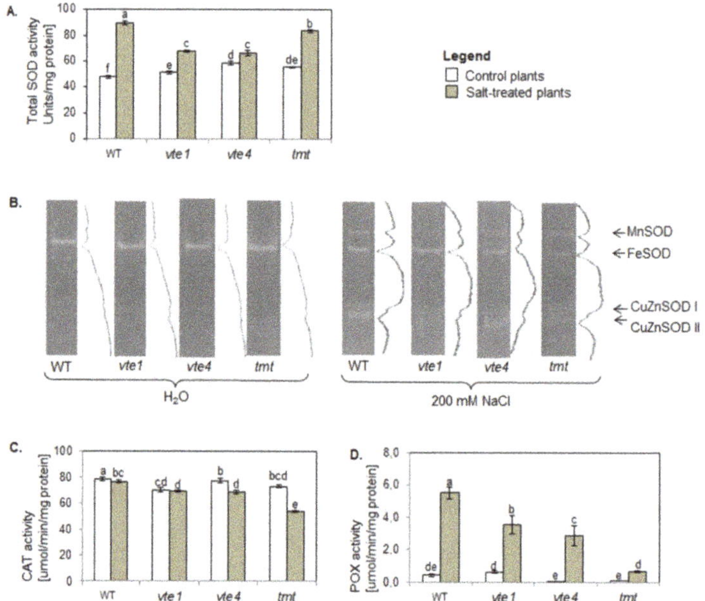

Figure 7. Changes in the activity of (**A**) total superoxide dismutase (SOD); (**B**) SOD isoforms (Mn-SOD, Fe-SOD, CuZn-SOD) visualized on native polyacrylamide gels, with 18 μg of soluble protein loaded on each well; (**C**) catalase (CAT); and (**D**) non-specific peroxidase (POX) in WT and α-TC deficient mutants (*vte1*, *vte4*) and α-TC enriched transgenic *tmt* line of *A. thaliana* irrigated with water (control) or with 200 mM NaCl solution after 10 days of the experiment. The plants grew at 100–120 μmol m^{-2} s^{-1} light intensity. The samples are described and marked as in Figure 5. The relative significance between samples is indicated as in Figure 5. For the PAGE (**B**) we showed a typical example of three repetitions with related densitograms corresponding to the activity of SOD isoforms.

The most pronounced activity among SOD isoforms was found for Fe-SOD and was similar in all tested plants (Figure 7B). The activity of Mn-SOD and CuZn-SOD was lower only in plants without α-TC. Also, the plants with altered TC composition had lower CAT activity than WT plants, with the lowest activity of this enzyme detected in *vte1* mutants (Figure 7C). Non-specific POX activity in plants with altered TC composition was comparable to WT plants (Figure 7D), and the plants overaccumulating α- or γ-TC (*tmt*, *vte4*) had lower total POX activity than *vte1* mutants.

NaCl treatment enhanced total SOD activity in all tested plants. The increase was the highest in WT plants, and higher in α-TC accumulating transgenic *tmt* line than in α-TC deficient plants (Figure 7A). NaCl treatment increased also the activity of Mn-SOD and two CuZn-SOD isoforms (CuZn-SODI, CuZn-SODII), more strongly in α-TC accumulating plants (WT, *tmt*), than in α-TC deficient plants (*vte1*, *vte4*; Figure 7B). Plants overaccumulating γ-TC showed the lowest intensity of Fe-SOD. Moreover, NaCl treatment resulted in a decline in CAT activity and an increase in non-specific POX activity in all studied plants as compared with their controls and WT plants (Figure 7C,D). NaCl treated plants can be ranked according to their CAT activity in the following order: WT> *vte1* = *vte4* = *tmt*, and according to their POX activity as: WT > *vte1* > *vte4* > *tmt*.

3. Discussion

Although a salinity of 100 – 200 mM has been proved toxic for most glycophytes [5,42], salt-induced lethality was not observed in plants growing under salt stress and low light intensity [21,43]. Our studies proved that the changed α- and γ-TC content is not a critical prerequisite for the survival of *A. thaliana* plants growing under salt stress (200 mM NaCl) and light intensity not exceeding 120 μmol photons m^{-2} s^{-1}. *A. thaliana* plants with altered TC composition were capable of adapting to NaCl stress through modulation of their defense mechanisms that include not only TCs as quenchers of singlet oxygen formed at PSII [16,21,35–37,44].

3.1. Tocopherol Composition Modulates the Level of Photosynthetic Pigments (Chlorophylls, Carotenoids) and Intensity of Vx De-Epoxidation

The Raman spectra showing changes in the number of chlorophylls and carotenoids that depended on TC composition in control, and salt stressed plants (Figures 1 and 2) revealed that mutual (α and γ) TC balance is important for the adjustment of the photosynthetically active pigments under both control and salt stress conditions. The changes in the level of photosynthetic pigments (chlorophylls, CARs) in *A. thaliana* plants with different TC composition subjected to salt stress (Figure 1) were accompanied by a slight stimulation of the maximum quantum yield of PSII [21]. We therefore suppose that a dynamic reorganization/recycling of metabolites related to TC enabled the plants to survive salt stress. The interdependence of TCs and chlorophylls is probably related to chlorophyll and protein degradation as well as *de novo* synthesis of TC precursors. These processes contribute to the increased demand for substrates used in TC synthesis during abiotic stress [22]. As showed by Ischebeck et al. [45] and Dörmann [46], free phytol from chlorophyll breakdown under stress might be directly used for TC biosynthesis. Moreover, α- and γ-TC dependent changes in chlorophyll and CAR accumulation (Figure 1) can be linked with various α- and γ-TC physical and chemical traits (e.g., stronger nucleophilic properties of γ-TC than α-TC), and their different membrane or cellular location [13,19]. We also found that under salt stress α-TC enabled more efficient modulation of signals related to photosynthetic pigment (chlorophylls, CARs) biosynthesis or accumulation than γ-TC. As the maximum quantum yield of PSII (Fv/Fm, stress indicator) in salt-stressed *A. thaliana* with altered TC composition was slightly stimulated [21], we speculated that TC dependent changes in CAR composition were related not only to the antioxidant mechanism of TCs [9,19,21] and CARs [24,47,48] protecting physical properties of the membranes, but probably also resulted from downregulation/alteration of CAR biosynthesis (as regulation of shikimate-phenylpropanoid and MEP pathways). The heterogeneity of plants with or without α-TC

(Figure 2) is due to their chemical composition, especially the content of CARs, chlorophylls, polysaccharides, and compounds with CH, CH_2, and CH_3 groups and C-C or C=C bonds.

Oxygenated derivatives of CARs, such as Vx, Ax, and Zx are involved in the violaxanthin cycle, which represents an important element of plant cell stress defense strategy, including high light intensity and salinity. The interdependent processes of epoxidation and particularly de-epoxidation (removal of two epoxide groups of Vx in two steps by VDE, forming first a monoepoxide Ax, and finally a completely epoxide free Zx) are crucial for photoprotection of chloroplasts. The formation of Zx and Ax was found to correlate with dissipation of excess excitation energy (non-photochemical quenching; NPQ) of chlorophyll fluorescence [11,25]. The intensity of Vx de-epoxidation, in addition to light stress, can be enhanced by other synergistically interacting environmental stress factors, including salinity [11,15]. Our results confirmed that initial rates of Vx de-epoxidation and Zx formation were higher in salt-stressed than in control WT plants (Figure 3A,B), and indicated that the naturally balanced level of α-/γ-TCs (WT plants) played an important role in optimizing the rate of Vx de-epoxidation and Zx formation (Figure 4A,B). Altered TC composition, particularly the lack of α-TCs modified various steps of Vx de-epoxidation, and caused a temporary change in ROS levels even if high light intensity was the only stress factor. Stronger differences in various steps of Vx de-epoxidation in salt-stressed α-TC deficient plants than in α-TC accumulating plants (Figures 3 and 4), indicated a more important role of α-TC than γ-TC in regulating Zx formation under high light and salinity (Figure 3A,B). In addition, the lowest CAR level and changes in Vx de-epoxidation in salt-stressed mutants overaccumulating γ-TC, may be related to the lack of α-TC and accumulation of intermediate metabolites formed by blocking α-TC from γ-TC synthesis pathway. Contrary to that, it seemed that double TC mutants (with knockout in tocopherol cyclase 1 gene catalyzing the penultimate step of TC synthesis [38]) presumably protected photosynthetic complexes from oxidative stress through tocotrienols [19], as tocopherol cyclase 1 is involved in the biosynthesis of both TCs and tocotrienols. Moreover, the lack of both TC forms, together with reduced level of CARs under salt stress (Figure 1), should facilitate de-epoxidation as TCs, tocotrienols, and CARs regulate the molecular dynamics and physical properties of the thylakoid membranes [10,19] in which Vx de-epoxidation occurs. Low amounts of TCs and CARs in salt-treated double TC mutants (Figures 1 and 2) can result in higher mobility of the membrane, increased diffusion rate of Vx, and higher initial rate of de-epoxidation followed by the most intensive accumulation of Ax (Figures 3C and 4C). In addition, our findings (Figures 3 and 4) confirmed that salinity affected xanthophyll cycle functioning independently of changes in the genes involved in TC biosynthesis. α- and γ-TCs seemed to selectively optimize Vx and Ax de-epoxidation (Figures 1–4), which might to a different degree affect the photoprotective and antioxidant functions of these molecules [11,23,24,47], their interactions with the components of antioxidant network governed by the ascorbate-glutathione cycle [13,49], and differentially modulate physical properties of lipid membranes [10,50]. Changes in Vx, Ax, and Zx profiles (Figures 3 and 4) may have reflected TC dependent modulations of metabolic pathways, including cytosolic mevalonate and plastid MEP pathways [11,48,51], (Scheme 1).

They could also indicate alterations in the biosynthesis of the precursors or derivatives of xanthophyls, including apocarotenoids (e.g., abscisic acid, ABA) [48] that can be involved in the regulation of TC biosynthesis in Arabidopsis exposed to abiotic stresses [52]. According to Ellouzi et al. [37], salt-stressed Arabidopsis plants overaccumulating γ-TC are not as efficient in modulating ABA accumulation as double TC mutants and WT plants. TC dependent reorganization of photoprotective pigment (carotenes, xanthophylls) accumulation in chloroplasts (Figures 1–4) seems to reflect their functional and metabolic relationships.

3.2. Tocopherols Affected Energy Use and Dissipation

We found that both the phenolics such as cinnamic and ferulic acids [21] and their derivatives such as CGAs [53] and RA [26] were influenced by TC composition in control and salt stress conditions (Figure 5A,B). TC related modification of phenylpropanoid accumulation and probably their interactions with other biocompounds (including CGA or RA precursors, derivatives, intermediates) (Scheme 1), might affect the dynamics of cellular antioxidant potential, adaptive signaling pathways, photochemical energy dissipation possibilities, and membrane properties [14]. TC dependent changes in CGA content (Figure 5A) might affect the mitochondrial tricarboxylic acid cycle (TCA) and amino acid metabolism [54,55].

3.3. Tocopherol Composition modulates the Pool of Primary and Secondary Metabolites under Salt Stress

Alterations in carbohydrate metabolism belong to salt stress response mechanisms [27,29,56]. Under salt stress, soluble sugar accumulation is influenced by TC composition [21,39] (Figures 1 and 6). Based on the Raman spectra, we stated that salt stress-triggered accumulation of polysaccharides localized in the cell wall or epidermis also depended on α- and γ-TC content (Figure 1), as α-TC containing plants adjusted cell wall or epidermis polysaccharide level more flexibly than γ-TC deficient plants. These results corresponded with α-TC dependent expression of several pectin and AGP epitopes in the leaves of salt-treated Arabidopsis [21]. Contrary to that, the fact that soluble oligosaccharides (hexose equivalents) were accumulated only in salt-stressed α-TC deficient plants (Figure 6A) suggested a reduced demand for soluble sugars. This was probably a consequence of shoot growth limitation [57], as well as various roles of α- and γ-TC in controlling the accumulation of osmotically active sugars involved in the regulation of membrane or cell wall potential [21] or sugars being used for stress-dependent cell wall reorganization.

TC composition affected the level of maltose (Figure 6B), the major product of starch degradation in chloroplasts. Maltose is also a source of glucosyl residue that is converted to hexose phosphate in the cytosol [27,56,58]. We only saw increased maltose content in salt-stressed plants overaccumulating α-TC, and its reduction in α-TC deficient plants (Figure 6B). These findings corresponded to those of Abbassi et al. [35], who reported on TC dependent changes in starch levels in salt treated transgenic (*vte2, vte4*) *Nicotiana tabacum*, and described modulation of maltose-related processes. Elevated maltose content in the plastids of freeze stressed *A. thaliana* was proposed as a mechanism protecting the photosynthetic electron transport chain, proteins, membranes [27], and biosynthesis of the compounds involved in cell walls reorganization [28].

A simultaneous reduction in inositol, particularly in γ-TC overaccumulating plants (Figure 6C), revealed different roles of α- and γ-TC in inositol-related processes (i.e., ascorbic acid biosynthesis) and signaling intricately tied to lipids [59].

Our study showed that under salt stress conditions, TC composition affected accumulation of polyamines [21] and of Pro and methylated amino acids derivatives (PB, GB) (Figure 6D–F). These are biocompounds involved in the Glu (glutamate)-Pro-Arg-PAs-GABA (γ-aminobutyric acid) pathway localized, at least partly, in chloroplasts and are responsible for (re)allocating of assimilated C and N [60]. This is in agreement with the studies showing that TC composition affects total amino acid and Pro levels in transgenic *N. tabacum* [35] and the accumulation of some non-aromatic amino acids in the leaves of TC deficient transgenic tomato [39]. As the accumulation of Pro, one of the first osmoprotectants activated under environmental stresses [30,31,61] was more intense in the salt-treated plants with altered TC composition (particularly without either TC form) than in WT plants (Figure 6D). We suggested that Pro may partially compensate the changes in TC composition and may reflect elevated carbon flux from the carboxylate into the amino acid pool. Pro also serves as a C, N, and energy source for the cellular recovery processes or metabolic pathways in which also TCs can be involved [30,31,60]. Pro biosynthesis in Arabidopsis subjected to salinity is photosynthetically (-light)-dependent and is inhibited by terpenoid or isoprenoid groups (e.g., brassinosteroids [62]). The biosynthesis or accumulation of Pro are mediated by signaling pathways dependent and independent of ABA, linked with glucose oxidation and oxidative pentose phosphate pathway. Under salinity, Pro biosynthesis is raised in the plastids, although under normal conditions proper Pro concentration is maintained in the cytosol [8,31,60,61]. According to Signorelli [30], an increased rate of Pro biosynthesis or accumulation through $NADP^+$ re-oxidation may prevent photosynthetic electron leakage and ROS generation. Also, through the generation of NADH/H^+, $FADH_2$ and ATP in the mitochondria Pro promotes cell survival under stress conditions. Pro can affect the initial stages of phenylpropanoid and protein biosynthesis [8,63,64], controls the expression of salt stress responsive genes (e.g., PRE, ACTCAT) [60] and the activity of some antioxidants, affects cellular redox buffering [64] and modulates cell wall architecture [63]. On the other hand, Pro accumulation may lead to protein denaturation [65] and may initiate programmed cell death [66].

Our study indicated that α-TC levels under salinity also affected the accumulation of methylated amino acid derivatives (PB or GB) and PB or GB related processes (Figure 6E,F). For instance, GB takes part in the regulation of ROS level and Na^+/K^+ homeostasis [31–33], depending on TC composition [37]. It seems that under salt stress, the accumulation pattern of Pro and methylated amino acid derivatives (GB, PB) was linked with the level of chromanol ring methylation in α-TC and γ-TC. The methylation reaction was shown as being involved in the regulation of gene expression, and/or biosynthesis of stress-response biomolecules, including TCs [9], Pro, betaines (e.g., GB), phenolic compounds, chlorophylls, or plastoquinones [31,33,67]. The association between salt stress tolerance and methylation was reported (e.g., by Kumar et al. [68], who found salt-induced tissue-specific cytosine methylation in *Triticum aestivum*, and by Karan et al. [69], who indicated salt-induced variation in DNA methylation pattern and gene expression in rice). In addition, nitrogen metabolism rearrangement was associated with the dissipation of the excess light energy through xanthophyll cycle [70], and the composition of the pool of xanthophyll cycle pigments [71], whose de-epoxidation under saline conditions depends on α-/γ-TC ratio (Figures 3 and 4). We suggest that α- and γ-TC dependent carbon and nitrogen biocompounds accumulation revealed differences in the regulation of C/N-related metabolic pathways (compartmentalized between chloroplast, cytoplasm and mitochondria) and the energy status of the cell.

The ROS balance and ROS-associated redox signals, described as crucial for harmonious metabolism and the establishment of adaptive signaling pathways, are maintained in the cells mainly through the mutual cooperation of non-enzymatic (e.g., TCs, CARs) and enzymatic (SOD, CAT, POX) antioxidants [6,14,19,20,34,72]. In agreement with our previous studies on Arabidopsis seedlings [16], we showed that changes in α- and γ-TC content were differentially compensated by alterations in SOD, CAT (Figure 7A–C), and POX (Figure 7D) activity in both control and salt stress conditions. It was shown previously that the exposure of TC deficient Arabidopsis mutants to high light [20,72], and salinity Ellouzi et al. [37] boosted the oxidative stress as compared with WT plants with naturally balanced TC levels. The changes in SOD activity can be related to α- and γ-TC content through their influence on the substrate level for SOD isoforms in different cellular organelles, and thus the modulation of ROS-dependent signal induction under salt stress conditions. The inactivation of CuZn-SOD observed in α-TC deficient plants could be explained by intensive NaCl induced ROS generation. This suggests an important role of synergistic interaction of α-TC and SODs in cell protection against oxidative damage.

Total SOD activity pattern did not closely correspond with the patterns of CAT and POX activity (Figure 7A,C,D), which may suggest α- and γ-TC dependent differences in the level of substrates for CAT and POX generation and specific functions of these antioxidant enzymes. CAT (widely used as a peroxisomal marker) is crucial in removing photorespiratory H_2O_2 [7] and may also reflect other peroxisomal metabolic pathways linked with nitrogen metabolism, β-oxidation of fatty acids and biosynthesis of phytohormones (e.g., indolilo-3-acetic acid (IAA) and jasmonic acid (JA) [73]). Munne-Bosch et al. [74] showed TC dependent synthesis of JA, a precursor of which, 12-oxophytodienoicacid, is generated in chloroplasts and subsequently synthesized in peroxisomes via β-oxidation. TC dependent changes in POX activity (Figure 7D), the level of Pro (Figure 6D), carbohydrates and phenolics [21] (Figures 5 and 6A) can be associated with TC related cell wall modification under salt stress [21]. Our results also suggested TC related involvement of CAT and POX in H_2O_2-dependent signaling pathway in the cellular processes compensating for altered TC balance and salinity response.

4. Materials and Methods

4.1. Plant Material

The seeds of *A. thaliana*: Columbia ecotype (Col-0) (1), homozygous mutant *vte1* (GABI_111D07) in the Col background with an insertion in the third intron of the open-reading frame (At4g32770) of the gene encoding tocopherol cyclase; deficient in tocopherol cyclase, totally devoid of TCs (2), the homozygous mutant *vte4* (SALK_036736) in the Col background with an insertion in the first intron of the open-reading frame (At1g64970) of the gene encoding γ-tocopherol methyltransferase; devoid of γ-tocopherol methyltransferase (*γ-TMT*, catalyses the conversion γ-TCs to α-TC) gene, accumulating γ- instead of α-TC (3), and transgenic *γ-TMT* plants overexpressing *γ-TMT* methyltransferase under the control of 35S CaMV promoter, thus overproducing α-TC (4), described by Desel et al. [75] and Fritscheet al. [9], were kindly provided by Prof. Karin Krupińska (Kiel University, Germany). The experimental plant line of *A. thaliana* used in our experiments has a well-documented origin Desel et al. [75] and was thoroughly characterized e.g., by Porfirova et al. [76], Bergmuller et al. [77], Shintani and DellaPenna [78], Collakova and DellaPenna [22], or Rosso et al. [79].

The seeds were germinated in the soil for three weeks and then each plant was transferred to a single pot. The six to seven-week-old plants were then divided into two groups. The first set of plants was subjected to rapid salinization by irrigation with 200 mM solution of NaCl for 10 days. The second group (i.e., the control plants) were irrigated with tap water. The plants grew under 100–120 μmol m^{-2} s^{-1} light intensity, at 18 °C, 12/12 h light/dark photoperiod, and 40–60% relative air humidity (RH). The plants were cultivated in a randomized design and rotated daily to minimize positional effects. Each group within the genotype consisted of at least 18 to 22 replicates (plants). For biochemical analyses the pooled samples were harvested from the representative rosette leaves of five to seven plants per genotype, growing under control or salt stress conditions. The samples were collected at the end of the light period (between 16.00 and 18.00) in three replicates. Three independent experiments were performed. Vx de-epoxidation was analyzed in 5 mm diameter leaf discs after 60 min dark incubation. The rest of the collected plant material was immediately frozen in liquid nitrogen (LN2) and stored at −80 °C until analysis.

The content of α- and γ-TC in the rosette leaves collected at the end of the experiment in both control and salt stress conditions was monitored according to a modified method of Surówka et al. [80] and presented in Table 1.

Table 1. The content of α- and γ-TC in the rosette leaves of *A. thaliana* with various TC composition growing in control and salt stress conditions; ND*, under the limit of detection.

A. thaliana with Changed TC Composition /Treatment	α-TC μg/g FW	Standard Deviation (SD)	γ-TC μg/g FW	Standard Deviation (SD)
WT H_2O	72.6	3.67	4.1	0.2
WT NaCl	83.5	6.05	6.5	0.8
vte1 H_2O	ND *	ND	ND	ND
vte1 NaCl	ND	ND	ND	ND
vte4 H_2O	ND	ND	58.0	3.7
vte4 NaCl	ND	ND	72.3	5.9
tmt H_2O	115.5	7.51	ND	ND
tmt NaCl	134.7	11.68	ND	ND

In WT plants, the levels of α-TC were comparable throughout the study under control and salt-stress treatment (Table 1). As expected, *vte4* mutant showed γ-TC but not α-TC accumulation in the rosette leaves, *vte1* mutant did not accumulate either α- or γ-TCs in the leaves, and *tmt* transgenic line showed α-TC, but not γ-TC accumulation in the rosette leaves in both control and salt stress conditions.

4.2. Fourier Transform Raman Spectroscopy Measurements and Chemometrics

The changes in chemical component profiles were assessed using a non-invasive technique—Fourier-transformation Raman spectroscopy (FT-Raman) [81,82]. The Raman spectra of *A. thaliana* leaves were recorded with a FT-Raman Spectrometer Nicolet NXR 9650 (Thermo Scientific, Walthman, MA, USA) equipped with a Nd:YAG 1064 nm laser and a germanium detector cooled with liquid nitrogen. The spectrometer was provided with a xy stage, a mirror objective and a prism slide for redirecting the laser beam. The spectra were collected in the range of 100 to 4000 cm^{-1} at 250 mW laser power with a 4 cm^{-1} resolution. Each spectrum included 128 scans. All spectra were registered by the Omnic/Thermo Scientific software. The leaves of *A. thaliana* were lyophilized. One leaf from each collected sample was used for the measurements. Five spectra were collected for the leaf, and then the baseline was corrected and averaged.

4.3. Violaxanthin De-Epoxidation

Leaf discs 5 mm in diameter were placed in the Petri dishes lined with wet filter paper. The discs were illuminated with 1700 μmol m^{-2} s^{-1} for 0; 5; 10; 20; 30 and 40 min. Then the material was frozen in LN2 and stored at −80 °C until analyzing by high performance liquid chromatography (HPLC, Agilent 1260 Infinity system, Waldbronn, Germany) as described by Latowski et al. [83].

4.4. Assays for ROS Scavengers and Osmoprotectants

4.4.1. Analysis of Phenolic Compounds

Phenylpropanoid compounds, such as chlorogenic acid (CGA) or rosmarinic acid (RA) were estimated according to Hura et al. [84] and Surówka et al. [21]. Two sets of dynamically modified excitation (Ex) and emission (Em) wavelengths, for chlorogenic acid (CGA; Ex 325 and Em 424 mn) and rosmarinic acid (RA; Ex 330 and Em 410 nm), were used for the fluorimetric detection. For further technical details, please see Golebiowska-Pikania et al. [85].

4.4.2. Carbohydrates and Sugar Alcohols

Maltose and oligosaccharides (i.e., iso-maltose, maltotriose, maltotetraose, maltosepentaose, maltohexaose, maltoseptaose) determined as hexose equivalents [nmol/mg FW]) and inositol were measured according to Hura et al. [84] and Surówka et al. [21].

4.4.3. Proline and Betaine Estimation

- *Proline*

Free proline was estimated by means of spectrophotometric method in 96-well-plate format (Synergy II, Biotek, Winoski, VT, USA), after derivatization with ninhydrine reagent as reported by Surówka et al. [80].

- *Betaines*

Glycine betaine and stachydrine were estimated in lyophilized samples (≤0.02 g FW) as detailed by Wiszniewska et al. [86] with some modifications. $[^2H_4]$1-amino-1- cyclopropanecarboxylic acid (D-ACC) was added as an internal standard. UHPLC separation was performed by hydrophilic interaction liquid chromatography (HILIC) on an Agilent 1260 Infinity system (Agilent, Waldbronn, Germany). Separation was achieved on a Kinetex HILIC column (75 mm × 2.1 mm, 2.6 μm, Phenomenex, Torrance, CA, USA) at 35 °C. For detection, an Agilent Technologies 6410 triple quadruple mass spectrometer (MS/MS) equipped with electrospray ionization (ESI) in the positive ionization mode was used. Mass hunter software was used to control the UHPLC–MS/MS system and for data analysis. Pure GB and PB were used as external standards, a surrogate was an internal standard for GB and PB.

4.5. Protein Extraction and Determination of Antioxidant Enzyme Activity

Soluble proteins were isolated from the leaves (0.1 g FW) homogenized in a cooled mortar in 2.5 mL of an extraction buffer (3 mM $MgSO_4$, 1 mM DTT, 3 mM EDTA, 100 mM Tricine, pH 8.0, TRIS). The supernatant obtained after centrifugation (20,000× *g*, 20 min) was used to determine the activity of the antioxidant enzymes.

Protein concentration was determined according to Bradford [87], with Bio-Rad Protein Assay (Bio-Rad, Hercules, CA, USA), and using bovine serum albumin (BSA) as a standard.

Total activity of SOD was determined according to Minami and Yoshikawa [88], with 50 mM TRIS-cacodylic buffer pH 8.2. The reaction mixture contained 0.1 mM EDTA, 1.4% (*v*/*v*) Triton X-100, 0.055 μM NBT, 16 μM pyrogallol and the plant extract. The reduced form of NBT was measured at 540 nm. A unit of enzyme activity [U] was defined as the enzyme activity that inhibits auto-oxidation of pyrogallol by 50% according to McCord and Fridovich [89].

CAT activity was evaluated according to Aebi [90], by monitoring the disappearance of H_2O_2 at 240 nm, in 50 mM phosphate buffer pH 7.0. The enzyme activity was determined in units [U] defined as 1 mmol of H_2O_2 degraded in 1 min per 1 mg of protein.

Non-specific peroxidase (POX) activity was measured according to Luck [91] by following the decomposition of p-phenylenediamine (pPD) H_2O_2-dependent for 2 min at 460 nm. The extinction coefficient of 1.545 × 103 M^{-1} cm^{-1} was used as described by Allgood and Perry [92]. Total peroxidase activity was described as nmol of pPD decomposed in 1 min per 1 mg of protein.

4.6. Analysis of SOD by Native PAGE

For determining SOD isoform activity, the fractions of soluble proteins were separated by native polyacrylamide gel electrophoresis (PAGE) on 12% gel at 4 °C, 180 V, in the Laemmli [93] buffer system without sodium dodecyl sulfate (SDS), as described previously by Miszalski et al. [94]. SOD bands were visualized using the staining procedure by Beauchamp and Fridovich [95]. SOD bands were analyzed densitometrically.

4.7. Statistical Analysis

Analysis of variance (ANOVA) and post-hoc Tukey's multiple range test were performed to determine significant differences between *A. thaliana* genotypes and treatments at the significance set at $p \leq 0.05$).

The similarities between the FT-Raman spectra were studied using a hierarchical cluster analysis (Statistica package 10). The spectra were baseline corrected. The cluster analysis was performed separately for WT, *vte1*, *vte4*, and *tmt* plants for the entire wavenumber range using Ward's algorithm. The spectral distances for WT and *tmt* plants were calculated with the standard algorithm with previously unprocessed data. For *vte1* and *vte4* plants, the cluster analysis was carried out with the factorization algorithm using the first two factors followed by vector normalization.

5. Conclusions and Challenges

Under salt stress, α- and γ-TC content differentially influenced the induced compensatory mechanisms, including the components of both non-enzymatic (e.g., CARs, phenols) and enzymatic (SOD, CAT, PODs) antioxidant systems, as well as osmoprotective (Pro, GB, PB, carbohydrates) networks (Scheme 1). The new cellular balance, achieved by α- and γ-TC dependent accumulation of metabolites of the shikimate-phenylpropanoid, MEP and the Glu (glutamate)-Pro-Arg-PAs-GABA (γ-aminobutyric acid) pathways involved reorganization of carbon and nitrogen metabolism and allowed for energy dissipation and ROS scavenging via alternative pathways.

Changes in the status of lipophilic α- and γ-TC influenced the level of CARs and the rate of de-epoxidation. It seems that not only the final level of accumulated xanthophylls but also the amount of transient xanthophylls and the interplay among these photosynthetic pigments at different stages of de-epoxidation depended on the α-/γ-TC ratio and might have important functions in plant signaling (including ABA and JA formation).

Under salt stress, α-TC appears to have a stronger regulatory effect on the pattern of accumulated biocompounds and de-epoxidation than γ-TC, which seems to inhibit some of the defense reactions. α-TC is responsible for key aspects of salt stress adaptation through modification of signals originating in the chloroplasts.

It seems that TCs function at the crossroad of ROS and methylation dependent processes, affecting carbon and nitrogen dependent metabolism. Altered content of differentially methylated α- and γ-TCs and modified TC biosynthetic pathways can affect cellular methylation processes linked probably with methionine cycle, but this statement requires further research.

Author Contributions: E.S. received the grant support, presented the concept of the study, and designed the biochemical experiments. D.L. designed the experiments with xanthophyll cycle in consultation with E.S. and M.D., conducted the HPLC analysis (in collaboration with E.S.), and collected the data. E.S. measured (in collaboration with A.M.) the activity of antioxidant enzymes. E.S. collected, analyzed and visualized the data from with biochemical assays. D.L. and M.O.-P. analyzed and visualized the data from xanthophyll cycle measurements. M.R. collected, analyzed and visualized the data regarding the Raman spectroscopy measurements. C.D. offered advice and experimental counseling. Z.M., I.Ż. and M.K. reviewed and edited the manuscript and study design. E.S. prepared the original manuscript draft in consultation with D.L. and M.R. E.S. bears the main responsibility for the final content of the manuscript. All authors have read and agreed to the published version of the manuscript.

Funding: This work was supported by National Science Center (NCN) in Poland [grant no. N N310 298639], the European Cooperation in Science and Technology (COST) Organization [grant no. 830/1/N-COST/2010/11], and partially from the institutional funding awarded to The *Franciszek Górski* Institute of Plant Physiology, Polish Academy of Sciences.

Institutional Review Board Statement: Not applicable.

Informed Consent Statement: Not applicable.

Acknowledgments: The authors thank Anna Nowicka for helpful comments and critical reading of the manuscript.

Conflicts of Interest: The authors declare no conflict of interest.

Abbreviations

ABA	abscisic acid
Ax	antheraxanthin
BSA	bovine serum albumin
CAR	carotenoids
Car	carotenes
CAT	catalase (EC 1.11.1.6)
CGA	chlorogenic acid
D-ACC	[$^2H_4$]1-amino-1-cyclopropanecarboxylic acid
DTT	dithiothreitol
EDTA	ethylenediaminetetraacetic acid
EGTA	ethylene glycol-*bis*(2-aminoethylether)-N,N,N′,N′-tetraacetic acid
ESI	electrospray ionization
FLD	fluorescence detector
FT-Raman	(Fourier-transformation Raman) spectroscopy
GB	glycine betaine
GGPP	geranylgeranyl pyrophosphate
Glu	(glutamate)-Pro-Arg-PAs-GABA (γ-aminobutyric acid) pathway
HGA	homogentisic acid/chorismate

HILIC	hydrophilic interaction liquid chromatography
HPLC	high performance liquid chromatography
MeOH/HCOOH	methanol/formic acid
MEP	2-C-Methyl-D-erythritol 4-phosphate pathway
MS/MS	mass spectrometer
NBT	nitro blue tetrazolium
PAGE	polyacrylamide gel electrophoresis
PAs	polyamines
POX	non-specific peroxidase
Pro	proline
PAR	photosynthetically active radiation
pPD	p-phenylenediamine
RA	rosmarinic acid
RH	relative humidity
ROS	reactive oxygen species
SOD	superoxide dismutase (EC 1.15.1.1)
TC	tocopherols
TCA	tricarboxylic acid cycle
TEMED	N,N,N′;N-tetramethylethylenediamine
TRICINE	N-tris[hydroxymethyl]methylglycine
TRIS	tris(hydroxymethyl)aminomethane
[U]	a unit of enzyme activity
VDE	violaxanthin de-epoxidase
Vx	violaxanthin
vte1	*A. thaliana* mutant deficient in tocopherol cyclase, totally devoid of TCs
vte4	*A. thaliana* homozygous mutant deficient in γ-tocopherol methyltransferase (γ-*TMT*, catalyses the conversion of γ-TC to α-TC) gene, accumulating γ-TC instead of α-TC
WT	wild type *Arabidopsis thaliana* (Columbia ecotype, Col-0);
γ-*TMT*	*A. thaliana* transgenic line (*tmt*) overexpressing γ-*TMT*, overproducing α-TC
Zx	zeaxanthin

References

1. Gorji, T.; Yildirim, A.; Sertel, E.; Tanik, A. Remote sensing approaches and mapping methods for monitoring soil salinity under different climate regimes. *Int. J. Geoinform.* **2019**, *6*, 33–49. [CrossRef]
2. Surówka, E.; Rapacz, M.; Janowiak, F. Climate change influences the interactive effects of simultaneous impact of abiotic and biotic stresses on plants. In *Plant Ecophysiology and Adaptation under Climate Change: Mechanisms and Perspectives*; Hasanuzzuman, M., Ed.; Springer: Singapore, 2020; pp. 1–50. [CrossRef]
3. Che-Othman, M.H.; Millar, A.H.; Taylor, N.L. Connecting salt stress signalling pathways with salinity-induced changes in mitochondrial metabolic processes in C3 plants. *Plant Cell Environ.* **2017**, *40*, 2875–2905. [CrossRef] [PubMed]
4. Suo, J.; Zhao, Q.; David, L.; Chen, S.; Dai, S. Salinity response in chloroplasts: Insights from gene characterization. *Int. J. Mol. Sci.* **2017**, *18*, 1011. [CrossRef] [PubMed]
5. Bose, J.; Munns, R.; Shabala, S.; Gilliham, M.; Pogson, B.; Tyerman, S.D. Chloroplast function and ion regulation in plants growing on saline soils: Lessons from halophytes. *J. Exp. Bot.* **2017**, *68*, 3129–3143. [CrossRef] [PubMed]
6. Dumanovic, J.; Nepovimova, E.; Natic, M.; Kuca, K.; Jacevic, V. The significance of reactive oxygen species and antioxidant defense system in plants: A concise overview. *Front. Plant Sci.* **2021**, *11*, 552969. [CrossRef]
7. Corpas, F.J.; del Rio, L.A.; Palma, J.M. Plant peroxisomes at the crossroad of NO and H_2O_2 metabolism. *J. Integr. Plant Biol.* **2019**, *61*, 803–816. [PubMed]
8. Caretto, S.; Linsalata, V.; Colella, G.; Mita, G.; Lattanzio, V. Carbon fluxes between primary metabolism and phenolic pathway in plant tissues under stress. *Int. J. Mol. Sci.* **2015**, *16*, 26378–26394. [CrossRef]
9. Fritsche, S.; Wang, X.; Jung, C. Recent advances in our understanding of tocopherol biosynthesis in plants: An overview of key genes, functions, and breeding of vitamin E improved crops. *Antioxidants* **2017**, *6*, 99. [CrossRef] [PubMed]
10. Grudziński, W.; Nierzwicki, L.; Welc, R.; Reszczyńska, E.; Luchowski, R.; Czub, J.; Gruszecki, W.I. Localization and orientation of xanthophylls in a lipid bilayer. *Sci. Rep.* **2017**, *7*, 9619. [CrossRef]
11. Sun, T.; Yuan, H.; Cao, H.; Yazdani, M.; Tadmor, Y.; Li, L. Carotenoid metabolism in plants: The role of plastids. *Mol. Plant* **2018**, *11*, 58–74. [CrossRef]
12. Sharma, A.; Shahzad, B.; Rehman, A.; Bhardwaj, R.; Landi, M.; Zheng, B. Response of phenylpropanoid pathway and the role of polyphenols in plants under abiotic stress. *Molecules* **2019**, *24*, 2452. [CrossRef] [PubMed]

13. Munoz, P.; Munne-Bosch, S. Vitamin E in plants: Biosynthesis, transport, and function. *Trends Plant Sci.* **2019**, *24*, 1040–1051. [CrossRef] [PubMed]
14. Marchiosi, R.; dos Santos, W.D.; Constantin, R.P.; de Lima, R.B.; Soares, A.R.; Finger-Teixeira, A.; Mota, T.R.; de Oliveira, D.M.; Foletto-Felipe, M.d.P.; Abrahao, J.; et al. Biosynthesis and metabolic actions of simple phenolic acids in plants. *Phytochem. Rev.* **2020**, *19*, 865–906. [CrossRef]
15. Misra, A.N.; Latowski, D.; Strzałka, K. The xanthophyll cycle activity in kidney bean and cabbage leaves under salinity stress. *Russ. J. Plant Physiol.* **2006**, *53*, 102–109. [CrossRef]
16. Surówka, E.; Kornaś, A.; Miszalski, Z. On the role of vitamin E in *Arabidopsis thaliana* seedlings growing at low light intensity. In Proceedings of the SFRR-E Meeting "Free Radicals, Health and Lifestyle: From Cell Signalling to Disease Prevention—Satellite Symposium on Vitamin E", Rome, Italy, 26 August 2009; pp. 131–136.
17. Farouk, S. Ascorbic acid and α-tocopherol minimize salt-induced wheat leaf senescence. *J. Stress Physiol. Biochem.* **2011**, *7*, 58–79.
18. Agarwal, P.K.; Shukla, P.S.; Gupta, K.; Jha, B. Bioengineering for salinity tolerance in plants: State of the art. *Mol. Biotechnol.* **2013**, *54*, 102–123. [CrossRef] [PubMed]
19. Kruk, J.; Szymańska, R.; Cela, J.; Munne-Bosch, S. Plastochromanol-8: Fifty years of research. *Phytochemistry* **2014**, *108*, 9–16. [CrossRef] [PubMed]
20. Rastogi, A.; Yadav, D.K.; Szymańska, R.; Kruk, J.; Sedlarova, M.; Pospisil, P. Singlet oxygen scavenging activity of tocopherol and plastochromanol in *Arabidopsis thaliana*: Relevance to photooxidative stress. *Plant Cell Environ.* **2014**, *37*, 392–401. [CrossRef]
21. Surówka, E.; Potocka, I.; Dziurka, M.; Wróbel-Marek, J.; Kurczynska, E.; Żur, I.; Maksymowicz, A.; Gajewska, E.; Miszalski, Z. Tocopherols mutual balance is a key player for maintaining *Arabidopsis thaliana* growth under salt stress. *Plant Physiol. Biochem.* **2020**, *156*, 369–383. [CrossRef]
22. Collakova, E.; DellaPenna, D. The role of homogentisate phytyltransferase and other tocopherol pathway enzymes in the regulation of tocopherol synthesis during abiotic stress. *Plant Physiol.* **2003**, *133*, 930–940. [CrossRef] [PubMed]
23. Cazzaniga, S.; Bressan, M.; Carbonera, D.; Agostini, A.; Dall'Osto, L. Differential roles of carotenes and xanthophylls in photosystem i photoprotection. *Biochemistry* **2016**, *55*, 3636–3649. [CrossRef]
24. Latowski, D.; Kuczyńska, P.; Strzałka, K. Xanthophyll cycle—A mechanism protecting plants against oxidative stress. *Redox Rep.* **2011**, *16*, 78–90. [CrossRef]
25. Jahns, P.; Holzwarth, A.R. The role of the xanthophyll cycle and of lutein in photoprotection of photosystem II. *BBA Bioenerg.* **2012**, *1817*, 182–193. [CrossRef] [PubMed]
26. Trocsanyi, E.; Gyorgy, Z.; Zamborine-Nemeth, E. New insights into rosmarinic acid biosynthesis based on molecular studies. *Curr. Plant Biol.* **2020**, *23*, 100162. [CrossRef]
27. Kaplan, F.; Sung, D.Y.; Guy, C.L. Roles of β-amylase and starch breakdown during temperatures stress. *Physiol. Plant.* **2006**, *126*, 120–128. [CrossRef]
28. Purdy, S.J.; Bussell, J.D.; Nunn, C.P.; Smith, S.M. Leaves of the *Arabidopsis maltose* exporter1 mutant exhibit a metabolic profile with features of cold acclimation in the warm. *PLoS ONE* **2013**, *8*, e79412. [CrossRef] [PubMed]
29. Zulfiqar, F.; Akram, N.A.; Ashraf, M. Osmoprotection in plants under abiotic stresses: New insights into a classical phenomenon. *Planta* **2020**, *251*, 3. [CrossRef] [PubMed]
30. Signorelli, S. The fermentation analogy: A point of view for understanding the intriguing role of proline accumulation in stressed plants. *Front. Plant Sci.* **2016**, *7*, 1339. [CrossRef] [PubMed]
31. Dikilitas, M.; Simsek, E.; Roychoudhury, A. Role of proline and glycine betaine in overcoming abiotic stresses. In *Protective Chemical Agents in the Amelioration of Plant Abiotic Stress: Biochemical and Molecular Perspectives, Roychoudhury*; Tripathi, D.K.A., Ed.; John Wiley & Sons Ltd: Chichester, UK, 2020; pp. 1–23. [CrossRef]
32. Mansour, M.M.F.; Ali, E.F. Glycinebetaine in saline conditions: An assessment of the current state of knowledge. *Acta Physiol. Plant.* **2017**, *39*, 56. [CrossRef]
33. Annunziata, M.G.; Ciarmiello, L.F.; Woodrow, P.; Dell'Aversana, E.; Carillo, P. Spatial and temporal profile of glycine betaine accumulation in plants under abiotic stresses. *Front. Plant Sci.* **2019**, *10*, 230. [CrossRef]
34. Smirnoff, N.; Arnaud, D. Hydrogen peroxide metabolism and functions in plants. *New Phytol.* **2019**, *221*, 1197–1214. [CrossRef] [PubMed]
35. Abbasi, A.-R.; Hajirezaei, M.; Hofius, D.; Sonnewald, U.; Voll, L.M. Specific roles of α-and γ-tocopherol in abiotic stress responses of transgenic tobacco. *Plant Physiol.* **2007**, *143*, 1720–1738. [CrossRef] [PubMed]
36. Cela, J.; Chang, C.; Munne-Bosch, S. Accumulation of γ- rather than α-tocopherol alters ethylene signaling gene expression in the *vte4* mutant of *Arabidopsis thaliana*. *Plant Cell Physiol.* **2011**, *52*, 1389–1400. [CrossRef] [PubMed]
37. Ellouzi, H.; Ben Hamed, K.; Cela, J.; Muller, M.; Abdelly, C.; Munne-Bosch, S. Increased sensitivity to salt stress in tocopherol-deficient Arabidopsis mutants growing in a hydroponic system. *Plant Signal. Behav.* **2013**, *8*, e23136. [CrossRef]
38. Kanwischer, M.; Porfirova, S.; Bergmuller, E.; Dormann, P. Alterations in tocopherol cyclase activity in transgenic and mutant plants of Arabidopsis affect tocopherol content, tocopherol composition, and oxidative stress. *Plant Physiol.* **2005**, *137*, 713–723. [CrossRef]
39. Asensi-Fabado, M.A.; Ammon, A.; Sonnewald, U.; Munne-Bosch, S.; Voll, L.M. Tocopherol deficiency reduces sucrose export from salt-stressed potato leaves independently of oxidative stress and symplastic obstruction by callose. *J. Exp. Bot.* **2015**, *66*, 957–971. [CrossRef] [PubMed]

40. Gajewska, E.; Surówka, E.; Kornaś, A.; Kuźniak, E. Nitrogen metabolism-related enzymes in *Mesembryanthemum crystallinum* after *Botrytis cinerea* infection. *Biol. Plant.* **2018**, *62*, 579–587. [CrossRef]
41. Naliwajski, M.R.; Skłodowska, M. The relationship between carbon and nitrogen metabolism in cucumber leaves acclimated to salt stress. *PeerJ* **2018**, *6*, e6043. [CrossRef]
42. Isayenkov, S.V.; Maathuis, F.J.M. Plant salinity stress: Many unanswered questions remain. *Front. Plant Sci.* **2019**, *10*, 80. [CrossRef]
43. Mitsuya, S.; Kawasaki, M.; Taniguchi, M.; Miyake, H. Light dependency of salinity-induced chloroplast degradation. *Plant Prod. Sci.* **2003**, *6*, 219–223. [CrossRef]
44. Skłodowska, M.; Gapińska, M.; Gajewska, E.; Gabara, B. Tocopherol content and enzymatic antioxidant activities in chloroplasts from NaCl-stressed tomato plants. *Acta Physiol. Plant.* **2009**, *31*, 393–400. [CrossRef]
45. Ischebeck, T.; Zbierzak, A.M.; Kanwischer, M.; Dormann, P. A salvage pathway for phytol metabolism in *Arabidopsis*. *J. Biol. Chem.* **2006**, *281*, 2470–2477. [CrossRef] [PubMed]
46. Dormann, P. Functional diversity of tocochromanols in plants. *Planta* **2007**, *225*, 269–276. [CrossRef] [PubMed]
47. Latowski, D.; Szymańska, R.; Strzałka, K. Carotenoids Involved in Antioxidant System of Chloroplasts. In *Oxidative Damage to Plants*; Parvaiz, A., Ed.; Academic Press: Cambrigde, MA, USA, 2014; pp. 289–319.
48. Brunetti, C.; Guidi, L.; Sebastiani, F.; Tattini, M. Isoprenoids and phenylpropanoids are key components of the antioxidant defense system of plants facing severe excess light stress. *Environ. Exp. Bot.* **2015**, *119*, 54–62. [CrossRef]
49. Latowski, D.; Surówka, E.; Strzałka, K. Regulatory Role of Components of Ascorbate–Glutathione Pathway in Plant Stress Tolerance. In *Ascorbate-Glutathione Pathway and Stress Tolerance in Plants*; Anjum, N.A., Chan, M.-T., Umar, S., Eds.; Springer: Dordrecht, The Netherlands, 2010; pp. 1–53.
50. Gruszecki, W.I.; Strzałka, K. Carotenoids as modulators of lipid membrane physical properties. *BBA Mol. Basis Dis.* **2005**, *1740*, 108–115. [CrossRef]
51. Lombard, J.; Moreira, D. Origins and early evolution of the mevalonate pathway of isoprenoid biosynthesis in the three domains of life. *Mol. Biol. Evol.* **2011**, *28*, 87–99. [CrossRef]
52. Kreszies, V. *ABA-Dependent and -Independent Regulation of Tocopherol (Vitamin E) Biosynthesis in Response to Abiotic Stress in Arabidopsis*; Rheinische Friedrich-Wilhelms-Universität Bonn: Bonn, Germany, 2019.
53. Lallemand, L.A.; Zubieta, C.; Lee, S.G.; Wang, Y.; Acajjaoui, S.; Timmins, J.; McSweeney, S.; Jez, J.M.; McCarthy, J.G.; McCarthy, A.A. A structural basis for the biosynthesis of the major chlorogenic acids found in coffee. *Plant Physiol.* **2012**, *160*, 249–260. [CrossRef]
54. Akazawa, T.; Uritani, I. Respiratory oxidation and oxidative phosphorylation by cytoplasmic particles of sweet potato. *J. Biochem.* **1954**, *41*, 631–638. [CrossRef]
55. Makovec, P.; Sindelar, L. The effect of phenolic compounds on the activity of respiratory chain enzymes and on respiration and phosphorylation activities of potato tuber mitochondria. *Biol. Plant.* **1984**, *26*, 415–422. [CrossRef]
56. Thalmann, M.; Santelia, D. Starch as a determinant of plant fitness under abiotic stress. *New Phytol.* **2017**, *214*, 943–951. [CrossRef]
57. Hummel, I.; Pantin, F.; Sulpice, R.; Piques, M.; Rolland, G.; Dauzat, M.; Christophe, A.; Pervent, M.; Bouteille, M.; Stitt, M.; et al. Arabidopsis plants acclimate to water deficit at low cost through changes of carbon usage: An integrated perspective using growth, metabolite, enzyme, and gene expression analysis. *Plant Physiol.* **2010**, *154*, 357–372. [CrossRef]
58. Kaplan, F.; Guy, C.L. β-amylase induction and the protective role of maltose during temperature shock. *Plant Physiol.* **2004**, *135*, 1674–1684. [CrossRef] [PubMed]
59. Williams, S.P.; Gillaspy, G.E.; Perera, I.Y. Biosynthesis and possible functions of inositol pyrophosphates in plants. *Front. Plant Sci.* **2015**, *6*, 67. [CrossRef]
60. Majumdar, R.; Barchi, B.; Turlapati, S.A.; Gagne, M.; Minocha, R.; Long, S.; Minocha, S.C. Glutamate, ornithine, arginine, proline, and polyamine metabolic interactions: The pathway is regulated at the post-transcriptional level. *Front. Plant Sci.* **2016**, *7*, 78. [CrossRef] [PubMed]
61. Meena, M.; Divyanshu, K.; Kumar, S.; Swapnil, P.; Zehra, A.; Shukla, V.; Yadav, M.; Upadhyay, R.S. Regulation of L-proline biosynthesis, signal transduction, transport, accumulation and its vital role in plants during variable environmental conditions. *Heliyon* **2019**, *5*, e02952. [CrossRef]
62. Abraham, E.; Rigo, G.; Szekely, G.; Nagy, R.; Koncz, C.; Szabados, L. Light-dependent induction of proline biosynthesis by abscisic acid and salt stress is inhibited by brassinosteroid in *Arabidopsis*. *Plant Mol. Biol.* **2003**, *51*, 363–372. [CrossRef]
63. Kishor, P.B.K.; Kumari, P.H.; Sunita, M.S.L.; Sreenivasulu, N. Role of proline in cell wall synthesis and plant development and its implications in plant ontogeny. *Front. Plant Sci.* **2015**, *6*, 544. [CrossRef]
64. Roychoudhury, A.; Banerjee, A.; Lahiri, V. Metabolic and molecular-genetic regulation of proline signaling and itscross-talk with major effectors mediates abiotic stress tolerance in plants. *Turk. J. Botany* **2015**, *39*, 887–910. [CrossRef]
65. Hayat, S.; Hayat, Q.; Alyemeni, M.N.; Wani, A.S.; Pichtel, J.; Ahmad, A. Role of proline under changing environments: A review. *Plant Signal. Behav.* **2012**, *7*, 1456–1466. [CrossRef] [PubMed]
66. Deuschle, K.; Funck, D.; Forlani, G.; Stransky, H.; Biehl, A.; Leister, D.; van der Graaff, E.; Kunzee, R.; Frommer, W.B. The role of Δ1-pyrroline-5-carboxylate dehydrogenase in proline degradation. *Plant Cell* **2004**, *16*, 3413–3425. [CrossRef] [PubMed]

67. Rahikainen, M.; Alegre, S.; Trotta, A.; Pascual, J.; Kangasjarvi, S. Trans-methylation reactions in plants: Focus on the activated methyl cycle. *Physiol. Plant.* **2018**, *162*, 162–176. [CrossRef] [PubMed]
68. Kumar, S.; Beena, A.S.; Awana, M.; Singh, A. Salt-induced tissue-specific cytosine methylation downregulates expression of HKT genes in contrasting wheat (*Triticum aestivum* L.) genotypes. *DNA Cell Biol.* **2017**, *36*, 283–294. [CrossRef]
69. Karan, R.; DeLeon, T.; Biradar, H.; Subudhi, P.K. Salt stress induced variation in DNA methylation pattern and its influence on gene expression in contrasting rice genotypes. *PLoS ONE* **2012**, *7*, e40203. [CrossRef] [PubMed]
70. Toth, V.R.; Meszaros, I.; Veres, S.; Nagy, J. Effects of the available nitrogen on the photosynthetic activity and xanthophyll cycle pool of maize in field. *J. Plant Physiol.* **2002**, *159*, 627–634. [CrossRef]
71. Cheng, L.L. Xanthophyll cycle pool size and composition in relation to the nitrogen content of apple leaves. *J. Exp. Bot.* **2003**, *54*, 385–393. [CrossRef]
72. Kruk, J.; Szymańska, R.; Nowicka, B.; Dłużewska, J. Function of isoprenoid quinones and chromanols during oxidative stress in plants. *New Biotechnol.* **2016**, *33*, 636–643. [CrossRef]
73. Mur, L.A.J.; Mandon, J.; Persijn, S.; Cristescu, S.M.; Moshkov, I.E.; Novikova, G.V.; Hall, M.A.; Harren, F.J.M.; Hebelstrup, K.H.; Gupta, K.J. Nitric oxide in plants: An assessment of the current state of knowledge. *AoB Plants* **2013**, *5*, pls052. [CrossRef] [PubMed]
74. Munne-Bosch, S.; Weiler, E.W.; Alegre, L.; Muller, M.; Duechting, P.; Falk, J. α-tocopherol may influence cellular signaling by modulating jasmonic acid levels in plants. *Planta* **2007**, *225*, 681–691. [CrossRef] [PubMed]
75. Desel, C.; Hubbermann, E.M.; Schwarz, K.; Krupińska, K. Nitration of gamma-tocopherol in plant tissues. *Planta* **2007**, *226*, 1311–1322. [CrossRef]
76. Porfirova, S.; Bergmuller, E.; Tropf, S.; Lemke, R.; Dormann, P. Isolation of an *Arabidopsis* mutant lacking vitamin E and identification of a cyclase essential for all tocopherol biosynthesis. *Proc. Natl. Acad. Sci. USA* **2002**, *99*, 12495–12500. [CrossRef]
77. Bergmuller, E.; Porfirova, S.; Dormann, P. Characterization of an *Arabidopsis* mutant deficient in gamma-tocopherol methyltransferase. *Plant Mol. Biol.* **2003**, *52*, 1181–1190. [CrossRef]
78. Shintani, D.; DellaPenna, D. Elevating the vitamin E content of plants through metabolic engineering. *Science* **1998**, *282*, 2098–2100. [CrossRef]
79. Rosso, M.G.; Li, Y.; Strizhov, N.; Reiss, B.; Dekker, K.; Weisshaar, B. An *Arabidopsis thaliana* T-DNA mutagenized population (GABI-Kat) for flanking sequence tag-based reverse genetics. *Plant Mol. Biol.* **2003**, *53*, 247–259. [CrossRef] [PubMed]
80. Surówka, E.; Dziurka, M.; Kocurek, M.; Goraj, S.; Rapacz, M.; Miszalski, Z. Effects of exogenously applied hydrogen peroxide on antioxidant and osmoprotectant profiles and the C-3-CAM shift in the halophyte *Mesembryanthemum crystallinum* L. *J. Plant Physiol.* **2016**, *200*, 102–110. [CrossRef]
81. Skoczowski, A.; Troć, M. Isothermal Calorimetry and Raman Spectroscopy to Study Response of Plants to Abiotic and Biotic Stresses. In *Molecular Stress Physiology of Plants*; Rout, G.R., Das, A.B., Eds.; Springer: New Delhi, India, 2013; pp. 263–288. [CrossRef]
82. Ryś, M.; Juhasz, C.; Surówka, E.; Janeczko, A.; Saja, D.; Tobias, I.; Skoczowski, A.; Barna, B.; Gullner, G. Comparison of a compatible and an incompatible pepper-tobamovirus interaction by biochemical and non-invasive techniques: Chlorophyll a fluorescence, isothermal calorimetry and FT-Raman spectroscopy. *Plant Physiol. Biochem.* **2014**, *83*, 267–278. [CrossRef] [PubMed]
83. Latowski, D.; Kruk, J.; Strzałka, K. Inhibition of zeaxanthin epoxidase activity by cadmium ions in higher plants. *J. Inorg. Biochem.* **2005**, *99*, 2081–2087. [CrossRef] [PubMed]
84. Hura, T.; Dziurka, M.; Hura, K.; Ostrowska, A.; Dziurka, K. Different allocation of carbohydrates and phenolics in dehydrated leaves of triticale. *J. Plant Physiol.* **2016**, *202*, 1–9. [CrossRef]
85. Gołębiowska-Pikania, G.; Dziurka, M.; Wąsek, I.; Wajdzik, K.; Dyda, M.; Wędzony, M. Changes in phenolic acid abundance involved in low temperature and *Microdochium nivale* (Samuels and Hallett) cross-tolerance in winter triticale (*x Triticosecale* Wittmack). *Acta Physiol. Plant* **2019**, *41*, 38. [CrossRef]
86. Wiszniewska, A.; Kosmińska, A.; Hanus-Fajerska, E.; Dziurka, M.; Dziurka, K. Insight into mechanisms of multiple stresses tolerance in a halophyte Aster tripolium subjected to salinity and heavy metal stress. *Ecotoxicol. Environ. Saf.* **2019**, *180*, 12–22. [CrossRef]
87. Bradford, M.M. A rapid and sensitive method for the quantitationof microgram quantities of protein utilizing the principle ofprotein-dye binding. *Anal. Biochem.* **1976**, *72*, 248–254. [CrossRef]
88. Minami, M.; Yoshikawa, H. A simplified assay method at superoxide dismutase activity for clinical use. *Clin. Chim. Acta* **1979**, *92*, 337–342. [PubMed]
89. McCord, J.M.; Fridovic, I. Superoxide dismutase: An enzymic function for erythrocuprein (hemocuprein). *J. Biol. Chem.* **1969**, *244*, 6049–6050. [CrossRef]
90. Aebi, H. Catalase In Vitro. In *Methods in Enzymology. Oxygen Radicals in Biological Systems*; Academic Press: Cambridge, MA, USA, 1984; Volume 105, pp. 121–126.
91. Luck, H. Peroxidase. In *Methods of Enzymatic Analysis*; Bergmeyer, H.U., Ed.; Academic Press Inc.: London, UK, 1963; pp. 895–897.
92. Allgood, G.S.; Perry, J.J. Oxygen defense systems in obligately thermophilic bacteria. *Can. J. Microbiol.* **1985**, *31*, 1006–1010. [CrossRef] [PubMed]

93. Laemmli, U.K. Cleavage of structural proteins during as-sembly of the head of bacteriophage T4. *Nature* **1970**, *227*, 680–685. [CrossRef]
94. Miszalski, Z.; Ślesak, I.; Niewiadomska, E.; Bączek-Kwinta, R.; Lüttge, U.; Ratajczak, R. Subcellular localization and stress responses of superoxide dismutase isoforms from leaves in the C-3-CAM intermediate halophyte *Mesembryanthemum crystallinum* L. *Plant Cell Environ.* **1998**, *21*, 169–179. [CrossRef]
95. Beauchamp, C.; Fridovic, I. Superoxide dismutase: Improved assays and an assay implicable to acrylamide gels. *Anal. Biochem.* **1971**, *44*, 276–287. [CrossRef]

International Journal of
Molecular Sciences

Article

Differential Regulation of NAPDH Oxidases in Salt-Tolerant *Eutrema salsugineum* and Salt-Sensitive *Arabidopsis thaliana*

Maria Pilarska [1,*], Dorothea Bartels [2] and Ewa Niewiadomska [1]

1 The Franciszek Górski Institute of Plant Physiology, Polish Academy of Sciences, Niezapominajek 21, 30-239 Kraków, Poland; e.niewiadomska@ifr-pan.edu.pl
2 Institute of Molecular Physiology and Biotechnology of Plants (IMBIO), University of Bonn, Kirschallee 1, 53115 Bonn, Germany; dbartels@uni-bonn.de
* Correspondence: m.pilarska@ifr-pan.edu.pl

Abstract: Reactive oxygen species (ROS) signalling is crucial in modulating stress responses in plants, and NADPH oxidases (NOXs) are an important component of signal transduction under salt stress. The goal of this research was to investigate whether the regulation of NOX-dependent signalling during mild and severe salinity differs between the halophyte *Eutrema salsugineum* and the glycophyte *Arabidopsis thaliana*. Gene expression analyses showed that salt-induced expression patterns of two NOX genes, *RBOHD* and *RBOHF*, varied between the halophyte and the glycophyte. Five days of salinity stimulated the expression of both genes in *E. salsugineum* leaves, while their expression in *A. thaliana* decreased. This was not accompanied by changes in the total NOX activity in *E. salsugineum*, while the activity in *A. thaliana* was reduced. The expression of the *RBOHD* and *RBOHF* genes in *E. salsugineum* leaves was induced by abscisic acid (ABA) and ethephon spraying. The in silico analyses of promoter sequences of *RBOHD* and *RBOHF* revealed multiple *cis*-acting elements related to hormone responses, and their distribution varied between *E. salsugineum* and *A. thaliana*. Our results indicate that, in the halophyte *E. salsugineum*, the maintenance of the basal activity of NOXs in leaves plays a role during acclimation responses to salt stress. The different expression patterns of the *RBOHD* and *RBOHF* genes under salinity in *E. salsugineum* and *A. thaliana* point to a modified regulation of these genes in the halophyte, possibly through ABA- and/or ethylene-dependent pathways.

Keywords: halophyte species; NADPH oxidases; NOX; respiratory burst oxidase homolog *RBOH* gene expression; saline adaptations

Citation: Pilarska, M.; Bartels, D.; Niewiadomska, E. Differential Regulation of NAPDH Oxidases in Salt-Tolerant *Eutrema salsugineum* and Salt-Sensitive *Arabidopsis thaliana*. *Int. J. Mol. Sci.* **2021**, *22*, 10341. https://doi.org/10.3390/ijms221910341

Academic Editors: Mirza Hasanuzzaman and Masayuki Fujita

Received: 29 August 2021
Accepted: 21 September 2021
Published: 25 September 2021

1. Introduction

Soil salinity is one of the major environmental factors that restrict crop productivity and the functioning of plants in natural ecosystems [1,2]. A common reaction of plants to salt-triggered osmotic stress and ionic imbalance is to increase the levels of reactive oxygen species (ROS) [3,4]. These reactive species include superoxide ($O_2^{\bullet -}$), hydroxyl radical ($OH^{\bullet}$), hydrogen peroxide (H_2O_2), and singlet oxygen ($^1O_2$) and are generated constantly as products of normal cell metabolism. For a long time, ROS were thought to have primarily negative effects on the cells, such as lipid peroxidation, protein denaturation, and DNA damage [5]. However, more recently, ROS have been perceived as universal signalling metabolites regulating plant growth, development, and defence against biotic and abiotic stresses [6–8]. ROS are generated in several cellular compartments, such as chloroplasts, mitochondria, and peroxisomes. In the plasma membrane, ROS are synthesised by NADPH oxidases (NOXs), also termed respiratory burst oxidase homologs (RBOHs) [9]. These enzymes with homology to the NADPH oxidase from mammalian phagocytes can use NADPH as an electron donor to reduce O_2 molecules and to generate $O_2^{\bullet -}$ in the apoplast. The resulting $O_2^{\bullet -}$ can be converted into H_2O_2 spontaneously or enzymatically by superoxide dismutase (SOD) [9,10]. NADPH oxidases in plants constitute a multigene family and the *Arabidopsis thaliana* genome contains 10 genes, *RBOHA-J*, which exhibit a different pattern of expression during ontogenesis and in response to stress stimuli [11,12]. The results of extensive studies confirmed that ROS produced by NOXs regulate many physiological processes, such as seed germination [13], stomatal opening [14], root growth [15], and pollen tube elongation [16]. RBOH-dependent ROS also appear to play an important role in signalling networks enabling acclimation to various stresses, including heat [17], wounding [18], drought [19], and salinity [20,21].

The regulatory mechanisms of NOXs are still under extensive research and their activity is controlled by different factors, such as calcium and protein kinases [22,23]. A link has been established between phytohormones and the regulation of NADPH oxidases. Abscisic acid (ABA) induced the expression of two *RBOH* genes in *A. thaliana* guard cells [14], and the overexpression of the ABA biosynthesis gene in tobacco resulted in salt-tolerance associated with NOX-dependent ROS production [24].

The two pleiotropic NOX genes in *A. thaliana*, *RBOHD* and *RBOHF*, are the main NADPH oxidases associated with acclimation to salinity [20]. It has been shown that an early response to salt stress is a calcium wave, which depends on the ROS produced by RBOHD and which then propagates the systemic response to salt [25]. The analyses of *A. thaliana* mutants enables linking the ROS produced by RBOHD and RBOHF with the regulation of Na^+/K^+ homeostasis under salinity, which relates to salt resistance [20,26]. The RBOHD- and RBOHF-related ROS also enhanced the accumulation of proline, an osmolyte associated with salinity tolerance [27].

Salt stress is harmful to most plants except for some species, termed halophytes, which can grow and reproduce in high salinity environments [28]. Halophytes activate protective mechanisms to prevent high Na^+ accumulation in the cytosol and to maintain photosynthesis [29–31]. The salt-tolerance of halophytes seems to be linked with their ability to control the redox balance [3,32,33]. The comparison of two close relatives, the glycophyte *Arabidopsis thaliana* and the halophyte *Eutrema salsugineum*, showed that the latter was capable of enhancing the production of H_2O_2 under control and stress conditions [34,35]. This suggests a different regulation of ROS formation in these two species. The effect of H_2O_2 signalling depends not only on its type but also on the site of its origin in the cell, as shown for *A. thaliana* [36]; therefore, precise control of ROS formation may contribute to the outcome of the plant stress responses.

The goal of this research was to investigate whether the regulation of NOX-dependent signalling under salinity differs between the glycophyte *A. thaliana* and the halophyte *E. salsugineum*. We aimed to characterise the expression patterns of the *RBOHD* and *RBOHF* genes and total NADPH oxidase activity in *A. thaliana* and *E. salsugineum* during five days of salinity treatment. To gain insight into possible hormonal regulation mechanisms of *RBOHD* and *RBOHF*, we monitored the gene expression in response to ABA and ethephon and performed an in silico search for *cis*-acting promoter elements.

2. Results

The *E. salsugineum* genome was searched using the available genome sequences of *RBOHD* and *RBOHF* in *A. thaliana*, and one homolog of each gene was found. The conserved domains characteristic of the RBOH family [37] were searched after aligning the amino acid sequences of RBOHD and RBOHF in *E. salsugineum* with the homologs in *A. thaliana*. The C-terminal region included the FAD-binding domain and ferric reductase NAD binding domain (Supplementary Figure S1). The N-terminal regions of the predicted EsRBOHD and EsRBOHF proteins contained the respiratory burst NADPH oxidase domain and putative Ca^{2+}-binding EF hands.

To establish whether the *RBOHD* and *RBOHF* genes in leaves of *E. salsugineum* respond to short (between 6 and 48 h) and prolonged (5 days) salinity conditions, the expression patterns were examined under moderate and severe NaCl stress and compared with the response of homologous genes in *A. thaliana*. The first significant changes in the expression of *RBOHD* in *E. salsugineum* were detected after 24 h of salt-stress. At this time, mild salinity suppressed the accumulation of the *EsRBOHD* transcripts (Figure 1A). Mild and severe salt treatment downregulated the *RBOHD* in this halophyte after 2 days, and the suppression was stronger under 300 mM NaCl. After 5 days of salinity conditions, the expression of the *RBOHD* gene increased (over 2-fold) regardless of the salt concentration used. Only under severe salt stress, 600 mM NaCl, was *RBOHF* in *E. salsugineum* upregulated at 12 h (less than 2-fold) and at 24 h (less than 2.5-fold) (Figure 1B). The *EsRBOHF* transcript also accumulated after 5 days of salinity but at a slightly higher level in response to 600 mM NaCl conditions (over 3.5-fold increase) than to 300 mM NaCl (over 2.5-fold increase).

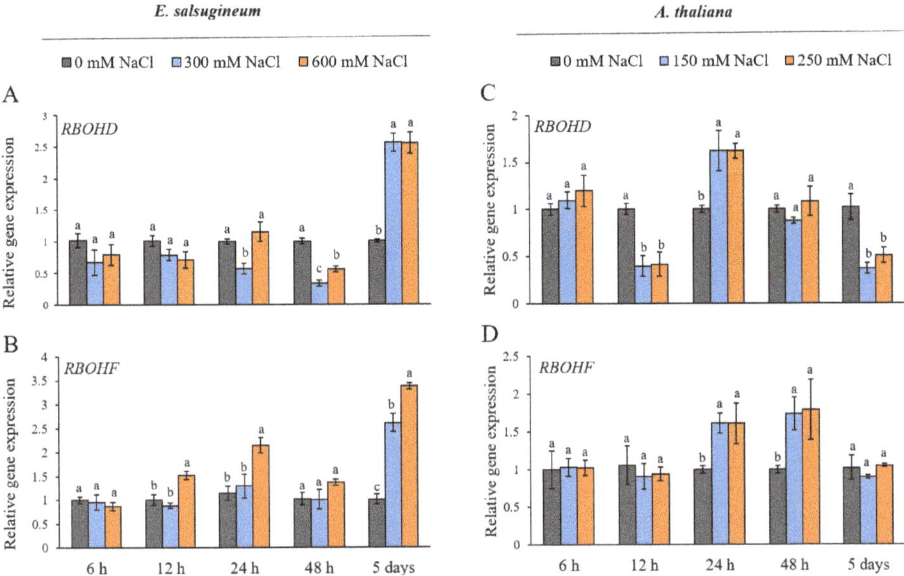

Figure 1. Gene expression patterns of *RBOHD* (**A**,**C**) and *RBOHF* (**B**,**D**) in leaves of *E. salsugineum* and *A. thaliana* under NaCl stress conditions. *E. salsugineum* plants were exposed to 0, 300, or 600 mM NaCl and *A. thaliana* plants were exposed to 0, 150, or 250 mM NaCl for 6, 12, 24, or 48 h or 5 days. Data represent mean ± SE (n = 3). Different letters illustrate significant differences at $p \leq 0.05$.

In *A. thaliana*, a 12-h exposure to moderate and severe salinity suppressed the expression of the *AtRBOHD* gene, but at 24 h, a slight upregulation was observed (less than 2-fold) (Figure 1C). After 5 days of mild and severe salt stress, the level of the *RBOHD* transcripts was significantly lower in relation to respective controls. Under both salinity treatments, the upregulation of *AtRBOHF* (less than 2-fold) was observed after 24 and 48 h (Figure 1D). The level of the *AtRBOHF* transcripts was not different from control samples after 5 days of salinity conditions.

Next, we aimed to verify whether changes in the transcript levels of the *RBOHD* and *RBOHF* genes in *E. salsugineum* and *A. thaliana* leaves under salinity translated to the activity of the enzymes; therefore, the total activity of NOXs in the leaf microsomal fraction was analysed. In *E. salsugineum*, the activity was decreased at 24 and 48 h of 300 mM NaCl conditions by 35.6 % and 11.3 %, respectively (Figure 2A). The activity of NADPH oxidase in the halophyte was not different from the control after 5 days of salinity, whereas in *A. thaliana*, changes in the activity of NOX were detected after 5 days of mild and severe salinity. The activity decreased by 27 % in plants treated with 150 mM NaCl and by 17.6 % in plants treated with 250 mM NaCl (Figure 2B).

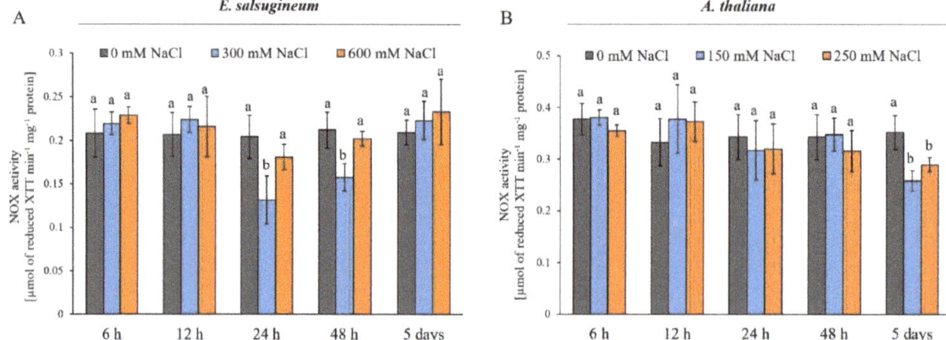

Figure 2. Total activity of NOX in leaves of *E. salsugineum* (**A**) and *A. thaliana* (**B**) plants under NaCl stress conditions. *E. salsugineum* plants were exposed to 0, 300, or 600 mM NaCl and *A. thaliana* were exposed to 0, 150, or 250 mM NaCl for 6, 12, 24, or 48 h or 5 days. Data represent mean ± SD ($n = 3$). Different letters illustrate significant differences at $p \leq 0.05$.

To evaluate the response of the *RBOHD* and *RBOHF* genes in *E. salsugineum* to exogenously applied stress hormones, the leaves were sprayed with ABA or ethephon, which quickly converts to ethylene [38]. As shown in Figure 3A, *RBOHD* was slightly induced by ABA after 7 h (less than 2-fold). The *RBOHD* transcripts were increased (less than 2-fold) at 3 and 7 h after the ethephon treatment. Both ABA and ethephon induced the expression of *RBOHF* by more than 2-fold at 3 h and by over 3-fold at 7 h after spraying of the leaves (Figure 3B).

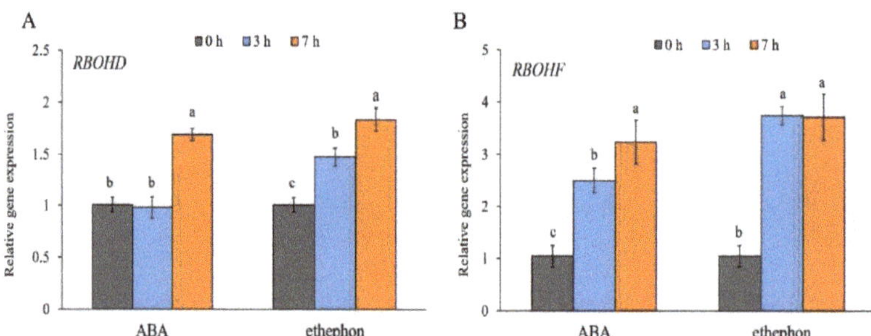

Figure 3. Gene expression patterns of *RBOHD* (**A**) and *RBOHF* (**B**) *in E. salsugineum* leaves sprayed with ABA or ethephon. Data represent mean ± SE ($n = 3$). Different letters illustrate significant differences at $p \leq 0.05$.

To gain an insight into possible transcriptional regulation of the *RBOHD* and *RBOHF* genes in *A. thaliana* and *E. salsugineum*, their promoter sequences were scanned using the PLACE database to predict possible *cis*-acting regulatory elements. Considering that the influence of stress-mediating hormones on the expression of the *RBOH* genes in plants has been reported earlier [14,39], we focused on the presence of *cis*-elements related to hormone signalling in 1500 bp upstream region from the transcription start site. The promoters of *RBOHD* and *RBOHF* were enriched in *cis*-elements responsive to salicylic acid, ethylene, cytokinins, auxins, gibberellins, and ABA (Table 1). However, the distribution of most identical sites varied between *E. salsugineum* and its homologs in *A. thaliana*. In all analysed promoters, multiple motifs involved in responses to ABA and gibberellins were localised. Conversely, only one motif was predicted for ethylene (ERELEE4) and cytokinins (CPBCSPOR) in all the promoters. The ABA responsive *cis*-element, LTRECOREATCOR15, was predicted only in *AtRBOHD* and *AtRBOHF*, while another ABA responsive motif,

RYREPEATVFLEB4, was limited to *EsRBOHF*. Additionally, one auxin responsive motif, CATATGGMSAUR, was observed only in *EsRBOHD*.

Table 1. Putative hormone *cis*-acting regulatory elements in *EsRBOHD*, *EsRBOHF*, *AtRBOHD*, and *AtRBOHF* promoters using the PLACE database.

Cis-Acting Regulatory Elements	Core Sequence	Hormone	Number of Elements			
			EsRBOHD	*EsRBOHF*	*AtRBOHD*	*AtRBOHF*
ERELEE4	AWTTCAAA	Ethylene	1 (+)	1 (−)	1 (+) 2 (−)	1 (−)
GT1CONSENSUS	GRWAAW	Salicylic acid	5 (+) 10 (−)	7 (+) 7 (−)	7 (+) 6 (−)	11 (+) 9 (−)
WBOXATNPR1	TTGAC	Salicylic acid	2 (+) 3 (−)	2 (−) 2 (−)	4 (+) 1 (−)	1 (+) 2 (−)
ASF1MOTIFCAMV	TGACG	Salicylic acid, Auxins	-	1 (−)	-	3 (−)
ARFAT	TGTCTC	Auxins	1 (+)	-	1 (−)	-
CATATGGMSAUR	CATATG	Auxins	1 (+) 1 (−)	-	-	-
CPBCSPOR	TATTAG	Cytokinins	1 (+)	1 (−)	2 (−)	1 (+) 3 (−)
WRKY71OS	TGAC	Gibberellins	6 (+) 6 (−)	4 (+) 4 (−)	8 (+) 6 (−)	2 (+) 5 (−)
MYBGAHV	TAACAAA	Gibberellins	1 (−)	1 (+)	2 (−)	-
GAREAT	TAACAAR	Gibberellins	1 (−)	1 (+)	3 (−)	1 (−)
PYRIMIDINEBOXHVEPB1	TTTTTTCC	Gibberellins, ABA	-	1 (+)	-	1 (−)
DPBFCOREDCDC3	ACACNNG	ABA	1 (+)	2 (+)	2 (+) 2 (−)	1 (+)
RYREPEATVFLEB4	CATGCATG	ABA	-	1 (+) 1 (−)	-	-
MYB1AT	WAACCA	ABA	1 (+) 1 (−)	1 (+)	1 (+) 2 (−)	1 (+)
MYCATRD22	CACATG	ABA	1 (+)	1 (+)	1 (+)	-
LTRECOREATCOR15	CCGAC	ABA	-	-	1 (+)	1 (+) 1 (−)

A + sign within brackets indicates the location of a motif on the presented promoter sequence; a - sign within brackets denotes the position of a motif on the complementary strand of the presented promoter sequence; N = A/C/G/T; R = G/A; W= A/T.

3. Discussion

Salt-sensitive and salt-tolerant plant species use generally the same basic mechanisms of adaptation to salinity; therefore, it has been postulated that the differences in salt tolerance are most probably based on specific regulatory mechanisms [40]. As ROS signalling is a crucial component modulating stress responses in plants [3,7,41], it is reasonable to speculate that differences in the regulation of its components, such as NOXs, might influence stress tolerance. To verify this hypothesis, we compared the expression profiles of two NOX genes important in ROS signalling under salt-stress, *RBOHD* and *RBOHF*, in the halophyte *E. salsugineum* and the glycophyte *A. thaliana*. Our results showed that short-time NaCl conditions, up to 48 h, were accompanied by the increased activity of *RBOHF* in both species. Thus, in *E. salsugineum*, similar to *A. thaliana*, *RBOHF* might be involved in the early acclimation of leaves to salinity. However, in our studies, in contrast with *A. thaliana*, the early transcriptional changes of the *RBOHF* gene in *E. salsugineum* were dependent on NaCl concentrations since only severe salinity triggered the gene expression at 12 and 24 h of exposure to salt. A similar observation was made for the expression of aldehyde dehydrogenase genes, where high salt concentrations were required to trigger the defence mechanisms in *E. salsugineum* compared with *A. thaliana* [42]. In *A. thaliana*, the *RBOHD* and *RBOHF* genes were previously assigned to play a role in the primary responses to NaCl, since the stimulation of their expression was observed in seedlings within a few hours of the stress treatment [20,43]. In agreement with this view, we also detected the salinity-induced activation of *AtRBOHD* after 24 h of the stress treatment. Salinity conditions lasting for 5 days stimulated an expression of *RBOH* genes in *E. salsugineum*, which stood in contrast to *A. thaliana*. This suggests that, in the leaves of the halophyte, the *RBOHD* and *RBOHF* genes might be involved in the late acclimation response to ionic and osmotic stress.

We demonstrated that the total activity of NADPH oxidases in *E. salsugineum* was not affected by severe salt stress. Prolonged mild salinity also did not trigger changes in the enzyme activity despite an initial decline. It may be concluded from the activity measurements that the acclimation to salinity was associated with maintaining the basal activity of NOXs in the halophytic *E. salsugineum*, while the activity declined in *A. thaliana*. These observations are in agreement with Srivastava et al. [44] who showed that the total NADPH oxidase activity in the halophyte *Sesuvium portulacastrum* was unaffected, while it decreased significantly in the glycophyte *Brassica juncea*. The discrepancy of the *EsRBOHD* and *EsRBOHF* transcript level and enzyme activities after 5 days of salinity may result from post-transcriptional regulation of the RBOH proteins. Such a phenomenon was reported earlier for NOXs in cucumber [45]. The molecular mechanisms of NOXs activation are complex, which might explain the lack of stimulation in the in vitro activity assays. It is known that these enzymes are subjected to activation depending on Ca^{2+} and phosphatidic acid and are modified by various kinases [22,23,46].

The ABA and ethylene signalling pathways have been associated with the acclimation of plants to salinity [47,48]. In our studies, the *EsRBOHD* and *EsRBOHF* genes were induced by ABA and ethephon, which indicates a regulation of NOXs in *E. salsugineum* through the ABA- and ethylene-dependent pathways. In *A. thaliana*, the ABA caused a suppression of the *RBOHD* gene, whereas *RBOHF* was suppressed by ethylene [12,49]. Our observations contrast these results. These differences between species in the regulation of the *RBOHD* and *RBOHF* genes might influence their responses to the salt treatment. Earlier studies documented that the ethylene precursor was increased due to salinity in *E. salsugineum* and *A. thaliana* [35,50]. A significant increase in ABA was detected only in *A. thaliana* leaves as a result of exposure to salinity, while only a slight or no increase in ABA levels was observed in *E. salsugineum* [35,51,52].

The in silico analyses of *cis*-acting elements in promoters of the *RBOHD* and *RBOHF* genes in *E. salsugineum* and *A. thaliana* indicated a possible regulation by stress-related hormones (ABA, ethylene, salicylic acid, and jasmonic acid) and by growth-stimulating hormones (auxins, cytokinins, and gibberellins). It is in agreement with earlier studies, where multiple *cis*-acting elements related to hormonal regulation were predicted in promoters of the *RBOH* genes in *A. thaliana* and rice, including ethylene, ABA, auxins, and salicylic acid [12]. Our studies showed that the number of predicted *cis*-acting elements varied between the analysed homologs, which points to differences in the promoter architecture and possibly promoter activity.

4. Methods

4.1. Plant Material, Growth Conditions, and Treatments

The seeds of *Eutrema salsugineum* (earlier *Thellungiella salsuginea*) ecotype Shandong and *Arabidopsis thaliana* ecotype Columbia were obtained from Nottingham Arabidopsis Stock Centre (University of Nottingham, Loughborough, UK). The plants were individually grown in 100 mL pots containing market available soil (pH 5.5–6.5, NaCl < 1.9 g dm^{-3}; Verve, Greenyard Horticulture, Pasłęk, Poland) and were irrigated with tap water (the water quality parameters are listed in Supplementary Table S1). Both species were cultivated in a growth chamber at photoperiod 10/14 h, an irradiance of about 120 μmol m^{-2} s^{-1}, temperatures of 23/20 °C day/night, and 55–65% relative humidity. Plants with fully developed rosette leaves were used for salt stress and hormone treatment. The salt stress was applied by daily irrigation with 20 ml of the NaCl solution (Sigma-Aldrich, St. Louis, MO, USA). Taking into consideration different salt sensitivities of the species used in this study, the concentration of NaCl was different for *E. salsugineum* and *A. thaliana*. To induce mild stress, 300 mM NaCl was used for the former and 150 mM NaCl was used for the latter species. To induce severe salt stress, 600 mM NaCl was used for *E. salsugineum* and 250 mM NaCl was used for *A. thaliana*. Plants irrigated with 20 ml of water were treated as the control. The leaves were collected at 6, 12, 24, 48 h, and 5 days after the onset of NaCl treatment, used immediately or frozen in liquid nitrogen, and kept at −80 °C for further analysis. For spraying the leaves of *E. salsugineum*, 400 mM ABA (Sigma-Aldrich) [49] or 7 mM ethephon (Sigma-Aldrich) [38] dissolved in ethanol was used. Mock-treated plants were sprayed with 1 % ethanol solution. The leaves were collected 3 and 7 h after the spraying, immediately frozen in liquid nitrogen, and kept at −80 °C for further analysis.

4.2. Database Search and Prediction of Cis-Acting Regulatory Elements

The sequences of the *EsRBOHD* and *EsRBOHF* genes were retrieved from the Phytozome v12.1 database (http://phytozome.jgi.doe.gov/; accessed on 23 February 2018) after the genome of *Eutrema salsugineum* was searched using the *Arabidopsis thaliana RBOHD* (At5g47910) and *RBOHF* (At1g64060) sequences as queries. The Phytozome ID were Thhalv10003619m for *EsRBOHD* and Thhalv100023240m for *EsRBOHF*.

The PLACE online tool [53] was used to predict *cis*-acting regulatory elements in the promoter region (1500 bp upstream region from transcription start site) of *RBOHD* and *RBOHF* in *A. thaliana* and *E. salsugineum*.

4.3. Gene Expression Analysis

Total RNA was extracted from the frozen leaf tissue according to Valenzuela-Avendaño et al. [54]. RNA purity and quantity were determined by Biospec-Nano (SHIMADZU, Kyoto, Japan). The integrity of the RNA samples was assessed on a 2.0 % (*w/v*) agarose gel. RNA samples were treated with DNase I (Thermo Fisher Scientific, Waltham, MA, USA) to remove any traces of DNA. To produce a single-stranded cDNA population, 2 μg of total RNA were reversely transcribed with a RevertAid First Strand cDNA Synthesis Kit (Thermo Fisher Scientific), using the oligo $(dT)_{18}$ primer technique according to the manufacturer's instructions. Relative quantitative real-time polymerase chain reaction (qRT-PCR) was performed with Maxima SYBR Green Master Mix (Thermo Fisher Scientific) using a CFX96 Touch Real-Time PCR Detection System (Bio-Rad, Hercules, CA, USA). Each qPCR analysis was performed for three samples of each variant and three technical replicates of each sample. The transcript levels were normalised to adenine phosphoribosyl transferase 1 (*APT1*) in the case of *E. salsugineum* and ubiquinol-cytochrome C reductase iron-sulphur subunit (*AT5G*) in the case of *A. thaliana* [21]. The gene sequences were obtained from the Phytozome database as described above. The primers used are listed in Supplementary Table S2. The probability of secondary structure folding in resulting target sequences was predicted with the M-fold webserver [55]. Reaction efficiency was tested by serial dilutions of cDNAs with gene-specific primers and the primer specificities were confirmed with the melting-curve analysis after amplification during the subsequent qPCR analysis. The expression was calculated according to Pfaffl [56], with water-treated plants serving as the calibrator.

4.4. Membrane Protein Extraction

Freshly harvested leaves were homogenised in a protein extraction buffer containing 50 mM Tris-HCl, 0.25 M sucrose, 2.5 mM dithiothreitol, and 0.1 mM $MgCl_2$ under chilled conditions. The homogenate was filtered through a cheesecloth and centrifuged at

10,000× *g* for 15 min at 4 °C. The microsomal fraction was separated from the supernatant by centrifugation at 80,000× *g* for 45 min at 4 °C, according to the procedure described by Janeczko et al. [57]. The protein content was measured according to the method of Bradford [58] using BSA as a standard. All chemicals used were purchased from Sigma-Aldrich.

4.5. NADPH Oxidase (NOX) Activity

The total NOX activity was determined by measuring the reduction of 2,3-*bis*(2-methoxy-4-nitro-5-sulfophenyl)-2H-tetrazolium-5-carboxanilide inner salt (XTT; BioShop, Burlington, ON, Canada) by $O_2^{\bullet-}$ radicals at 470 nm [59,60]. The reaction mixture contained 50 mM Tris-HCl (Sigma-Aldrich) pH 7.5, 0.5 mM XTT, 0.6 mM NADPH (Sigma-Aldrich), and 5 µg of membrane proteins. The reaction was started by the addition of the NADPH solution. Measurements of absorbance changes in the presence and absence of 50 U SOD (Sigma-Aldrich) were carried out at 470 nm (for XTT $\varepsilon = 2.16 \times 10^4$ M^{-1} cm^{-1}) using an xMark™ Microplate Absorbance Spectrophotometer (Bio-Rad). Enzyme activity was defined as 1 µmol of XTT reduced by 1 mg of membrane proteins per minute.

4.6. Statistical Analysis

All data presented were expressed as mean ± standard error (SE) or standard deviation (SD). The differences between means ($p \leq 0.05$) were determined by one-way ANOVA followed by Duncan test post hoc using SigmaPlot 12 (Systat Software, Inc, Palo Alto, CA, USA).

5. Conclusions

Our results suggest that the maintenance of the basal activity of NOXs in the leaves of the halophytic *E. salsugineum* plays a role in late acclimation responses to salt stress. The different expression patterns of the *RBOHD* and *RBOHF* genes under salinity in *E. salsugineum* and *A. thaliana* point to a modified regulation of these genes in the halophytic *E. salsugineum*, possibly through the ABA- and/or ethylene-dependent pathways.

Supplementary Materials: The following are available online at https://www.mdpi.com/article/10.3390/ijms221910341/s1, Figure S1: The multiple alignment of RBOHD and RBOHF amino acids in *Arabidopsis thaliana* and *Eutrema salsugineum* constructed using the EMBL-EBI Clustal OMEGA tool (www.ebi.ac.uk/Tools/msa/clustalo/; accessed on 20 March 2020). Table S1: Selected parameters of tap water used for irrigation according to the information provided by the Kraków waterworks. Table S2: PCR primers used for quantitative RT-PCR.

Author Contributions: Conceptualisation, funding acquisition, investigation, formal analysis, data curation, visualisation, writing—original draft, writing—review and editing, M.P.; conceptualisation, and writing—review and editing, D.B.; writing—review and editing, E.N. All authors have read and approved the manuscript.

Funding: MP acknowledges a fellowship from the Polish National Science Centre (MINIATURA, Project No. 2017/01/X/NZ3/00293), which supported her research stay at the Institute of Molecular Physiology and Biotechnology of Plants (IMBIO), University of Bonn, Germany.

Institutional Review Board Statement: Not applicable.

Informed Consent Statement: Not applicable.

Data Availability Statement: All relevant data of this article are available within the manuscript and its Supplementary Materials.

Conflicts of Interest: The authors declare that they have no conflicts of interest.

References

1. Pan, T.; Liu, M.; Kreslavski, V.D.; Zharmukhamedov, S.K.; Nie, C.; Yu, M.; Kuznetsov, V.V.; Allakhverdiev, S.I.; Shabala, S. Non-stomatal limitation of photosynthesis by soil salinity. *Crit. Rev. Environ. Sci. Technol.* **2020**, *51*, 791–825. [CrossRef]
2. Zhao, C.; Zhang, H.; Song, C.; Zhu, J.K.; Shabala, S. Mechanisms of plant responses and adaptation to soil salinity. *Innovation* **2020**, *1*, 100017. [CrossRef]

3. Bose, J.; Rodrigo-Moreno, A.; Shabala, S. ROS homeostasis in halophytes in the context of salinity stress tolerance. *J. Exp. Bot.* **2014**, *65*, 1241–1257. [CrossRef] [PubMed]
4. Kumar, V.; Khare, T.; Sharma, M.; Wani, S.H. ROS-induced signaling and gene expression in crops under salinity stress. In *Reactive Oxygen Species and Antioxidant Systems in Plants: Role and Regulation under Abiotic Stress*; Khan, M.I.R., Khan, N.A., Eds.; Springer: Singapore, 2017; pp. 159–184.
5. Møller, I.M.; Jensen, P.E.; Hansson, A. Oxidative modifications to cellular components in plants. *Annu. Rev. Plant Biol.* **2007**, *58*, 459–481. [CrossRef]
6. Foyer, C.H.; Ruban, A.V.; Noctor, G. Viewing oxidative stress through the lens of oxidative signaling rather than damage. *Biochem. J.* **2017**, *474*, 877–883. [CrossRef] [PubMed]
7. Czarnocka, W.; Karpiński, S. Friend or foe? Reactive oxygen species production, scavenging and signaling in plant response to environmental stresses. *Free Radic. Biol. Med.* **2018**, *122*, 4–20. [CrossRef] [PubMed]
8. Fichman, Y.; Mittler, R. Rapid systemic signaling during abiotic and biotic stresses: Is the ROS wave master of all trades? *Plant J.* **2020**, *102*, 887–896. [CrossRef]
9. Kaur, G.; Guruprasad, K.; Temple, B.R.; Shirvanyants, D.G.; Dokholyan, N.V.; Pati, P.K. Structural complexity and functional diversity of plant NADPH oxidases. *Amino Acids.* **2018**, *50*, 79–94. [CrossRef]
10. Suzuki, N.; Miller, G.; Morales, J.; Shulaev, V.; Torres, M.A.; Mittler, R. Respiratory burst oxidases: The engines of ROS signaling. *Curr. Opin. Plant Biol.* **2011**, *14*, 691–699. [CrossRef]
11. Sagi, M.; Fluhr, R. Production of reactive oxygen species by plant NADPH oxidases. *Plant Physiol.* **2006**, *141*, 336–340. [CrossRef]
12. Kaur, G.; Pati, P.K. Analysis of cis-acting regulatory elements of respiratory burst oxidase homolog (Rboh) gene families in Arabidopsis and rice provides clues for their diverse functions. *Comput Biol. Chem.* **2016**, *62*, 104–118. [CrossRef]
13. Müller, K.; Carstens, A.C.; Linkies, A.; Torres, M.A.; Leubner-Metzger, G. The NADPH-oxidase AtrbohB plays a role in *Arabidopsis* seed after-ripening. *New Phytol.* **2009**, *184*, 885–897. [CrossRef] [PubMed]
14. Kwak, J.M.; Mori, I.C.; Pei, Z.M.; Leonhardt, N.; Torres, M.A.; Dangl, J.L.; Bloom, R.E.; Bodde, S.; Jones, J.D.; Schroeder, J.I. NADPH oxidase AtrbohD and AtrbohF genes function in ROS-dependent ABA signaling in Arabidopsis. *EMBO J.* **2003**, *22*, 2623–2633. [CrossRef] [PubMed]
15. Monshausen, G.B.; Bibikova, T.N.; Messerli, M.A.; Shi, C.; Gilroy, S. Oscillations in extracellular pH and reactive oxygen species modulate tip growth of Arabidopsis root hairs. *Proc. Natl. Acad. Sci. USA* **2007**, *104*, 20996–21001. [CrossRef]
16. Potocký, M.; Jones, M.A.; Bezvoda, R.; Smirnoff, N.; Žárský, V. Reactive oxygen species produced by NADPH oxidase are involved in pollen tube growth. *New Phytol.* **2007**, *174*, 742–751. [CrossRef] [PubMed]
17. Larkindale, J.; Hall, J.D.; Knight, M.R.; Vierling, E. Heat stress phenotypes of Arabidopsis mutants implicate multiple signaling pathways in the acquisition of thermotolerance. *Plant Physiol.* **2005**, *138*, 882–897. [CrossRef]
18. Miller, G.; Schlauch, K.; Tam, R.; Cortes, D.; Torres, M.A.; Shulaev, V.; Dangl, J.L.; Mittler, R. The plant NADPH oxidase RbohD mediates rapid systemic signaling in response to diverse stimuli. *Sci. Signal.* **2009**, *2*, ra45. [CrossRef] [PubMed]
19. Wang, X.; Zhang, M.M.; Wang, Y.J.; Gao, Y.T.; Li, R.; Wang, G.F.; Li, V.Q.; Liu, W.T.; Chen, K.M. The plasma membrane NADPH oxidase OsRbohA plays a crucial role in developmental regulation and drought-stress response in rice. *Physiol. Plant.* **2016**, *156*, 421–443. [CrossRef]
20. Ma, L.; Zhang, H.; Sun, L.; Jiao, Y.; Zhang, G.; Miao, C.; Hao, F. NADPH oxidase AtrbohD and AtrbohF function in ROS dependent regulation of Na+/K+ homeostasis in *Arabidopsis* under salt stress. *J. Exp. Bot.* **2012**, *63*, 305–317. [CrossRef]
21. Rejeb, K.B.; Benzarti, M.; Debez, A.; Bailly, C.; Savouré, A.; Abdelly, C. NADPH oxidase-dependent H_2O_2 production is required for salt-induced antioxidant defense in *Arabidopsis thaliana. J. Plant Physiol.* **2015**, *174*, 5–15. [CrossRef]
22. Ogasawara, Y.; Kaya, H.; Hiraoka, G.; Yumoto, F.; Kimura, S.; Kadota, Y.; Hishinuma, H.; Senzaki, E.; Yamagoe, S.; Nagata, K.; et al. Synergistic activation of the Arabidopsis NADPH oxidase AtrbohD by Ca^{2+} and phosphorylation. *J. Biol. Chem.* **2008**, *283*, 8885–8892. [CrossRef]
23. Li, L.; Li, M.; Yu, L.; Zhou, Z.; Liang, X.; Liu, Z.; Cai, G.; Gao, L.; Zhang, X.; Wang, Y.; et al. The FLS2-associated kinase BIK1 directly phosphorylates the NADPH oxidase RbohD to control plant immunity. *Cell Host Microbe.* **2014**, *15*, 329–338. [CrossRef]
24. Zhang, Y.; Tan, J.; Guo, Z.; Lu, S.; He, S.; Shu, W.; Zhou, B. Increased abscisic acid levels in transgenic tobacco over-expressing 9 cis-epoxycarotenoid dioxygenase influence H_2O_2 and NO production and antioxidant defences. *Plant Cell Environ.* **2009**, *32*, 509–519. [CrossRef]
25. Evans, M.J.; Choi, W.G.; Gilroy, S.; Morris, R.J. A ROS-assisted calcium wave dependent on the AtRBOHD NADPH oxidase and TPC1 cation channel propagates the systemic response to salt stress. *Plant Physiol.* **2016**, *171*, 1771–1784. [CrossRef] [PubMed]
26. Chung, J.S.; Zhu, J.K.; Bressan, R.A.; Hasegawa, P.M.; Shi, H. Reactive oxygen species mediate Na+-induced SOS1 mRNA stability in Arabidopsis. *Plant J.* **2008**, *53*, 554–565. [CrossRef] [PubMed]
27. Ben Rejeb, K.B.; Vos, D.L.; Disquet, I.L.; Leprince, A.; Bordenave, M.; Maldiney, R.; Jdey, A.; Abdelly, C.; Savouré, A. Hydrogen peroxide produced by NADPH oxidases increases proline accumulation during salt or mannitol stress in *Arabidopsis thaliana. New Phytol.* **2015**, *208*, 138–1148. [CrossRef] [PubMed]
28. Grigore, M.N. Defining halophytes: A conceptual and historical approach in an ecological frame. In *Halophytes and Climate Change: Adaptive Mechanisms and Potential Uses*; Hasanuzzaman, M., Shabala, S., Fujita, M., Eds.; CABI: Boston, MA, USA, 2019; pp. 3–18.
29. Niewiadomska, E.; Wiciarz, M. Adaptations of chloroplastic metabolism in halophytic plants. In *Progress in Botany*; Lüttge, U., Beyschlag, W., Eds.; Springer: Cham, Switzerland, 2015; pp. 177–193.

30. Bose, J.; Munns, R.; Shabala, S.; Gilliham, M.; Pogson, B.; Tyerman, S.D. Chloroplast function and ion regulation in plants growing on saline soils: Lessons from halophytes. *J. Exp. Bot.* **2017**, *68*, 3129–3143. [CrossRef] [PubMed]
31. Kazachkova, Y.; Eshel, G.; Pantha, P.; Cheeseman, J.M.; Dassanayake, M.; Barak, S. Halophytism: What have we learnt from *Arabidopsis* thaliana relative model systems? *Plant Physiol.* **2018**, *178*, 972–988. [CrossRef]
32. Ozgur, R.; Uzilday, B.; Sekmen, A.H.; Turkan, I. Reactive oxygen species regulation and antioxidant defence in halophytes. *Funct. Plant Biol.* **2013**, *40*, 832–847. [CrossRef]
33. Surówka, E.; Latowski, D.; Libik-Konieczny, M.; Miszalski, Z. ROS Signalling, and Antioxidant Defence Network in Halophytes. In *Halophytes and Climate Change: Adaptive Mechanisms and Potential Uses*; Hasanuzzaman, M., Shabala, S., Fujita, M., Eds.; CABI: Boston, MA, USA, 2019; pp. 179–195.
34. Wiciarz, M.; Gubernator, B.; Kruk, J.; Niewiadomska, E. Enhanced chloroplastic generation of H_2O_2 in stress-resistant *Thellungiella salsuginea* in comparison to *Arabidopsis thaliana*. *Physiol. Plant.* **2015**, *153*, 467–476. [CrossRef]
35. Pilarska, M.; Wiciarz, M.; Jajić, I.; Kozieradzka-Kiszkurno, M.; Dobrev, P.; Vanková, R.; Niewiadomska, E. A different pattern of production and scavenging of reactive oxygen species in halophytic Eutrema salsugineum (*Thellungiella salsuginea*) plants in comparison to *Arabidopsis thaliana* and its relation to salt stress signaling. *Front. Plant Sci.* **2016**, *7*, 1179. [CrossRef] [PubMed]
36. Sewelam, N.; Jaspert, N.; Van Der Kelen, K.; Tognetti, V.B.; Schmitz, J.; Frerigmann, H.; Stahl, E.; Zeier, J.; Van Breusegem, F.; Maurino, V.G. Spatial H_2O_2 signaling specificity: H_2O_2 from chloroplasts and peroxisomes modulates the plant transcriptome differentially. *Mol. Plant* **2014**, *7*, 1191–1210. [CrossRef]
37. Torres, M.A.; Onouchi, H.; Hamada, S.; Machida, C.; Hammond-Kosack, K.E.; Jones, J.D. Six *Arabidopsis thaliana* homologues of the human respiratory burst oxidase (gp91phox). *Plant J.* **1998**, *14*, 365–370. [CrossRef]
38. Zhang, W.; Wen, C.K. Preparation of ethylene gas and comparison of ethylene responses induced by ethylene, ACC, and ethephon. *Plant Physiol. Bioch.* **2010**, *48*, 45–53. [CrossRef]
39. Cheng, C.; Xu, X.; Gao, M.; Li, J.; Guo, C.; Song, J.; Wang, X. Genome-wide analysis of respiratory burst oxidase homologs in grape (*Vitis vinifera* L.). *Int. J. Mol. Sci.* **2013**, *14*, 24169–24186. [CrossRef]
40. Zhu, J.-K. Plant salt tolerance. *Trends Plant Sci.* **2001**, *6*, 66–71. [CrossRef]
41. Hasan, M.; Rahman, A.; Skalicky, M.; Alabdallah, N.; Waseem, M.; Jahan, M.; Ahammed, G.; El-Mogy, M.; El-Yazied, A.; Ibrahim, M.; et al. Ozone Induced Stomatal Regulations, MAPK and Phytohormone Signaling in Plants. *Int. J. Mol. Sci.* **2021**, *22*, 6304. [CrossRef]
42. Hou, Q.; Bartels, D. Comparative study of the aldehyde dehydrogenase (ALDH) gene superfamily in the glycophyte *Arabidopsis thaliana* and Eutrema halophytes. *Ann. Bot.* **2014**, *115*, 465–479. [CrossRef] [PubMed]
43. Xie, Y.J.; Xu, S.; Han, B.; Wu, M.Z.; Yuan, X.X.; Han, Y.; Gu, Q.; Xu, D.K.; Yang, Q.; Shen, W.B. Evidence of Arabidopsis salt acclimation induced by up-regulation of HY1 and the regulatory role of RbohD-derived reactive oxygen species synthesis. *Plant J.* **2011**, *66*, 280–292. [CrossRef]
44. Srivastava, A.K.; Srivastava, S.; Lokhande, V.H.; D'souza, S.F.; Suprasanna, P. Salt stress reveales differential antioxidant and energetics responses in, glycophyte (*Brassica juncea* L.) and halophyte (*Sesuvium portulacastrum* L.). *Front. Environ. Sci.* **2015**, *3*, 19. [CrossRef]
45. Jakubowska, D.; Janicka-Russak, M.; Kabała, K.; Migocka, M.; Reda, M. Modification of plasma membrane NADPH oxidase activity in cucumber seedling roots in response to cadmium stress. *Plant Sci.* **2015**, *234*, 50–59. [CrossRef]
46. Zhang, Y.; Zhu, H.; Zhang, Q.; Li, M.; Yan, M.; Wang, R.; Wang, L.; Welti, R.; Zhang, W.; Wang, X. Phospholipase Da1 and phosphatidic acid regulate NADPH oxidase activity and production of reactive oxygen species in ABA-mediated stomatal closure in Arabidopsis. *Plant Cell* **2009**, *21*, 2357–2377. [CrossRef]
47. Zhang, M.; Smith, J.A.C.; Harberd, N.P.; Jiang, C. The regulatory roles of ethylene and reactive oxygen species (ROS) in plant salt stress responses. *Plant Mol. Biol.* **2016**, *91*, 651–659. [CrossRef]
48. Yu, Z.; Duan, X.; Luo, L.; Dai, S.; Ding, Z.; Xia, G. How plant hormones mediate salt stress responses. *Trends Plant Sci.* **2020**, *25*, 1117–1130. [CrossRef]
49. Sewelam, N.; Kazan, K.; Thomas-Hall, S.R.; Kidd, B.N.; Manners, J.M.; Schenk, P.M. Ethylene response factor 6 is a regulator of reactive oxygen species signaling in Arabidopsis. *PLoS ONE* **2013**, *8*, e70289. [CrossRef]
50. Ellouzi, H.; Ben Hamed, K.; Hernández, I.; Cela, J.; Müller, M.; Magné, C.; Abdelly, C.; Munné-Bosch, S. A comparative study of the early osmotic, ionic, redox and hormonal signaling response in leaves and roots of two halophytes and a glycophyte to salinity. *Planta* **2014**, *240*, 1299–1317. [CrossRef] [PubMed]
51. Arbona, V.; Argamasilla, R.; Gómez-Cadenas, A. Common and divergent physiological, hormonal and metabolic responses of *Arabidopsis thaliana* and *Thellungiella halophila* to water and salt stress. *J. Plant. Physiol.* **2010**, *167*, 1342–1350. [CrossRef] [PubMed]
52. Prerostova, S.; Dobrev, P.I.; Gaudinova, A.; Hosek, P.; Soudek, P.; Knirsch, V.; Vankova, R. Hormonal dynamics during salt stress responses of salt-sensitive *Arabidopsis thaliana* and salt-tolerant *Thellungiella salsuginea*. *Plant Sci.* **2017**, *264*, 188–198. [CrossRef] [PubMed]
53. Higo, K.; Ugawa, Y.; Iwamoto, M.; Korenaga, T. Plant cis-acting regulatory DNA elements (PLACE) database: 1999. *Nucleic Acids Res.* **1999**, *27*, 297–300. [CrossRef]
54. Valenzuela-Avendaño, J.; Mota, I.E.; Uc, G.; Perera, R.; Valenzuela-Soto, E.; Aguilar, J.Z. Use of a simple method to isolate intact RNA from partially hydrated *Selaginella lepidophylla* plants. *Plant Mol. Biol. Rep.* **2005**, *23*, 199–200. [CrossRef]

55. Zuker, M. Mfold web server for nucleic acid folding and hybridization prediction. *Nucleic Acids Res.* **2003**, *31*, 3406–3415. [CrossRef]
56. Pfaffl, M.W. A new mathematical model for relative quantification in real-time RT–PCR. *Nucleic Acids Res.* **2001**, *29*, e45. [CrossRef] [PubMed]
57. Janeczko, A.; Budziszewska, B.; Skoczowski, A.; Dybała, M. Specific binding sites for progesterone and 17ß-estradiol in cells of *Triticum aestivum* L. *Acta Biochem. Pol.* **2008**, *55*, 708–711.
58. Bradford, M.M. A rapid and sensitive method for quantitation of microgram quantities of protein utilizing the principle of proteindye- binding. *Anal. Biochem.* **1975**, *72*, 248–254. [CrossRef]
59. Able, A.; Guest, D.; Sutherland, M. Use of a new tetrazolium-based assay to study the production of superoxide radicals by tobacco cell cultures challenged with avirulent zoospores of Phytophthora parasitica var nicotianae. *Plant Physiol.* **1998**, *117*, 491–499. [CrossRef] [PubMed]
60. Potocký, M.; Pejchar, P.; Gutkowska, M.; Jiménez-Quesada, M.J.; Potocká, A.; de Dios, A.J.; Kost, B.; Žárský, V. NADPH oxidase activity in pollen tubes is affected by calcium ions, signaling phospholipids and Rac/Rop GTPases. *J. Plant Physiol.* **2012**, *169*, 1654–1663. [CrossRef] [PubMed]

International Journal of *Molecular Sciences*

MDPI

Article

At the Edges of Photosynthetic Metabolic Plasticity—On the Rapidity and Extent of Changes Accompanying Salinity Stress-Induced CAM Photosynthesis Withdrawal

Michał Nosek [1,*], Katarzyna Gawrońska [1], Piotr Rozpądek [2], Marzena Sujkowska-Rybkowska [3], Zbigniew Miszalski [4] and Andrzej Kornaś [1]

1 Institute of Biology, Pedagogical University of Krakow, Podchorążych 2, 30-084 Kraków, Poland; katarzyna.gawronska@up.krakow.pl (K.G.); andrzej.kornas@up.krakow.pl (A.K.)
2 Małopolska Centre of Biotechnology, Jagiellonian University in Kraków, Gronostajowa 7a, 30-387 Kraków, Poland; piotr.rozpadek@uj.edu.pl
3 Department of Botany, Institute of Biology, Warsaw University of Life Sciences (SGGW), Nowoursynowska 159, 02-776 Warsaw, Poland; marzena_sujkowska@sggw.pl
4 W. Szafer Institute of Botany, Polish Academy of Sciences, Lubicz 46, 31-512 Kraków, Poland; z.miszalski@botany.pl
* Correspondence: michal.nosek@up.krakow.pl

Citation: Nosek, M.; Gawrońska, K.; Rozpądek, P.; Sujkowska-Rybkowska, M.; Miszalski, Z.; Kornaś, A. At the Edges of Photosynthetic Metabolic Plasticity—On the Rapidity and Extent of Changes Accompanying Salinity Stress-Induced CAM Photosynthesis Withdrawal. *Int. J. Mol. Sci.* **2021**, *22*, 8426. https://doi.org/10.3390/ijms22168426

Academic Editors: Mirza Hasanuzzaman and Masayuki Fujita

Received: 24 June 2021
Accepted: 3 August 2021
Published: 5 August 2021

Abstract: The common ice plant (*Mesembryanthemum crystallinum* L.) is a facultative crassulacean acid metabolism (CAM) plant, and its ability to recover from stress-induced CAM has been confirmed. We analysed the photosynthetic metabolism of this plant during the 72-h response period following salinity stress removal from three perspectives. In plants under salinity stress (CAM) we found a decline of the quantum efficiencies of PSII (Y(II)) and PSI (Y(I)) by 17% and 15%, respectively, and an increase in nonphotochemical quenching (NPQ) by almost 25% in comparison to untreated control. However, 48 h after salinity stress removal, the PSII and PSI efficiencies, specifically Y(II) and Y(I), elevated nonphotochemical quenching (NPQ) and donor side limitation of PSI (Y_{ND}), were restored to the level observed in control (C_3 plants). Swelling of the thylakoid membranes, as well as changes in starch grain quantity and size, have been found to be components of the salinity stress response in CAM plants. Salinity stress induced an over 3-fold increase in average starch area and over 50% decline of average seed number in comparison to untreated control. However, in plants withdrawn from salinity stress, during the first 24 h of recovery, we observed chloroplast ultrastructures closely resembling those found in intact (control) ice plants. Rapid changes in photosystem functionality and chloroplast ultrastructure were accompanied by the induction of the expression (within 24 h) of structural genes related to the PSI and PSII reaction centres, including *PSAA*, *PSAB*, *PSBA* (D_1), *PSBD* (D_2) and *cp43*. Our findings describe one of the most flexible photosynthetic metabolic pathways among facultative CAM plants and reveal the extent of the plasticity of the photosynthetic metabolism and related structures in the common ice plant.

Keywords: C_3–CAM intermediate; common ice plant; *Mesembryanthemum crystallinum*; osmotic stress

1. Introduction

Facultative CAM (Crassulacean acid metabolism) plants are plants that can induce or upregulate CAM photosynthesis in response to water-related environmental stresses (drought, high salinity). This was previously recognised as a unique trait but has been identified in a large and constantly growing group of plant families, including *Aizoaceae*, *Bromeliaceae*, *Cactaceae*, *Didiereaceae*, *Lamiaceae*, *Montiaceae* and *Vitaceae*. For most facultative CAM plants, the reversibility of stress-induced CAM photosynthesis has been confirmed [1].

Mesembryanthemum crystallinum L. (the common ice plant), in the *Aizoaceae* family, perfectly reflects the distinctive features of facultative CAM plants and for years was identified as a plant model in studies of osmotic stress-induced CAM photosynthesis. Elevated salinity generates a plant response that involves a wide variety of modifications resulting primarily from Na^+ sensing and secondarily from the build-up of a toxic concentration of ions in the aerial plant parts [2–4]. The presence of osmoprotective mechanisms (e.g., proline synthesis and Na^+ and Cl^- accumulation in bladder cells of the aerial plant parts) and the induction of antioxidative system components allows the maintenance of the main metabolic pathways in a minimally disturbed fashion, even under high-salinity conditions. In addition to the above mentioned traits, β-carboxylation as a part of stress-induced CAM photosynthesis greatly enhances water use efficiency. This is often recognised as a key factor allowing normal growth and development under water-related stresses. A recent study regarding the symptoms of stress-related CAM induction in the common ice plant confirmed the occurrence of β-carboxylation after 6 days of salt treatment [5]. The reorganisation of the CO_2 metabolism is accompanied by substantial modifications at the chloroplast ultrastructure, photosystem reactive centre organisation and photochemical efficiency levels [6]. It was shown that in the common ice plant, salinity stress is also responsible for the enhancement of linear electron transport and the extension of sinks for reducing power [7]. In addition to these modifications of photosynthetic pathways, redox homeostasis and the osmoprotective $C_3 \rightarrow$CAM transition are associated with changes in the structural organisation of tissues and organs [8,9].

As previously mentioned, recovery from stress-induced CAM to C_3 photosynthesis seems to be a common trait in all known facultative CAM plants, including the ice plant, however, the rate, as well as the range of withdrawal-related changes, have been suggested to differ among them. It was shown that members of the *Talinum* and *Clusia* genera can fully stop overnight carbon fixation (distinguished as a hallmark of CAM photosynthesis) at up to 4 days following drought stress withdrawal [10]. On the other hand, for members of the *Portulacaceae* and *Aizoaceae* families, the full development of light phase-related CO_2 fixation without overnight acidification was confirmed as early as 24 h after re-watering [10]. Our earlier study involving the withdrawal of stress-induced CAM in common ice plants showed that removal of osmotic stress inhibited nocturnal malate synthesis within 24 h. This sudden retreat from β-carboxylation was accompanied by a rapid decline in *PEPC1* expression [11]. The recovery of C_3 photosynthesis was also accompanied by the reversion of the activity of the main antioxidant enzymes to a level resembling that in unstressed plants within 48 h after osmotic stress removal. The rapidity of these modifications unequivocally puts the common ice plant's photosynthetic metabolism among those with the highest possible plasticity. According to the previously mentioned studies, photosynthetic metabolic plasticity allows for the quick adjustment to the presence and absence of stressful conditions and must play an important role in acclimation. Understanding the roles of and connection between photosynthesis and acclimation processes seems to be a top priority, especially if we take into account the fact that the ranges of many stress factors, such as high salinity, are covering an increasing area. This study aimed to demonstrate not only the speed and rate of changes occurring in response to osmotic stress withdrawal in our model facultative plant but also the extent to which these changes are related, including the changes in photochemical apparatus efficiency, chloroplast ultrastructure and the regulation of photosystem I and II structural gene expression.

2. Results

2.1. Removal of Osmotic Stress Results in Fast Recovery of PSI and PSII Efficiency

Our expectations about the time needed for stress recovery-related photosystem efficiency changes were based on our earlier study showing the full reversal of functional CAM within hours and modifications of enzymatic antioxidative system components (gene expression, protein activity) within days after osmotic stress removal. As we could not precisely predict the exact timing of such modifications in the context of photosynthetic apparatus efficiency, we decided to extend the experiment to 72 h. We performed our analyses at three consecutive time points, namely, 24, 48 and 72 h after stressor removal.

Earlier studies showed a clear effect of osmotic stress on photosynthetic apparatus efficiency in ice plants, including modifications of the maximal quantum yield of PSII (F_v/F_m), the PSII and PSI electron transport chains (ETRII, ETRI), quanta yield (Y(II), Y(I)), nonphotochemical quenching (NPQ) and donor side limitation (Y_{ND}). F_v/F_m was shown to be affected differently in ice plants under salinity stress (from photoinhibition to no changes at all). Therefore, we decided to concentrate our efforts on the remaining parameters, including the analysis of photochemical quenching according to the lake model (qL), as this model shows the share of photochemical processes in the energy sink. We found a decline by 18% and 12% in both the Y(II) and qL, respectively, of ice plants exposed to salt stress (CAM) in comparison to control (C_3) at 24 h after stress removal. These changes were accompanied in salt-stressed (CAM) plants by a 50% increase in NPQ in comparison to control (Figure 1A) and were sustained in salt-stressed plants 48 and 72 h after stressor removal (Figure 1B,C). We found no substantial changes

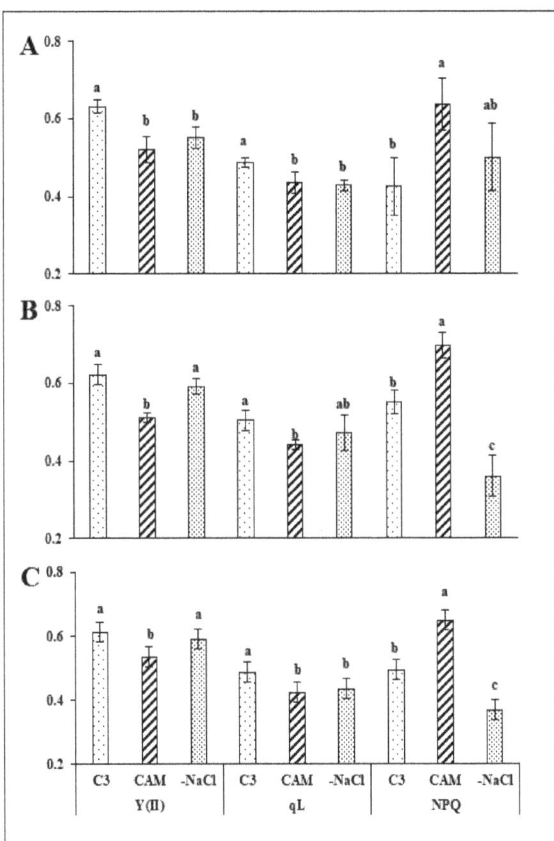

Figure 1. Quantum efficiency of PSII—Y(II), photochemical quenching coefficient—qL and non-photochemical quenching—NPQ in the leaves of unstressed control (C_3), NaCl-treated (CAM) and salt-stress withdrawn (-NaCl) *Mesembryanthemum crystallinum* L. plants measured in the middle of the light phase 24 (**A**), 48 (**B**) and 72 (**C**) hours after osmotic stress removal. Bars represent mean values (±SD) for n = 5. Different letters indicate statistically significant differences according to Tukey's HSD test at $p \leq 0.05$.

In the analysed PSII parameters of stress-withdrawn plants following 24 h after stressor removal. Forty-eight hours after osmotic stress removal, the PSII quantum efficiency Y(II)

reached the level measured in control plants, while the NPQ dropped below the value found in the control by almost 35%. Salt stress removal did not affect the qL of PSII. The changes observed in the Y(II) and NPQ of stress-withdrawn plants were sustained up to 72 h after stressor removal (Figure 1C). At all analysed time points in the experiment, the electron transport in photosystem II (ETRII) of plants affected with salinity stress (CAM) was inhibited in comparison to that in the control (Figure 2A–C). Removal of the stressor (-NaCl) resulted in more rapid induction of ETRII; however, the rate of induction measured in the unstressed control was achieved only 48 h after removal of the osmotic stress and was sustained further (Figure 2B,C).

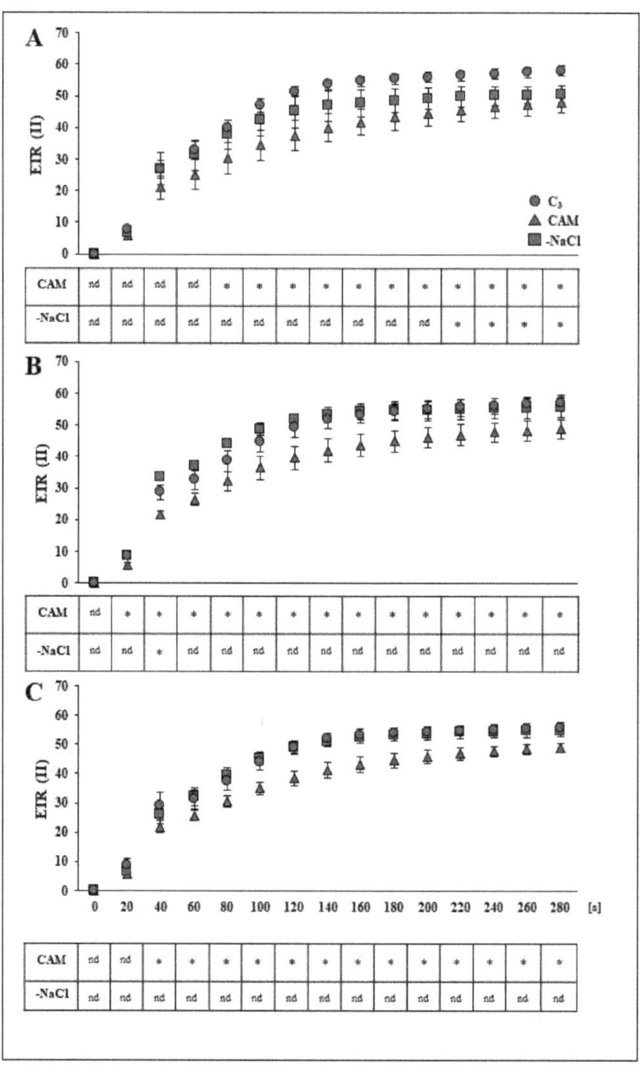

Figure 2. Induction curves of the PSII electron transport rate in the leaves of unstressed control (C_3), NaCl-treated (CAM) and salt-stress-withdrawn (-NaCl) *Mesembryanthemum crystallinum* L. plants measured in the middle of the light phase 24 (**A**), 48 (**B**) and 72 (**C**) h after osmotic stress removal. Asterisks indicate a statistically significant difference in comparison to the unstressed control (C_3) plants according to Dunnett's test for $n = 5$; nd, no differences; [s], seconds.

To assess whether the effects of osmotic stress removal extended to PSI efficiency, we employed a dual PAM measurement of the quantum yield Y(I) and electron transport rate of PSI (ETRI). Unlike in the PSII analysis, we found no evidence of a detrimental effect of salt stress on the analysed parameters of PSI at up to 48 h after stressor removal (Figure 3A). However, at the last time point (72 h), we found a lower by 15% PSI quantum yield and increased by 84% PSI donor side limitation (Y_{ND}) in salt-stressed (CAM) plants in comparison to unstressed plants. Salt stress had no visible effect on acceptor side limitation (Y_{NA}) at any of the experimental time points. As in the PSII analysis, no substantial changes were observed in the analysed parameters (Y(II), Y_{ND}) in stress-withdrawn plants 24 h after stressor removal. However, during the 48-h time frame, plants withdrawn from salt stress (-NaCl) achieved the PSI quantum yield level of the control (Figure 3B). This was accompanied by a substantial, precisely 42% decrease in PSI donor side limitation (Y_{ND}) of the desalinated plants in comparison to salt-stressed plants. We found that, as in the case of PSII, the described changes were sustained for 72 h (Figure 3C). On the other hand, salt stress removal (-NaCl) did not modify the PSI acceptor side limitation Y_{NA} compared with that in the unstressed control. Similar to its effect on ETRII, salt stress substantially retarded the induction of ETRI (Figure 4A) when compared with the control plants, and this effect was sustained at subsequent experimental time points. However, contrary to ETRII PSI, electron transport in stress-withdrawn plants achieved the rate observed in the control as early as 24 h after removal of osmotic stress (Figure 4A); this effect was also sustained at 48 and 72 h after osmotic stress removal (Figure 4B,C). The trends described for the analysed PSI and PSII parameters were confirmed in most cases during the second replication of the experiment (supplementary Figures S1–S3).

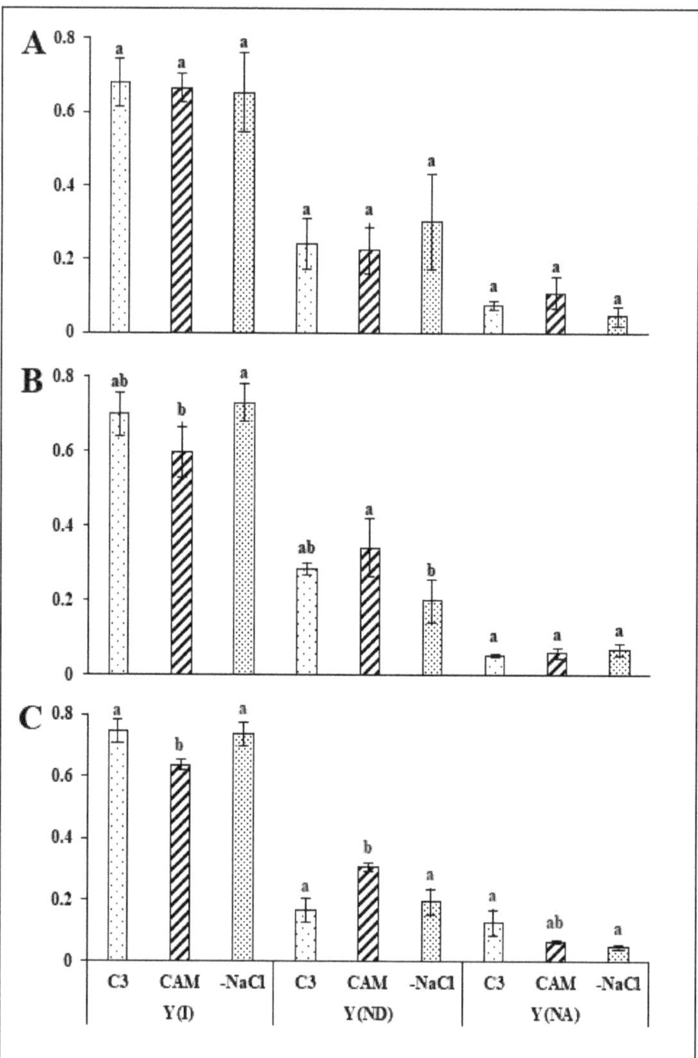

Figure 3. Quantum efficiency of PSI—Y(I), donor side limitation of PSI—Y_{ND} and acceptor side limitation—Y_{NA} in the leaves of unstressed control (C_3), NaCl-treated (CAM) and salt-stress withdrawn (-NaCl) *Mesembryanthemum crystallinum* L. plants measured in the middle of the light phase 24 (**A**), 48 (**B**) and 72 (**C**) h after osmotic stress removal. Bars represent mean values (±SD) for $n = 5$. Different letters indicate statistically significant differences according to Tukey's HSD test at $p \leq 0.05$.

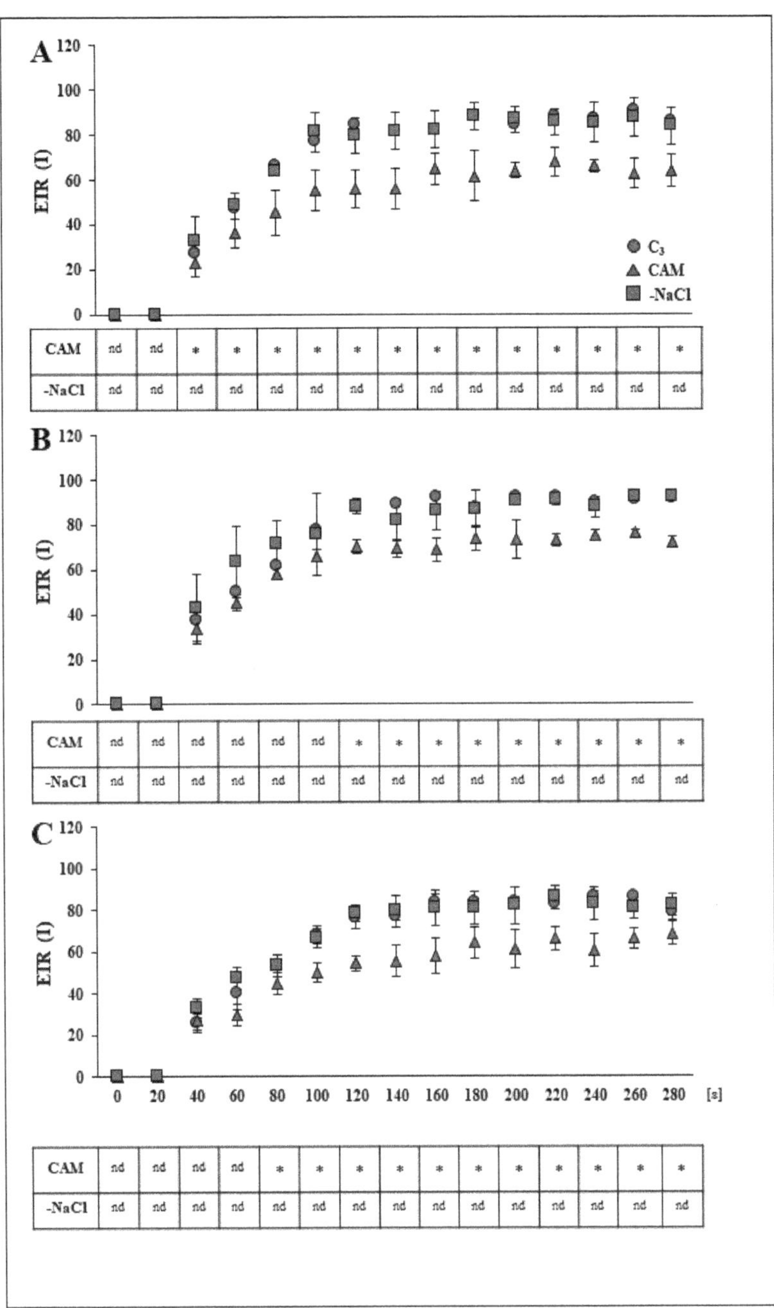

A	0	20	40	60	80	100	120	140	160	180	200	220	240	260	280
CAM	nd	nd	*	*	*	*	*	*	*	*	*	*	*	*	*
-NaCl	nd	nd	nd	nd	nd	nd	nd	nd	nd	nd	nd	nd	nd	nd	nd

B	0	20	40	60	80	100	120	140	160	180	200	220	240	260	280
CAM	nd	nd	nd	nd	nd	nd	*	*	*	*	*	*	*	*	*
-NaCl	nd	nd	nd	nd	nd	nd	nd	nd	nd	nd	nd	nd	nd	nd	nd

C	0	20	40	60	80	100	120	140	160	180	200	220	240	260	280
CAM	nd	nd	nd	nd	*	*	*	*	*	*	*	*	*	*	*
-NaCl	nd	nd	nd	nd	nd	nd	nd	nd	nd	nd	nd	nd	nd	nd	nd

Figure 4. Induction curves of the PSI electron transport rate in the leaves of unstressed control (C_3), NaCl-treated (CAM) and salt-stress withdrawn (-NaCl) *Mesembryanthemum crystallinum* L. plants measured in the middle of the light phase 24 (**A**), 48 (**B**) and 72 (**C**) h after osmotic stress removal. Asterisks indicate a statistically significant difference in comparison to unstressed control (C_3) plants according to Dunnett's test for $n = 5$; nd, no differences; [s], seconds.

2.2. *Rapid Recovery of PSII and PSI Efficiency Is Combined with Induced Expression of Structural Genes for the Reaction Centres of Both Photosystems*

To assess how salt stress removal affects the expression of the structural genes associated with the reaction centres of PSI and PSII, we analysed the abundance of their respective transcripts with *q*PCR during a 72-h recovery period following stress removal. In all studied genes, substantial modifications related to osmotic stress withdrawal were observed during the first 24 h of recovery. We found over 2-fold and over 3-fold increases in *PSAA* (PSI-A core protein of PS I) and *PSBD* (D_2 protein of PS II) expression in salt-stressed (CAM) plants in comparison to intact plants (Figure 5A,D). Except for *PSAA* and *PSBD*, the expression of the analysed structural genes of the photosystem centres remained mostly unaffected during salt stress (Figure 5B,C,E). On the other hand, we found that *PSAA* was the gene mostly affected by osmotic stressor withdrawal, with expression over 3-fold and over 8-fold higher than that in intact (C_3) and stressed (CAM) plants, respectively (Figure 5A). The expression of *CP43* was upregulated in a similar manner, with an over 2-fold increase compared with that in both intact (C_3) and salt-stressed (CAM) plants (Figure 5E). The expression of *PSAB* (PSI-B core subunit of PS I) and *PSBA* (D_1 protein of PS II) was modified in the same direction; however, in both cases, the upregulation related to osmotic stress removal was less intense than that of *PSAA*, *PSBD* and *cp43* (Figure 5B,C). The upregulated expression of *PSBD* due to salt stress was enhanced even further with osmotic stress removal, resulting in an over 1,5-fold increase in comparison to that in salt-stressed (CAM) plants (Figure 5D). All modifications described here at the 24-h experimental time point were sustained at 48 and 72 h after stressor removal (supplementary Figure S4).

2.3. *Withdrawal from Osmotic Stress Is Accompanied by the Rapid Reorganisation of Chloroplast Ultrastructure*

To determine the range of modifications resulting from osmotic stress withdrawal, we analysed the chloroplast ultrastructure of salt-stressed and salt stress-withdrawn plants. In the chloroplasts of intact (C_3) plants, we found densely packed, intact thylakoids that were unstacked or organised in irregular grana stacks located between small- and medium-sized starch grains (Figure 6A–C). In response to osmotic stress, the thylakoid membrane system was reorganised. In the chloroplasts of salt-stressed plants, we found swollen thylakoids that were mostly unstacked and localised between large starch grains (Figure 6D).

We found that osmotic stress increased the size and reduced the average starch grain quantity over 3-fold and 2-fold, respectively (Figure 6F,G). In plants withdrawn from osmotic stress, thylakoid swelling was barely visible. This effect was accompanied by the reorganisation of the membrane system, which began to resemble the one observed in the untreated control, with densely packed thylakoids organised mostly in irregular grana (Figure 6E). The changes resulting from osmotic stress removal were also expressed in the reduced area and increased number of starch grains, similar to those found in unstressed (C_3) plants. The ultrastructural modifications observed 24 h after osmotic stress removal were sustained at 48 and 72 h (Supplementary Figure S5).

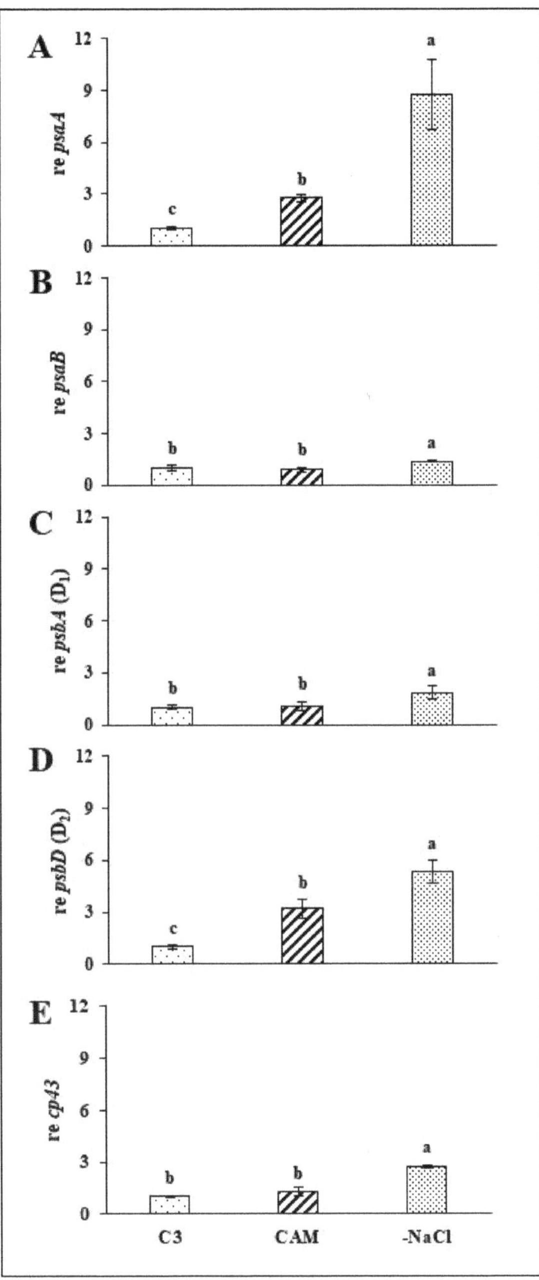

Figure 5. Relative expression of the PSI-A core protein of photosystem I—*PSAA* (**A**), PSI-B core subunit of photosystem I—*PSAB* (**B**), D_1 protein of photosystem II—*PSBA* (**C**), D_2 protein of photosystem II—*PSBD* (**D**) and cp43 protein of photosystem II—*cp43* (**E**) in the leaves of unstressed control (C_3), NaCl-treated (CAM) and salt-stress withdrawn (-NaCl) *Mesembryanthemum crystallinum* L. plants measured 24 h after osmotic stress removal. Bars represent mean values (±SD) for $n = 5$. Different letters indicate statistically significant differences according to Tukey's HSD test at $p \leq 0.05$.

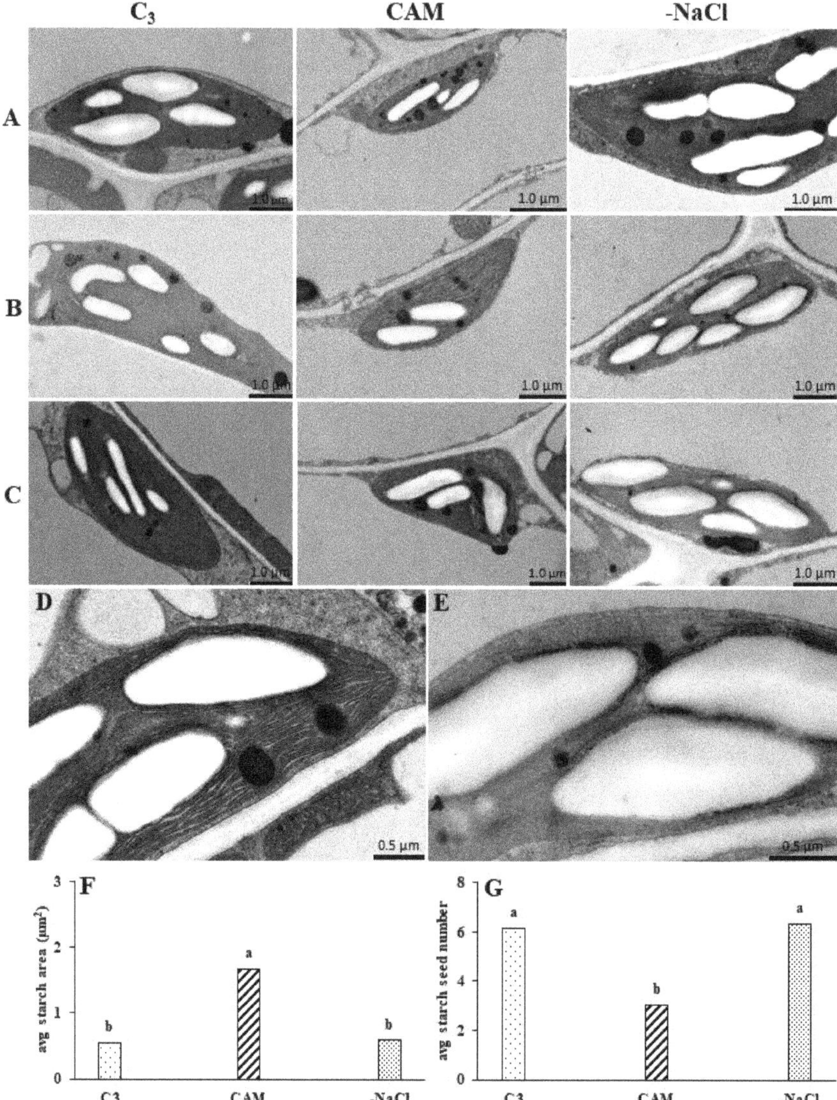

Figure 6. Electron micrographs of chloroplasts in unstressed control (C_3), NaCl-treated (CAM) and salt-stress withdrawn (-NaCl) *Mesembryanthemum crystallinum* L. plants taken in the middle of the light phase 24 (**A**), 48 (**B**) and 72 (**C**) h after osmotic stress removal. To clearly display the ultrastructural changes, sections (**D**,**E**) of the panel show the thylakoid membranes of NaCl-treated (CAM) and salt-stress withdrawn (-NaCl) plants, respectively, taken at higher magnification. The presented micrographs are representative examples of at least 23 repetitions. Sections (**F**,**G**) show the average area (μm^2) and the number of starch grains, respectively, assessed through an image analysis of chloroplast micrographs of unstressed control (C_3), NaCl-treated (CAM) and salt-stress withdrawn (-NaCl) *Mesembryanthemum crystallinum* L. plants 24 h after osmotic stress removal. Bars represent mean values ($\pm$SD) for $n = 23$. Different letters indicate statistically significant differences according to Tukey's HSD test at $p \leq 0.05$.

3. Discussion

3.1. Rapid Modifications in PSII and PSI Functionality during Recovery from Osmotic Stress Confirm the Great Flexibility of the Common Ice Plant Photosynthetic Apparatus

An earlier study on the functionality of the photosynthetic apparatus in *M. crystallinum* showed higher and similar quantum efficiencies of PSII and PSI, respectively, in salt-stressed (CAM) plants than in control (C_3) plants [5]. Our analysis revealed quite a different picture of PSII and PSI functionality during a similar salt treatment, with lower PSII and PSI quantum efficiencies, elevated nonphotochemical quenching (NPQ), decreased photochemical quenching (qL) and elevated acceptor side limitation (Y_{ND}) of PSI detected at midday in salt-stressed plants. We also found retarded photosynthetic electron transport (PET) in both photosystems in salt-stressed plants. Most of these modifications of PSII and PSI functionality were previously found to be a part of salt stress-induced CAM photosynthesis [12,13]. Moreover, it was suggested that despite the high salt tolerance of *M. crystallinum*, the plants still experience salt stress, and the resultant CAM induction was reported to be a source of photoinhibition [12–15]. Although we could not confirm the occurrence of photoinhibition in our study, we suggest that salt-induced CAM plants struggled with excess energy that could not be quenched with photochemical processes. We believe that the substantially lower amount of RubisCO (ribulose-1,5-bisphosphate carboxylase-oxygenase) previously detected in salt-stressed plants than in C_3 plants [6,9] may to some extent explain the retardation of the photochemical processes of PSII and PSI. Photochemical quenching is at its highest at midday due to elevated RubisCO activity resulting from the abundance of CO_2. An osmotic stress-induced decline in the amount of RubisCO results in a substantially lower capacity for electron sinkage as well as a lower energy demand, which may be responsible for the declines observed in the functionalities of both PSII and PSI and the necessity of excess energy dissipation through nonphotochemical mechanisms. It was recently confirmed that the largest share of NPQ build-up accompanying salinity-induced CAM photosynthesis belonged to heat dissipation through antennae [13]. Among CAM facultative species, those with increasingly flexible photosynthetic metabolisms can be distinguished. It was shown that the return from nocturnal (CAM photosynthesis) to diel (C_3 photosynthesis) CO_2 fixation may be re-established in different facultative CAM plants as a response to stress withdrawal within as early as 24 h (*Portulaca oleracea, M. crystallinum* L.) and even up to 96 h (*Talinum triangulare, Clusia pratensis*) [10]. Our previous study involving salinity stress withdrawal in *M. crystallinum* confirmed these findings. We discovered substantial downregulation of *pepc1* (an osmotic stress-related member of the PEPC gene family) expression during the first hours after osmotic stress removal. This effect was accompanied by the re-established Δ malate, a hallmark of functional CAM photosynthesis that, in salt stress-withdrawn plants, rapidly (24 h) returned to the values typical for C_3 plants [11]. Here, we showed that the PSII and PSI functionality of intact (C_3) plants was quickly achieved by salt-stressed withdrawn plants. Most of the disturbances found in PSII and PSI functionality disappeared within 48 h after osmotic stress removal. It can be hypothesised that one of the reasons for the rapid return of the functionality of both photosystems was the restoration of the content and activity of RubisCO characteristic of C_3 plants. Altogether, our findings show that rapid modifications of carbon metabolism-related genetic (*PEPC1* expression) and biochemical (CO_2 fixation, malate concentration build-up, tissue acidification) pathways resulting from the withdrawal of osmotic stress are coupled with modifications of photosynthetic apparatus performance.

3.2. Expression of PSII and PSI Structural Genes Is Rapidly Modified in Response to Osmotic Stress Absence

The regulation of photosynthesis-related genes during salinity stress remains a poorly understood phenomenon in both salt-sensitive (glycophyte) and salt-tolerant (halophyte) plants. It has been shown that in glycophytes, salt stress is responsible mostly for photosynthesis-related gene downregulation [16]. Large-scale analysis of the common ice plant transcriptome under salt stress revealed an increase in the abundance of genes related to CAM photosynthesis [17]. Our analysis showed that in addition to inducing transcripts of *PSAA* (PSI-A core protein of PS I) and *PSAD* (D_2 protein of PS II), the regulation of the main structural genes of PSII and PSI remained unaffected by salt stress. To some extent, this result correlates with an earlier study [6], and the regulation of photosystem gene expression during the salt stress response seems to rely on a combination of different signals. Two of these signals may have opposite effects; to meet the high demand for de novo synthesis of photosystem proteins resulting from stress-related protein degradation, the upregulation of their respective genes is required. On the other hand, it was previously proposed that a decline in PET is responsible for photosystem gene downregulation [18]. Our results show that salt stress had a rather minimal impact, however, the withdrawal of salt stress strongly induced the expression of all analysed genes. According to the previously mentioned studies, it can be suggested that the rapid induction of these genes is stimulated not only by the recovered efficiency of both photosystems and PET but also to supply the high demand for de novo synthesis of photosystem proteins.

3.3. Rapid Changes in PSII and PSI Functionality Are Accompanied by Chloroplast Ultrastructure Modification during Osmotic Stress Recovery

Ion homeostasis disorder accompanying salt stress is responsible for alterations in chloroplast ultrastructure, including thylakoid membrane swelling, grana thylakoid lamellae disorder, the fracturing of stroma thylakoid lamellae and chloroplast membrane disintegration; these effects have been confirmed in both glycophytes and halophytes of different plant species [6,19–21]. To present the extent of changes resulting from osmotic stress withdrawal, we analysed the chloroplast ultrastructure of plants that had recovered from salinity stress. Our analysis of the salt-stressed (CAM) chloroplast ultrastructure confirmed the presence of swollen thylakoids organised mostly in unstacked grana [22]. It is possible that the salt-induced decline in photochemical efficiency observed, especially in PSII, was related not only to the decreased amount of RubisCO but also to the detrimental effects of salt stress on the thylakoid ultrastructure. Although it was previously suggested that thylakoid swelling might not be related to salt stress occurrence [6], the strong reduction in swelling observed in our experiments during the first 24 h of recovery indicates a relationship with osmotic stress. Modifications of the thylakoid membrane system during the salt stress response in common ice plants are accompanied by starch remobilisation. It was previously shown in the common ice plant that increased daily starch turnover resulting from the induced activity of α- and β-amylases, the main starch-degrading enzymes, was unequivocally related to salt stress [23]. In CAM-performing plants, the products of transitory starch degradation supply the cytosol with the carbohydrates required for the synthesis of phospho*enol*pyruvate (PEP), a substrate in primary CO_2 fixation, by phospho*enol*pyruvate carboxylase (PEPC). In isolated chloroplasts supplied with oxaloacetic acid (OAA), a substrate of malate synthesis by plastidic malate dehydrogenase (MDH), the rate of starch degradation was increased [24]. Our analysis confirmed the salinity-induced degradation of starch. Additionally, stress-related starch degradation was confirmed by the increased number of starch grains found in the chloroplasts recovered (24 h) from osmotic stress. Taken together, our results show that the flexibility of ice plants' carbon metabolism also extends to the chloroplast ultrastructure.

4. Materials and Methods

4.1. Plant Material

Mesembryanthemum crystallinum L. seeds (provided by the Technical University, Darmstadt, Germany) were sown onto a soil substrate in a greenhouse under 300–350 µmol photons $m^{-2} \cdot s^{-1}$ of photosynthetically active radiation (PAR), a 16/8 h day/night photoperiod, a day/night thermoperiod of 25/17 °C and 55–65% relative humidity (RH), as previously described [25]. The substrate implemented in the experiment was made of market available soil ("Athena" Bio-Products, Golczewo, Poland; pH 6.75; d = 0.24 kg dm^{-3})

and sand (grain size in the range of 1–2 mm) mixed in a 4:1 *v*/*v* ratio. Two weeks after sowing, each seedling with a fully developed 2nd leaf pair was transferred to an individual 0.4-L pot with 360 ± 0.1 g of mentioned substrate applied per each pot. After 6 weeks, the plants were divided into two groups: the first group was irrigated with tap water (C_3, intact control plants), and the second group was irrigated with 0.4 M NaCl (CAM, salt-stressed plants). After 14 days, CAM development in the salt-stressed plants was confirmed by the measurement of the diurnal Δ malate, a hallmark of functional CAM photosynthesis expressed as the difference in cell sap malate concentration between the beginning and the end of the light phase. Δ malate was measured according to the method previously described for *Clusia hilariana* Schltdl [26]. The next day, half of the CAM-performing plants were subsequently subjected desalinisation process by continuous rinsing of the soil substrate with tap water for 2 h (-NaCl, salt stress-withdrawn plants). Fluorometric measurements, as well as the collection of C_3, CAM and NaCl plant leaves for ultrastructure and molecular analysis, were performed at midday at 24, 48 and 72 h after osmotic stress removal. For the analysis of gene expression in leaves, frozen (LN_2) leaf tissue was ground to a fine powder in LN_2 and then stored at −80 °C until further analysis.

4.2. Quantum Efficiencies of PSII and PSI

PSII and PSI photochemistry was analysed simultaneously with a Dual PAM 100 (Heinz Walz GmbH, Effeltrich, Germany) fluorescence system. Induction curves were obtained from dark-adapted (20 min) plants, and each experimental variant was measured in 5 repetitions during two independent experiments. The minimal fluorescence yield (F_0) was measured at less than 1 µmol photons $m^{-2} \cdot s^{-1}$ intensity, whereas the maximum fluorescence yield (F_m) was measured after the application of a 1 s. saturating pulse of 2500 µmol photons $m^{-2} \cdot s^{-1}$. After dark relaxation (45 s), the centres were oxidised under red-orange actinic light with an irradiance of 126 µmol photons $m^{-2} \cdot s^{-1}$. After 260 s of light adaptation, the current (F_t) and maximum ($F_{m'}$) light-adapted fluorescence were measured. The PSII parameters were calculated using the following Equations (1–4).

$$Y(II) = (F_{m'} - F)/F_{m'} \quad (1)$$

$$qL = (F_{m'} - F)/(F_{m'} - F_{0'}) \times F_{0'}/F = qP \times F_{0'}/F \quad (2)$$

$$NPQ = (F_m - F_{m'})/F_{m'} \quad (3)$$

$$F_{0'} = F_0/(F_v/F_m + F_0/F_{m'}) \quad (4)$$

Y(II) was calculated according to [27]. Calculation of photochemical quenching (qL) coefficient was delivered by PAM 100 (Heinz Walz GmbH, Effeltrich, Germany) user manual. Non-photochemical quenching (NPQ) and F_0 calculations were performed according to [28] and [29], respectively.

The parameters describing PSI efficiency were calculated as described in the Dual-PAM 100 (Walz, Germany) user manual. The photochemical quantum yield of PSI-Y(I) was calculated from the complementary PSI quantum yields using the following Equation (5), namely, the nonphotochemical energy dissipation, Y_{ND} and Y_{NA}.

$$Y(I) = 1 - Y_{ND} - Y_{NA} \quad (5)$$

The donor side limitation of PSI-Y(ND), was calculated from the reduced P700 using the following Equation (6), according to the manufacturer's manual.

$$Y_{ND} = 1 - P_{700\ red} \quad (6)$$

Similarly, Y_{NA}, representing the acceptor side limitation of PSI, was determined as the fraction of the P700 centres that could not be oxidised with a saturation pulse calculated according to the following Equation (7).

$$Y_{NA} = (P_m - P_{m'})/P_m \quad (7)$$

where P_m and $P_{m'}$ represent the maximal change in the P700 signal upon the application of a saturation pulse in the dark-adapted state and light state, respectively.

4.3. RNA Preparation

Total RNA was isolated with an Aurum™ Total RNA Mini Kit (Bio-Rad, Hercules, CA, USA) according to the method previously described [30]. For the removal of DNA contamination, digestion with DNase I (DNA I Amplification Grade, Merck, Darmstadt, Germany) was used. RNA purity and quantity were determined using a Biospec-Nano (Shimadzu, Japan). To assess the integrity and purity of the RNA, the extracted RNA was separated by electrophoresis on agarose (1.5%) gels stained with EtBr. The bands were visualised on a Molecular Imager® ChemiDoc™ XRS+ Imaging System (Bio-Rad, Hercules, CA, USA).

4.4. qPCR

Reverse transcription was carried out on 1000 ng of total RNA with an iScript cDNA Synthesis Kit (Bio-Rad, Hercules, CA, USA). During qPCR, the samples were labelled with iQ™ SYBR® Green Supermix (Bio-Rad, Hercules, CA, USA) fluorescent dye. For a single reaction, 10–15 ng of cDNA and 150 nM of gene-specific primers were used. To test the amplification specificity, a dissociation curve was acquired by heating samples from 60 °C to 95 °C. Polyubiquitin was used as a housekeeping reference gene. The reaction efficiency was tested by serial dilutions of cDNAs with gene-specific primers (Supplementary Table S1). The expression was calculated from at least three reactions with unstressed control (C_3) plants after 24 h as calibrators according to a previously described method [31].

4.5. Chloroplast Ultrastructure—TEM Analysis

For (ultra)structural studies, collected leaf fragments were fixed in a mixture of 4% paraformaldehyde and 5% glutaraldehyde in 0.1 M sodium cacodylate buffer, pH 7.3, for 2 h at room temperature and air pressure (0.3 hPa), washed with 0.1 M cacodylate buffer, and postfixed for 2 h at 4 °C in 1% osmium tetroxide in 0.1 M cacodylate buffer. Then, the samples were dehydrated in an ethanol series at room temperature, rinsed three times in propylene oxide, embedded in epoxy resin Epon 812 (Fluka, Buchs, Switzerland) and polymerised for 24 h at 60 °C. Ultrathin sections collected on formvar-coated grids were briefly stained with uranyl acetate and lead citrate and examined under an FEI 268D 'Morgagni' transmission electron microscope (FEI-Thermo Fisher Scientific, Waltham, MA, USA) operating at 80 kV and equipped with a 'Morada' digital camera (Olympus-SIS, Tokyo, Japan). The collected digital microscopic images were saved as.jpg files and, if necessary, processed in Photoshop CS 8.0 (Adobe Systems, Mountain View, CA, USA) software with non-destructive tools (contrast and/or levels).

4.6. Image Analysis of Electron Micrographs

Image analysis of the chloroplast micrographs was performed with ImageJ 2 (under a GPL licence; NIH, Bethesda, MD, USA).

4.7. Statistical Analysis

The results were analysed with Statistica 13.3 (TIBCO, Palo Alto, CA, USA) statistical software. For the determination of statistically significant differences in the electron transport rates of PSII and PSI between experimental variants and the control, Dunnett's test was applied. One-way ANOVA followed by a post hoc Tukey HSD test was used to evaluate individual treatment effects at $p \leq 0.05$.

5. Conclusions

Here presented results show that the presence of salinity stress is required not only for the induction of stress-dependent CAM photosynthesis but also for maintaining its functioning. The rapid shutdown of the energy-demanding functional CAM seems to be one small component of the flexibility features. As we showed here, the metabolic flexibility of the common ice plant includes rapid and far-reaching changes. This includes photosystem high-performance recovery, the induction of the expression of reaction centre structural genes and the reorganisation of the chloroplast ultrastructure, and most of the mentioned modifications are completed within 24 h after osmotic stress removal. Here, we confirmed that the photosynthetic metabolism of the common ice plant may be one of the most highly plastic processes documented among facultative CAM plants.

Supplementary Materials: The following are available online at https://www.mdpi.com/article/10.3390/ijms22168426/s1.

Author Contributions: M.N.—conceptualisation, formal analysis, funding acquisition, investigation, visualisation, writing—original draft; K.G.—formal analysis, investigation, methodology, writing-review & editing; P.R.—conceptualisation, writing—original draft; M.S.-R.—formal analysis, methodology; Z.M.—writing—review & editing; A.K.—funding acquisition, supervision, visualisation, writing—review & editing. All authors have read and agreed to the published version of the manuscript.

Funding: This research was supported by the National Science Centre (NCN, Poland) MINIATURA 3 Project No. 2019/03/X/NZ9/01191 and Project Faculty of Exact and Natural Sciences, Pedagogical University of Cracow No. 610-469/PBU.2020.

Institutional Review Board Statement: Not applicable.

Informed Consent Statement: Not applicable.

Data Availability Statement: Data is contained within the article or supplementary material.

Conflicts of Interest: The authors declare no conflict of interest.

Abbreviations

CAM	Crassulacean acid metabolism;
ETRI	Electron transport chain of PSI;
ETRII	Electron transport chain of PSII;
F_0	Minimal fluorescence yield;
F_m	Maximum fluorescence yield;
$F_{m'}$	Maximum light-adapted fluorescence;
F_v/F_m	Maximum quantum yield of PSII;
MDH	Malate dehydrogenase;
NPQ	Non-photochemical quenching;
OAA	Oxaloacetic acid;
PAR	Photosynthetically active radiation;
PEP	Phospho*enol*pyruvate;
PEPC	Phospho*enol*pyruvate carboxylase;
PET	Photosynthetic electron transport;
PSAA	PSI-A core protein of PS I;
PSBD	D2 protein of PS II;
PSI	Photosystem I;
PSII	Photosystem II;
qP, qL	Photochemical quenching calculated based on the puddle and lake model, respectively;

RuBisCO	Ribulose-1,5-bisphosphate carboxylase-oxygenase;
Y(I)	Quantum yield of (PSI);
Y(II)	Quantum yield of (PSII);
Y_{NA}	Quantum yield of energy dissipation due to acceptor side limitation in PSI;
Y_{ND}	Quantum yield of energy dissipation due to donor side limitation in PSI.

References

1. Winter, K. Ecophysiology of constitutive and facultative CAM photosynthesis. *J. Exp. Bot.* **2019**, *70*, 6495–6508. [CrossRef] [PubMed]
2. Munns, R.; Tester, M. Mechanisms of salinity tolerance. *Ann. Rev. Plant. Biol.* **2008**, *59*, 651–681. [CrossRef] [PubMed]
3. Gilroy, S.; Suzuki, N.; Miller, G.; Choi, W.-G.; Toyota, M.; Devireddy, A.R.; Mittler, R. A tidal wave of signals: Calcium and ROS at the forefront of rapid systemic signalling. *Trend Plant. Sci.* **2014**, *19*, 623–630. [CrossRef]
4. Acosta-Motos, J.R.; Ortuno, M.F.; Bernal-Vicente, A.; Diaz-Vivancos, P.; Sanchez-Blanco, M.J.; Hernandez, J.A. Plant response to salt stress: Adaptive mechanisms. *Agronomy* **2017**, *7*, 18. [CrossRef]
5. Guan, Q.; Bowen, T.; Kelley, T.M.; Jingkui, T.; Chen, S. Physiological changes in *Mesembryanthemum crystallinum* during C_3 to CAM transition induced by salt stress. *Front. Plant Sci.* **2020**, *11*, 283. [CrossRef]
6. Niewiadomska, E.; Bilger, W.; Gruca, M.; Mulisch, M.; Miszalski, Z.; Krupinska, K. CAM-related changes in chloroplastic metabolism of *Mesembryanthemum crystallinum* L. *Planta* **2011**, *233*, 275–285. [CrossRef]
7. Niewiadomska, E.; Pilarska, M. Acclimation to salinity in halophytic ice plant prevents a decline of linear electron transport. *Environ. Exp. Bot.* **2021**, *184*, 104401. [CrossRef]
8. Adams, P.; Nelson, D.; Yamada, S.; Chmara, W.; Jensen, R.; Bohnert, H.; Griffiths, H. Growth and development of *Mesembryanthemum crystallinum* (*Aizoaceae*). *New Phytol.* **1998**, *138*, 171–190. [CrossRef]
9. Kuźniak, E.; Kornas, A.; Kaźmierczak, A.; Rozpądek, P.; Nosek, M.; Kocurek, M.; Zellnig, G.; Müller, M.; Miszalski, Z. Photosynthesis-related characteristics of the midrib and the interveinal lamina in leaves of the C_3–CAM intermediate plant *Mesembryanthemum crystallinum*. *Ann. Bot.* **2016**, *117*, 1141–1151. [CrossRef]
10. Winter, K.; Holtum, J.A.M. Facultative crassulacean acid metabolism (CAM) plants: Powerful tools for unravelling the functional elements of CAM photosynthesis. *J. Exp. Bot.* **2014**, *65*, 3425–3441. [CrossRef] [PubMed]
11. Nosek, M.; Gawrońska, K.; Rozpądek, P.; Szechyńska-Hebda, M.; Kornaś, A.; Miszalski, Z. Withdrawal from functional Crassulacean acid metabolism (CAM) is accompanied by changes in both gene expression and activity of antioxidative enzymes. *J. Plant Physiol.* **2018**, *229*, 151–157. [CrossRef]
12. Keiller, D.R.; Slocombe, S.P.; Cockburn, W. Analysis of chlorophyll a fluorescence in C_3 and CAM forms of *Mesembryanthemum crystallinum*. *J. Exp. Bot.* **1994**, *45*, 325–334. [CrossRef]
13. Matsuoka, T.; Onozawa, A.; Sonoike, K.; Kore-eda, S. Crassulacean acid metabolism induction in *Mesembryanthemum crystallinum* can be estimated by non-photochemical quenching upon actinic illumination during the dark period. *Plant Cell Physiol.* **2018**, *59*, 1966–1975. [CrossRef]
14. Schöttler, M.A.; Kirchhoff, H.; Siebke, K.; Weis, E. Metabolic control of photosynthetic electron transport in Crassulacean acid metabolism-induced *Mesembryanthemum crystallinum*. *Funct. Plant Biol.* **2002**, *29*, 697–705. [CrossRef]
15. Broetto, F.; Duarte, H.M.; Lüttge, U. Responses of chlorophyll fluorescence parameters of the facultative halophyte and C_3–CAM intermediate species *Mesembryanthemum crystallinum* to salinity and high irradiance stress. *J. Plant Physiol.* **2007**, *164*, 904–912. [CrossRef] [PubMed]
16. Chaves, M.M.; Flexas, J.; Pinheiro, C. Photosynthesis under drought and salt stress: Regulation mechanisms from whole plant to cell. *Ann. Bot.* **2009**, *103*, 551–560. [CrossRef]
17. Cushman, J.C.; Tillett, R.L.; Wood, J.A.; Branco, J.M.; Schlauch, K.A. Large-scale mRNA expression profiling in the common iceplant, *Mesembryanthemum crystallinum*, performing C_3 photosynthesis and Crassulacean acid metabolism (CAM). *J. Exp. Bot.* **2008**, *59*, 1875–1894. [CrossRef] [PubMed]
18. Pfannschmidt, T. Chloroplast redox signals: How photosynthesis controls its own genes. *Trends Plant Sci.* **2003**, *8*, 33–41. [CrossRef]
19. Mitsuya, S.; Takeoka, Y.; Miyake, H. Effects of sodium chloride on foliar ultrastructure of sweet potato (*Ipomoea batatas* Lam.) plantlets grown under light and dark conditions *in vitro*. *J. Plant Physiol.* **2000**, *157*, 661–667. [CrossRef]
20. Yamane, K.; Hayakawa, K.; Kawasaki, M.; Taniguchi, M.; Miyake, H. Bundle sheath chloroplasts of rice are more sensitive to drought stress than mesophyll chloroplasts. *J. Plant Physiol.* **2003**, *160*, 1319–1327. [CrossRef]
21. Shu, S.; Yuan, Y.; Chen, J.; Sun, J.; Zhang, W.; Tang, Y.; Zhong, M.; Guo, S. The role of putrescine in the regulation of proteins and fatty acids of thylakoid membranes under salt stress. *Sci. Rep.* **2015**, *5*, 14390. [CrossRef] [PubMed]
22. Paramanova, N.V.; Shevyakova, N.I.; Kuznetsov, V. Ultrastructure of chloroplasts and their storage inclusions in the primary leaves of *Mesembryanthemum crystallinum* affected by putrescine and NaCl. *Russ. J. Plant Physiol.* **2004**, *51*, 86–96. [CrossRef]
23. Paul, M.J.; Loos, K.; Stitt, M.; Ziegler, P. Starch-degrading enzymes during the induction of CAM in *Mesembryanthemum crystallinum*. *Plant Cell Environ.* **1993**, *16*, 531–538. [CrossRef]
24. Neushaus, E.H.; Schulte, N. Starch degradation in chloroplasts isolated from C_3 or CAM (Crassulacean acid metabolism)-induced *Mesembryanthemum crystallinum* L. *Biochem. J.* **1996**, *318*, 945–953. [CrossRef]

25. Nosek, M.; Kaczmarczyk, A.; Jędrzejczyk, R.J.J.; Supel, P.; Kaszycki, P.; Miszalski, Z. Expression of genes involved in heavy metal trafficking in plants exposed to salinity stress and elevated cd concentrations. *Plants* **2020**, *9*, 475. [CrossRef]
26. Miszalski, Z.; Kornas, A.; Rozpadek, P.; Fischer-Schliebs, E.; Lüttge, U. Independent fluctuations of malate and citrate in the CAM species *Clusia hilariana* Schltdl. under low light and high light in relation to photoprotection. *J. Plant Physiol.* **2013**, *170*, 453–458. [CrossRef] [PubMed]
27. Genty, B.; Briantais, J.-M.; Baker, N.R. The relationship between the quantum yield of photosynthetic electron transport and quenching of chlorophyll fluorescence. *Biochim. Biophys. Acta* **1989**, *990*, 87–92. [CrossRef]
28. Kramer, D.M.; Johnson, G.; Kiirats, O.; Edwards, G.E. New flux parameters for the determination of QA redox state and excitation fluxes. *Photosynth. Res.* **2004**, *79*, 209–218. [CrossRef]
29. Oxborough, K. Imaging of chlorophyll a fluorescence: Theoretical and practical aspects of an emerging technique for the monitoring of photosynthetic performance. *J. Exp. Bot.* **2004**, *55*, 1195–1205. [CrossRef]
30. Nosek, M.; Rozpadek, R.; Kornas, A.; Kuzniak, E.; Schmitt, A.; Miszalski, Z. Veinal-mesophyll interaction under biotic stress. *J. Plant Physiol.* **2015**, *185*, 52–56. [CrossRef]
31. Pfaffl, M.W. A new mathematical model for relative quantification in real-time RT-PCR. *Nucleic Acid Res.* **2001**, *29*, 16–21. [CrossRef] [PubMed]

International Journal of *Molecular Sciences*

MDPI

Article

Transcriptomic Analysis of Salt-Stress-Responsive Genes in Barley Roots and Leaves

Rim Nefissi Ouertani [1], Dhivya Arasappan [2], Ghassen Abid [3], Mariem Ben Chikha [1], Rahma Jardak [1], Henda Mahmoudi [4], Samiha Mejri [1], Abdelwahed Ghorbel [1], Tracey A. Ruhlman [5] and Robert K. Jansen [5,6,*]

1 Laboratory of Plant Molecular Physiology, Center of Biotechnology of Borj Cedria, B.P. 901, Hammam-Lif 2050, Tunisia; rimnefissi@gmail.com (R.N.O.); mariam.tn@gmx.fr (M.B.C.); rjardak@yahoo.fr (R.J.); mejrisamiha2018@gmail.com (S.M.); wahidghorbel@yahoo.fr (A.G.)
2 Center for Biomedical Research Support, University of Texas at Austin, Austin, TX 78712, USA; darasappan@austin.utexas.edu
3 Laboratory of Legumes and Sustainable Agrosystems, Center of Biotechnology of Borj Cedria, B.P. 901, Hammam-Lif 2050, Tunisia; abidghassen@gmail.com
4 International Center for Biosaline Agriculture, Dubai 00000, United Arab Emirates; HMJ@biosaline.org.ae
5 Department of Integrative Biology, University of Texas at Austin, Austin, TX 78712, USA; truhlman@austin.utexas.edu
6 Biotechnology Research Group, Department of Biological Sciences, Faculty of Science, King Abdulaziz University (KAU), Jeddah 21589, Saudi Arabia
* Correspondence: jansen@austin.utexas.edu

Citation: Nefissi Ouertani, R.; Arasappan, D.; Abid, G.; Ben Chikha, M.; Jardak, R.; Mahmoudi, H.; Mejri, S.; Ghorbel, A.; Ruhlman, T.A.; Jansen, R.K. Transcriptomic Analysis of Salt-Stress-Responsive Genes in Barley Roots and Leaves. *Int. J. Mol. Sci.* **2021**, *22*, 8155. https://doi.org/10.3390/ijms22158155

Academic Editors: Masayuki Fujita and Mirza Hasanuzzaman

Received: 21 May 2021
Accepted: 23 July 2021
Published: 29 July 2021

Abstract: Barley is characterized by a rich genetic diversity, making it an important model for studies of salinity response with great potential for crop improvement. Moreover, salt stress severely affects barley growth and development, leading to substantial yield loss. Leaf and root transcriptomes of a salt-tolerant Tunisian landrace (Boulifa) exposed to 2, 8, and 24 h salt stress were compared with pre-exposure plants to identify candidate genes and pathways underlying barley's response. Expression of 3585 genes was upregulated and 5586 downregulated in leaves, while expression of 13,200 genes was upregulated and 10,575 downregulated in roots. Regulation of gene expression was severely impacted in roots, highlighting the complexity of salt stress response mechanisms in this tissue. Functional analyses in both tissues indicated that response to salt stress is mainly achieved through sensing and signaling pathways, strong transcriptional reprograming, hormone osmolyte and ion homeostasis stabilization, increased reactive oxygen scavenging, and activation of transport and photosynthesis systems. A number of candidate genes involved in hormone and kinase signaling pathways, as well as several transcription factor families and transporters, were identified. This study provides valuable information on early salt-stress-responsive genes in roots and leaves of barley and identifies several important players in salt tolerance.

Keywords: *Hordeum vulgare* L.; salinity; RNA-seq analysis; differentially expressed genes; tolerance; candidate genes

1. Introduction

Salinity is one of the most pressing abiotic stressors threatening plant growth and agricultural production worldwide. Saline conditions are increasing rapidly along with the alarming rise of global warming, particularly in arid and semiarid regions [1]. Given these severe conditions, understanding the molecular mechanisms underlying salinity stress response in plants could contribute to the development of salt-tolerant crops in order to sustain productivity and quality.

Salinity is often recognized as an excessive accumulation of sodium ions in the soil [2], leading to osmotic stress and ion toxicity [3,4]. These two main effects of salt damage result in decreased photosynthetic efficiency, redistribution of cell wall constituents, reduction of cell expansion and division, and oxidative damage from reactive oxygen species

(ROS) [2,5,6]. Hence, salinity stress generates deleterious effects on plant growth and productivity.

In response to salt stress, plants activate several tolerance mechanisms, including physiological, biochemical, and molecular changes. These diverse mechanisms allow the accumulation of osmoprotectants, regulation of ion homeostasis, and detoxification by the activation of ROS scavengers via efficient signal transduction networks [1]. Several important mechanisms have been characterized, including the osmoprotectant pathway and scavengers that regulate ROS homeostasis [7]. Both antioxidant enzymes and nonenzymatic compounds play critical roles in detoxifying ROS induced by salinity stress [8]. In addition, membrane transporters and ion channels, namely, the high-affinity potassium transporter and salt overly sensitive families involved in Na^+-specific transport, play a crucial role in Na^+ homeostasis through the regulation of K^+/Na^+ and H^+/Na^+ balances, respectively [4,9,10]. Further, several transcription factor families, such as the dehydration-responsive element-binding protein (CBF/DREB) family [11] and the mitogen-activated protein kinase family, function in pathways that regulate the expression of stress-related genes [12,13] along with other factors that regulate abiotic stress responses. Finally, the interaction of several plant hormones, such as abscisic acid (ABA), cytokinins, auxins, ethylene, salicylic acid, and jasmonic acid, plays vital roles in salt stress signaling and response [14,15].

High-throughput RNA sequencing (RNA-seq) is an important approach to study the expression of a large number of genes in a given tissue at a given time point [16]. RNA transcript profiling is a powerful technology for genomewide transcript characterization, differential gene expression analysis, variant detection, and gene-specific expression. These features are facilitating a deeper understanding of the genetic variation in complex phenotypic traits, such as salt tolerance, and allowing the enrichment of salt stress response pathways [17]. Several RNA sequencing studies have examined salt stress responses in different plants, such as barley (*Hordeum vulgare* L.) [18], sweet potato (*Ipomoea batatas* (L.) Lam.) [19], rapeseed (*Brassica napus* L.) [20], and *Arabidopsis* [21]. These studies suggest the involvement of a substantial number of salt tolerance genes encoding oxidation–reduction processes and osmoprotectant metabolism, ion transport, heat shock proteins (HSPs), and hormone signaling. Furthermore, differentially expressed genes (DEGs) encoding several transcription factors and signal transduction components associated with salinity tolerance have been identified. However, a paucity of information is available regarding differences between expression profiles of shoot and root in salt-tolerant barley cultivars.

Barley is an important food, feed, and industry crop with economic significance worldwide [22]. Barley yields are seriously threatened by escalating levels of salinization due to the overall reduction of root and leaf growth. Indeed, salt stress first impacts the root system of plants by inducing osmotic stress, leading to ion toxicity effects due to nutrient imbalance in cytosol, decreasing turgor due to limits in leaf gas exchange and stomata closure, and increasing oxidative damage, all of which interfere with normal cell division and expansion, leading to lower growth and yield rates [23–26]. Barley is still considered a relatively salt-tolerant crop and an important model for investigations of plant responses to changes in salinity [24]. Barley nuclear genomes are characterized by robust genetic diversity, making it attractive for stress tolerance breeding. Barley landrace accessions harbor novel genetic resources; Boulifa is a Tunisian accession with high salinity tolerance [27,28]. Therefore, investigating salt tolerance mechanisms remains important for barley breeding programs and helps to identify key genes involved in salt tolerance. The identified candidate genes represent valuable resources for future genetic engineering studies in cereals as well as in other crops towards the development of new varieties with more salt-tolerant characters and would be exploited to establish efficient applied breeding plans.

In order to identify candidate genes, molecular functions, and biological processes involved in response to salinity, stress high-throughput RNA-seq was performed on the salt-tolerant Tunisian accession Boulifa. Leaves and roots were examined separately and at different time points following exposure to severe salt stress (200 mM NaCl) in order to deepen our understanding of the specific response of these tissues under different stress durations. The results improve the current understanding of salt stress response mechanisms in barley leaves and roots and can be applied to developing salt-tolerant cereals.

2. Results

2.1. Analyses of RNA-Seq Datasets

An average of 23.9 million high-quality reads were obtained for each sample (an average total count of 24.14 million reads per sample before filtering low-quality reads (Table S1)). On average, 92.5% of the reads were mapped to the barley genome, indicating that the samples were comparable. On average, 62.5% of the reads were pseudoaligned to the barley transcriptome, and 32,587 genes were detected (Table S1). Furthermore, the time point clustering of replicates in 21 of the 24 samples, shown in Figure 1, indicates the high quality of sampling and RNA-seq analysis.

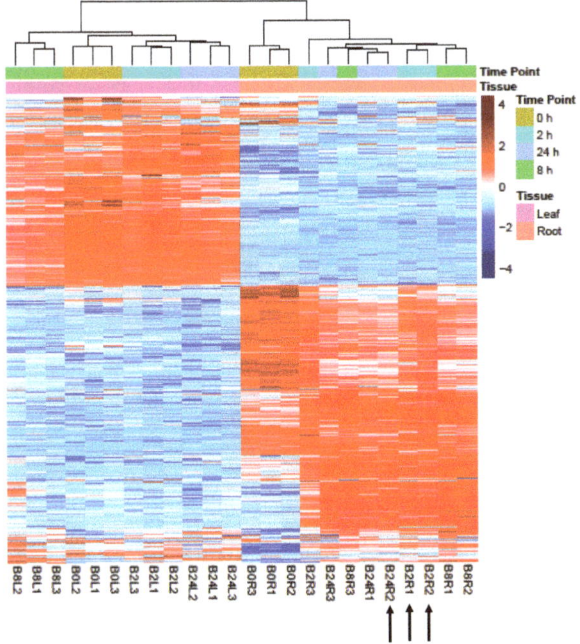

Figure 1. Cluster analysis using 25% highly variable genes. Gene expression in leaf (pink) and root (orange) was assessed under salt (200 mM NaCl) treatment for 2, 8, and 24 h relative to untreated plants (0 h). Samples and biological replicates (L1, L2, and L3 for leaves and R1, R2, and R3 for roots) are shown. The color gradient (upper right) indicates the pattern of expression from lowly (blue) to highly (red) expressed genes and ranges from −4- to +4-fold changes in expression. Sampling time point and tissue type are shown at the top of the cluster plot and defined on the right. The dendrogram at the top shows the sample clustering, and the black arrows indicate the samples that did not cluster by time point in the analysis.

2.2. Differentially Expressed Genes in Leaves and Roots under Salt Stress

Principal component analysis (PCA), a dimensionality reduction technique that projects high-dimensional data on the principal components that represent the largest variation in the data, was conducted in order to assess the largest source of variation among data. The samples were projected onto principal component 1 (PC1) as the *X*-axis and principal component 2 (PC2) as the *Y*-axis, representing, respectively, the first and the second largest sources of variation in the data. In Figure S2, PC1 (*X*-axis), which represents 86.5% of the variation in data, separates out samples by tissue, indicating that tissue is likely the largest source of variation in our data. Therefore, differential expression analysis was performed separately for each tissue. Differentially expressed genes were assessed in leaves and roots of plants exposed to short (2 h), intermediate (8 h), and long-term (24 h) salt stress, and responses were compared with pretreatment plants. The pattern of differential expression in untreated and salt-treated barley seedlings is shown in Figure 1. Red and blue signify overexpressed and underexpressed genes, respectively. The numbers of DEGs are depicted as volcano plots (Figure 2). For all salt treatment durations, differences in total DEGs were observed between leaves and roots. In leaves, the numbers of DEGs were 1290, 4338, and 3546 compared with 6449, 8915, and 8414 in roots after 2, 8, and 24 h salt treatment, respectively. The intermediate term response (8 h) showed the highest number of DEGs, followed by the long-term (24 h) and the short-term (2 h) (Figure S1). In both leaves and roots across time points, the greatest numbers of shared DEGs were between 8 and 24 h. Stress-responsive DEGs were either common to both tissue types or specific to each type (Figure 3). Tissue-specific DEGs were much more prominent in roots at all time points, particularly in short-term stress response. At 2 h of salt treatment, 6033 DEGs (representing 82% of the total DEGs in this tissue/time of treatment) were specific to roots, whereas only 874 DEGs (12%) were specific to leaves, and 415 DEGs (6%) were shared by both tissues. Intermediate- and long-term responses shared similar proportions of tissue-specific DEGs, with 63% and 66% DEGs in roots, 14% and 13% shared DEGs, and 23% and 21% DEGs in leaves at 8 and 24 h of salt stress, respectively.

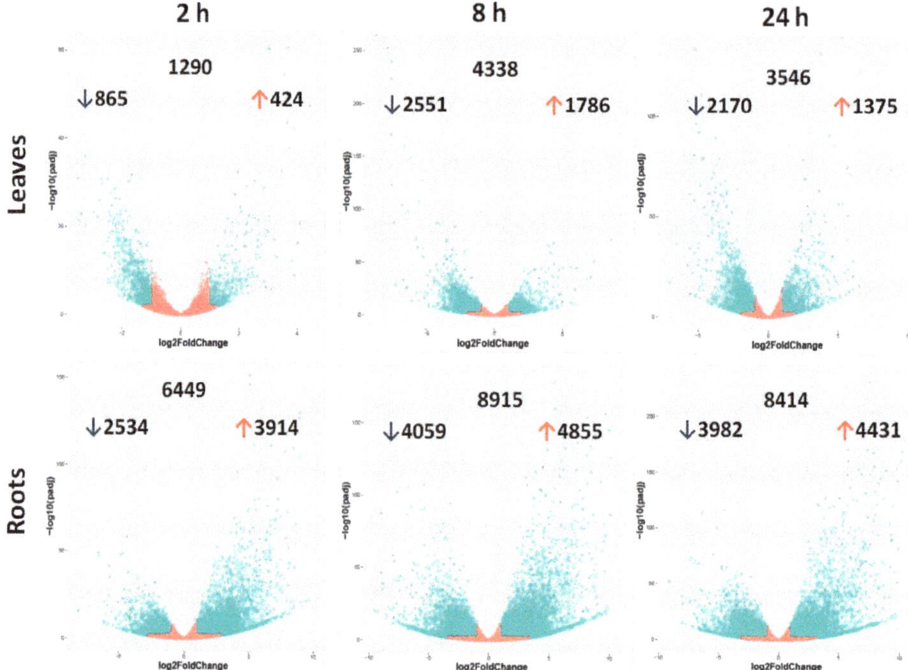

Figure 2. Volcano plots depict differentially expressed genes. Salt-treated plants were sampled at 2, 8, and 24 h of salt treatment, and differentially expressed genes (DEGs) were assessed relative to untreated plants (0 h) in leaves and roots. For each plot, the *X*-axis shows a log base 2-fold change, and the *Y*-axis indicates the adjusted *p*-values for the differences in expression. For each time point, the total DEG numbers are shown in the upper middle of each graph, upregulated DEGs are indicated by a red up arrow, and downregulated DEGs by a black down arrow. Blue dots and red dots equate to significant DEGs and nonsignificant genes, respectively.

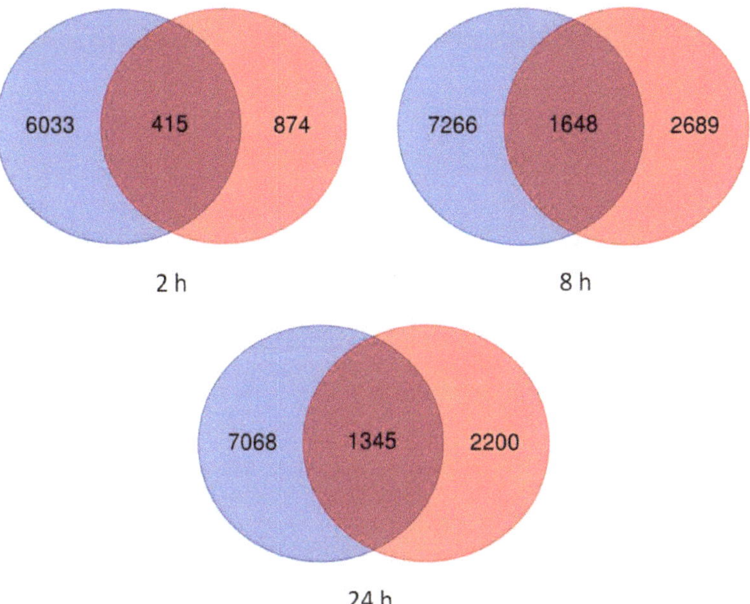

Figure 3. Shared and unique differentially expressed genes in barley seedlings. Differentially expressed genes (DEGs) exposed to 2, 8, and 24 h salt treatment were evaluated relative to untreated plants (0 h). Blue and red circles indicate roots and leaves, respectively, and shared DEGs are indicated by overlap. The number of affected genes is given for each segment of the Venn diagram.

At all time points, fewer up- than downregulated DEGs were detected in leaves (3585 up and 5586 down) (Figure 2). However, in roots, more DEGs were up- than downregulated (13,200 up and 10,575 down) (Figure 2). Across time points, only 921 DEGs (representing 0.36% of the total DEGs in leaves and roots) were oppositely modulated in the two tissues. The highest degree of differential modulation was observed at 8 h of salt treatment (501 DEGs) compared with only 69 at 2 h and 351 at 24 h of salt stress.

2.3. *Gene Ontology Enrichment Analysis of Differentially Expressed Genes*

In leaf samples, the most enriched biological processes were biosynthetic and metabolic processes (Figure 4). These categories were consistent across time points, although the largest DEG numbers for both GO terms were detected at 8 h. In roots, the predominant processes were metabolic, cellular metabolic, and response to stress (Figure 4a).

Enrichment of GO terms involved in metabolism, such as biosynthetic, small-molecule, cellular, organic substance, nitrogen compound, and primary metabolic processes, was detected in both leaves and roots. Although these categories were enriched in both tissues, roots also included more enriched GO terms and several other metabolic processes, including cellular protein, lignin, and polysaccharide, along with protein modification and phosphorylation. Biosynthetic processes, such as alpha amino acid, cellular amino acid compound, lysine, and organic substance biosynthetic, as well as the chlorophyll metabolic process, were enriched only in leaves mainly after 8 and 24 h salt treatments. Response to stimulus, enriched predominantly in roots, included response to oxidative stress and response to biotic stress.

Catalytic activity was over-represented in both leaves and roots at all time points (Figure 4b). While this was the main molecular function highly enriched in leaves, several other GO categories were ascribed to roots, such as binding, antioxidant, and kinase activities. In both tissues, catalytic activity was mainly represented by GO terms for oxidoreductase, ligase, transferase, hydrolase, and decarboxylase. In roots, GO terms associated with binding affinity were mainly represented, such as nucleotide, protein, ion, organic cyclic compound, anion, carbohydrate, and ATP (Figure 4b).

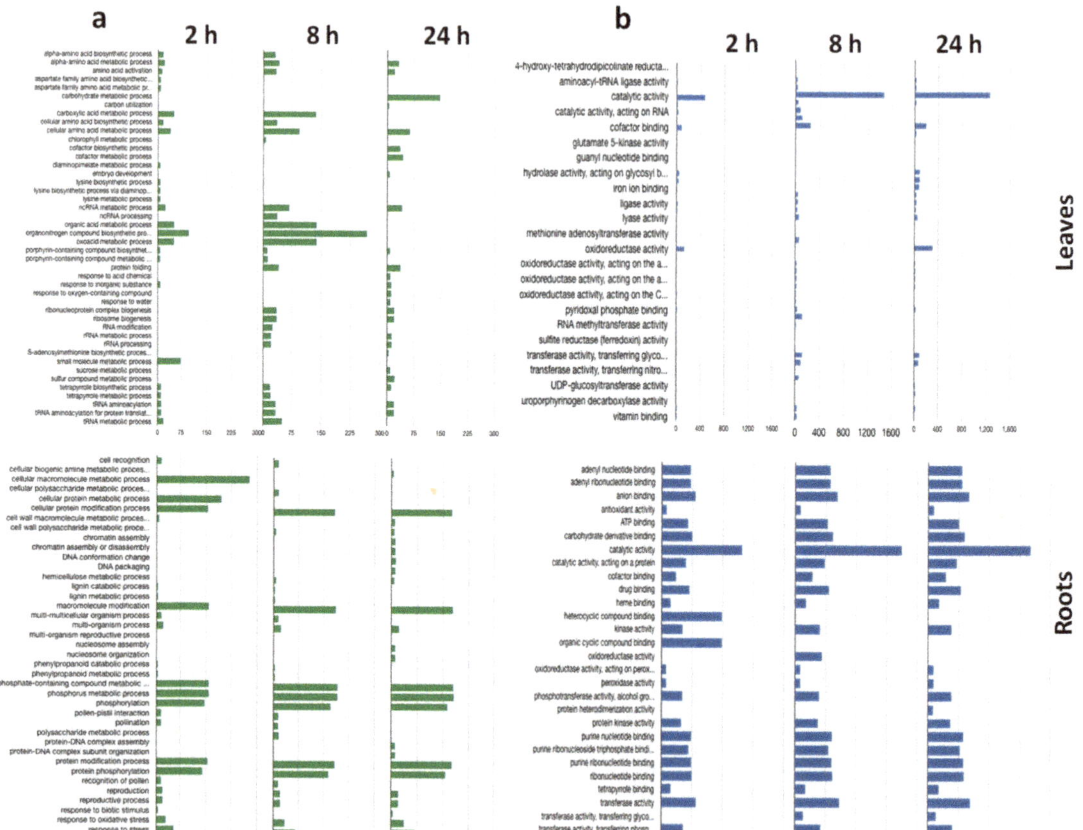

Figure 4. Gene ontology analysis. The gene ontology analysis of biological processes (**a**) and molecular functions (**b**) for the functional groups most significantly enriched among the differentially expressed genes (DEGs) under 2, 8, and 24 h salt stress in both leaf and root tissues. Gene ontology term categories are indicated on the *Y*-axis. Green (**a**) and blue (**b**) histograms indicate numbers (*X*-axis) of DEGs enriched in each category.

KEGG (Kyoto Encyclopedia of Genes and Genomes) pathway analysis showed that the pathways enriched in leaves and roots in response to salt stress were different, but were conserved across time points for each tissue (Table 1). In leaves, porphyrin and chlorophyll metabolism, biosynthesis, and metabolism of various amino acids (lysine, alanine, aspartate, and glutamate), as well as biosynthesis of aminoacyl-tRNA, antibiotics, and several secondary metabolites, were identified. The over-represented pathways in roots were drug metabolism of various enzymes (oxidoreduction and phospholipase) and biosynthesis of phenylpropanoids, characterized by their antioxidant activity (such as phenylalanine and flavonoids).

Table 1. Over-represented KEGG pathways identified in barley leaves and roots under salt stress.

Leaves		Roots	
KEGG ID	**Enriched Pathway**	**KEGG ID**	**Enriched Pathway**
Ko00860	Porphyrin and chlorophyll metabolism	ko00980	Metabolism of xenobiotics by cytochrome P450
Ko00970	Aminoacyl-tRNA biosynthesis	ko00982	Drug metabolism—cytochrome P450
Ko00261	Monobactam biosynthesis (glutamate dehydrogenase (NAD(P)+))	ko00983	Drug metabolism—other enzymes (phospholipase)
Ko00300	Lysine biosynthesis	ko00940	Phenylpropanoid biosynthesis
Ko00250	Alanine, aspartate, and glutamate metabolism	ko00460	Cyanoamino acid metabolism
Ko00997	Biosynthesis of various secondary metabolites	ko00480	Glutathione metabolism
Ko00270	Cysteine and methionine metabolism	ko00400	Phenylalanine, tyrosine, and tryptophan biosynthesis
Ko00941	Flavonoid biosynthesis	ko00140	Steroid hormone biosynthesis
Ko00332	Carbapenem biosynthesis (NADH-quinone oxidoreductase subunit C)	ko00943	Isoflavonoid biosynthesis
Ko00906	Carotenoid biosynthesis		

2.4. Candidate Salt-Responsive DEGs

Based on functional annotation, several candidate genes were found to be differentially regulated in both leaves and roots at short- (2 h), intermediate- (8 h), and long-term (24 h) salt stress treatments. The most prevalent salt-responsive genes were categorized in different families, including ion-transporter-related, antioxidant, hormone-related, abiotic stress-responsive, transcription factors, and signal transduction (Table S2).

Several components of the ABA signaling pathway, such as ABA sensor pyrabactin resistance 1, ABA-independent SNF1-related protein kinase 2 (*SnRK2*), and 2C-type protein phosphatases (*PP2C*), as well as ABA-responsive elements GRAM-domain-containing protein and ABC transporter G family member 3, were differentially regulated under all stress durations in both tissues. Auxin-signal-transduction-related genes were also differentially expressed, including several auxin-responsive factors, auxin-induced proteins, and dormancy/auxin-associated family protein. Additionally, a number of both ethylene- and jasmonic-acid-mediated signaling pathways were differentially regulated in barley seedlings (Table S2).

Several other gene families involved in signaling were differentially regulated in both leaf and root tissues under all salt stress durations (2, 8, and 24 h), including calcium signaling, leucine-rich repeats receptor-like kinase (*LRR-RLK*), and protein kinases. Differential expression of several others was observed only in roots, such as the NTPase-domain- (*NACHT*) and the pyrin-domain (*PYD*)-containing protein. Among protein kinases, histidine kinase, histidine-tRNA ligase, and hybrid signal transduction histidine kinase I were identified. Compared with leaves, more differentially expressed kinases were detected in roots. Various transcription factor families were differentially expressed in both tissues, among them a basic helix-loop-helix DNA-binding superfamily protein, *WRKY* DNA-binding protein, kinase interacting (*KIP1-like*) family protein, homeobox-leucine zipper protein 3, and heat shock transcription factor C1.

Several oxidoreductase, glutathione S-transferase, and peroxidase families were differentially expressed, particularly in roots after 8 and 24 h salt exposure. Superoxide dismutase (SOD) and catalase antioxidant enzymes were also differentially regulated by salt stress. Catalase genes, including catalases 1 and 3, were upregulated in both leaves and roots under all stress durations except 24 h in roots, while Fe superoxide dismutase 2 was downregulated only in roots. Chalcone synthase 2, with a crucial role in ROS detoxification, was differentially expressed in both roots and leaves. Genes involved in the biosynthesis of proline, sugars, and glycine betaine, all of which are major osmoprotectants, were also differentially expressed in both tissues, particularly in leaves after 8 and 24 h salt stress. The ATP-binding cassette (*ABC*) transporter family, such as ABC transporter G family members, and solute transporters, including those for sugars, amino acids, and peptides, were differentially regulated in both tissues under all salt stress durations.

Transcript expression of various membrane transporters and ion channels, including glutamate receptor-like, cyclic nucleotide-gated, high-affinity K^+ transporters (*HKTs*), salt overly sensitive Na^+/H^+ exchanger (*SOS*), and two-pore-domain K^+ channel (TPK), were differentially expressed in both tissues mainly after 8 and 24 h salt treatments.

Components of photosystems I (*PSI*) and II (*PSII*) and light-induced proteins were also differentially expressed in both leaves and roots under all salt stress durations. Furthermore, plant–pathogen interaction genes, such as thaumatin superfamily proteins, and defensin genes were up- or downregulated.

2.5. Validation of RNA-Seq Data by Quantitative Real-Time qRT-PCR

Eight genes, including two up- and two downregulated genes in leaves and two up- and two downregulated genes in roots, were selected for confirmation of RNA-seq data by qRT-PCR. The expression fold changes for all six transcripts were in agreement with RNA-seq regardless of salt stress duration (Figure 5).

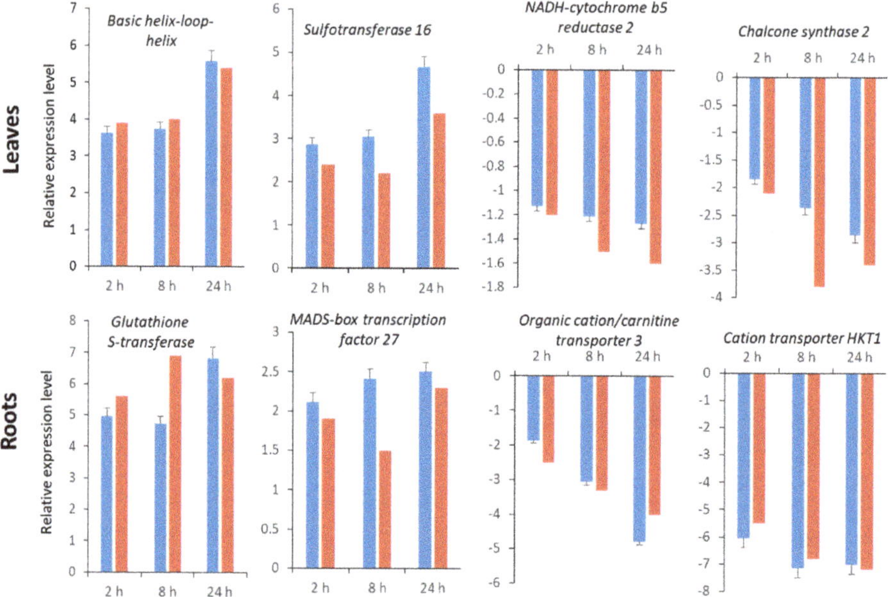

Figure 5. Expression pattern validation of eight randomly selected genes in leaves and roots by qRT-PCR. Red bar indicates transcript abundance changes of RNA-seq data calculated by the Log_2 fold change method. Blue bar, mean with associated standard error bar (n = 3), represents the relative expression level determined by qRT-PCR using the $2^{-\Delta\Delta CT}$ method.

3. Discussion

Salt stress tolerance is determined by several interconnecting effects of different molecular, cellular, metabolic, and physiological mechanisms [8]. Understanding the networks that underlie the barley salt stress response will be of great interest for the identification of possible breeding targets in order to improve barley stress tolerance under a future scenario of global climate change.

Transcriptomic approaches can provide relevant information to elucidate the complex molecular and genetic mechanisms involved in barley salt tolerance response [17,29]. Furthermore, comparing transcriptomes of stressed vs. nonstressed barley plants across different time points supplies important key markers to support salt tolerance breeding programs. Tunisian local barely accessions may hold genes of high value for salinity tolerance due to their potential to grow under adverse conditions [28].

Barley leaf and root transcription profiles after 2, 8, and 24 h of high salt stress treatments (200 mM NaCl) revealed an array of up- and downregulations in various biological processes involved in the overall salt stress response (Figures 4 and 6). Based on pairwise comparisons between control and salt-stressed samples, the number of DEGs was high and varied with the duration of salt treatment (Figure 2). Elevation of DEGs at all time points suggests important changes in the Boulifa seedling gene expression in response to salt stress. This extensive genetic regulation could be responsible for salt tolerance in the Boulifa genotype by affecting several physiological and biochemical processes. High salt induced the greatest number of DEGs following 8 h of exposure (Figure 2 and Figure S1). In agreement with a previous study, higher DEGs in wild barley leaves were found after 12 h of salt treatment compared with 24 h [30]. The salt-induced increase and subsequent decrease of DEGs may be attributed to the increasing stress of extended salt exposure, followed by possible recovery after 24 h. Based on these findings, 8 to 12 h of salt treatment should be the most appropriate duration to elucidate the genes involved salt stress responses. In contrast, a positive correlation between the duration of imposed stress and the number of DEGs in barley roots exposed to 6 and 24 h salt stress was demonstrated [15]. This discordance could be attributed to the differences in salt exposure times and/or imposed salt concentrations. Indeed, in our experiment the severe salt stress applied (200 mM) could induce a rapid and strong response compared with 150 mM used by Osthoff et al. [15].

Comparison of DEGs between root and leaf tissues revealed significant differences in expression in response to salt stress (Figure 1). For all treatment durations, the most severe impact on gene expression regulation was observed in roots compared with leaves (Figures 2 and 3 and Figure S1), emphasizing the more prominent role of roots in sensing salinity and responding through regulation of very complex transcriptional processes [31]. Early (2 h) salt-responsive DEGs were seven times more abundant in roots than in leaves; however, after 8 and 24 h, root DEGs were around three times greater than those of leaves (Figure S1), indicating a rapid salt stress response in roots, the primary organ of exposure [31]. At all time points, downregulated genes were more abundant than upregulated genes only in leaves. These results are consistent with previous reports on transcriptional responses in root and leaf tissues of different plant species subjected to abiotic stresses. Baldoni et al. [32] detected a higher number of DEGs in the roots (6007 genes) of a tolerant rice genotype (Eurosis) subjected to osmotic stress compared with leaves (3065) after 3 h treatment; however, the number of DEGs in both roots and leaves were similar after 24 h. Additionally, they detected a higher number of upregulated genes (61.6% of all DEGs) than downregulated genes (38.4%) only in roots. Furthermore, Luo et al. [19] detected more DEGs in roots than in leaves of salt-stress-treated sweet potato with also a greater number of upregulated DEGs than downregulated DEGs (544 up and 392 down) in roots and more downregulated DEGs than upregulated DEGs (75 up and 145 down) in leaves. Even in quinoa and peach, similar trends of DEG distribution between roots and leaves subjected to salt stress were reported [33,34]. The high number of downregulated genes in leaves under high salinity could be attributed to the efficiency in conserving resources and energy under stress conditions by repressing the transcriptional process of genes mainly associated with oxidative activities and cell wall compartment, which could be constitutively active. This may have contributed to the salt tolerance phenotype of Boulifa [28].

To gain further insight into the mechanisms underlying barley salt stress tolerance at an early seedling stage, DEGs in both leaves and roots were annotated, GO-enriched, and categorized into different functional groups (biological processes), including sensing and signaling pathways, transcriptional reprograming, hormone and ion homeostasis regulation, and metabolic changes as summarized in Figure 6.

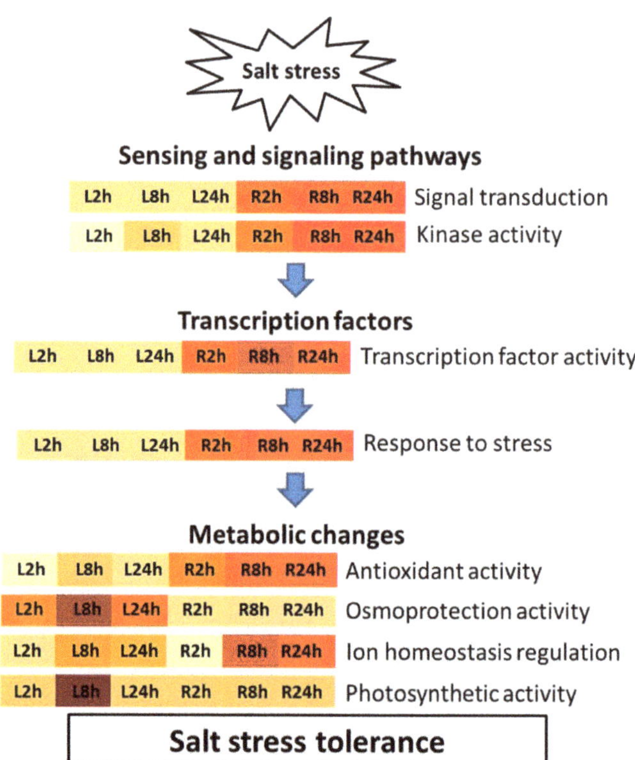

Figure 6. Flow chart of the salt stress tolerance mechanism in barley. The induced stress response begins with the activation of several sensing and signaling pathways, followed by transcriptional reprograming, resulting in the activation of diverse cellular homeostasis mechanisms for ROS detoxification, osmoprotection, ion homeostasis, and photosystem protection. Samples L2h, L8h, and L24h and R2h, R8h, and R24h for leaves and roots, respectively, under different stress durations are shown. The color gradient indicates the pattern of GO term enrichment level from low (yellow) to high (brown).

To avoid salinity damage, plants have evolved sensors to detect stress and activate signal transduction for the modification of cellular traits through transcriptional regulation [35,36]. Therefore, sensing and signaling are crucial for salt stress response [37–39]. Several salt-induced signaling pathways have been previously reported, including abscisic acid (ABA) [40], hormone [41], calcium [42], and receptor-like kinase pathways [43] (Table S2).

The current study suggests the involvement of several signal transduction genes, including differentially expressed ABA signaling pathway genes. The ABA receptors, which are critical for plant growth and development under abiotic stress [44], were differentially expressed particularly in roots. Indeed, the ABA sensor pyrabactin resistance 1, a negative regulator of the ABA-independent *SnRK2* and a selective inhibitor of the *PP2C* [45], and both *PP2C* and *SnRK2*, major negative regulators of ABA signaling [46], were differentially expressed at all time points in both tissues. Furthermore, a number of ABA-responsive elements were differentially regulated under all stress durations.

Several other hormones play important roles in barley salt signaling, including differentially expressed hormone-related genes, such as auxins, ethylene, and jasmonic acid.

Auxins play an important role in determining plant architecture and contribute mainly to cell elongation and division. Several auxin-responsive protein family members, auxin signaling pathway constituents, and dormancy/auxin-associated family proteins, involved in defense against virulent bacterial pathogens [47], were differentially expressed, suggesting their important roles in barley salt stress response.

Ethylene is involved in ion homeostasis, ROS detoxification, and salt stress tolerance in plants [48]. Several genes encoding ethylene signaling pathway proteins were differentially expressed in both tissues under all stress durations. Relative to leaves, the majority of ethylene-signaling-associated DEGs were found in roots. Moreover, after 2 h salt exposure, the ethylene-signaling-associated DEGs were found only in roots (Table S2), demonstrating the rapid salt stress response in these tissues.

A number of jasmonic-acid-mediated signaling pathways that regulate many developmental and defense mechanisms, including root growth inhibition and activation of antioxidant enzymes upon exposure to high salinity [37], were upregulated under all stress durations mainly in roots. In addition, several calcium–calmodulin signal transduction proteins were differentially expressed in barley seedlings, particularly in roots, emphasizing the importance of calcium signaling and related mechanisms in regulating transcriptional activity in response to salt stress [37].

Various transcripts for proteins involved in signal transduction, such as the LRR-RLK protein and scaffold protein families were differentially expressed under all salt stress durations. Additionally, several transcripts encoding signal transduction histidine kinases were salt-regulated in barley seedlings. Relative to leaves, more kinases were differentially expressed in roots. These results are in agreement with previous studies [29,30,34] and confirm the involvement of signal transduction in salinity tolerance in barley and highlight the more complex signaling regulation in roots, supporting the notion that roots are the primary sensors of salt stress.

At 8 h of salt exposure, there were higher numbers of kinase DEGs in both leaves and roots. Furthermore, several transcription factors regulating gene expression in different signaling pathways [49] were identified under severe salt stress. Similar transcription factor families were identified by RNA-seq analysis in a mutant barley line exposed to salt stress [18].

Several metabolic pathways that include numerous proteins were differentially expressed in barley leaves and roots under salt stress. Indeed, in response to ROS, synthesis of stress-related metabolites (hormone, multiple osmolytes, and cell wall components) and photosynthesis and metabolite and water transport systems were upregulated (Table S2).

The antioxidant defense system was strongly affected in barley seedlings under salt stress with over-represented DEGs corresponding to antioxidant and oxidoreductase activities (Figure 4). Antioxidant enzymes such as SOD and catalase were represented among DEGs. These genes, widely described as active in ROS homeostasis [37], were also indicated through RNA-seq analyses of barley roots, mutant barley, and wild barley subjected to salt stress [15,18,30]. Chalcone synthase 2, which plays diverse roles in cell protection and detoxification as ROS scavengers and osmoregulators [1,15,31], was differentially expressed in both roots and leaves.

Several genes involved in the biosynthesis of major osmotic components, including proline, sugars, glycine betaine, and polyamines, were differentially expressed. These osmoprotectants, which have very important roles in maintaining water uptake, membrane and protein protection, and stabilization against abiotic stresses [50], were also identified in mutant barley exposed to salt stress [18]. These results highlight the involvement of osmotic and oxidative homeostasis maintenance in preventing stress-induced ROS accumulation, resulting in high growth performance previously detected in the tolerant genotype Boulifa [27,28].

Several proteins belonging to the ATP-binding cassette (ABC) transporter family involved in the modulation of stomatal response to CO_2 [51] were upregulated in leaves and differentially expressed in roots under all stress treatments. Additionally, solute transporters were highly differentially expressed at all time points in both roots and leaves, suggesting their involvement in the re-establishment of cellular osmotic homeostasis.

Cellular ionic homeostasis, which is an essential process for growth during salt stress [52], was also highly differentially expressed in barley seedlings under salt stress. Two major classes of nonselective cation channels permeable to potassium and calcium were involved in re-establishing ionic balance after defense action [53]. The *HKT* transporters were upregulated in leaves and roots at all time points, while the *SOS*, which contributes to ion homeostasis by transporting Na^+ out of the cell [54], was downregulated only in leaves after 24 h of salt treatment. The *TPK* channel, activated by calcium with strong selectivity for K^+ over Na^+, is involved in intracellular K^+ redistribution between different tissues and stomatal regulation by monitoring turgor pressure [55]. The expression of this transporter was downregulated in both tissues at all stress time points. Other transporters active in the root tissue, such as the stelar K^+ outward rectifier (*SKOR*) that mediates the delivery of K^+ to the xylem for root–shoot potassium allocation [56], and the transient receptor potential cation channel subfamily V member 6 (*Trpv6*) that mediates Na^+ and Ca^{2+} influx [57] were differentially expressed. In roots, expression of both genes was upregulated after 24 h of salt exposure, emphasizing the importance of root function in water and nutrient uptake during salt stress.

Salt-stressed plants must respond to not only ionic and osmotic disruptions but also impaired photosynthesis [37]. A large number of genes involved in the protection of photosystems were differentially expressed in leaf samples mainly after 8 h of exposure to salt stress. The regulation of these photosystem genes could be the origin of the sustained high growth rate, maintenance of water use efficiency, and photosynthetic potential in the tolerant genotype Boulifa under severe salt stress [27,28]. Significantly upregulated were components of PSI and PSII, including PSI assembly protein and PSII reaction center proteins required for stability and/or assembly, PSI P700 chlorophyll apoproteins, PSI iron–sulfur center, PSII D2 protein, and PSII 10 kDa polypeptide essential for photochemical activities [58]. Almost all of these genes were downregulated in roots under all stress durations. Earlier observations based on transcriptome analysis are consistent with this finding [32,59]. Both studies found downregulation of several photosynthesis-related genes in rice roots under osmotic stress and wheat roots under low-phosphorus stress. It was suggested that the repression of these genes in roots may be related to energy conservation. Furthermore, some transcripts related to carbohydrates, fatty acids, protein metabolism, and biosynthetic processes were differentially expressed in both leaves and roots in order to mobilize an alternative source of energy because of the impairment of photosynthesis during salt stress [1].

A number of pathogenesis-related (*PR*) genes were also differentially expressed in barley seedlings following salt stress. Upregulation of the *PR* gene expression was noted in 3-day-old barley seedlings geminated and grown in 100 mM NaCl [34]. At 8 h of salt stress, the leaf expression of a pathogenesis-related thaumatin superfamily protein was downregulated, while the expression in roots was downregulated at all treatment durations. The expression of another PR protein, *Solanum tuberosum* pathogenesis (STH-2), was upregulated only in roots, and defensin genes 1 and 2 were downregulated at all time points in roots but not at the 2 h time point in leaves. These observations support the hypothesis that pathogenesis-related proteins are involved in plant responses to environmental stresses.

4. Conclusions

In this study, a comprehensive comparison of the gene expressions of barley leaves and roots after 2, 8, and 24 h high salt stress was performed. The results suggest cross-talk among diverse pathways in barley in response to salt stress. Differential response included a rapid regulation of several candidate genes related to hormone and kinase sensing and signaling, such as ABA-responsive elements, calcium signaling, LRR-RLK, and protein kinases, and several transcription factors mainly belong to the MYB, bHLH, HD-ZIP, WRKY, MADS-box, and NAC families. Moreover, differential regulation of antioxidant genes, genes involved in the biosynthesis of osmolytes, and transporter genes was observed. Both common and tissue-specific salt-responsive candidate genes identified here constitute valuable resources for plant breeders and for further omics studies in barley and other crops. Future research, such transgenic assay, complementation assay, and subcellular localization, is needed to further validate the functions of the identified genes in providing salinity tolerance to plants and the physiological mechanisms in which they are involved.

5. Material and Methods

5.1. Plant Material and Hydroponic Salt Stress Treatment

Seeds of a salt-tolerant Tunisian barley cultivar (Boulifa) [27,28] were surface-sterilized with 5% sodium hypochlorite solution for 5 min, then thoroughly rinsed with distilled water and germinated in the dark at 25 °C in Petri dishes with distilled water. After 5 days, germinated seedlings were transferred to an aerated hydroponic system containing half-strength modified Hoagland's solution [60] under 16 h light at 22 °C. After 3 days of acclimatization, gradual salt stress was applied. NaCl concentrations were brought up to 200 mM by increments of 50 mM NaCl on the first and second day and 100 mM on the third day. Root and shoot tissues were sampled at 0 h (before adding the first 50 mM NaCl), then again at 2, 8, and 24 h after reaching a final concentration of 200 mM NaCl. Five plants in each time point were harvested, pooled, washed thoroughly and separated into roots and shoots, frozen in liquid nitrogen, and stored at −80 °C for RNA isolation. All experiments were performed in triplicate.

5.2. Total RNA Isolation and DNase Treatment

Total RNA was isolated from shoots and roots representing each time point using the ZR Plant RNA MiniPrep™ Kit (Zymo Research, Irvine, CA, USA). The quality and quantity of isolated RNAs were checked by agarose gel electrophoresis and spectrophotometrically using a BioPhotometer (Eppendorf BioPhotometer plus, Hamburg, Germany). Residual DNA was eliminated using a TURBO DNA-free™ Kit (Promega, Madison, WI, USA).

5.3. Sequencing

Library construction and sequencing were carried out at the Beijing Genomics Institute (BGI, Shenzhen, China) for the three replicates of each treatment using the Illumina NextSeq 500 platform. Single-end reads 50 bp in length were generated for each sample at an average of 24.14 million reads per sample using the oligoDT selection method. Low-quality reads and adaptor sequences were removed from all samples (clean reads).

All clean reads were deposited in the Sequence Read Archive (SRA) database in NCBI with accession number PRJNA715166.

5.4. Pseudoalignment and Transcript Abundance Analysis

The reads were pseudoaligned to the barley transcriptome [61] (barley reference from PGSB barley genome database 2017) using kallisto [62], and gene-level abundances were obtained. The abundances were normalized using DESeq2 [63] and principal component analysis (PCA, Figure S2), and hierarchical clustering was performed using the top 25% highly varying genes in order to examine the underlying structure of the data and to identify the largest sources of variance.

5.5. Differential Expression Analysis

Differential expression analysis was performed separately for each tissue (root vs. leaf). Within each tissue, DESeq2 [63] was used to model the gene abundances as a negative binomial distribution, and three pairwise contrasts were performed (2 vs. 0 h, 8 vs. 0 h, and 24 vs. 0 h). Genes with adjusted p-value (after Benjamini–Hochberg correction) ≤ 0.01 and absolute $\log_2$ fold changes ≥ 1 were reported as significantly differentially expressed in each contrast.

5.6. GO Enrichment Analysis and Visualization

For each contrast, gene ontology (GO) terms enriched among the differentially expressed genes were identified using topGO [64], an R package that provides various algorithms for calculating the statistical enrichment of GO terms among a list of genes. A classic Fisher's enrichment test was performed using each list of differentially expressed genes to identify significantly enriched GO terms in the molecular function and biological process domains.

All barley genes were annotated using Blast2go (OmicsBox 2.0.10) [65] to identify KEGG pathway information. Fisher's enrichment test was run, and over-represented pathways were identified using FDR cutoff 0.1.

5.7. Validation of RNA-Seq Findings by Real-Time PCR

Validation of RNA-seq findings was performed by reverse transcription PCR. Quantitative real-time PCR (qRT-PCR) was performed using Applied Biosystems Power SYBR Green qPCR Master Mix (Life technologies, Carlsbad, CA, USA) on 96-well plates using specific primers designed by the Primer 3 software (http://bioinfo.ut.ee/primer3-0.4.0/) [66]. The expression levels of eight randomly selected genes and the internal control alpha tubulin (TUB2) were checked using an Applied Biosystems thermal cycler. The primer names and sequences used for primer design are in Table S3. Each qRT-PCR reaction mixture (20 µL) contained 1 µL of fourfold diluted cDNA, 10 µL of PCR mixture (2 × SYBR Green buffer), 7 µL of water, and 2 µL of primers (10 ppM). The following thermal profile was used: 95 °C (10 min), 40 amplification cycles of 95 °C (30 s), 60 °C (60 s). Melting curves were obtained by slow heating from 65 to 95 °C at 0.5 °C/s and continuous monitoring of the fluorescence signal. Results were analyzed by the StepOne™ software, v2.2.2 (Applied Biosystems, Foster City, CA, USA). Transcript abundance quantification was performed according to the $2^{-\Delta\Delta CT}$ method [67]. Fold change was calculated for the salt-treated seedlings relative to the untreated plants (0 h).

Total RNA isolation, quantification, and DNase treatment were performed as previously described. cDNA synthesis was performed from 1 µg of RNA using a GoScript™ Reverse Transcription System Kit (Promega) following the manufacturer's instructions.

Supplementary Materials: The following are available online at https://www.mdpi.com/article/10.3390/ijms22158155/s1. Figure S1. Differentially expressed genes in leaves (a) and roots (b). Differentially expressed genes (DEGs) at 2, 8, and 24 h of salt treatments relative to untreated plants (0 h) are shown. The specific DEG numbers for each time point and the shared DEGs between the different time points are shown by dots and lines, respectively. Figure S2. Principal component analysis (PCA) of the data showing the variation due to tissue. Table S1. Statistics of RNA-seq numerical data analysis in barley seedlings. B0L (control leaves), B0R (control roots), B2L (leaves salt-stressed for 2 h), B2R (roots salt-stressed for 2 h), B8L (leaves salt-stressed for 8 h), B8R (roots salt-stressed for 8 h), B24L (leaves salt-stressed for 24 h), B24R (roots salt-stressed for 24 h). Table S2. Candidate genes of barley in response to salt stress, yellow: downregulated; red: upregulated; blue: up- and downregulated genes (multiple genes). Table S3. Primers used for qRT-PCR to validate RNA-seq findings.

Author Contributions: Conceptualization, R.N.O., R.K.J., D.A., T.A.R. and H.M.; methodology, R.N.O., R.K.J., D.A. and H.M.; software, D.A.; validation, R.K.J., G.A. and A.G.; formal analysis, R.J. and M.B.C.; investigation, M.B.C., R.J. and S.M.; resources, M.B.C., R.J., S.M. and D.A.; data curation, M.B.C., R.J. and D.A.; writing—original draft preparation, R.N.O.; writing—review and editing, T.A.R., R.K.J., D.A. and A.G.; visualization, D.A.; supervision, R.K.J. and A.G.; project administration, R.N.O. and R.K.J.; funding acquisition, R.N.O. and R.K.J. All authors have read and agreed to the published version of the manuscript.

Funding: This research was funded by the "International Center for Biosaline Agriculture (ICBA)", grant number JRCAFS-23396, and the "CRDF Global", grant number #62795.

Institutional Review Board Statement: Not applicable.

Informed Consent Statement: Not applicable.

Data Availability Statement: The data presented in this study are openly available in the Sequence Read Archive (SRA) database in NCBI under accession number PRJNA715166.

Acknowledgments: The authors are thankful to Aida Bouajila and Badra Bouamama for prospecting barley accessions.

Conflicts of Interest: The authors declare no conflict of interest.

References

1. Hernández, J.A. Salinity tolerance in plants: Trends and perspectives. *Int. J. Mol. Sci.* **2019**, *20*, 2408. [CrossRef]
2. Zhu, J.K. Plant salt tolerance. *Trends Plant Sci.* **2001**, *6*, 66–71. [CrossRef]
3. Acosta-Motos, J.R.; Ortuño, M.F.; Bernal-Vicente, A.; Diaz-Vivancos, P.; Sanchez-Blanco, M.J.; Hernández, J.A. Plant responses to salt stress: Adaptive mechanisms. *Agronomy* **2017**, *7*, 18. [CrossRef]
4. Isayenkov, S.V.; Maathuis, F.J.M. Plant salinity stress: Many unanswered questions remain. *Front. Plant Sci.* **2019**, *10*, 80–91. [CrossRef]
5. Krasensky, J.; Jonak, C. Drought, salt, and temperature stress-induced metabolic rearrangements and regulatory networks. *J. Exp. Bot.* **2012**, *63*, 1593–1608. [CrossRef] [PubMed]
6. Kwon, O.K.; Mekapogu, M.; Kim, K.S. Effect of salinity stress on photosynthesis and related physiological responses in carnation (*Dianthus caryophyllus*). *Hortic. Environ. Biotechnol.* **2019**, *60*, 831–839. [CrossRef]
7. Bohnert, H.J.; Sheveleva, E. Plant stress adaptations—Making metabolism move. *Curr. Opin. Plant Biol.* **1998**, *1*, 267–274. [CrossRef]
8. Gupta, B.; Huang, B. Mechanism of salinity tolerance in plants: Physiological, biochemical, and molecular characterization. *Int. J. Genom.* **2014**, *2014*, 701596–701614. [CrossRef]
9. Blumwald, E. Sodium transport and salt tolerance in plants. *Curr. Opin. Cell Biol.* **2000**, *12*, 431–434. [CrossRef]
10. Munns, R.; Tester, M. Mechanisms of salinity tolerance. *Annu. Rev. Plant Biol.* **2008**, *59*, 651–681. [CrossRef]
11. Li, H.; Zhang, D.; Li, X.; Guan, K.; Yang, H. Novel DREB A-5 subgroup transcription factors from desert moss (*Syntrichia caninervis*) confers multiple abiotic stress tolerance to yeast. *J. Plant Physiol.* **2016**, *194*, 45–53. [CrossRef]
12. Shinozaki, K.; Yamaguchi-Shinozaki, K. Gene expression and signal transduction in water-stress response. *Plant Physiol.* **1997**, *115*, 327–334. [CrossRef]
13. Nakagami, H.; Pitzschke, A.; Hirt, H. Emerging MAPkinase pathways in plant stress signaling. *Trends Plant Sci.* **2005**, *10*, 339–346. [CrossRef] [PubMed]
14. Ma, S.; Gong, Q.; Bohnert, H.J. Dissecting salt stress pathways. *J. Exp. Bot.* **2006**, *57*, 1097–1107. [CrossRef] [PubMed]
15. Osthoff, A.; Rose, P.D.D.; Baldauf, J.A.; Piepho, H.P.; Hochholdinger, F. Transcriptomic reprogramming of barley seminal roots by combined water deficit and salt stress. *BMC Genom.* **2019**, *20*, 325–339. [CrossRef] [PubMed]
16. Wilhelm, B.T.; Marguerat, S.; Watt, S.; Schubert, F.; Wood, V.; Goodhead, I.; Penkett, C.J.; Rogers, J.; Bähler, J. Dynamic repertoire of a eukaryotic transcriptome surveyed at single-nucleotide resolution. *Nature* **2008**, *453*, 1239–1243. [CrossRef] [PubMed]
17. Han, Y.; Gao, S.; Muegge, K.; Zhang, W.; Zhou, B. Advanced applications of RNA sequencing and challenges. *Bioinform. Biol. Insights* **2015**, *9*, 29–46. [CrossRef]
18. Yousefirad, S.; Soltanloo, H.; Ramezanpour, S.S.; Nezhad, K.Z.; Shariati, V. The RNA-seq transcriptomic analysis reveals genes mediating salt tolerance through rapid triggering of ion transporters in a mutant barley. *PLoS ONE* **2020**, *15*, e0229513. [CrossRef]
19. Luo, Y.; Reid, R.; Freese, D.; Li, C.; Watkins, J.; Shi, H.; Zhang, H.; Loraine, A.; Song, B.H. Salt tolerance response revealed by RNA-Seq in a diploid halophytic wild relative of sweet potato. *Sci. Rep.* **2016**, *7*, 9624–9637. [CrossRef]
20. Yong, H.Y.; Zou, Z.; Kok, E.P.; Kwan, B.H.; Chow, K.; Nasu, S.; Nanzyo, M.; Kitashiba, H.; Nishio, T. Comparative transcriptome analysis of leaves and roots in response to sudden increase in salinity in *Brassica napus* by RNA-seq. *BioMed Res. Int.* **2014**, *6*, 467395–467415.
21. Shen, X.Y.; Wang, Z.L.; Song, X.F.; Xu, J.J.; Jiang, C.Y.; Zhao, Y.X.; Ma, C.L.; Zhang, H. Transcriptomic profiling revealed an important role of cell wall remodeling and ethylene signaling pathway during salt acclimation in *Arabidopsis*. *Plant Mol. Biol.* **2014**, *86*, 303–317. [CrossRef]
22. Feriani, W.; Rezgui, S.; Cherif, M. Grain yield assessment of genotype by environment interaction of Tunisian doubled-haploid barley lines. *J. New Sci.* **2016**, *27*, 1507–1512.
23. Adem, G.D.; Roy, S.J.; Zhou, M.; Bowman, J.P.; Shabala, S. Evaluating contribution of ionic, osmotic and oxidative stress components towards salinity tolerance in barley. *BMC Plant Biol.* **2014**, *14*, 113–126. [CrossRef] [PubMed]
24. Witzel, K.; Matros, A.; Strickert, M.; Kaspar, S.; Peukert, M.; Mühling, K.H.; Börner, A.; Mock, H.P. Salinity stress in roots of contrasting barley genotypes reveals time-distinct and genotype-specific patterns for defined proteins. *Mol. Plant* **2014**, *7*, 336–355. [CrossRef] [PubMed]
25. Hasanuzzaman, M.D.; Nahar, K.; Alam, M.M.; Roychowdhury, R.; Fujita, M. Physiological, biochemical, and molecular mechanisms of heat stress tolerance in plants. *Int. J. Mol. Sci.* **2013**, *14*, 9643–9684. [CrossRef] [PubMed]
26. Negrao, S.; Schmöckel, S.M.; Tester, M. Evaluating physiological responses of plants to salinity stress. *Ann. Bot.* **2017**, *119*, 1–11. [CrossRef]

27. Ben Chikha, M. Variability of Tolerance to Salt Stress in Local Genotypes of Barley (*Hordeum vulgare* L.) Depending on the Stage of Development. Ph.D. Thesis, Faculty of Sciences of Tunis, University of Tunis EL Manar, Tunis, Tunisia, 2017; p. 162.
28. Ben Chikha, M.; Hessini, K.; Ourteni, R.N.; Ghorbel, A.; Zoghlami, N. Identification of barley landrace genotypes with contrasting salinity tolerance at vegetative growth stage. *Plant Biotechnol.* **2016**, *33*, 287–295. [CrossRef]
29. Ziemann, M.; Kamboj, A.; Hove, R.M.; Loveridge, S.; El-Osta, A.; Bhave, M. Analysis of the barley leaf transcriptome under salinity stress using mRNA-Seq. *Acta Physiol. Plant* **2013**, *35*, 1915–1924. [CrossRef]
30. Bahieldin, A.; Atef, A.; Sabir, J.S.; Gadalla, N.O.; Edris, S.; Alzohairy, A.M.; Radhwan, N.A.; Baeshen, M.N.; Ramadan, A.M.; Eissa, H.F.; et al. RNA-Seq analysis of the wild barley (*H. spontaneum*) leaf transcriptome under salt stress. *C. R. Biol.* **2015**, *338*, 285–297. [CrossRef]
31. Hill, B.; Cassin, A.; Keeble-Gagnère, G.; Doblin, M.S.; Bacic, A.; Roessner, U. De novo transcriptome assembly and analysis of differentially expressed genes of two barley genotypes reveal root-zone-specific responses to salt exposure Camilla. *Sci. Rep.* **2016**, *6*, 31558–31572. [CrossRef]
32. Baldoni, E.; Bagnaresi, P.; Locatelli, F.; Monica Mattana, M.; Genga, A. Comparative leaf and root transcriptomic analysis of two rice japonica cultivars reveals major differences in the root early response to osmotic stress. *Rice* **2016**, *9*, 25–45. [CrossRef]
33. Ruiz, K.B.; Maldonado, J.; Biondi, S.; Silva, H. RNA-seq analysis of salt-stressed versus non salt-stressed transcriptomes of *Chenopodium quinoa* landrace R49. *Genes* **2019**, *10*, 1042. [CrossRef] [PubMed]
34. Ksouri, N.; Jiménez, S.; Wells, C.E.; Contreras-Moreira, B.; Gogorcena, Y. Transcriptional responses in root and leaf of *Prunus persica* under drought stress using RNA sequencing. *Front. Plant Sci.* **2016**, *7*, 1715–1734. [CrossRef] [PubMed]
35. Ghosh, D.; Xu, J. Abiotic stress responses in plant roots: A proteomics perspective. *Front. Plant Sci.* **2014**, *5*, 6–19. [CrossRef]
36. Agarwal, P.K.; Jha, B. Transcription factors in plants and ABA dependent and independent abiotic stress signaling. *Biol. Plant.* **2010**, *54*, 201–212. [CrossRef]
37. Zhao, C.; Zhang, H.; Song, C.; Zhu, J.K.; Shabala, S. Mechanisms of plant responses and adaptation to soil salinity. *Innovation* **2020**, *1*, 1–42. [CrossRef]
38. Golldack, D.; Li, C.; Mohan, H.; Probst, N. Tolerance to drought and salt stress in plants: Unraveling the signaling networks. *Front. Plant Sci.* **2014**, *5*, 151–162. [CrossRef]
39. Seyfferth, C.; Tsuda, K. Salicylic acid signal transduction: The initiation of biosynthesis, perception and transcriptional reprogramming. *Front. Plant Sci.* **2014**, *9*, 697–707. [CrossRef]
40. Osakabe, Y.; Yamaguchi-Shinozaki, K.; Shinozaki, K.; Tran, L.S.P. ABA control of plant macroelement membrane transport systems in response to water deficit and high salinity. *New Phytol.* **2014**, *202*, 35–49. [CrossRef]
41. Zhang, L.; Li, Z.; Quan, R.; Li, G.; Wang, R.; Huang, R. An AP2 domain-containing gene, *ESE1*, targeted by the ethylene signaling component EIN3 is important for the salt response in *Arabidopsis*. *Plant Physiol.* **2011**, *157*, 854–865. [CrossRef]
42. Dodd, A.N.; Kudla, J.; Sanders, D. The language of calcium signaling. *Annu. Rev. Plant Biol.* **2010**, *61*, 593–620. [CrossRef]
43. Wolf, S. Plant cell wall signaling and receptor-like kinases. *Biochem. J.* **2017**, *474*, 471–492. [CrossRef] [PubMed]
44. Sah, S.K.; Reddy, K.R.; Li, J. Abscisic acid and abiotic stress tolerance in crop plants. *Front. Plant Sci.* **2016**, *7*, 571–596. [CrossRef] [PubMed]
45. Zhao, Y.; Zhang, Z.; Gao, J.; Wang, P.; Hu, T.; Wang, Z.; Hou, Y.J.; Wan, Y.; Liu, W.; Xie, S.; et al. *Arabidopsis* duodecuple mutant of PYL ABA receptors reveals PYL repression of ABA-independent SnRK2 Activity. *Cell. Rep.* **2018**, *23*, 3340–3351. [CrossRef] [PubMed]
46. Hirayama, T.; Umezawa, T. The PP2C-SnRK2 complex—The central regulator of an abscisic acid signaling pathway. *Plant Signal. Behav.* **2010**, *5*, 160–163. [CrossRef] [PubMed]
47. Roy, S.; Saxena, S.; Sinha, A.; Nandi, A.K. Dormancy/auxin associated family protein 2 of *Arabidopsis thaliana* is a negative regulator of local and systemic acquired resistance. *J. Plant Res.* **2020**, *133*, 409–417. [CrossRef] [PubMed]
48. Peng, J.; Li, Z.; Wen, X.; Li, W.; Shi, H.; Yang, L.; Zhu, H.; Guo, H. Salt-induced stabilization of EIN3/EIL1 confers salinity tolerance by deterring ROS accumulation in Arabidopsis. *PLoS Genet.* **2014**, *10*, e1004664. [CrossRef]
49. Chew, W.; Hrmova, M.; Lopato, S. Role of homeodomain leucine zipper (HD-Zip) IV transcription factors in plant development and plant protection from deleterious environmental factors. *Int. J. Mol. Sci.* **2013**, *14*, 8122–8147. [CrossRef]
50. Khan, M.S.; Ahmad, D.; Khan, M.A. Utilization of genes encoding osmoprotectants in transgenic plants for enhanced abiotic stress tolerance. *Electron. J. Biotechnol.* **2015**, *18*, 257–266. [CrossRef]
51. Borghi, L.; Kang, J.; Francisco, R.B. Filling the Gap: Functional clustering of ABC proteins for the investigation of hormonal transport in planta. *Front. Plant Sci.* **2019**, *10*, 422–442. [CrossRef]
52. Hasegawa, P.M. Sodium (Na+) homeostasis and salt tolerance of plants. *Environ. Exp. Bot.* **2013**, *92*, 19–31. [CrossRef]
53. Demidchik, V. ROS-activated ion channels in plants: Biophysical characteristics, physiological functions and molecular nature. *Int. J. Mol. Sci.* **2018**, *19*, 1263. [CrossRef]
54. Shi, H.; Quintero, F.J.; Pardo, J.M.; Zhu, J.K. The putative plasma membrane Na^+/H^+ antiporter SOS1 controls long-distance Na^+ transport in plants. *Plant Cell.* **2002**, *14*, 465–477. [CrossRef]
55. Bihler, H.; Eing, C.; Hebeisen, S.; Roller, A.; Czempinski, K.; Bertl, A. TPK1 is a vacuolar ion channel different from the slow-vacuolar cation channel. *Plant Physiol.* **2005**, *139*, 417–424. [CrossRef] [PubMed]

56. Johansson, I.; Wulfetange, K.; Porée, F.; Michard, E.; Gajdanowicz, P.; Lacombe, B.; Sentenac, H.; Thibaud, J.B.; Mueller-Roeber, B.; Blatt, M.R.; et al. External K^+ modulates the activity of the *Arabidopsis potassium* channel SKOR via an unusual mechanism. *Plant J.* **2006**, *46*, 269–281. [CrossRef] [PubMed]
57. Stoerger, C.; Flockerzi, V. The transient receptor potential cation channel subfamily V member 6 (TRPV6): Genetics, biochemical properties, and functions of exceptional calcium channel proteins. *Biochem. Cell. Biol.* **2014**, *92*, 441–448. [CrossRef] [PubMed]
58. Gao, J.; Wang, H.; Yuan, Q.; Feng, Y. Structure and function of the photosystem supercomplexes. *Front. Plant Sci.* **2018**, *20*, 357–364. [CrossRef]
59. Wang, J.; Qin, Q.; Pan, J.; Sun, L.; Sun, Y.; Xue, Y.; Song, K. Transcriptome analysis in roots and leaves of wheat seedlings in response to low-phosphorus stress. *Sci. Rep.* **2019**, *9*, 19802–19814. [CrossRef] [PubMed]
60. Hoagland, D.R.; Arnon, D.I. The water culture method for growing plants without soil. *Calif. Agric. Exp. Stn.* **1950**, *347*, 32.
61. Mascher, M.; Gundlach, H.; Himmelbach, A.; Beier, S.; Twardziok, S.O.; Wicker, T.; Radchuk, V.; Dockter, C.; Hedley, P.E.; Russell, J.; et al. A chromosome conformation capture ordered sequence of the barley genome. *Nature* **2017**, *544*, 427–433. [CrossRef]
62. Bray, N.; Pimentel, H.; Melsted, P.; Pachter, L. Near-optimal probabilistic RNA-Seq quantification. *Nat. Biotechnol.* **2016**, *34*, 525–527. [CrossRef] [PubMed]
63. Love, M.; Huber, W.; Anders, S. Moderated estimation of fold change and dispersion for RNA-seq data with DESeq2. *Genom. Biol.* **2014**, *15*, 550–571. [CrossRef]
64. Alexa, A.; Rahnenfuhrer, J. *topGO: Enrichment Analysis for Gene Ontology*; R Package Version 2.44.0. 2021. Available online: https://bioconductor.org/packages/release/bioc/html/topGO.html (accessed on 1 July 2021).
65. Conesa, A.; Götz, S.; García-Gómez, J.M.; Terol, J.; Talón, M.; Roble, M. Blast2GO: A universal tool for annotation, visualization and analysis in functional genomics research. *Bioinformatics* **2005**, *21*, 3674–3676. [CrossRef] [PubMed]
66. Untergasser, A.; Cutcutache, I.; Koressaar, T.; Ye, J.; Faircloth, B.C.; Remm, M.; Rozen, S.G. Primer3-new capabilities and interfaces. *Nucleic Acids Res.* **2012**, *40*, 115–127. [CrossRef] [PubMed]
67. Schmittgen, T.D.; Livak, K.J. Analyzing real-time PCR data by the comparative C_T method. *Nat. Protoc.* **2008**, *3*, 1101–1108. [CrossRef]

International Journal of
Molecular Sciences

Article

Comparative Proteomic Analysis of Tolerant and Sensitive Varieties Reveals That Phenylpropanoid Biosynthesis Contributes to Salt Tolerance in Mulberry

Tiantian Gan, Ziwei Lin, Lijun Bao, Tian Hui, Xiaopeng Cui, Yanzhen Huang, Hexin Wang, Chao Su, Feng Jiao, Minjuan Zhang * and Yonghua Qian *

The Sericultural and Silk Research Institute, College of Animal Science and Technology, Northwest A&F University, Yangling 712100, China; gantiantian1987@163.com (T.G.); linxiaoliu510@163.com (Z.L.); baolijun@nwafu.edu.cn (L.B.); ht190322@163.com (T.H.); cuixpssyjs@163.com (X.C.); whyiggy@126.com (Y.H.); wanghexin1111@163.com (H.W.); suchao503@126.com (C.S.); fjiao@nwsuaf.edu.cn (F.J.)
* Correspondence: mjzhang1008@nwsuaf.edu.cn (M.Z.); qyh@nwafu.edu.cn (Y.Q.)

Abstract: Mulberry, an important woody tree, has strong tolerance to environmental stresses, including salinity, drought, and heavy metal stress. However, the current research on mulberry resistance focuses mainly on the selection of resistant resources and the determination of physiological indicators. In order to clarify the molecular mechanism of salt tolerance in mulberry, the physiological changes and proteomic profiles were comprehensively analyzed in salt-tolerant (Jisang3) and salt-sensitive (Guisangyou12) mulberry varieties. After salt treatment, the malondialdehyde (MDA) content and proline content were significantly increased compared to control, and the MDA and proline content in G12 was significantly lower than in Jisang3 under salt stress. The calcium content was significantly reduced in the salt-sensitive mulberry varieties Guisangyou12 (G12), while sodium content was significantly increased in both mulberry varieties. Although the Jisang3 is salt-tolerant, salt stress caused more reductions of photosynthetic rate in Jisang3 than Guisangyou12. Using tandem mass tags (TMT)-based proteomics, the changes of mulberry proteome levels were analyzed in salt-tolerant and salt-sensitive mulberry varieties under salt stress. Combined with GO and KEGG databases, the differentially expressed proteins were significantly enriched in the GO terms of amino acid transport and metabolism and posttranslational modification, protein turnover up-classified in Guisangyou12 while down-classified in Jisang3. Through the comparison of proteomic level, we identified the phenylpropanoid biosynthesis may play an important role in salt tolerance of mulberry. We clarified the molecular mechanism of mulberry salt tolerance, which is of great significance for the selection of excellent candidate genes for saline-alkali soil management and mulberry stress resistance genetic engineering.

Keywords: mulberry; salt stress; TMT proteomics; phenylpropanoid metabolism

Citation: Gan, T.; Lin, Z.; Bao, L.; Hui, T.; Cui, X.; Huang, Y.; Wang, H.; Su, C.; Jiao, F.; Zhang, M.; et al. Comparative Proteomic Analysis of Tolerant and Sensitive Varieties Reveals That Phenylpropanoid Biosynthesis Contributes to Salt Tolerance in Mulberry. *Int. J. Mol. Sci.* **2021**, 22, 9402. https://doi.org/10.3390/ijms22179402

Academic Editors: Mirza Hasanuzzaman and Masayuki Fujita

Received: 31 May 2021
Accepted: 23 August 2021
Published: 30 August 2021

Publisher's Note: MDPI stays neutral with regard to jurisdictional claims in published maps and institutional affiliations.

1. Introduction

Plants need to adapt with the constantly changing environment, including frequent stress environments like drought, salinity, chilling, cold, and heavy metal stress that are not conducive to plant growth and development. In particular, the frequent extreme weather in recent years has caused huge losses to agricultural production [1,2]. Therefore, the understanding of how plants respond to environmental stress is a fundamental question in plant biology [3]. Water deficit and high salinity induced osmotic stress severely restricts plant growth and productivity. Salinity has caused great loss to crop production around the world [4]. The area of saline-alkali land in the world is 950 million hectares, of which the area of saline-alkali land in China is about 99 million hectares. One fifth of arable land is threatened by salinity stress, especially in coastal areas where it is eroded by seawater. Salt stress disrupts many physiological and biochemical processes in plant cell and then induces iron toxicity, osmotic stress, and nutritional deficiencies [5]. Plants evolved to survive in adverse environmental stresses through a variety of strategies, such as developing serious sensors and signaling pathways, stress induced organelles (chloroplasts, mitochondria, peroxisomes, nuclei, and cell walls) response. The stress signals generated from all organelles are integrated to regulate the expression of stress-related genes and other cell activities, thereby restoring cell homeostasis [3]. In addition, a series of physiological processes including the synthesis of osmolytes (e.g., proline, glutathione, mannitol, carbohydrate, glycine betaine, and polyamines) and the activation of antioxidant enzymes (e.g., catalase, glutathione peroxidase, superoxide dismutase, and peroxidase) enhance the tolerance to salt stress [6,7].

Plants have established elaborate mechanisms to cope with salt stress. In recent years, the exploration of salt tolerance mechanisms has been conducted on herbaceous plants like rice, Arabidopsis, wheat and maize, while infrequently in wood plant like cotton and mulberry trees [8–10]. High-throughput biological technology such as metabolomics, transcriptomics and proteomics, have been used in the discovery of genetic resources to salt tolerance [11]. The molecular mechanism underlying salt-tolerant rice cultivar (sea rice) has been explored through whole genome sequencing and comparative transcriptome analysis [12]. A comparative transcriptional profiling was performed to explore salt tolerance rice cultivate (FL478) and its sensitive parent (IR29), and more than two thousand genes were found to respond to salt stress [13]. Another comparative transcriptome analysis of salt-tolerant rice under salt stress found that ABA signal transduction is highly related to salt stress [14]. Isobaric tags for relative and absolute quantitation (iTRAQ)-based proteomic analysis of salinity-tolerant rice (*sd58*) and wild-type *Kitaake* revealed that heterotrimeric G protein alpha subunit (RGA1) participates in salt response through ROS scavenging [15]. RNA-seq analysis of cotton found that 14,172 alternative splicing (AS) events were changed under salt stress, and Ser/Arg-rich (SR) proteins related to AS regulation may be crucial in salt stress [16]. Comprehensive RNA sequencing and SWATH-MS-based quantitative proteomics analyses of maize uncovered splicing factors and transcription factors have an important role in salt stress [17].

Mulberry (Rosales: Moraceae), a perennial woody plant, has remarkable nutritional, medicinal value, and economic value, as well as ecological functions. The fruit of mulberry is nutritious for human beings and the leaves are important food for silkworms. Mulberry is also a valuable genetic resource for abiotic stress (e.g., salinity, heavy metal ions, cold and drought) and can be used for drought relief and remediation of polluted water [18]. Genes involved in stress signals were explored in various studies in mulberry. For example, the polyphenol oxidase 1 gene (*MnPPO1*) is regulated during plant stress responses; ABA pathway-related genes *RD29A*, *RD29B*, *RD22*, *ABI3* and *ABI5* were significantly higher in drought tolerance [19]; the receptor for activated C kinase 1 (RACK1) protein is regulated in drought and salt tolerance. DNA methyltransferase genes (*MnCMT2*, *MnCMT3*, *MnMET1*, *MnDRM1*, and *MnDRM3*) are responsive to biotic stresses [20]. The activities of superoxide dismutase (SOD), catalase (CAT), peroxidase (POD) and ascorbate peroxidase (APX) in mulberry can eliminate the production of oxidative stress induced by Cd stress [21]. MiR397a and its target copper-assisted laccase (LAC) can respond to copper regulatory response in mulberry [22]. Furthermore, its popularization is expected to improve saline-alkali land and gradually turn it into fertile land. Through the release of mulberry (*Morus notablilis*) genome sequence [23], the biological techniques including transcriptome and proteome were used in understanding of the mechanisms of mulberry response to abiotic stress [24–26]. However, the comparative proteomic studies on salt tolerance of mulberry are seldom reported. Over the past decade, chemical labeling with isobaric tandem mass tags, such as isobaric tags for relative and absolute quantification reagents (iTRAQ) and tandem mass tag (TMT) reagents, have been employed in a wide range of different studies to explore the signal transduction pathways and protein interaction networks. In this study, we conducted physiological analysis and TMT-based proteomics to elucidate mulberry response to salt stress, which will provide a scientific basis for further understanding of abiotic stress response in mulberry and cultivation of salt-tolerant crops.

2. Results

2.1. Physiological Responses of Mulberry to Salt Stress

The salt-tolerant mulberry variety Jisang3 and sensitive variety Guisangyou12 were used in this study. After treatment with 200 mM NaCl for 10 days, compared to the Jisang3 seedlings with no obvious stress phenotype, the Guisangyou12 seedlings were wilted and rumpled (Figure 1A). Proline and MDA contents [27] in roots of both varieties were significantly increased after salt stress, while the antioxidant ability of Jisang3 and Guisangyou12 seedlings were significantly decreased (Figure 1B–D). The ion content including potassium, sodium and calcium were also determined. The contents of potassium and calcium were significantly decreased, and the content of sodium was significantly increased (Figure 1E–G). Net photosynthetic rate was further analyzed in Jisang3 and Guisangyou12. It was revealed that the photosynthetic efficiency is significantly reduced in the two varieties during salt stress, but the salt-sensitive variety Guisangyou12 keeps a relatively higher photosynthetic efficiency after salt treatment (Figure 2A). Consistent with plant phenotype, the water content was not decreased in salt-tolerant variety Jisang3 but was significantly reduced in salt-sensitive variety Guisangyou12 (Figure 2B). Intercellular carbon dioxide concentration was significantly increased in Jisang3 but no obvious change in Guisangyou12 (Figure 2C). Stomatal conductance reflecting the degree of stoma opening directly affects the demand for photosynthesis in tree species, and then affects photosynthesis rate. Stomatal conductance has a great influence on the CO_2 concentration in the intercellular space. As the salt increases, the measured values of stomatal conductance and intercellular concentration decrease sequentially, and the daily change amplitude gradually decreases (Figure 2D). This revealed that the harmony of all varieties is seriously inhibited with the increase of salt stress intensity. These results indicated that the Jisang3 seedlings have a better adaptability to salt stress environment.

2.2. Differentially Regulated Proteins (DRPs) in Mulberry Roots

A differential regulated protein (DRP) expressions analysis was performed to determine the responses of mulberry to salt stress and to identify specific strategies of mulberry survival in saline-alkali soil. In total, 1,166,253 mass spectra were produced by mass spectrometer and 162,897 mass spectra were matched with alignment protein, 82,899 peptides in which spectra hits were found and 7044 proteins were detected by spectrum search analysis (Table S1).

In salt-sensitive variety Guisangyou12 after salt stress, there were 255 upregulated DRPs and 532 downregulated DRPs with the fold-change higher than 1.2 (Table 1). With the fold-change higher than 1.5, there were 71 upregulated DRPs and 117 downregulated DRPs in Guisangyou12. These data indicate that a great number of proteins were depressed in salt-sensitive variety by salt stress. In salt-tolerant variety Jisang3, there were 412 upregulated DRPs and 288 downregulated DRPs after salt stress, with the fold-change higher than 1.2 (Table 1). With the fold-change higher than 1.5, there were 68 upregulated DRPs and 64 downregulated DRPs. These data indicate that a large number of proteins were activated in salt-tolerant variety by salt stress. Furthermore, a comparative analysis between Jisang3 and Guisangyou12 showed that 600 DRPs were upregulated and 619 DRPs were downregulated under salt stress, with fold-change higher than 1.2 (Table 1). With the fold-change up to 1.5, there were 169 upregulated DRPs and 165 downregulated DRPs under salt stress.

In order to analyze the differential responsive pattern between the salt-sensitive variety and salt-tolerant variety, we created a Venn diagram of the DRPs with fold-change higher than 1.3. As shown in Figure 3, 94 upregulated proteins were unique in G12 group during salt stress and 59 upregulated proteins were unique in G12 group. After salt treatment, 221 proteins were upregulated between two varieties. Among the upregulated proteins, there were only 2 DRPs shared between the two varieties. For the downregulated proteins, there were 135 unique proteins in G12 group and 83 unique proteins in J group, and only one DRP shared between the two varieties. These results indicated that the salt-sensitive mulberry variety and salt-tolerant mulberry variety have different protein profiles in response to salt stress.

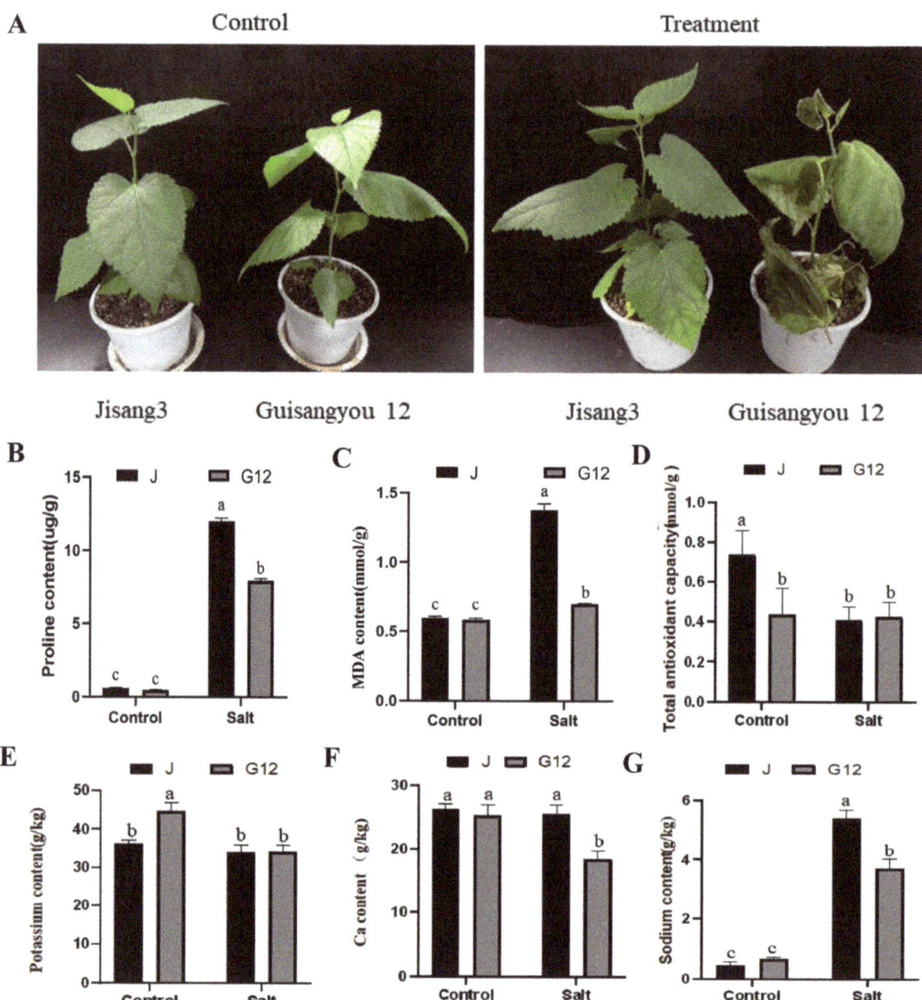

Figure 1. Phenotypic and physiological characteristics of mulberry varieties Jisang3 and Guisangyou12 under control and saline conditions. (**A**) Performance of Jisang3 and Guisangyou12 plants under control and saline conditions. (**B**) Proline content of root in both J and G12 groups. (**C**) MDA content of root in both J and G12 groups. (**D**) Total antioxidant capacity of root of the two varieties in saline condition. (**E–G**) Ion content (potassium, calcium, sodium) of root of mulberry varieties Jisang3 and Guisangyou12 under control and saline conditions. Data are means ± SD with three biological replicates. Different letters represent statistically significant differences between control and salt treated plants by Duncan's multiple range test ($p < 0.05$). Jck, JS, G12ck and G12S represent the two states of Jisang3 and Guisangyou12 under control and salt stress, respectively.

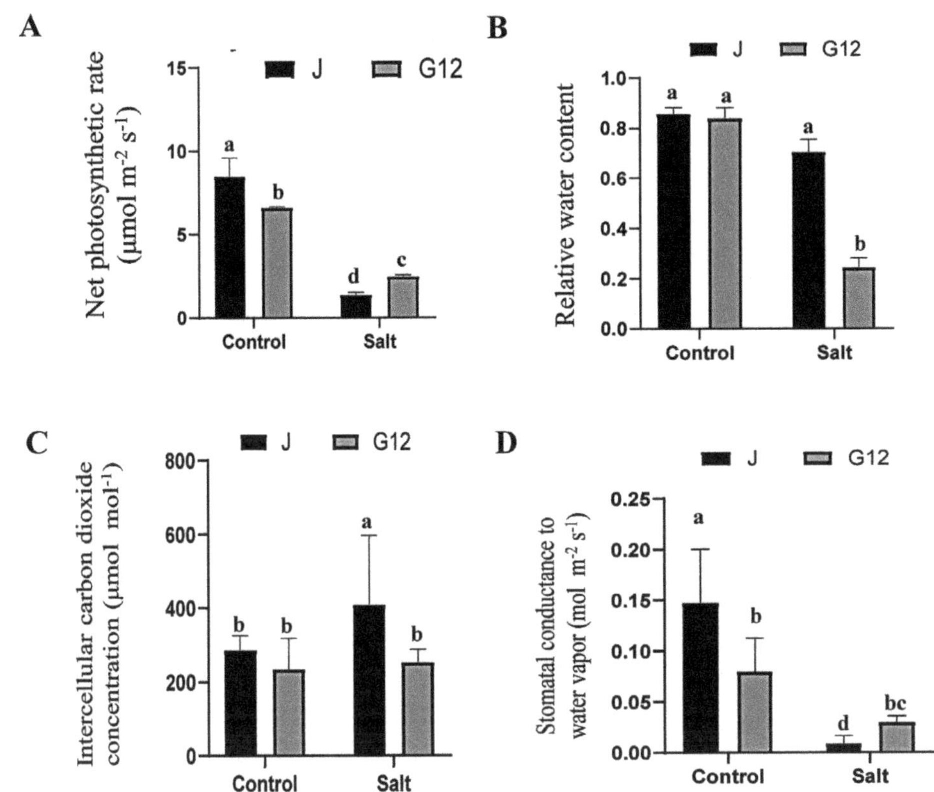

Figure 2. Physiological responses of mulberry to salt stress. (**A**) Net photosynthetic rate of mulberry leaves in both J and G12 groups. (**B**) Relative water content in both J and G12 groups. (**C**) Intercellular carbon dioxide concentration in both J and G12 groups. (**D**) Stomatal conductance to water vapor in both J and G12 groups. Data are means ± SD with three biological replicates. Different letters represent statistically significant differences between control and salt treated plants by Duncan's multiple range test ($p < 0.05$). Jck, JS, G12ck and G12S represent the two states of Jisang3 and Guisangyou12 under control and salt stress, respectively.

Table 1. Differentially expressed protein summary (filtered with threshold value of expression fold change and p value < 0.05). Jck, JS, G12ck and G12S represent the two states of Jisang3 and Guisangyou12 under control and salt stress, respectively.

Compare Group	Regulated Type	Fold Change > 1.2	Fold Change > 1.3	Fold Change > 1.5	Fold Change > 2
G12S/G12ck	Up-regulated	255	147	71	24
	Down-regulated	532	320	117	15
JS/G12S	Up-regulated	600	402	169	29
	Down-regulated	619	396	165	51
JS/Jck	Up-regulated	412	218	68	11
	Down-regulated	288	162	64	21
Jck/G12ck	Up-regulated	380	232	113	35
	Down-regulated	686	375	135	21

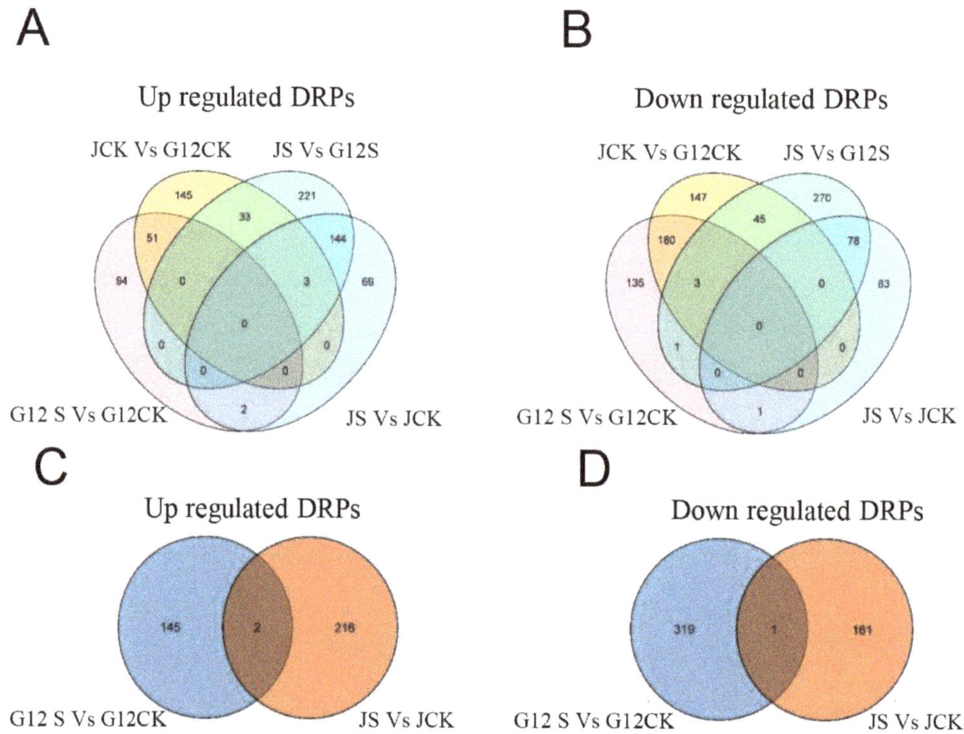

Figure 3. Analysis of Venn diagram. (**A**,**B**) Venn diagram of differentially expressed proteins identified at different treatments. (**C**,**D**) Venn diagram of differentially expressed proteins identified between Jisang3 and Guisangyou12 in the saline condition.

We further selected DRPs with fold-change higher than 1.5 to build a heatmap (Figure 4 and Table S2). Most of them were enzymes that participated in the metabolism. Additionally, these proteins were mostly downregulated in Guisangyou12 but upregulated in Jisang3 after salt stress.

2.3. GO Terms and KEGG Pathway in Mulberry Response to Salt Tolerance

To further understand the function of DRPs, these DRPs were enriched in the clusters of orthologous groups of proteins (COG/KOG) category (Figure S2). On the whole, DRPs involved in salt stress were classified into processes including posttranslational modification, protein turnover, chaperones, amino acid transport and metabolism, lipid transport and metabolism, carbohydrate transport and metabolism. The upregulated proteins were classified into processes such as signal transduction mechanisms, intracellular trafficking, secretion, vesicular transport cytoskeleton and cytoskeleton, while the downregulated proteins were classified into processes such as amino acid transport and metabolism, posttranslational modification, protein turnover, chaperones, secondary metabolites biosynthesis, transport and catabolism, lipid transport and metabolism. In Guisangyou12, the upregulated proteins were enriched in the processes including energy production and conversion, posttranslational modification, protein turnover, chaperones, amino acid transport and metabolism, while the downregulated proteins were enriched in the processes including carbohydrate transport and metabolism, intracellular trafficking, secretion, and catabolism. In Jisang3, the upregulated proteins were enriched in processes including signal transduction mechanisms, intracellular trafficking secretion and vesicular transport, cytoskeleton, while downregulated proteins were enriched in processes such as amino acid transport and metabolism, posttranslational modification, protein turnover, chaperones, secondary metabolites biosynthesis transport and catabolism.

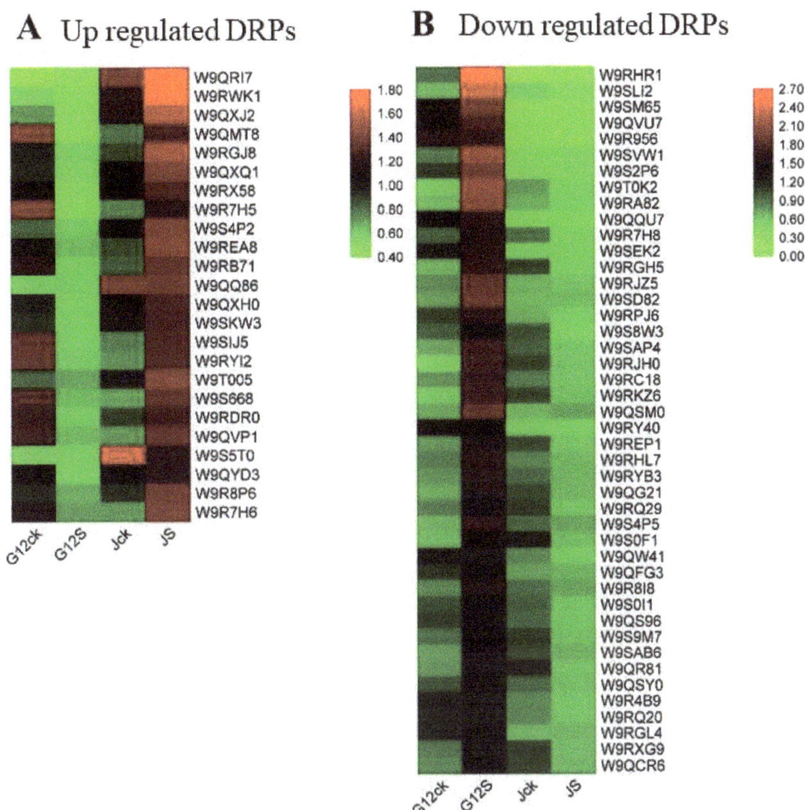

Figure 4. Heatmap of differentially expressed proteins identified in mulberry roots after NaCl treatment. Fold change >1.5. Red represents a high gene expression level and green represents a low gene expression level.

To further understand the function of these proteins from a pathway-specific perspective, we subsequently subjected the data to the Kyoto Encyclopedia of Genes and Genomes (KEGG) enrichment (Figure S3). In Guisangyou12, the glucosinolate biosynthesis had significantly upregulated expression under salt stress. The synthesis and degradation of amino acids, beta-alanine metabolism, lysing degradation, valine, leucine and isoleucine degradation and fatty acid degradation were also changed under salt stress. The downregulated DRPs were mostly enriched in phenylpropanoid biosynthesis and steroid biosynthesis. In Jisang3, the upregulated DRPs were enriched in phagosome, pyrimidine and endocytosis, while the downregulated DRPs were enriched in the carbon metabolism, glycolysis/gluconeogenesis and arginine and proline metabolism. In sum, a large number of proteins involved in phenylpropanoid biosynthesis were upregulated while DRPs involved in the carbon metabolism were downregulated. This provides important clues to elucidate the salt tolerant mechanism in mulberry.

Subcellular localization analysis was also performed to understand the localization of the DRPs. As shown in Figure S4, most of the DRPs were located in the nucleus, cytoplasm, plasma membrane and chloroplast, some proteins were located in mitochondria, cytoskeleton, and vacuolar membrane, were also participated in the salt stress response. In the salt-tolerant variety Jisang3, protein transport into nucleus occurred when cytoplasm proteins and chloroplast proteins were degraded. However, in the salt-sensitive variety Guisangyou12, proteins were still processing in cytoplasm and chloroplast, and large amounts of proteins in chloroplast, nucleus and cytoplasm were downregulated. Combined with the physiological and biochemical data (Figure S4), we inferred that DRPs in the salt-tolerant variety were degraded for the synthesis of osmotic stress resistant materials while the salt-sensitive variety could keep a high level of photosynthesis and therefore exhibit salt-stressed phenotype quickly.

2.4. Biological Process in Mulberry Response to Salt Stress

After the DRPs in different comparison groups were classified by GO and KEGG pathway, we further performed cluster analysis on these proteins to find the correlation of the DRPs in the comparison groups (Figures S5 and S6). According to the *p*-value of Fisher's exact test obtained by enrichment analysis, the relevant functions in different groups are grouped together using the hierarchical clustering method and drawn as a heatmap. Different groups of DRPs and color blocks corresponding to the functional description indicate the degree of enrichment. Together with these results, we found some interesting pathways including amino acid synthesis, reactive oxygen scavenging activity and phenylpropanoid biosynthesis, may contribute to salt tolerance. We further analyzed DRPs involved in these pathways, listed in Tables S3–S5. In Guisangyou12, 37 DRPs were involved in amino acid synthesis such as arginine, glycine, serine, valine and tryptophan metabolism and 8 DRPs were involved in glutathione metabolism which were all upregulated, only 10 DRPs related to phenylpropanoid biosynthesis were downregulated during salt stress. In Jisang3, only downregulated DRPs were enriched in these pathways, and 31 DRPs involved in amino acid synthesis and 4 DRPs involved in glutathione metabolism. For comparison of the two varieties after salt stress, 10 DRPs involved in phenylpropanoid biosynthesis were highly upregulated and 90 DRPs involved in amino acid metabolism were downregulated, and another 9 DRPs involved in glutathione metabolism, 14 DRPs involved in peroxisome metabolism and 12 DRPs involved in phenylpropanoid biosynthesis were all downregulated. In a comparison of the two varieties, we found that fatty acid metabolic process (e.g., long-chain fatty acid metabolic process, fatty acid beta-oxidation, fatty acid oxidation, fatty acid catabolic process, monocarboxylic acid catabolic process and carboxylic acid catabolic process) and hyperosmotic salinity response were significantly enriched (Figure S6). Taken together, these results suggest that phenylpropanoid biosynthesis may contribute to salt tolerance in mulberry.

2.5. Phenylpropanoid Biosynthesis in Mulberry Response to Salt Stress

Phenylpropanoids contribute to all aspects of plant responses to biotic and abiotic stimuli [28]. We found that phenylpropanoid biosynthesis pathway is responsive to salt stress (Figure 5A). Aldehyde dehydrogenase were downregulated in Jisang3 and upregulated in Guisangyou12; caffeic acid 3-O-methyltransferase and quinate hydroxycinnamoyl transferase were induced in Jisang3 and depressed in Guisangyou12, while the peroxidase showed a diametrically exposed trend (Figure 5A). The protein abundance of glycosyl transferase was changed irregularly, while reticuline oxidase and vinorine synthase showed a stable expression pattern under salt stress (Figure 5A). As shown in Figure 5B, the DRPs involved in phenylpropanoid biosynthesis were further analyzed. The expressions of aldehyde dehydrogenase, glycosyltransferase, peroxidase, cinnamyl alcohol dehydrogenase were all depressed in Jisang3 but induced in Guisangyou12. Phenylpropanoids including phenol, phenylpropanol, phenylpropionic acid are natural ingredients. Its condensates, leggins, lignans, lignins, and phenylpropanoids are mostly biosynthesized through the shikimic acid pathway [29]. We further measured the contents of coumarin, lignins and flavonoid through high performance liquid chromatography (HPLC). As shown in Figure S7, the coumarin and flavonoid were all induced in Guisangyou12 under salt stress and were correlated with our inference. These results indicated that phenylpropanoids may be involved in salt response and can be used as a target to improve salt resistance in mulberry.

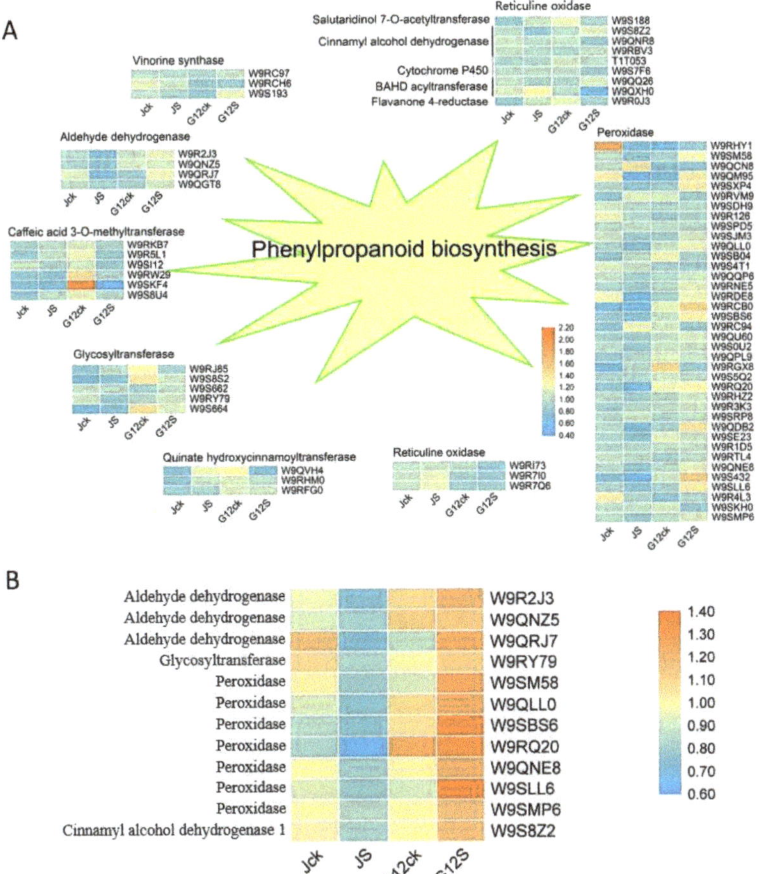

Figure 5. Heatmap of proteins involved in phenylpropanoid biosynthesis pathway. (**A**) Heatmap of proteins identified in phenylpropanoid biosynthesis pathway. (**B**) Heatmap of DRPs involved in phenylpropanoid biosynthesis pathway. Red represents a high gene expression level and blue represents a low gene expression level.

2.6. Post-Transcription and Translation Show a Different Pattern in Salt-Tolerant and Salt-Sensitive Mulberry during Early Salt Responsiveness

To further characterize the transcriptome profiles under salt stress, genes involved in salt stress in the two varieties were further screened through RNA-Seq. There were 24,448 genes identified and 4009 genes were both annotated in transcriptome and proteome (Figure S8). The transcriptomic and proteomic data describe the expression of genes at the transcription and translation levels, respectively. In addition to the one-to-one correspondence between transcriptome and proteome, there are other more complex regulatory relationships. Shown in Figure 6A is a scatter plot between the sample's transcript and its corresponding protein expression. By comparing the quantitative correlation between the two omics, we found that there was a positive regulatory relationship between the protein and the transcript in Guisangyou12 under salt stress while there was a negative regulatory relationship in Jisang3. We further performed the GSEA analysis (Figure 6B) based on KEGG pathway through the above overall analysis of the quantitative correlation between the transcriptome and the proteome. The result showed that there was obvious positive regulatory relationship between the transcriptome and the proteome such as plant hormone signal transduction and phenylpropanoid biosynthesis. Additionally, there were still some negative regulatory relationships between the protein and the transcript such as oxidative phosphorylation, carbon metabolism, phosphatidylinositol signaling system and proteasome. In order to explore the potential relationship between gene transcription level and protein level under multiple experimental conditions, the protein group and transcriptome expression data sets were merged. Then, we used the hclust "ward.D" method to cluster the expression of genes in two dimensions (Figures S9 and S10). The abscissa is the sample in the quantitative study of the proteome and transcriptome, and the ordinate is the protein or transcript quantified in the two omics at the same time. Through the hclust clustering method, these proteins or transcripts were divided into six categories. There is a specific relationship between protein and transcript expression in each category. Preparation of whole transcriptome libraries and deep sequencing were performed by the Annoroad Gene Technology Corporation (Beijing, China).

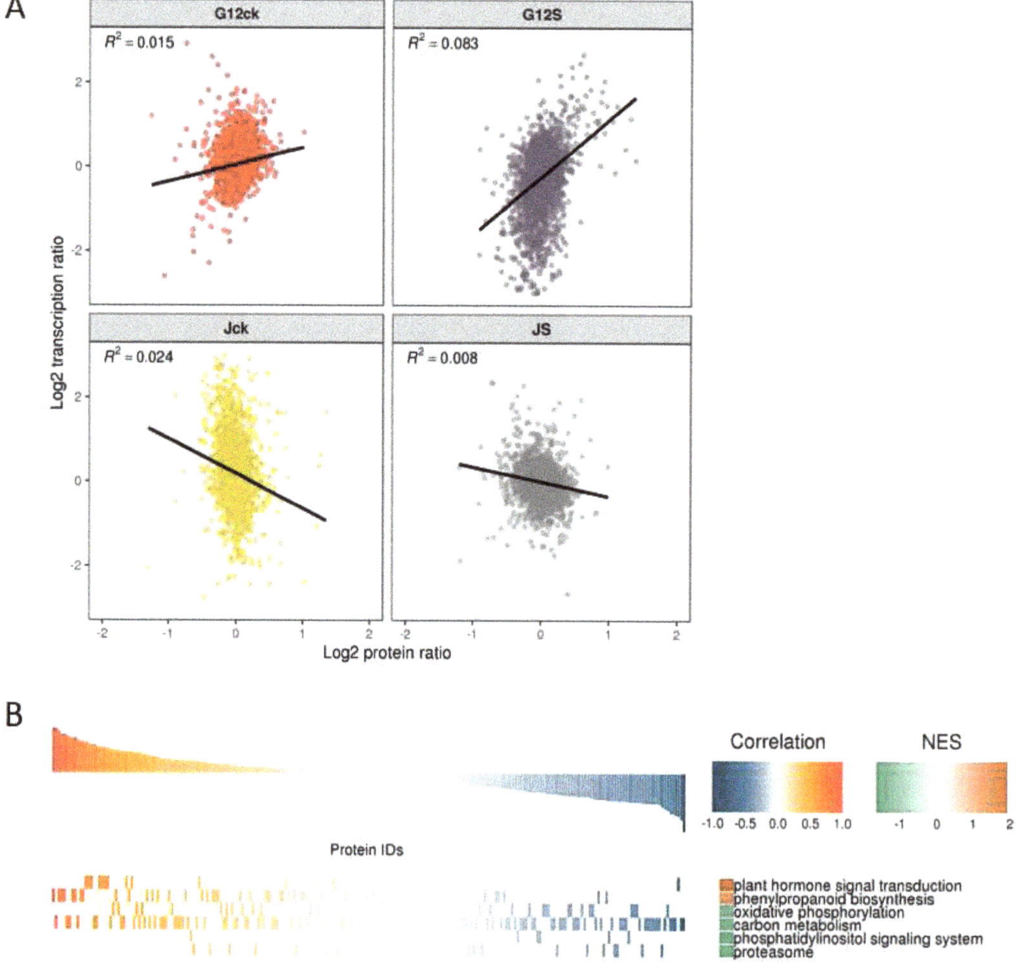

Figure 6. Combined analysis of proteome and transcriptome. (**A**) Scatter plot of transcripts and their corresponding proteins. (**B**) GSEA analysis of KEGG pathway based on quantitative correlation coefficients of transcriptome and proteome.

3. Discussion

Soil salinity is a complex environmental issue threat to plant growth and crop production. Mulberry has the characteristics of drought tolerance, salt tolerance, barren tolerance, and strong adaptability to soil [18]. Proteins have a crucial role in stress response and modulate physiological characteristics to develop different phenotypes. In this study, we selected salt-sensitive variety Guisangyou12 and salt-tolerant variety Jisang3 for physiological and proteomic analyses to uncover the mechanism of salt stress adaptability. We found 787 DRPs in Guisangyou12 and 700 DRPs in Jisang3, through the analysis of proteomics data. We found that phenylpropanoid biosynthesis pathway may play important roles in salt tolerance of mulberry. These results may contribute to the genetic improvement of plants for saline-alkali soil management and mulberry stress resistance in genetic engineering.

One important strategy for plants to improve salt tolerance is to re-establish the ionic homeostasis [30]. Stomata are another factor involved in photosynthesis, gas exchange of leaf and microbe interaction [31–33]. Studies have also shown that salt stress inhibits the growth and development of mulberry trees [18,34,35]. Proline responds to oxidative processes, protects plants from damage, and acts as a protein stabilizer [36–38]. MDA naturally produced form lipid peroxidation and MDA content is used as a parameter to measure the degree of plant cell damage [39,40]. Proline, MDA, were evaluated and upregulated after salt treatment while antioxidant ability reduced in Jisang3 and was unchanged in Guisangyou12. Under salt stress, the content of osmotic adjustment substances such as proline and soluble sugar in mulberry cells increases, which is conducive to its adaptation to the external environment. Ionic homeostasis including potassium, sodium and calcium were detected and sodium content was significantly increased during salt stress. The mulberry Na^+/H^+ antiporter NHX1 can have a positive role in salt stress response [41]. Transformation of the *NHX1* gene and the pyrophosphatase gene *AVP1* into Arabidopsis can obtain high salt-tolerant transgenic lines [42]. Net photosynthetic rate was significantly reduced, and the photosynthetic efficiency was higher in the salt-sensitive cultivar (Guisangyou12). Stomata guard gas exchange and enable CO_2 entry into the leaf for photosynthesis [43,44]. Stomata have a role in water transportation [32,45]. Stomatal conductance balances water–gas exchange and photosynthesis [46]. CO_2 concentration is also directly related to saline stress and photosynthesis [47]. Stomatal conductance and net photosynthetic rate decreased with the increase of salt content. The intercellular carbon dioxide concentration was increased in Jisang3 while changed little in Guisangyou12. Mulberry trees are salt-tolerant varieties have higher water use efficiency and intercellular CO_2 concentration than salt-sensitive cultivars [48], but net photosynthetic rate and stomatal conductance of mulberry are lower. Studies have shown that photosynthesis in mulberry trees shows a downregulated trend with the increase of salt concentration [48]. The salt-tolerant mulberry Jisang3 has lower photosynthesis and stomatal conductance, elevates CO_2 to improve water use efficiency thus avoiding the harm of salt stress. In barley (*Hordeum vulgare*) leaves, elevated CO_2 partially reduces the impact of salinity on photosynthesis [49]. Comparing the two materials, we found that sodium transfer and ROS scavenge activity were involved in salt stress response. These observations indicate that Jisang3 has a better adaptation to saline conditions than Guisangyou12 plants.

To further elucidate the molecular network response to salt stress between these two varieties, the proteomic data of Jisang3 and Guisangyou12 plants under salt stress was analyzed. We found that 532 of 787 DRPs in Guisangyou12 were downregulated while 412 of 700 DRPs in Jisang3 were upregulated, this indicated that proteins in salt-sensitive plants were depressed while in salt-tolerant plants were induced during salt stress. In Guisangyou12, DRPs were enriched in the glucosinolate biosynthesis, beta-alanine metabolism, lysing degradation, valine, leucine and isoleucine degradation and fatty acid degradation; the detoxification processes such as peroxisome and glutathione metabolism were upregulated in the salt condition. In Jisang3, DRPs related to phagosome, pyrimidine, endocytosis, carbon metabolism, glycolysis/gluconeogenesis and arginine and proline metabolism were enriched after salt treatment. Comparing the two materials, we found that large amounts of proteins involved in much of secondary metabolic pathways were upward expressed including phenylpropanoid biosynthesis, isoflavonoid biosynthesis, cutin, suberine and wax biosynthesis, porphyrin and chlorophyII metabolism, flavonoid biosynthesis, glycerophospholipid metabolism, stilbenoid, diarylheptanoid and gingerol biosynthesis. The pathways such as alanine, aspartate and glutamate metabolism, alpha-linolenic acid metanolism, valine, leucine and isoleucine degradation, 2-oxocarboxylic acid metabolism were enriched in downregulated DRPs. The metabolism related to phenylpropanoid biosynthesis was also marked in Figures S5 and S6. The degradation of fatty acids was in accordance with the increased MDA content. In the previous studies, Luo et al. compared proteomics of two maize sister lines after salt treatment and found that proteins related to phenylpropanoid biosynthesis, phagosome, endocytosis, galactose metabolism, starch and sucrose metabolism, and oxidative phosphorylation were downregulated in the salt sensitive line and oxygen-dependent pentose phosphate pathway, glutathione metabolism and nitrogen metabolism were enhanced [50]. Meng et al. also explored the different responses in sweet potato through high-throughput sequencing in a saline condition and found that ion accumulation, stress signaling, transcriptional regulation, redox reactions, plant hormone signal transduction, and secondary metabolite accumulation may be the response of salt tolerance genotypes [51]. In the analysis of salt treatment to two varieties of *Origanum majorana*, the Tunisian *O. majorana* plants developed tolerance to salinity by improved the process of galactosylation of quercetin into quercetin-3-galactoside and quercetin-3-rhamnoside [52]. Quantitative proteomic analyses of seedling roots from salt-sensitive and salt-tolerant maize were performed by using the iTRAQ method, phenylpropanoid biosynthesis, starch and sucrose metabolism, and the mitogen-activated protein kinase (MAPK) signaling pathway were enriched in salt-tolerant maize while only the nitrogen metabolism pathway was enriched in salt-sensitive maize [53]. Taken together, we found that synthesis and degradation of amino acids, ROS response, carbon metabolism and phenylpropanoid biosynthesis were highly related to salt response. On account of the phenylpropanoids being indicators and key mediators of plant responses to biotic and abiotic stimuli [28,29,54,55], we further analyzed the DRPs involved in phenylpropanoid biosynthesis pathway.

Phenylpropane metabolism is one of the most important plant secondary metabolic pathways, producing more than 8000 metabolites, which play an important role in plant growth and development and plant environmental interaction. Terrestrial plants have evolved a cultivar of branch pathways of phenylpropane metabolism, producing various metabolites such as flavonoids, lignin, lignans, and cinnamic acid amides [50]. Phenylpropane metabolism starts with the phenylalanine produced by the shikimate pathway and undergoes a series of enzymatic reactions to produce secondary metabolites. It is composed of Phenylalanine ammonia-lyase (PAL), Cinnamate 4-hydroxylase (C4H), and 4-coumarate: CoA ligase (4CL) catalyzes to form *p*-coumaric acid: Coenzyme A, which provides precursors for different branches of downstream metabolic pathways (Figure 7). In this study, Aldehyde dehydrogenase was downregulated in Jisang3 and upregulated in Guisangyou12; Caffeic acid 3-*O*-methyltransferase and quinate hydroxycinnamoyltransferase were induced in Jisang3 and degraded in Guisangyou12, while the peroxidase showed a diametrically opposite trend; Glycosyltransferase changes irregularly; Reticuline oxidase and vinorine synthase showed a stable expression pattern under salt stress. Aldehyde dehydrogenase, glycosyltransferase, peroxidase, cinnamyl alcohol dehydrogenase in DRPs were all degraded in Jisang3 and induced in Guisangyou12. Glycosyltransferase including O-glycosyltransferase (OGT) and C-glycosyltransferase (CGT) are related to the production of flavones [56,57]. Cinnamyl alcohol dehydrogenase (CAD) is the final step of monolignol to become caffeyl alcohol, coniferyl alcohol, 5H coniferyl alcohol, and sinapyl alcohol [54]. Peroxidase catalyzes the oxidation of phenylpropanoids to their phenoxyl radicals, and the subsequent nonenzymatic coupling controls the pattern and extent of polymerization [58]. Secondary metabolites explored by HPLC also confirm our conclusion.

Abundant metabolites produced from phenylpropanoid biosynthesis pathway coupled with high antioxidant capacity endow mulberry trees with salt tolerance peculiarity.

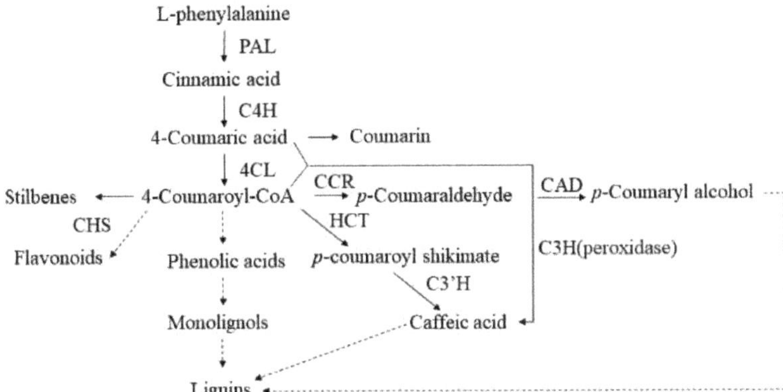

Figure 7. A part scheme of phenylpropanoid metabolism. 4CL, 4-coumarate-CoA ligase; C3H, coumarate 3-hydroxylase; C3′H, p-coumaroyl shikimate 3′ hydroxylase; C4H, cinnamic acid 4-hydroxylase; CAD, cinnamyl alcohol dehydrogenase; HCT, Hydroxycinnamoyl-CoA shikimate/quinate hydroxycinnamoyl transferase; CCR, cinnamoyl-CoA reductase; CHS, chalcone synthase; PAL, phenylalanine ammonia-lyase.

Comprehensive transcriptome and proteome analyses showed a different expression pattern. A large number of genes showed consistency between the transcript and protein levels in salt-sensitive mulberry Guisangyou12 while salt-tolerant mulberry Jisang3 showed an opposite effect. The integrative transcriptomic and proteomic data are important in deciphering the molecular processes involved in salt stress. Genes and proteins involved in plant hormone signal transduction and phenylpropanoid biosynthesis showed a positive correlation in salt response. Similar to previous studies [53,59,60], the metabolic pathways of plant hormone signal transduction, carotenoid biosynthesis, flavonoid biosynthesis, and starch and sucrose metabolism were involved in salt stress.

4. Materials and Method

4.1. Plant Material

Two mulberry varieties, Jisang3 (J) and Guisangyou12 (G12) seeds were used in this study. Mulberry seeds were disinfected in 1% $HgCl_2$ for 15 min and then soaked in distilled water for 24 h for germination. The mulberry young seedlings were transplanted to sterilized soil substrate for 2 months with a 14 h/10 h (day/night), 25/22 °C (day/night) and 75% air humidity condition. The 21d plants were treated with 200 mM NaCl for two days to impose salinity stress according to the previous study [18,61]. The control group was watered normally. After 10 days, the leaves and roots from each mulberry plant were sampled for physiological analysis and part of them were then stored at −80 °C for TMT-labeled proteomics, respectively.

4.2. Physiological Analysis

Photosynthesis of the leaf and the water content was tested. Net photosynthetic rate, intercellular carbon dioxide concentration and stomatal conductance were measured by a portable photosynthesis system (LI-6800; LI-COR, Lincoln, NE, USA). For leaf chamber environment, chamber temperature, relative humidity and CO_2 concentration were set at 25 °C, 70%, and 400 μmol $mol^{-1}{\cdot}s^{-1}$, each experiment was repeated 10 times. MDA content was analyzed through the TAB method (Banga & Lengyel, 1980) with MDA assay kit (sigma). Proline content, antioxidant capacity and ion content were measured as previously described [62] with test kits (Jiancheng Bioengineering Institute, Nanjing,

China). Element content analysis was performed using inductively coupled plasma-optical emission spectrophotometry ICP-OES (ICAP 7000, Thermo Scientific, Waltham, MA, USA).

4.3. Protein Extraction, Digestion, and TMT Labeling

Samples of the roots from each mulberry plant were used for protein extraction (n = 3), trypsin digestion and TMT labeling (Thermo Scientific, Waltham, MA, USA). The protein solution was reduced with 5 mM dithiothreitol for 30 min at 56 °C and alkylated with 11 mM iodoacetamide for 15 min at room temperature in darkness. The protein sample was then diluted by adding 100 mM Triethylamonium bicarbonate (TEAB) to urea concentration less than 2 M. Finally, trypsin was added at 1:50 trypsin-to-protein mass ratio for the first digestion overnight and 1:100 trypsin-to-protein mass ratio for a second 4 h digestion. After trypsin digestion, peptide was desalted by Strata X C18 SPE column (Phenomenex) and vacuum dried. Peptide was reconstituted in 0.5 M TEAB and processed according to the manufacturer's protocol for TMT kit. Briefly, one unit of TMT reagent was thawed and reconstituted in acetonitrile. The peptide mixtures were then incubated for 2 h at room temperature and pooled, desalted and dried by vacuum centrifugation. The tryptic peptides were fractionated into fractions by high pH reverse-phase HPLC using Thermo Betasil C18 column (5 μm particles, 10 mm ID, 250 mm length). The tryptic peptides were treated and subjected to NSI source followed by tandem mass spectrometry (MS/MS) in Q Exactive™ Plus (Thermo Scientific, Waltham, MA, USA) coupled online to the UPLC. A data-dependent procedure alternated between one MS scan followed by 20 MS/MS scans with 15.0 s dynamic exclusion.

4.4. Data Processing, Protein Identification, and Quantification

GO Annotation: The Gene Ontology, or GO, is a major bioinformatics initiative to unify the representation of gene and gene product attributes across all species. More specifically, the project aims to maintain and develop its controlled vocabulary of gene and gene product attributes; annotate genes and gene products, and assimilate and disseminate annotation data; provide tools for easy access to all aspects of the data provided by the project. Gene Ontology (GO) annotation proteome was derived from the UniProt-GOA database (http://www.ebi.ac.uk/GOA/ accessed on 20 July 2020). Identified proteins domain functional description were annotated by InterProScan (a sequence analysis application) based on protein sequence alignment method, and the InterPro domain database was used. InterPro (http://www.ebi.ac.uk/interpro/ accessed on 20 July 2020) is a database that integrates diverse information about protein families, domains, and functional sites, and makes it freely available to the public via Web-based interfaces and services. Central to the database are diagnostic models, known as signatures, against which protein sequences can be searched to determine their potential function. InterPro has utility in the large-scale analysis of whole genomes and meta-genomes, as well as in characterizing individual protein sequences.

KEGG Pathway Annotation: KEGG connects known information on molecular interaction networks, such as pathways and complexes (the "Pathway" database), information about genes and proteins generated by genome projects (including the gene database) and information about biochemical compounds and reactions (including compound and reaction databases). These databases are different networks, known as the "protein network", and the "chemical universe", respectively. There are efforts in progress to add to the knowledge of KEGG, including information regarding ortholog clusters in the KEGG Orthology database. There, we used wolfpsort, a subcellular localization predication tool, to predict subcellular localization. Wolfpsort is an updated version of PSORT/PSORT II for the prediction of eukaryotic sequences. Special for protokaryon species, subcellular localization prediction tool CELLO was used.

4.5. Functional Enrichment

Enrichment of Gene Ontology analysis: Proteins were classified by GO annotation into three categories: biological process, cellular compartment, and molecular function. For each category, a two-tailed Fisher's exact test was employed to test the enrichment of the differentially expressed protein against all identified proteins. The GO with a corrected p-value < 0.05 is considered significant. Enrichment of pathway analysis: Encyclopedia of Genes and Genomes (KEGG) database was used to identify enriched pathways by a two-tailed Fisher's exact test to test the enrichment of the differentially expressed protein against all identified proteins. The pathway with a corrected p-value < 0.05 was considered significant. These pathways were classified into hierarchical categories according to the KEGG website. Enrichment of protein domain analysis: For each category proteins, InterPro (a resource that provides functional analysis of protein sequences by classifying them into families and predicting the presence of domains and important sites) database was researched and a two-tailed Fisher's exact test was employed to test the enrichment of the differentially expressed protein against all identified proteins. Protein domains with a corrected p-value < 0.05 were considered significant.

4.6. Enrichment-Based Clustering

For further hierarchical clustering based on differentially expressed protein functional classification (such as GO, Domain, Pathway, Complex). We first collated all the categories obtained after enrichment along with their p-values, and then filtered for those categories which were at least enriched in one of the clusters with p-value < 0.05. This filtered p-value matrix was transformed by the function x = −log10 (p value). Finally, these x values were z-transformed for each functional category. These z scores were then clustered by one-way hierarchical clustering (Euclidean distance, average linkage clustering) in Genesis. Cluster membership was visualized by a heat map using the "heatmap.2" function from the "gplots" R-package.

5. Conclusions

Collectively, in order to clarify the molecular mechanism of mulberry salt tolerance, comparative proteomic analyses were performed based on the phenotypic, physiological differences in the roots of salt-tolerant and sensitive mulberry varieties after salt treatment. The tolerant and sensitive mulberry genotypes respond differently to salt stress, large amounts DRPs were detected in the two cultivars. Further, our results showed that phenylpropanoid biosynthesis and ROS scavenging system may facilitate the salt tolerance and can be used as the target of genetic breeding. In summary, our results provide a reference for the molecular mechanism for salt condition and reveal the different response pattern between genotypes.

Supplementary Materials: The following are available online at https://www.mdpi.com/article/10.3390/ijms22179402/s1.

Author Contributions: Conceptualization, T.G., C.S. and Y.Q.; methodology, T.G. and M.Z.; formal analysis, T.G., Z.L. and L.B.; investigation, T.H., X.C., Y.H. and H.W.; resources, C.S. and F.J.; writing—original draft preparation, T.G.; writing—review and editing, M.Z. and Y.Q.; funding acquisition, C.S. and Y.Q. All authors have read and agreed to the published version of the manuscript.

Funding: This research was funded by Science and Technology Innovation Project of Shaanxi Academy of Forestry (No.SXLK2020-0211), Modern Agricultural Industry Technology System (CARS-18), and Key Promotion Project of Northwest A& F University (XTG2018-29).

Institutional Review Board Statement: Not applicable.

Informed Consent Statement: Not applicable.

Data Availability Statement: The data presented in this study are available in the article and supplementary material.

Conflicts of Interest: The authors declare no conflict of interest.

References

1. Fedoroff, N.V.; Battisti, D.S.; Beachy, R.N.; Cooper, P.J.; Fischhoff, D.A.; Hodges, C.N.; Knauf, V.C.; Lobell, D.; Mazur, B.J.; Molden, D.; et al. Radically rethinking agriculture for the 21st century. *Science* **2010**, *327*, 833–834. [CrossRef] [PubMed]
2. Shabala, S.; Bose, J.; Fuglsang, A.T.; Pottosin, I. On a quest for stress tolerance genes: Membrane transporters in sensing and adapting to hostile soils. *J. Exp. Bot.* **2016**, *67*, 1015–1031. [CrossRef] [PubMed]
3. Zhu, J.K. Abiotic Stress Signaling and Responses in Plants. *Cell* **2016**, *167*, 313–324. [CrossRef] [PubMed]
4. Chaves, M.M.; Pereira, J.S.; Maroco, J.; Rodrigues, M.L.; Ricardo, C.P.; Osório, M.L.; Carvalho, I.; Faria, T.; Pinheiro, C. How plants cope with water stress in the field. Photosynthesis and growth. *Ann. Bot.* **2002**, *89*, 907–916. [CrossRef] [PubMed]
5. Conde, A.; Chaves, M.M.; Gerós, H. Membrane transport, sensing and signaling in plant adaptation to environmental stress. *Plant Cell Physiol.* **2011**, *52*, 1583–1602. [CrossRef] [PubMed]
6. Yamaguchi, T.; Hamamoto, S.; Uozumi, N. Sodium transport system in plant cells. *Front. Plant Sci.* **2013**, *4*, 410. [CrossRef] [PubMed]
7. Deinlein, U.; Stephan, A.B.; Horie, T.; Luo, W.; Xu, G.; Schroeder, J.I. Plant salt-tolerance mechanisms. *Trends Plant Sci.* **2014**, *19*, 371–379. [CrossRef]
8. Zhao, C.; Zhang, H.; Song, C.; Zhu, J.K.; Shabala, S. Mechanisms of Plant Responses and Adaptation to Soil Salinity. *Innovation* **2020**, *1*, 100017. [CrossRef]
9. Wani, S.H.; Kumar, V.; Khare, T.; Guddimalli, R.; Parveda, M.; Solymosi, K.; Suprasanna, P.; Kavi, K.P. Engineering salinity tolerance in plants: Progress and prospects. *Planta* **2020**, *251*, 76. [CrossRef]
10. Sharif, I.; Aleem, S.; Farooq, J.; Rizwan, M.; Younas, A.; Sarwar, G.; Chohan, S.M. Salinity stress in cotton: Effects, mechanism of tolerance and its management strategies. *Physiol. Mol. Biol. Plants* **2019**, *25*, 807–820. [CrossRef]
11. Qin, H.; Li, Y.; Huang, R. Advances and challenges in the breeding of salt-tolerant rice. *Int. J. Mol. Sci.* **2020**, *21*, 8385. [CrossRef]
12. Chen, R.; Cheng, Y.; Han, S.; Van Handel, B.; Dong, L.; Li, X.; Xie, X. Whole genome sequencing and comparative transcriptome analysis of a novel seawater adapted, salt-resistant rice cultivar-sea rice 86. *BMC Genom.* **2017**, *18*, 655. [CrossRef]
13. Mirdar, M.R.; Shobbar, Z.S.; Babaeian, J.N.; Ghaffari, M.R.; Nematzadeh, G.A.; Asari, S. Dissecting molecular mechanisms underlying salt tolerance in rice: A comparative transcriptional profiling of the contrasting genotypes. *Rice (N.Y.)* **2019**, *12*, 13. [CrossRef]
14. Sun, B.R.; Fu, C.Y.; Fan, Z.L.; Chen, Y.; Chen, W.F.; Zhang, J.; Jiang, L.Q.; Lv, S.; Pan, D.J.; Li, C. Genomic and transcriptomic analysis reveal molecular basis of salinity tolerance in a novel strong salt-tolerant rice landrace Changmaogu. *Rice (N.Y.)* **2019**, *12*, 99. [CrossRef]
15. Peng, P.; Gao, Y.; Li, Z.; Yu, Y.; Qin, H.; Guo, Y.; Huang, R.; Wang, J. Proteomic analysis of a rice mutant sd58 possessing a novel d1 allele of heterotrimeric G protein alpha subunit (RGA1) in salt stress with a focus on ROS scavenging. *Int. J. Mol. Sci.* **2019**, *20*, 167. [CrossRef]
16. Zhu, G.; Li, W.; Zhang, F.; Guo, W. RNA-seq analysis reveals alternative splicing under salt stress in cotton, Gossypium davidsonii. *BMC Genom.* **2018**, *19*, 73. [CrossRef] [PubMed]
17. Chen, M.; Lu, C.; Sun, P.; Nie, Y.; Tian, Y.; Hu, Q.; Das, D.; Hou, X.; Gao, B.; Chen, X.; et al. Comprehensive transcriptome and proteome analyses reveal a novel sodium chloride responsive gene network in maize seed tissues during germination. *Plant Cell Environ.* **2021**, *44*, 88–101. [CrossRef] [PubMed]
18. Liu, Y.; Ji, D.; Turgeon, R.; Chen, J.; Lin, T.; Huang, J.; Luo, J.; Zhu, Y.; Zhang, C.; Lv, Z. Physiological and proteomic responses of Mulberry trees (*Morus alba* L.) to combined salt and drought stress. *Int. J. Mol. Sci.* **2019**, *20*, 2486. [CrossRef] [PubMed]
19. Li, R.; Hu, F.; Li, B.; Zhang, Y.; Chen, M.; Fan, T.; Wang, T. Whole genome bisulfite sequencing methylome analysis of mulberry (*Morus alba*) reveals epigenome modifications in response to drought stress. *Sci. Rep.-UK* **2020**, *10*, 8013. [CrossRef] [PubMed]
20. Xin, Y.; Ma, B.; Zeng, Q.; He, W.; Qin, M.; He, N. Dynamic changes in transposable element and gene methylation in mulberry (*Morus notabilis*) in response to Botrytis cinerea. *Hortic. Res.* **2021**, *8*, 154. [CrossRef] [PubMed]
21. Guo, Z.; Zeng, P.; Xiao, X.; Peng, C. Physiological, anatomical, and transcriptional responses of mulberry (*Morus alba* L.) to Cd stress in contaminated soil. *Environ. Pollut.* **2021**, *284*, 117387. [CrossRef]
22. Du, Q.; Guo, P.; Shi, Y.; Zhang, J.; Zheng, D.; Li, Y.; Acheampong, A.; Wu, P.; Lin, Q.; Zhao, W. Genome-wide identification of copper stress-regulated and novel microRNAs in mulberry leaf. *Biochem. Genet.* **2021**, *59*, 589–603. [CrossRef] [PubMed]
23. He, N.; Zhang, C.; Qi, X.; Zhao, S.; Tao, Y.; Yang, G.; Lee, T.H.; Wang, X.; Cai, Q.; Li, D. Draft genome sequence of the mulberry tree Morus notabilis. *Nat. Commun.* **2013**, *4*, 2445. [CrossRef] [PubMed]
24. Dhanyalakshmi, K.H.; Naika, M.B.; Sajeevan, R.S.; Mathew, O.K.; Shafi, K.M.; Sowdhamini, R.; Nataraja, K.N. An approach to function annotation for proteins of unknown function (PUFs) in the transcriptome of indian Mulberry. *PLoS ONE* **2016**, *11*, e151323. [CrossRef] [PubMed]
25. Dhanyalakshmi, K.H.; Nataraja, K.N. Mulberry (Morus spp.) has the features to treat as a potential perennial model system. *Plant Signal. Behav.* **2018**, *13*, e1491267. [CrossRef] [PubMed]
26. Li, R.; Chen, D.; Wang, T.; Wan, Y.; Li, R.; Fang, R.; Wang, Y.; Hu, F.; Zhou, H.; Li, L.; et al. High throughput deep degradome sequencing reveals microRNAs and their targets in response to drought stress in mulberry (*Morus alba*). *PLoS ONE* **2017**, *12*, e172883. [CrossRef]

27. Per, T.S.; Khan, N.A.; Reddy, P.S.; Masood, A.; Hasanuzzaman, M.; Khan, M.; Anjum, N.A. Approaches in modulating proline metabolism in plants for salt and drought stress tolerance: Phytohormones, mineral nutrients and transgenics. *Plant Physiol. Biochem.* **2017**, *115*, 126–140. [CrossRef]
28. Zhang, X.; Liu, C.-J. multifaceted regulations of gateway enzyme phenylalanine ammonia-lyase in the biosynthesis of phenylpropanoids. *Mol. Plant* **2015**, *8*, 17–27. [CrossRef]
29. Vogt, T. Phenylpropanoid biosynthesis. *Mol. Plant* **2010**, *3*, 2–20. [CrossRef]
30. Yang, Y.; Guo, Y. Elucidating the molecular mechanisms mediating plant salt-stress responses. *New Phytol.* **2018**, *217*, 523–539. [CrossRef]
31. Chen, Z.H.; Chen, G.; Dai, F.; Wang, Y.; Hills, A.; Ruan, Y.L.; Zhang, G.; Franks, P.J.; Nevo, E.; Blatt, M.R. Molecular evolution of grass stomata. *Trends Plant Sci.* **2017**, *22*, 124–139. [CrossRef]
32. Lawson, T.; Vialet-Chabrand, S. Speedy stomata, photosynthesis and plant water use efficiency. *New Phytol.* **2019**, *221*, 93–98. [CrossRef]
33. Aung, K.; Jiang, Y.; He, S.Y. The role of water in plant-microbe interactions. *Plant J.* **2018**, *93*, 771–780. [CrossRef]
34. Liu, C.; Fan, W.; Zhu, P.; Xia, Z.; Hu, J.; Zhao, A. Mulberry RGS negatively regulates salt stress response and tolerance. *Plant. Signal. Behav.* **2019**, *14*, 1672512. [CrossRef] [PubMed]
35. Zhang, H.; Li, X.; Guan, Y.; Li, M.; Wang, Y.; An, M.; Zhang, Y.; Liu, G.; Xu, N.; Sun, G. Physiological and proteomic responses of reactive oxygen species metabolism and antioxidant machinery in mulberry (*Morus alba* L.) seedling leaves to NaCl and $NaHCO_3$ stress. *Ecotoxicol. Environ. Saf.* **2020**, *193*, 110259.
36. Hu, H.; Xiong, L. Genetic engineering and breeding of drought-resistant crops. *Annu. Rev. Plant Biol.* **2014**, *65*, 715–741. [CrossRef] [PubMed]
37. Hong, Y.; Zhang, H.; Huang, L.; Li, D.; Song, F. Overexpression of a stress-responsive NAC transcription factor gene *ONAC022* improves drought and salt tolerance in rice. *Front. Plant Sci.* **2016**, *7*, 4. [CrossRef]
38. Thakur, M.; Sharma, A.D. Salt-stress-induced proline accumulation in germinating embryos: Evidence suggesting a role of proline in seed germination. *J. Arid Environ.* **2005**, *62*, 517–523. [CrossRef]
39. Ma, J.; Du, G.; Li, X.; Zhang, C.; Guo, J. A major locus controlling malondialdehyde content under water stress is associated with *Fusarium* crown rot resistance in wheat. *Mol. Genet. Genom.* **2015**, *290*, 1955–1962. [CrossRef]
40. Morales, M.; Munné-Bosch, S. Malondialdehyde: Facts and Artifacts. *Plant Physiol.* **2019**, *180*, 1246–1250. [CrossRef]
41. Sandhu, D.; Cornacchione, M.V.; Ferreira, J.F.; Suarez, D.L. Variable salinity responses of 12 alfalfa genotypes and comparative expression analyses of salt-response genes. *Sci. Rep.* **2017**, *7*, 42958. [CrossRef]
42. Zhao, W.T.; Feng, S.J.; Li, H.; Faust, F.; Kleine, T.; Li, L.N.; Yang, Z.M. Salt stress-induced *FERROCHELATASE 1* improves resistance to salt stress by limiting sodium accumulation in *Arabidopsis thaliana*. *Sci. Rep.* **2017**, *7*, 14737. [CrossRef]
43. Lawson, T. Guard cell photosynthesis and stomatal function. *New Phytol.* **2009**, *181*, 13–34. [CrossRef]
44. Blatt, M.R.; Brodribb, T.J.; Torii, K.U. Small pores with a big impact. *Plant Physiol.* **2017**, *174*, 467–469. [CrossRef] [PubMed]
45. Duan, H.; O'Grady, A.P.; Duursma, R.A.; Choat, B.; Huang, G.; Smith, R.A.; Jiang, Y.; Tissue, D.T. Drought responses of two gymnosperm species with contrasting stomatal regulation strategies under elevated [CO_2] and temperature. *Tree Physiol.* **2015**, *35*, 756–770. [CrossRef] [PubMed]
46. Urban, J.; Ingwers, M.; McGuire, M.A.; Teskey, R.O. Stomatal conductance increases with rising temperature. *Plant Signal. Behav.* **2017**, *12*, e1356534. [CrossRef] [PubMed]
47. Eller, F.; Lambertini, C.; Nguyen, L.X.; Brix, H. Increased invasive potential of non-native *Phragmites australis*: Elevated CO_2 and temperature alleviate salinity effects on photosynthesis and growth. *Glob. Chang. Biol.* **2014**, *20*, 531–543. [CrossRef] [PubMed]
48. Kumar, S.G.; Madhusudhan, K.V.; Sreenivasulu, N.; Sudhakar, C. Stress responses in two genotypes of mulberry (*Morus alba* L.) under NaCl salinity. *Ind. J. Exp. Biol.* **2000**, *38*, 192–195.
49. Pérez-López, U.; Robredo, A.; Lacuesta, M.; Mena-Petite, A.; Muñoz-Rueda, A. Elevated CO2 reduces stomatal and metabolic limitations on photosynthesis caused by salinity in *Hordeum vulgare*. *Photosynth. Res.* **2012**, *111*, 269–283. [CrossRef]
50. Luo, M.; Zhao, Y.; Wang, Y.; Shi, Z.; Zhang, P.; Zhang, Y.; Song, W.; Zhao, J. Comparative proteomics of contrasting maize genotypes provides insights into salt-stress tolerance mechanisms. *J. Proteome Res.* **2018**, *17*, 141–153. [CrossRef]
51. Meng, X.; Liu, S.; Dong, T.; Xu, T.; Zhu, M. Comparative transcriptome and proteome analysis of salt-tolerant and salt-sensitive sweet potato and overexpression of IbNAC7 confers salt tolerance in Arabidopsis. *Front. Plant Sci.* **2020**, *11*, 572540. [CrossRef]
52. Baâtour, O.; Mahmoudi, H.; Tarchoun, I.; Nasri, N.; Trabelsi, N.; Kaddour, R.; Zaghdoudi, M.; Hamdawi, G.; Ksouri, R.; Lachaâl, M.; et al. Salt effect on phenolics and antioxidant activities of *Tunisian* and *Canadian sweet marjoram* (*Origanum majorana* L.) shoots. *J. Sci. Food Agric.* **2013**, *93*, 134–141. [CrossRef]
53. Chen, F.; Fang, P.; Peng, Y.; Zeng, W.; Ren, B. Comparative proteomics of salt-tolerant and salt-sensitive maize inbred lines to reveal the molecular mechanism of salt tolerance. *Int. J. Mol. Sci.* **2019**, *20*, 4725. [CrossRef] [PubMed]
54. Dong, N.Q.; Lin, H.X. Contribution of phenylpropanoid metabolism to plant development and plant-environment interactions. *J. Integr. Plant Biol.* **2021**, *63*, 180–209. [CrossRef]
55. Silva, R.R.D.; Câmara, C.A.G.D.; Almeida, A.V.; Ramos, C.S. Biotic and abiotic stress-induced phenylpropanoids in leaves of the mango (*Mangifera indica* L., Anacardiaceae). *J. Braz. Chem. Soc.* **2012**, *23*, 206–211. [CrossRef]
56. Falcone Ferreyra, M.L.; Rodriguez, E.; Casas, M.I.; Labadie, G.; Grotewold, E.; Casati, P. Identification of a bifunctional maize C- and O-Glucosyltransferase. *J. Biol. Chem.* **2013**, *288*, 31678–31688. [CrossRef] [PubMed]

57. Casas, M.I.; Ferreyra, M.L.F.; Jiang, N.; Mejía-Guerra, M.K.; Rodríguez, E.; Wilson, T.; Engelmeier, J.; Casati, P.; Grotewold, E. Identification and characterization of maize salmon silks genes involved in insecticidal maysin biosynthesis. *Plant Cell* **2016**, *28*, 1297–1309. [CrossRef]
58. Russell, W.R.; Burkitt, M.J.; Scobbie, L.; Chesson, A. EPR investigation into the effects of substrate structure on peroxidase-catalyzed phenylpropanoid oxidation. *Biomacromolecules* **2006**, *7*, 268–273. [CrossRef] [PubMed]
59. Lai, Y.; Zhang, D.; Wang, J.; Wang, J.; Ren, P.; Yao, L.; Si, E.; Kong, Y.; Wang, H. Integrative transcriptomic and proteomic analyses of molecular mechanism responding to salt stress during seed germination in Hulless Barley. *Int. J. Mol. Sci.* **2020**, *21*, 359. [CrossRef] [PubMed]
60. Geng, G.; Lv, C.; Stevanato, P.; Li, R.; Liu, H.; Yu, L.; Wang, Y. Transcriptome analysis of salt-sensitive and tolerant genotypes reveals salt-tolerance metabolic pathways in sugar beet. *Int. J. Mol. Sci.* **2019**, *20*, 5910. [CrossRef]
61. Zheng, L.; Meng, Y.; Ma, J.; Zhao, X.; Cheng, T.; Ji, J.; Chang, E.; Meng, C.; Deng, N.; Chen, L.; et al. Transcriptomic analysis reveals importance of ROS and phytohormones in response to short-term salinity stress in *Populus tomentosa*. *Front. Plant Sci.* **2015**, *6*, 678. [CrossRef] [PubMed]
62. Liu, C.; Xu, Y.; Feng, Y.; Long, D.; Cao, B.; Xiang, Z.; Zhao, A. Ectopic expression of mulberry G-proteins alters drought and salt stress tolerance in Tobacco. *Int. J. Mol. Sci.* **2018**, *20*, 89. [CrossRef] [PubMed]

International Journal of *Molecular Sciences*

MDPI

Article

BolTLP1, a Thaumatin-like Protein Gene, Confers Tolerance to Salt and Drought Stresses in Broccoli (*Brassica oleracea* L. var. *Italica*)

Lixia He [1,†], Lihong Li [1,†], Yinxia Zhu [1], Yu Pan [1], Xiuwen Zhang [1], Xue Han [1], Muzi Li [2], Chengbin Chen [1], Hui Li [2,*] and Chunguo Wang [1,3,4,*]

1. Department of Genetics and Cell Biology, College of Life Sciences, Nankai University, Tianjin 300071, China; helx@mail.nankai.edu.cn (L.H.); llh348536673@126.com (L.L.); 18322712771@163.com (Y.Z.); Pan1997666@126.com (Y.P.); zxw20150316@163.com (X.Z.); hx18322106151@126.com (X.H.); chencb@nankai.edu.cn (C.C.)
2. College of Horticulture and Landscape, Tianjin Agricultural University, Tianjin 300384, China; mz5872991001@163.com
3. State Key Laboratory of Tree Genetics and Breeding, Northeast Forestry University, Harbin 150040, China
4. State Key Laboratory of Vegetable Germplasm Innovation, Tianjin Academy of Agricultural Sciences, Tianjin 300381, China

* Correspondence: lihui@tjau.edu.cn (H.L.); wangcg@nankai.edu.cn (C.W.)

† Lixia He and Lihong Li contributed equally to this work.

Citation: He, L.; Li, L.; Zhu, Y.; Pan, Y.; Zhang, X.; Han, X.; Li, M.; Chen, C.; Li, H.; Wang, C. *BolTLP1*, a Thaumatin-like Protein Gene, Confers Tolerance to Salt and Drought Stresses in Broccoli (*Brassica oleracea* L. var. *Italica*). *Int. J. Mol. Sci.* **2021**, *22*, 11132. https://doi.org/10.3390/ijms222011132

Academic Editor: Juan Manuel Ruiz Lozano

Received: 2 September 2021
Accepted: 11 October 2021
Published: 15 October 2021

Abstract: Plant thaumatin-like proteins (TLPs) play pleiotropic roles in defending against biotic and abiotic stresses. However, the functions of TLPs in broccoli, which is one of the major vegetables among the *B. oleracea* varieties, remain largely unknown. In the present study, *bolTLP1* was identified in broccoli, and displayed remarkably inducible expression patterns by abiotic stress. The ectopic overexpression of *bolTLP1* conferred increased tolerance to high salt and drought conditions in *Arabidopsis*. Similarly, *bolTLP1*-overexpressing broccoli transgenic lines significantly improved tolerance to salt and drought stresses. These results demonstrated that *bolTLP1* positively regulates drought and salt tolerance. Transcriptome data displayed that *bolTLP1* may function by regulating phytohormone (ABA, ethylene and auxin)-mediated signaling pathways, hydrolase and oxidoreductase activity, sulfur compound synthesis, and the differential expression of histone variants. Further studies confirmed that RESPONSE TO DESICCATION 2 (RD2), RESPONSIVE TO DEHYDRATION 22 (RD22), VASCULAR PLANT ONE-ZINC FINGER 2 (VOZ2), SM-LIKE 1B (LSM1B) and MALATE DEHYDROGENASE (MDH) physically interacted with bolTLP1, which implied that bolTLP1 could directly interact with these proteins to confer abiotic stress tolerance in broccoli. These findings provide new insights into the function and regulation of bolTLP1, and suggest potential applications for bolTLP1 in breeding broccoli and other crops with increased tolerance to salt and drought stresses.

Keywords: thaumatin-like proteins (TLPs); bolTLP1; broccoli; salt stress; drought stress

1. Introduction

Thaumatin-like proteins (TLPs) are widely distributed in plants, animals and fungi [1], and are named for their high sequence similarity to the thaumatin protein initially identified in shrub *Thaumatococcus daniellii* Benth [2]. Most TLPs contain about 200 amino acids and have a molecular mass ranging from 21 to 26 kDa, except the poplar TLPs, most of which are 24 to 34 kDa in size [1,3,4]. In plants, TLPs are classified as pathogenesis-related protein family 5 (PR-5), one out of the 17 defense-related PR protein families [5,6]. Examination of genome databases revealed at least 37 *TLPs* in *Oryza sativa*, 33 in *Zea mays*, 18 in *Vitis vinifera*, 25 in *Arabidopsis thaliana*, 28 in *Prunus persica* and 42 in *Populus trichocarpa* [4,7]. However, the functions of the majority of TLPs are still unknown. Previous reports demonstrated that a few plant TLPs play important roles in defending against pathogen infection, particularly fungal infection. A 20 kDa TLP from the French bean legumes displays considerable resistance against *Fusarium oxysporum, Pleurotus ostreatus* and *Coprinus comatus* [8]. In *Cassia didymobotrya*, a 23 kDa TLP was identified, which exerts antifungal activity toward some Candida species [9]. *TaLr35PR5*, a *TLP* from wheat, was involved in *Lr35*-mediated adult wheat defense in response to leaf rust attack [10]. The recombinant expression of *Mald2*, a *TLP* identified in apple, conferred antifungal activity against *Fusarium oxysporum* and *Penicillium expansum* in *Nicotiana benthamiana* [11]. In tobacco, the ectopic overexpression of *PpTLP* from *Pyrus pyrifolia* or *AdTLP* from *Arachis diogoi* can enhance resistance to *Rhizoctonia solani* [12,13]. Transgenic canola, wheat and banana overexpressing *TLPs* from rice all exhibit increased resistance to diverse pathogens [14–16]. In addition to resistance against biotic stress, some *TLPs* in plants also respond to abiotic stress. Transgenic tobacco plants ectopically expressing *AdTLP* not only have enhanced resistance to *Rhizoctonia solani*, but also have improved tolerance to salt and oxidative stresses [13]. The ectopic expression of *ObTLP1* isolated from *Ocimum basilicum* enhances tolerance to dehydration and salt stresses in *Arabidopsis* [17]. Under drought stress, *TLPs* showing inducible expression patterns were also identified in carrot and tea, suggesting that these *TLPs* could be involved in drought response [18,19]. In addition, *TLPs* have also been demonstrated to play roles in regulating plant growth and development including fruit maturation, lentinan degradation and seed development [20–24]. These investigations indicate that *TLPs* have undergone functional diversification in plant species, and play pleiotropic roles in response to biotic and abiotic stresses as well as in the regulation of growth and development [4,7].

Broccoli (*Brassica oleracea* L. var. *italica*) is one of the major vegetables among the *Brassica oleracea* varieties, and is widely planted in Asia, Europe and North America [25]. Soil salinization and drought are two potential threats in planting broccoli, which can result in curd yield reduction or even no harvest [26]. Breeding broccoli varieties with high tolerance to drought and salt is an important target for broccoli breeders. Several studies related to the responses to these two abiotic stresses were reported in broccoli and other species of *Brassica*. Broccoli pre-treating with SA and chitosan showed the highest drought stress recovery in a dose-dependent manner [27]. Drought-tolerant broccoli cultivars presented higher levels of methionine and ABA than that of the drought-sensitive cultivars [28]. 42 putative known and 39 putative candidate miRNAs displayed differentially expressed patterns between control and salt-stressed broccoli [29]. *Bra-botrytis-ERF056* from cauliflower was demonstrated to increase tolerance to salt and drought stresses in overexpressing *bra-botrytis-ERF056 Arabidopsis* [30]. *BrEXLB1* is involved in drought tolerance in *Brassica rapa* [31]. *BnaABF2* from *Brassica napus* confers enhanced drought and salt tolerance in transgenic *Arabidopsis* [32]. *BolOST1* was dramatically induced by drought and high-salt stress, and the ectopic expression of *BolOST1* restored the drought-sensitive phenotype of *ost1* [33]. However, natural response to salt and drought stresses in broccoli is still largely unknown. In our previous transcriptome data analysis, a *TLP* gene named *bolTLP1* was identified in broccoli which exhibited a significant positive response to salt stress. However, the role of *bolTLP1* is unknown. In the present study, the full-length cDNA of *bolTLP1* was cloned, and its expression patterns under salt stress and drought stress were analyzed. Transgenic *Arabidopsis* and broccoli overexpressing *bolTLP1* were generated. The phenotypes of these *bolTLP1* overexpressing transgenic plants under salt and drought stresses were observed. The genes and regulatory processes involved in *bolTLP1*-mediated tolerance to abiotic stress were explored.

2. Results

2.1. BolTLP1 Was Identified in Broccoli

In previous comparative transcriptome sequence analysis to identify genes involved in salt stress response in broccoli, *bolTLP1* was found to be rapidly induced under salt stress condition (Figure 1A). In the present study, the full-length cDNA of *bolTLP1* was identified based on the ESTs of this gene from broccoli transcriptome data. Sequence analysis indicated that the coding region of *bolTLP1*comprises 696 bp and encodes a predicted

231-amino acid protein, which contained a conserved TLP domain. The phylogenetic analysis indicated that *bolTLP1* and 1 out of the 29 *TLP* genes identified in *Brassica oleracea* var. *oleracea* (XP_013621285), are classified into a separate group, and they appear to show far genetic distance with other *TLP* genes (Figure 1B).

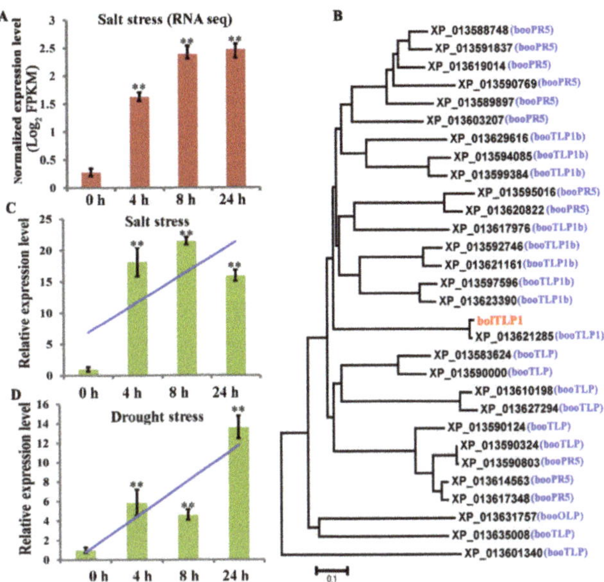

Figure 1. Phylogenetic tree and expression profiles of *bolTLP1* under salt and drought stresses. (**A**): Expression profiles of *bolTLP1* at 0 h (control), 4 h, 8 h and 24 h after salt stress (200 mM NaCl treatment) detected by RNA-seq. (**B**): Phylogenetic analysis of *bolTLP1* and genes with a TLP domain from *Brassica oleracea* var. *oleracea* (Bootstrap value = 1000). The bar indicated the genetic distance of 0.1. The red font showed the position of bolTLP1. (**C**): Expression profiles of *bolTLP1* at 0 h (control), 4 h, 8 h and 24 h after salt stress (200 mM NaCl treatment) detected by qRT-PCR. (**D**): Expression profiles of *bolTLP1* at 0 h (control), 4 h, 8 h and 24 h after mimetic drought stress (300 mM mannitol treatment). The blue lines indicated the expression trends of *bolTLP1* under salt and drought stresses. ** indicated the significantly differential expression levels of *bolTLP1* under salt and drought stresses ($p < 0.01$).

2.2. *BolTLP1 Positively Responded to Salt and Drought Stresses in Broccoli*

The expression pattern of *bolTLP1* under salt treatment in broccoli was further confirmed by qRT-PCR. Consistent with the expression trend identified by comparative transcriptome sequence analysis, the expression level of *bolTLP1* was increased significantly in 4 h after exposure to salt stress (Figure 1C). In addition, the expression pattern of *bolTLP1* in the leaves of broccoli treated with 300 mM mannitol, which could mimic drought condition, was explored. Similar to observations under salt stress, the expression level of *bolTLP1* was significantly higher under drought stress than that observed under control condition (0 h) (Figure 1D).

2.3. *Overexpression of bolTLP1 Increased Resistance to Salt and Drought Stresses in Arabidopsis*

To further uncover the roles of *bolTLP1*, a transformation vector containing the *bolTLP1* under the control of the enhanced CaMV 35S promoter was constructed, and used to transform *Arabidopsis*. A total of 26 independent 35S::*bolTLP1* transgenic lines were obtained (Figure S1). Under normal growth conditions, there were no phenotypic differences between the 35S::*bolTLP1* transgenic lines and the vector controls, thereby indicating that the overexpression of *bolTLP1* did not affect the growth and development of the transgenic lines in *Arabidopsis* (Figure 2A). To determine if overexpression of *bolTLP1* could improve resistance to salt and/or drought stress, 35S::*bolTLP1 Arabidopsis* lines were planted in soil, and the 25-day-old seedlings were subjected to 125 and 200 mM NaCl treatments, respectively. At 12 days after 200 mM NaCl treatment, the aerial organs of the vector controls were wilted, growth was retarded, and leaves were yellow and senescent. Compared with the vector controls, the 35S::*bolTLP1 Arabidopsis* was only slightly affected by the high salt condition (Figure 2B). At 18 days after salt stress, the growth of the remaining vector control plants was substantially inhibited. The survival rate of 35S::*bolTLP1 Arabidopsis* was significantly higher than that of the vector controls, and the aerial organs of the 35S::*bolTLP1 Arabidopsis* still showed normal growth. (Figure 2C,G). Under 125 mM NaCltreatment, although over 45% of the vector control plants died, almost all individual plants of the 35S::*bolTLP1 Arabidopsis* survived (Figure 2G). At both low (125 mM) and high (200 mM) NaCl treatments, the primary and lateral roots of the 35S::*bolTLP1 Arabidopsis* were stronger than those of the vector controls (Figure 2D–F).

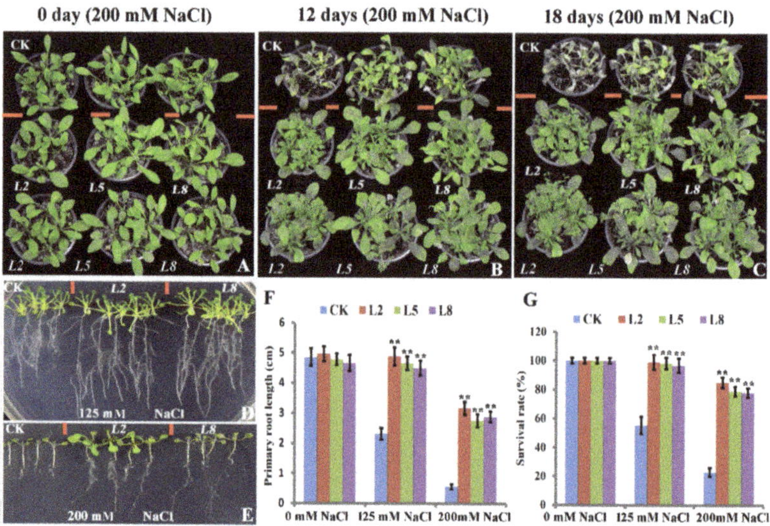

Figure 2. Phenotypes of transgenic *Arabidopsis* plants overexpressing *bolTLP1* under salt stress. (**A–C**): Phenotypes of 35S::*bolTLP1 Arabidopsis* at 0 day, 12 days and 18 days after irrigated with 200 mM NaCl in nutrient soil, respectively. (**D**,**E**): Phenotypes of 35S::*bolTLP1 Arabidopsis* seedlings at 15 days after growing in 1/2 MS medium with 125 mM and 200 mM NaCl, respectively. (**F**): Primary root length of 35S::*bolTLP1 Arabidopsis* at 15 days after growing in 1/2 MS medium with 125 mM and 200 mM NaCl. (**G**): Survival rates of 35S::*bolTLP1 Arabidopsis* at 18 days after irrigated with 125 mM and 200 mM NaCl in nutrient soil. (Student's *t*-test, ** $p < 0.01$; data are means ± SD ($n \geq 15$)). The top row of (**A–C**) showed the vector controls (CK), and the middle and bottom rows showed the independent 35S::*bolTLP1 Arabidopsis* lines (*L2*, *L5* and *L8*).

The 35S::*bolTLP1 Arabidopsis* plants were treated with water deficit. At 15 days after the water deficit, the leaves of the vector controls were seriously withered, and plant growth was inhibited. The growth and development of the 35S::*bolTLP1 Arabidopsis* were much better than those of the vector controls. In brief, the aerial organs of the 35S::*bolTLP1 Arabidopsis* were stronger than those of the vector controls, and their leaves were normal, although some of them also became senescent. The survival rate of 35S::*bolTLP1 Arabidopsis* was over 75%, whereas only 10% of the vector control plants survived after rewatering (Figure 3A–C,G). To further confirm the roles of *bolTLP1* under drought stress, the 35S::*bolTLP1 Arabidopsis* plants were treated with different concentrations of mannitol. At 200 mM mannitol, the aerial organs of the 35S::*bolTLP1 Arabidopsis* were larger than those of the vector controls. Similarly, at 300 mM mannitol, the growth and development of the aerial organs of the 35S::*bolTLP1 Arabidopsis* were still stronger than those of the vector controls (Figure 3D,E). Nevertheless, the roots, especially the primary roots, were not significantly affected by the mimic drought stresses (Figure 3F).

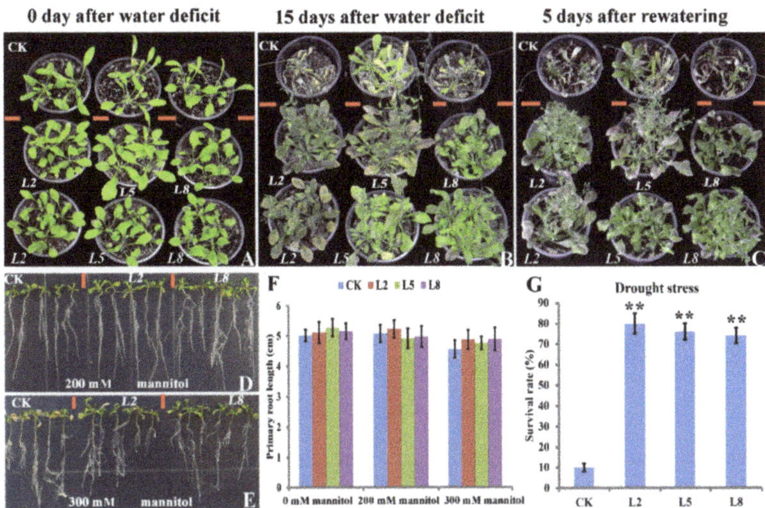

Figure 3. Phenotypes of transgenic *Arabidopsis* plants overexpressing *bolTLP1* under drought stress. (**A**,**B**): Phenotypes of 35S::*bolTLP1 Arabidopsis* at 0 day and 15 days after the water deficit. (**C**): Phenotypes of 35S::*bolTLP1 Arabidopsis* at 5 days after rewatering. (**D**,**E**): Phenotypes of 14-day-old 35S::*bolTLP1 Arabidopsis* seedlings growing in 1/2 MS medium with 200 mM and 300 mM mannitol, respectively. (**F**): Primary root length of 35S::*bolTLP1* transgenic *Arabidopsis* at 14 days after growing in 1/2 MS medium with 200 mM and 300 mM mannitol. (**G**): Survival rate of 35S::*bolTLP1 Arabidopsis* at 5 days after rewatering. (Student's *t*-test, ** $p < 0.01$; data are means $\pm$ SD ($n \geq 20$).) The top row of A, B and C showed the vector controls (CK), and the middle and bottom rows showed the independent 35S::*bolTLP1 Arabidopsis* lines (*L2*, *L5* and *L8*).

2.4. *Transgenic Broccoli Overexpressing bolTLP1 Exhibited High Salt and Drought Tolerance*

BolTLP1 was also overexpressed in broccoli. A total of 16 independent 35S::*bolTLP1* broccoli transgenic lines were obtained (Figure S1). Similar to the 35S::*bolTLP1 Arabidopsis*, the growth and development of the 35S::*bolTLP1* broccoli were unaffected under the normal growth conditions (Figure 4A). The 30-day-old 35S::*bolTLP1* broccoli seedlings were irrigated with 200 mM NaCl. At 13 days after the salt treatment, the growth of the 35S::*bolTLP1* broccoli was normal. Under these conditions, the vector controls grew slower and the first leaves showed signs of senescence (Figure 4B). As time went on, the 35S::*bolTLP1* broccoli showed slight salt damage, but their growth vigor was much better than that of the vector controls (Figure 4C–E). The growth and development of the 35S::*bolTLP1* broccoli under drought stress were also examined. The phenotypic data demonstrated that at 12 days after the water deficit, the aerial organs of the vector controls became withered, and their first and second leaves were senescing. Interestingly, at this same stage of water deficit, the growth of the 35S::*bolTLP1* broccoli was not considerably affected, and their aerial organs were larger and stronger than those of the vector controls. At 15 days after the water deficit, although the leaves of the 35S::*bolTLP1* broccoli and vector controls both showed withering, the growth vigor of the vector controls was inhibited to a greater degree than the 35S::*bolTLP1* broccoli. After rewatering, the majority of the 35S::*bolTLP1* broccoli (over 78%) survived, whereas almost all vector controls died (Figure 4F–J).

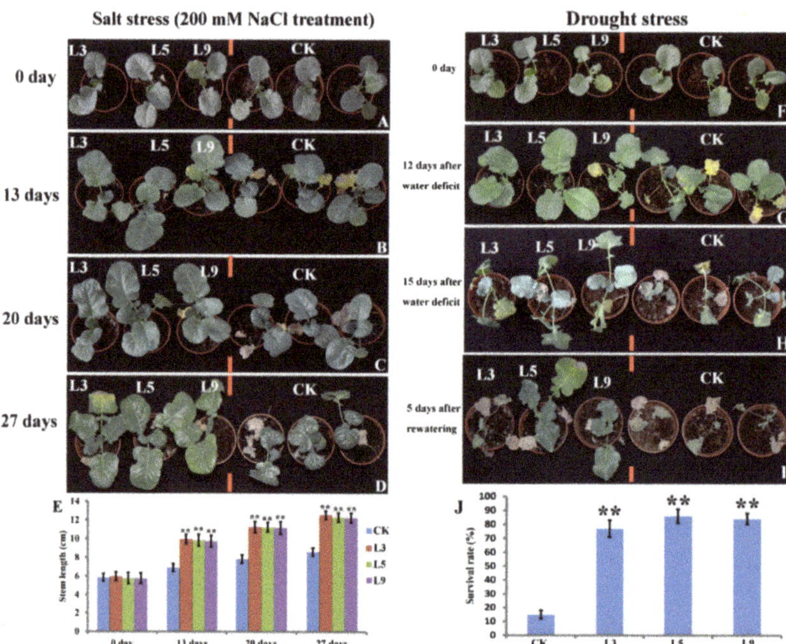

Figure 4. Phenotypes of transgenic broccoli plants overexpressing *bolTLP1* under salt and drought stresses. (**A–D**): Phenotypes of 35S::*bolTLP1* broccoli at 0 day, 13 days, 20 days and 27 days after irrigated with 200 mM NaCl, respectively. (**E**): Stem length of 35S::*bolTLP1* broccoli at 0 day, 13 days, 20 days and 27 days after irrigated with 200 mM NaCl. (Student's *t*-test, ** $p < 0.01$; data are means ± SD ($n \geq 18$)). (**F–H**): Phenotypes of 35S::*bolTLP1* broccoli at 0 day, 12 days and 15 days after water deficit, respectively. (**I**): Phenotypes of 35S::*bolTLP1* broccoli at 5 days after rewatering. (**J**): Survival rate of 35S::*bolTLP1* broccoli at 5 days after rewatering. (Student's *t*-test, ** $p < 0.01$; data are means ± SD ($n \geq 15$)). L3, L5 and L9 indicated the three independent 35S::*bolTLP1* broccoli lines. CK indicated the vector control plants.

2.5. DEGs Were Confirmed between the 35S::bolTLP1 Broccoli and Vector Control

Comparative transcriptome analysis was conducted to identify the DEGs between the 35S::*bolTLP1* broccoli and the vector controls. In total, 34,258 genes showing transcriptional expression were detected, among which 3284 genes showed significantly differential expression levels (corrected *p*-value < 0.01). The expression levels of 2202 out of the 3284 DEGs were up-regulated in the 35S::*bolTLP1* broccoli, and the other approximately 33% of the DEGs displayed lower expression levels in the 35S::*bolTLP1* broccoli compared to the vector control plants (Table S1). GO functional annotations indicated that these DEGs were mapped to 933 GO terms in the biological process, 154 GO terms in the cellular component and 343 GO terms in the molecular function, among which 33, 1 and 1 GO terms were significantly enriched in biological process, molecular function and cellular component, respectively (corrected *p*-value < 0.01) (Figure 5, Table S2). The significantly enriched GO terms in the biological process were mainly involved in stress responses, glycosinolate and sulfur compound synthetic and metabolic process, and peptidase, proteolysis and hydrolase activity. The GO terms involved in oxidoreductase activity and chromatin organization were significantly enriched in the molecular function and the cellular components, respectively (Figure 5, Table S2).

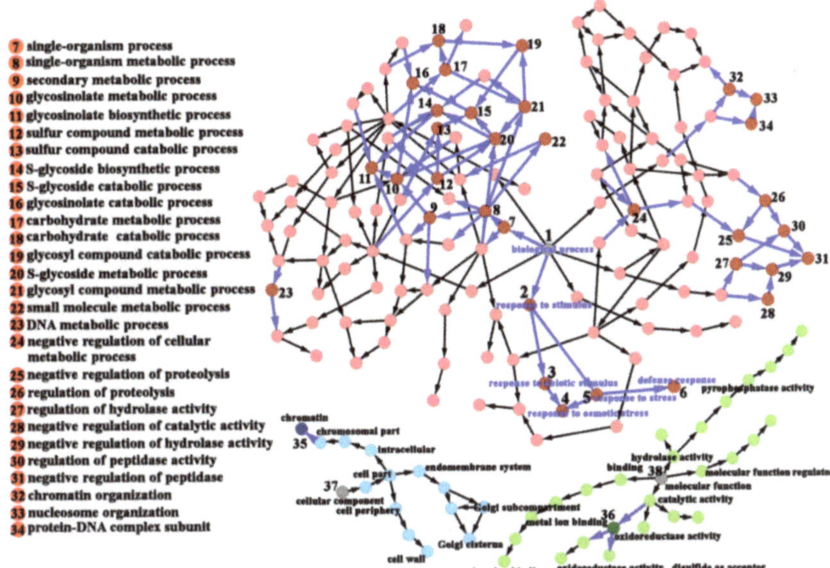

Figure 5. Regulatory network relationships of GO terms of the differentially expressed genes between 35S::*bolTLP1* broccoli and the vector control plants. Each node represents a GO term. GO terms in biological process, cellular component and molecular function were distinguished by red, blue and green color, respectively. The significantly enriched GO terms (corrected *p*-value < 0.01) are further marked by dark color, and their inclusion relations are highlighted by thick and blue arrows. The inclusion relations of other GO terms are marked by black arrows. Only those significantly enriched GO terms are serially numbered and their functional annotations were showed. The functional annotations of other enriched GO terms can be found in Table S2.

2.6. Expression Profiles of Genes Involved in ABA, Ethylene and Auxin-Mediated Signaling Pathways Were Significantly Altered by Overexpression of bolTLP1

According to the functional annotations of all DEGs by GO analysis, genes involved in the biotic and abiotic stress response processes were obviously predominant in the DEGs. At least 685 out of the 3284 DEGs (>21%) were functionally associated with stress response processes. Among the stress response-associated DEGs, approximately 70% of them displayed higher expression levels in the 35S::*bolTLP1* broccoli than those in the vector controls (Table S3). A large proportion of these DEGs were key regulators in phytohormone-mediated signaling pathways. In brief, twenty-six DEGs were involved in the auxin signaling pathway. Twenty-four out of the 26 DEGs displayed higher expression levels in the 35S::*bolTLP1* broccoli than those in the vector controls. Thirty-one genes involved in the ABA signaling pathway displayed differential expression patterns between the 35S::*bolTLP1* broccoli and the vector controls, among which 14 genes, such as *RD22*, *HVA22*, *RCI2A* and *ABSCISIC ACID-INSENSITIVE 5* (*ABI 5*), displayed significantly up-regulated expression patterns in the 35S::*bolTLP1* broccoli. At least 11 genes involved in the ethylene signaling pathway displayed differential expression patterns between the 35S::*bolTLP1* broccoli and the vector controls (Figure 6A–C). The expression levels of some of these DEGs were further verified by qRT-PCR (Figure 7).

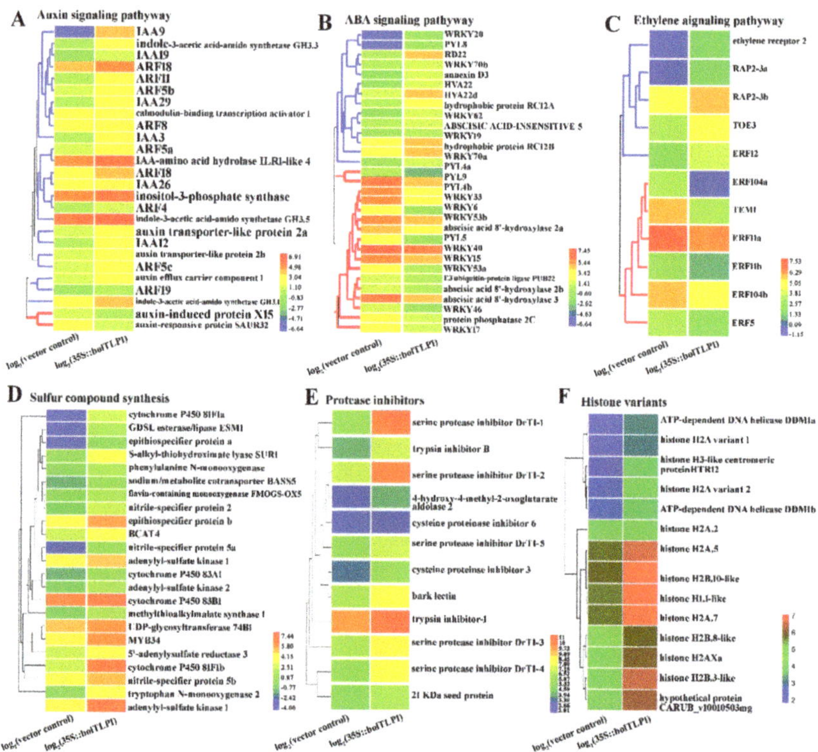

Figure 6. Transcriptional expression patterns of the differentially expressed genes involved in phytohormone-mediated signaling pathways (**A–C**), sulfur compound synthesis (**D**), protease inhibitors (**E**) and histone variants (**F**). The blue and red branches in (**A–C**) indicated the genes displayed significantly up-regulated and down-regulated expression in the 35S::*bolTLP1* broccoli compared with the vector control (corrected *p*-value < 0.01), respectively. Log_2 (35S::bolTLP1) and log_2 (vector control) indicated the log_2FPKM of *bolTLP1* in 35S::*bolTLP1* broccoli and vector control plants, respectively, which represents the normalized expression level of *bolTLP1* detected by RNA-seq.

2.7. Genes Involved in Sulfur Compound Synthesis and Hydrolase/Oxidoreductase Activity Were Significantly Inductively Expressed by the Overexpression of bolTLP1

Besides the DEGs involved in stress response, at least 62 DEGs were confirmed to map the GO terms associated with the glycosinolate and sulfur compound synthesis and metabolism, which displayed significant enrichment in the biological process. Forty-seven out of the 62 DEGs, such as genes encoding cytochrome P450 81F1, epithiospecifier protein, S-alkyl-thiohydroximate lyase, adenylyl-sulfate kinase, nitrile-specifier protein, flavin-containing monooxygenase, 5′-adenylylsulfate reductase and cytochrome P450 83B1, had significantly higher expression levels in the 35S::*bolTLP1* broccoli compared to the vector control plants. The majority of these genes play crucial roles in sulfur compound synthesis and metabolism (Figure 6D, Table S3).

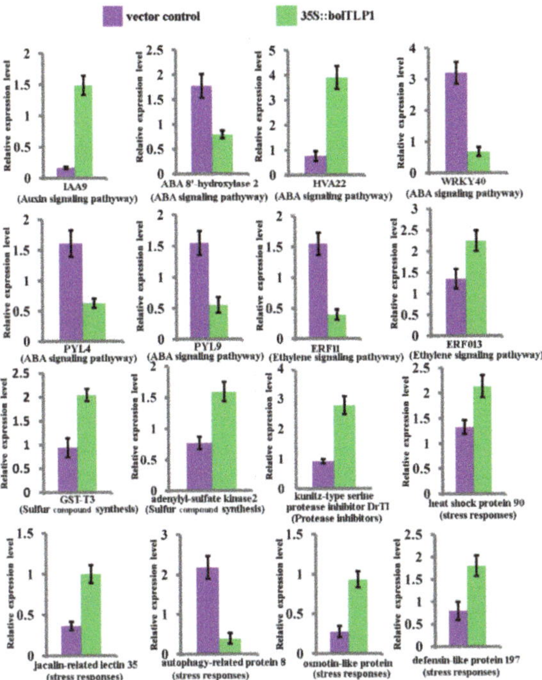

Figure 7. Identification of the relative expression levels of the representative genes by qRT-PCR. GST: glutathione S-transferase. PYL: abscisic acid receptor PYR/PYL/RCAR. ERF: ethylene-responsive transcription factor. IAA9: indol-yl-3-acetic acid 9. RD22: dehydration-responsive protein 22. 35S::bolTLP1: 35S::*bolTLP1* broccoli. Vector control: vector control plants.

The expression patterns of genes related to significantly enriched GO terms involved in the activity of diverse enzymes, such as the proteolysis, peptidase, hydrolase and oxidoreductase, were also analyzed. A total of 12 genes including genes encoding kunitz-type serine protease inhibitor, cysteine proteinase inhibitor, trypsin inhibitor and 4-hydroxy-4-methyl-2-oxoglutarate aldolase were associated with the activity of proteolysis and hydrolase, all of which exhibited significantly increased expression levels in the 35S::*bolTLP1* broccoli (Figure 6E). At least 192 DEGs were confirmed to be associated with the oxidoreductase activity. The majority of them (146 out of 192) also exhibited higher expression levels in the 35S::*bolTLP1* broccoli than those in the vector controls (Table S3). Moreover, 15 genes associated with the significantly enriched GO terms involved in chromatin organization, especially histone variants, were identified. All these genes showed increased expression levels in the 35S::*bolTLP1* broccoli (Figure 6F, Table S3).

2.8. Five Proteins Involved in Abiotic Stress Responses Were Confirmed to Interact with bolTLP1

Yeast two-hybrid screening was conducted to identify proteins that could interact with bolTLP1. In total, 266 positive clones were identified and sequenced. These sequences were annotated as 20 diverse genes, among which *RD2*, *RD22*, *VOZ2*, *LSM1B* and *MDH* were functionally associated with stress responses in plants. The interaction of these five proteins with bolTLP1 was further confirmed by using bolTLP1 as the prey and RD2, RD22, VOZ2, LSM1B and MDH as the baits in the yeast two-hybrid assay (Figure 8A, Table S4). Quantitative expression analysis revealed that the expression levels of these stress response-associated genes were significantly higher in 35S::*bolTLP1* broccoli than those in the vector controls (Figure 8B).

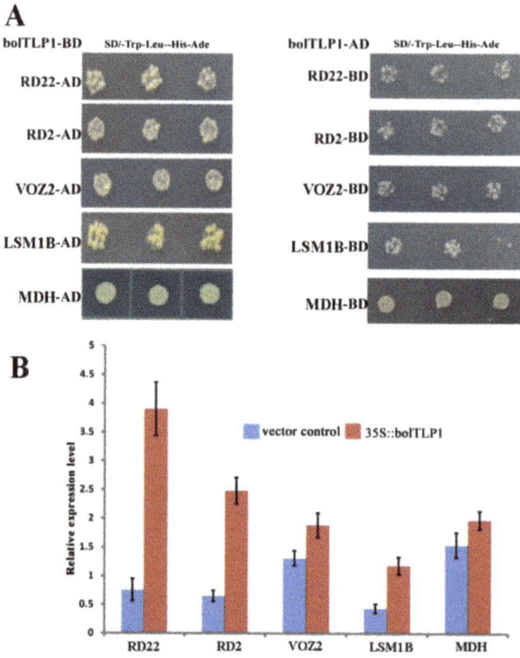

Figure 8. Identification and expression profiles of proteins interacting with bolTLP1. (**A**): Identification of proteins interacting with bolTLP1 by yeast two-hybrid assay. Three spots from left to right in each row indicated the 1:10 gradient dilution of yeast clone. (**B**): Relative expression levels of the five candidate genes, which encode proteins that interact with bolTLP1, in 35S::*bolTLP1* broccoli (35S::bolTLP1) and the vector control plants (vector control).

3. Discussion

TLPs are present as a large group of PR-5 gene family in plants, and most *TLPs* were demonstrated to play a role in defending against pathogen [1,9,11,14,16]. Only a few *TLPs* have been demonstrated to function in response to abiotic stress. Nevertheless, in the present study, *bolTLP1*, a member of *TLP* subfamily, was first identified in broccoli. The expression profiles assay confirmed that *bolTLP1* displayed inducible expression patterns under salt and drought stresses. Overexpression of *bolTLP1* in broccoli, as well as in *Arabidopsis*, significantly increased the salt and drought tolerance of transgenic plants. These findings demonstrated that *bolTLP1* is an important positive response factor that plays a role in combatting salt and drought stress in broccoli, and suggested the potential applications of *bolTLP1* in breeding crops with high salt and drought tolerance via genetic engineering.

The molecular mechanisms by which the TLPs respond to abiotic stress remain largely unknown, although investigations have confirmed the roles of several *TLP* genes in cold temperature, salt and drought stresses [13,17–19]. In the present study, comparative transcriptome analysis revealed that the DEGs detected between the 35S::*bolTLP1* broccoli and the vector controls were mainly involved in the regulatory processes including "stress responses", "glycosinolate and sulfur compound metabolic process", "peptidase, proteolysis and hydrolase activity", "oxidoreductase activity" and "chromatin organization" (Figure 5, Table S2). These results suggested that these regulatory processes should participate in *bolTLP1*-mediated tolerance to salt and drought stresses. Furthermore, the functional annotation analysis of the DEGs enriched in these regulatory processes confirmed that over 21% DEGs were involved in stress response processes, which represented the largest gene group in the DEGs. Remarkably, a large proportion of these DEGs were associated with phytohormone-mediated signaling pathways. For example, *RD2*, *RD22*,

ABI5, WRKY15/17/20/33/40/46/53/62/70, abscisic acid 8′-hydroxylase 2, annexin, PUB23, HVA22, RCI2A, RCI2B and *PYL4/5/8/9* are the important members of ABA-mediated signaling pathway, which all displayed significantly differential expression levels between the 35S::*bolTLP1* broccoli and the vector controls (Figure 6B). Most of these genes were confirmed to play crucial roles in response to abiotic stresses in diverse plant species [34–38]. Interestingly, RD2 and RD22 were confirmed to directly interact with bolTLP1 (Figure 8). *RD2* and *RD22* are up-regulated by drought stress, salinity stress and exogenously supplied ABA. Their inducible expression has been used as a marker for abiotic stress [39,40]. In addition, at least 26 DEGs, such as auxin response factor (*ARF*) *4/5/8/11/14/18/19, auxin transporter 2, IAA3/9/12/19/26/29, inositol-3-phosphate synthase, indole-3-aceticacid-amido synthetase GH3, calmodulin-binding transcription activator 1* (*CAMTA 1*) and auxin-induced gene *X15* and *SAUR32* are involved in auxin-mediated regulatory processes (Figure 6A). Similarly, some of these genes such as *CAMTA 1* [41,42] and *inositol-3-phosphate synthase* [43,44] were identified to function in response to drought or salt stress. On the other hand, a few DEGs, such as *Ethylene receptor 2* and *ethylene responsive factor*s (*ERFs*) including *ERF04, ERF05, ERF09, ERF12, RAP2-3, RAP2-10, ERF11, ERF104, TEM1* and *TOE3*, are involved in ethylene signaling pathway (Figure 6C). Inducible expression of *RAP2-3*, as well as *RAP2-12* and *RAP2-2*, confers tolerance to anoxia, oxidative and osmotic stresses [45]. The *ERF6* overexpressing transgenic lines are hypersensitive to osmotic stress, while the growth of *erf5erf6* loss-of-function mutants is less affected by stress [46]. These results demonstrated that the overexpression of *bolTLP1* significantly affects the expression levels of a set of genes involved in the ABA, ethylene and auxin signaling pathways. Moreover. The homologues of a large proportion of these genes have been demonstrated to function in defending against abiotic stress in diverse plants. Our results also support the role of these genes in the *bolTLP1*-mediated tolerance to salt and drought stresses. This finding provided new insight into the roles of phytohormone-associated genes in response to abiotic stress in broccoli. Moreover, beside RD2 and RD22, three other proteins VOZ2, LSM1B and MDH were confirmed to directly interact with bolTLP1, and their genes displayed differently expressed pattens between the 35S::*bolTLP1* broccoli and vector controls (Figure 8). The investigations demonstrated that *VOZ* transcription factors act as positive regulators of salt tolerance in *Arabidopsis* [47]. *LSM1-7* are the components of the 5′–3′ mRNA decay pathway, which play a crucial role in regulating ABA signaling and modulating ABA-dependent expression of stress related transcription factors from the AP2/ERF/DREB family [48]. Overexpression of plastidic maize NADP-malate dehydrogenase (*ZmNADP-MDH*) increases tolerance to salt stress in *Arabidopsis* [49]. It suggested that these three abotic-stress associated genes could also play important roles in *bolTLP1*-mediated regulatory network.

Another remarkably overrepresented group of DEGs was associated with various enzymes, particularly proteolysis, hydrolase and oxidoreductase. The DEGs involved in proteolysis and hydrolase activity mainly function as lectin and kunitz-type protease inhibitors such as serine protease inhibitor, cysteine proteinase inhibitor and trypsin inhibitor (Figure 6E). Plant lectins and protease inhibitors constitute a class of proteins which play a crucial role in plant defense against insect and pathogen attacks [50,51]. Previous investigations confirmed that protease inhibitors also function in abiotic stress. The constitutive expression of a trypsin protease inhibitor confers tolerance to multiple stresses in transgenic tobacco [52]. *Oryzacystatin-I* (*OCI*) is a rice cysteine proteinase inhibitor. Ectopic overexpression of *OCI* can enhance drought stress tolerance in soybean and *Arabidopsis* [53]. *AtCYSa* and *AtCYSb* are two cysteine proteinase inhibitors from *Arabidopsis*. Overexpression of these inhibitors in transgenic yeast and *Arabidopsis* increases their resistance to high salt, drought, oxidative and cold stresses [54]. The present study confirmed that all these DEGs involved in proteolysis and hydrolase activity displayed significantly up-regulated expression in 35S::*bolTLP1* broccoli (Figure 6E). These findings indicated that the increased expression levels of these genes could also play important roles in promoting salt and drought tolerance of the *bolTLP1* transgenic plants.

A series of histone variants including *H1.1, H2A.1, H2A.2, H2A.5, H2B.3, H2A.7, H2B.8* and *DDM1* were also significantly overrepresented in the DEGs (Figure 6F). These genes are important epigenetic regulators, and display significantly higher expression levels in the 35S::*bolTLP1* broccoli than those in the vector controls (Figure 6F). *TaH2A.7* in wheat can enhance drought tolerance and promote stomatal closure when overexpressed in *Arabidopsis* [55]. In *Arabidopsis*, *DDM1* is an epigenetic link between salicylic acid metabolism and heterosis, in which salicylic acid can protect plants from pathogens and abiotic stress [56]. Consistently, the mutants of *DDM1* show higher sensitivity to NaCl stress than the wild type plants [57]. In addition, the data confirmed that over 60 genes involved in glycosinolate and sulfur compound synthetic and metabolic processes showed significant differential expression patterns in the *bolTLP1* overexpressing transgenic plants and the vector controls. Genes associated with sulfur compound synthesis were up-regulated expression in 35S::*bolTLP1* broccoli (Figure 6D). Sulfur compound such as glucosinolates are best known for their roles in plant defense against herbivores and pathogens as well as their cancer-preventive properties [58–60]. A few investigations confirmed that glucosinolates also contributed to abiotic stress [61–64]. These results suggested that the histone variants and these sulfur compound-associated genes may also play an important role in *bolTLP1*-mediated tolerance to abiotic stress.

4. Materials and Methods

4.1. Plant Materials and Stress Treatment

Homozygous broccoli seeds of KRJ-012 were kindly provided by Dr. Hanmin Jiang, Tianjin Kernel Vegetable Research Institute, Tianjin, China. The seeds were planted in a greenhouse with a 16 h/8 h light/dark cycle at 25 °C and 22 °C, respectively. The 25-day-old seedlings were subjected to salt and drought stresses. For the salt-stress treatment, at least 20 individual plants were irrigated with 200 mM NaCl. To mimic the drought stress, 30 individual plants were irrigated with 300 mM mannitol. The leaves of each plant were harvested at 0, 4, 8 and 24 h after these treatments, frozen immediately in liquid nitrogen, and then stored at −80 °C. To evaluate the salt tolerance of 35S::*bolTLP1* broccoli, at least 18 individual plants from three independent transgenic lines were continually irrigated with 200 mM NaCl (Tianjin Kemiou Chemical Reagent Co., Ltd., Tianjin, China), and three such replicated treatments were conducted. Similarly, a total of 48 individual plants from three independent 35S::*bolTLP1* broccoli lines were divided into three groups, and unirrigated to evaluate their drought tolerance. In addition, at least three 35S::*bolTLP1 Arabidopsis* lines in 1/2 MS medium or in soil were treated with 125 and 200 mM NaCl to mimic salt stress. Three 35S::*bolTLP1 Arabidopsis* lines in 1/2 MS medium were treated with 200 and 300 mM mannitol to mimic drought stress. The 35S::*bolTLP1 Arabidopsis* planted in soil were also unirrigated to further evaluate their drought tolerance. To validate the reliability of the assessments of salt and drought tolerance in 35S::*bolTLP1 Arabidopsis*, three replicates were performed for each treatment.

4.2. BolTLP1 Cloning and Phylogenetic Analysis

The ESTs of *bolTLP1* were identified from the broccoli transcriptome data, and the specific primers (*bolTLP1*-F/*bolTLP1*-R) were designed to amplify the full coding sequences of *bolTLP1* based on these expressed sequence tags (ESTs) (Table S5). Total RNAs were isolated from the 25-day-old broccoli seedlings by using TRIzol reagent (Invitrogen, Carlsbad, CA, USA) according to the manufacturer's instructions. The first-strand cDNA was then synthesized using M-MLV reverse transcriptase (Promega, Madison, WI, USA), and the coding sequences of *bolTLP1* were amplified by PCR. Because the genome data of broccoli is still limited, to elucidate the sequence homology of *bolTLP1* with other *TLP* genes, those genes containing a TLP domain were identified from the genome database of *Brassica oleracea* var. *oleracea*, which is genetically closely related with broccoli (Table S6). The deduced amino acid sequences of *bolTLP1* and its homologous genes from *B. oleracea* var. *oleracea* were used to construct the phylogenetic tree using the neighbor-joining method by the MEGA 6.0 program with the following parameters: bootstrap value of 1000, poisson correction and pairwise deletion [65].

4.3. *BolTLP1* Expression Profile Analysis

Total RNAs from the leaves of individual plants, which were harvested at different time points after imposing salt or drought stress, were isolated and reverse transcribed to cDNAs. Specific primers of *bolTLP1* (*qbolTLP1*-F/*qbolTLP1*-R) was designed by Primer premier 5.0 software (Table S5). The *bolActin* gene from broccoli was selected as the internal control and Faststart Universal SYBR Green Master (Roche, Basel, Switzerland) was used in quantitative real-time RT-PCR (qRT-PCR) assay. The relative expression levels of *bolTLP1* under salt or drought stress were calculated by the comparative $2^{-\Delta\Delta CT}$ method. Three biological replicates and three technological replicates were performed to ensure the reliability of quantitative analysis.

4.4. Expression Vector Construction and Genetic Transformation

The full coding sequence of *bolTLP1* with incorporated *Nco* I and *Bst*E II restriction sites was amplified (Table S5) and cloned into the pEASY-T1 vector. The insert released from the pEASY-T1 vector with *Nco* I/*Bst*E II double digestion was sub-cloned into the pCAMBIA3301 binary vector. The recombinant 35S::*bolTLP1* plasmids and vectors without exogenous gene insertion (empty vectors) were transformed into *A. tumefaciens* strain LBA4404. *Agrobacterium*-mediated *Arabidopsis* genetic transformation was performed via the floral dip method. The transgenic *Arabidopsis* lines were screened by spraying 1:10,000 dilute Basta solution and further identified by PCR using specific primers (*bolTLP1*-F/*bolTLP1*-R) and combined primers (*35S*-F/*bolTLP1*-R). Simultaneously, the plants with empty vector were identified by PCR using specific primer pair (*35S*-F/*NOS*-R). The transgenic *Arabidopsis* plants were further verified by qRT-PCR using unique primers (*qbolTLP1*-F/*qbolTLP1*-R) and *atActin* as an internal control (Table S5). The 35S::*bolTLP1* plasmids and empty vectors were also transformed into broccoli using *Agrobacterium*-mediated method. In brief, broccoli seeds were rinsed with 75% ethanol for 2 min; 2% NaClO (Tianjin Fengchuan Chemical Reagent Co., Ltd., Tianjin, China) was then further used for surface sterilization of the seeds for 10 min. The sterilized seeds were planted on Murashige and Skoog (MS) medium with a 16 h/8 h light/dark cycle at 22 °C. The hypocotyls of 7-day-old broccoli seedlings were cut to 0.5 cm and pre-cultured on MS medium containing 1.5 mg/L 6-BA and 0.15 mg/L NAA for 2 days. Then, the truncated hypocotyls were dipped into a suspension of *Agrobacterium* containing the 35S::*bolTLP1* expression vector or an empty vector for 1 min. The infected explants were transferred onto co-cultivation medium with 1 mg/L 6-BA, 0.1 mg/L NAA and 100 μmol/L acetosyringone, and cultured for 2 days in the dark. Subsequently, the explants were washed with sterile water and transferred onto MS medium containing 1.6 mg/L 6-BA, 0.2 mg/L NAA and 200 mg/L cefotaxime for 10 days for callus induction. The differentiated explants were transferred into screening medium supplemented with 3 mg/L Basta for 2 weeks. Finally, buds that continued to differentiate were transferred to the 1/2 MS medium containing 1 mg/L IBA for root initiation. The transgenic broccoli plants were further verified by PCR using prime pair (*35S*-F/*bolTLP1*-R) and qRT-PCR using unique primers (*qbolTLP1*-F/*qbolTLP1*-R) (Table S5).

4.5. Transcriptome Sequencing and Data Analysis

Leaves from six 30-day-old individual plants of per independent 35S::*bolTLP1* broccoli line were equally mixed and used to perform three batches of independent RNA isolation. Equal amounts of RNA from each independent 35S::*bolTLP1* broccoli line were then used to construct sequencing library. In total, two such sequencing libraries from two independent 35S::*bolTLP1* broccoli lines and one sequencing library from the vector controls were constructed. The sequencing reaction was conducted by the Illumina HiSeq^{TM} 2500 sequencing platform (Beijing Genomics Institute, Shenzhen, China) with three technological replicates. The clean reads were annotated and mapped to the reference genome of *B. oleracea* var. *oleracea*. The expression levels of genes were calculated by the fragments per kilobase of transcript sequence per million base pairs sequenced (FPKM). The significantly expression levels of genes between the 35S::*bolTLP1* broccoli and the vector controls were identified based on the thresholds: $|\log_2$ (fold-change (35S::*bolTLP1* broccoli/vector control))$| > 1$ and corrected p-value < 0.05. Gene Ontology (GO) analysis of the differentially expressed genes (DEGs) was performed by the agriGO platform. Available online: http://bioinfo.cau.edu.cn/agriGO/ (accessed on 4 May 2018).

4.6. Differentially Expressed Gene Identification by qRT-PCR

The gene expression profiles detected by comparative transcriptome analysis were verified by qRT-PCR by using the specific primer pairs (Table S5). Three independent 35S::*bolTLP1* broccoli lines (Line 3, Line 5 and Line 9) were used to perform three biological replicates. Similar to RNA isolation in transcriptome sequencing, RNAs from leaves of six randomly selected 30-day-old plants per 35S::*bolTLP1* broccoli line were isolated and reverse transcribed to cDNAs. The comparative $2^{-\Delta\Delta CT}$ method was conducted to calculate the relative expression levels of genes using *bolActin* as an internal control. To further ensure the reliability of qRT-PCR, three technological replicates were carried out.

4.7. Yeast Two-Hybrid Screening and Assays

The proteins which interact with bolTLP1 were screened using Matchmaker™ Gold Yeast Two-Hybrid System (Clontech, Mountain View, CA, USA) according to the manufacturer's instructions. In brief, the full-length coding sequences of *bolTLP1* with the *Nco* I and *Bam*H I restriction enzyme sites were constructed into the bait vector pGBKT7 (Table S5). The recombinant bait vector pGBKT7-*bolTLP1* was transformed into the Y2H Gold yeast train, and the autoactivation and toxicity of bolTLP1 were detected. The positive pGBKT7-*bolTLP1* yeast then was mixed and mated with the universal *Arabidopsis* Mate & Plate library (Clontech, Mountain View, CA, USA). The positive mating yeast was screened on SD/-Trp-Leu-His-Ade medium with 0.5 mM 3-Amino-1, 2, 4-triazole (3-AT), which could effectively inhibit the autoactivation of bolTLP1. The plasmids from the mating yeast were isolated and used as a template to amplify the candidate genes using the universal primers (T7-F/3′AD-R) (Table S5). To further confirm the relationship of bolTLP1 and its candidate interacting proteins, the broccoli homologs of the positive *Arabidopsis* genes were cloned and inserted into the bait vector pGBKT7. In this case, bolTLP1 was inserted into the prey vector pGADT7 (Table S5). Subsequently, the yeast two-hybrid experiments were conducted according to the manufacturer's instructions. Transcript expression levels of the broccoli genes encoding proteins interacting with bolTLP1 were analyzed by qRT-PCR as mentioned above (Table S5).

5. Conclusions

Overexpression of *bolTLP1* significantly increased the salt and drought tolerance in both *Arabidopsis* and broccoli. BolTLP1 may directly interact with stress response-associated proteins RD2, RD22, VOZ2, LSM1B and MDH to regulate a series of genes involved in phytohormone (ABA, ethylene and auxin)-mediated signaling pathways, hydrolase/oxidoreductase activity, sulfur compound synthesis, and histone variants, which could play important roles in *bolTLP1*-mediated tolerance to salt and drought stresses in broccoli (Figure 9). *BolTLP1* is a potential candidate gene in breeding crops with high tolerance to abiotic stress via genetic engineering.

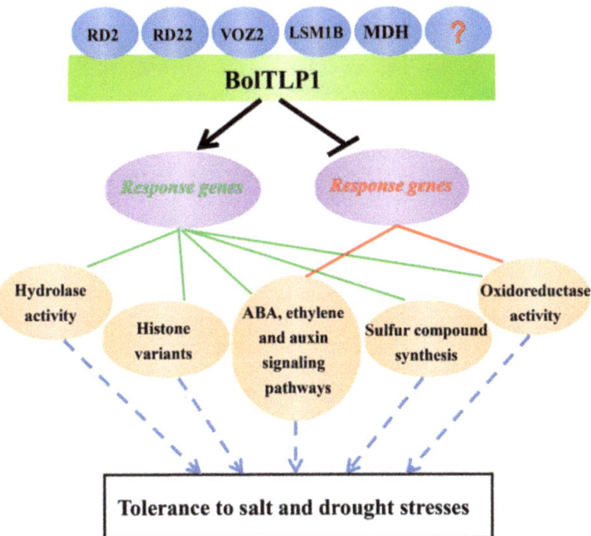

Figure 9. A proposed model indicating the regulatory processes of *bolTLP1* in increasing the tolerance to salt and drought stresses in broccoli. This model indicated that in response to salt and drought stresses, bolTLP1 may directly interact with RD2, RD22, VOZ2, LSM1B and MDH to positively regulate a series of genes involved in phytohormone (ABA, ethylene and auxin)-mediated signaling pathways, hydrolase activity, oxidoreductase activity, sulfur compound synthesis and histone variants. Meanwhile, the expression of several genes involved in phytohormone (ABA, ethylene and auxin)-mediated signaling pathways and oxidoreductase activity were inhibited by the overexpression of *bolTLP1*. The green font and lines showed the positive regulation of response genes. The red font and lines showed the negative regulation of response genes.

Supplementary Materials: The following are available online at https://www.mdpi.com/article/10.3390/ijms222011132/s1.

Author Contributions: L.L. and L.H. performed the molecular biological experiments in *Arabidopsis* and broccoli; Y.Z. and Y.P. performed the data analysis; X.Z., X.H. and M.L. performed the yeast two-hybrid assay; C.C. and Y.Z. conducted the phenotypic assay; H.L. and C.W. designed the project and wrote the manuscript. All authors have read and agreed to the published version of the manuscript.

Funding: This work was funded by grants from the Natural Science Foundation of China (No. 31872115 and No. 31470669), the State Key Laboratory of Tree Genetics and Breeding, Northeast Forestry University, China (NFUZD2015), the National science and technology major special project of transgenes (2018ZX08020003-001-003) and the Science and Technology of Tianjin, China (No. 18ZXZYNC00160). The funding body was not involved in the design of the study and analysis or interpretation of the data and in writing of the manuscript.

Institutional Review Board Statement: Not applicable.

Informed Consent Statement: Not applicable.

Data Availability Statement: Data are contained in Supplementary Materials.

Acknowledgments: We are grateful to Hanmin Jiang of Tianjin Kernel Vegetable Research Institute, Tianjin, China, for kindly providing the homozygous broccoli seeds.

Conflicts of Interest: The authors declare no competing interests.

References

1. Liu, J.J.; Sturrock, R.; Ekramoddoullah, A.K. The superfamily of thaumatin-like proteins: Its origin, evolution, and expression towards biological function. *Plant Cell Rep.* **2010**, *29*, 419–436. [CrossRef] [PubMed]
2. Cornelissen, B.J.; Hooft, V.; Bol, J.F. A tobacco mosaic virus-induced tobacco protein is homologous to the sweet-testing protein thaumatin. *Nature* **1986**, *32*, 1531–1532.
3. Ghosh, R.; Chakrabarti, C. Crystal structure analysis of NP24-I, A thaumatin-like protein. *Planta* **2008**, *228*, 883–890. [CrossRef]
4. Petre, B.; Major, I.; Rouhier, N.; Duplessis, S. Genome-wide analysis of eukaryote thaumatin-like proteins (TLPs) with an emphasis on poplar. *BMC Plant Biol.* **2011**, *11*, 33. [CrossRef] [PubMed]
5. Christensen, A.B.; Cho, B.H.; Næsby, M.; Gregersen, P.L.; Brandt, J.; Madriz-Ordeñana, K.; Collinge, D.B.; Thordal-Christensen, H. The molecular characterization of two barley proteins establishes the novel PR-17 family of pathogenesis-related proteins. *Mol. Plant Pathol.* **2002**, *3*, 135–144. [CrossRef] [PubMed]
6. Van Loon, L.C.; Rep, M.; Pieterse, C.M.J. Significance of inducible defense-related proteins in infected plants. *Annu. Rev. Phytopathol.* **2006**, *44*, 135–162. [CrossRef]
7. Iqbal, I.; Tripathi, R.K.; Wilkins, O.; Singh, J. Thaumatin-Like Protein (TLP) gene family in barley: Genome-wide exploration and expression analysis during germination. *Genes* **2020**, *11*, 1080. [CrossRef] [PubMed]
8. Ye, X.Y.; Wang, H.X.; Ng, T.B. First chromatographic isolation of an antifungal thaumatin-like protein from French bean legumes and demonstration of its antifungal activity. *Biochem. Biophys. Res. Commun.* **1999**, *263*, 130–134. [CrossRef]
9. Vitali, A.; Pacini, L.; Bordi, E.; De Mori, P.; Pucillo, L.; Maras, B.; Botta, B.; Brancaccio, A.; Giardina, B. Purification and characterization of an antifungal thaumatin-like protein from Cassia didymobotrya cell culture. *Plant Physiol. Biochem.* **2006**, *44*, 604–610. [CrossRef]
10. Zhang, J.; Wang, F.; Liang, F.; Zhang, Y.; Ma, L.; Wang, H.; Liu, D. Functional analysis of a pathogenesis-related thaumatin-like protein gene *TaLr35PR5* from wheat induced by leaf rust fungus. *BMC Plant Biol.* **2018**, *18*, 76. [CrossRef]
11. Krebitz, M.; Wagner, B.; Ferreira, F.; Peterbauer, C.; Campillo, N.; Witty, M.; Kolarich, D.; Steinkellner, H.; Scheiner, O.; Breiteneder, H. Plant-based heterologous expression of *Mald2*, a thaumatin-like protein and allergen of apple (*Malus domestica*), and its characterization as an antifungal protein. *J. Mol. Biol.* **2003**, *329*, 721–730. [CrossRef]
12. Liu, D.Q.; He, X.; Li, W.X.; Chen, C.Y.; Ge, F. Molecular cloning of a thaumatin-like protein gene from *Pyrus pyrifolia* and overexpression of this gene in tobacco increased resistance to pathogenic fungi. *Plant Cell Tissue Organ Cult.* **2012**, *111*, 29–39. [CrossRef]
13. Singh, N.K.; Kumar, K.R.; Kumar, D.; Shukla, P.; Kirti, P.B. Characterization of a pathogen induced thaumatin-like protein gene *AdTLP* from *Arachis diogoi*, a wild peanut. *PLoS ONE* **2013**, *8*, e83963. [CrossRef] [PubMed]
14. Aghazadeh, R.; Zamani, M.; Motallebi, M.; Moradyar, M. *Agrobacterium*-mediated transformation of the *Oryza sativa* thaumatin-like protein to canola (R Line Hyola308) for enhancing resistance to *Sclerotinia sclerotiorum*. *Iran. J. Biotechnol.* **2017**, *15*, 201–207. [CrossRef] [PubMed]
15. Chen, W.P.; Chen, P.D.; Liu, D.J.; Kynast, R.; Friebe, B.; Velazhahan, R.; Muthukrishnan, S.; Gil, B.S. Development of wheat scab symptoms is delayed in transgenic wheat plants that constitutively express a rice thaumatin-like protein gene. *Theor. Appl. Genet.* **1999**, *99*, 755–760. [CrossRef]
16. Mahdavi, F.; Sariah, M.; Maziah, M. Expression of rice thaumatin-like protein gene in transgenic banana plants enhances resistance to fusarium wilt. *Appl. Biochem. Biotechnol.* **2012**, *166*, 1008–1019. [CrossRef]
17. Misra, R.C.; Sandeep; Kamthan, M.; Kumar, S.; Ghosh, S. A thaumatin-like protein of *Ocimum basilicum* confers tolerance to fungal pathogen and abiotic stress in transgenic *Arabidopsis*. *Sci. Rep.* **2016**, *6*, 25340. [CrossRef]
18. Jung, Y.C.; Lee, H.J.; Yum, S.S.; Soh, W.Y.; Cho, D.Y.; Auh, C.K.; Lee, T.K.; Soh, H.C.; Kim, Y.S.; Lee, S.C. Drought-inducible-but ABA-independent-thaumatin-like protein from carrot (*Daucus carota* L.). *Plant Cell Rep.* **2005**, *24*, 366–373. [CrossRef]
19. Muoki, R.C.; Paul, A.; Kumar, S. A shared response of thaumatin like protein, chitinase, and late embryogenesis abundant protein 3 to environmental stresses in tea *Camellia sinensis* (L.) O. *Kuntze. Funct. Integr. Genom.* **2012**, *12*, 565–571. [CrossRef] [PubMed]
20. Salzman, R.A.; Tikhonova, I.; Bordelon, B.P.; Hasegawa, P.M.; Bressan, R.A. Coordinate accumulation of antifungal proteins and hexoses constitutes a developmentally controlled defense response during fruit ripening in grape. *Plant Physiol.* **1998**, *117*, 465–472. [CrossRef]
21. Sakamoto, Y.; Watanabe, H.; Nagai, M.; Nakade, K.; Takahashi, M.; Sato, T. *Lentinula edodes tlg1* encodes a thaumatin-like protein that is involved in lentinan degradation and fruiting body senescence. *Plant Physiol.* **2006**, *141*, 793–801. [CrossRef]
22. Damme, E.; Charels, D.; Menu-Bouaouiche, L.; Proost, P.; Barre, A.; Rougé, P.; Peumans, W.J. Biochemical, molecular and structural analysis of multiple thaumatin-like proteins from the elderberry tree (*Sambucus nigra* L.). *Planta* **2002**, *214*, 853–862. [CrossRef]
23. Skadsen, R.W.; Sathish, P.; Kaeppler, H.F. Expression of thaumatin-like permatin PR-5 genes switches from the ovary wall to the aleurone in developing barley and oat seeds. *Plant Sci.* **2000**, *156*, 11–22. [CrossRef]
24. Singh, S.; Tripathi, R.K.; Lemaux, P.G.; Buchanan, B.B.; Singh, J. Redox-dependent interaction between thaumatin-like protein and β-glucan influences malting quality of barley. *Proc. Natl. Acad. Sci. USA* **2017**, *114*, 7725–7730. [CrossRef] [PubMed]
25. Ilahy, R.; Tlili, I.; Pék, Z.; Montefusco, A.; Siddiqui, M.W.; Homa, F.; Hdider, C.; R'Him, T.; Lajos, H.; Lenucci, M.S. Pre- and Post-harvest Factors Affecting Glucosinolate Content in Broccoli. *Front. Nutr.* **2020**, *7*, 147. [CrossRef]
26. Zhu, J.K. Abiotic Stress Signaling and Responses in Plants. *Cell* **2016**, *167*, 313–324. [CrossRef] [PubMed]

27. Venegas-Molina, J.; Proietti, S.; Pollier, J.; Orozco-Freire, W.; Ramirez-Villacis, D.; Leon-Reyes, A. Induced tolerance to abiotic and biotic stresses of broccoli and *Arabidopsis* after treatment with elicitor molecules. *Sci. Rep.* **2020**, *10*, 10319. [CrossRef] [PubMed]
28. Chevilly, S.; Dolz-Edo, L.; López-Nicolás, J.M.; Morcillo, L.; Vilagrosa, A.; Yenush, L.; Mulet, J.M. Physiological and molecular characterization of the differential response of broccoli (*Brassica oleracea* var. *Italica*) cultivars reveals limiting factors for broccoli tolerance to drought stress. *J. Agric. Food Chem.* **2021**, *69*, 10394–10404. [CrossRef] [PubMed]
29. Tian, Y.; Tian, Y.; Luo, X.; Zhou, T.; Huang, Z.; Liu, Y.; Qiu, Y.; Hou, B.; Sun, D.; Deng, H.; et al. Identification and characterization of microRNAs related to salt stress in broccoli, using high-throughput sequencing and bioinformatics analysis. *BMC Plant Biol.* **2014**, *14*, 226. [CrossRef]
30. Li, H.; Wang, Y.; Wu, M.; Li, L.H.; Li, C.; Han, Z.P.; Yuan, J.Y.; Chen, C.B.; Song, W.Q.; Wang, C.G. Genome-wide identification of AP2/ERF transcription factors in cauliflower and expression profiling of the erf family under salt and drought stresses. *Front. Plant Sci.* **2017**, *8*, 946. [CrossRef]
31. Muthusamy, M.; Kim, J.Y.; Yoon, E.K.; Kim, J.A.; Lee, S.I. *BrEXLB1*, a *Brassica rapa* expansin-like B1 gene is associated with root development, drought stress response, and seed germination. *Genes* **2020**, *11*, 404. [CrossRef]
32. Zhao, B.Y.; Hu, Y.F.; Li, J.J.; Yao, X.; Liu, K.D. *BnaABF2*, a bZIP transcription factor from rapeseed (*Brassica napus* L.), enhances drought and salt tolerance in transgenic *Arabidopsis*. *Bot. Stud.* **2016**, *57*, 12. [CrossRef] [PubMed]
33. Wang, M.; Yuan, F.; Hao, H.; Zhang, Y.; Zhao, H.; Guo, A.; Hu, J.; Zhou, X.; Xie, C.G. *BolOST1*, an ortholog of Open Stomata 1 with alternative splicing products in *Brassica oleracea*, positively modulates drought responses in plants. *Biochem. Biophys. Res. Commun.* **2013**, *442*, 214–220. [CrossRef] [PubMed]
34. Samota, M.K.; Sasi, M.; Awana, M.; Yadav, O.P.; Amitha Mithra, S.V.; Tyagi, A.; Kumar, S.; Singh, A. Elicitor-induced biochemical and molecular manifestations to improve drought tolerance in rice (*Oryza Sativa*, L.) through seed-priming. *Front. Plant Sci.* **2017**, *8*, 934. [CrossRef] [PubMed]
35. Harshavardhan, V.T.; Van Son, L.; Seiler, C.; Junker, A.; Weigelt-Fischer, K.; Klukas, C.; Altmann, T.; Sreenivasulu, N.; Bäumlein, H.; Kuhlmann, M. AtRD22 and AtUSPL1, members of the plant-specific BURP domain family involved in *Arabidopsis thaliana* drought tolerance. *PLoS ONE* **2014**, *9*, e110065. [CrossRef]
36. Ijaz, R.; Ejaz, J.; Gao, S.; Liu, T.; Imtiaz, M.; Ye, Z.; Wang, T. Overexpression of annexin gene *AnnSp2*, enhances drought and salt tolerance through modulation of ABA synthesis and scavenging ROS in tomato. *Sci. Rep.* **2017**, *7*, 12087. [CrossRef]
37. Luo, X.; Bai, X.; Sun, X.; Zhu, D.; Liu, B.; Ji, W.; Cai, H.; Cao, L.; Wu, J.; Hu, M.; et al. Expression of wild soybean *WRKY20* in *Arabidopsis* enhances drought tolerance and regulates ABA signaling. *J. Exp. Bot.* **2013**, *64*, 2155–2169. [CrossRef]
38. Takeuchi, J.; Okamoto, M.; Mega, R.; Kanno, Y.; Ohnishi, T.; Seo, M.; Todoroki, Y. Abscinazole-E3M, a practical inhibitor of abscisic acid 8′-hydroxylase for improving drought tolerance. *Sci. Rep.* **2016**, *6*, 37060. [CrossRef]
39. Ju, Y.L.; Yue, X.F.; Min, Z.; Wang, X.H.; Fang, Y.L.; Zhang, J.X. *VvNAC17*, a novel stress-responsive grapevine (*Vitis vinifera* L.) NAC transcription factor, increases sensitivity to abscisic acid and enhances salinity, freezing, and drought tolerance in transgenic *Arabidopsis*. *Plant Physiol. Biochem.* **2020**, *146*, 98–111. [CrossRef]
40. Wang, H.; Zhou, L.; Fu, Y.; Cheung, M.Y.; Wong, F.L.; Phang, T.H.; Sun, Z.; Lam, H.M. Expression of an apoplast-localized BURP-domain protein from soybean (*GmRD22*) enhances tolerance towards abiotic stress. *Plant Cell Environ.* **2012**, *35*, 1932–1947. [CrossRef]
41. Galon, Y.; Aloni, R.; Nachmias, D.; Snir, O.; Feldmesser, E.; Scrase-Field, S.; Boyce, J.M.; Bouché, N.; Knight, M.R.; Fromm, H. Calmodulin-binding transcription activator 1 mediates auxin signaling and responds to stresses in *Arabidopsis*. *Planta* **2010**, *232*, 165–178. [CrossRef]
42. Pandey, N.; Ranjan, A.; Pant, P.; Tripathi, R.K.; Ateek, F.; Pandey, H.P.; Patre, U.V.; Sawant, S.V. *CAMTA 1* regulates drought responses in *Arabidopsis thaliana*. *BMC Genom.* **2013**, *14*, 216. [CrossRef]
43. Kusuda, H.; Koga, W.; Kusano, M.; Oikawa, A.; Saito, K.; Hirai, M.Y.; Yoshida, K.T. Ectopic expression of myo-inositol 3-phosphate synthase induces a wide range of metabolic changes and confers salt tolerance in rice. *Plant Sci.* **2015**, *232*, 49–56. [CrossRef]
44. Das-Chatterjee, A.; Goswami, L.; Maitra, S.; Dastidar, K.G.; Ray, S.; Majumder, A.L. Introgression of a novel salt-tolerant L-myo-inositol 1-phosphate synthase from *Porteresia coarctata* (Roxb.) *Tateoka* (PcINO1) confers salt tolerance to evolutionary diverse organisms. *FEBS. Lett.* **2006**, *580*, 3980–3988. [CrossRef] [PubMed]
45. Papdi, C.; Pérez-Salamó, I.; Joseph, M.P.; Giuntoli, B.; Bögre, L.; Koncz, C.; Szabados, L. The low oxygen, oxidative and osmotic stress responses synergistically act through the ethylene response factor VII genes *RAP2.12*, *RAP2.2* and *RAP2.3*. *Plant J.* **2015**, *82*, 772–784. [CrossRef] [PubMed]
46. Dubois, M.; Skirycz, A.; Claeys, H.; Maleux, K.; Dhondt, S.; De Bodt, S.; Vanden Bossche, R.; De Milde, L.; Yoshizumi, T.; Matsui, M.; et al. Ethylene Response Factor 6 acts as a central regulator of leaf growth under water-limiting conditions in *Arabidopsis*. *Plant Physiol.* **2013**, *162*, 319–332. [CrossRef] [PubMed]
47. Prasad, K.; Xing, D.; Reddy, A. Vascular plant one-zinc-finger (VOZ) transcription factors are positive regulators of salt tolerance in *Arabidopsis*. *Int. J. Mol. Sci.* **2018**, *19*, 3731. [CrossRef]
48. Wawer, I.; Golisz, A.; Sulkowska, A.; Kawa, D.; Kulik, A.; Kufel, J. mRNA Decapping and 5′-3′ decay contribute to the regulation of aba signaling in *Arabidopsis thaliana*. *Front. Plant Sci.* **2018**, *9*, 312. [CrossRef]
49. Kandoi, D.; Mohanty, S.; Tripathy, B.C. Overexpression of plastidic maize NADP-malate dehydrogenase (*ZmNADP-MDH*) in *Arabidopsis thaliana* confers tolerance to salt stress. *Protoplasma* **2018**, *255*, 547–563. [CrossRef]

50. Koiwa, H.; Bressan, R.A.; Hasegawa, P.M. Regulation of protease inhibitors and plant defense. Trends. *Plant. Sci.* **1997**, *2*, 379–384. [CrossRef]
51. Breitenbach Barroso Coelho, L.C.; Marcelino Dos Santos Silva, P.; Felix de Oliveira, W.; de Moura, M.C.; Viana Pontual, E.; Soares Gomes, F.; Guedes Paiva, P.M.; Napoleão, T.H.; Dos Santos Correia, M.T. Lectins as antimicrobial agents. *J. Appl. Microbiol.* **2018**, *125*, 1238–1252. [CrossRef]
52. Srinivasan, T.; Kumar, K.R.; Kirti, P.B. Constitutive expression of a trypsin protease inhibitor confers multiple stress tolerance in transgenic tobacco. *Plant Cell Physiol.* **2009**, *50*, 541–553. [CrossRef] [PubMed]
53. Quain, M.D.; Makgopa, M.E.; Márquez-García, B.; Comadira, G.; Fernandez-Garcia, N.; Olmos, E.; Schnaubelt, D.; Kunert, K.J.; Foyer, C.H. Ectopic phytocystatin expression leads to enhanced drought stress tolerance in soybean (*Glycine max*) and *Arabidopsis thaliana* through effects on strigolactone pathways and can also result in improved seed traits. *Plant Biotechnol. J.* **2014**, *12*, 903–913. [CrossRef]
54. Zhang, X.; Liu, S.; Takano, T. Two cysteine proteinase inhibitors from *Arabidopsis thaliana*, *AtCYSa* and *AtCYSb*, increasing the salt, drought, oxidation and cold tolerance. *Plant Mol. Biol.* **2008**, *68*, 131–143. [CrossRef]
55. Xu, W.; Li, Y.; Cheng, Z.; Xia, G.; Wang, M. A wheat histone variant gene *TaH2A.7* enhances drought tolerance and promotes stomatal closure in *Arabidopsis*. *Plant Cell Rep.* **2016**, *35*, 1853–1862. [CrossRef]
56. Zhang, Q.; Li, Y.; Xu, T.; Srivastava, A.K.; Wang, D.; Zeng, L.; Yang, L.; He, L.; Zhang, H.; Zheng, Z.; et al. The chromatin remodeler DDM1 promotes hybrid vigor by regulating salicylic acid metabolism. *Cell Discov.* **2016**, *2*, 16027. [CrossRef]
57. Yao, Y.; Bilichak, A.; Golubov, A.; Kovalchuk, I. *ddm1* plants are sensitive to methyl methane sulfonate and NaCl stresses and are deficient in DNA repair. *Plant Cell Rep.* **2012**, *31*, 1549–1561. [CrossRef]
58. Bednarek, P.; Osbourn, A. Plant–microbe interactions: Chemical diversity in plant defense. *Science* **2009**, *324*, 746–748. [CrossRef] [PubMed]
59. Clay, N.K.; Adio, A.M.; Denoux, C.; Jander, G.; Ausubel, F.M. Glucosinolate metabolites required for an *Arabidopsis* innate immune response. *Science* **2009**, *323*, 95–101. [CrossRef] [PubMed]
60. Zhang, Y.; Kensler, T.W.; Cho, C.G.; Posner, G.H.; Talalay, P. Anticarcinogenic activities of sulforaphane and structurally related synthetic norbornyl isothiocyanates. *Proc. Natl. Acad. Sci. USA* **1994**, *91*, 3147–3150. [CrossRef] [PubMed]
61. Martínez-Ballesta, M.; Moreno-Fernández, D.A.; Castejón, D.; Ochando, C.; Morandini, P.A.; Carvajal, M. The impact of the absence of aliphatic glucosinolates on water transport under salt stress in *Arabidopsis thaliana*. *Front. Plant. Sci.* **2015**, *6*, 524. [CrossRef]
62. Seo, M.S.; Jin, M.; Sohn, S.H.; Kim, J.S. Expression profiles of *BrMYB* transcription factors related to glucosinolate biosynthesis and stress response in eight subspecies of *Brassica rapa*. *FEBS Open. Bio.* **2017**, *7*, 1646–1659. [CrossRef] [PubMed]
63. Zhao, L.; Wang, C.; Zhu, F.; Li, Y. Mild osmotic stress promotes 4-methoxy indolyl-3-methyl glucosinolate biosynthesis mediated by the MKK9-MPK3/MPK6 cascade in *Arabidopsis*. *Plant Cell Rep.* **2017**, *36*, 543–555. [CrossRef]
64. Martínez-Ballesta, M.D.C.; Moreno, D.A.; Carvajal, M. The physiological importance of glucosinolates on plant response to abiotic stress in *Brassica*. *Int. J. Mol. Sci.* **2013**, *14*, 11607–11625. [CrossRef] [PubMed]
65. Tamura, K.; Stecher, G.; Peterson, D.; Filipski, A.; Kumar, S. MEGA6: Molecular evolutionary genetics analysis version 6.0. *Mol. Biol. Evol.* **2013**, *30*, 2725–2729. [CrossRef] [PubMed]

International Journal of
Molecular Sciences

MDPI

Review

Gene Mapping, Cloning and Association Analysis for Salt Tolerance in Rice

Xiaoru Fan [1], Hongzhen Jiang [2], Lijun Meng [3,4,*] and Jingguang Chen [2,*]

1 School of Chemistry and Life Science, Anshan Normal University, Anshan 114007, China; 2017203043@njau.edu.cn
2 School of Agriculture, Shenzhen Campus of Sun Yat-sen University, Shenzhen 518107, China; jianghzh23@mail.sysu.edu.cn
3 Shenzhen Branch, Guangdong Laboratory of Lingnan Modern Agriculture, Genome Analysis Laboratory of the Ministry of Agriculture and Rural Affairs, Agricultural Genomics Institute at Shenzhen, Chinese Academy of Agricultural Sciences, Shenzhen 518120, China
4 Kunpeng Institute of Modern Agriculture at Foshan, Foshan 528200, China
* Correspondence: menglijun@caas.cn (L.M.); chenjg28@mail.sysu.edu.cn (J.C.)

Abstract: Soil salinization caused by the accumulation of sodium can decrease rice yield and quality. Identification of rice salt tolerance genes and their molecular mechanisms could help breeders genetically improve salt tolerance. We studied QTL mapping of populations for rice salt tolerance, period and method of salt tolerance identification, salt tolerance evaluation parameters, identification of salt tolerance QTLs, and fine-mapping and map cloning of salt tolerance QTLs. We discuss our findings as they relate to other genetic studies of salt tolerance association.

Keywords: salt tolerance; quantitative trait locus (QTL); association analysis; marker-assisted selection (MAS); rice (*Oryza sativa* L.)

Citation: Fan, X.; Jiang, H.; Meng, L.; Chen, J. Gene Mapping, Cloning and Association Analysis for Salt Tolerance in Rice. *Int. J. Mol. Sci.* **2021**, 22, 11674. https://doi.org/10.3390/ijms222111674

Academic Editors: Mirza Hasanuzzaman and Masayuki Fujita

Received: 10 October 2021
Accepted: 27 October 2021
Published: 28 October 2021

1. Introduction

Land clearing, excessive irrigation, salt intrusion into coastal zones and sea-level rise has increased soil salinity, and this is now a significant abiotic stress affecting crop production and quality [1]. A total of 6% of the world's land area and 20% of irrigated agriculture have been affected by soil salinity. Salinity also poses a serious threat to irrigated agriculture [2,3]. The salinity problem in crop production will likely worsen due to the increasing human population [4].

Rice (*Oryza sativa* L.) is a staple food for much of the global population [4,5]. Rice is a salt-sensitive crop and yield can be greatly reduced (by over 50%) when soil salinity exceeds 6 dS/m [6]. Salt tolerance in rice varies as the growth stage does. Rice is salt-sensitive at the seedling stage, moderately salt-tolerant at the vegetative stage, and highly sensitive at the reproductive stage [7].

Salt tolerance in rice is controlled by multiple physiological and biochemical reactions, including osmotic stress and ionic stress [3]. Therefore, it is difficult to improve the salt tolerance of rice using traditional breeding methods [7]. Marker-assisted selection (MAS) and genetic engineering technology can accelerate the process of selecting for salt-tolerant rice varieties, but it is difficult to obtain salt-tolerant varieties for crop production by the insertion of single genes [8]. Therefore, it is necessary to simultaneously introduce multiple key genes to improve many pathways in the salt-tolerant regulatory network [8]. It is important to understand the molecular mechanisms and to identify the quantitative trait loci (QTL) and key genes of rice salt tolerance [1,9,10].

Genome-wide QTL analysis has been used to identify salt tolerance-related sites, and this has identified many QTLs related to rice salt tolerance. These studies have provided a foundation for the cloning of salt tolerance genes. The location and cloning of salt-tolerant genes, or QTLs, have promoted molecular-assisted selection breeding in rice. This review summarizes research on rice salt tolerance gene mapping, cloning, and breeding applications to aid breeding of salt-tolerant varieties.

2. QTL Analysis of Salt Tolerance in Rice

2.1. QTL Mapping Population for Salt Tolerance

Mapping QTLs provides insights in the inheritance mechanisms of the quantitative traits in plants and animals [11]. The mapping populations used for QTL analysis could be divided into permanent populations and temporary populations [11]. In the QTL analysis of salt tolerance in rice, the permanent populations included recombinant inbred lines (RILs) and introgression lines (ILs). RIL population–parent combinations included Kolajoha × Ranjit [12], Jiucaiqing × IR26 [13,14], Changbai10 × Dongnong425 [15], Tesanai 2 × CB, (Nona Bokra × Pokkali) × (IR4630-22-2-5-1-3 × IR10167-129-3-4) [16], IR4630 × IR15324 [17], Co39 × Moroberekan [18], Milyang23 × Gihobyeo [19,20], H359 × Acc8558 [21], IR29 × Pokkali B [22], Yiai1 × Lishuinuo [23], CSR11 × MI48 [24], CSR27 × MI48 [25], and Dongxiang × NJ16 [7]. IL population–parent combinations included IR64 × Tarom Molaii [26], Ilpumbyeo × Moroberekan [27], Minghui86 × ZDZ057, Minghui86 × Teqing Shuhui527 × ZDZ057, Shuhui527 × Teqing [28], Lemont × Teqing [29], Pokkali × IR29 [30,31], Teqing × Oryza rufipogon [32], Ce258 × IR758 62 [33], Tarome-Molaei × Tiqing [34], Xiushui 09 × IR2061 [35], IR64 × Binam [36], and Nipponbare × Kasalath [37]. In addition, there are doubled haploid (DH) groups that include IR64 × Aucena [38] and Zhaiyeqing 8 × Jingxi 17 [39,40]. Some studies also used a set of chromosome segment substitution lines (CSSLs) to detect salt tolerance in seedlings [41]. Mapped salt-tolerant QTLs that have permanent populations could analyze phenotypic variation at multiple points over multiple years. In this way, the identified salt-tolerant QTLs are more stable and not affected by the environment, which was of benefit to map-based cloning and molecular breeding applications. However, most of the permanent populations in the studies were not used for salt tolerance analysis. There was a lack of highly salt-tolerant or salt-sensitive parental varieties. The salt tolerance difference between the parents was small, which was not conducive to the identification of major salt-tolerant sites. Only a few populations were constructed that had salt-tolerant varieties as their parents and used for salt tolerance research, such as Kolajoha × Ranjit [12], Jiucaiqing × IR26 [13,14], (Nona Bokra × Pokkali) × (IR4630-22-2-5-1-3 × IR10167-129-3-4) [16], IR29 × Pokkali [22], CSR11 × MI48 [24], and CSR27 × MI48 [25].

Most of the salt-tolerant QTL mapping of rice has used temporary populations. Most of these populations were F_2 and F_3 populations, and a few were F_4, BC_1F_1, $BC_1F_{2:3}$, and $BC_2F_{2:3}$ populations. The parent populations included Gharib × Sepidroud [42,43], Nona Bokra × Koshihikari [44], Tarommahali × Khazar [45,46], Pokkali × Shaheen Basmati [47], BRRI Dhan40 × IR61920-3B-22-2-1 [48], Dongnong425 × Changbai10 [15,49], Jiucaiqing × IR36 [50], Sadri × FL478 [51], NERICA-L-19 × Hasawi, Sahel 108 × Hasawi, and BG90-2 × Hasawi [52], IR36 × Pokkali [53,54], CSR27 × MI48 [55], Cheriviruppu × Pusa Basmati1 [56], and Peta × Pokkali [57]. These populations were used for QTL analysis of salt-tolerant materials such as Gharib, Nona Bokra, Tarommahali, Pokkali, Jiucaiqing, FL478, Hasawi, IR61920-3B-22-2-1, Cheriviruppu, Changbai10, and CSR27 [42–52,55,56].

Some studies used two or more populations simultaneously for salt tolerance QTL analysis. Tiwari et al. [24] identified the salt-tolerant QTLs that had two RIL populations CSR11 × MI48 and CSR27 × MI48; Cheng et al. and Yang et al. [29,35] used the two-way combination of Xiushui09 × IR2061 and Lemont × Teqing; Qian et al. [28] selected Shuhui 527 × ZDZ057, Minghui 86 × ZDZ057, Shuhui 527 × teqing, and Minghui 86 × teqing for salt tolerance QTL analysis; Sun et al. [49] used F_3 and $BC_1F_{2:3}$ populations of Dongnong 425 × Changbai 10 to analyze the dynamic QTL that controls the ion content in rice roots; Bimpong et al. [52] used F_2 populations of NERICA-L-19 × Hasawi, Sahel 108 × Hasawi and BG90-2 × Hasawi to identify QTLs for salt tolerance in Hasawi. QTL analysis and comparison with multiple mapped populations were conducive to finding salt-tolerant sites that could be stably expressed and less affected by genetic background.

2.2. Period and Method of Salt Tolerance Identification

Rice has different tolerances to salt stress at different growth stages [7]. The seedling stage and the reproductive growth stage are salt-sensitive, while the seed germination stage and the vegetative growth stage are more salt-tolerant [7]. Therefore, most of the studies of salt tolerance QTL analysis in rice have been conducted during the seedling and reproductive growth stages [58].

More than half of the QTL studies on rice salt tolerance have used the seedling stage. The methods used for the identification of salt tolerance at the seedling stage were uniform. Rice seedlings were cultivated by hydroponics, and treated with salt at, or near, the three-leaf stage [13,14,16,17,19–23,26,28–30,34–36,40,43–46]. For the reproductive growth stage, most rice studies used plants in artificial salt ponds. A small number of studies used rice planted in soil and treated with salt water [24,25,40,51–56]. The initial and final salt treatments were different in different studies. Most of the studies transplanted rice to salt ponds in the seedling or tillering stage, where they were grown to maturity. The plants were then scored for agronomic traits and physiological indicators of salt tolerance [24,25,40,51–54,57]. A few studies analyzed salt tolerance QTL in the seed germination stage, and conducted the germination in a medium with salt as a treatment [13,38,42].

Some studies simultaneously analyzed salt tolerance QTL in two or more growth and development stages. Gu et al. and Pandit et al. [25,57] identified the salt tolerance QTL in the vegetative and reproductive growth stages of rice; Zang et al. [36] identified tolerance in the seedling stage and vegetative growth stage; Ammar et al. [55] analyzed salt tolerance QTLs in seedling, vegetative growth, and reproductive growth stages. These studies helped to identify the genes that control salt tolerance in multiple growth and development stages of rice.

To analyze the influence of plant developmental differences on salt tolerance, some studies used a control group. They analyzed the salt tolerance of the mapping population under the salt and the control treatments at the same time [13,14,24,27,28,36–38,40,49,52–54,57]. Most of the studies used permanent populations that are homozygous for each strain. A few studies used different tillers from F_2 populations for different treatments.

2.3. Salt Tolerance Evaluation Parameter

The salt tolerance of rice is a complex and comprehensive trait that has various evaluations that differ between development stages. In QTL analysis of rice salt tolerance, the evaluation parameters at the seedling stage can be divided into three categories: morphological, growth and physiological. Morphological parameter analysis evaluates the salt tolerance of the seedlings (score of salt tolerance, SST) by observing the blade tips, leaves, tillers, and the growth inhibition and death of plants after salt stress, and also investigating the survival days of seedling (SDS) after salt stress [14,15,18–20,22,23,26–30,33,35,36,43,44,46–48,55,59]. Most studies have used the standard evaluation system (SES) proposed by the International Rice Research Institute (IRRI) to evaluate the salt damage level [60]. Some studies modified the evaluation criteria based on experimental materials and experimental design [19,22,26,28–30,33,35,36,43,46,47,55,59]. The growth indicators used to evaluate the salt tolerance during the seedling stage include plant height and the fresh and dry weight of shoots and roots [14,17,22,27,43,45–47]. There are many physiological parameters for evaluating the salt tolerance of rice, and the indicators for QTL analysis include plant ion content, the concentration of shoot Na^+ (SNC) and K^+ (SKC), shoot Na^+/K^+ ratio (SNKR), the concentrations of root Na^+ (RNC) and K^+ content (RKC), and root Na^+/K^+ ratio (RNKR) [14–17,21,22,26,29,33–35,43–47,59]. Some studies also analyzed QTL with the chlorophyll content of seedlings after salt stress [22,43,45].

The evaluation parameters for the salt tolerance of rice seeds during germination include germination rate and germination vigor. Some studies further analyzed growth of the embryo and the radicle of seedlings after germination [13,38,42]. The evaluation parameters during the vegetative growth stage included plant growth and physiological indicators. Most studies analyzed the growth and ion content of the shoot rather than the root [18,25,36,49,50,55,57]. The evaluation during reproductive growth included yield-related agronomic traits, such as the heading date, plant height, tiller number panicles per plant, grains per panicle, seed setting rate, 1000-seed weight, and yield per plant [24,25,40,51,52,55–57]. Some studies analyzed the content of Na^+, K^+, Ca^{2+}, and Cl^- in rice leaves or straw after salt treatment in the reproductive growth stage [25,51,53,55]. Some studies included control groups, and they used the absolute value of each evaluation parameter for QTL analysis between the control and comparison groups. They also used the relative value of each salt tolerance trait (treatment/control) or decrease rate ((control–treatment)/control) as an indicator, which was beneficial in reducing the influence of individual plant differences [24,27,49,53,54,57].

2.4. Salt Tolerance QTL

We found 52 salt tolerance QTL studies in rice, as shown in Table 1. More than half of the salt-tolerant QTLs were in the seedling stage. Salt-tolerant QTLs at each growth stage were distributed on the 12 rice chromosomes.

The phenotypic contribution rate of a single QTL ranged from 0.02% to 81.56%. A total of 167 QTLs had a contribution greater than 20% and these occupied 22.0% of the total QTLs (Table 1). Salt-tolerant QTLs that have a large contribution to the phenotype were found in the studies that follow. Thomson et al. [22] detected 16 salt-tolerant QTLs which explained more than 20% of the phenotypic variation in the seedling stage. Of these, five QTLs had a contribution exceeding 50%. Qian et al. [28] used four mapping populations and detected 43 QTLs that control SST or SDS in seedlings. The contributions of 12 QTLs were more than 20%. Sabouri et al. [45] identified 32 QTLs that control different growth and physiological indicators of salt tolerance in rice seedlings. Among them, 14 QTLs explained more than 20% of the phenotypic variation. Bimpong et al. [52] detected 75 salt tolerance QTLs in three mapping populations during the reproductive stage, of which about half of the QTLs (37) explained more than 20% of the phenotypic variation. Ammar et al. [55] detected 25 QTLs that had a contribution greater than 10% in the seedling, vegetative growth, or reproductive growth stage, and 22 QTLs among these had a contribution rate over 20%. In these studies, there were 101 salt-tolerant QTLs that had a phenotypic contribution rate exceeding 20%. There were only a few QTLs that had large effects in other studies, and 13 studies had no salt-tolerant QTLs that exceeded a 20% variation. [5,15,17,25,27,30,33,38,47,50,59,61,62].

Table 1. Identified QTL for salt tolerance in rice.

Stage	Parents for Cross	Population Type	Evaluation Parameter for Salt Tolerance	PVE%	QTL	High-PVE QTL	Reference
Germination stage	IR64 × Azucena	DH	GR, seedling root length, seedling dry mass, seedling vigor	13.5–19.5	7	0	[38]
	Jiucaiqing × IR26	RIL	GR, RL, SH	6.5–43.7	7	4	[13]
	Gharib × Sepidroud	$F_2/F_{2:4}$	GR, germination percentage, radicle length, plumule length, coleoptile length, radicle fresh weight, plumule fresh weigh, radicle dry weight, plumule dry weight, coleoptile fresh weight, coleoptile dry weight	10.0–21.9	17	2	[42]
	9311 × japonica	CSSL	Survival rate	5.1–93.2	4	-	[41]
Seedling stage	Dongnong425 × Changbai10	$BC_2F_2/BC_2F_{2:3}$	SST, SNC, SKC, RNC, RKC	6.45–17.95	13	0	[63]
	O. rufipogon × O. Sative	ILs	SDS, STT	2–8	10	-	[64]
	(Nona Bokra × Pokkali) × (IR4630-22-2-5-1-3 × IR10167-129-3-4)	RIL	SNC, SKC, SNKR	-	4	-	[16]
	IR4630 × IR15324	RIL	SNC, SKC, SNKR, total Na^+ and K^+, SDW	6.4–19.6	11	0	[17]
	Milyang 23 × Gihobyeo	RIL	SST	9.2–27.8	2	1	[19]
	Milyang 23 × Gihobyeo	RIL	SST	9.1–27.8	2	1	[20]
	H359 × Acc 8558	RIL	SNC	1.68–45.39	13	3	[21]
	IR29 × Pokkali	RIL	SNC, SKC, RKC, RNKC, SH, chlorophyll content, seedling survival rate, initial and final SST	6–67	27	16	[22]
	Yiai1 × Lishuinuo	RIL	Dead rate of leaf and seedling	8.65–27.20	6	1	[23]
	IR64 × Tarom Molaii	IL	SST, SDS, SKC, SNC, RKC, RNC	-	23	-	[26]
	Ilpumbyeo × Moroberekan	IL	The reduction rate of fresh and dry weight, leaf area and SH	10.2–13.9	8	0	[27]
	Shuhui527 × ZDZ057, Minghui86 × Teqing, Minghui86 × ZDZ057, Shuhui527 × Teqing	IL	SST, SDS	8.17–42.18	43	12	[28]
	Lemont × Teqing	IL	SST, SDS, SKC, SNC	-	36	-	[29]
	Pokkali × IR29	IL	SST	4.00-18.42	6	0	[30]
	Ce258 × IR75862, ZGX1 × IR75862	IL	SST, SDS, SKC, SNC	5.13–13.75/3.73–8.26 *	18/2 *	0	[33]
	Tarome-Molaei × Tiqing	IL	SNC, SKC, SNKR, RNC, RKC, RNKR	9.0–30.0	14	5	[34]
	Xiushui 09 × IR2061-520-6-9	IL	SST, SDS, SKC, SNC, SKNR	5.14–18.89/2.60–14.30 *	26/21 *	0	[35]
	Zaiyeqing8 × Jingxi17	DH	SDS	10.2–38.4	10	2	[39]
	Nona Bokra × Koshihikari	F_2/F_3	SDS, SNC, SKC, RNC, RKC, Na^+ and K^+ in root, SDW	12.4–48.5	11	3	[44]

Table 1. *Cont.*

Stage	Parents for Cross	Population Type	Evaluation Parameter for Salt Tolerance	PVE%	QTL	High-PVE QTL	Reference
	Tarommahalli × Khazar	F_2/F_3	Survival rate, chlorophyll content, SH, RL, leaf area, the weight of stem and root, total Na^+ and K^+ in shoot, SNKR	9.03–38.22	32	14	[45]
	Tarommahali × Khazar	F_2/F_3	STR, DM, Na^+ content, K^+ content, Na^+/K^+	9.03–20.90	14	1	[46]
	Pokkali × Shaheen Basmati	F_2/F_3	SST, SH, SDW, SFW, SNC, SKC, SNKR, RNC, RKC, RNKR	4.89–10.55	22	0	[47]
	BRRI Dhan40 × IR61920-3B-22-2-1	F_2	SST	12.5–29.0	3	2	[48]
	Jiucaiqing × IR26	RIL	RNKR, SH, SDW, RDW	7.8–23.9/- *	15/5 *	2	[14]
	Jiucaiqing × IR26	RIL	RKC, SNC, SKC, SST	8.5–18.9/- *	13/9 *	0	[59]
	Tesanai 2 × CB	RIL	SDS	1.5–11.6	4	0	[61]
	Tesanai 2 × CB	RIL	SDS, SDW, RDW, SNC, SKC, SKNR	4.4–15.0	31	0	[62]
	Teqing × *Oryza rufipogon*	IL	SST, relative SDW, RDW and total plant dry weight	8–26	15	3	[32]
Vegetative growth stage	Co39 × Moroberekan	RIL	Content of Na^+ in shoot, SNKR, fresh weight of stem, moisture content of leaf	11.0–26.3	14	3	[18]
	Nipponbare × Kasalath	IL	SH, SDW, number of tillers	12–41	31	11	[37]
	Dongnong425 × Changbai10	$BC_1F_2/BC_1F_{2:3}$, F_2/F_3	RNC, RKC, RNKR, relative RNC, relative RKC, relative RNKR	3.61–27.9	50	4	[49]
	CSR10 × Taraori Basmati	F_3	Relative growth rate, SNKR, visual salt-injury symptoms	25.6–31.3	14	-	[65]
	Jiucaiqing × IR26	F_2	SST, SNKR, SDW	6.7–19.3	7	0	[50]
Reproductive growth stage	CSRll × MI48, CSR27 × MI48	RIL	Sensitivity index of grain yield stress	-	55	-	[24]
	Zhaiyeqing 8 × Jingxi 17	DH	Effective tiller number, thousand-grain weight, PH, heading date, number of grains per panicle	7.9–40.1	24	3	[40]
	Sadri × FL478	F_2	Heading date, PH, length and number of panicles, dry weight of straw, number of fertile and sterile spikelets per plant, total number of spikelets per plant, yield per plant, spikelet fertility, thousand grain weight	4.2–30.0	37	1	[51]
	NERICA-L-19 × Hasawi, Sahel108 × Hasawi, BG90-2 × Hasawi	F_2	SST, PH, TN, heading date, panicle number per plant, panicle sterility rate, grain number per ear, thousand-grain weight, yield per plant	6.5–49.5	75	37	[52]
	IR36 × Pokkali	F_2	Content of Na^+ and Ca^{2+}, absorption rate of Ca^{2+}, relative content of Na^+, K^+ and Ca^{2+}, relative ion content, relative absorption rate of Na^+, K^+, Ca^{2+} and Na^+/K^+	7.69–26.33	14	3	[53]

Table 1. *Cont.*

Stage	Parents for Cross	Population Type	Evaluation Parameter for Salt Tolerance	PVE%	QTL	High-PVE QTL	Reference
	IR36 × Pokkali	F_2	PH, TN, number of effective tillers, panicle weight, panicle length, number of spikelets panicle, number of unfilled grains panicle, number of grains panicle, panicle fertility, days of 50% flowering, days to maturity, grain length, grain width, grain length–width ratio, grain yield, thousand-grain weight, straw yield, harvest index	11.52–81.56	6	1	[54]
	Cheriviruppu × Pusa Basmati 1	F_2	PH, TN, panicle length, yield, biomass, pollen fertility, Na^+ content in flag leaf, Na^+/K^+	3.8–48.7	24	5	[56]
	HHZ × Budda, HHZ × Gang46B	BC_2F_5	Grain weight, spikelet number, thousand-grain weight, seed fertility	4.7–90.6	22	1	[66]
	Sahel108 × Hasawi, NERICA-L-19 × Hasawi, BG90-2 × Hasawi	F_2	Days to flowering/heading, PH, TN, panicle sterility, grain yield, yield per plant, yield-component data for each plot, salt tolerance score	7.3–31.9	75	-	[52]
	Tarommahalli × Khazar	F_2/F_3	PH, TN, number of full grains, number of empty grains, length and number of panicle, biomass	8.76–26.83	12	3	[56]
Multiple growth stages	CSR27 × MI48	RIL	Vegetative growth period: content of Na+ in stem, content of K^+ and Cl^- content in leaf	5.86–8.55	4	0	[25]
			Reproductive growth period: content of Na^+, K^+ in straw, Na^+/K^+ in straw, spikelet fertility stress sensitivity index	7.22–14.05	5	0	
	IR64 × Binam	IL	Seedling stage: SST, SDS, SKC, SNC	-	13	-	[36]
			Vegetative growth period: PH, panicle number, fresh weight	-	22	-	
	CSR27 × MI48	F_2/F_3	Seedling stage: SST	14.38	1	0	[55]
		F_2	Vegetative growth period: Na^+, Cl^- content in leaf and stem, K^+ content in stem, Na^+/K^+ in leaf and stem	11.13–55.72	17	15	
		F_2	Reproductive growth period: content of Na^+, K^+ and Cl^- in leaf, Na^+/K^+ in leaf	26.26–52.63	7	7	

Table 1. *Cont.*

Stage	Parents for Cross	Population Type	Evaluation Parameter for Salt Tolerance	PVE%	QTL	High-PVE QTL	Reference
	Peta×Pokkali	BC_1F_1	Vegetative growth stage: SST, SFW, SDW, Na^+ content	-	4	-	[57]
			Reproductive growth stage: weight of stem and leaf, PH, TL, effective panicle, number, panicle weight, main panicle length, grain weight, seed setting rate	-	11	-	

Note: DH: double haploid. RIL: recombinant inbred lines. IL: introgression line. CSSL: chromosome segment substitution line. F_2: second filial generation. F_2/F_3: second and third filial generation. $F_2/F_{2:4}$: F_2 generation and F_2 derived fourth filial generation. BC_1F_1: first backcross generation. BC_2F_5: twice backcross and four selling generation. $BC_1F_2/BC_1F_{2:3}$: twice backcross and one selling generation, and one backcross and one selling derived third filial generation. $BC_2F_2/BC_2F_{2:3}$: twice backcross and one selling generation, and twice backcross and one selling generation derived fourth filial generation. High-PVE QTL: the number of QTL with contribution > 20%. GR: germination rate. SDS: survival days of seedling. SST: score of salt tolerance. SNC and SKC: the concentrations of Na^+ and K^+ in shoots. SNKR: shoot Na^+/K^+ ratio. RNC and RKC: the concentrations of Na^+ and K^+ in roots. RNKR: root Na^+/K^+ ratio. SH: shoot height, RL: root length, SDW: shoot dry weight. SFW: shoot fresh weight. RDW: root dry weight. PH: plant height. TN: tiller number, STR: standard tolerance ranking, DM: dry matter weight. * indicate major QTL/epistatic QTL.

We constructed a framework genetic map using 70 QTLs with high PVE in the reports [13,14,19–21,34,39,40,42,44,45,48,51] (Figure 1). Figure 1 showed that QTLs related to salt tolerance are distributed on 12 chromosomes, but less on chromosomes 11 and 12.

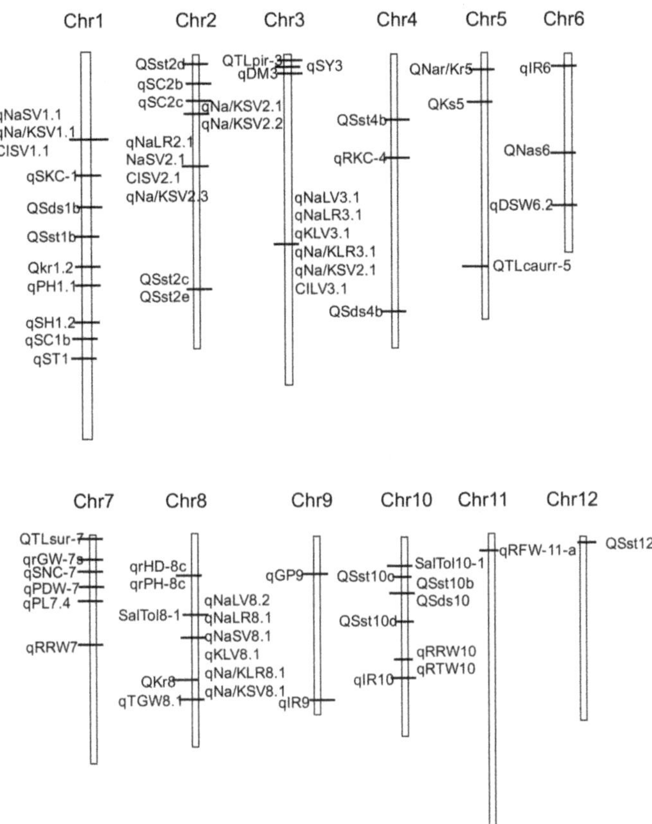

Figure 1. Genetic linkage map showing the location of QTLs for salt tolerance-related traits detected in reports.

2.5. Fine Mapping and Map-Based Cloning of QTLs for Salt Tolerance in Rice

Because many salt-tolerant rice QTLs have a low phenotypic contribution rate and are difficult to fine-map and clone, relevant research has progressed slowly. However, two QTLs located on the first chromosome, *qSKC-1* and *Saltol*, are suitable for fine-mapping or map-based cloning.

qSKC-1 is a major QTL that controls the K^+ content in the shoot. It was detected in the F_2 population that was constructed by the salt-tolerant variety Nona Bokra and the salt-sensitive variety Koshihikari, and it explained 40.1% of the total phenotypic variation [44]. Ren et al. [67] used the map-based cloning method, followed by fine-mapping of the BC_2F_2 population and high-precision linkage analysis of the BC_3F_2 population, and they restricted *qSKC-1* within the 7.4 kb chromosome interval and isolated the *qSKC-1* gene. This gene encoded an ion transporter (OsHKT1;5) of the HKT (high-affinity K^+ transporter) family, which exists in the parenchyma cells of the xylem of rice roots and has the function of specifically transporting Na^+. This transporter could transport Na^+ out of the xylem, and transport Na^+ from the phloem back to the root where it was excreted from the plant through the action of other Na^+ transporters. This process reduced the Na^+ content in the shoot, regulated the Na^+/K^+ balance in the shoots, and improved rice salt tolerance [68].

Gregorio [69] used AFLP markers to analyze the salt tolerance QTL of the F_8 recombinant inbred line population of the Pokkali/IR29 combination, and detected a major QTL on rice chromosome 1 that simultaneously controls the Na^+, K^+ content and Na^+/K^+ ratio in rice. This QTL was named *Saltol*. In the population, the LOD value of the *Saltol* site was greater than 14.5, and the phenotypic contribution rate was 64.3–80.2%. Subsequently, Bonilla et al. [70] used the same population to map *Saltol* to the chromosome between SSR markers RM23 and RM140, and they found that the contribution rates of *Saltol* sites to the Na^+, K^+ content and Na^+/K^+ ratio were 39.2%, 43.9%, and 43.2%, respectively. Niones and Thomson et al. [22,71] used the near-isogenic lines BC_3F_4 and BC_3F_5 with IR29 as the background and Pokkali as the donor to confirm the position of the *Saltol* locus. Since the positions of *Saltol* and *qSKC-1* were nearby on the chromosome, and both were responsible for regulating the Na^+/K^+ balance of rice under salt stress, Thomson et al. [22] speculated that *Saltol* and *qSKC-1* may encode the same gene (OsHKT1;5).

Some studies conducted fine mapping and cloning on salt-tolerant and salt-sensitive mutants. Lan et al. [72] fine-mapped the seedling salt-tolerant mutant gene *SST* to the 17 kb interval on chromosome 6, and the only predicted gene in this interval is *OsSPL10*, which might be a candidate gene for *SST*. Ogawa et al. [73] and Toda et al. [74] used the salt-sensitive mutants *rss1* and *rss3* to clone the salt-tolerant-related genes *RSS1* and *RSS3*, respectively. *RSS1* participated in the regulation of the cell cycle and was an important factor for maintaining the viability and vigor of meristematic cells under salt stress; *RSS3* regulated the expression of the jasmonic acid-responsive gene, and was involved in maintaining root cell elongation at an appropriate rate under salt stress. Deng et al. [75,76] analyzed the salt-tolerant and salt-sensitive mutants *rst1*, *rss2* and *rss4*. They detected two QTLs (*qSNC-1* and *qSNC-6*) that control the Na^+ content of aerial parts on chromosomes 1 and 6, which explained 14.5% and 53.3% of the phenotypic variation, respectively. The synergistic alleles were derived from *rss2*.

3. Association Analysis of Rice Salt Tolerance

Traditional QTL mapping usually uses markers to perform linkage analysis on the segregating populations derived from the parental cross F_1. This requires construction of a mapping population with a long cycle, and it has limited mapping accuracy and detectable alleles. Therefore, association analysis based on linkage disequilibrium is more widely used to analyze quantitative traits of plants [77,78]. Association analysis is now commonly applied to the identification of rice salt tolerance genes.

3.1. Association Analysis of Salt Tolerance Candidate Genes

To identify salt tolerance QTLs or candidate genes in 180 japonica rice from the European Rice Core Collection (ERCC), Ahmadi et al. [79] used 124 SNP and 52 SSR markers to associate 14 salt tolerance QTLs and 65 salt tolerance candidate genes. They identified 19 gene loci that were significantly associated with one or more salt stress traits. Negrão et al. [80] analyzed 392 rice germplasm resources by EcoTILLING technology, and they found the allele polymorphisms of five salt tolerance candidate genes, which were related to Na^+/K^+ balance, signal cascade and stress protection. There were 40 new alleles in the coding sequences of these genes, and 11 SNPs related to salt tolerance in rice were identified by association analysis.

Association analysis identified QTLs, candidate genes, and alleles related to the salt tolerance of rice, and it also revealed different salt tolerance mechanisms of different genotypes of rice. However, none of the rice varieties carried favorable alleles at all salt tolerance loci [79,80].

3.2. SSR Association Analysis of Salt Tolerance

SSR markers were used on 300 rice resources to analyze the association between salt tolerance in the seedling stage and tolerance in the whole growth period. Zheng et al. [63] identified the salt tolerance of 342 japonica rice with seedling survival days and shoot

Na^+/K^+ as evaluation indicators at the seedling stage. They used 160 pairs of SSR markers for salt tolerance association analysis. A total of twelve SSR markers were significantly associated with salt tolerance. A total of nine of the markers were close to the positions of reported salt tolerance QTLs and four markers were in the same position of known salt tolerance-related genes (*OsEREBP1*, *OsABF2*, *HKT1;5* and *OsAHP1*). Cui et al. [81] planted 347 japonica rice on coastal tidal flats, and examined agronomic traits such as heading date, plant height, effective panicle number, grain number per panicle, spikelet fertility, and thousand-grain weight. These traits were used as evaluation parameters for salt tolerance association analysis with 148 SSR markers. The study identified 25 SSR markers linked to rice salt tolerance. These markers are located on 10 chromosomes, except for the fifth and sixth chromosomes, and explained from 4.58% to 31.65% of the phenotypic variation. Among these loci, 10 markers were consistent with, or close to, the positions of reported salt tolerance QTLs on the chromosome.

3.3. Salt Tolerance Genome-Wide Association Analysis

Genome-wide association analysis (GWAS) has recently played an important role in discovering genes that regulate plant salt tolerance [82]. Through phenotypic variation analysis, GWAS can analyze the interactions between traits that previously seemed independent, and this can help explain the interactions among potential genes [83–85].

Kumar et al. [86] identified rice salt tolerance gene loci with 220 rice materials, and they performed an association analysis on 12 agronomic traits related to salt tolerance during the reproductive growth stage and the accumulation of Na^+ and K^+ in leaves. They identified 20 SNPs significantly related to leaf Na^+/K^+, and 44 SNPs related to other salt tolerance traits. These gene loci explained 5–18% of the phenotypic variation. Zhang et al. [87] used a multiparent advanced generation intercross population, DC1, and identified salt tolerance QTLs by GWAS. There were 7 QTLs delineated from 186 associations that were detected on chromosomes 1, 2, 5 and 9, which explained 7.42–9.38% of the total phenotypic variation. Liu et al. [88] found five known genes (*OsSUT1*, *OsCTR3*, *OsMYB6*, *OsHKT1;4*, and *OsGAMYB*) and two novel genes (*LOC_Os02g49700* and *LOC_Os03g28300*) that were associated with grain yield under salinity stress. Batayeva et al. [89] performed GWAS on 9 seedling salt tolerance traits of 191 japonica rice, and they detected 26 significant loci. Neang et al. [90] used 296 accessions of rice to identify salt-related traits and used 36,901 SNPs to conduct GWAS. They found 13 candidate genes. Yu et al. [91] used 295 accessions and identified 93 candidate genes with high association peaks of salt stress. Cui et al. [92] reported six multi-locus GWAS methods (mrMLM, FASTmrMLM, FASTmrEMMA, pLARmEB, pKWmEB, and ISIS EM-BLASSO), and identified 162,529 SNPs at the seed germination stage for salt tolerance traits with 478 rice accessions. Lekklar et al. [93] conducted a GWAS of salt tolerance using Thai rice accessions, and they found 164 genes co-localized with reported salt quantitative trait loci. These accounted for 73% of the identified loci. Yu et al. [94] performed a GWAS of salt-tolerance-related phenotypes in rice during the germination stage with 295 accessions. They found *OsMADS31*, one of the MADS-box family transcription factors, had down-regulated expression and was predicted to participate in salt stress at the germination stage. Rohila et al. [95] conducted a GWAS of early vigor traits under salt stress with the natural genetic variation in the United States Department of Agriculture rice mini-core collection. They identified 14 salt-tolerant accessions, 6 new loci, and 16 candidate genes that could contribute to salt tolerance breeding. Warraich et al. [96] evaluated 180 rice accessions for salinity tolerance at the reproductive stage by GWAS and 19 associations were identified for Na^+, K^+ and Na^+/K^+ uptake in leaves and stems. Based on 6,361,920 single nucleotide polymorphisms in 478 rice accessions, Shi et al. [97] identified 22 salt tolerance-associated SNPs based on salt tolerance-related traits. There were seven loci on chromosomes 1, 5, 6, 11, and 12 that were close to six previously identified quantitative gene loci/genes related to salinity tolerance. These studies showed that there were some genes expected to be involved in salt resistance in rice, including a nitrate transporter gene *OsNRT2.1* [97], a MADS-box family transcription factor gene *OsMADS31* [94], a sucrose transport protein gene *OsSUT1*, a transcript factor gene *OsGAMYB*, and some function genes *OsCTR3*, *OsMYB6* and *OsHKT1;4* [88], etc. With the development of variety resequencing, salt tolerance GWAS has rapidly developed [98–105]. GWAS has exploited the natural variation in root architecture remodeling under salt tolerance to uncover the genetic controls underlying plant responses [83,106]. The studies provide insight into the genetic structure of salt tolerance and are important resources for breeding programs.

In Table 2, we summarize 25 association analyses of salt tolerance with more than 600 genetic sites related to salt tolerance in rice

Table 2. Association analysis of salt tolerance in rice.

Stage	Population Size	Maker Type	QTL	Traits	Reference
Generation stage	478	SNP (6.36M)	11	GR, germination index, vigor index, germination time, and imbibition rate	[97]
	295	SNP (1.65M)	12	GR, germination energy, germination index, SH, RL	[94]
	184	SNP (788K)	8	RL under control condition, alkaline stress and relative RL	[101]
Seedling stage	32	SSR (64)	28	Salt tolerance level	[107]
	342	SSR (160)	12	SST, SNC, SKC, RNC and RKC	[15]
	533	SNP (700K)	20	Relative growth rate, transpiration use efficiency and transpiration rates	[108]
	295	SNP (1.65M)	25	Leaf width, SH, RL, total dry weight	[91]
	235	SNP (30K)	27	Tiller number, SH, RL, SDW, RDW, RDW/SDW, leaf area, SNKR	[79]
	306	SNP (200K)	58	SNC, SKC	[103]
	203	SNP (68K)	26	Shoot Na^+ and K^+ content, standard evaluation score, percentage of damage, SDW	[89]
	708	SNP (3.45M)	41	SDS, SSI	[88]
	162	SNP (3.2M)	9	SSI, SDW, RDW, SDS	[95]
	176	SSR (154)	13	Salinization damage grade, SNC, SKC, SNKR	[109]
	221	SNP (55K)	7	SES, SDS, SH and RL under salt treatment, relative SH and RL, SDW and RDW after salt treatment, relative SDW and RDW, relative biomass.	[87]
	181	SNP (32K)	54	SH, SFW and SDW under control, salt stress conditions, relative SH, SFW and SDW	[98]
	295	SNP (788K)	8	SST, SNC, SKC, SNKR	[100]
	664	SNP (3M)	21	SH, RL, SFW, SDW, RFW, RDW, salt tolerance level	[105]
Vegetative growth stage	104	SNP (112K)	200	Photosynthetic parameters and cell membrane stability	[93]
	296	SNP (44K)	11	Na^+ and Cl^- of leaf blades, Na^+ and Cl^- sheath:blade ratios, SES	[90]
	179	SNP (21K)	26	SES, chlorophyll content, water content, Na^+ and K^+ contents, SNKR	[99]
	96	SNP (50K)	23	SH, RL, SFW, SDW, RFW, RDW, RNC, SNC, RKC, SKC, RNK, SNKR, RNKR	[104]
Reproductive growth stage	220	SNP (6K)	64	SNKR, PH, TN, spikelet fertility, unfilled or filled grains, yield	[86]
	347	SSR (148)	25	Salinity tolerance index	[81]
Multiple growth stages	180	SSR (150)	28	Na^+, K^+, Ca^{2+}, Mg^{2+} content in stem and leaves, grain yield and SSI	[96]
	208	SNP (395K)	20	Generation stage: GR.Seedling stage: SH, RL	[102]

Note: GR: germination rate. SDS: survival days of seedling. SSI: salt stress injury score. SST: score of salt tolerance. SNC and SKC: the concentrations of Na^+ and K^+ in shoots. SNKR: shoot Na^+/K^+ ratio. RNC and RKC: the concentrations of Na^+ and K^+ in roots. RNKR: root Na^+/K^+ ratio. SES: standard evaluation system score. SH: shoot height, RL: root length, SDW: shoot dry weight. SFW: shoot fresh weight. RDW: root dry weight. RFW: root fresh weight. PH: plant height. TN: tiller number.

4. Issues and Prospects

Most studies of rice salt tolerance gene mapping and cloning evaluated the salt tolerance of a specific growth and development stage such as the seedling stage. Few studies have evaluated salt tolerance in multiple growth stages. Due to the differences in the salt tolerance of rice at different growth and development stages, it is necessary to conduct QTL analysis or association mapping for salt tolerance in the different growth stages of rice, especially the seedling stage and reproductive growth stage. This will enable identification of genes that simultaneously control salt tolerance in multiple growth stages. This research could begin by screening salt-tolerant rice germplasm resources during the whole growth period, and then identifying materials that complete their life cycle under salt stress with less impact on yield traits. This approach could be useful for salt-tolerant gene mapping, cloning, and breeding.

Hundreds of QTLs related to salt tolerance have been identified but the progress of follow-up work on fine-mapping and map-based cloning of genes has been slow. One reason is the lack of salt-tolerant rice varieties in the parental combinations. A small difference in salt tolerance between the two parents may only result in a small contribution to the phenotype of the identified QTLs and disturb the fine-mapping by the genetic background. However, most of salt-tolerant QTL mapping studies involved a single time period or a single-year phenotypic identification, and there was a lack of QTL stability. Some salt-tolerant QTLs with a high phenotypic contribution rate (above 20%) could be difficult to fine-map and clone due to poor genetic stability. Therefore, it is necessary to select strong salt-tolerant rice varieties for salt-tolerant QTL mapping, and also to test the stability of salt-tolerant QTLs multiple times or by multi-year multi-point experiments for gene cloning and breeding.

For mining salt-tolerant genes for salt-tolerant germplasm resources, traditional QTL analysis methods could be combined with mutant construction, screening, and association analysis. Association analysis, especially GWAS, is now commonly used for the analysis of complex traits of plants, and it aids understanding of the genetic basis and differentiation of salt tolerance in rice. Using QTL mapping and GWAS, the genetic basis of many complex quantitative traits has been analyzed and many QTL segments or loci have been located. With the development of genome sequencing technology and bioinformatics, using the reference genome sequence information of the corresponding species could help determine the candidate genes related to the target trait and narrow the range of candidate genes [110,111]. Moreover, GWAS is useful for marker-assisted selection (MAS) of rice varieties suitable for cultivation in salinized fields.

With the development of molecular marker technology, MAS technology is now widely used in crop breeding, thus providing a new way to accelerate the genetic improvement of rice salt tolerance. MAS could select target traits in early generations, accelerate the breeding process, and aggregate multiple beneficial genes at the same time to improve breeding efficiency [112]. A QTL for salt tolerance in rice that is frequently used in MAS breeding is *Saltol*, which is located on chromosome 1 [113]. *Saltol* QTL is a major QTL associated with the Na^+/K^+ ratio and salinity tolerance at the seedling stage in rice. Several genes have been reported in the *Saltol* QTL (*LEA*, *CaMBP*, *V-ATPase*, *GST*, *OSAP1* zing finger protein and transcription factor *HBP1b*) that were salinity sensitive and regulated between the genotypes [114–116]. Most studies were performed using marker-assisted backcrossing (MABC) technology in India, the Philippines, Bangladesh, Thailand, Vietnam and Senegal [117–119]. Some *Saltol* introgression lines cultivated by MAS, such as BR11-SalTol and BRRI dhan28-SalTol, have been tested in salt damaged coastal areas in the Philippines, Bangladesh, India, and Vietnam [113,120]. Bimpong et al. [113] found that compared with traditional breeding, MAS breeding can shorten the germplasm improvement time by four to seven years.

Author Contributions: Conceptualization, L.M. and J.C.; writing-original draft preparation, X.F., H.J. and J.C.; data curation, X.F., H.J. and J.C.; supervision, L.M. and J.C. All authors have read and agreed to the published version of the manuscript.

Funding: This research was financially supported by the National Natural Science Foundation of China (31902103), Doctoral Scientific Research Foundation of Anshan Normal University (21b04),

Postgraduate Research and Practice Innovation Program of Jiangsu Province (KYCX19_0546), and Shenzhen Science and Technology Projects (No. JCYJ20210324124409027 to J.C.).

Institutional Review Board Statement: Not applicable.

Informed Consent Statement: Not applicable.

Data Availability Statement: All of the data generated or analyzed during this study are included in this published article.

Conflicts of Interest: The authors declare no conflict of interest.

References

1. Ismail, A.M.; Horie, T. Genomics, Physiology, and Molecular Breeding Approaches for Improving Salt Tolerance. *Annu. Rev. Plant Biol.* **2017**, *68*, 405–434. [CrossRef] [PubMed]
2. Chinnusamy, V.; Jagendorf, A.; Zhu, J.K. Understanding and Improving Salt Tolerance in Plants. *Crop Sci.* **2005**, *45*, 437–448. [CrossRef]
3. Munns, R.; Tester, M. Mechanisms of salinity tolerance. *Annu. Rev. Plant Biol.* **2008**, *59*, 651–681. [CrossRef]
4. Zeng, L.; Shannon, M.C.; Grieve, C.M. Evaluation of salt tolerance in rice genotypes by multiple agronomic parameters. *Euphytica* **2002**, *127*, 235–245. [CrossRef]
5. Ponce, K.S.; Meng, L.; Guo, L.; Leng, Y.; Ye, G. Advances in Sensing, Response and Regulation Mechanism of Salt Tolerance in Rice. *Int. J. Mol. Sci.* **2021**, *22*, 2254. [CrossRef] [PubMed]
6. Al-Tamimi, N.; Oakey, H.; Tester, M.; Negrão, S. Assessing Rice Salinity Tolerance: From Phenomics to Association Mapping. In *Rice Genome Engineering and Gene Editing: Methods and Protocols*; Bandyopadhyay, A., Thilmony, R., Eds.; Springer: New York, NY, USA, 2021; pp. 339–375.
7. Quan, R.; Wang, J.; Hui, J.; Bai, H.; Lyu, X.; Zhu, Y.; Zhang, H.; Zhang, Z.; Li, S.; Huang, R. Improvement of Salt Tolerance Using Wild Rice Genes. *Front. Plant Sci.* **2017**, *8*, 2269. [CrossRef] [PubMed]
8. Deinlein, U.; Stephan, A.B.; Horie, T.; Luo, W.; Xu, G.; Schroeder, J.I. Plant salt-tolerance mechanisms. *Trends Plant Sci.* **2014**, *19*, 371–379. [CrossRef] [PubMed]
9. Morton, M.J.L.; Awlia, M.; Al-Tamimi, N.; Saade, S.; Pailles, Y.; Negrão, S.; Tester, M. Salt stress under the scalpel-dissecting the genetics of salt tolerance. *Plant J.* **2019**, *97*, 148–163. [CrossRef] [PubMed]
10. Zelm, E.v.; Zhang, Y.; Testerink, C. Salt Tolerance Mechanisms of Plants. *Annu. Rev. Plant Biol.* **2020**, *71*, 403–433. [CrossRef]
11. Shan, T.; Pang, S.; Wang, X.; Li, J.; Li, Q.; Su, L.; Li, X. A method to establish an "immortalized F-2" sporophyte population in the economic brown alga Undaria pinnatifida (Laminariales: Alariaceae). *J. Phycol.* **2020**, *56*, 1748–1753. [CrossRef] [PubMed]
12. Mazumder, A.; Rohilla, M.; Bisht, D.S.; Krishnamurthy, S.L.; Barman, M.; Sarma, R.N.; Sharma, T.R.; Mondal, T.K. Identification and mapping of quantitative trait loci (QTL) and epistatic QTL for salinity tolerance at seedling stage in traditional aromatic short grain rice landrace Kolajoha (Oryza sativa L.) of Assam, India. *Euphytica* **2020**, *216*, 1211–1226. [CrossRef]
13. Wang, Z.; Wang, J.; Bao, Y.; Wu, Y.; Zhang, H. Quantitative trait loci controlling rice seed germination under salt stress. *Euphytica* **2010**, *178*, 297–307. [CrossRef]
14. Wang, Z.; Cheng, J.; Chen, Z.; Huang, J.; Bao, Y.; Wang, J.; Zhang, H. Identification of QTLs with main, epistatic and QTL x environment interaction effects for salt tolerance in rice seedlings under different salinity conditions. *Theor. Appl. Genet.* **2012**, *125*, 807–815. [CrossRef] [PubMed]
15. Zheng, H.; Zhao, H.; Liu, H.; Wang, J.; Zou, D. QTL analysis of Na+ and K+ concentrations in shoots and roots under NaCl stress based on linkage and association analysis in japonica rice. *Euphytica* **2014**, *201*, 109–121. [CrossRef]
16. Flowers, T.J.; Koyama, M.L.; Flowers, S.A.; Sudhakar, C.; Singh, K.P.; Yeo, A.R. QTL: Their place in engineering tolerance of rice to salinity. *J. Exp. Bot.* **2000**, *51*, 99–106. [CrossRef]
17. Koyama, M.L.; Levesley, A.; Koebner, R.; Flowers, T.; Yeo, A.R. Quantitative Trait Loci for Component Physiological Traits Determining Salt Tolerance in Rice. *Plant Physiol.* **2001**, *125*, 406–422. [CrossRef]
18. Haq, U.T.; Gorham, J.; Akhtar, J.; Akhtar, N.; Steele, K.A. Dynamic quantitative trait loci for salt stress components on chromosome 1 of rice. *Funct. Plant Biol.* **2010**, *37*, 634–645. [CrossRef]
19. Lee, S.; Ahn, J.; Cha, Y.; Yun, D.; Lee, M.; Ko, J.; Lee, K.; Eun, M. Mapping of quantitative trait loci for salt tolerance at the seedling stage in rice. *Mol. Cells* **2006**, *21*, 192–196. [PubMed]
20. Lee, S.Y.; Ahn, J.H.; Cha, Y.S.; Yun, D.W.; Lee, M.C.; Ko, J.C.; Lee, K.S.; Eun, M.Y. Mapping QTLs related to salinity tolerance of rice at the young seedling stage. *Plant Breed.* **2007**, *126*, 43–46. [CrossRef]
21. Wang, B.; Lan, T.; Wu, W. Mapping of QTLs for Na+ content in rice seedlings under salt stress. *Chin. J. Rice Sci.* **2007**, *21*, 585–590.
22. Thomson, M.J.; de Ocampo, M.; Egdane, J.; Rahman, M.A.; Sajise, A.G.; Adorada, D.L.; Tumimbang-Raiz, E.; Blumwald, E.; Seraj, Z.I.; Singh, R.K.; et al. Characterizing the Saltol Quantitative Trait Locus for Salinity Tolerance in Rice. *Rice* **2010**, *3*, 148–160. [CrossRef]
23. Liang, J.; Qu, Y.; Yang, C.; Ma, X.; Cao, G.; Zhao, Z.; Zhang, S.; Zhang, T.; Han, L. Identification of QTLs associated with salt or alkaline tolerance at the seedling stage in rice under salt or alkaline stress. *Euphytica* **2014**, *201*, 441–452. [CrossRef]

24. Tiwari, S.; Sl, K.; Kumar, V.; Singh, B.; Rao, A.R.; Mithra SV, A.; Rai, V.; Singh, A.K.; Singh, N.K. Mapping QTLs for Salt Tolerance in Rice (Oryza sativa L.) by Bulked Segregant Analysis of Recombinant Inbred Lines Using 50K SNP Chip. *PLoS ONE* **2016**, *11*, e0153610. [CrossRef]
25. Pandit, A.; Rai, V.; Bal, S.; Sinha, S.; Kumar, V.; Chauhan, M.; Gautam, R.K.; Singh, R.; Sharma, P.C.; Singh, A.K.; et al. Combining QTL mapping and transcriptome profiling of bulked RILs for identification of functional polymorphism for salt tolerance genes in rice (Oryza sativa L.). *Mol. Genet. Genom.* **2010**, *284*, 121–136. [CrossRef] [PubMed]
26. Sun, Y.; Zang, J.; Wang, Y.; Zhu, L.; Mohammadhosein, F.; Xu, J.; Li, Z. Mining Favorable Salt-tolerant QTL from Rice Germplasm Using a Backcrossing Introgression Line Population. *Acta Agron. Sin.* **2007**, *33*, 1611–1617.
27. Kim, D.M.; Ju, H.G.; Kwon, T.R.; Oh, C.S.; Ahn, S.N. Mapping QTLs for salt tolerance in an introgression line population between Japonica cultivars in rice. *J. Crop Sci. Biotech.* **2009**, *12*, 121–128. [CrossRef]
28. Qian, Y.; Wang, H.; Chen, M.; Zhang, L.; Chen, B.; Cui, J.; Liu, H.; Zhu, L.; Shi, Y.; Gao, Y.; et al. Detection of Salt-tolerant QTL Using BC2F3 Yield Selected Introgression Lines of Rice (Oryza sativa L.). *Mol. Plant Breed.* **2009**, *7*, 224–232.
29. Yang, J.; Sun, Y.; Cheng, L.R.; Zhou, Z.; Wang, Y.; Zhu, L.H.; Cang, J.; Xu, J.L.; Li, Z.K. Genetic Background Effect on QTL Mapping for Salt Tolerance Revealed by a Set of Reciprocal Introgression Line Populations in Rice. *Acta Agron. Sin.* **2009**, *35*, 974–982. [CrossRef]
30. Alam, R.; Rahman, M.S.; Seraj, Z.I.; Thomson, M.J.; Ismail, A.M.; Tumimbang-Raiz, E.; Gregorio, G.B. Investigation of seedling-stage salinity tolerance QTLs using backcross lines derived from *Oryza sativa* L. Pokkali. *Plant Breed.* **2011**, *130*, 430–437. [CrossRef]
31. Li, Y.F.; Zheng, Y.; Vemireddy, L.R.; Panda, S.K.; Jose, S.; Ranjan, A.; Panda, P.; Govindan, G.; Cui, J.; Wei, K.; et al. Comparative transcriptome and translatome analysis in contrasting rice genotypes reveals differential mRNA translation in salt-tolerant Pokkali under salt stress. *BMC Genom.* **2018**, *19*, 935. [CrossRef] [PubMed]
32. Tian, L.; Tan, L.; Liu, F.; Cai, H.; Sun, C. Identification of quantitative trait loci associated with salt tolerance at seedling stage from Oryza rufipogon. *J. Genet. Genom.* **2011**, *38*, 593–601. [CrossRef]
33. Qiu, X.; Yuan, Z.; Liu, H.; Xiang, X.; Yang, L.; He, W.; Du, B.; Ye, G.; Xu, J.; Xing, D.; et al. Identification of salt tolerance-improving quantitative trait loci alleles from a salt-susceptible rice breeding line by introgression breeding. *Plant Breed.* **2015**, *134*, 653–660. [CrossRef]
34. Ahmadi, J.; Fotokian, M.H. Identification and mapping of quantitative trait loci associated with salinity tolerance in rice (Oryza Sativa) using SSR markers. *Iran. J. Biotechnol.* **2011**, *9*, 21–30.
35. Cheng, L.; Wang, Y.; Meng, L.; Hu, X.; Cui, Y.; Sun, Y.; Zhu, L.; Ali, J.; Xu, J.; Li, Z. Identification of salt-tolerant QTLs with strong genetic background effect using two sets of reciprocal introgression lines in rice. *Genome* **2012**, *55*, 45–55. [CrossRef] [PubMed]
36. Zang, J.; Sun, Y.; Wang, Y.; Yang, J.; Li, F.; Zhou, Y.; Zhu, L.; Jessica, R.; Mohammadhosein, F.; Xu, J.; et al. Dissection of genetic overlap of salt tolerance QTLs at the seedling and tillering stages using backcross introgression lines in rice. *Sci. China C Life Sci.* **2008**, *51*, 583–591. [CrossRef] [PubMed]
37. Takehisa, H.; Shimodate, T.; Fukuta, Y.; Ueda, T.; Yano, M.; Yamaya, T.; Kameya, T.; Sato, T. Identification of quantitative trait loci for plant growth of rice in paddy field flooded with salt water. *Field Crop. Res.* **2004**, *89*, 85–95. [CrossRef]
38. Prasad, S.; Bagali, P.; Hittalmani, S.; Shashidhar, H. Molecular mapping of quantitative trait loci associated with seedling tolerance to salt stress in rice (Oryza sativa L). *Curr. Sci.* **2000**, *78*, 162–164.
39. Gong, J.; He, P.; Qian, Q.; Shen, L.; Zhu, L.; Chen, S. Identification of salt-tolerance QTL in rice (Oryza sativa L.). *Chin. Sci. Bull.* **1999**, *41*, 68–71. [CrossRef]
40. Gong, J.; Zheng, X.; Du, B.; Qian, Q.; Chen, S.; Zhu, L.; He, P. Comparative study of QTLs for agronomic traits of rice (Oryza sativa L.) between salt stress and nonstress environment. *Sci. China C Life Sci.* **2001**, *44*, 73–82. [CrossRef] [PubMed]
41. Zhao, C.; Zhang, S.; Zhao, Q.; Zhou, L.; Zhao, L.; Yao, S.; Zhang, Y.; Wang, C. Mapping of QTLs for bud-stage salinity tolerance based on chromosome segment substitution line in rice. *Acta Agric. Boreali-Sin.* **2017**, *32*, 106–111.
42. Mardani, Z.; Rabiei, B.; Sabouri, H.; Sabouri, A.; Virk, P. Identification of molecular markers linked to salt-tolerant genes at germination stage of rice. *Plant Breed.* **2014**, *133*, 196–202. [CrossRef]
43. Ghomi, K.; Rabiei, B.; Sabouri, H.; Sabouri, A. Mapping QTLs for traits related to salinity tolerance at seedling stage of rice (Oryza sativa L.): An agrigenomics study of an Iranian rice population. *OMICS* **2013**, *17*, 242–251. [CrossRef]
44. Lin, H.X.; Zhu, M.Z.; Yano, M.; Gao, J.P.; Liang, Z.W.; Su, W.A.; Hu, X.H.; Ren, Z.H.; Chao, D.Y. QTLs for Na+ and K+ uptake of the shoots and roots controlling rice salt tolerance. *Theor. Appl. Genet.* **2004**, *108*, 253–260. [CrossRef] [PubMed]
45. Sabouri, H.; Sabouri, A. New evidence of QTLs attributed to salinity tolerance in rice. *Afr. J. Biotechnol.* **2008**, *7*, 4376–4383.
46. Sabouri, H.; Rezai, A.; Moumeni, A.; Kavousi, A.; Katouzi, M.; Sabouri, A. QTLs mapping of physiological traits related to salt tolerance in young rice seedlings. *Biol. Plant.* **2009**, *53*, 657–662. [CrossRef]
47. Javed, M.A.; Huyop, F.Z.; Wagiran, A.; Salleh, F.M. Identification of QTLs for Morph-Physiological Traits Related to Salinity Tolerance at Seedling Stage in Indica Rice. *Procedia Environ. Sci.* **2011**, *8*, 389–395. [CrossRef]
48. Islam, M.; Salam, M.; Hassan, L.; Collard, B.Y.; Singh, R.; Gregorio, G. QTL mapping for salinity tolerance at seedling stage in rice. *Emir. J. Food Agric.* **2011**, *23*, 137. [CrossRef]
49. Sun, J.; Zou, D.T.; Luan, F.S.; Zhao, H.W.; Wang, J.G.; Liu, H.L.; Xie, D.W.; Su, D.Q.; Ma, J.; Liu, Z.L. Dynamic QTL analysis of the Na+ content, K+ content, and Na+ /K+ ratio in rice roots during the field growth under salt stress. *Biol. Plant.* **2014**, *58*, 689–696. [CrossRef]

50. Yao, M.; Wang, J.; Chen, H.; Zhai, H.; Zhang, H. Inheritance and QTL mapping of salt tolerance in rice. *Rice Sci.* **2005**, *12*, 25–32.
51. Mohammadi, R.; Mendioro, M.; Diaz, G.; Gregorio, G.; Singh, R. Mapping quantitative trait loci associated with yield and yield components under reproductive stage salinity stress in rice (Oryza sativa L.). *J. Genet.* **2013**, *93*, 433–443. [CrossRef] [PubMed]
52. Bimpong, I.K.; Manneh, B.; Diop, B.; Ghislain, K.; Sow, A.; Amoah, N.K.A.; Gregorio, G.; Singh, R.K.; Ortiz, R.; Wopereis, M. New quantitative trait loci for enhancing adaptation to salinity in rice from Hasawi, a Saudi landrace into three African cultivars at the reproductive stage. *Euphytica* **2014**, *200*, 45–60. [CrossRef]
53. Khan, M.S.K.; Saeed, M.; Iqbal, J. Identification of quantitative trait loci for Na+, K+ and Ca++ accumulation traits in rice grown under saline conditions using F2 mapping population. *Braz. J. Bot.* **2015**, *38*, 555–565. [CrossRef]
54. Khan, M.S.K.; Saeed, M.; Iqbal, J. Quantitative trait locus mapping for salt tolerance at maturity stage in indica rice using replicated F2 population. *Braz. J. Bot.* **2016**, *39*, 641–650. [CrossRef]
55. Ammar, M.; Pandit, A.; Singh, R.; Sameena, S.; Chauhan, M.; Singh, A.; Sharma, P.; Gaikwad, K.; Sharma, T.; Mohapatra, T.; et al. Mapping of QTLs controlling Na+, K+ and Cl− ion concentrations in salt tolerant indica rice variety. *J. Plant Biochem. Biot.* **2009**, *18*, 139–150. [CrossRef]
56. Hossain, H.; Rahman, M.A.; Alam, M.S.; Singh, R.K. Mapping of Quantitative Trait Loci Associated with Reproductive-Stage Salt Tolerance in Rice. *J. Agron. Crop Sci.* **2015**, *201*, 17–31. [CrossRef]
57. Gu, X.; Mei, M.; Yan, X.; Zheng, S.; Lu, Y. Preliminary detection of quantitative trait loci for salt tolerance in rice. *Chin. J. Rice Sci.* **2000**, *14*, 65–70. [CrossRef]
58. Negrão, S.; Courtois, B.; Ahmadi, N.; Abreu, I.; Saibo, N.; Oliveira, M.M. Recent Updates on Salinity Stress in Rice: From Physiological to Molecular Responses. *Crit. Rev. Plant Sci.* **2011**, *30*, 329–377. [CrossRef]
59. Wang, Z.; Chen, Z.; Cheng, J.; Lai, Y.; Wang, J.; Bao, Y.; Huang, J.; Zhang, H. QTL analysis of Na+ and K+ concentrations in roots and shoots under different levels of NaCl stress in rice (Oryza sativa L.). *PLoS ONE* **2012**, *7*, e51202. [CrossRef] [PubMed]
60. IRRI. *Standard Evaluation System for Rice*, 4th ed.; International Rice Research Institute: Manila, Philippines, 1996; Volume 52.
61. Lin, H.; Yanagihara, S.; Zhuang, J.; Senboku, T.; Yashima, S. Identification of QTL for salt tolerance in rice via molecular markers. *Chin. J. Rice Sci.* **1998**, *12*, 72–78. [CrossRef]
62. Masood, M.; Seiji, Y.; Shinwari, Z.; Anwar, R. Mapping quantitative trait loci (QTLs) for salt tolerance in rice (Oryza sativa) using RFLPs. *Pak. J. Bot.* **2004**, *36*, 825–834.
63. Zheng, H.; Wang, J.; Zhao, H.; Liu, H.; Sun, J.; Guo, L.; Zou, D. Genetic structure, linkage disequilibrium and association mapping of salt tolerance in japonica rice germplasm at the seedling stage. *Mol. Breed.* **2015**, *35*, 152. [CrossRef]
64. Wang, S.; Cao, M.; Ma, X.; Chen, W.; Zhao, J.; Sun, C.; Tan, L.; Liu, F. Integrated RNA Sequencing and QTL Mapping to Identify Candidate Genes from Oryza rufipogon Associated with Salt Tolerance at the Seedling Stage. *Front. Plant Sci.* **2017**, *8*, 1427. [CrossRef] [PubMed]
65. Poonam, R.; Sunita, J.; Sheetal, Y.; Navinder, S.; Jain, R.K. Identification of SSR Markers for Salt-tolerance in Rice Variety CSR10 by Selective Genotyping. *J. Plant Biochem. Biotechnol.* **2009**, *18*, 87–91.
66. Chai, L.; Zhang, J.; Pan, X.; Zhang, F.; Zheng, T.; Zhao, X.; Wang, W.; Jauhar, A.; Xu, J.; Li, Z. Advanced Backcross QTL Analysis for the Whole Plant Growth Duration Salt Tolerance in Rice (Oryza sativa L.). *J. Integr. Agric.* **2014**, *13*, 1609–1620. [CrossRef]
67. Ren, Z.H.; Gao, J.P.; Li, L.G.; Cai, X.L.; Huang, W.; Chao, D.Y.; Zhu, M.Z.; Wang, Z.Y.; Luan, S.; Lin, H.X. A rice quantitative trait locus for salt tolerance encodes a sodium transporter. *Nat. Genet.* **2005**, *37*, 1141–1146. [CrossRef] [PubMed]
68. Riedelsberger, J.; Miller, J.K.; Valdebenito-Maturana, B.; Pineros, M.A.; Gonzalez, W.; Dreyer, I. Plant HKT Channels: An Updated View on Structure, Function and Gene Regulation. *Int. J. Mol. Sci.* **2021**, *22*, 1892. [CrossRef] [PubMed]
69. Gregorio, G.B. Tagging salinity tolerance genes in rice using amplified fragment length polymorphism (AFLP). Ph.D. Thesis, University of the Philippines, Los Banos, CA, USA, 1997; 118p.
70. Bonilla, P.; Dvorak, J.; Mackill, D.; Deal, K.; Gregorio, G. RFLP and SSLP mapping of salinity tolerance genes in chromosome 1 of rice (Oryza sativa L.) using recombinant inbred lines. *Philipp. Agric. Sci.* **2002**, *65*, 68–76.
71. Niones, J.; Gregorio, G.; Tumimbang, E. Fine mapping of the salinity tolerance gene on chromosome 1 of rice (Oryza sativa L.) using near isogenic lines. *Crop Sci. Soc. Philipp.* **2004**, *31*, 46.
72. Lan, T.; Zhang, S.; Liu, T.; Wang, B.; Guan, H.; Zhou, Y.; Duan, Y.; Wu, W. Fine mapping and candidate identification of SST, a gene controlling seedling salt tolerance in rice (Oryza sativa L.). *Euphytica* **2015**, *205*, 269–274. [CrossRef]
73. Ogawa, D.; Abe, K.; Miyao, A.; Kojima, M.; Sakakibara, H.; Mizutani, M.; Morita, H.; Toda, Y.; Hobo, T.; Sato, Y.; et al. RSS1 regulates the cell cycle and maintains meristematic activity under stress conditions in rice. *Nat. Commun.* **2011**, *2*, 278. [CrossRef]
74. Toda, Y.; Tanaka, M.; Ogawa, D.; Kurata, K.; Kurotani, K.; Habu, Y.; Ando, T.; Sugimoto, K.; Mitsuda, N.; Katoh, E.; et al. RICE SALT SENSITIVE3 forms a ternary complex with JAZ and class-C bHLH factors and regulates jasmonate-induced gene expression and root cell elongation. *Plant Cell* **2013**, *25*, 1709–1725. [CrossRef] [PubMed]
75. Deng, P.; Jiang, D.; Dong, Y.; Shi, X.; Jing, W.; Zhang, W. Physiological characterisation and fine mapping of a salt-tolerant mutant in rice (Oryza sativa). *Funct. Plant Biol.* **2015**, *42*, 1026–1035. [CrossRef]
76. Deng, P.; Shi, X.; Zhou, J.; Wang, F.; Dong, Y.; Jing, W.; Zhang, W. Identification and Fine Mapping of a Mutation Conferring Salt-Sensitivity in Rice (Oryza sativa L.). *Crop Sci.* **2015**, *55*, 219–228. [CrossRef]
77. Huang, X.; Feng, Q.; Qian, Q.; Zhao, Q.; Wang, L.; Wang, A.; Guan, J.; Fan, D.; Weng, Q.; Huang, T.; et al. High-throughput genotyping by whole-genome resequencing. *Genome Res.* **2009**, *19*, 1068–1076. [CrossRef] [PubMed]

78. Yano, K.; Yamamoto, E.; Aya, K.; Takeuchi, H.; Lo, P.C.; Hu, L.; Yamasaki, M.; Yoshida, S.; Kitano, H.; Hirano, K.; et al. Genome-wide association study using whole-genome sequencing rapidly identifies new genes influencing agronomic traits in rice. *Nat. Genet.* **2016**, *48*, 927–934. [CrossRef] [PubMed]
79. Ahmadi, N.; Negrão, S.; Katsantonis, D.; Frouin, J.; Ploux, J.; Letourmy, P.; Droc, G.; Babo, P.; Trindade, H.; Bruschi, G.; et al. Targeted association analysis identified japonica rice varieties achieving Na+/K+ homeostasis without the allelic make-up of the salt tolerant indica variety Nona Bokra. *Theor. Appl. Genet.* **2011**, *123*, 881–895. [CrossRef] [PubMed]
80. Negrao, S.; Almadanim, M.C.; Pires, I.S.; Abreu, I.A.; Maroco, J.; Courtois, B.; Gregorio, G.B.; McNally, K.L.; Oliveira, M.M. New allelic variants found in key rice salt-tolerance genes: An association study. *Plant Biotechnol. J.* **2013**, *11*, 87–100. [CrossRef]
81. Cui, D.; Xu, C.; Yang, C.; Zhang, Q.; Zhang, J.; Ma, X.; Qiao, Y.; Cao, G.; Zhang, S.; Han, L. Association mapping of salinity and alkalinity tolerance in improved japonica rice (Oryza sativa L. subsp. japonica Kato) germplasm. *Genet. Resour. Crop Evol.* **2014**, *62*, 539–550. [CrossRef]
82. Li, B. Identification of Genes Conferring Plant Salt Tolerance using GWAS: Current Success and Perspectives. *Plant Cell Physiol.* **2020**, *61*, 1419–1426. [CrossRef]
83. Deolu-Ajayi, A.O.; Meyer, A.J.; Haring, M.A.; Julkowska, M.M.; Testerink, C. Genetic Loci Associated with Early Salt Stress Responses of Roots. *iScience* **2019**, *21*, 458–473. [CrossRef] [PubMed]
84. Cao, Y.; Zhang, M.; Liang, X.; Li, F.; Shi, Y.; Yang, X.; Jiang, C. Natural variation of an EF-hand Ca(2+)-binding-protein coding gene confers saline-alkaline tolerance in maize. *Nat. Commun.* **2020**, *11*, 186. [CrossRef] [PubMed]
85. Zhang, M.; Liang, X.; Wang, L.; Cao, Y.; Song, W.; Shi, J.; Lai, J.; Jiang, C. A HAK family Na+ transporter confers natural variation of salt tolerance in maize. *Nat. Plants* **2019**, *5*, 1297–1308. [CrossRef] [PubMed]
86. Kumar, V.; Singh, A.; Mithra, S.V.A.; Krishnamurthy, S.L.; Parida, S.K.; Jain, S.; Tiwari, K.K.; Kumar, P.; Rao, A.R.; Sharma, S.K.; et al. Genome-wide association mapping of salinity tolerance in rice (Oryza sativa). *DNA Res.* **2015**, *22*, 133–145. [CrossRef] [PubMed]
87. Zhang, Y.; Ponce, K.S.; Meng, L.; Chakraborty, P.; Zhao, Q.; Guo, L.; Gao, Z.; Leng, Y.; Ye, G. QTL identification for salt tolerance related traits at the seedling stage in indica rice using a multi-parent advanced generation intercross (MAGIC) population. *Plant Growth Regul.* **2020**, *92*, 365–373. [CrossRef]
88. Liu, C.; Chen, K.; Zhao, X.; Wang, X.; Shen, C.; Zhu, Y.; Dai, M.; Qiu, X.; Yang, R.; Xing, D.; et al. Identification of genes for salt tolerance and yield-related traits in rice plants grown hydroponically and under saline field conditions by genome-wide association study. *Rice (NY)* **2019**, *12*, 88. [CrossRef] [PubMed]
89. Batayeva, D.; Labaco, B.; Ye, C.; Li, X.; Usenbekov, B.; Rysbekova, A.; Dyuskalieva, G.; Vergara, G.; Reinke, R.; Leung, H. Genome-wide association study of seedling stage salinity tolerance in temperate japonica rice germplasm. *BMC Genet.* **2018**, *19*, 2. [CrossRef] [PubMed]
90. Neang, S.; de Ocampo, M.; Egdane, J.A.; Platten, J.D.; Ismail, A.M.; Seki, M.; Suzuki, Y.; Skoulding, N.S.; Kano-Nakata, M.; Yamauchi, A.; et al. A GWAS approach to find SNPs associated with salt removal in rice leaf sheath. *Ann. Bot.* **2020**, *126*, 1193–1202. [CrossRef]
91. Yu, J.; Zao, W.; He, Q.; Kim, T.S.; Park, Y.J. Genome-wide association study and gene set analysis for understanding candidate genes involved in salt tolerance at the rice seedling stage. *Mol. Genet. Genom.* **2017**, *292*, 1391–1403. [CrossRef] [PubMed]
92. Cui, Y.; Zhang, F.; Zhou, Y. The Application of Multi-Locus GWAS for the Detection of Salt-Tolerance Loci in Rice. *Front. Plant Sci.* **2018**, *9*, 1464. [CrossRef] [PubMed]
93. Lekklar, C.; Pongpanich, M.; Suriya-Arunroj, D.; Chinpongpanich, A.; Tsai, H.; Comai, L.; Chadchawan, S.; Buaboocha, T. Genome-wide association study for salinity tolerance at the flowering stage in a panel of rice accessions from Thailand. *BMC Genom.* **2019**, *20*, 76. [CrossRef] [PubMed]
94. Yu, J.; Zhao, W.; Tong, W.; He, Q.; Yoon, M.Y.; Li, F.P.; Choi, B.; Heo, E.B.; Kim, K.W.; Park, Y.J. A Genome-Wide Association Study Reveals Candidate Genes Related to Salt Tolerance in Rice (Oryza sativa) at the Germination Stage. *Int. J. Mol. Sci.* **2018**, *19*, 3145. [CrossRef]
95. Rohila, J.S.; Edwards, J.D.; Tran, G.D.; Jackson, A.K.; McClung, A.M. Identification of Superior Alleles for Seedling Stage Salt Tolerance in the USDA Rice Mini-Core Collection. *Plants* **2019**, *8*, 472. [CrossRef]
96. Warraich, A.S.; Krishnamurthy, S.L.; Sooch, B.S.; Vinaykumar, N.M.; Dushyanthkumar, B.M.; Bose, J.; Sharma, P.C. Rice GWAS reveals key genomic regions essential for salinity tolerance at reproductive stage. *Acta Physiol. Plant.* **2020**, *42*, 134. [CrossRef]
97. Shi, Y.; Gao, L.; Wu, Z.; Zhang, X.; Wang, M.; Zhang, C.; Zhang, F.; Zhou, Y.; Li, Z. Genome-wide association study of salt tolerance at the seed germination stage in rice. *BMC Plant Biol.* **2017**, *17*, 92. [CrossRef] [PubMed]
98. An, H.; Liu, K.; Wang, B.; Tian, Y.; Ge, Y.; Zhang, Y.; Tang, W.; Chen, G.; Yu, J.; Wu, W.; et al. Genome-wide association study identifies QTLs conferring salt tolerance in rice. *Plant Breed.* **2019**, *139*, 73–82. [CrossRef]
99. Le, T.D.; Gathignol, F.; Vu, H.T.; Nguyen, K.L.; Tran, L.H.; Vu, H.T.T.; Dinh, T.X.; Lazennec, F.; Pham, X.H.; Very, A.A.; et al. Genome-Wide Association Mapping of Salinity Tolerance at the Seedling Stage in a Panel of Vietnamese Landraces Reveals New Valuable QTLs for Salinity Stress Tolerance Breeding in Rice. *Plants (Basel)* **2021**, *10*, 1088. [CrossRef]
100. Li, N.; Zheng, H.; Cui, J.; Wang, J.; Liu, H.; Sun, J.; Liu, T.; Zhao, H.; Lai, Y.; Zou, D. Genome-wide association study and candidate gene analysis of alkalinity tolerance in japonica rice germplasm at the seedling stage. *Rice (NY)* **2019**, *12*, 24. [CrossRef] [PubMed]

101. Li, X.; Zheng, H.; Wu, W.; Liu, H.; Wang, J.; Jia, Y.; Li, J.; Yang, L.; Lei, L.; Zou, D.; et al. QTL mapping and candidate gene analysis for alkali tolerance in Japonica rice at the bud stage based on linkage mapping and genome-wide association study. *Rice (NY)* **2020**, *13*, 48. [CrossRef] [PubMed]
102. Naveed, S.A.; Zhang, F.; Zhang, J.; Zheng, T.Q.; Meng, L.J.; Pang, Y.L.; Xu, J.L.; Li, Z.K. Identification of QTN and candidate genes for salinity tolerance at the germination and seedling stages in rice by genome-wide association analyses. *Sci. Rep.* **2018**, *8*, 6505. [CrossRef]
103. Patishtan, J.; Hartley, T.N.; De Carvalho, R.F.; Maathuis, F.J.M. Genome-wide association studies to identify rice salt-tolerance markers. *Plant Cell Environ.* **2018**, *41*, 970–982. [CrossRef] [PubMed]
104. Yadav, A.K.; Kumar, A.; Grover, N.; Ellur, R.K.; Bollinedi, H.; Krishnan, S.G.; Bhowmick, P.K.; Vinod, K.K.; Nagarajan, M.; Singh, A.K. Genome-Wide association study reveals marker-trait associations for early vegetative stage salinity tolerance in rice. *Plants (Basel)* **2021**, *10*, 559. [CrossRef] [PubMed]
105. Yuan, J.; Wang, X.; Zhao, Y.; Khan, N.U.; Zhao, Z.; Zhang, Y.; Wen, X.; Tang, F.; Wang, F.; Li, Z. Genetic basis and identification of candidate genes for salt tolerance in rice by GWAS. *Sci. Rep.* **2020**, *10*, 9958. [CrossRef] [PubMed]
106. Julkowska, M.M.; Koevoets, I.T.; Mol, S.; Hoefsloot, H.; Feron, R.; Tester, M.A.; Keurentjes, J.J.B.; Korte, A.; Haring, M.A.; de Boer, G.J.; et al. Genetic components of root architecture remodeling in response to salt stress. *Plant Cell* **2017**, *29*, 3198–3213. [CrossRef] [PubMed]
107. Neeraja, C.N.; Mishra, B.; Rao, K.S.; Singh, R.K.; Padmavati, G.; Shenoy, V.V. Linkage disequilibrium in salt tolerant genotypes of rice (Oryza sativaL). *J. Plant Biochem. Biotechnol.* **2008**, *17*, 65–68. [CrossRef]
108. Al-Tamimi, N.; Brien, C.; Oakey, H.; Berger, B.; Saade, S.; Ho, Y.S.; Schmockel, S.M.; Tester, M.; Negrao, S. Salinity tolerance loci revealed in rice using high-throughput non-invasive phenotyping. *Nat. Commun.* **2016**, *7*, 13342. [CrossRef]
109. Li, X.; Zheng, H.; Zhao, H.; Wang, J.; Liu, H.; Sun, J.; Li, N.; Lei, L.; Zou, D. Association analysis of saline-alkali related traits and SSR markers in Japonica rice under soda saline-alkali stress. *Acta Agric. Boreali-Sin.* **2018**, *33*, 139–148.
110. Huang, X.; Wei, X.; Sang, T.; Zhao, Q.; Feng, Q.; Zhao, Y.; Li, C.; Zhu, C.; Lu, T.; Zhang, Z.; et al. Genome-wide association studies of 14 agronomic traits in rice landraces. *Nat. Genet.* **2010**, *42*, 961–967. [CrossRef]
111. Xie, L.; Zheng, C.; Li, W.; Pu, M.; Zhou, G.; Sun, W.; Wu, X.; Zhao, X.; Xie, X. Mapping and identification a salt-tolerant QTL in a salt-resistant rice landrace, Haidao86. *J. Plant Growth Regul.* **2021**. [CrossRef]
112. Das, P.; Nutan, K.K.; Singla-Pareek, S.L.; Pareek, A. Understanding salinity responses and adopting 'omics-based' approaches to generate salinity tolerant cultivars of rice. *Front. Plant Sci.* **2015**, *6*, 712. [CrossRef] [PubMed]
113. Bimpong, I.K.; Manneh, B.; Sock, M.; Diaw, F.; Amoah, N.K.A.; Ismail, A.M.; Gregorio, G.; Singh, R.K.; Wopereis, M. Improving salt tolerance of lowland rice cultivar 'Rassi' through marker-aided backcross breeding in West Africa. *Plant Sci.* **2016**, *242*, 288–299. [CrossRef]
114. Lakra, N.; Nutan, K.K.; Das, P.; Anwar, K.; Singla-Pareek, S.L.; Pareek, A. A nuclear-localized histone-gene binding protein from rice (OsHBP1b) functions in salinity and drought stress tolerance by maintaining chlorophyll content and improving the antioxidant machinery. *J. Plant Physiol.* **2015**, *176*, 36–46. [CrossRef] [PubMed]
115. Soda, N.; Kushwaha, H.R.; Soni, P.; Singla-Pareek, S.L.; Pareek, A. A suite of new genes defining salinity stress tolerance in seedlings of contrasting rice genotypes. *Funct. Integr. Genom.* **2013**, *13*, 351–365. [CrossRef] [PubMed]
116. Kumari, S.; Nee Sabharwal, V.P.; Kushwaha, H.R.; Sopory, S.K.; Singla-Pareek, S.L.; Pareek, A. Transcriptome map for seedling stage specific salinity stress response indicates a specific set of genes as candidate for saline tolerance in *Oryza sativa* L. *Funct. Integr. Genom.* **2008**, *9*, 109–123. [CrossRef] [PubMed]
117. Singh, R.; Singh, Y.; Xalaxo, S.; Verulkar, S.; Yadav, N.; Singh, S.; Singh, N.; Prasad, K.S.N.; Kondayya, K.; Rao, P.V.R.; et al. From QTL to variety-harnessing the benefits of QTLs for drought, flood and salt tolerance in mega rice varieties of India through a multi-institutional network. *Plant Sci.* **2016**, *242*, 278–287. [CrossRef] [PubMed]
118. Singh, A.K.; Gopalakrishnan, S.; Singh, V.P.; Prabhu, K.V.; Mohapatra, T.; Singh, N.K.; Sharma, T.R.; Nagarajan, M.; Vinod, K.K.; Singh, D.; et al. Marker Assisted Selection: A paradigm shift in Basmati breeding. *Indian J. Genet. Plant Breed.* **2011**, *71*, 120–128.
119. Linh le, H.; Linh, T.H.; Xuan, T.D.; Ham le, H.; Ismail, A.M.; Khanh, T.D. Molecular breeding to improve salt tolerance of rice (Oryza sativa L.) in the red river delta of Vietnam. *Int. J. Plant Genom.* **2012**, *2012*, 949038. [CrossRef]
120. Gregorio, G.; Islam, R.; Vergara, G.; Thirumeni, S. Recent advances in rice science to design salinity and other abiotic stress tolerant rice varietie. *SABRAO J. Breed. Genet.* **2013**, *45*, 31–41.

International Journal of
Molecular Sciences

Article

Genome-Wide Identification and Functional Characterization of the Cation Proton Antiporter (CPA) Family Related to Salt Stress Response in Radish (*Raphanus sativus* L.)

Yan Wang †, Jiali Ying †, Yang Zhang, Liang Xu, Wanting Zhang, Meng Ni, Yuelin Zhu * and Liwang Liu *

National Key Laboratory of Crop Genetics and Germplasm Enhancement, Key Laboratory of Horticultural Crop Biology and Genetic Improvement (East China) of MOAR, College of Horticulture, Nanjing Agricultural University, Nanjing 210095, China; wangyanhs@njau.edu.cn (Y.W.); 2019204031@njau.edu.cn (J.Y.); radishlab@njau.edu.cn (Y.Z.); nauxuliang@njau.edu.cn (L.X.); 2018104061@njau.edu.cn (W.Z.); 2019104066@njau.edu.cn (M.N.)

* Correspondence: ylzhu@njau.edu.cn (Y.Z.); nauliulw@njau.edu.cn (L.L.)

† These authors have contributed equally to this work.

Received: 1 October 2020; Accepted: 2 November 2020; Published: 4 November 2020

Abstract: The CPA (cation proton antiporter) family plays an essential role during plant stress tolerance by regulating ionic and pH homeostasis of the cell. Radish fleshy roots are susceptible to abiotic stress during growth and development, especially salt stress. To date, *CPA* family genes have not yet been identified in radish and the biological functions remain unclear. In this study, 60 *CPA* candidate genes in radish were identified on the whole genome level, which were divided into three subfamilies including the Na^+/H^+ exchanger (NHX), K^+ efflux antiporter (KEA), and cation/H^+ exchanger (CHX) families. In total, 58 of the 60 *RsCPA* genes were localized to the nine chromosomes. RNA-seq. data showed that 60 *RsCPA* genes had various expression levels in the leaves, roots, cortex, cambium, and xylem at different development stages, as well as under different abiotic stresses. RT–qPCR analysis indicated that all nine *RsNHXs* genes showed up regulated trends after 250 mM NaCl exposure at 3, 6, 12, and 24h. The *RsCPA31 (RsNHX1)* gene, which might be the most important members of the RsNHX subfamily, exhibited obvious increased expression levels during 24h salt stress treatment. Heterologous over-and inhibited-expression of *RsNHX1* in *Arabidopsis* showed that *RsNHX1* had a positive function in salt tolerance. Furthermore, a turnip yellow mosaic virus (TYMV)-induced gene silence (VIGS) system was firstly used to functionally characterize the candidate gene in radish, which showed that plant with the silence of endogenous *RsNHX1* was more susceptible to the salt stress. According to our results we provide insights into the complexity of the *RsCPA* gene family and a valuable resource to explore the potential functions of *RsCPA* genes in radish.

Keywords: *CPA* gene family; *RsNHX1*; over-expression; virus-induced gene silence; salt resistance; radish

1. Introduction

Plants respond to salt stress by regulating the cell ion and pH balance through a variety of mechanisms, which are mainly dependent on ion transporters in cell membranes and organelle membranes [1,2]. Cation proton antiporters (CPAs) are mainly involved in the exchange and transport of monovalent cations in plants, which not only reverses the transport of Na^+ with H^+ but also exchanges and transports monovalent cations, such as K^+ and Li^+ [3–5]. The CPA family is divided into two main superfamilies, named CPA1 and CPA2 [6]. The CPA1 superfamily contains the Na^+/H^+ exchanger family (NHX), which is predicted to have 10–12 membrane-spanning domains in plants and it has been confirmed with an important effect on salt tolerance of plants [2,7]. CPA2 consists of the K^+ efflux reverse transporter family (KEA) and cation/H^+ exchanger family (CHX), which is predicted to have 8–14 transmembrane domains [2,7], all of which contain the Na^+/H^+ exchanger domain (PF00999) [8]. In addition, CPA proteins are mainly localized on the plasma, vacuole membrane and organelle membrane of plant cells [2,9], which can play essential roles in plants responding to environmental stress and maintaining the homeostasis of pH and ions [8,10].

Identification and characterization of the CPA family has been extensively reported in several plant species, including *Arabidopsis thaliana* [11], rice (*Oryza sativa*) [12], grape (*Vitis vinifera*) [13], and pear (*Pyrus bretschneideri*) [14]. In recent years, increasing evidence has indicated that some *CPA*s, especially *NHX*s, respond to salt stress, cell expansion, pH and ion balance regulation, osmotic regulation, and vesicle transport, as well as protein processing and floral organ development [15–19]. The previous studies showed that the NHX subfamily consisted of eight members (AtNHX1–AtNHX8) in *Arabidopsis*, which were mainly involved in the exchange and transport of Na^+ and H^+ [20–22]. Recent reports further revealed that the *NHX1* gene in *Vigna unguiculate*, *Malus domestica*, *Sesuvium portulacastrum*, *Arachis hypogaea*, and *Helianthus tuberosus* can regulate salt tolerance [12,23–26]. For example, *VuNHX1* displayed higher salt tolerance through over-expression in *Arabidopsis* [23]. Over-expression of *SpNHX1* yeast cells grew better and accumulated more Na^+ than control, indicating that *SpNHX1* was a key response gene to salt stress [25]. Furthermore, over-expression of *AtNHX7* (*AtSOS1*) limited Na^+ accumulation in xylem and stem, and improved salt tolerance in transgenic *Arabidopsis thaliana* [27]. Interestingly, the *AtCHX17* mutant accumulated less K^+ than the control in salt stress and K^+ deficient environment, indicating that the *AtCHX* gene was involved in the exchange and transport of K^+ rather than Na^+ [28]. In addition, K^+/H^+ reverse transporters AtKEA1–AtKEA3 played an important role in chloroplast function, osmoregulation, photosynthesis, and pH regulation [29,30].

Radish (*Raphanus sativus* L.) belongs to the Brassicaceae family and is one of the most economically important annual or biennial root vegetable crops, which is widely cultivated all over the world with high nutritional and medicinal value. The fleshy taproot is the edible organ of the radish, which has different sensitivities to salt stress [31,32]. Although some *CPA* genes have been proved to be related to salt stress response in other plant species, such as *Arabidopsis*, pear (*Pyrus bretschneideri*), grape (*Vitis vinifera*) and *Helianthus tuberosus*, the information of *CPA* genes identification at the whole genome level in radish is still limited [11–14]. The radish genomes were released to provide a helpful resource to identify the *CPA* gene family at the whole genome level [33]. Our study aimed to systematically identify CPA family members from the radish genome, map *RsCPAs* onto chromosomes, and investigate gene structure and conserved motifs. Moreover, the expression patterns of *RsCPAs* in different developmental stages and tissues were analyzed and also explored for differentially sensitive genes under abiotic stress, especially in salt stress. Furthermore, the biological function of *RsNHX1* was validated for heterologous over-expression and inhibited-expression, as well as through turnip yellow mosaic virus (TYMV)-mediated silencing (VIGS) in radish. The outcomes of our study lay the foundation for further characterization of these *CPA* genes for roles in radish salt-tolerance processes.

2. Results

2.1. Identification and Classification of RsCPA Members in Radish

The Hidden Markov Model (HMM) profile was firstly performed to search the whole genome protein sequence of radish with the Na^+/H^+ exchanger domain (PF00999), and a total of 61 putative CPA proteins were obtained. Following, the CDD and InterPro tools were employed for detecting the completeness of the Na^+/H^+ exchanger domain and then one was excluded. Finally, 60 non-redundant and complete RsCPA members were identified among the radish genome, which were correspondingly named as RsCPA01–RsCPA60 (Table S1).

Through the physical and chemical properties analysis, the protein sizes of RsCPA ranged from 231 to 1172 amino acids (AAs) with molecular weight (MWs) from 25.97 to 126.11 kDa and the theoretical isoelectric point (pI) varied from 4.98 to 9.21. In addition, the instability coefficient reached from 27.56 to 46.90 and 42 members were <40.00, which were considered as stable proteins. The aliphatic index varied from 95.31 to 127.56, indicating that most RsCPA proteins contained a lot of aliphatic amino acids. The grand average of hydropathicity (GRAVY) ranged from 0.048 to 0.798, suggesting that all RsCPA proteins were hydrophobic proteins (Table S1).

2.2. Phylogenetic Analysis of RsCPA Members

To investigate the classification of the CPA subfamily of radish and the evolutionary relationship with other species, full-length CPA protein sequences of radish, *Arabidopsis* and *Brassica rapa* were extracted and aligned to construct a neighbor-joining (NJ) phylogenetic tree (Figure 1). A total of 166 CPA protein members in these three species (containing 60 radish, 42 *Arabidopsis*, and 64 *Brassica rapa*) were categorized into three subfamilies, namely NHX, KEA, and CHX. Among them, the NHX group had 28 members containing nine, eight, and 11 members of radish (15%), *Arabidopsis* (19.05%), and *Brassica rapa* (17.19%), respectively. The KEA group included 29 members with ten, six, and 13 members of radish (16.67%), *Arabidopsis* (14.29%), and *Brassica rapa* (20.31%), respectively. The CHX group was the most abundant subfamily and had 109 members with 41, 28, and 40 members in radish (68.33%), *Arabidopsis* (66.67%), and *Brassica rapa* (62.5%), respectively (Table 1). The phylogenetic relationships indicated that the CPA proteins in radish had stronger homology with *Brassica rapa* than *Arabidopsis*.

Table 1. The number of cation proton antiporter (*CPA*) genes in 17 plant species.

Species	CHX Number	NHX Number	KEA Number	Total Gene Number
Arabidopsis thaliana	28	8	6	42
Raphanus sativus	41	9	10	60
Brassica rapa	40	11	13	64
Pyrus bretschneideri	27	14	12	53
Malus domestica	42	12	7	61
Prunus persica	26	6	5	37
Fragaria vesca	24	6	5	35
Prunus mume	23	7	4	34
Vitis vinifera	17	8	4	29
Oryza sativa	18	8	4	30
Zea mays	16	11	6	33
Sorghum bicolor	17	7	4	28
Selaginella	3	7	4	14
Ostreococcus	0	6	4	10
Chlorophyta reinhardtii	0	9	3	12
Physcomitrella patens	5	10	7	22
Populus trichocarpa	29	8	7	44

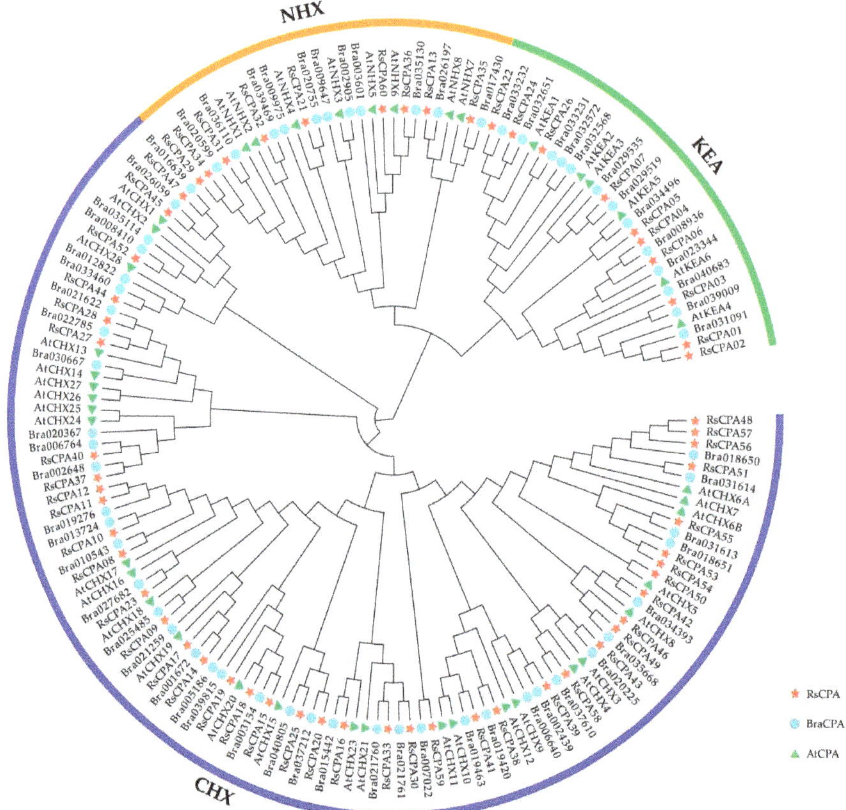

Figure 1. Phylogenetic relationship of RsCPA, BraCPA, and AtCPA members.

Meanwhile, compared with several dicotyledon and monocotyledon crops, the number of CPA gene in radish was closer to *Malus domestica* and *Pyrus bretschneideri*, while it was significantly different from that in *Prunus persica*, *Fragaria vesca*, *Prunus mume*, and *Vitis vinifera*. Especially compared with monocotyledonous plants, such as *Oryza sativa*, *Zea mays*, and *Sorghum bicolor*, the difference of CPA gene numbers was very significant (Table 1).

2.3. Gene Structure and Motif Composition Analysis

All the 60 RsCPA members were divided into three subfamilies, including 9 RsNHXs, 10 RsKEAs, and 41 RsCHXs (Figure 2a). The distributions of RsCPA protein motifs were conducted by Multiple Em for Motif Elicitation (MEME) and 20 conserved motifs were generated (Figure 2b, Figure S1). Most RsCPA members in the same subfamily had similar motif compositions, suggesting that these proteins might have conservative functions. Among them, motif 1 and 12 were found in the CHX, KEA, and NHX subfamilies, indicating that were highly conserved in all RsCPA proteins. Additionally, some motifs were distributed in two subfamilies. For instance, motif 5, 11, and 17 were distributed in the CHX and NHX subfamilies, while motif 14 was distributed in the CHX and KEA subfamilies. Intriguingly, several motifs were only detected in specific RsCPA subfamilies. For example, the RsCHX subfamily independently contained diverse motifs, such as motif 3, 4, 6, 7, 8, 10, 13, 18, and 19. However, the KEA subfamily exclusively contained the motif 15 and 20, as well as the RsNHX subfamily that specifically contained motif 16 (Figure 2b).

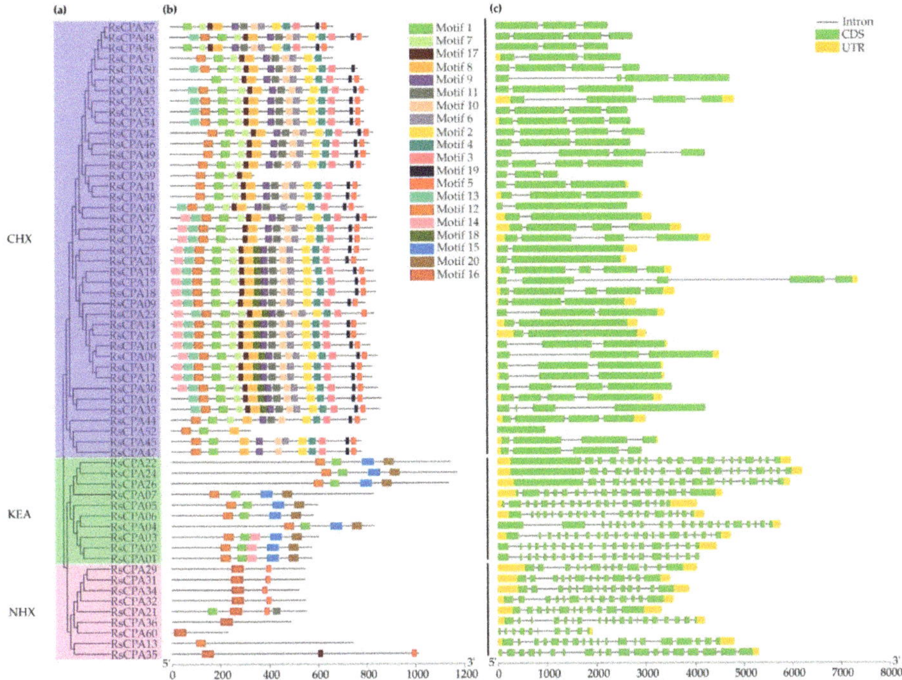

Figure 2. Conserved motif and gene structure distribution of RsCPA proteins. (**a**) Phylogenetic tree of RsCPA proteins. The scale bar indicates 200 aa; (**b**) Conserved motif distribution of RsCPA proteins; (**c**) Exon-intron structure of *CPA* genes in radish. The scale bar indicates 1 kb.

Furthermore, exon-intron analysis was investigated to obtain the structure information of *RsCPA* genes. As shown in Figure 2c, the gene structures involved in the same subfamily were similar, while the lengths of the exon and intron were different. Compared with the *RsCHXs* subfamily, the gene structures of *RsNHXs* and *RsKEAs* were more complex. Among them, the exon numbers of *RsCHXs* were generally one to five. However, the exon numbers of *RsNHXs* and *RsKEAs* ranged from 10 to 19 and 17 to 21, respectively. Additionally, all members in *RsNHX* subfamily as well as most *RsKEA* ones (except *RsCPA01*) had UTR (untranslated region), whereas several members of the *RsCHX* subfamily had no UTR region (Figure 2c).

2.4. Promoter Elements and Transmembrane Region Analysis

The putative promoter sequence (2000 bp upstream region of transcription initiation site) of *RsCPA* genes was submitted to PlantCARE to search for *cis*-acting elements. A total of 103 *cis*-acting elements were identified. Except basic promoter elements, such as the CAAT box and TATA box, 18 other important *cis*-acting elements related to plant growth and development and various stresses were explored. In Figure 3, it was shown that the distribution pattern of *cis*-acting elements were diverse, indicating that the expression of *RsCPA* genes might be regulated by various factors. In the aspect of hormone regulation, Abscisic Acid (ABA), auxin, Gibberellin A_3 (GA_3), and Methyl Jasmonate (MeJA) responsiveness elements frequently existed. While in terms of stress, anaerobic induction, defense and stress, low temperature, and wound-responsive elements were resided. Moreover, the zein metabolism regulation element existed in 14 promoters of *RsCPA* genes, suggesting that *RsCPA* genes might participate in the process of metabolic regulation. Intriguingly, there were 12 promoters of the *RsCPA* gene involved in endosperm expression, among them, *RsCPA27* and *RsCPA38* participated in endosperm specific negative expression (Figure 3 and Figure S2). Additionally, the transmembrane regions of RsCPA proteins showed that all RsCPA proteins, except RsCPA07, contained transmembrane regions that varied from three to 14 (Table S2).

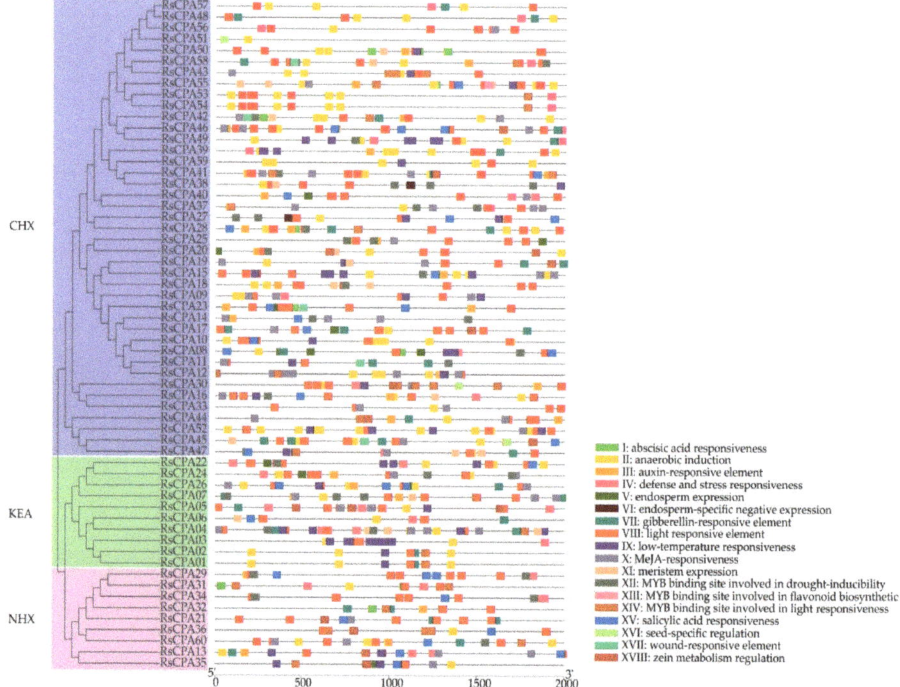

Figure 3. *Cis*-acting elements on promoters of *RsCPA* genes. Distribution of *cis*-acting elements on promoters of *RsCPA* genes.

Futhermore, an interaction of CPA orthologs co-regulatory network was constructed based on the stress-inducible RsCPA orthologs in *Arabidopsis* (Figure S3). It was found that the combination score of the ATCHX1 and KEA4 protein was highest at 0.868, suggesting that it was involved in a stronger relationship between some specific RsCPA proteins, such as RsCPA45, RsCPA47, RsCPA03, RsCPA01, and RsCPA02.

2.5. Chromosomal Localization and Gene Distribution Analysis

A total of 58 *RsCPA* genes (96.67%) were successfully mapped to the R1–R9 chromosomes of radish by TBtools, except *RsCPA12* and *RsCPA51* (Figure 4a). R1 and R4 harbored the most *RsCPA* genes (Ten, 17.67%), followed by R5 and R6 (Eight, 13.37%), while R3 and R8 contained the least *RsCPA* genes (Two, 3.33%). Genome duplication events have facilitated the expansion of plant gene families, including whole-genome duplication (WGD)/segmental duplication, dispersed duplication (DD), tandem duplication (TD), proximal duplication (PD), and transposed duplication (TRD) [34–36]. The duplication types driving expansion of the *RsCPA* gene family was explored by Multiple Collinearity Scan toolkit (MCScanX). Each *RsCPA* gene was mapped on the radish genome based on the position coordinates to deduce the evolutionary relationship (Figure 4b). Totally, 24 pairs of *CPA* genes in radish had collinear relationships. Of these, 36 (60%) *RsCPA* genes were duplicated and retained in the WGD event, indicating that the WGD/segmental duplication type played an important role in expansion of the *RsCPA* gene family.

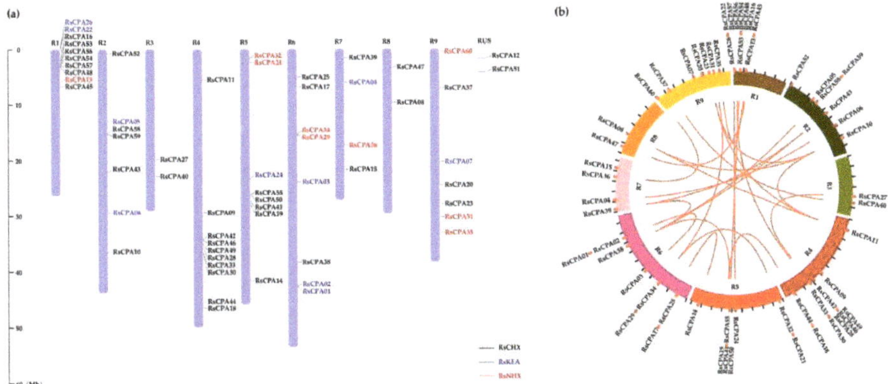

Figure 4. Chromosomal distribution and chromosomal relationships of *RsCPA* genes. (**a**) Chromosomal distribution of *RsCPA* genes; (**b**) Genome distribution and collinearity of the RsCPA family. Red lines indicate the collinear relationship among genes.

2.6. Evolution Analysis of the RsCPA Genes

The possible evolution mechanism of the RsCPA family was investigated by using synteny blocks. To further infer the origin and evolutionary history of CPA members, the synthetic regions between radish and *Arabidopsis* were analyzed and compared (Figure 5). According to the synthetic map, there were 57 pairs of collinear genes in radish and *Arabidopsis*, containing 12 *NHXs*, six *KEAs*, and 39 *CHXs*. Five pairs of syntenic orthologous genes (one to one) were identified, including *RsCPA06AtKEA6*, *RsCPA20–AtCHX15*, *RsCPA26–AtKEA2*, *RsCPA44–AtCHX28*, and *RsCPA45–AtCHX2*. These genes could be traced back to a common ancestor in *Arabidopsis* and radish. Among the two synthetic orthologous gene pairs, one radish gene corresponded to multiple *Arabidopsis* genes, such as *RsCPA26–AtKEA1/2*, *RsCPA27–AtCHX13/14/21/25/26*, and *RsCPA32–AtNHX1/2/6*. Accordingly, there also existed syntenic orthologous gene pairs with one *Arabidopsis* gene corresponding to multiple radish genes, for instance, *AtCHX6B–RsCPA39/42/43/53/55*, *AtNHX6–RsCPA32/36/60*, *AtKEA1–RsCPA24/26*, among others. In addition, the gene pairs of two or three *Arabidopsis* genes corresponding to the same two radish gene pairs were also found, containing *AtCHX3/4/6B–RsCPA39/43*, *AtCHX13/14/26–RsCPA27/28*, *AtCHX6B/7–RsCPA42/53*, and *AtNHX1/2–RsCPA31/32*. A series of synteny events indicated that many *CPA* genes appeared before the divergence of the *Arabidopsis* and radish lineages (Figure 5, Table S3).

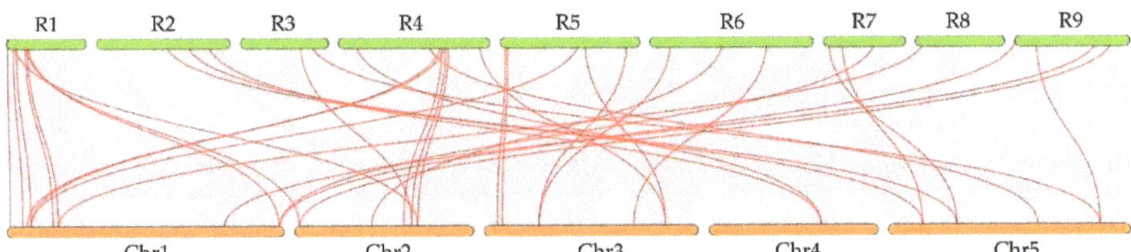

Figure 5. Evolution analysis of the *RsCPA* gene family. Synteny blocks of *CPA* genes between radish and *Arabidopsis*. Colored lines connecting two chromosomal regions indicate syntenic regions between *Arabidopsis* (Chr1–5) and radish (R1–9) chromosomes.

The non-synonymous/synonymous substitution ratio (Ka/Ks) for the 24 gene pairs was calculated to determine the selection pressure among duplicated *RsCPA* genes. Most of the *RsCPA* duplication genes (except *RsCPA11–RsCPA23* and *RsCPA21–RsCPA34*) had a Ka/Ks < 1, indicating that they had experienced strong purifying selective pressure (Table S4).

2.7. Spatial and Temporal Expression Profiles of RsCPA Genes

According to the reads per kilobase per million (RPKM) values, the heatmap was generated to characterize the divergence in expression patterns of *RsCPA* genes among special tissues (cortical, cambium, xylem, root tip, and leaf) and different development stages (40, 60, and 90 d) (Figure 6). In general, the RPKM value varied from 0 to 103.06 and all *RsCPA* genes exhibited diverse expression patterns. Most of the *RsNHX* and *RsKEA* genes showed high expression levels among the five tissues, while the *RsCHX* genes had extremely low expression or were hardly expressed, as well as a few that showed tissue-specific expression (Figure 6a). It was found that *RsCPA58* (*RsCHX*) was only highly expressed in leaves rather than other tissues, indicating that it might be a leaf-specific gene. Moreover, *RsCPA35* was highly expressed in roots after 7 days, while it was down-regulated in roots at other development stages, suggesting it might be a spatiotemporal-specific gene. Furthermore, 68.33, 43.33, 28.33, 30, and 83.33% of the *RsCPA* genes showed a higher transcriptional abundance value (RPKM value > 10) in the five tissues, respectively (Figure 6b). Notably, *RsCPA24*, *RsCPA29*, *RsCPA34*, and *RsCPA35* had abundant expression (RPKM value > 10) in five tissues (Figure 6c). These *RsCPA* genes might play various functions in the development of different tissues during various stages.

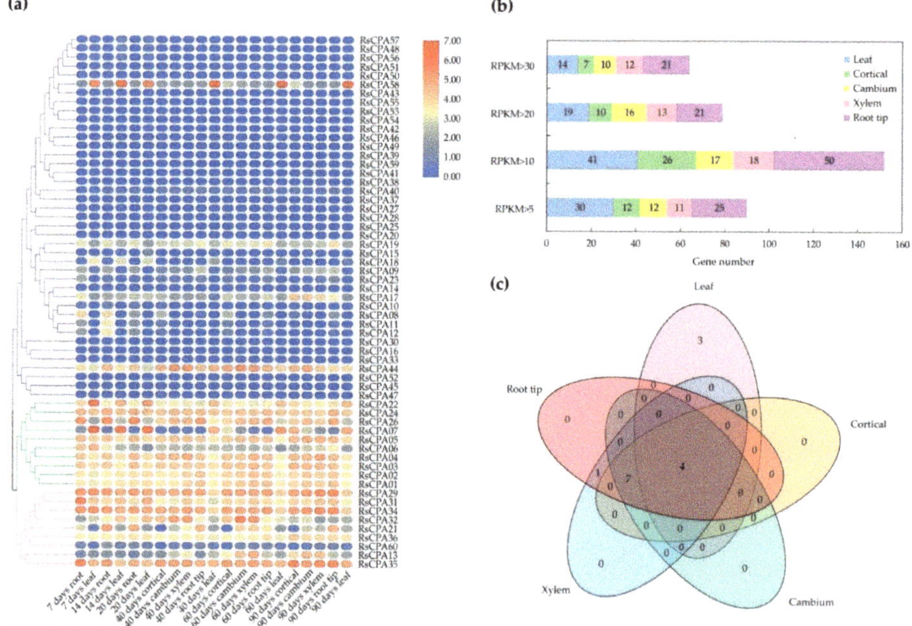

Figure 6. Expression profile of *RsCPA* genes in different stages and tissues. (**a**) *RsCPA* genes expression heatmap in six stages (7, 14, 20, 40, 60, and 90 days) and five tissues (cortical, cambium, xylem, root tip, and leaf). Expression values were calculated by (reads per kilobase per million) RPKM. The scale represents relative expression values; (**b**) Number of *RsCPAs* with a high transcriptional abundance level (RPKM > 10) in each tissue; (**c**) Venn diagram of overlapping *RsCPAs* that are abundantly expressed (RPKM > 10) in different tissues.

2.8. The RsCPA Genes Expression Levels under Abotic Stresses

Based on our previous RNA-Seq. data in radish taproots, the differential expression levels of *RsCPA* genes under various abiotic stresses were investigated, including heavy metal (HM, such as Cadmium (Cd), Chromium (Cr), and lead (Pb), temperature, and salt exposure. As a result, a total of 35, 38, 35, 33, and 29 *RsCPA* genes were differentially expressed during Cd, Pb, Cr, heat, and salt stress (Fold change >1, *p*-value < 0.05), respectively (Figure S4). For instance, *RsCPA13* was up-regulated in response to Cd, Pb, Cr, and salt stresses, while it was down-regulated under high temperatures. *RsCPA08* was up-regulated in response to Cd and Pb, whereas it was down-regulated under Cr and salt stress. Additionally, *RsCPA35* was up-regulated in response to Cd, Pb, high temperature, and salt stresses, but it was down-regulated under Cr stress.

Furthermore, RT–qPCR was conducted to explore the expression levels of *RsNHXs* subfamily genes under the stress of salt exposure. On the whole, all of the *RsNHX* genes were significantly up-regulated under the 250 mM salt stress treatment. The expression level of *RsCPA13*, *RsCPA29*, *RsCPA31*, and *RsCPA35* were highly increased during the 24 h salt exposure. Moreover, several genes were significantly up-regulated after 12 h salt exposure, such as *RsCPA21* and *RsCPA34*. Notably, the expression level of *RsNHXs* recovered to normal levels after 96 h salt exposure, implying that *RsNHXs* might play a crucial role in the process of salt stress response (Figure S5).

2.9. Ectopic Expression of the RsNHX1 Gene in Arabidopsis Can Influence Salt Tolerance

To confirm the biologic function of *RsNHX1* gene in the salt stress response of plant, over-expression (OE–*RsNHX1*) and amiRNA-induced inhibit-expression (amiR–*RsNHX1*) constructs were introduced into wild-type *Arabidopsis* (WT). Four and six independent transgenic lines were respectively generated for OE–*RsNHX1* (#2, 4, 5 and 6) and amiR–*RsNHX1* vectors (#1, 2, 3, 4, 5, and 6) (Figure S6). Following, each of the six OE–*RsNHX1* and amiR–*RsNHX1* transgenic lines were compared with nine WT lines under the 200 mM NaCl stress. It was found that the OE–*RsNHX1* transgenic lines were slightly yellowed and grew better than the control lines, while most of the amiR–*RsNHX1* transgenic lines turned markedly yellow and their growth was inferior to control lines (Figure 7a). For salt stress assays at the seedling stage, the germination ratio of two transgenic seedlings and WT seedlings were counted under 0, and 100 mM NaCl treatment (Figure 7b). As shown in Figure 7b,c, the germination rate of OE–*RsNHX1* seedlings treated with 50mM NaCl was higher than the WT, while amiR–*RsNHX1* seedlings treated with 100 mM NaCl was significantly inferior to the WT. Furthermore, the root length of transgenic seedlings and the WT were measured under 0 and 100 mM NaCl treatment (Figure 7d). The root of OE–*RsNHX1* seedlings exhibited continuing elongation, whereas the amiR–*RsNHX1* seedlings were significantly inhibited under 100 mM NaCl exposure (Figure 7e). These results indicated that *RsNHX1* might play a positive role in the salt tolerance of radish.

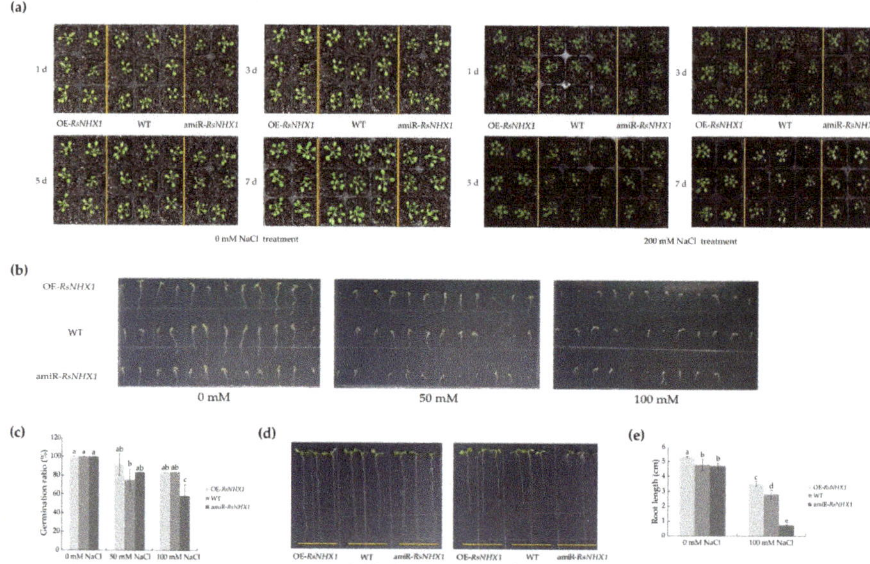

Figure 7. Phenotypic identification of over-expression and inhibit-expression *RsNHX1* transgenic lines responding to salt stress in *Arabidopsis*. (**a**) Morphological comparison between wild-type (WT), OE–*RsNHX1* and amiR–*RsNHX1* transgenic lines; (**b**) Morphological comparison of germination between WT, OE–*RsNHX1* and amiR–*RsNHX1* transgenic seedlings with different concentrations of NaCl; (**c**) Statistical analysis of germination ratio after salt stress. Each bar shows the mean ± SD of the double assay. Values with different letters indicate significant differences at $p < 0.05$ according to Duncan's multiple range tests; (**d**) Morphological comparison of root length between WT, OE–*RsNHX1*, and amiR–*RsNHX1* transgenic seedlings with different concentrations of NaCl; (e) Statistical analysis of root length after salt stress. Each bar shows the mean ± SD of the triplicate assay. Values with different letters indicate a significant difference at $p < 0.05$ according to Duncan's multiple range tests.

2.10. Functional Analysis of RsNHX1 in Radish Confirms that It Can Positively Regulates Salt Tolerance

We further identify the function of *RsNHX1* gene responding to salt stress in radish by using turnip yellow mosaic virus (TYMV)-induced gene silencing (VIGS) technique to silence *RsNHX1* expression. Phytoene desaturase (PDS) was employed as a reporter gene to test whether the TYMV–derived vector can silence the endogenous gene of radish. Plants treated with the pTY–S virus vector designed to silence *RsNHX1* expression (pTY–*RsNHX1*), with the wide type (WT), empty pTY–S silencing vector as well as the pTY–*RsPDS* gene served as mock, empty vector, and positive controls, respectively. Three weeks after particle gun bombardment in radish seedlings, typical phenotype of chlorophyll photobleaching and TYMV spots were separately observed on the leaves of pTY–*RsPDS*, pTY–*RsNHX1*, and pTY–S plants, indicating that TYMV–VIGS system was effective in radish (Figure 8a).

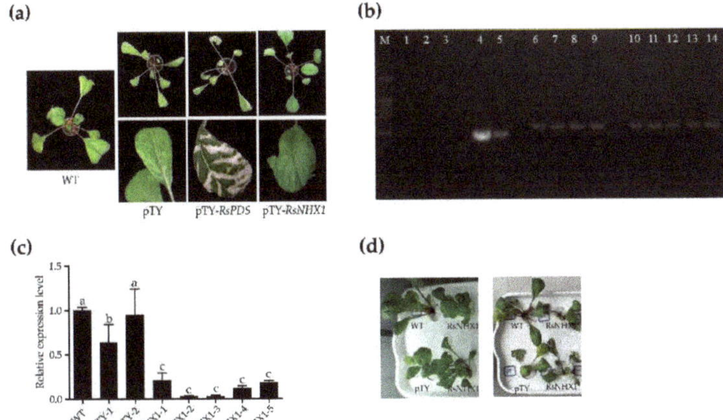

Figure 8. Functional analysis of *RsNHX1* gene in radish during the salt stress. (**a**) Phenotype of virus symptoms on leaves of the entire plant; (**b**) Electrophoresis identification of pTY–S. M–marker, WT (*line 1–3*), pTY–S (*line 4* and 5), pTY–*RsPDS* (*line 6–9*), pTY–*RsNHX1* (*line 10–14*); (**c**) Relative expression levels of the *RsNHX*1 gene. The data represented are means of the triplicate assay and error bars represent the standard deviation of means. Different letters above bars indicate significant differences ($p < 0.05$) between plants; (**d**) Phenotype of plants after salt stress for seven days.

The total RNA was extracted from the diseased leaves of the suspected radish. Gel electrophoresis showed that the 488 bp PCR products were amplified in two pTY–S plants, the 522bp PCR products were amplified in four pTY–*RsPDS* plants and five pTY–*RsNHX1* plants, indicating that they were successfully silenced in radish plants (Figure 8b). Furthermore, RT–qPCR revealed that the expression of positive pTY–*RsNHX1* was significantly decreased compared to the WT and positive pTY–S plants (Figure 8c). Subsequently, the positive pTY–*RsNHX1* plants showed more severe yellowing and wilting than positive pTY–S plants after 7 days of 250mM NaCl exposure (Figure 8d), showing that *RsNHX1* might be a salt sensitive gene.

3. Discussion

3.1. Genome-Wide Identification and Phylogenetic Analysis of CPA Genes in Radish

The *CPA* gene encoded a conserved Na^+/H^+ domain and played an important role in diverse biology processes, including salt stress, cell expansion, pH and ion balance regulation, osmotic regulation, vesicle transport, protein processing, and floral organ development [15–19]. With the completion of genome sequencing in many plant species, CPA family members have been reported in various plant species, including model plant *Arabidopsis* (42) and *Oryza sativa* (30), as well as several horticulture plants, such as *Vitis vinifera* (29) and *Pyrus bretschneideri* (53) [11–14]. Herein, 60 *RsCPA* members were identified from the radish genome, which showed more members than other reported species indicating that the *CPA* gene family in radish may be duplicated and expanded.

Phylogenetic analysis indicated that a total of 166 CPAs among the three species containing radish, *Brassica rapa*, and *Arabidopsis* were divided into three subfamilies: CHX (109, 65.66%), KEA (29, 17.47%), and NHX (28, 16.87%), which was largely consistent with previous studies in *Arabidopsis*, *Vitis vinifera*, and *Pyrus bretschneideri* [11–14]. Through the phylogenetic relationships analysis, it was showed that RsCPAs exhibited closer relations to BraCPAs than AtCPAs, demonstrating that the CPA proteins in radish had stronger homology with the *Brassica rapa* rather than *Arabidopsis* (Figure 1). Furthermore, gene structure and motif analysis indicated that the RsCPAs family harbored similar exon–intron structure and shared motif composition with other species, such as *Vitis vinifera* and *Pyrus bretschneideri* [13,14] (Figure 2).

3.2. Evolutionary Characterization of the RsCPA Family

The expansion of the gene family was mainly caused by gene duplication [34,35]. In the process of plant evolution, duplicated genes could obtain new functions or segment existing functions to improve the adaptability of plants [34]. For instance, expansion of the *RsHSF* gene family was primarily driven by WGD or segmental duplication, which might be largely related with gene duplication [36]. It was previously reported that WGD and PDs event were mainly involved in the expansion of the CPA family in *Pyrus bretschneideri* [14]. In the present study, the predicted gene duplication was also found in radish among the *CPA* genes, indicating that the WGD event played an important role in the expansion of the *CPA* gene family (Figure 4b). In addition, Ka/Ks could be used to identify whether selective pressure existed on the *RsCPA* gene family, including positive, negative, and neutral selection. Herein, except *RsCPA11–RsCPA23* and *RsCPA21–RsCPA34*, all of the *RsCPA* duplication genes displayed a Ka/Ks < 1, suggesting that they had experienced strong purifying selective pressure. A similar Ka/Ks was also reported in the *Pyrus bretschneideri* and cotton, further confirming that the evolutionary pattern of *CPA* genes was very conservative (Table S4) [14,37].

Combined with evolutionary classification and synteny analysis, a large number of the *RsCPAs* were identified as orthologous genes in *Arabidopsis*. For example, eleven pairs seemed to be single radish-to-*Arabidopsis* pairs, presuming that these genes might exist in the genome of the last common ancestor of the two species. There also existed more complex relationships, such as single radish and multiple *Arabidopsis* genes, one *Arabidopsis* and multiple radish genes, and two *Arabidopsis* and multiple radish genes, etc. This phenomenon was similar to the evolutionary relationship of *bZIP* genes in radish and *Arabidopsis* [38]. The close relationship of orthologous genes may exhibit similar functions in different species. For instance, *AtNHX1* gene was reported high salt sensitivity [24], accordingly in our study its orthologues comparising *RsCPA31*, *RsCPA32*, and *RsCPA34* also exhibited high induction under 250 Mm NaCl salt treatments (Figure S5). The collinearity-orthologues analysis of radish and *Arabidopsis* could provide a valuable reference for further exploring the functions of these highly homologous genes in radish.

3.3. Roles of RsCPA Genes in Response to Different Abiotic Stresses

Increasing evidences indicated that *CPA* genes played vital roles in a variety of abiotic stresses, including HM, temperature, and salt exposure [39–42]. For instance, *AtCHX17* mutant accumulated less K^+ than wide type under the salt stress and K^+ deficiency environment, indicating that AtCHX17 was involved in the absorption and transport of K^+ [28]. Among the *CPA* genes in other plants, it was *NHX1* that could decrease the salt-tolerance ability of kallar grass at the concentration of 100 and 150 μM cadmium concentrations [43]. Moreover, *NHX1* regulated Cd^{2+} and H^+ flow during short-term Cd^{2+} shock and confirmed that it could enhance tolerance during Cd^{2+} stress [44]. Recent studies have shown that *NHX2* homologues had a high expression under salinity stress at higher time intervals in *G. barbadense* and *G. hirsutum* [45]. The expressions of *NHX* were up-regulated in root tissues of wheat under salinity stress [46]. In this study, the transcriptome data of the radish taproot showed that nearly one-half of *RsCPA* genes displayed diverse expression profiles under HM, heat, and salt stress, indicating that they might play important roles in the plant response to abiotic stress. For example, *RsCPA09* was significantly up-regulated under HM and heat stress, while exhibited down-regulated under salt stress, indicating that the expression of *RsCPA09* might be a repress factor during salt stress. While *RsCPA13* and *RsCPA31* were all up-regulated under the HM and salt stress, indicating these two genes might play positive roles in response the various abiotic stresses of radish. A recent study in genome-wide identification of the *Gossypium hirsutum NHX* genes showed that most of the *GhNHX* genes were affected by salinity through salt-induced expression patterns analysis [47]. Here, according to the RT–qPCR analysis of *RsNHX* subfamily genes, we found all of them were significantly up-regulated under the 250 mM salt treatment, which indicated that the *RsNHX* genes may be the critical characters for the salt response of radish.

3.4. Potential Functions of RsNHXs Genes in Salt Stress

Emerging evidence indicates the Plant NHX proteins play critical roles for salt tolerance through biological function verification. For instance, over-expression of the soybean gene *GmNHX1* in *Arabidopsis thaliana* could enhance salt tolerance through maintaining higher Na^+ efflux rate and K^+/Na^+ ratio, while silencing it may cause soybean plants became more susceptible to salt stress [48]. Similar, over-expression of wheat *TaNHX2* gene in transgenic sunflower improved salinity stress tolerance and growth performance, which showed better growth performance and accumulated higher Na^+,K^+ contents in leaves and roots under 200 mM NaCl salt stress [49]. Addtionally, a yeast functional complementation test proved that *GhNHX4A* can partially restore the salt tolerance of the salt-sensitive yeast mutant *AXT3*, while silencing it decreased the resistance of cotton [47]. In the present study, ectopic over-expression and inhibited-expression of the *RsNHX1* gene in *Arabidopsis* significantly affected salt tolerance. Soil culture experiments showed that the growth of *Arabidopsis* OE–*RsNHX1* responded more positively and amiR–*RsNHX1* responded more negatively than the non-transgenic control plants. Furthermore, a TYMV-based VIGS system was used to functionally characterize the *RsNHX1* gene, which was the first time to be employed for silencing the endogenous gene of radish. The expression levels of *RsNHX1* gene were successfully silenced in pTY–*RsNHX1* lines and the seedlings showed more salt damage than controls, which clarified that *RsNHX1* may be a potential regulator in response to salt stress of radish.

According to our results, 60 *CPA* candidate genes of radish were firstly identified on the whole genome level. These genes could be clustered into three subfamilies, including nine *RsNHXs*, ten *RsKEAs*, and 41 *RsCHXs*. 58 genes were mapped to the nine chromosomes based on radish genome sequences. All the 60 *RsCPA* genes had various expression levels in the leaves, roots, cortex, cambium, and xylem at different development stages, as well as under various abiotic stresses. RT–qPCR analysis indicated that all nine *RsNHXs* genes showed upregulated trends after 250 mM NaCl exposure. The *RsCPA31* (*RsNHX1*) gene, which might be the most important members of the RsNHX subfamily, exhibited obvious increased expression levels during 24h salt stress treatment. Heterologous over-expression and inhibited expression of *RsNHX1* in *Arabidopsis* showed that *RsNHX1* had a positive function in salt tolerance. Meanwhile, TYMV-based gene silence system was firstly used to functionally characterize the candidate gene in radish, and the silence of endogenous *RsNHX1* in radish was more susceptible to the salt stress. The results would be useful to understand the complexity of the *RsCPA* gene family and could provide a valuable resource to explore the potential functions of *RsCPA* genes in radish.

4. Materials and Methods

4.1. Sequence Collection, CPA Identification, and Phylogenetic Analysis

Whole genome sequences of CPA were obtained from the radish genome database (RGD, http://radish-genome.org/) [33]. The CPA protein sequence of *Arabidopsis* was downloaded from TAIR10 (http://www.arabidopsis.org) [50,51]. The CPA protein sequence of *Brassica rapa* was available from *Brassica* database (BRAD, http://brassicadb.org/brad/) [52]. *CPA* family candidate genes with the Na^+/H^+ exchanger domain (PF00999) were obtained from the Pfam database (http://pfam.xfam.org) [53]. Then, the Hidden Markov Model (HMM) search was carried out by HMMER 3.0 [54,55]. Subsequently, CDD (https://www.ncbi.nlm.nih.gov/cdd) and InterPro (http://www.ebi.ac.uk/x/pfa/iprscan/) were employed to verify the integrity of the Na^+/H^+ exchanger domain [56,57]. In addition, the ExPASy ProtParam (https://www.expasy.org/) was performed to predict the physical and chemical properties of RsCPA proteins, such as the number of amino acids (AA), MW, pI, and instability index [58].

All CPA protein sequences in radish, *Brassica rapa* and *Arabidopsis* were imported to generate the phylogenetic tree using MEGA 6.0 with the neighbor-joining (NJ) and bootstrap value set to 1000 replicates [59,60]. The phylogenetic tree was visualized using Evolview (http://www.evolgenius.info/evolview/) [61].

4.2. Gene Structure, Motif Composition, and Promoter Element Analysis

The structure information of *RsCPA* genes were displayed by TBtools (https://github.com/CJ-Chen/TBtools) and the conserved motifs were identified by MEME (http://meme.nbcr.net/meme/tools/meme) [62,63]. Moreover, the 2000 bp upstream sequence (putative promoter region) of all *RsCPA* genes were extracted by TBtools [62]. The *cis*-acting regulatory elements were predicted through PlantCARE (http://bioinformatics.psb.ugent.be/webtools/plantcare/html/) [64]. Additionally, TransMembrane prediction was analyzed by Hidden Markov Models Server v.2.0 (TMHMM, http://www.cbs.dtu.dk/services/TMHMM/) [65] and the protein–protein relationships of stress-inducible *RsCPA* orthologs was evaluated by STRING 9 (https://string-db.org) [66].

4.3. Synteny Analysis and Chromosomal Localization

Synteny analysis was performed by the method described in the Plant Duplicate Gene Database (PlantDGD, http://pdgd.njau.edu.cn:8080/) [67]. The collinear block was identified by RsCPA duplication events in the MCScanX [68]. The data were integrated and plotted by using Circos [69]. Based on the annotation information from RGD and duplications of the *RsCPA* genes, the corresponding location distributions of *RsCPA* genes in chromosomes were displayed by TBtools [62].

4.4. Expression Analysis of RsCPAs Based on the RNA-Seq. Data

Illumina RNA-Seq. data were downloaded from the NODAI Radish Genome Database and used for the transcriptional profiling of *RsCPA* genes in five tissues (cortical, cambium, xylem, root tip, and leaf) and six stages (7, 14, 20, 40, 60, and 90 days after sowing) [70]. The RPKM method was used to analyze the expression level for each *RsCPA* gene and the heatmap was displayed by TBtools [62]. Furthermore, the expression patterns of *RsCPA* genes during abiotic stress, including heavy metal (HM, such as Cd, Cr and Pb), high temperature, and salt stress were extracted and analyzed from the transcriptome data of radish taproots [71–75].

4.5. Plant Material, Salt Stress Treatment, and RT-qPCR Expression Analysis

The radish variety 'NAU–XBC' was used, which is an advanced inbred line with white flesh and white skin in taproot, and sensitive to salt stress. The seeds were germinated at 25 °C in the dark for 3 days and then cultivated in the growth chamber with a day/night temperature of 25/18 °C (16/8 h), 60% humidity, and 12,000 lx light. The three-week-old radish seedlings were grown in a plastic container with half-strength Hoagland nutrient solution, as previously described [76]. One week later, seedlings were treated with 250 mM NaCl and the NaCl-free solution was used as a control. Leaves were collected in triplicate at 0, 3, 6, 12, 24, 48, and 96 h after NaCl treatment and immediately frozen in liquid nitrogen and stored at −80 °C. Total RNA was isolated using the TRIzol reagent RNAsimple total RNA kit (Tiangen, Beijing, China) and reverse transcribed into cDNA using the PrimeScript™ RT reagent kit (Takara, Dalian, China) according to the instructions. RT–qPCR analysis was carried out on a LightCycler® 480 System (Roche, Mannheim, Germany). The $2^{-\Delta\Delta CT}$ formula was used to calculate the relative expression level [77]. *RsActin* was regarded as the internal reference gene. The primers used for RT–qPCR are listed in Table S5.

Arabidopsis thaliana (ecotype: Col–0) plants were used for heterologous over-expression and inhibited expression *RsNHX1*. In addition, the advanced inbred line 'NAU–YH' seedlings were used to perform the VIGS experiment [78].

4.6. Genetic Transformation and Generation of RsNHX1 Transgenic Lines in Arabidopsis

To generate over-expression lines, the ORF sequence of *RsNHX1* was cloned into the pCAMBIA2301 vector, using with *Bam*HI/*Kpn*I restriction sites, and driven by a 35S promoter [79]. For amiRNA construction, the natural miR319a sequence as the backbone was exchanged with the *amiRRsNHX1* via the overlapping PCR method [80]. The 418 bp specific sequence was inserted with the *Xba*I and *Sac*I digestion sites and then transferred into the pCAMBIA2301 vector [76]. These fusion plasmids were introduced into *Arabidopsis* using *A. tumefaciens*-mediated transformation with the floral dip method [81,82]. Transgenic lines were screened by $\frac{1}{2}$ Murashige and Skoog ($\frac{1}{2}$ MS) solid medium with 100 mg·L^{-1} kanamycin. Following, the 14-day-old wide type and transgenic line seedlings were planted in sterilized soil one week later and treated with 200 mM NaCl solution every other day [83]. Additionally, the seeds were sown on $\frac{1}{2}$ MS solid media containing 0 and 100 mM NaCl to calculate the germination ratio as well as the root length after two weeks [24].

4.7. VIGS-Mediated Silencing of RsNHX1 in Radish

The VIGS experiment was conducted to functionally characterize *RsCPA,* as previously described [78]. The 40 bp specific sequence and its reverse complementary fragment of *RsNHX1* and *RsPDS* were separately synthesized and phosphorylated, and inserted into the pTY–S vector by digestion with *SnaBI* and transformed to obtain positive clones. The primers used for vector construction to identify *RsNHX1*-silenced plants are listed in Table S6. The pTY–S empty vector and self-hybridized palindromic oligonucleotide of *RsPDS*-silencing vector were regarded as negative and positive controls, respectively. Particle bombardment was performed on the two-four fully expanded leaves from 'NAU–YH' plants using the PDS–1000/He bio-gun (Bio–Rad, Hercules, CA, USA) to trigger *RsNHX1* silencing [84]. Five plants were bombarded in each experiment. Three weeks later, the inoculated plant leaf phenotype was observed and sampled to analyze the downstream gene level and silencing efficiency. RT–qPCR was used to further confirm *RsNHX1*-silencing. The primers used for RT–qPCR are listed in Table S5. Afterwards, the two positive plants were treated with 250 mM NaCl solution to observe the phenotypes.

Supplementary Materials: Supplementary Materials can be found at http://www.mdpi.com/1422-0067/21/21/8262/s1. Figure S1. LOGO of 20 amino acid motifs in RsCPA proteins. Figure S2. Number of *cis*–acting elements on promoters of *RsCPA* genes. Figure S3. Functional interaction networks of 60 RsCPA proteins. Figure S4. RNA–Seq of *RsCPA* genes under different treatments in the radish taproot. Figure S5. The expression levels of *RsNHX* genes at different times under 250 mM NaCl treatment. Figure S6. PCR analysis of over–expression and inhibited–expression in T_3 transgenic *Arabidopsis* plants. Table S1. Basic data for the CPA proteins in radish. Table S2. Number of transmembrane regions in RsCPA proteins. Table S3. Synteny blocks of *CPA* genes between radish and *Arabidopsis* genomes. Table S4. Ka/Ks analysis of 24 *RsCPA* duplicated gene pairs. Table S5. Specific primers of *RsCPA* genes for qRT–PCR. Table S6. Specific primers of *RsNHX1* for vector construction.

Author Contributions: Conceptualization, Y.W., J.Y. and Y.Z. (Yang Zhang); Writing—original draft preparation, Y.W. and J.Y.; Writing—review and editing, L.X. and L.L.; Visualization, J.Y., Y.Z. (Yang Zhang) and L.X.; Validation, J.Y., W.Z. and M.N.; Project administration, Y.Z. (Yuelin Zhu) and L.L. All authors have read and agreed to the published version of the manuscript.

Funding: This research was funded by grants from the Jiangsu Agricultural S&T Innovation Fund [CX(19)3045], China Postdoctoral Science Foundation (2016M590466), the Fundamental Research Funds for the Central Universities (KYZZ201910), key laboratory of biology and genetic improvement of horticultural crops, ministry of agriculture, China (IVF201706).

Conflicts of Interest: The authors declare no conflict of interest. The funders had no role in the design of the study; in the collection, analyses, or interpretation of data; in the writing of the manuscript, or in the decision to publish the results.

Abbreviations

AA	Amino Acids
ABA	Abscisic Acid
AtCPA	*Arabidopsis thaliana Cation proton antiporter* gene
BraCPA	*Brassica rapa Cation proton antiporter* gene
BRAD	Brassica database
Cd	Cadmium
Chr(s)	Chromosome(s)
CHX	Cation/H^+ exchanger
CPA	Cation proton antiporter
Cr	Chromium
DD	Dispersed duplication
GA_3	Gibberellin A_3
GRAVY	Grand average of hydropathicity
HM	Heavy metal
HMM	Hidden Markov Model
KEA	K^+ efflux antiporter
MCScanX	Multiple Collinearity Scan toolkit
MeJA	Methyl Jasmonate
MEME	Multiple Em for Motif Elicitation
MW	Molecular weight
NHX	Na^+/H^+ exchanger family
NJ	Neigbor-joining
ORF	Open reading frame
Pb	Lead
PD	Proximal duplicatio
pI	Theoretical isoelectric point
PlantDGD	Plant Duplicate Gene Database
RGD	Radish Genome Database
RPKM	Reads per kilobase per million reads
RsCPA	*Raphanus sativus Cation proton antiporter* gene
RT–qPCR	Real-Time Quantitative Polymerase Chain Reaction
TD	Tandem duplication
TMHMM	TransMembrane prediction was analyzed by Hidden Markov Models
TRD	Transposed duplication
TYMV	Turnip yellow mosaic virus
UTR	Untranslated region
VIGS	Virus-induced gene silencing
WGD	Whole-genome duplication
WT	Wide type

References

1. Zhu, J.K. Abiotic stress signaling and responses in plants. *Cell* **2016**, *16*, 313–324. [CrossRef] [PubMed]
2. Chanroj, S.; Wang, G.Y.; Venema, K.; Zhang, M.W.; Delwiche, C.F.; Sze, H. Conserved and diversified gene families of monovalent cation/H^+ antiporters from algae to flowering plants. *Front. Plant Sci.* **2012**, *9*, 50–55. [CrossRef] [PubMed]
3. Ye, C.Y.; Yang, X.; Xia, X.; Yin, W.L. Comparative analysis of cation/proton antiporter superfamily in plants. *Gene* **2013**, *521*, 245–251. [CrossRef] [PubMed]
4. An, R.; Chen, Q.J.; Chai, M.F.; Lu, P.L.; Su, Z.; Qin, Z.X.; Chen, J.; Wang, X.C. *AtNHX8*, a member of the monovalent cation: Proton antiporter-1 family in *Arabidopsis thaliana*, encodes a putative Li^+/H^+ antiporter. *Plant J.* **2007**, *49*, 718–728. [CrossRef]
5. Véry, A.A.; Sentenac, H. Molecular mechanisms and regulation of K^+ transport in higher plants. *Annu. Rev. Plant Biol.* **2003**, *54*, 575–603. [CrossRef]

6. Inês, S.P.; Sónia, N.; Melissa, M.P.; Isabel, A.A.; Margarida, M.O.; Michael, D.P. Different evolutionary histories of two cation/proton exchanger gene families in plants. *BMC Plant Biol.* **2013**, *13*, 97.
7. Jia, Q.; Zheng, C.; Sun, S.; Amjad, H.; Liang, K.; Lin, W. The role of plant cation/proton antiporter gene family in salt tolerance. *Biol. Plant.* **2018**, *62*, 617–629. [CrossRef]
8. Rodríguez-Rosales, M.P.; Gálvez, F.J.; Huertas, R.; Aranda, M.N.; Baghour, M.; Cagnac, O.; Venema, K. Plant NHX cation/proton antiporters. *Plant Signal. Behav.* **2009**, *4*, 265–276. [CrossRef]
9. Brett, C.L.; Donowitz, M.; Rao, R. Evolutionary origins of eukaryotic sodium/proton exchangers. *Am. J. Physiol. Cell Physiol.* **2005**, *288*, C223. [CrossRef]
10. Dong, W.; Li, D.L.; Qiu, N.W.; Song, Y.G. The functions of plant cation/proton antiporters. *Biol. Plant.* **2018**, *62*, 421–427. [CrossRef]
11. Mäser, P.; Thomine, S.; Schroeder, J.I.; Ward, J.M.; Hirschi, K.; Sze, H.; Talke, I.N.; Amtmann, A.; Maathuis, F.J.M.; Sanders, D.; et al. Phylogenetic relationships within cation transporter families of Arabidopsis. *Plant Physiol.* **2001**, *126*, 1646–1667. [CrossRef]
12. Zeng, Y.; Li, Q.; Wang, H.Y.; Zhang, J.L.; Du, J.; Feng, H.M.; Blumwald, E.; Yu, L.; Xu, G.H. Two NHX-type transporters from *Helianthus tuberosus* improve the tolerance of rice to salinity and nutrient deficiency stress. *Plant Biotechnol. J.* **2017**, *16*, 310–321. [CrossRef]
13. Ma, Y.; Wang, J.; Zhong, Y.; Cramer, G.R.; Cheng, Z.M.M. Genome-wide analysis of the cation/proton antiporter (CPA) super family genes in grapevine (*Vitis vinifera* L.). *Plant Omics* **2015**, *8*, 300–311.
14. Zhou, H.S.; Qi, K.J.; Liu, X.; Yin, H.; Wang, P.; Chen, J.Q.; Wu, J.Y.; Zhang, S.L. Genome-wide identification and comparative analysis of the cation proton antiporters family in pear and four other Rosaceae species. *Mol. Genet. Genom.* **2016**, *291*, 1727–1742. [CrossRef]
15. Banjara, M.; Zhu, L.F.; Shen, G.X.; Payton, P.; Zhang, H. Expression of an *Arabidopsis* sodium/proton antiporter gene (*AtNHX1*) in peanut to improve salt tolerance. *Plant Biotechnol. Rep.* **2012**, *6*, 59–67. [CrossRef]
16. Bassil, E.; Tajima, H.; Liang, Y.C.; Ohto, M.A.; Ushijima, K.; Nakano, R.; Esumi, T.; Coku, A.; Belmonte, M.; Blumwald, E. The *Arabidopsis* Na^+/H^+ antiporters NHX1 and NHX2 control vacuolar pH and K^+ homeostasis to regulate growth, flower development, and reproduction. *Plant Cell* **2011**, *23*, 3482–3497. [CrossRef] [PubMed]
17. Bassil, E.; Ohto, M.; Esumi, T.; Tajima, H.; Zhu, Z.; Cagnac, O.; Belmonte, M.; Peleg, Z.; Yamaguchi, T.; Blumwald, E. The *Arabidopsis* intracellular Na^+/H^+ antiporters NHX5 and NHX6 are endosome associated and necessary for plant growth and development. *Plant Cell* **2011**, *23*, 224–239. [CrossRef] [PubMed]
18. Sze, H.; Padmanaban, S.; Cellier, F.; Honys, D.; Cheng, N.H.; Bock, K.W.; Conéjéro, G.; Li, X.Y.; Twell, D.; Ward, J.M.; et al. Expression patterns of a novel *AtCHX* gene family highlight potential roles in osmotic adjustment and K^+ homeostasis in pollen development. *Plant Physiol.* **2004**, *136*, 2532–2547. [CrossRef] [PubMed]
19. Bassil, E.; Coku, A.; Blumwald, E. Cellular ion homeostasis: Emerging roles of intracellular NHX Na^+/H^+ antiporters in plant growth and development. *J. Exp. Bot.* **2012**, *63*, 5727–5740. [CrossRef]
20. Apse, M.P.; Aharon, G.S.; Snedden, W.A.; Blumwald, E. Salt tolerance conferred by overexpression of a vacuolar Na^+/H^+ antiport in *Arabidopsis*. *Science* **1999**, *285*, 1256–1258. [CrossRef]
21. Yokoi, S.; Quintero, F.J.; Cubero, B.; Ruiz, M.T.; Bressan, R.A.; Hasegawa, P.M.; Pardo, J.M. Differential expression and function of *Arabidopsis thaliana* NHX Na^+/H^+ antiporters in the salt stress response. *Plant J.* **2002**, *30*, 529–539. [CrossRef] [PubMed]
22. Li, H.T.; Liu, H.; Gao, X.S.; Zhang, H.X. Knock-out of Arabidopsis *AtNHX4* gene enhances tolerance to salt stress. *Biochem. Biophys. Res. Commun.* **2009**, *382*, 637–641. [CrossRef]
23. Mishra, S.; Alavilli, H.; Lee, B.; Panda, S.K.; Sahoo, L. Cloning and characterization of a novel vacuolar Na^+/H^+ antiporter gene (*VuNHX*1) from drought hardy legume, cowpea for salt tolerance. *Plant Cell Tissue Organ* **2015**, *120*, 19–33. [CrossRef]
24. Sun, M.H.; Ma, Q.J.; Liu, X.; Zhu, X.P.; Hu, D.G.; Hao, Y.J. Molecular cloning and functional characterization of *MdNHX1* reveals its involvement in salt tolerance in apple calli and *Arabidopsis*. *Sci. Hortic.* **2017**, *215*, 126–133. [CrossRef]
25. Zhou, Y.; Yang, C.; Hu, Y.; Zhu, X.P.; Hu, D.G.; Hao, Y.J. The novel Na^+/H^+ antiporter gene *SpNHX1* from *Sesuvium portulacastrum* confers enhanced salt tolerance to transgenic yeast. *Acta Physiol. Plant.* **2018**, *40*, 61. [CrossRef]

26. Zhang, W.W.; Meng, J.J.; Xing, J.Y.; Yang, S.; Guo, F.; Li, X.G.; Wan, S.B. The K^+/H^+ antiporter *AhNHX1* improved tobacco tolerance to NaCl stress by enhancing K^+ retention. *J. Plant Biol.* **2017**, *60*, 259–267. [CrossRef]
27. Qiu, Q.S.; Guo, Y.; Dietrich, M.A.; Schumaker, K.S.; Zhu, J.K. Regulation of SOS1, a plasma membrane Na^+/H^+ exchanger in *Arabidopsis thaliana*, by SOS2 and SOS3. *Proc. Natl. Acad. Sci. USA* **2002**, *99*, 8436–8441. [CrossRef] [PubMed]
28. Cellier, F.; Conéjéro, G.; Ricaud, L.; Luu, D.T.; Lepetit, M.; Gosti, F.; Casse, F. Characterization of *AtCHX17*, a member of the cation/H^+ exchangers, CHX family, from *Arabidopsis thaliana* suggests a role in K^+ homeostasis. *Plant J.* **2004**, *39*, 834–846. [CrossRef]
29. Kunz, H.H.; Gierth, M.; Herdean, A.; Satoh-Cruzd, M.; Kramerd, D.M.; Speteac, C.; Schroeder, J.I. Plastidial transporters KEA-1, -2, and-3 are essential for chloroplast osmoregulation, integrity, and pH regulation in *Arabidopsis*. *Proc. Natl. Acad. Sci. USA* **2014**, *111*, 7480–7485. [CrossRef]
30. Wang, L.; Jun, L.; Shi, W.M.; Yan, H.; Han, L. Identification and localized expression of putative K^+/H^+ antiporter genes in *Arabidopsis*. *Acta Physiol. Plant.* **2015**, *37*, 101.
31. Sun, X.C.; Xu, L.; Wang, Y.; Yu, R.G.; Zhu, X.W.; Luo, X.B.; Gong, Y.Q.; Wang, R.H.; Limera, C.; Zhang, K.Y.; et al. Identification of novel and salt-responsive miRNAs to explore miRNA-mediated regulatory network of salt stress response in radish (*Raphanus sativus* L.). *BMC Genom.* **2015**, *16*, 1–16. [CrossRef]
32. Sun, X.C.; Wang, Y.; Xu, L.; Li, C.; Zhang, W.; Luo, X.B.; Jiang, H.Y.; Liu, L.W. Unraveling the root proteome changes and its relationship to molecular mechanism underlying salt stress response in radish (*Raphanus sativus* L.). *Front. Plant Sci.* **2017**, *8*, 1192. [CrossRef]
33. Jeong, Y.M.; Kim, N.; Ahn, B.O.; Oh, M.; Chung, W.H.; Chung, H.; Jeong, S.; Lim, K.B.; Hwang, Y.J.; Kim, G.B.; et al. Elucidating the triplicated ancestral genome structure of radish based on chromosome-level comparison with the *Brassica* genomes. *Theor. Appl. Genet.* **2016**, *129*, 1357–1372. [CrossRef]
34. De, G.A.; Lanave, C.; Saccone, C. Genome duplication and gene-family evolution: The case of three *OXPHOS* gene families. *Gene* **2008**, *421*, 1–6.
35. Airoldi, C.A.; Davies, B. Gene duplication and the evolution of plant MADS-box transcription factors. *J. Genet. Genom.* **2012**, *39*, 157–165. [CrossRef] [PubMed]
36. Tang, M.J.; Xu, L.; Wang, Y.; Cheng, W.W.; Luo, X.B.; Xie, Y.; Fan, L.X.; Liu, L.W. Genome-wide characterization and evolutionary analysis of heat shock transcription factors (HSFs) to reveal their potential role under abiotic stresses in radish (*Raphanus sativus* L.). *BMC Genom.* **2019**, *20*, 772. [CrossRef]
37. Fu, X.; Lu, Z.; Wei, H.; Zhang, J.; Yang, X.; Wu, A.; Ma, L.; Kang, M.; Lu, J.; Wang, H.; et al. Genome-Wide identification and expression analysis of the NHX (Sodium/Hydrogen Antiporter) gene family in cotton. *Front. Genet.* **2020**, *11*, 964. [CrossRef]
38. Fan, L.X.; Xu, L.; Wang, Y.; Tang, M.J.; Liu, L.W. Genome- and Transcriptome-wide characterization of *bZIP* gene family identifies potential members involved in abiotic stress response and anthocyanin biosynthesis in radish (*Raphanus sativus* L.). *Int. J. Mol. Sci.* **2019**, *20*, 6334. [CrossRef]
39. Chen, H.T.; Chen, X.; Wu, B.Y.; Yuan, X.X.; Zhang, H.M.; Cui, X.Y.; Liu, X.Q. Whole-Genome identification and expression analysis of KEA and NHX antiporter family under abiotic stress in soybean. *J. Integr. Agric.* **2015**, *14*, 1171–1183. [CrossRef]
40. Jha, A.; Joshi, M.; Yadav, N.S.; Agarwal, P.; Jha, B. Cloning and characterization of the *Salicornia brachiata* Na^+/H^+ antiporter gene *SbNHX1* and its expression by abiotic stress. *Mol. Biol. Rep.* **2011**, *38*, 1965–1973. [CrossRef]
41. Hamed, A.; Hamid, R.K.; Hadi, H.; Iraj, T. Assessment of the vacuolar Na^+/H^+ antiporter (NHX1) transcriptional changes in *Leptochloa fusca* L. in response to salt and cadmium stresses. *Mol. Biol. Res. Commun.* **2015**, *4*, 133–142.
42. Sharma, H.; Taneja, M.; Upadhyay, S.K. Identification, characterization and expression profiling of *cation-proton antiporter* superfamily in *Triticum aestivum* L. and functional analysis of *TaNHX4-B*. *Genomics* **2020**, *112*, 356–370. [CrossRef]
43. Yang, L.; Han, Y.J.; Wu, D.; Yong, W.; Liu, M.M.; Wang, S.T.; Liu, W.X.; Lu, M.Y.; Wei, Y.; Sun, J.S. Salt and cadmium stress tolerance caused by overexpression of the *Glycine Max* Na^+/H^+ antiporter (*GmNHX1*) gene in duckweed (*Lemna turionifera 5511*). *Aquat. Toxicol.* **2017**, *192*, 127–135. [CrossRef] [PubMed]

44. Yang, L.; Wei, Y.; Li, N.; Zeng, J.Y.; Han, Y.J.; Zuo, Z.J.; Wang, S.T.; Zhu, Y.R.; Zhang, Y.; Sun, J.S.; et al. Declined cadmium accumulation in Na^+/H^+ antiporter (NHX1) transgenic duckweed under cadmium stress. *Ecotox. Environ. Saf.* **2019**, *182*, 109397. [CrossRef] [PubMed]
45. Akram, U.; Song, Y.H.; Liang, C.Z.; Abid, M.A.; Askari, M.; Myat, A.A.; Abbas, M.; Malik, W.; Ali, Z.; Guo, S.D.; et al. Genome-wide characterization and expression analysis of *NHX* gene family under salinity stress in *Gossypium barbadense* and its comparison with *Gossypium Hirsutum*. *Genes* **2020**, *11*, 803. [CrossRef]
46. Tiwari, B.K.; Aquib, A.; Anand, R. Analysis of physiological traits and expression of NHX and SOS3 genes in bread wheat (*Triticum aestivum* L.) under salinity stress. *J. Pharmacogn. Phytochem.* **2020**, *9*, 362–366.
47. Ma, W.; Ren, Z.; Zhou, Y.; Zhao, J.; Zhang, F.; Feng, J.; Liu, W.; Ma, X. Genome-wide identification of the *Gossypium hirsutum* NHX genes reveals that the endosomal-type *GhNHX4A* is critical for the salt tolerance of cotton. *Int. J. Mol. Sci.* **2020**, *21*, 7712. [CrossRef]
48. Sun, T.J.; Fan, L.; Yang, J.; Cao, R.Z.; Yang, C.Y.; Zhang, J.; Wang, D. A glycine max sodium/hydrogen exchanger enhances salt tolerance through maintaining higher Na^+ efflux rate and K^+/Na^+ ratio in *Arabidopsis*. *BMC Plant Biol.* **2019**, *19*, 469. [CrossRef]
49. Mushke, R.; Rajesh, Y.; Kirti, P.B. Improved salinity tolerance and growth performance in transgenic sunflower plants via ectopic expression of a wheat antiporter gene (*TaNHX2*). *Mol. Biol. Rep.* **2019**, *46*, 5941–5953. [CrossRef] [PubMed]
50. Lamesch, P.; Berardini, T.Z.; Li, D.H.; Swarbreck, D.; Wilks, C.; Sasidharan, R.; Muller, R.; Dreher, K.; Alexander, D.L.; Garcia-Hernandez, M.; et al. The Arabidopsis Information Resource (TAIR): Improved gene annotation and new tools. *Nucleic Acids Res.* **2012**, *40*, D1202–D1210. [CrossRef]
51. Berardini, T.Z.; Reiser, L.; Li, D.H.; Mezheritsky, Y.; Muller, R.; Strait, E.; Huala, E. The arabidopsis information resource: Making and mining the "gold standard" annotated reference plant genome. *Genesis* **2015**, *53*, 474–485. [CrossRef]
52. Cheng, F.; Liu, S.Y.; Wu, J.; Fang, L.; Sun, S.L.; Liu, B.; Li, P.X.; Hua, W.; Wang, X.W. BRAD, the genetics and genomics database for Brassica plants. *BMC Plant. Biol.* **2011**, *11*, 136. [CrossRef]
53. Finn, R.D.; Tate, J.; Mistry, J.; Coggill, P.C.; Sammut, S.J.; Hotz, H.R.; Ceric, G.; Forslund, K.; Eddy, S.R.; Sonnhammer, E.L.L. The Pfam protein families database. *Nucleic Acids Res.* **2008**, *32*, D138.
54. Sun, Y.N.; Buhler, J. Designing patterns for profile HMM search. *Bioinformatics* **2007**, *23*, e36. [CrossRef]
55. Eddy, S.R.; Pearson, W.R. Accelerated Profile HMM Searches. *PLoS Comput. Biol.* **2011**, *7*, e1002195. [CrossRef]
56. Marchler-Bauer, A.; Anderson, J.B.; Cherukuri, P.F.; DeWweese-Scott, C.; Geer, L.Y.; Gwadz, M.; He, S.Q.; Hurwitz, D.I.; Jackson, J.D.; Ke, Z.X. CDD: A Conserved Domain Database for protein classification. *Nucleic Acids Res.* **2005**, *33*, D192–D196. [CrossRef]
57. Zdobnov, E.M.; Apweiler, R. InterProScan—An integration platform for the signature-recognition methods in InterPro. *Bioinformatics* **2001**, *17*, 847–848. [CrossRef]
58. Elisabeth, G.; Alexandre, G.; Christine, H.; Ivan, I.; Ron, D.A.; Amos, B. ExPASy: The proteomics server for in-depth protein knowledge and analysis. *Nucleic Acids Res.* **2003**, *31*, 3784–3788.
59. Tamura, K.; Stecher, G.; Peterson, D.; Filipski, A.; Kumar, S. MEGA6: Molecular evolutionary genetics analysis version 6.0. *Mol. Biol. Evol.* **2013**, *30*, 2725–2729. [CrossRef]
60. Olivier, G.; Mike, S. Neighbor-Joining Revealed. *Mol. Biol. Evol.* **2006**, *23*, 1997–2000.
61. Zhang, H.K.; Gao, S.H.; Lercher, M.J.; Hu, S.N.; Chen, W.H. EvolView, an online tool for visualizing, annotating and managing phylogenetic trees. *Nucleic Acids Res.* **2012**, *40*, 569–572. [CrossRef]
62. Chen, C.J.; Xia, R.; Chen, H.; He, Y.H. TBtools, a toolkit for biologists integrating various biological data handling tools with a user-friendly interface. *BioRxiv* **2018**, 289660. [CrossRef]
63. Bailey, T.L. MEME-ChIP: Motif analysis of large DNA datasets. *Bioinformatics* **2011**, *27*, 1696–1697.
64. Stephane, R.; Patrice, D.; Marc, V.M.; Pierre, R. PlantCARE, a plant *cis*-acting regulatory element database. *Nucleic Acids Res.* **1999**, *27*, 295–296.
65. Chen, Y.J.; Yu, P.; Luo, J.C.; Jiang, Y. Secreted protein prediction system combining CJ-SPHMM, TMHMM, and PSORT. *Mamm. Genome* **2003**, *14*, 859–865. [CrossRef]
66. Damian, S.; Andrea, F.; Michael, K.; Milan, S.; Alexander, R.; Pablo, M.; Tobias, D.; Manuel, S.; Jean, M.; Peer, B. The STRING database in 2011: Functional interaction networks of proteins, globally integrated and scored. *Nucleic Acids Res.* **2011**, *39*, 561–568.

67. Qiao, X.; Li, Q.H.; Yin, H.; Qi, K.; Li, L.; Wang, R.; Zhang, S.L.; Paterson, A.H. Gene duplication and evolution in recurring polyploidization-diploidization cycles in plants. *Genome Biol.* **2019**, *20*, 38. [CrossRef]
68. Wang, Y.P.; Tang, H.B.; Debarry, J.D.; Tan, X.; Li, J.P.; Wang, X.Y.; Lee, T.H.; Jin, H.Z.; Marler, B.; Guo, H.; et al. MCScanX: A toolkit for detection and evolutionary analysis of gene synteny and collinearity. *Nucleic Acids Res.* **2012**, *40*, e49. [CrossRef]
69. Krzywinski, M.; Schein, J.; Birol, I.; Connors, J.; Gascoyne, R.; Horsman, D.; Jones, S.J.; Marra, M.A. Circos: An information aesthetic for comparative genomics. *Genome Res.* **2009**, *19*, 1639–1645. [CrossRef]
70. Mitsui, Y.; Shimomura, M.; Komatsu, K.; Namiki, N.; Shibata-Hatta, M.; Imai, M.; Katayose, Y.; Mukai, Y.; Kanamori, H.; Kurita, K.; et al. The radish genome and comprehensive gene expression profile of tuberous root formation and development. *Sci. Rep.* **2015**, *5*, 10835. [CrossRef]
71. Xu, L.; Wang, Y.; Liu, W.; Wang, J.; Zhu, X.W.; Zhang, K.Y.; Yu, R.G.; Wang, R.H.; Xie, Y.; Zhang, W.; et al. De novo sequencing of root transcriptome reveals complex cadmium-responsive regulatory networks in radish (*Raphanus sativus* L.). *Plant Sci.* **2015**, *236*, 313–323. [CrossRef]
72. Xie, Y.; Ye, S.; Wang, Y.; Xu, L.; Zhu, X.W.; Yang, J.; Feng, H.Y.; Yu, R.G.; Karanja, B.; Gong, Y.Q.; et al. Transcriptome-based gene profiling provides novel insights into the characteristics of radish root response to Cr stress with next-generation sequencing. *Front. Plant Sci.* **2015**, *6*, 202. [CrossRef]
73. Wang, Y.; Xu, L.; Chen, Y.L.; Shen, H.; Gong, Y.Q.; Limera, C.; Liu, L.W. Transcriptome profiling of radish (*Raphanus sativus* L.) root and identification of genes involved in response to Lead (Pb) stress with next generation sequencing. *PLoS ONE* **2013**, *8*, e66539. [CrossRef]
74. Wang, R.H.; Mei, Y.; Xu, L.; Zhu, X.W.; Wang, Y.; Guo, J.; Liu, L.W. Genome-wide characterization of differentially expressed genes provides insights into regulatory network of heat stress response in radish (*Raphanus sativus* L.). *Funct. Integr. Genom.* **2018**, *18*, 225–239. [CrossRef]
75. Sun, X.C.; Xu, L.; Wang, Y.; Luo, X.B.; Zhu, X.W.; Kinuthia, K.B.; Nie, S.S.; Feng, H.Y.; Li, C.; Liu, L.W. Transcriptome-based gene expression profiling identifies differentially expressed genes critical for salt stress response in radish (*Raphanus sativus* L.). *Plant Cell Rep.* **2016**, *35*, 329–346. [CrossRef] [PubMed]
76. Xu, L.; Wang, Y.; Zhai, L.L.; Xu, Y.Y.; Wang, L.J.; Zhu, X.W.; Gong, Y.Q.; Yu, R.G.; Limera, C.; Liu, L.W. Genome-wide identification and characterization of cadmium-responsive microRNAs and their target genes in radish (*Raphanus sativus* L.) roots. *J. Exp. Bot.* **2013**, *14*, 4271–4287. [CrossRef] [PubMed]
77. Livak, K.J.; Schmittgen, T.D. Analysis of relative gene expression data using real-time quantitative PCR and the $2^{-\Delta\Delta CT}$ method. *Methods* **2001**, *25*, 402–408. [CrossRef]
78. Yu, J.; Yang, X.D.; Wang, Q.; Gao, L.W.; Yang, Y.; Xiao, D.; Liu, T.K.; Li, Y.; Hou, X.L.; Zhang, C.W. Efficient virus-induced gene silencing in *Brassica rapa* using a turnip yellow mosaic virus vector. *Biol. Plant.* **2018**, *62*, 826–834. [CrossRef]
79. Topping, J.F.; Wei, W.B.; Clarke, M.C.; Muskett, P.; Lindsey, K. *Agrobacterium*-mediated transformation of *Arabidopsis thaliana*: Application in T-DNA tagging. *Methods Mol. Biol.* **1995**, *49*, 63–76.
80. Schwab, R.; Ossowski, S.; Riester, M.; Warthmann, N.; Weigel, D. Highly specific gene silencing by artificial microRNAs in Arabidopsis. *Plant Cell* **2006**, *18*, 1121–1133. [CrossRef] [PubMed]
81. Clough, S.J.; Bent, A.F. Floral dip: A simplified method for *Agrobacterium*-mediated transformation of *Arabidopsis thaliana*. *Plant J.* **2010**, *16*, 735–743. [CrossRef] [PubMed]
82. Zhang, X.R.; Henriques, R.; Lin, S.S.; Niu, Q.W.; Chua, N.H. *Agrobacterium*-mediated transformation of *Arabidopsis thaliana* using the floral dip method. *Nat. Protoc.* **2006**, *2*, 641–646. [CrossRef]
83. Li, M.R.; Lin, X.J.; Li, H.Q.; Pan, X.P.; Wu, G.J. Overexpression of *AtNHX5* improves tolerance to both salt and water stress in rice (*Oryza sativa* L.). *Plant Cell Tissue Org.* **2011**, *107*, 283–293. [CrossRef]
84. Simmonds, D.H.; Donaldson, P.A. Genotype screening for proliferative embryogenesis and biolistic transformation of short-season soybean genotypes. *Plant Cell Rep.* **2000**, *19*, 485–490. [CrossRef]

International Journal of
Molecular Sciences

MDPI

Article

Genome-Wide Analysis of the Apple CBL Family Reveals That Mdcbl10.1 Functions Positively in Modulating Apple Salt Tolerance

Peihong Chen †, Jie Yang †, Quanlin Mei, Huayu Liu, Yunpeng Cheng, Fengwang Ma * and Ke Mao *

State Key Laboratory of Crop Stress Biology for Arid Areas/Shaanxi Key Laboratory of Apple, College of Horticulture, Northwest A & F University, Xianyang 712100, China; cph0219@163.com (P.C.); yangjie320@163.com (J.Y.); mqlyes@163.com (Q.M.); huayuliu93@163.com (H.L.); cyp2020055284@163.com (Y.C.)
* Correspondence: fwm64@nwsuaf.edu.cn (F.M.); maoke2002@nwsuaf.edu.cn (K.M.)
† Peihong Chen and Jie Yang contributed equally to this work.

Abstract: Abiotic stresses are increasingly harmful to crop yield and quality. Calcium and its signaling pathway play an important role in modulating plant stress tolerance. As specific Ca^{2+} sensors, calcineurin B-like (CBL) proteins play vital roles in plant stress response and calcium signaling. The CBL family has been identified in many plant species; however, the characterization of the CBL family and the functional study of apple MdCBL proteins in salt response have yet to be conducted in apple. In this study, 11 *MdCBL* genes were identified from the apple genome. The coding sequences of these *MdCBL* genes were cloned, and the gene structure and conserved motifs were analyzed in detail. The phylogenetic analysis indicated that these MdCBL proteins could be divided into four groups. The functional identification in Na^+-sensitive yeast mutant showed that the overexpression of seven *MdCBL* genes could confer enhanced salt stress resistance in transgenic yeast. The function of *MdCBL10.1* in regulating salt tolerance was also verified in cisgenic apple calli and apple plants. These results provided valuable insights for future research examining the function and mechanism of CBL proteins in regulating apple salt tolerance.

Keywords: *Malus domestica*; calcium; calcineurin B-like proteins; Na^+ accumulation; salt tolerance

Citation: Chen, P.; Yang, J.; Mei, Q.; Liu, H.; Cheng, Y.; Ma, F.; Mao, K. Genome-Wide Analysis of the Apple CBL Family Reveals That Mdcbl10.1 Functions Positively in Modulating Apple Salt Tolerance. *Int. J. Mol. Sci.* **2021**, *22*, 12430. https://doi.org/10.3390/ijms222212430

Academic Editors: Masayuki Fujita and Mirza Hasanuzzaman

Received: 30 October 2021
Accepted: 16 November 2021
Published: 18 November 2021

Publisher's Note: MDPI stays neutral with regard to jurisdictional claims in published maps and institutional affiliations.

1. Introduction

Plants are inevitably exposed to a variety of adverse environmental conditions due to their sessile lifestyle. They have evolved a series of signal transduction mechanisms to fine-regulate the body's adaptation to the environment. Calcium ion (Ca^{2+}) is a nutrient element crucial for plant growth and development. It also acts as the ubiquitous intracellular secondary messenger initiating the Ca^{2+} signaling pathway, which is a fine regulatory mechanism responsible for the acquisition, perception, transformation, transmission, and decryption of external stimuli [1]. Ca^{2+} signals are an important regulator of growth, development, and biotic and abiotic stresses in plants. Refs. [2–5]. Under stress conditions, Ca^{2+} signals induced by the stimulus are first perceived, decoded, and transmitted by Ca^{2+} sensors [6].

Based on the protein structural characteristics, Ca^{2+} sensors are divided into two types of sensors: sensor relays and sensor responders [7,8]. Sensor relays include calmodulin (CaM)-like proteins (CMLs) and calcineurin B-like (CBL) proteins, which do not have kinase activity [8,9]. They specifically target downstream proteins to transfer the perceived calcium signals. Sensor responder proteins, such as CaMs and Ca^{2+}-dependent protein kinases (CDPKs), have all the functions of Ca^{2+} sensor relay proteins as well as the kinase activity [8,10,11]. As a result, CaMs, CMLs, CDPKs, and CBLs constitute sensors in the Ca^{2+} signal transduction pathway [12]. CaM is a ubiquitous conserved Ca^{2+}-binding protein found in both animals and plants. The CML family was identified as encoding proteins that contain the CaM-like EF-hand structures and share at least 16% homology with CaM in amino acid residues [13–16]. CDPK comprises a kinase domain and a CaM-like domain (four EF-hands) in a single protein; thus, it acts as not only a Ca^{2+} sensor but also an effector [17–19]. The CBL family belongs to a unique group of calcium sensors in plants. Ca^{2+} can bind to the elongation factor (EF) hand domains of the CBL proteins, which changes their phosphorylation status. The change in the phosphorylation status of Ca^{2+} sensors activates several protein kinases, which sometimes lead to a protein phosphorylation cascade [6,20,21].

Decades of research have revealed that the CBL family proteins play important roles in plant stress response and resistance regulation. For example, CBL1 functions under drought, high salt, and hyperosmotic stresses in plants [22,23]. CBL2 and CBL3 regulate ion homeostasis across the vacuolar membranes under salt stress. Studies also proved that CBLs were involved in K^+ regulation, which indirectly regulated the Na^+ homeostasis. For example, AtCBL1/AtCBL9 regulates plant K^+ homeostasis and salt tolerance by regulating the K^+ channel AKT1 and K^+ transporter HAK5 [3,24,25], while CBL4 modulates the activity of the plasma membrane K^+ channel AKT2 [26]. The involvement of CBL in regulating plant salt stress response has been widely reported in a CIPK-dependent manner. The first identified CBL-CIPK pathway was the salt overly sensitive (SOS) pathway in *Arabidopsis*. Under salt stress, SOS3 (CBL4) interacts with SOS2, and the SOS2–SOS3 complex activates the transport properties of the cell membrane-located SOS1 to promote the Na^+ efflux, thereby enhancing plant salt tolerance [27,28]. Other studies also found that the SOS3 homolog SOS3-like calcium-binding protein8 (SCABP8)/CBL10 interacted with SOS2 and enhanced plant salt tolerance by activating SOS1 [28–31]. Because of the crucial roles of CBL family proteins in plant growth and stress response, this family has been identified at the genome-wide level in many plant species [1,8,32–34]. For example, ten CBLs are present in *Arabidopsis* and rice [35], eight in pineapples [32], nine in peppers [10], and five in eggplant [36]. However, no detailed identification and characterization of the apple CBL family have been reported till now.

Apple (*Malus domestica*) is one of the most widely grown and economically valuable fruit crops globally. Abiotic stresses, such as high salinity, severely restrict its global yield and quality. Although CBL proteins play an important role in regulating plant salt tolerance, little is known regarding the function of apple CBL proteins in salt stress response. Genome-wide identification and gene cloning of CBL family genes were performed in apples in this study. The collinearity, phylogenetic relationship, gene structure, and conserved motifs of these MdCBLs were analyzed in detail. The functional identification in the Na^+-sensitive yeast mutant showed that several *MdCBLs* played positive roles in modulating salt response. The function of MdCBL10.1 to enhance apple salt tolerance was verified in cisgenic apple calli and apple plants. These results provided valuable insights for subsequent research on the functions and regulatory mechanisms of MdCBLs in apples.

2. Materials and Methods

2.1. Sequence Retrieval and Identification of Apple CBL Family Proteins

The apple proteome file (GDDH13_1-1_prot.fasta) was downloaded from the GDR database (Genome Database for Rosaceae; https://www.rosaceae.org/, accessed on 6 April 2021), and the protein sequences of 10 AtCBLs were downloaded from the TAIR (The Arabidopsis Information Resource) database (https://www.arabidopsis.org/, accessed on 6 April 2021). The HMM file EF-hand_7.hmm (PF13499.8) was downloaded from the Pfam database and used as a query to search the apple proteome. Phylogenetic analysis was subsequently performed with the protein sequences of the HMMER screening results and the 10 AtCBLs. Further, 11 MdCBL family proteins were identified from the phylogenetic tree (Figure S1). This result was also verified with the local BLASTp search using the 10 AtCBLs as queries, which was conducted with the BioEdit software (version 7.0.9.0).

2.2. Collinearity Analysis and Characterization of Apple CBL Family Genes

The GFF file (gene_models_20170612.gff3) that contained location data for apple CBL family genes was downloaded from the GDR database. The collinearity analysis between different apple chromosomes was performed with MCScanX software, and the results were visualized using TBtools software. The protein length, mass weight, p*I* (isoelectric point) values, and charge at pH 7.0 were determined with the DNAstar software (version 7.1.0). The best hits in *Arabidopsis* for MdCBL proteins were determined by the local BLASTp search.

2.3. Phylogenetic Relationships, Gene Structure, and Conserved Motif Analysis

Phylogenetic analyses were constructed with the MEGA-X software (version 10.0.5) using the neighbor-joining method (bootstrap method, 1000 replicates, Poisson model, pairwise deletion). The intron–exon schematic structures of *MdCBL* genes were drawn with the TB tools. The online MEME software (version 5.3.3; https://meme-suite.org/meme/tools/meme, accessed on 26 April 2021) was used to identify conserved motifs in the protein sequences of these MdCBLs. Detailed methods refer to previous studies [37,38].

2.4. Cis-Acting Elements in the Promoters of MdCBL Genes

The upstream regions (1500 bp) of MdCBL genes were obtained from the apple genome file (GDDH13_1-1_formatted.fasta) downloaded from the GDR database. Abiotic stress– or hormone response–related regulatory elements were identified using the PlantCARE software (http://bioinformatics.psb.ugent.be/webtools/plantcare/html/, accessed on 13 May 2021).

2.5. Vector Construction, Genetic Transformation, and Stress Treatment

For gene cloning of MdCBLs, total RNA was extracted from the leaves of "Golden Delicious" apple using a plant RNA isolation kit [Wolact, Vicband Life Sciences Company (HK) Limited] and reverse transcribed using a PrimeScript First Strand cDNA Synthesis Kit (Takara, Dalian, China). Gene-specific primers were designed based on the predicted coding sequences (CDSs) of these *MdCBLs* obtained from the GDR database.

For the transformation of yeast mutants, CDSs of *MdCBLs* were cloned into the pDR196 vector. The recombinant MdCBLs–pDR196 vector was transformed into the Na^+-sensitive yeast mutant Δ*ena1-4* using the LiAc/ss carrier DNA/PEG method. After growth selection on selective medium (synthetic defined medium minus the appropriate amino acids) and PCR confirmation of transgene presence, three single colonies of each strain were selected for subsequent experiments. Detailed methods of yeast transformation and salt treatment referred to previous studies [39,40].

For the overexpression of the *MdCBL10.1* gene in the apple callus, the CDSs of *MdCBL10.1* were cloned into the pBI121 vector under the control of the 35S promoter. Detailed methods of genetic transformation and salt stress treatment on the apple calluses refer to previous studies [39,40].

To obtain the composite apple plants with *MdCBL10.1* expression whose expression increased or was interfered in roots, the CDSs and a 300-bp fragment of *MdCBL10.1* were cloned into the overexpression vector pCambia2300-GFP and the RNAi vector pK7GWIWG2D, respectively. Methods of *Agrobacterium rhizogenes* K599-mediated genetic transformation referred to previous studies [39,40]. For NaCl stress treatment, the plants in the treatment group were irrigated with 150 mM NaCl solution for 10 days at 5-day intervals. The same number of plants in the control group were irrigated with distilled water.

2.6. Measurement of Stress-Related Physiological Parameters

Relative electrolyte leakage (REL) and Na^+ content were measured with a flame photometer (M410; Sherwood Scientific, Cambridge, UK) as described previously [39,40]. Malondialdehyde (MDA) content, H_2O_2 and O_2^- content, and plant root activity were measured using Suzhou Comin Biotechnology test kits (Suzhou Comin Biotechnology Co., Ltd., Suzhou, China).

2.7. Statistical Analysis

IBM SPSS Statistics software (version 26; SPSS Inc., Chicago, IL, USA) was used for statistical analysis. Significant differences ($p < 0.05$) were determined using the Student *t* test or Tukey's multiple range test.

3. Results

3.1. Identification, Characterization, and Gene Duplication of Apple CBL Family Genes

The HMM file (PF13499.8) was used as a query to search the apple proteome using the HMMER software (hmmsearch, version 3.1b2) so as to screen the CBL family proteins in apple. With default inclusion threshold, 230 protein sequences were obtained (Supplementary File S1). Then, the protein sequences of the ten *Arabidopsis* CBL family members were downloaded from the TAIR database. These ten AtCBL proteins, along with the 230 proteins in HMMER screening results, were used for phylogenetic analysis. Based on the phylogenetic tree (Figure S1), 11 proteins were finally identified as apple CBL family members and named based on their orthlogs in *Arabidopsis* (Table 1). The predicted protein sequences of these 11 MdCBLs were 187–259 amino acids (aa) in length, with predicted mass weight, isoelectric point (pI), and charge at pH 7.0 ranging from 21.38 to 29.53, 4.42 to 4.89, and −10.78 to −17.30, respectively (Table 1).

Based on the genomic location information obtained from the GDR database, 10 of the 11 *MdCBL* genes were randomly distributed on 8 of the 17 chromosomes of the apple, and the remaining one (*MdCBL1.2*) was localized to unassembled genomic scaffolds (Table 1 and Figure 1). Segmental and tandem duplications are the main causes of gene family expansion in plants. The collinear analyses between different apple chromosomes were performed with MCScanX software to investigate the gene duplication events among these *MdCBL* genes. The results showed complex patterns of collinearity between different chromosomes. The collinear analysis also revealed two segmental duplication events (*MdCBL3.1* and *MdCBL3.2*, and *MdCBL5* and *MdCBL8*) and one tandem duplication event (*MdCBL10.1* and *MdCBL10.2*) among these *MdCBL* genes (Figure 1).

3.2. Phylogenetic Analysis, Gene Structure Display, Prediction of Conserved Motifs, and Cloning of Apple CBL Family Genes

Previous studies showed that the members of the CBL family in plants could be divided into four groups. A phylogenetic analysis based on the protein sequences of the 11 MdCBLs and 10 AtCBLs was performed to investigate the relationship between these 11 CBL family members in apples. The phylogenetic tree showed that the 11 MdCBL proteins could be divided into four groups, similar to the grouping results in *Arabidopsis* and many other plant species (Figure 2A).

Table 1. Characterization of the CBL family genes in apple.

Group	Gene Name	Gene Locus (GDDH13)	Genomic Location (GDDH13)	Deduced Polypeptide				Best Hits
				Length (aa)	Mass Weight (kDa)	pI	Charge at PH 7.0	
A	MdSOS3	MD07G1288800	Chr07: 34980079–34983632	212	24.39	4.52	−16.44	AT5G24270
	MdSOS3.2	MD01G1081000	Chr01: 18815406–18816403	**187**	**21.38**	**4.53**	**−9.88**	AT5G24270
	MdCBL5	MD14G1130200	Chr14: 20784655–20786608	214	24.53	4.42	−15.87	AT4G26570
	MdCBL8	MD06G1109200	Chr06: 24877949–24881781	212	24.42	4.60	−13.48	AT1G64480
B	**MdCBL10.1**	MD09G1194600	Chr09: 17493583–17495782	**187 (266)**	**21.57 (30.51)**	**4.48 (4.88)**	**−14.55 (−11.27)**	AT4G33000
	MdCBL10.2	MD09G1194500	Chr09: 17485706–17488584	**259 (245)**	**29.53 (27.88)**	**4.54 (4.63)**	**−17.30 (−14.31)**	AT4G33000
	MdCBL10.3	MD08G1043100	Chr08: 3254626–3256803	246	28.28	4.65	−17.03	AT4G33000
C	MdCBL1.1*	MD06G1046100	Chr06: 6240265–6244670	213	24.52	4.63	−10.78	AT4G17615
	MdCBL1.2	MD00G1132600	Chr00: 28662543–28667386	213	24.52	4.63	−10.78	AT4G17615
D	MdCBL3.1	MD03G1036500	Chr03: 2901761–2905532	226	26.06	4.69	−14.17	AT4G26570
	MdCBL3.2	MD11G1037200	Chr11: 3199328–3202169	**210 (226)**	**24.20 (25.96)**	**4.89 (4.75)**	**−10.34 (−13.17)**	AT4G26570

Best hits in *Arabidopsis* were determined by local blastp method. Sequence length, mass weight, pI, and charge at PH 7.0 of the revised proteins (MdCBL10.1, MdCBL10.2, and MdCBL3.2) were listed in parentheses.

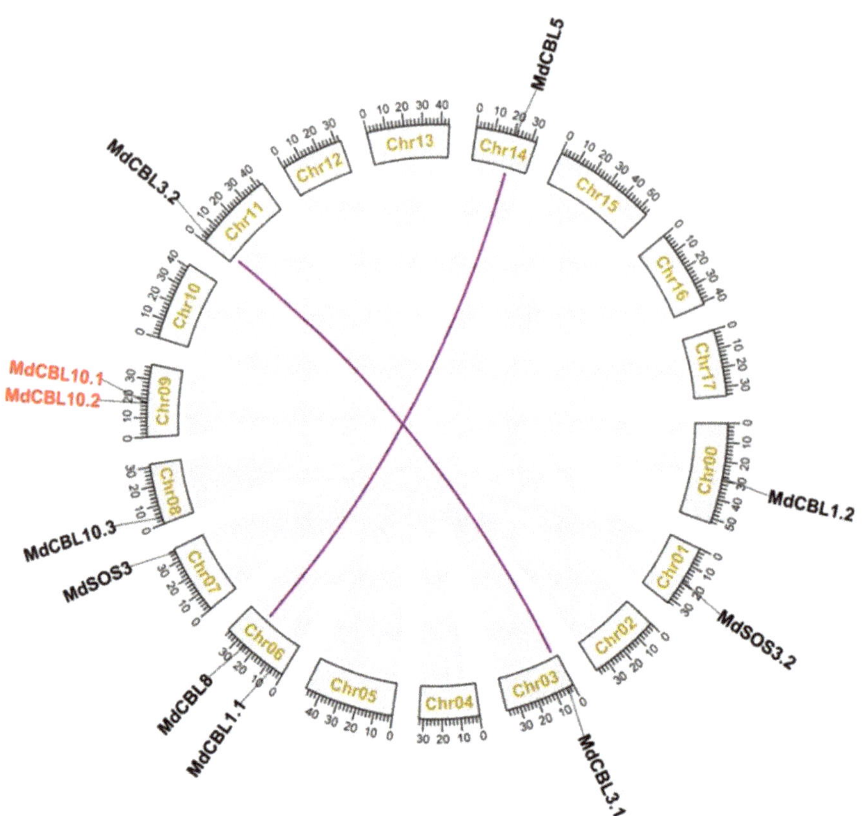

Figure 1. Genome locations of CBL family genes in apple. Purple lines and red font indicate the segmental and tandem duplication genes, respectively. Collinear blocks are represented by grayish lines. Chr00 represents the unassembled genomic scaffolds.

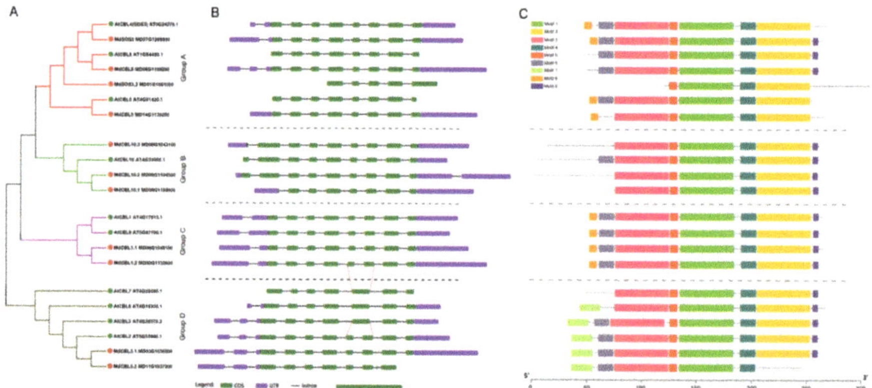

Figure 2. Phylogenic relationships (**A**), gene structure (**B**), and conserved motifs (**C**) for CBL family proteins in apples and *Arabidopsis*. Green and red dots indicate CBL proteins in *Arabidopsis* and apple, respectively. Red lines show that the fifth exon of *AtCBL7* and *AtCBL3* could be further differentiated into two shorter exons. Introns are represented by black line segments of the same length to facilitate the comparison of the exon–intron composition patterns of these genes.

Gene structure is one of the factors that reflects the evolution of a multigene family. Thus, the exon–intron composition patterns of these *MdCBL* and *AtCBL* genes were analyzed. Although the length of introns varied significantly (Figure S2), genes that belonged to the same group exhibited the same exon–intron composition patterns, with the largest variation in the N-terminal region of these genes (Figure 2B). Based on gene structure analysis, three *MdCBL* genes with partial exons missing compared with other *CBL* genes in the same group, namely *MdSOS3.2*, *Md10.1*, and *MdCBL3.2*, were also identified. This result suggested that the predicted coding sequences of these three genes in apple genome (GDDH13) might be wrong. The conserved motifs of the CBL family were further explored using the online software MEME, and nine conserved motifs were found (Figure S3). The motifs 1–5 were conserved in almost all CBL proteins except MdSOS3.2 and MdCBL3.2 (Figure 2C). Besides, MdCBL10.1 was the missing part of the N-terminal compared with other proteins in group B (Figure 2C). These results further supported the presence of errors in the prediction of coding sequences of these three genes.

Gene-specific primers were designed based on the predicted sequences in the apple genome (GDDH13) and used for gene cloning to confirm the coding sequence of *MdCBL* genes. Using total RNA extracted from the leaves of "Golden Delicious" apple as the template, the coding sequences of these *MdCBL* genes were finally obtained by polymerase chain reaction (PCR) amplification, except for *MdSOS3.2* (Supplementary File S2). The sequence alignment showed that the predicted protein sequences of MdCBL10.1 (MD09G1194600) and MdCBL3.2 (MD11G1037200) were missing a segment of N-terminal and C-terminal, respectively (Figure S4A). Although the coding sequence of *MdSOS3.2* was not obtained, the sequence comparison between MD01G1081000 and MdSOS3 indicated an incorrect fragment deletion and an incorrect fragment insertion in the N- and C-terminals of MD01G1081000, respectively (Figure S4B).

3.3. Promoter Analysis of MdCBL Genes

CBL family genes are involved in plant response to various environmental stresses. The promoter regions (upstream 1500 bp of the start codon ATG) of these genes were obtained from the apple genome (GDDH13) and submitted to the online software PlantCARE for *cis*-acting element analysis to explore the possible response of *MdCBL* genes to abiotic stresses. Various *cis*-elements related to abiotic stresses (hypoxia, cold, and drought) and plant hormone (ABA, auxin, MeJA, ethylene, GA, and SA) responsiveness were found (Figure 3 and Supplementary File S3), suggesting that these *MdCBL* genes played an important role in apple stress response. Many of these *cis*-elements appeared multiple times in the promoter region of the same gene (Figure 3). Besides, more hormone response–related *cis*-elements were present in the promoter regions of *MdCBLs* in groups A and B, especially *cis*-elements related to ABA and MeJA, while the *cis*-elements in the promoter regions of *MdCBLs* in groups C and D were mostly related to abiotic stress response (Figure 3). This indicated the functional differentiation of *MdCBL* genes between different groups.

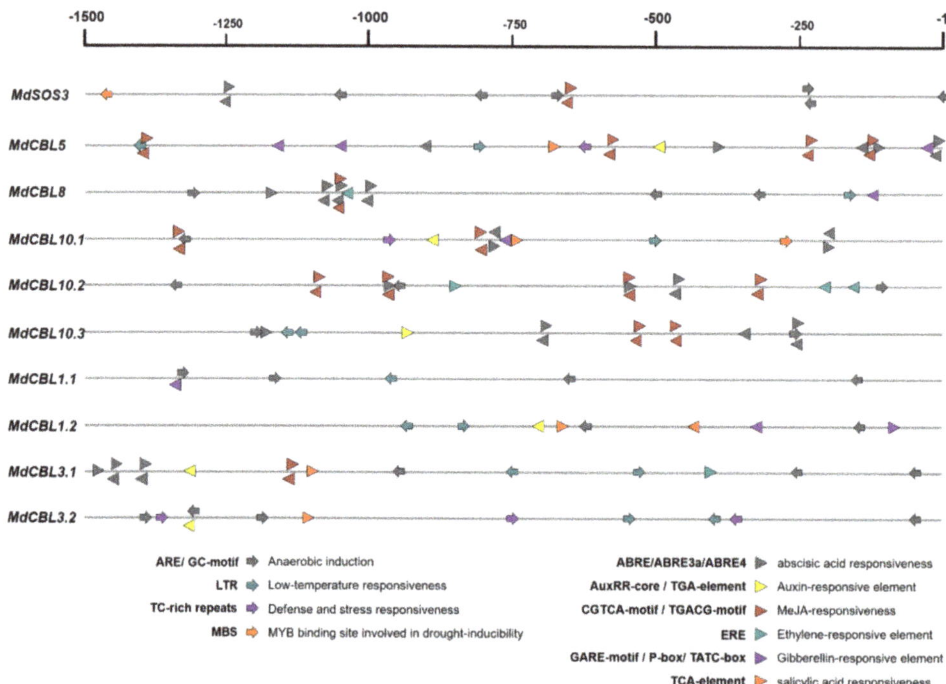

Figure 3. *Cis*-element analysis of the *MdCBL* gene promoter regions in apple. Arrows and triangles indicate the *cis*-elements related to abiotic stress response and hormone response, respectively. Positive and negative directions indicate whether the motif existed in the plus or minus strand of the *cis*-acting elements, respectively.

3.4. Functional Identification of Mdcbls in Regulating Salt Tolerance in Yeast

Previous studies demonstrated that CBL proteins affected plant salt tolerance through the SOS pathway. The yeast mutant strain Δ*ena*1−4 that lacked Na^+-ATPase and was sensitive to high [Na^+] was used to identify which CBL proteins in apples had a significant regulatory effect on salt tolerance. The full-length coding sequences of the ten *MdCBL* genes were cloned into the pDR196 vector and then transformed into the yeast mutant Δ*ena*1−4. The yeast strain W313-1B and positive transformants of the empty vector pDR196 were used as positive and negative controls. All of these strains were cultured in a YPD medium containing different concentrations of NaCl for three days. The growth of all strains was almost the same in a normal YPD medium (0 NaCl) (Figure 4). The addition of NaCl significantly inhibited the growth of these strains. The growth of the yeast mutant Δ*ena*1−4 was almost completely inhibited at a 200 mM NaCl concentration. The overexpression of any one of the seven *MdCBL* genes (MdCBL5, MdCBL10.1, MdCBL10.2, MdCBL1.1, MdCBL1.2, MdCBL3.1, and MdCBL3.2) could significantly inhibit the sensitivity of Δ*ena*1−4 to a high NaCl concentration (Figure 4). These results indicated that these seven *MdCBL* genes played a positive role in modulating salt tolerance in yeast. The growth of *MdCBL5* transgenic yeast was also completely inhibited with the further increase in the NaCl concentration (Figure 4), indicating a stronger capacity of the other six *MdCBL* genes than *MdCBL5* in enhancing yeast salt tolerance.

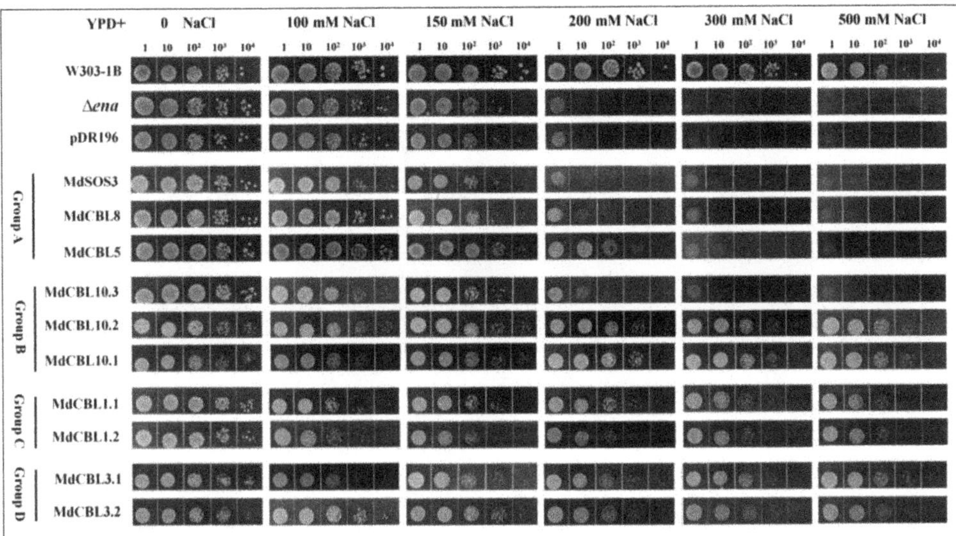

Figure 4. Functional identification of *MdCBLs* in the Na^+-sensitive yeast mutant. Aliquots (10 μL) of serial dilutions (10^0, 10^1, 10^2, 10^3, and 10^4) were dotted onto the YPD medium in the presence of 0, 100, 150, 200, 300, and 500 mM NaCl and grown for 3 days. W303-1B was the positive control. Δ*ena1−4* and Δ*ena1−4* transformed with the pDR196 empty vector were used as negative controls.

3.5. Overexpression of MdCBL10.1 Improved Salt Tolerance of Cisgenic Apple Calli

The SOS3/CBL10-SOS2-SOS1 and CBL10-CIPK8-SOS1 signaling pathways are the paramount regulatory mechanisms for facilitating Na^+ extrusion and are critical to the ability of plants to adapt to high salinity [31,41]. Based on the functional identification results in yeast, *MdCBL10.1* was selected for the subsequent transgene and functional identification in apple. Full-length CDS of *MdCBL10.1* was cloned into the pBI121 vector and transformed into apple calli. PCR identification and quantitative reverse transcription (qRT)-PCR expression analysis demonstrated that several cisgenic lines with high *MdCBL10.1* expression were obtained (Figure S5). Three lines (OE-3, OE-4, and OE-7) with high *MdCBL10.1* expression levels were selected for NaCl treatment.

After 20 days of culture, no significant difference was found between OE lines and wild type (WT) on the normal MS medium. The growth of all lines was inhibited under NaCl treatment, with significantly reduced fresh weight compared with the apple calli cultured in MS medium (Figure 5A,B). However, the growth of *MdCBL10.1* cisgenic lines was significantly better, and the fresh weight was significantly higher than that of the WT under salt treatment. These results suggested that the overexpression of *MdCBL10.1* could significantly improve the salt tolerance of cisgenic apple calli (Figure 5A,B). In *Arabidopsis*, CBL10 participates in the SOS pathway and promotes the Na^+ efflux [31,41]. Therefore, the Na^+ content of the apple calli was measured. No significant difference was observed between different lines cultured in the normal MS medium. Under NaCl treatment, the Na^+ contents of three cisgenic lines were significantly lower than that of the WT (Figure 5C), suggesting that the overexpression of *MdCBL10.1* could inhibit the excessive accumulation of Na^+ in cisgenic apple calli.

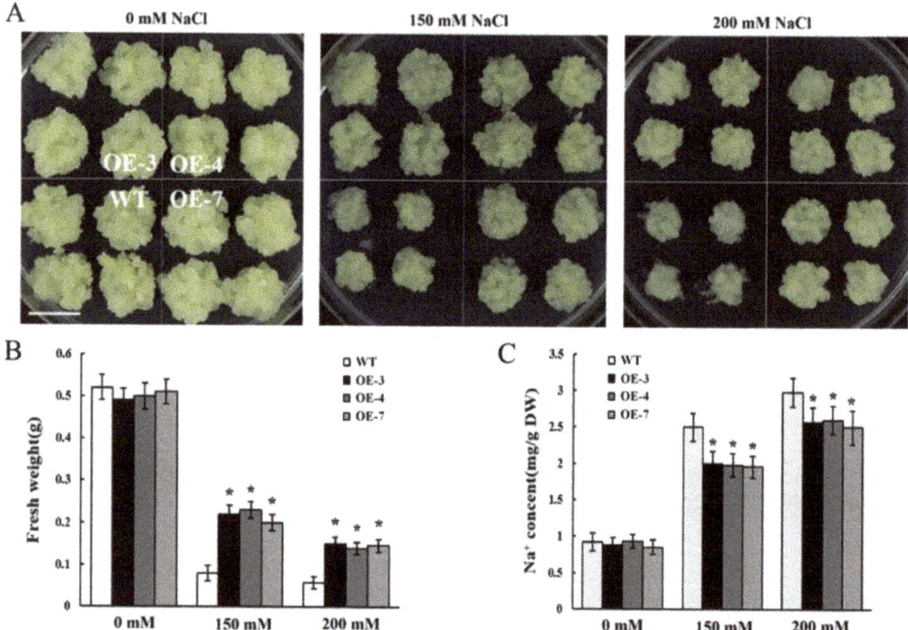

Figure 5. Functional identification of *MdCBL10.1* in cisgenic apple calli. (**A**) Phenotypes of the cisgenic (OE) and wild-type (WT) apple calli treated with NaCl stress. The scale bar represents 1 cm. Fresh weight (**B**) and Na^+ content (**C**) in cisgenic and WT apple calli. For (**B**) and (**C**), error bars represent the SD of three independent biological replicates, with each biological repeat having at least three dishes. Bars labeled with * in each panel are significantly different from the WT ($p < 0.05$, Student t test).

3.6. Overexpression of MdCBL10.1 in Roots Enhanced the Salt Tolerance of Apple Plants

The use of *A. rhizogenes* K599 to obtain plants with cisgenic roots provides a convenient way for studying the function of *MdCBL10.1* in apple plants because of the low genetic transformation efficiency [39,40]. Full-length CDS of *MdCBL10.1* was cloned into the pCAMBIA2300 vector fused with a GFP tag, and a selected inhibitory fragment of *MdCBL10.1* was cloned into the RNA-interference vector pK7GWIWG2D. Through GFP fluorescence detection and expression analysis (Figure S6), 20 plants with high *MdCBL10.1* expression (MdCBL10.1-OE) and 20 plants with *MdCBL10.1* expression significantly inhibited (MdCBL10.1-RNAi) were selected for subsequent salt treatment. Plants with their roots transformed with empty vectors pCAMBIA2300-GFP (OE-EV) and pK7GWIWG2D (RNAi-EV) were used as controls.

Under normal conditions (control group), no significant difference was observed between different types of cisgenic plants. After ten days of 150 mM NaCl irrigation, the RNAi plants were severely damaged by salt treatment, with their leaves showing obvious yellowing, wilting, and even death. Compared with the RNAi and control lines, the OE plants exhibited a better growth state, with their leaves still bright green and vigorous (Figure 6A). The measurements of relative ion leakage (REL) and MDA content in the leaves of these plants also indicated that the OE plants suffered less, whereas the RNAi plants suffered more stress damage caused by NaCl treatment (Figure 6B,C).

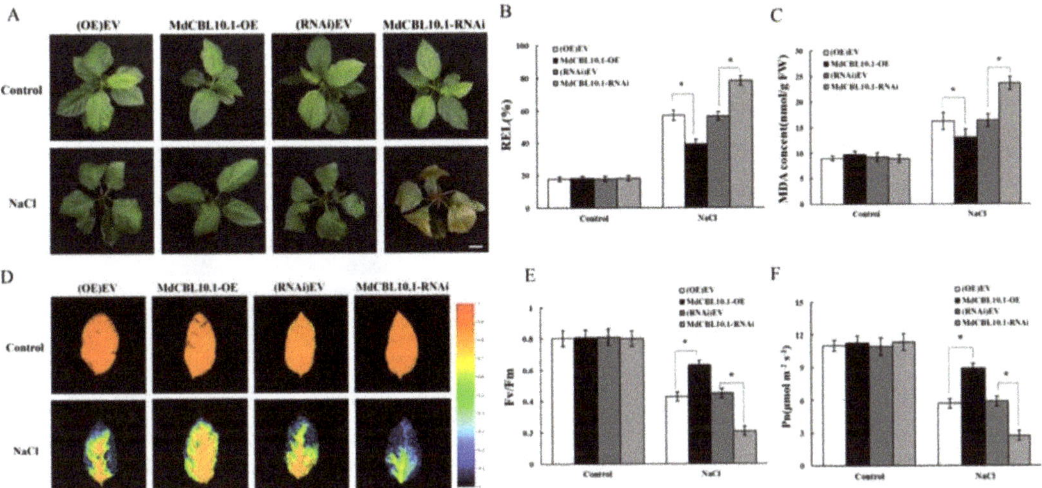

Figure 6. Phenotypic analysis of *MdCBL10.1* cisgenic apple plants under 150 mM NaCl treatment. (**A**) Growth phenotypes of cisgenic apple plants after NaCl treatment. Scale bars represent 3 cm. (**B**) Relative electrolyte leakage (REL) of apple leaves. (**C**) Malondialdehyde (MDA) content in apple leaves. Representative chlorophyll fluorescence images (**D**), Fv/Fm ratios (**E**), and net photosynthetic rate (Pn) (**F**) of cisgenic apple plants under normal and NaCl stress conditions. (OE)-EV and (RNAi)-EV represent roots transformed with the empty vector pCambia2300 and pK7GWIWG2D, respectively. The values of each index for cisgenic plants of the same type are the average values from all lines. Values are means of 20 replicates ± SD (each plant acts as a biological replicate). * in each panel denotes values significantly different from the corresponding control lines ($p < 0.05$, Student *t* test).

Besides REL and MDA content, the maximum quantum yield of PSII (Fv/Fm) is also an appropriate indicator for the early identification of the degree of damage in plants. Under normal conditions, the leaves of all lines were healthy and maintained high Fv/Fm ratios (Figure 6D,E). After NaCl treatment, the Fv/Fm of leaves of the RNAi plants decreased significantly, while that of the OE plants remained high (Figure 6D,E). The damage to photosynthetic units directly affected photosynthesis. After NaCl treatment, the net photosynthetic rate (Pn) of RNAi plants was significantly lower, while that of the OE plants was significantly higher than that of the control lines (Figure 6F). This was consistent with the performance of Fv/Fm and further supported that the overexpression of *MdCBL10.1* alleviated the stress damage to apple plants caused by NaCl treatment.

Salt stress triggers the accumulation of reactive oxygen species (ROS), which is harmful to plant growth. The accumulation of ROS in the roots of these NaCl-treated apple plants was determined to evaluate the damage caused by salt stress to root systems. The results showed that the roots of the OE lines accumulated less H_2O_2 and O_2^-, while the roots of the RNAi plants accumulated more ROS than controls (Figure 7A,B). Since *MdCBL10.1* overexpression could inhibit Na^+ accumulation in apple calli, the Na^+ content in these apple plants after NaCl treatment was measured. As expected, the Na^+ content in the roots and leaves of OE plants was significantly lower than that of the controls, while the RNAi plants showed the opposite trend (Figure 7C,D). These results indicated that *MdCBL10.1* overexpression in roots could reduce the Na^+ content in both roots and leaves, thereby alleviating salt stress-induced damage to apple plants.

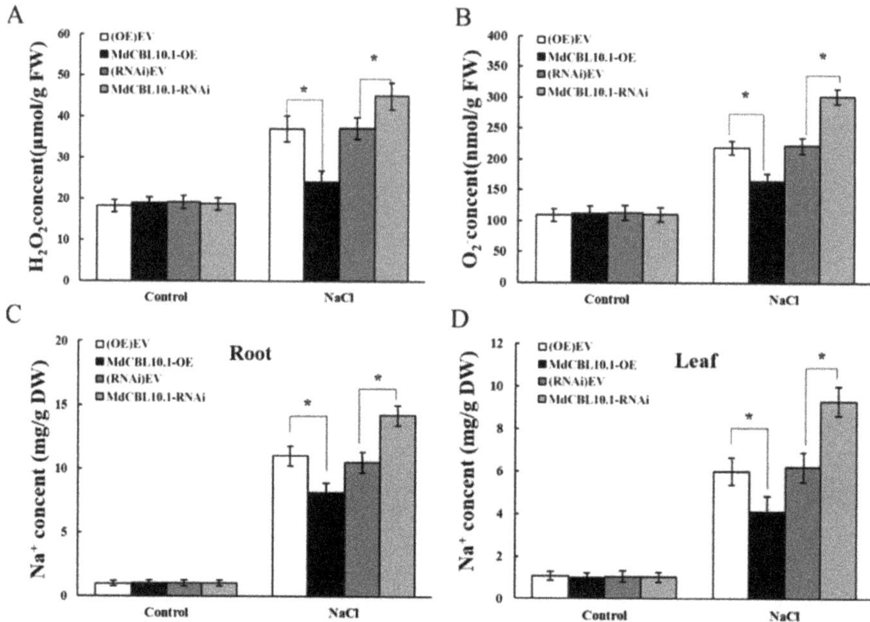

Figure 7. Overexpression of *MdCBL10.1* in apple roots inhibited the ROS and Na^+ accumulation in cisgenic apple plants. (**A**) Hydrogen peroxide (H_2O_2) content in roots. (**B**) Superoxide anion (O_2^-) content in roots. (**C**) Na^+ content in roots. (**D**) Na^+ content in leaves. The values of each index are the average values of all lines in cisgenic plants of the same type. Values are means of 20 replicates ± SD. * in each panel denotes values significantly different from the corresponding control lines ($p < 0.05$, Student t test).

4. Discussion

Plants are inevitably exposed to a variety of adverse environmental conditions, such as water shortage, low temperature, high salinity, and so forth, due to the sessile lifestyle. These abiotic stresses are increasingly harmful to crop yield and quality. Calcium and its signaling pathway play an important role in plant stress response. As specific Ca^{2+} sensors, CBL proteins play vital roles in calcium signaling and stress resistance regulation [7]. Apple is one of the most economically important fruits in the world. Its cultivation and extension are restricted by various abiotic stresses. The study of CBL proteins is thus important for the resistance breeding of apple. To date, the CBL family proteins in many plant species have been identified at the genome-wide level. However, no detailed characterization of apple CBL family proteins has been reported, and little is known about their functions in abiotic stress response.

Ca^{2+} is an essential element for plant growth and survival, and Ca^{2+} signals are an important regulator of growth, development, and biotic and abiotic stresses in plants [7,38,42]. Studies in various fruit tree crops also have shown that Ca^{2+} plays an important role in regulating fruit development, ripening, quality, and storage [43–47]. The CBL family has been identified and systematically studied in many plant species, such as dicotyledons *Arabidopsis* (10 AtCBLs) [33,48], eggplant (5 SmCBLs) [36], cotton (13 GaCBLs, 13 GrCBLs, 22 GhCBLs) [1], Cassava (9 MeCBLs) [12], pepper (9 CaCBLs) [10], tea plant (7 CsCBLs) [32], and pigeon pea (9 CcCBLs) [49], as well as the monocotyledon rice (10 OsCBLs) [33,48], due to the important role of CBL proteins in Ca^{2+} signaling. The CBL family has also been identified in some fruit trees such as grapevine (8 VvCBLs) [8], banana (11 MaCBLs) [50], and pineapple (8 AcCBLs) [11]. Although the genomes of these species are significantly larger than that of *Arabidopsis*, the number of *CBL* genes in most of these species is not significantly higher than that in *Arabidopsis* [48]. In this study, 11 *MdCBL* genes were identified from apples (Table 1), only one more than that from *Arabidopsis*, suggesting that the CBL family in apples had not expanded significantly during evolution. This was different from other gene families that were previously identified in apple, such as the bHLH [37], Lhc [51], and CaCA [38] families. Tandem and segmental duplication events are fundamental mechanisms of gene family expansion. Considering the two genome-wide duplication events that occurred during apple evolution [52], collinear analysis between these *MdCBLs* was performed. The results showed that only two segmental duplication events occurred (Figure 1), indicating that the CBL family did not greatly expand during apple evolution. The evolutionary analysis of many plant species also suggested that the number of CBL members was independent of their genome size [48].

Although the number of CBL members varied, the phylogenetic analysis of CBL proteins in various plant species indicated that this family should be divided into four groups. In this study, the MdCBL proteins were divided into four groups (groups A to D), as in *Arabidopsis* (Figure 2A). This grouping result was also supported by the gene structure analysis and conserved motif prediction. The intron/exon and motif composition patterns of genes within the same group were consistent, whereas significant differences were observed between different groups (Figure 2B,C). In addition, based on the comparison of gene structure and conserved motifs of genes within the same group, three *MdCBL* genes (*MdSOS3.2*, *MdCBL10.1*, and *MdCBL3.2*) that might have errors in their predicted CDS in the apple genome were identified. The gene cloning results confirmed this speculation (Table 1 and Figure S4) and further suggested that this comparative analysis method could help people identify genes in a gene family whose coding sequences were incorrectly predicted. On the contrary, the high similarity of gene structure, conserved motifs, and *cis*-acting elements (Figures 2 and 3) suggested functional redundancy among *MdCBL* genes, especially *MdCBL* genes in the same group. Similar results were also found in *Arabidopsis*, such as SOS3 and CBL10 [31].

Stress stimulation causes a transient increase in the intracellular Ca^{2+} concentration. CBL proteins can sense and interact with intracellular increased Ca^{2+}. The binding of Ca^{2+} promotes the interaction between CBL and CIPK proteins, which is crucial for the activation of the kinase activity of CIPKs. Then, the activated CIPKs phosphorylate downstream substrates, further triggering a range of response mechanisms [7,33,42]. Decades of research have revealed extensive and complex interaction networks between CBL and CIPK family proteins. As Ca^{2+} sensors, each CBL has three or more EF-hand domains and Ca^{2+}-binding sites. Besides, the CBL proteins harbor a conserved FPSF domain in their C-terminal, which is the target of phosphorylation by CIPK. Moreover, many of the CBL proteins also contain conserved MGCXXS/T motifs in their N-terminal, which contribute to the anchorage of CBLs in the membrane to transduce a Ca^{2+} signal [48–50,53]. Several studies proved that different CBLs could be localized to the plasma or vacuole membrane and could regulate plant salt tolerance by promoting the Na^+ efflux or sequestration [7,42]. In this study, all MdCBL proteins contained the FPSF domain in their C-terminal (Figures 2, S3 and S4), suggesting complex interactions between MdCBLs and CIPK family proteins in apple. For example, four MdCBL proteins were found to interact with the CIPK family protein MdSOS2L1 [54]. Many of these MdCBLs contained the conserved MGCXXS/T domain in their N-terminal. Further, the distribution of this domain showed obvious group specificity, which was found only in groups A and C of the MdCBL family (Figures 2, S3 and S4). These results suggested that MdCBL proteins in these two groups might be more likely to function on the membrane and also indicated the functional and subcellular differentiation of CBL proteins between different groups. More studies on the subcellular localization and functional identification of apple CBL proteins are needed to confirm this hypothesis.

The CBL-CIPK model is reported to characterize many forms of abiotic stress responses. The first identified CBL-CIPK pathway was the SOS pathway, which plays vital roles in plant salt tolerance regulation [55]. It contains three key components: SOS1 (Na^+/H^+ antiporter), SOS2 (CIPK24), and SOS3 (CBL4). Under salt stress, the SOS2–SOS3 complex activates the transport properties of SOS1 to promote the Na^+ efflux, thus enhancing plant salt tolerance. Subsequent studies found that CBL10 (SCABP8) could also interact with SOS2 and activate SOS1, suggesting the functional redundancy between SOS3 and CBL10 [31]. Studies in apples also showed that MdCBL10 could interact with MdSOS2L1, an SOS2-like protein that played positive roles in regulating plant salt tolerance [54]. In addition, a recent study found that the CBL10–CIPK8 complex regulated plant salt tolerance through its interaction with SOS1 [41]. These results, combined with the functional identification results in yeast (Figure 4), led to the selection of *MdCBL10.1* for further genetic transformation and functional identification in apple. The phenotypic comparison and measurement of stress-related physiological indicators showed that *MdCBL10.1* overexpression significantly enhanced the salt tolerance of cisgenic apple materials (Figures 5–7). Moreover, the significantly reduced Na^+ content in transgenic and cisgenic materials indicated that *MdCBL10.1* overexpression could significantly inhibit Na^+ accumulation under salt stress. These results suggested that MdCBL10.1 enhanced plant salt tolerance by inhibiting Na^+ accumulation, and this was probably achieved through the SOS pathway.

In conclusion, 11 *MdCBL* genes were identified from apple. CDSs of these genes were cloned, and the gene structure and conserved motifs were analyzed. Several *MdCBLs* that played positive roles in salt tolerance were identified using the Na^+-sensitive yeast mutant. The function of *MdCBL10.1* in regulating salt tolerance was also identified in cisgenic apple materials in detail. This study provided a foundation for future research examining the function and mechanism of CBL proteins in regulating apple salt tolerance.

Supplementary Materials: The following are available online at https://www.mdpi.com/article/10.3390/ijms222212430/s1. Figure S1: Phylogenetic analysis of proteins in HMMER screening results and CBL family members in Arabidopsis; Figure S2: Comparison of gene structure of the CBL family genes in apples and Arabidopsis; Figure S3: Putative conserved motifs identified in the sequences of apple CBL family proteins. Red dots indicate the conserved "FPSF" and "MGCxxS/T" domains; Figure S4: Sequence comparison between the predicted MdCBLs in the apple genome and the MdCBLs that were actually cloned in this study; Figure S5: Identification of cisgenic apple calli. (**A**) PCR identification of the MdCBL10.1 transgene based on genomic DNA extracted from apple calli. (**B**) Expression level of MdCBL10.1 in cisgenic apple calli. MdMDH served as an internal reference gene. Bars labeled with different letters in each panel are significantly different ($p < 0.05$, one-way ANOVA and Duncan's test); Figure S6: Identification of the MdCBL10.1 transgene in the roots of apple plants. (**A**) GFP expression in cisgenic apple roots. (**B**) qRT-PCR determination of MdCBL10.1 expression in cisgenic apple roots. (OE)-EV and (RNAi)-EV represent roots transformed with the empty vector pCambia2300 and pK7GWIWG2D, respectively. MdMDH served as an internal reference gene. Bars labeled with different letters in each panel are significantly different ($p < 0.05$, one-way ANOVA and Duncan's test). Supplementary File S1: HMMER screening results. Supplementary File S2: CDS and protein sequences of the cloned MdCBL genes. Supplementary File S3: Cis-elements identified in the promoter regions of MdCBL genes. Plus and minus signs indicate whether the cis-element was located in the plus or minus strand.

Author Contributions: P.C., J.Y., K.M. and F.M. conceived the project; P.C., J.Y. and K.M. designed the research plan; P.C., J.Y., Q.M., H.L. and Y.C. carried out the experiments; P.C., J.Y. and Q.M. analyzed the data; J.Y., K.M. and F.M. wrote the manuscript. All authors have read and agreed to the published version of the manuscript.

Funding: This work was supported by National Key Research and Development Program of China (2018YFD1000300/2018YFD1000301), the National Natural Science Foundation of China (31701894), the Key S&T Special Projects of Shaanxi Province (2020zdzx03–01-02), and the China Agriculture Research System of MOF and MARA (CARS-27).

Institutional Review Board Statement: No applicable.

Informed Consent Statement: No applicable.

Data Availability Statement: All data supporting the findings of this study are available within the paper and within its supplementary data published online.

Acknowledgments: We thank Huazhong Shi (Texas Tech University) and Jiafu Jiang (Nanjing Agricultural University) for providing the Na^+-sensitive yeast mutants. Apple calli ('Orin') and the 'Gala' (GL-3) seedlings were kindly provided by Yujin Hao (Shandong Agricultural University) and Zhinghong Zhang (Shenyang Agricultural University), respectively.

Conflicts of Interest: The authors declare no conflict of interest.

References

1. Lu, T.T.; Zhang, G.F.; Sun, L.R.; Wang, J.; Hao, F.S. Genome-wide identification of CBL family and expression analysis of CBLs in response to potassium deficiency in cotton. *PeerJ* **2017**, *5*, e3653. [CrossRef]
2. Luan, S.; Kudla, J.; Rodriguez-Concepcion, M.; Yalovsky, S.; Gruissem, W. Calmodulins and calcineurin B-like proteins: Calcium sensors for specific signal response coupling in plants. *Plant Cell* **2002**, *14*, S389–S400. [CrossRef]
3. Cheong, Y.H.; Pandey, G.K.; Grant, J.J.; Batistic, O.; Li, L.; Kim, B.G.; Lee, S.C.; Kudla, J.; Luan, S. Two calcineurin B-like calcium sensors, interacting with protein kinase CIPK23, regulate leaf transpiration and root potassium uptake in Arabidopsis. *Plant J. Cell Mol. Biol.* **2007**, *52*, 223–239. [CrossRef]
4. Mazars, C.; Bourque, S.; Mithofer, A.; Pugin, A.; Ranjeva, R. Calcium homeostasis in plant cell nuclei. *New Phytol.* **2009**, *181*, 261–274. [CrossRef] [PubMed]
5. Gilroy, S.; Bethke, P.C.; Jones, R.L. Calcium homeostasis in plants. *J. Cell Sci.* **1993**, *106*, 453–462. [CrossRef]
6. Kumar, M.; Sharma, K.; Yadav, A.K.; Kanchan, K.; Baghel, M.; Kateriya, S.; Pandey, G.K. Genome-wide identification and biochemical characterization of calcineurin B-like calcium sensor proteins in *Chlamydomonas reinhardtii*. *Biochem. J.* **2020**, *477*, 1879–1892. [CrossRef] [PubMed]
7. Sanyal, S.K.; Pandey, A.; Pandey, G.K. The CBL-CIPK signaling module in plants: A mechanistic perspective. *Physiol. Plant.* **2015**, *155*, 89–108. [CrossRef]
8. Xi, Y.; Liu, J.Y.; Dong, C.; Cheng, Z.M. The CBL and CIPK gene family in grapevine (*Vitis vinifera*): Genome-wide analysis and expression profiles in response to various abiotic stresses. *Front. Plant Sci.* **2017**, *8*, 978. [CrossRef] [PubMed]
9. Sun, W.N.; Zhang, B.; Deng, J.W.; Chen, L.; Ullah, A.; Yang, X.Y. Genome-wide analysis of CBL and CIPK family genes in cotton: Conserved structures with divergent interactions and expression. *Physiol. Mol. Biol. Plant* **2021**, *27*, 359–368. [CrossRef]
10. Ma, X.; Gai, W.X.; Qiao, Y.M.; Ali, M.; Wei, A.M.; Luo, D.X.; Li, Q.H.; Gong, Z.H. Identification of CBL and CIPK gene families and functional characterization of CaCIPK1 under *Phytophthora capsici* in pepper (*Capsicum annuum* L.). *BMC Genom.* **2019**, *20*, 775. [CrossRef]
11. Aslam, M.; Fakher, B.; Jakada, B.H.; Zhao, L.H.; Cao, S.J.; Cheng, Y.; Qin, Y. Genome-wide identification and expression profiling of CBL-CIPK gene family in pineapple (*Ananas comosus*) and the role of AcCBL1 in abiotic and biotic stress response. *Biomolecules* **2019**, *9*, 293. [CrossRef] [PubMed]
12. Hu, W.; Yan, Y.; Tie, W.W.; Ding, Z.H.; Wu, C.L.; Ding, X.P.; Wang, W.Q.; Xia, Z.Q.; Guo, J.C.; Peng, M. Genome-wide analyses of calcium sensors reveal their involvement in drought stress response and storage roots deterioration after harvest in Cassava. *Genes* **2018**, *9*, 221. [CrossRef] [PubMed]
13. Tang, M.F.; Xu, C.; Cao, H.H.; Shi, Y.; Chen, J.; Chai, Y.; Li, Z.G. Tomato calmodulin-like protein SlCML37 is a calcium (Ca^{2+}) sensor that interacts with proteasome maturation factor SlUMP1 and plays a role in tomato fruit chilling stress tolerance. *J. Plant Physiol.* **2021**, *258*, 153373. [CrossRef]
14. Jung, H.; Chung, P.J.; Park, S.H.; Redillas, M.C.F.R.; Kim, Y.S.; Suh, J.W.; Kim, J.K. Overexpression of OsERF48 causes regulation of OsCML16, a calmodulin-like protein gene that enhances root growth and drought tolerance. *Plant Biotechnol. J.* **2017**, *15*, 1295–1308. [CrossRef] [PubMed]
15. La Verde, V.; Trande, M.; D'Onofrio, M.; Dominici, P.; Astegno, A. Binding of calcium and target peptide to calmodulin-like protein CML19, the centrin 2 of Arabidopsis thaliana. *Int. J. Biol. Macromol.* **2018**, *108*, 1289–1299. [CrossRef]
16. Trande, M.; Pedretti, M.; Bonza, M.C.; Di Matteo, A.; D'Onofrio, M.; Dominici, P.; Astegno, A. Cation and peptide binding properties of CML7, a calmodulin-like protein from *Arabidopsis thaliana*. *J. Inorg. Biochem.* **2019**, *199*, 110796. [CrossRef]
17. Zhao, P.C.; Liu, Y.J.; Kong, W.Y.; Ji, J.Y.; Cai, T.Y.; Guo, Z.F. Genome-wide identification and characterization of calcium-dependent protein kinase (CDPK) and CDPK-related kinase (CRK) gene families in *Medicago truncatula*. *Int. J. Mol. Sci.* **2021**, *22*, 1044. [CrossRef]
18. Bi, Z.Z.; Wang, Y.H.; Li, P.C.; Sun, C.; Qin, T.Y.; Bai, J.P. Evolution and expression analysis of CDPK genes under drought stress in two varieties of potato. *Biotechnol. Lett.* **2021**, *43*, 511–521. [CrossRef]
19. Crizel, R.L.; Perin, E.C.; Vighi, I.L.; Woloski, R.; Seixas, A.; Pinto, L.D.; Rombaldi, C.V.; Galli, V. Genome-wide identification, and characterization of the CDPK gene family reveal their involvement in abiotic stress response in Fragaria × ananassa. *Sci. Rep.* **2020**, *10*, 11040. [CrossRef]
20. Chu, L.C.; Offenborn, J.N.; Steinhorst, L.; Wu, X.N.; Xi, L.; Li, Z.; Jacquot, A.; Lejay, L.; Kudla, J.; Schulze, W.X. Plasma membrane calcineurin B-like calcium-ion sensor proteins function in regulating primary root growth and nitrate uptake by affecting global phosphorylation patterns and microdomain protein distribution. *New Phytol.* **2021**, *229*, 2223–2237. [CrossRef]
21. Lu, L.; Chen, X.Y.; Zhu, L.M.; Li, M.J.; Zhang, J.B.; Yang, X.Y.; Wang, P.K.; Lu, Y.; Cheng, T.L.; Shi, J.S.; et al. NtCIPK9: A calcineurin B-like protein-interacting protein kinase from the Halophyte *Nitraria tangutorum*, enhances arabidopsis salt tolerance. *Front. Plant Sci.* **2020**, *11*, 1112. [CrossRef]

22. Albrecht, V.; Weinl, S.; Blazevic, D.; D'Angelo, C.; Batistic, O.; Kolukisaoglu, U.; Bock, R.; Schulz, B.; Harter, K.; Kudla, J. The calcium sensor CBL1 integrates plant responses to abiotic stresses. *Plant J. Cell Mol. Biol.* **2003**, *36*, 457–470. [CrossRef] [PubMed]
23. Gao, H.; Wang, C.; Li, L.; Fu, D.; Zhang, Y.; Yang, P.; Zhang, T.; Wang, C. A novel role of the calcium sensor CBL1 in response to phosphate deficiency in *Arabidopsis thaliana*. *J. Plant Physiol.* **2020**, *253*, 153266. [CrossRef]
24. Yan, Y.; He, M.; Guo, J.; Zeng, H.; Wei, Y.; Liu, G.; Hu, W.; Shi, H. The CBL1/9-CIPK23-AKT1 complex is essential for low potassium response in cassava. *Plant Physiol. Biochem. PPB* **2021**, *167*, 430–437. [CrossRef] [PubMed]
25. Xu, J.; Li, H.D.; Chen, L.Q.; Wang, Y.; Liu, L.L.; He, L.; Wu, W.H. A protein kinase, interacting with two calcineurin B-like proteins, regulates K+ transporter AKT1 in Arabidopsis. *Cell* **2006**, *125*, 1347–1360. [CrossRef] [PubMed]
26. Held, K.; Pascaud, F.; Eckert, C.; Gajdanowicz, P.; Hashimoto, K.; Corratge-Faillie, C.; Offenborn, J.N.; Lacombe, B.; Dreyer, I.; Thibaud, J.B.; et al. Calcium-dependent modulation and plasma membrane targeting of the AKT2 potassium channel by the CBL4/CIPK6 calcium sensor/protein kinase complex. *Cell Res.* **2011**, *21*, 1116–1130. [CrossRef] [PubMed]
27. Qiu, Q.S.; Guo, Y.; Dietrich, M.A.; Schumaker, K.S.; Zhu, J.K. Regulation of SOS1, a plasma membrane Na+/H+ exchanger in *Arabidopsis thaliana*, by SOS_2 and SOS_3. *Proc. Natl. Acad. Sci. USA* **2002**, *99*, 8436–8441. [CrossRef]
28. Liu, J.; Zhu, J.K. A calcium sensor homolog required for plant salt tolerance. *Science* **1998**, *280*, 1943–1945. [CrossRef]
29. Zhu, J.K. Abiotic stress signaling and responses in plants. *Cell* **2016**, *167*, 313–324. [CrossRef]
30. Kim, B.G.; Waadt, R.; Cheong, Y.H.; Pandey, G.K.; Dominguez-Solis, J.R.; Schultke, S.; Lee, S.C.; Kudla, J.; Luan, S. The calcium sensor CBL10 mediates salt tolerance by regulating ion homeostasis in Arabidopsis. *Plant J. Cell Mol. Biol.* **2007**, *52*, 473–484. [CrossRef]
31. Quan, R.; Lin, H.; Mendoza, I.; Zhang, Y.; Cao, W.; Yang, Y.; Shang, M.; Chen, S.; Pardo, J.M.; Guo, Y. SCABP8/CBL10, a putative calcium sensor, interacts with the protein kinase SOS2 to protect Arabidopsis shoots from salt stress. *Plant Cell* **2007**, *19*, 1415–1431. [CrossRef]
32. Liu, H.; Wang, Y.X.; Li, H.; Teng, R.M.; Wang, Y.; Zhuang, J. Genome-wide identification and expression analysis of calcineurin B-like protein and calcineurin B-like protein-interacting protein kinase family genes in tea plant. *DNA Cell Biol.* **2019**, *38*, 824–839. [CrossRef] [PubMed]
33. Kolukisaoglu, U.; Weinl, S.; Blazevic, D.; Batistic, O.; Kudla, J. Calcium sensors and their interacting protein kinases: Genomics of the Arabidopsis and rice CBL-CIPK signaling networks. *Plant Physiol.* **2004**, *134*, 43–58. [CrossRef] [PubMed]
34. Zhang, K.; Yue, D.; Wei, W.; Hu, Y.; Feng, J.; Zou, Z. Characterization and functional analysis of calmodulin and calmodulin-like genes in *Fragaria vesca*. *Front. Plant Sci.* **2016**, *7*, 1820. [CrossRef] [PubMed]
35. Hu, W.; Hou, X.; Xia, Z.; Yan, Y.; Wei, Y.; Wang, L.; Zou, M.; Lu, C.; Wang, W.; Peng, M. Genome-wide survey and expression analysis of the calcium-dependent protein kinase gene family in cassava. *Mol. Genet. Genom. MGG* **2016**, *291*, 241–253. [CrossRef]
36. Li, J.; Jiang, M.M.; Ren, L.; Liu, Y.; Chen, H.Y. Identification and characterization of CBL and CIPK gene families in eggplant (*Solanum melongena* L.). *Mol. Genet. Genom. MGG* **2016**, *291*, 1769–1781. [CrossRef]
37. Mao, K.; Dong, Q.; Li, C.; Liu, C.; Ma, F. Genome wide identification and characterization of apple bHLH Transcription factors and expression analysis in response to drought and salt stress. *Front. Plant Sci.* **2017**, *8*, 480. [CrossRef]
38. Mao, K.; Yang, J.; Wang, M.; Liu, H.; Guo, X.; Zhao, S.; Dong, Q.; Ma, F. Genome-wide analysis of the apple CaCA superfamily reveals that MdCAX proteins are involved in the abiotic stress response as calcium transporters. *BMC Plant Biol.* **2021**, *21*, 81. [CrossRef]
39. Yang, J.; Guo, X.; Li, W.; Chen, P.; Cheng, Y.; Ma, F.; Mao, K. MdCCX2 of apple functions positively in modulation of salt tolerance. *Environ. Exp. Bot.* **2021**, *192*, 104663. [CrossRef]
40. Yang, J.; Li, W.; Guo, X.; Chen, P.; Cheng, Y.; Mao, K.; Ma, F. Cation/Ca^{2+} Exchanger 1 (MdCCX1), a plasma membrane-localized Na+ transporter, enhances plant salt tolerance by inhibiting excessive accumulation of Na+ and reactive oxygen species. *Front. Plant Sci.* **2021**, *12*, 46189. [CrossRef]
41. Yin, X.; Xia, Y.; Xie, Q.; Cao, Y.; Wang, Z.; Hao, G.; Song, J.; Zhou, Y.; Jiang, X. The protein kinase complex CBL10-CIPK8-SOS1 functions in Arabidopsis to regulate salt tolerance. *J. Exp. Bot.* **2020**, *71*, 1801–1814. [CrossRef]
42. Tang, R.J.; Wang, C.; Li, K.L.; Luan, S. The CBL-CIPK Calcium signaling network: Unified paradigm from 20 years of discoveries. *Trends Plant Sci.* **2020**, *25*, 604–617. [CrossRef] [PubMed]
43. Michailidis, M.; Karagiannis, E.; Tanou, G.; Samiotaki, M.; Tsiolas, G.; Sarrou, E.; Stamatakis, G.; Ganopoulos, I.; Martens, S.; Argiriou, A.; et al. Novel insights into the calcium action in cherry fruit development revealed by high-throughput mapping. *Plant Mol. Biol.* **2020**, *104*, 597–614. [CrossRef] [PubMed]
44. Michailidis, M.; Karagiannis, E.; Tanou, G.; Karamanoli, K.; Lazaridou, A.; Matsi, T.; Molassiotis, A. Metabolomic and physico-chemical approach unravel dynamic regulation of calcium in sweet cherry fruit physiology. *Plant Physiol. Biochim.* **2017**, *116*, 68–79. [CrossRef]
45. Yu, J.; Zhu, M.T.; Wang, M.J.; Xu, Y.S.; Chen, W.T.; Yang, G.S. Transcriptome analysis of calcium-induced accumulation of anthocyanins in grape skin. *Sci. Hortic.* **2020**, *260*, 108871. [CrossRef]
46. Martins, V.; Billet, K.; Garcia, A.; Lanoue, A.; Geros, H. Exogenous calcium deflects grape berry metabolism towards the production of more stilbenoids and less anthocyanins. *Food Chem.* **2020**, *313*, 126123. [CrossRef] [PubMed]
47. Michailidis, M.; Karagiannis, E.; Tanou, G.; Sarrou, E.; Stavridou, E.; Ganopoulos, I.; Karamanoli, K.; Madesis, P.; Martens, S.; Molassiotis, A. An integrated metabolomic and gene expression analysis identifies heat and calcium metabolic networks underlying postharvest sweet cherry fruit senescence. *Planta* **2019**, *250*, 2009–2022. [CrossRef]

48. Mohanta, T.K.; Mohanta, N.; Mohanta, Y.K.; Parida, P.; Bae, H.H. Genome-wide identification of Calcineurin B-Like (CBL) gene family of plants reveals novel conserved motifs and evolutionary aspects in calcium signaling events. *BMC Plant Biol.* **2015**, *15*, 189. [CrossRef]
49. Song, Z.H.; Dong, B.Y.; Yang, Q.; Niu, L.L.; Li, H.H.; Cao, H.Y.; Amin, R.; Meng, D.; Fu, Y.J. Screening of CBL genes in pigeon pea with focus on the functional analysis of CBL4 in abiotic stress tolerance and flavonoid biosynthesis. *Environ. Exp. Bot.* **2020**, *177*, 104102. [CrossRef]
50. Xiong, Y.; Li, R.M.; Lin, X.J.; Zhou, Y.J.; Tang, F.L.; Yao, Y.; Liu, J.; Wang, L.X.; Yin, X.M.; Liu, Y.X.; et al. Genome-wide analysis and expression tendency of banana (*Musa acuminata* L.) calcineurin B-like (MaCBL) genes under potassium stress. *Horticulturae* **2021**, *7*, 70. [CrossRef]
51. Zhao, S.; Gao, H.; Luo, J.; Wang, H.; Dong, Q.; Wang, Y.; Yang, K.; Mao, K.; Ma, F. Genome-wide analysis of the light-harvesting chlorophyll a/b-binding gene family in apple (Malus domestica) and functional characterization of MdLhcb4.3, which confers tolerance to drought and osmotic stress. *Plant Physiol. Biochem. PPB* **2020**, *154*, 517–529. [CrossRef]
52. Velasco, R.; Zharkikh, A.; Affourtit, J.; Dhingra, A.; Cestaro, A.; Kalyanaraman, A.; Fontana, P.; Bhatnagar, S.K.; Troggio, M.; Pruss, D.; et al. The genome of the domesticated apple (Malus × domestica *Borkh.*). *Nat. Genet.* **2010**, *42*, 833–839. [CrossRef] [PubMed]
53. Jiang, M.; Zhao, C.L.; Zhao, M.F.; Li, Y.Z.; Wen, G.S. Phylogeny and evolution of calcineurin B-like (CBL) gene family in grass and functional analyses of rice CBLs. *J. Plant Biol.* **2020**, *63*, 117–130. [CrossRef]
54. Hu, D.G.; Ma, Q.J.; Sun, C.H.; Sun, M.H.; You, C.X.; Hao, Y.J. Overexpression of MdSOS2L1, a CIPK protein kinase, increases the antioxidant metabolites to enhance salt tolerance in apple and tomato. *Physiol. Plant.* **2016**, *156*, 201–214. [CrossRef]
55. Mahajan, S.; Tuteja, N. Cold, salinity and drought stresses: An overview. *Arch. Biochem. Biophys.* **2005**, *444*, 139–158. [CrossRef] [PubMed]

International Journal of
Molecular Sciences

Article

Genome-Wide Identification and Characterization of Wheat 14-3-3 Genes Unravels the Role of TaGRF6-A in Salt Stress Tolerance by Binding MYB Transcription Factor

Wenna Shao [1,2,†], Wang Chen [3,†], Xiaoguo Zhu [2,†], Xiaoyi Zhou [1,2], Yingying Jin [1], Chuang Zhan [1], Gensen Liu [1], Xi Liu [1], Dongfang Ma [1,*] and Yongli Qiao [2,*]

[1] Engineering Research Center of Ecology and Agricultural Use of Wetland, Ministry of Education, Hubei Collaborative Innovation Center for Grain Industry, College of Agriculture, Yangtze University, Jingzhou 434000, China; 201871379@yangtzeu.edu.cn (W.S.); 201871384@yangtzeu.edu.cn (X.Z.); 202072787@yangtzeu.edu.cn (Y.J.); 202072788@yangtzeu.edu.cn (C.Z.); 202071677@yangtzeu.edu.cn (G.L.); 202072786@yangtzeu.edu.cn (X.L.)

[2] Shanghai Key Laboratory of Plant Molecular Sciences, College of Life Sciences, Shanghai Normal University, Shanghai 200234, China; zhuxiaoguo1990@126.com

[3] Oil Crops Research Institute, Chinese Academy of Agricultural Sciences, Wuhan 430062, China; chenwangchw@163.com

* Correspondence: madf@yangtzeu.edu.cn (D.M.); qyl588@gmail.com (Y.Q.)

† These authors contributed equally to this work.

Citation: Shao, W.; Chen, W.; Zhu, X.; Zhou, X.; Jin, Y.; Zhan, C.; Liu, G.; Liu, X.; Ma, D.; Qiao, Y. Genome-Wide Identification and Characterization of Wheat 14-3-3 Genes Unravels the Role of TaGRF6-A in Salt Stress Tolerance by Binding MYB Transcription Factor. *Int. J. Mol. Sci.* **2021**, *22*, 1904. https://doi.org/10.3390/ijms22041904

Academic Editors: Mirza Hasanuzzaman and Masayuki Fujita

Received: 29 January 2021
Accepted: 9 February 2021
Published: 14 February 2021

Abstract: 14-3-3 proteins are a large multigenic family of general regulatory factors (GRF) ubiquitously found in eukaryotes and play vital roles in the regulation of plant growth, development, and response to stress stimuli. However, so far, no comprehensive investigation has been performed in the hexaploid wheat. In the present study, A total of 17 potential 14-3-3 gene family members were identified from the Chinese Spring whole-genome sequencing database. The phylogenetic comparison with six 14-3-3 families revealed that the majority of wheat *14-3-3* genes might have evolved as an independent branch and grouped into ε and non-ε group using the phylogenetic comparison. Analysis of gene structure and motif indicated that 14-3-3 protein family members have relatively conserved exon/intron arrangement and motif composition. Physical mapping showed that wheat *14-3-3* genes are mainly distributed on chromosomes 2, 3, 4, and 7. Moreover, most *14-3-3* members in wheat exhibited significantly down-regulated expression in response to alkaline stress. VIGS assay and protein-protein interaction analysis further confirmed that TaGRF6-A positively regulated slat stress tolerance by interacting with a MYB transcription factor, TaMYB64. Taken together, our findings provide fundamental information on the involvement of the wheat 14-3-3 family in salt stress and further investigating their molecular mechanism.

Keywords: 14-3-3 gene family; *Triticum aestivum* L.; bioinformatics analysis; salt tolerance; protein-protein interactions

1. Introduction

General regulatory factor (GRF) proteins, also known as 14-3-3 proteins, are found among all eukaryotic organisms [1]. The 14-3-3 protein, first discovered in the bovine brain as a soluble acidic protein, acquired its name based on the fraction number in diethylaminoethyl (DEAE) cellulose chromatography and migration position on starch-gel electrophoresis [2]. Initially, 14-3-3 proteins were believed to be unique to animals. However, further research showed that these proteins are found in almost all eukaryotes [1,3,4]. The 14-3-3 proteins are a class of highly conserved regulatory proteins that can form homo- or heterodimers, and each monomer in the dimer can interact with a separate target protein. This dimeric property of 14-3-3s allows them to serve as scaffolds for bringing together different regions of a protein or two different proteins in close proximity [5,6]. Therefore,

14-3-3 proteins regulate activities of numerous target proteins via protein-protein interactions, specifically by binding to the phosphoserine/phosphothreonine residues of target proteins. Evidence suggests that 14-3-3s are the most important phosphopeptide-binding proteins that play a regulatory role in various biological processes including the regulation of cell cycle, metabolism, apoptosis, development, protein trafficking, stress response, and gene transcription [7–15].

In plants, the first plant 14-3-3 protein was isolated from *Arabidopsis thaliana* as a component of the G-box complex and was named GF14 [16]. Subsequently, 14-3-3 proteins were detected in many plant species and were named GF or GRF, followed by Arabic numerals [17]. With the development of new high-throughput sequencing technologies and modern bioinformatics tools, more and more plant genomes were sequenced, which dramatically accelerated the genome-wide survey of 14-3-3 proteins in plants. To date, the 14-3-3 gene family has been identified in *Arabidopsis* (13), rice (*Oryza sativa* L.) (8), tomato (*Solanum lycopersicum* L.) (12), soybean (*Glycine max* L.) (18), cotton (*Gossypium hirsutum* L.) (25), banana (*Musa acuminata* L.) (25), grape (*Vitis vinifera* L.) (11), and many other plant species [18].

Wheat (*Triticum aestivum* L.) is the most widely cultivated crop in the world and one of the primary grains consumed by humans [19]. Drought, extreme temperatures, and salinity are the major abiotic stresses that reduce wheat production throughout the growing season [20]. The 14-3-3 proteins act as key regulators of signaling networks and abiotic stress responses [21]. For example, in rice, OsGF14e interacts with OsCPK21 to promote the salt stress response [22]. Additionally, studies show that OsGF14f has a negative effect on grain development and filling, while OsGF14e negatively impacts cell death and disease resistance [23,24]. In tomato, four 14-3-3 genes (*TFT1*, *TFT4*, *TFT7*, and *TFT10*) are significantly up-regulated under salt stress [25]. Ectopic expression of the wheat *TaGF14b* gene in tobacco (*Nicotiana tabacum* L.) enhanced the tolerance of mature tobacco plants to drought and salt stresses, which are related to the abscisic acid (ABA) signaling pathway, and improved their growth and survival compared with the control [26]. Overexpression of the soybean *GsGF14o* gene in *Arabidopsis* showed that GsGF14o regulates stomata size, root hair development, and drought tolerance [27].

Recently, Guo et al. [28] analyzed wheat *TaGF14* genes based on the TGACv1 reference genome and found that five of these genes were up-regulated under drought stress, and all of the analyzed *TaGF14s* were down-regulated during heat stress. These results suggest that some of the *TaGF14* genes play a vital role in combating drought stress, and all *TaGF14s* are negatively associated with heat stress. Additionally, the study showed that TaGF14s might be involved in starch biosynthesis. However, no information is available on the structure of 14-3-3 genes and their role in response to salt tolerance in wheat.

In this study, we conducted an in-depth analysis of the wheat 14-3-3 gene family members based on the newly released Chinese Spring reference genome (IWGSC RefSeq v1.1, https://urgi.versailles.inra.fr/download/iwgsc/IWGSC_RefSeq_Annotations/v1.1/ (accessed on 13 February 2021)). We performed a comprehensive analysis of the structure, evolutionary relationship, *cis*-acting elements, Gene Ontology (GO) annotations, and expression profiles of the *14-3-3* genes, and deciphered the biological roles of these genes in salt stress tolerance using the virus-induced gene silencing (VIGS) assay and protein-protein interaction analysis. It provided an updated view of the wheat 14-3-3 family and lays the foundation for studying the molecular mechanism of *TaGRF6-A* in regulating salt tolerance in wheat.

2. Results

2.1. Genome-Wide Identification and Characterization of 14-3-3 Family Members in Wheat

A total of 32 14-3-3 protein sequences of wheat were identified using the basic local alignment search tool, BLASTp, and validated using the Hidden Markov Model (HMM) search tool, HMMER. These sequences were further confirmed using SMART and NCBI-CDD online tools, which revealed that all sequences contained conserved 14-3-3 protein domains. These protein sequences were encoded by 17 genes, including 12 genes showing alternative splicing, and splice variants with complete domains were chosen as representatives (Table 1). Multiple sequence alignment and secondary structure analysis revealed that the 14-3-3 family members contained nine typical α-helices. But the *TraesCS4A02G167100* gene lacked the first α-helix, making it the first 14-3-3 family member that does not contain nine α-helices (Figure S1). The 14-3-3 amino acid sequences were highly conserved, indicating that these proteins may perform similar functions to the 14-3-3 proteins of other plant species. The C-terminal end of 14-3-3 proteins was relatively divergent, which explains the functional diversity observed among these proteins. Genes with a 1:1:1 correspondence in all three sub-genomes (A, B, and D) of wheat are called triads [29]. We identified five triads based on the results of Ramírez-González et al. (Table S1). Because of the peculiarity of *TraesCS4A02G167100*, this gene was named *TaGRF-like1*. The remaining 16 genes were named according to their order on various homologous chromosomes (Table 1).

A detailed description of *TaGRFs* is summarized in Table 1. The deduced wheat 14-3-3 proteins contained 244–282 amino acid residues, and their molecular weight (MW) ranged from 27.25 to 31.93 kDa. The predicted isoelectric point (pI) of these proteins ranged from 4.67 to 6.32, implying that these proteins were acidic in nature. Three-dimensional modeling clearly showed the presence of α-helices and indicated that higher structures of these proteins were very similar (Figure S2).

2.2. Gene Structure and Conserved Motif Analysis

Phylogenetic analysis revealed that *TaGRFs* were divided into two groups (ε and non-ε) and three corresponding members of each triad were closely clustered together (Figure 1A). To examine the structural diversity of *TaGRFs*, we manually extracted the exon-intron structure from the GFF3 annotation file (Table S2). All *TaGRF* genes, especially those within the same phylogenetic group, shared a similar exon-intron structure (Figure 1B). The number of exons varied from 4 to 6 and the genes within each phylogenetic group exhibited nearly identical exon lengths. To understand the motif composition, a search was implemented using the MEME website, with the maximum number of motifs set at 10 (Figure 1C). Functional analysis using SMART revealed that 3 of these 10 motifs (motifs 1–3) were annotated as 14-3-3, which were included in all members (Table S3). This result suggests that motifs 1–3 represent the main domains that determine the function of 14-3-3 proteins. At the same time, this result also indicates the functional conservation of the *14-3-3* genes during evolution. In addition, motif composition was highly similar within each phylogenetic group but was relatively divergent between groups. All members of the non-ε group contained six main motifs (1–6), whereas members of the ε group contained two unique motifs (9 and 10), which indicates the potentially different roles during plant growth and response to various stresses.

Table 1. Detailed information of putative TaGRFs.

Gene	Sequence ID	AS [a]	Genomic Position	CDS [b]	Protein[c]	MW [d]	pI [e]	Information
TaGRF1-A	TraesCS2A02G337300.1	2	Chr[f]2:570782629–57078596 (−)[g]	792	263	29.95	4.69	TaGF14a [#]
TaGRF1-B	TraesCS2B02G344600.1	2	Chr2:491180649–491184008 (+)[h]	792	263	29.97	4.73	TaGF14e [#]
TaGRF1-D	TraesCS2D02G325600.1	2	Chr2:419039525–419043028 (+)	792	263	29.95	4.69	TaGF14k[#], TaGF14r [#]
TaGRF2-A	TraesCS3A02G055600.1	2	Chr3:32150045–32153332 (−)	789	262	29.69	4.67	TaGF14b [#]
TaGRF2-B	TraesCS3B02G068000.1	3	Chr3:40243799–40247161 (−)	849	282	31.93	4.98	Identified in this study
TaGRF2-D	TraesCS3D02G055500.1	1	Chr3:23061692–23065089 (−)	789	262	29.75	4.71	TaGF14m [#], TaGF14s [#]
TaGRF3-A	TraesCS4A02G268700.1	2	Chr4:580905848–580908549 (−)	786	261	29.26	4.83	TaGF14d [#]
TaGRF3-B	TraesCS4B02G045500.2	2	Chr4:32805467–32808180 (+)	786	261	29.26	4.83	TaGF14h [#]
TaGRF3-D	TraesCS4D02G046400.2	2	Chr4:21886194–21888840 (−)	786	261	29.26	4.83	TaGF14j [#]
TaGRF4-A	TraesCS4A02G151000.2	3	Chr4:299599404–299602584 (−)	813	270	29.90	4.76	TaGF14c [#]
TaGRF4-B	TraesCS4B02G159900.1	3	Chr4:310973441–310976945 (+)	822	273	30.22	4.75	TaGF14g [#]
TaGRF4-D	TraesCS4D02G155900.1	2	Chr4:209095178–209098724 (+)	801	266	29.49	4.76	TaGF14n [#]
TaGRF5-B	TraesCS4B02G148900.1	2	Chr4:217846208–217849422 (+)	738	245	27.56	5.90	TaGF14i [#]
TaGRF6-A	TraesCS7A02G295500.1	1	Chr7:386346364–386357777 (−)	780	259	28.70	4.80	TaGF14o [#]
TaGRF6-B	TraesCS7B02G183800.1	1	Chr7:292828452–292841187 (+)	786	261	28.80	4.80	Identified in this study
TaGRF6-D	TraesCS7D02G292400.1	1	Chr7:356896682–356911158 (+)	786	261	28.80	4.74	TaGF14q [#], TaGF14t [#]
TaGRF-like1	TraesCS4A02G167100.1	1	Chr4:412260859–412267803(+)	735	244	27.25	6.32	Identified in this study

AS [a] (Number of alternative splices); CDS [b] (Length of the coding sequence, bp); Protein [c] (Amino acid length, aa); MW [d] (Molecular weight, kD); pI [e] (Isoelectric point); Chr [f] (represents Chromosome); (−) [g] (represents antisense strand); (+) [h] (represents sense strand); reference [#].

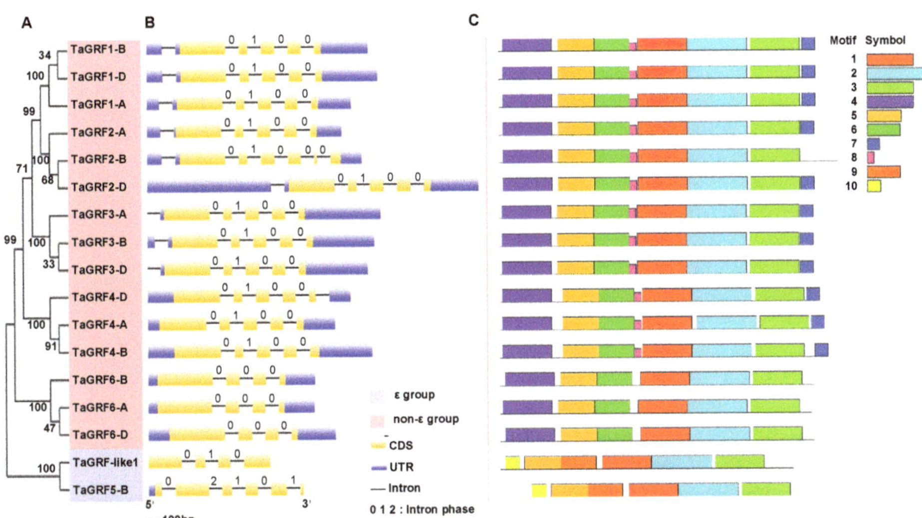

Figure 1. Structural analysis of *TaGRFs*. (**A**) The phylogenetic tree of *TaGRFs*. (**B**) Gene structure analysis of *TaGRFs* was conducted using the GSDS database. All introns are shown the same length. (**C**) Motif analysis of *TaGRFs* was identified by MEME. Motif components in the amino acid sequence of *TaGRFs* are displayed using MAST. Different motifs were represented in different color blocks.

2.3. Chromosomal Distribution and Evolutionary Analysis of TaGRF Genes

To clarify the chromosome position of *TaGRFs*, a schematic diagram was constructed. The *TaGRF* genes were mainly distributed on homoeologous chromosomes 4A/4B/4D, and only one gene was located on each of the remaining chromosomes (Figure 2A).

To trace the evolutionary origin of wheat *TaGRFs*, we performed a synteny analysis of family members in common wheat (AABBDD) and its progenitor species *Aegilops tauschii* L. (DD), *Triticum urartu* L. (AA), and *T. dicoccoides* L. (AABB) (Figure 2B). Given that the B sub-genome donor has not been sequenced, wild emmer wheat (*T. dicoccoides* L.) was used as the source of the B sub-genome. The results showed that *TaGRFs* had seven, twelve, and five orthologs in *T. urartu*, *T. dicoccoides*, and *Ae. tauschii*, respectively (Table S4). Interestingly, the origin of *TaGRF* genes was traceable. Our results indicated that the *TaGRF* genes in common wheat were entirely derived from the sub-genome donors and have the same number on the corresponding chromosomes. Wheat chromosomes 4A, 4B, and 4D harbored three, three, and two *TaGRF* genes, respectively. The same number of *GRF* homologs were detected on chromosomes 4A and 4B of *T. dicoccoides* and on chromosome 4D of *Ae. tauschii*. However, the corresponding gene was not detected on chromosome 2A of *T. urartu*. We speculated that the genes on chromosome 2A of common wheat evolved from *T. dicoccoides* or the corresponding gene on *T. urartu* used to exist but was later lost. This implied that members of the *TaGRF* gene family did not experience large fluctuations. The only polyploidization doubles the number of genes twice. In addition, the non-synonymous-to-synonymous substitution ratio (Ka/Ks) of *TaGRF* gene duplicates was <1 (Figure 2C, Table S3), indicating that these genes are under negative selection pressure to maintain the protein sequence.

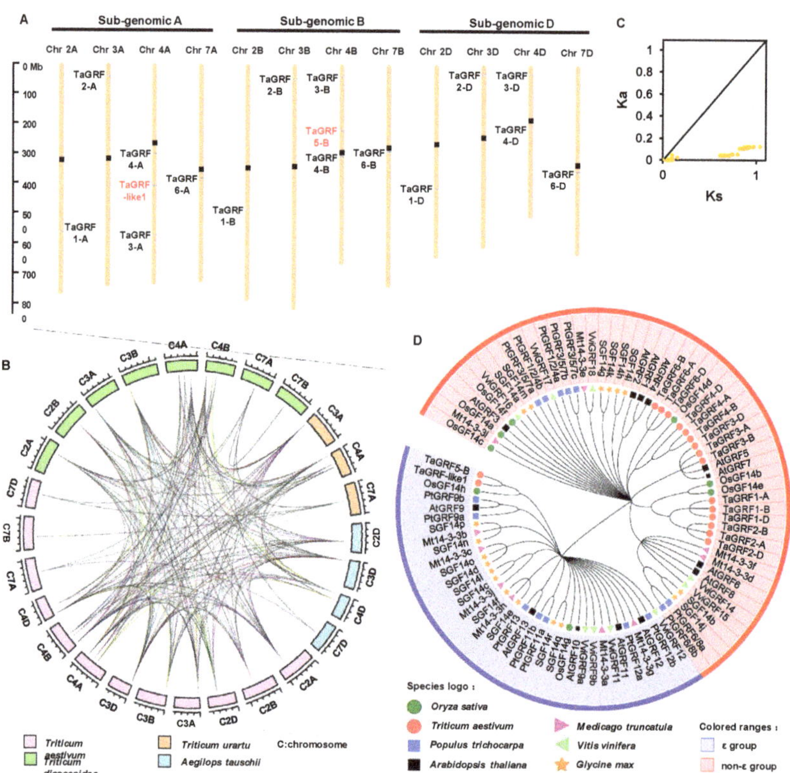

Figure 2. Evolutionary analysis of TaGRFs (**A**) Chromosomal locations and repeated events of TaGRFs. Orange bars represent chromosomes. The black rectangular dots above is the centromere. The scale indicates the corresponding physical distance. Chr 2A is the second chromosome on sub-genome A. The ε group genes are shown in red font, and non-ε group genes are shown in black font. The segmentally duplicated homologous genes were marked with blue lines. (**B**) Orthologous relationships analysis among T. aestivum (Ta, AABBDD, pink box) and progenitor species Ae. tauschii (Ae, DD, blue box), T. urartu (Tu, AA, orange box), and T. dicoccoides (Td, AABB, green box). The lines represent collinearity between chromosomes. (**C**) Ka/Ks value of TaGRF repeat gene pair. (**D**) Phylogenetic analysis of TaGRF proteins and other species such as *A. thaliana*, *O. sativa*, *G. max*, *P. trichocarpa*, *M. truncatula*, and *V. vinifera*. ClustalW2 was used to align the protein sequences, and the phylogenetic tree was constructed by the Neighbor-Joining (NJ) method of MEGA 7 with 1000 bootstrap replicates. Then the tree was modified by ITOL online tool. The two major groups were marked with different colors.

To understand the evolutionary relationships of wheat 14-3-3 proteins with those of other plant species, a phylogenetic tree of full-length 14-3-3 amino acid sequences from seven plant species, including *T. aestivum*, *A. thaliana*, *O. sativa*, *G. max*, *Populus trichocarpa*, *Medicago truncatula*, and *V. vinifera*, was constructed using the maximum likelihood (ML) method (Figure 2D, Table S5). Consistent with the previous classification [30], all 14-3-3 proteins were divided into two main evolutionary branches, namely, ε and non-ε. Only *TaGRF-like1* and *TaGRF5-B* were grouped into the ε group, and the remaining *TaGRFs* clustered in the non-ε group. The proteins from wheat, *Arabidopsis*, and rice were closely related, suggesting that these proteins may perform similar functions, which provides clues for functional analysis.

2.4. Cis-Element Analysis and Functional Annotation of TaGRF Genes

To examine *cis*-acting elements in *TaGRF* promoter sequences, we searched 1.5 kb sequence upstream of the start codon (ATG) of all *TaGRF* genes. A total of 1789 *cis*-acting elements were identified (Table S6). These included elements involved in biotic and abiotic stress responses, plant growth and development, and phytohormone response processes. The number of elements involved in the first two processes was significantly more (Figure 3A,B). Motifs such as TGACG and CGTCA (involved in responding to methyl jasmonate [MeJA]) and ABA-responsive element (ABRE; involved in the response to ABA) were commonly found in *TaGRF* gene promoters, suggesting that these genes are involved in hormone regulation. In addition, stress-related *cis*-elements, such as Myb, STRE, Sp1, G-box, Myc, as-1, LTR, and GC-motif, were also identified, implying that *TaGRFs* are important players in response to adverse conditions.

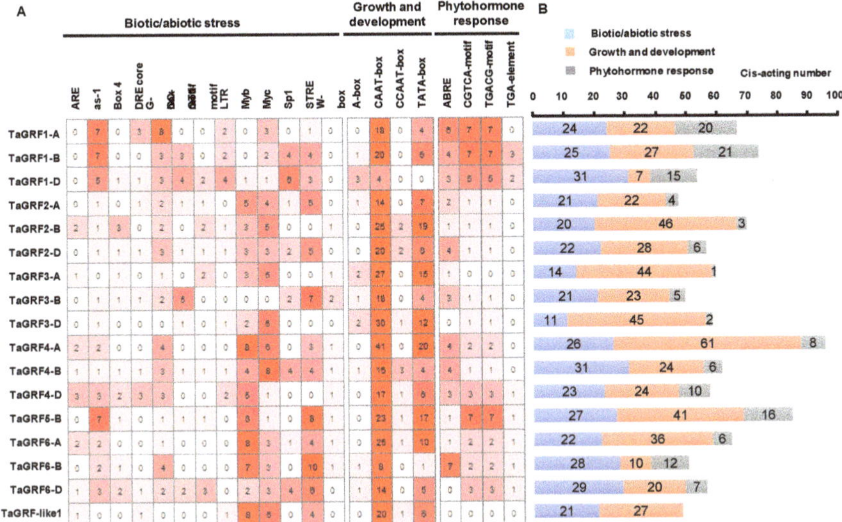

Figure 3. Analysis of *cis*-acting elements in the promoter of *TaGRFs*. (**A**) Different numbers and shades of color in the grid indicated the number of different promoter elements. (**B**) The histogram showed the sum of the cis-acting elements in different classes of each gene.

To further predict the function of *TaGRFs*, we performed GO annotation analysis. The results showed that *TaGRFs* were divided into three categories (molecular function, cell composition, and biological process) (Figure S3). In the molecular function category, GO terms such as "ATP binding" and "protein domain specific binding" were highly enriched, indicating that TaGRF proteins may bind to other proteins to perform various cellular functions. In the cell component category, GO terms such as nucleus, cytosol, plasma membrane, and organelles, such as mitochondrion, were highly enriched, suggesting that *TaGRFs* perform a wide range of functions. In the biological process category, *TaGRFs* play a prominent role in the response to cadmium ions. Both *TaGRF5-B* and *TaGRF-like1* were annotated only under the GO term "protein domain specific binding", suggesting that these proteins play a limited role in wheat.

2.5. Expression Profiling of TaGRF Genes

The results of *cis*-element and GO annotation analyses suggested that *TaGRFs* are potentially involved in abiotic stress responses. To better understand the function of *TaGRFs*, we examined their expression patterns using RNA-seq data. The expression profiles of 16 *TaGRFs* in different wheat varieties under salt stresses are shown in Figure 4A. In general, the expression of *TaGRFs* decreased after the salt treatment. Furthermore, the expression levels of *TaGRF4-A*, *TaGRF4-B*, and *TaGRF5-B* were extremely low. In the wheat variety Kharchia Local, *TaGRFs* showed different expression profiles in different tissues; for example, *TaGRF1-B*, *TaGRF1-D*, and *TaGRF2-A* showed low expression in roots, while *TaGRF6-A*, *TaGRF6-B*, and *TaGRF6-D* showed high expression in leaves, which indicated that these genes may perform a function mainly in leaves. Compared with Kharchia Local, QM6 and Chinese Spring showed greater differences in gene expression after the salt treatment, suggesting that genotype is an important factor affecting the differences in abiotic stress responses.

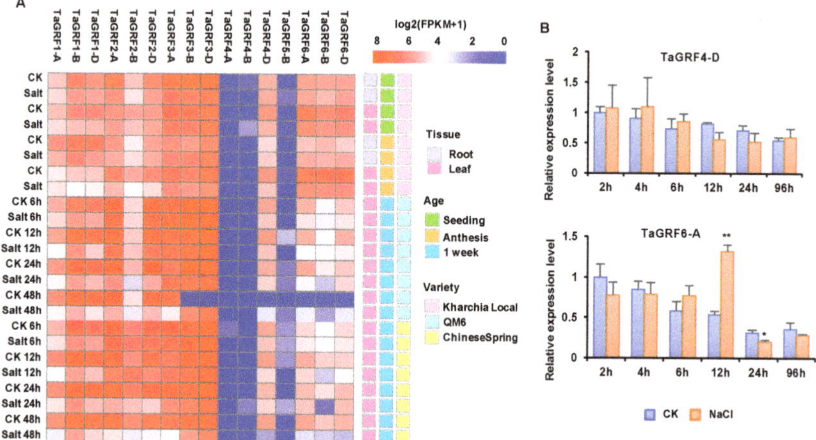

Figure 4. Expression patterns of *TaGRFs*. (**A**) Transcriptome analysis of *TaGRFs* across salt stress. The expression data were generated from NCBI and viewed in RStudio software. The relative expression level of a particular gene in each column was normalized against the mean value by log2 transformation. The color scale represents relative expression levels. (**B**) qRT-PCR analyses of two *TaGRF* genes in leaves under NaCl treatment. The asterisks indicate statistical significance between CK and NaCl treatments (**, $p < 0.01$; *, $p < 0.05$; Student's *t*-test).

To verify the reliability of the transcriptome data, we analyzed the expression of two genes (*TaGRF4-D* and *TaGRF6-A*) in 1-week-old salt-treated wheat seedlings by quantitative real-time PCR. Both the *TaGRFs* responded to salt stress, but with different rates and intensities (Figure 4B). Compared with the control, the expression of *TaGRF4-D* did not change significantly, while the expression of *TaGRF6-A* was significantly up-regulated at 12 h post salt treatment, and then decreased sharply at 24 h post salt treatment. This trend suggests that *TaGRF6-A* can be induced by salt stress at a certain stage, and may play a role during this period.

2.6. TaGRF6-A Positively Regulates Salt Tolerance in Wheat

The VIGS assay is used to silence specific genes in plants, thus enabling rapid characterization of gene function [31]. To further understand the role of *TaGRF6-A* in salt stress, we performed a virus-induced gene silencing assay. Two fragments of the *TaGRF6-A* gene were chosen for silencing (Figure S3). At 10 days post-inoculation (dpi), wheat leaves inoculated with BSMV showed mild chlorotic mosaic symptoms while leaves inoculated with BSMV: *TaPDS* showed bleaching symptoms, indicating that the BSMV-induced gene silencing system was working (Figure 5A). When treated with 300 mM NaCl for 12 days, both fragments of *TaGRF6-A* were well silenced by qRT-PCR (Figure 5C), and at this time, the leaf curling degree of plants carrying BSMV: *TaGRF6-A-1/2* was stronger than that of the control group carrying the empty vector (Figure 5B). According to the statistical results, the number of leaf curls of plants carrying BSMV: *TaGRF6-A-1/2* was significantly more than that of the empty carrier group (Figure 5D), indicating that *TaGRF6-A* contributes to salt tolerance in wheat.

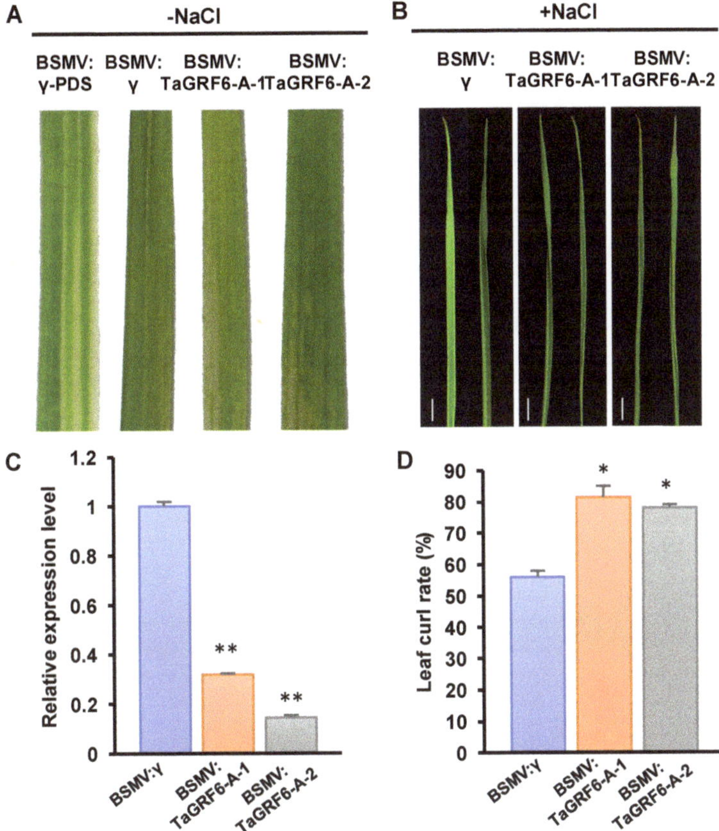

Figure 5. The silencing of *TaGRF6-A* decreases salt tolerance in wheat. (**A**) Phenotypic characteristics of wheat leaves after *TaGRF6-A* silenced by VIGS technology. (**B**) The curling phenotype of wheat leaves after 12 days of 300 mM NaCl treatment. Bars, 1 cm. (**C**) Silencing efficiency assessment of *TaGRF6-A* in the *TaGRF6-A*-knockdown plants treated with 12 days of 300 mM NaCl. (**D**) Statistics of leaf curls rate of *TaGRF6-A* in the *TaGRF6-A*-knockdown plants treated with 12 days of 300 mM NaCl. Error bars represent the SD of three biological replicates. The asterisks indicate statistical significance using Student's *t*-tests. (**, $p < 0.01$; *, $p < 0.05$).

2.7. *TaGRF6-A* Interacts with *TaMYB64* In Vitro and In Vivo

Since *TaGRF6-A* expressed salt tolerance, we started to look for its internal mechanism. We focused on the expression levels of stress-related genes and their potential targets (Figure 6A). It is worth noting that *TaMYB64* is significantly down-regulated, suggesting that there may be a synergistic effect to enhance abiotic stress tolerance together with *TaGRF6-A*. Previous studies in soybean and rice showed that interactions between MYB transcription factors and 14-3-3 proteins enhance plant growth and stress tolerance [32,33]. These findings, combined with our qRT-PCR results, prompt us to investigate the association between TaMYB64 and TaGRF6-A. Before that, we analyzed the subcellular localization of TaGRF6-A and TaMYB64 by generating a C-terminal fusion of gene and yellow fluorescent protein (YFP) gene. After transient expression in *N. benthamiana* leaves using *Agrobacterium*-mediated transformation, compared with the YFP empty vector, confocal microscopy showed that TaGRF6-A has a similar subcellular localization and was detected in both the nucleus and the cytoplasm (Figure 6B), and TaMYB64 was detected in the nucleus, indicating the place where it functions. When exploring the interaction of TaMYB64 and TaGRF6-A, we first performed yeast two-hybrid (Y2H) assays. Although yeast transformants expressing both TaMYB64 and TaGRF6-A could grow on high stringency media, the number of yeast cells was less (Figure 6C). To further verify the interaction between TaMYB64 and TaGRF6-A, we performed bimolecular fluorescence complementation (BiFC) and co-immunoprecipitation (Co-IP) assay. After transiently co-expressing TaGRF6-A labeled with the N-terminal half of YFP (nYFP), together with TaMYB64 labeled with the C-terminal half of YFP (cYFP) in *N. benthamiana* leaves, the YFP signal was detected in the nucleus and cytoplasm of *N. benthamiana* leaf epidermal cells. While no signal was detected in the control combinations (Figure 6D). The TaMYB64-YFP-3×FLAG fusion proteins were co-expressed with TaGRF6-A-YFP-HA or YFP-HA in *N. benthamiana*. After immunoprecipitation using the anti-HA antibody, Western blot was performed using the anti-FLAG antibody to detect TaMYB64 proteins. The TaMYB64-YFP-3×FLAG protein could be coimmunoprecipitated by TaGRF6-A-YFP-HA, but not the YFP-HA control, revealing a specific interaction between TaMYB64 and TaGRF6-A in vivo (Figure 6D). These results proved that TaGRF6-A could physically interact with TaMYB64.

2.8. TaGRF6-A and TaMYB64 Work Together to Cope with Salt Stress

To clarify whether TaGRF6-A interacts with TaMYB64 to respond to salt stress, we chose to silence the *TaMYB64* gene. Ten days post-inoculation (dpi), wheat leaves inoculated with BSMV showed mild symptoms of chlorosis, while leaves inoculated with BSMV: *TaPDS* showed symptoms of bleaching stripes, indicating that BSMV-induced gene silencing was effective (Figure 7A). qRT-PCR results revealed that both fragments of *TaMYB64* were successfully silenced (Figure 7C). After treated with 300 mM NaCl for 12 days, the plants carrying BSMV: *MYB64-1/2* had stronger curls and a significantly higher number of rolled leaves compared with the control group carrying the empty vector (Figure 7B,D). These results indicated that *TaMYB64* was also involved in response to salt stress.

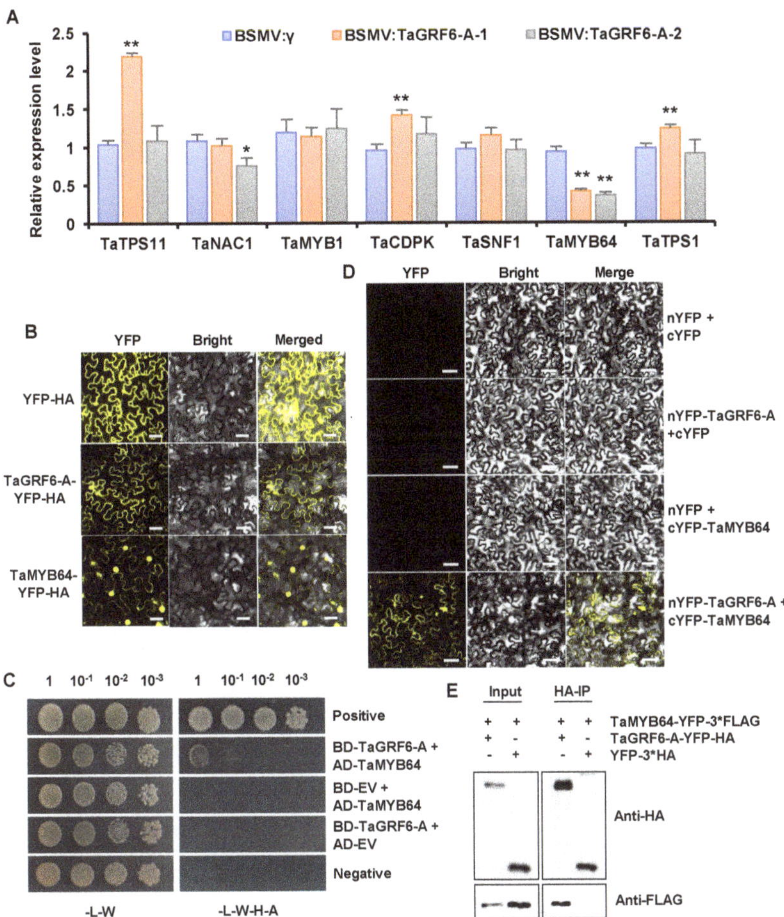

Figure 6. Analysis of protein interaction between TaGRF6-A and TaMYB64. (**A**) Relative expression levels of potential targets and stress-related genes in the leaves of *TaGRF6-A*-knockdown plants treated with 12 days of 300 mM NaCl. Error bars represent the SD of three biological replicates. The asterisks indicate statistical significance using Student's *t*-tests. (**, $p < 0.01$; *, $p < 0.05$). (**B**) Subcellular localization of TaGRF6-A in *N. benthamiana*. Free YFP and TaGRF6-A-YFP and TaMYB64-YFP fusion proteins were transiently injected into *N. benthamiana* leaves by *Agrobacterium* (GV3101) transformation. Subcellular localization observed by laser confocal microscopy after 48 h post-infiltration. Scale bars, 40 μm. (**C**) Yeast two-hybrid interaction results of TaGRF6-A and TaMYB64 proteins. -L-W refers to media lacking leucine and tryptophan, -L-W-H-A refers to media lacking leucine, tryptophan, histidine, and adenine. (**D**) BiFC assay showing interactions of TaGRF6-A and TaMYB64 proteins. Scale bars, 40 μm. (**E**) Co-immunoprecipitation experiments show that TaGRF6-A interacts with TaMYB64 in planta. The total proteins of TaMYB64 with FLAG tag and TaGRF6-A with HA tag were extracted from *N. benthamiana* leaves. The immune complex was pulled down using anti-HA agarose gel, and the co-precipitation of TaGRF6-A and TaMYB64 was detected by Western blotting.

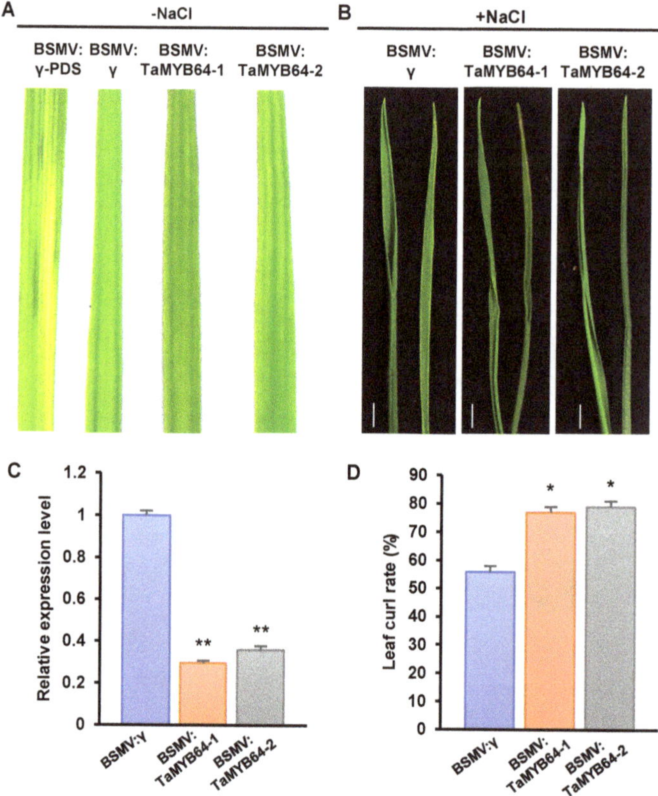

Figure 7. Silencing of *TaMYB64* decreases salt tolerance in wheat. (**A**) Phenotypic characteristics of wheat leaves after *TaMYB64* silenced by VIGS technology. (**B**) The curling phenotype of wheat leaves after 12 days of 300 mM NaCl treatment. Bars, 1 cm. (**C**) Silencing efficiency assessment of *TaMYB64* in the *TaMYB64*-knockdown plants treated with 12 days of 300 mM NaCl. (**D**) Statistics of leaf curls rate of *TaMYB64* in the *TaMYB64*-knockdown plants treated with 12 days of 300 mM NaCl. Error bars represent the SD of three biological replicates. The asterisks indicate statistical significance using Student's *t*-tests. (**, $p < 0.01$; *, $p < 0.05$).

3. Discussion

14-3-3 genes are ubiquitous in many species, and the study of its gene function is very necessary. Recent studies have shown that the four 14-3-3 proteins (MdGF14a, MdGF14d, MdGF14i, and MdGF14j) of apples interact with the flower integrator TFL1/FT and participate in flowering regulation [34]. Under the condition of nitrate deficiency, the interaction of MdGRF11 and MdBT2 in apples can induce the accumulation of anthocyanins and provide a new mechanism for anthocyanin biosynthesis [35]. Under sugar starvation, MYBS2 interacted with 14-3-3 protein to inhibit the expression of α-amylase, thereby to improve rice plant growth, stress resistance, and grain weight [32]. As the hub of different signal pathways, 14-3-3 protein can transmit and integrate different hormone signals in plant signal transduction [36]. 14-3-3 interacts with AtWRI1, and the biosynthesis of oil increased in transgenic tobacco leaves co-expressing these two genes [37]. 14-3-3λ and k are important salt tolerance regulators. When there is no salt stress, 14-3-3λ and k interact with SOS2 and inhibit the kinase activity of SOS2; while in the presence of salt stress, the interaction of them weakens and activates SOS2 kinase activity [38].

In the present study, we performed a genome-wide characterization and in silico analysis of the *14-3-3* gene family in wheat. A total of 17 *14-3-3* genes were identified in wheat based on the genome sequence of Chinese Spring. However, this result is inconsistent with previous studies, which identified 20 *14-3-3* genes based on the TGACv1 wheat genome assembly [28]. We found that previous results contained duplicate genes and several other genes that do not exist in our newly released genome sequencing data. For example, Traes_2DL_1D912517E and Traes_2DL_21639EB55, which were previously identified as two distinct genes, represent the same gene in the new genome. In addition, TraesCS2D02G325600 and Traes_3B_3BF024700350CFD_c1 are not identified in the present study. We speculate that this difference between studies is caused by differences in genome sequencing depth, genome analysis, and gene annotations. We also cannot rule out the possibility of wheat germplasm resource differences.

The 14-3-3 proteins are highly conserved and usually contain nine α-helices, which constitute the conserved core region of each monomer in the dimer [13,39]. In the current study, we showed that *TaGRF-like1* lacks the first α-helix, which may affect the binding of phosphorylated target proteins. while the secondary structure of TaGRF proteins was conserved except in the C-terminal region (Figure S1). Previous studies have shown that the variable C-terminus is the key to dimer formation, and it can interact with different ligands to show the target specificity of 14-3-3 protein [40,41].

Phylogenetic analysis of 89 14-3-3 proteins belonging to wheat and six other plant species showed that these proteins were clustered into two major groups (ε and non-ε), which is consistent with previous reports [18,34]. The result indicates that the formation of these two isoforms is a basic and ancient difference (Figure 2D). As *TaGRF6-A* was adjacent to *OsGF14d*, which is expressed under salt, heat, and cold stress [42], the function of TaGRF6-A may be similar to that of *OsGF14d*. According to previous studies, the ε group genes usually contain a greater number of exons and motifs than the non-ε genes. However, this finding was not in agreement with our results; although the ε group genes contained more motifs, not all ε group genes contained more exons than non-ε group genes (Figure 1B,C). For instance, the ε group gene *TaGRF-like1* contained four exons, whereas the non-ε group gene *TaGRF2-B* contained six exons. This difference may be due to the insertion and deletion of introns over the long-term evolution of wheat. Nevertheless, gene structure and motif distribution were diverse between the two groups but similar within a group, which supports the results of the phylogenetic analysis (Figure 1B,C).

Common wheat is an allohexaploid with a large and complex genome, which makes its genome research difficult. Genome sequencing of common wheat and its sub-genome donors has been completed, which provides an important reference for the evolutionary analysis of wheat. The chromosome map of common wheat shows that the distance between genes is much greater than 200 kb (Figure 2A), which means that these genes were not generated by tandem duplication [43]. Collinearity analysis of common wheat and its ancestors revealed the direct source of the wheat 14-3-3 sub-genomic donor (Figure 2B), indicating that the entire family of wheat 14-3-3 proteins originated by polyploidization. The Ka/Ks ratio of 14-3-3 genes was <1 (Figure 2C), implying that wheat 14-3-3 proteins may remain unchanged during the process of long-term domestication. Analysis of *cis*-acting elements in the promoter region of *TaGRFs* indicated that these genes may be involved in biotic and abiotic stress responses, growth and development, and phytohormone response (Figure 3A). Previously, there have been some reports showing the involvement of 14-3-3 proteins in abiotic stress tolerance [44,45]. Considering the potential function of 14-3-3 proteins in abiotic stress, we performed transcriptome analysis of wheat plants under salt stress. Different *TaGRF* genes showed different expression profiles under salt stress. Most *TaGRF* genes were down-regulated after the salt treatment, suggesting that these genes may negatively regulate the response to salt stress (Figure 4A). However, in rice, tomato, *Brachypodium distachyon*, and other plant species, members of the 14-3-3 family are involved in salt stress tolerance [22,25,46]. In addition, wheat *TaGF14b* (named as *TaGRF1-B* in this article) also enhanced salt stress tolerance when ectopically expressed in tobacco [26]. In the current study, we selected two genes for the salt treatment and found that *TaGRF6-A* was changed greatly responding to salt stress (Figure 4B). To further verify the role of 14-3-3 family genes in salt stress, we used VIGS technology to silence the *TaGRF6-A* gene. The results showed that the level of leaf curling and number of BMSV: *TaGRF6-A* were more than the wild type (Figure 5B–D), indicating that *TaGRF6-A* plays a positive role in response to salt stress.

To elucidate the possible mechanism of action of *TaGRF6-A* under salt stress, we examined the expression of seven stress-related genes or potential interacting partners of *TaGRF6-A*. Interestingly, the expression level of *TaMYB64* was significantly down-regulated after *TaGRF6-A* silencing (Figure 6A). Since the interaction between MYBS2 and 14-3-3 protein has been reported in rice and soybean [32,33,47], we cloned the homolog of OsMYBS2 in wheat and performed Y2H, BiFC, and CoIP assays. The results showed that TaMYB64 and TaGRF6-A did interact in wheat (Figure 6C–E). Thus, we have the question of whether they use this interaction to improve salt tolerance. Next, we used the same method to silence *TaMYB64* in wheat, and the results proved our deduction. Compared with the control, the silenced plants also have a higher degree and amount of curling (Figure 7B,D). In summary, the results in the present study revealed the function of *TaGRF6-A* in salt stress, and initially elucidated its salt tolerance mechanism. Moreover, our study improved our understanding of the biological functions of the wheat *14-3-3* gene family.

4. Materials and Methods

4.1. Plant Material and Salt Treatment

Wheat (*T. aestivum* L.) cultivar Emai 170 was used in this study. Seeds of uniform size were selected and soaked in clear water for 24 h. Then, the seeds were sown in a small pot filled with nutrient-rich soil, and the pots were placed in a greenhouse maintained at 23 ± 2 °C, 16 h light/8 h dark photoperiod, and 200 lux light intensity. At the 2–3-leaf stage, wheat plants were treated with either double-distilled water (control) or 300 mM NaCl solution. Leaves were collected at 2, 4, 6, 12, 24, and 96 h after the treatment, frozen in liquid nitrogen and stored at −80 °C.

4.2. Mining and Phylogenetic Analysis of Wheat 14-3-3 Genes

To identify potential members of the wheat 14-3-3 gene family, published 14-3-3 amino acid sequences of *A. thaliana*, *O. sativa*, and *G. max* were used as queries in BLASTp searches against the wheat genome IWGSC RefSeq v.1.1, with an *E*-value cut-off $<10^{-10}$. The HMM profile of the conserved domain of 14-3-3 (PF00244) was downloaded from the PFAM 32.0 database [48] and used to examine all wheat protein sequences with the HMMER search tool [49]. According to the sequence ID of the target candidate 14-3-3 protein, genomic sequences were isolated. Then, all matching sequences were validated using the SMART website (Simple Modular Architecture Research Tool, http://smart.embl-heidelberg.de/ (accessed on 13 February 2021)) and NCBI-CDD (National Center for Biotechnology Information-Conserved domain database, https://www.ncbi.nlm.nih.gov/Structure/bwrpsb/bwrpsb.cgi (accessed on 13 February 2021)) search to identify the conserved 14-3-3 protein domain with default parameters, and proteins lacking the 14-3-3 domain were discarded [50,51]. Previously known 14-3-3 protein sequences from *A. thaliana* (13), *O. sativa* (8), and *G. max* (18) used in this study were retrieved from the Phytozome v.12 database (http://phytozome.jgi.doe.gov (accessed on 13 February 2021)) [52].

4.3. Amino Acid Sequence Alignment and Characterization of TaGRF Proteins

Multiple sequence alignments were performed using DNAMAN software (version 7.212, Lynnon Corp., Quebec, QC, Canada). Secondary structures and three-dimensional models were constructed using Protein Homology/analogY Recognition Engine v.2.0 (Phyre2, http://www.sbg.bio.ic.ac.uk/phyre2/html/ (accessed on 13 February 2021)) [53]. The basic physicochemical properties of the 14-3-3 proteins, including MW, pI, and the number of amino acid residues, were examined using ProtParam on the ExPASy website (https://www.expasy.org/vg/index/protein (accessed on 13 February 2021)) [54].

4.4. Gene Structure Analysis and Conserved Motif Prediction

The GFF3 annotation file was obtained from the wheat reference genome IWGSC RefSeq v.1.1. Gene structure analysis was conducted using the Gene Structure Display Server 2.0 (GSDS, http://gsds.cbi.pku.edu.cn (accessed on 13 February 2021)) [55]. Conserved motifs in protein sequences were analyzed by MEME (http://meme-suite.org/tools/meme (accessed on 13 February 2021)) [56], with the maximum number of motifs set to 10.

4.5. Chromosome Distribution, Synteny, Ka/Ks, and Phylogenetic Analysis of TaGRFs

Information about the start and end of *TaGRFs* was extracted from the GFF3 file. A physical map of *TaGRFs* was constructed using MapInspect software Version 1.0 (http://www.softsea.com/review/MapInspect.html (accessed on 13 February 2021)) [57]. The TBtools software (https://github.com/CJChen/TBtools/ (accessed on 13 February 2021)) was used to determine the Ka and Ks values of *TaGRFs*, based on their coding sequences (CDSs) [58]. Reference genome sequences of wheat sub-genome donors were downloaded from NCBI, and *14-3-3* genes of each species were identified using the same methods as those used for determining the *TaGRFs*. To determine the paralogous or orthologous relationship between wheat *TaGRFs* and *14-3-3* genes of its sub-genome donors, the general tool "all against all BLAST searches" was used, with an E-value of 1×10^{-10} and sequence similarity > 75% [59]. The "circlize" package of the R program was used to draw the relationship between wheat *TaGRFs* and *14-3-3* genes of its sub-genome donors [60]. Phylogenetic analysis was conducted using the ML method of MEGA7, based on the aligned *14-3-3* sequences of *T. aestivum*, *A. thaliana*, *O. sativa*, *G. max*, *P. trichocarpa*, *M. truncatula*, and *V. vinifera*, with 1000 bootstrap replications [61]. The phylogenetic tree file was then uploaded to the Interactive Tree of Life (https://itol.embl.de/ (accessed on 13 February 2021)) for adjustment and modified [62].

4.6. Cis-Acting Element Analyses

The 1.5 kb genomic DNA sequences upstream of the start codon (ATG) of each *TaGRFs* genes were extracted from the wheat genome sequence. *Cis*-regulatory elements in the promoters were identified using the PlantCARE database (http://bioinformatics.psb.ugent.be/webtools/plantcare/html/ (accessed on 13 February 2021)) [63].

4.7. Expression Analysis of TaGRF Genes under Abiotic Stress Conditions

To determine the expression patterns of 14-3-3 genes in wheat under salt stress, wheat transcriptome data were downloaded from the NCBI Short Read Archive (SRA) database and mapped onto the reference genome of wheat using hisat2. The FPKM (fragments per kilobase of transcript per million) values obtained after "cufflinks" assembly were log-transformed, and a heatmap was drawn using the RStudio software "pheatmap" to display the expression profiles of *TaGRFs* [64,65].

4.8. RNA Isolation and qRT-PCR Analysis

Total RNA was extracted from leaf and root tissues of wheat plants treated with salt or water (control) using the RNAprep Pure Plant Kit (Invitrogen). The isolated total RNA was reverse transcribed to synthesize cDNA using the HiScript® II 1st Strand cDNA Synthesis Kit (Vazyme) for qRT-PCR analysis. The cDNA was diluted to 100 ng/μL with RNase-free water, and qRT-PCR was performed in a 10 μL reaction volume containing 5 μL ChamQ Universal SYBR qPCR Master Mix, 0.5 μL each forward and reverse primer (10 μM), and 4 μL cDNA template. The following conditions were used for PCR: initial denaturation at 95 °C for 3 min, followed by 40 cycles of denaturation at 95 °C for 10 s, and annealing at 60 °C for 30 s. Fluorescence signals were collected after each cycle, and the temperature was increased from 60 °C to 95 °C after each cycle for melting curve analysis. The EF-1α gene

(GeneBank accession: BT009129.1) was used as a reference gene. The relative expression level of genes was calculated using the $2^{-\Delta\Delta CT}$ method [66]. Three technical repeats were performed for each sample, and three independent replicates were carried out. Primers used for qRT-PCR are listed in Table S7.

4.9. Virus-Induced Gene Silencing (VIGS) Assay in Wheat

There are four kinds of vectors involved in the VIGS test: α, β, γ, and γ-PDS. The gene fragments were cloned into the γ vector to obtain a recombinant vector. The vector (α, β, γ, γ-PDS, and recombinant vector) were linearized, and the linearized plasmid was treated with RiboMAX ™ Large Scale RNA Production System-T7 and the Ribom7G Cap Analog (Promega) to obtain capped in vitro transcription products. VIGS inoculation was carried out at the 3-leaf stage [67]. Steps are as follows: Mix equal volumes of in vitro transcription products α, β, γ (or γ-PDS /recombinant γ), dilute with DEPC water, add 1 × FES buffer (0.1 M glycine, 0.06 M K2HPO4, 1% *w*/*v* tetrasodium pyrophosphate, 1% *w*/*v* bentonite, and 1% *w*/*v* celite, pH 8.5), and then rubbed onto wheat leaves. BSMV: γ-PDS (PDS: wheat phytoene desaturase gene) and BSMV: γ were used as controls for BSMV infection. After inoculation, when BSMV: γ-PDS showed bleaching and yellowing phenomenon (about 10 days later), ddH_2O and 300 mmol NaCl solution were used for irrigation. After 12 days of irrigation, the curl phenotype of the fourth leaf was recorded and the curl rate was counted. Curl rate is the percentage of curled leaves in all leaves.

4.10. Subcellular Localization of the TaGRF6-A Protein

The CDS of *TaGRF6-A* was cloned into the pQBV3 Gateway entry vector and then cloned into the pEarlyGate101 destination expression vector [68]. The resulting TaGRF6-A-YFP fusion construct was transformed into *Agrobacterium tumefaciens* strain GV3101, which was grown on LB (add antibiotics: kanamycin, rifampicin, gentamicin) solid medium for 2 days. The positive colonies were verified by PCR and transferred to LB (add antibiotics: kanamycin, rifampicin, gentamicin) liquid medium. The culture was grown for 16 h at 28 °C on a shaker until the optical density of the culture (measured at 600 nm absorbance; OD600) reached 1.5–1.8. The cells were harvested by centrifugation at 4000× g for 15 min and resuspended in acetosyringone (AS) culture solution. Then, 1 mL culture (OD600 = 0.8) was injected into the abaxial surface of the leaves of 3–4-week-old *N. benthamiana* plants using a needleless syringe, followed by incubation in the dark for 4 h. At 48 h post-inoculation, the distribution of the YFP signal in leaf epidermal cells was observed under a confocal laser scanning microscope (Zeiss LSM710) [69].

4.11. Y2H, BiFC, and CoIP Assays

The CDSs of *TaGRF6-A* and *TaMYB64* were cloned into pGBKT7 (BD) and pGADT7 (AD) vectors, respectively. According to the Yeast Protocols Handbook (Clontech, Mountain View, CA, USA), the recombinant plasmids were transformed into the yeast AH109 strain (*Saccharomyces cerevisiae*) and plated on an SD/-LW selection medium. The plates were incubated at 30 °C for 3–5 days until the appearance of colonies. Single colonies were picked using an inoculation ring and streaked onto SD/-LW and SD/-LWHA solid media. Plates were incubated at 30 °C for 3–5 days, and photographs were taken to record the growth of yeast colonies. T + P53 and T + lam served as positive and negative controls, respectively. Full-length cDNA sequences of *TaGRF6-A* and *TaMYB64* minus the stop codon were PCR amplified using the Pfu polymerase (NEB). The PCR products were ligated into the pQBV3 vector and then cloned into pEarleyGate201-YN and pEarleyGate202-YC vectors using the LR enzyme (Gateway LR Clonase II Enzyme mix, Invitrogen). The resulting plasmids were transformed into *A. tumefaciens* strain GV3101 and then transiently expressed in *N. benthamiana* leaves using the method described above. For Co-IP analysis, the PCR products were constructed on pEarleyGate100 and pEarleyGate104 vectors. Using the method of Qiao et al. [69], the protein was transiently expressed in *N. benthamiana* and the total protein was extracted, and then incubated with HA magnetic beads (MBL, Tokyo, Japan) at 4 °C and enriched with magnetic beads on ice. The precipitated protein was then separated by SDS-PAGE electrophoresis. Use anti-FLAG and anti HA antibody (MBL, Japan) to detect the coprecipitation signal of TaMYB64-YFP-3FLAG and TaGRF6-A-YFP-HA.

4.12. Statistical Analysis

Student's *t*-test were applied to test differences among treatments.

Supplementary Materials: The following are available online at https://www.mdpi.com/1422-0067/22/4/1904/s1.

Author Contributions: X.Z. (Xiaoguo Zhu), Y.Q., and D.M. conceived this research. W.S. and W.C. designed the experiment and wrote the manuscript. Y.J., C.Z., G.L., and X.L. collected and analyzed public data sets, and W.S., X.Z. (Xiaoyi Zhou) conducted bioinformatics analysis and experiments. All authors have read and agreed to the published version of the manuscript.

Funding: This work was supported by National Key R&D Program of China [2018YFD0200506]; the Key Project of Hubei Provincial Department of Education [D20191305].

Institutional Review Board Statement: Not applicable.

Informed Consent Statement: Not applicable.

Data Availability Statement: Data is contained within the article or supplementary material.

Acknowledgments: The authors thank Zhensheng Kang (Northwest A & F University, Yangling, China) for providing vector of VIGS.

Conflicts of Interest: The authors declare no conflict of interest.

References

1. Ferl, R.J. 14-3-3 Proteins and signal transduction. *Annu. Rev. Plant Biol.* **1996**, *47*, 49–73. [CrossRef]
2. Moore, B.W.; Perez, V.J. Specific Acidic Proteins of the Nervous System. *Physiol. Biochem. Asp. Neverous Integr.* **1967**, 343–359.
3. Ichimura, T.; Isobe, T.; Okuyama, T.; Yamauchi, T.; Fujisawa, H. Brain 14-3-3 protein is an activator protein that activates tryptophan 5-monooxygenase and tyrosine 3-monooxygenase in the presence of Ca2+, calmodulin-dependent protein kinase II. *FEBS Lett.* **1987**, *219*, 79–82. [CrossRef]
4. Robinson, K.; Jones, D.; Patel, Y.; Martin, H.; Madrazo, J.; Martin, S.; Howell, S.; Elmore, M.; Finnen, M.J.; Aitken, A. Mechanism of inhibition of protein kinase C by 14-3-3 isoforms. 14-3-3 isoforms do not have phospholipase A2 activity. *Biochem. J.* **1994**, *299*, 853–861. [CrossRef] [PubMed]
5. Braselmann, S.; McCormick, F. Bcr and Raf form a complex in vivo via 14-3-3 proteins. *EMBO J.* **1995**, *14*, 4839–4848. [CrossRef]
6. Yaffe, M.B. How do 14-3-3 proteins work?–Gatekeeper phosphorylation and the molecular anvil hypothesis. *FEBS Lett.* **2002**, *513*, 53–57. [CrossRef]
7. Aitken, A. 14-3-3 Proteins: A historic overview. *Semin. Cancer Biol.* **2006**, *16*, 162–172. [CrossRef]
8. Darling, D.L.; Ling, Y.J.; Boris, W.A. Role of 14-3-3 proteins in eukaryotic signaling and development. *Curr. Top. Dev. Biol.* **2005**, *68*, 281–315.
9. Fu, H.; Subramanian, R.R.; Masters, S.C. 14-3-3 PROTEINS: Structure, Function, and Regulation. *Annu. Rev. Pharmacol. Toxicol.* **2000**, *40*, 617–647. [CrossRef]
10. Hermeking, H.; Benzinger, A. 14-3-3 proteins in cell cycle regulation. *Semin. Cancer Biol.* **2006**, *16*, 183–192. [CrossRef]
11. Muslin, A.J.; Tanner, J.W.; Allen, P.M.; Shaw, A.S. Interaction of 14-3-3 with signaling proteins is mediated by the recognition of phosphoserine. *Cell* **1996**, *84*, 889–897. [CrossRef]
12. Obsilová, V.; Silhan, J.; Boura, E.; Teisinger, J.; Obsil, T. 14-3-3 Proteins: A Family of Versatile Molecular Regulators. *Physiol. Res.* **2008**, *57*, S11–S21. [PubMed]
13. Yaffe, M.B.; Volinia, S.; Leffers, H.; Rittinger, K.; Caron, P.R.; Aitken, A.; Gamblin, S.J.; Smerdon, S.J.; Cantley, L.C. The structural basis for 14-3-3: Phosphopeptide binding specificity. *Cell* **1997**, *91*, 961–971. [CrossRef]
14. Yang, X.; Lee, W.H.; Sobott, F.; Papagrigoriou, E.; Robinson, C.V.; Grossmann, J.G.; Sundström, M.; Doyle, D.A.; Elkins, J.M. Structural basis for protein-protein interactions in the 14-3-3 protein family. *Proc. Natl. Acad. Sci. USA* **2006**, *103*, 17237–17242. [CrossRef]
15. Bian, Y.; Lv, D.; Cheng, Z.; Gu, A.; Cao, H.; Yan, Y. Integrative proteome analysis of Brachypodium distachyon roots and leaves reveals a synergetic responsive network under H2O2 stress. *J. Proteom.* **2015**, *128*, 388–402. [CrossRef]
16. Lu, G.; Delislet, A.J.; Vetten, N.C.D.; Ferlo, R.J. Brain proteins in plants: An Arabidopsis homolog to neurotransmitter pathway activators is part of a DNA binding complex. *Proc. Natl. Acad. Sci. USA* **1992**, *89*, 11490–11494. [CrossRef]
17. Tian, F.; Wang, T.; Xie, Y.; Zhang, J.; Hu, J. Genome-wide identification, classification, and expression analysis of 14-3-3 gene family in Populus. *PLoS ONE* **2015**, *10*, 3225. [CrossRef]

18. Cheng, C.; Wang, Y.; Chai, F.; Li, S.; Xin, H.; Liang, Z. Genome-wide identification and characterization of the 14–3-3 family in *Vitis vinifera* L. during berry development and cold- and heat-stress response. *BMC Genom.* **2018**, *19*, 579. [CrossRef]
19. Yin, J.; Fang, Z.; Sun, C.; Zhang, P.; Zhang, X.; Lu, C.; Wang, S.; Ma, D.F.; Zhu, Y. Rapid identification of a stripe rust resistant gene in a space-induced wheat mutant using specific locus amplified fragment (SLAF) sequencing. *Sci. Rep.* **2018**, *8*, 3086. [CrossRef]
20. Tester, M.; Bacic, A. Abiotic stress tolerance in grasses. From model plants to crop plants. *Plant Physiol.* **2005**, *137*, 791–793. [CrossRef]
21. Oecking, C.; Jaspert, N. Plant 14-3-3 proteins catch up with their mammalian orthologs. *Curr. Opin. Plant Biol.* **2009**, *12*, 760–765. [CrossRef]
22. Chen, Y.; Zhou, X.; Chang, S.; Chu, Z.; Wang, H.; Han, S.; Wang, Y. Calcium-dependent protein kinase 21 phosphorylates 14-3-3 proteins in response to ABA signaling and salt stress in rice. *Biochem. Biophys. Res. Commun.* **2017**, *493*, 1450–1456. [CrossRef] [PubMed]
23. Zhang, Z.; Zhao, H.; Huang, F.; Long, J.; Song, G.; Lin, W. The 14-3-3 protein GF14f negatively affects grain filling of inferior spikelets of rice (*Oryza sativa* L.). *Plant J.* **2019**, *99*, 344–358. [PubMed]
24. Manosalva, P.M.; Bruce, M.; Leach, J.E. Rice 14-3-3 protein (GF14e) negatively affects cell death and disease resistance. *Plant J.* **2011**, *68*, 777–787. [CrossRef]
25. Xu, W.; Shi, W. Expression profiling of the 14-3-3 gene family in response to salt stress and potassium and iron deficiencies in young tomato (*Solanum lycopersicum*) roots: Analysis by real-time RT-PCR. *Ann. Bot.* **2006**, *98*, 965–974. [CrossRef]
26. Zhang, Y.; Zhao, H.; Zhou, S.; He, Y.; Luo, Q.; Zhang, F.; Qiu, D.; Feng, J.; Wei, Q.; Chen, L. Expression of TaGF14b, a 14-3-3 adaptor protein gene from wheat, enhances drought and salt tolerance in transgenic tobacco. *Planta* **2018**, *248*, 117–137. [CrossRef]
27. Sun, X.; Luo, X.; Sun, M.; Chen, C.; Ding, X.; Wang, X.; Yang, S.; Yu, Q.; Jia, B.; Ji, W. A Glycine soja 14-3-3 protein GsGF14o participates in stomatal and root hair development and drought tolerance in Arabidopsis thaliana. *Plant Cell Physiol.* **2014**, *55*, 99–118. [CrossRef]
28. Guo, J.; Dai, S.; Li, H.; Liu, A.; Liu, C.; Cheng, D.; Cao, X.; Chu, X.; Zhai, S.; Liu, J. Identification and Expression Analysis of Wheat *TaGF14* Genes. *Front. Genet.* **2018**, *9*, 12. [CrossRef] [PubMed]
29. González, R.R.H.; Borrill, P.; Lang, D.; Harrington, S.A.; Brinton, J.; Venturini, L.; Davey, M.; Jacobs, J.; Ex, F.V.; Pasha, A. The transcriptional landscape of polyploid wheat. *Science* **2018**, *361*, 6089. [CrossRef]
30. De Lille, J.M.; Sehnke, P.C.; Ferl, R.J. The Arabidopsis 14-3-3 Family of Signaling Regulators. *Plant Physiol.* **2001**, *126*, 35–38. [CrossRef] [PubMed]
31. Kumar, S.M.; Mysore, K.S. New dimensions for VIGS in plant functional genomics. *Trends Plant Sci.* **2011**, *16*, 656–665. [CrossRef]
32. Chen, Y.; Ho, T.D.; Liu, L.; Lee, D.H.; Lee, C.; Chen, Y.; Lin, S.; Lu, C.; Yu, S. Sugar starvation-regulated MYBS2 and 14-3-3 protein interactions enhance plant growth, stress tolerance, and grain weight in rice. *Proc. Natl. Acad. Sci. USA* **2019**, *116*, 21925–21935. [CrossRef] [PubMed]
33. Li, X.; Dhaubhadel, S. 14-3-3 proteins act as scaffolds for GmMYB62 and GmMYB176 and regulate their intracellular localization in soybean. *Plant Signal Behav.* **2012**, *7*, 965–968. [CrossRef] [PubMed]
34. Zuo, X.; Wang, S.; Xiang, W.; Yang, H.; Tahir, M.M.; Zheng, S.; An, N.; Han, M.; Zhao, C.; Zhang, D. Genome-wide identification of the 14-3-3 gene family and its participation in floral transition by interacting with TFL1/FT in apple. *BMC Genom.* **2021**, *22*, 41. [CrossRef]
35. Ren, Y.R.; Zhao, Q.; Yang, Y.Y.; Zhang, T.E.; Wang, X.F.; You, C.X.; Hao, Y.J. The apple 14-3-3 protein MdGRF11 interacts with the BTB protein MdBT2 to regulate nitrate deficiency-induced anthocyanin accumulation. *Hortic. Res.* **2021**, *8*, 22. [CrossRef]
36. Camoni, L.; Visconti, S.; Aducci, P.; Marra, M. 14-3-3 Proteins in Plant Hormone Signaling: Doing Several Things at Once. *Front. Plant. Sci.* **2018**, *9*, 297. [CrossRef] [PubMed]
37. Ma, W.; Kong, Q.; Mantyla, J.J.; Yang, Y.; Ohlrogge, J.B.; Benning, C. 14-3-3 protein mediates plant seed oil biosynthesis through interaction with AtWRI1. *Plant J.* **2016**, *88*, 228–235. [CrossRef] [PubMed]
38. Zhou, H.; Lin, H.; Chen, S.; Becker, K.; Yang, Y.; Zhao, J.; Kudla, J.; Schumaker, K.S.; Guo, Y. Inhibition of the Arabidopsis salt overly sensitive pathway by 14-3-3 proteins. *Plant Cell* **2014**, *26*, 1166–1182. [CrossRef] [PubMed]
39. Benzinger, A.; Popowicz, G.M.; Joy, J.K.; Majumdar, S.; Holak, T.A.; Hermeking, H. The crystal structure of the non-liganded 14-3-3σ protein: Insights into determinants of isoform specific ligand binding and dimerization. *Cell Res.* **2005**, *15*, 219–227. [CrossRef]
40. Truong, A.B.; Masters, S.C.; Yang, H.; Fu, H. Role of the 14-3-3 C-terminal loop in ligand interaction. *Proteins* **2002**, *49*, 321–325. [CrossRef]
41. Börnke, F. The variable C-terminus of 14-3-3 proteins mediates isoform-specific interaction with sucrose-phosphate synthase in the yeast two-hybrid system. *J. Plant Physiol.* **2005**, *162*, 161–168. [CrossRef] [PubMed]
42. Yao, Y.; Du, Y.; Jiang, L.; Liu, J. Molecular analysis and expression patterns of the 14-3-3 gene family from *Oryza sativa*. *J. Biochem. Mol. Biol.* **2007**, *40*, 349–357. [CrossRef] [PubMed]
43. Holub, E.B. The arms race is ancient history in Arabidopsis, the wildflower. *Nat. Rev. Genet.* **2001**, *2*, 516–527. [CrossRef] [PubMed]
44. Chen, F.; Li, Q.; Sun, L.; He, Z. The rice 14-3-3 gene family and its involvement in responses to biotic and abiotic stress. *DNA Res.* **2006**, *13*, 53–63. [CrossRef]
45. Roberts, M.R.; Salinas, J.; Collinge, D.B. 14-3-3 proteins and the response to abiotic and biotic stress. *Plant Mol. Biol.* **2002**, *50*, 1031–1039. [CrossRef]
46. He, Y.; Zhang, Y.; Chen, L.; Wu, C.; Luo, Q.; Zhang, F.; Wei, Q.; Li, K.; Chang, J.; Yang, G. A Member of the 14-3-3 Gene Family in Brachypodium distachyon, BdGF14d, Confers Salt Tolerance in Transgenic Tobacco Plants. *Front. Plant Sci.* **2017**, *8*, 340. [CrossRef]
47. Li, X.; Chen, L.; Dhaubhadel, S. 14-3-3 proteins regulate the intracellular localization of the transcriptional activator GmMYB176 and affect isoflavonoid synthesis in soybean. *Plant J.* **2012**, *71*, 239–250. [CrossRef]

48. Finn, R.D.; Mistry, J.; SchusterBockler, B.; GriffithsJones, S.; Hollich, V.; Lassmann, T.; Moxon, S.; Marshall, M.; Khanna, A.; Durbin, R. Pfam: Clans, web tools and services. *Nucleic Acids Res.* **2006**, D247–D251. [CrossRef] [PubMed]
49. Wheeler, T.J.; Eddy, S.R. nhmmer: DNA homology search with profile HMMs. *Bioinformatics* **2013**, *29*, 2487–2489. [CrossRef] [PubMed]
50. Yang, M.; Derbyshire, M.K.; Yamashita, R.A.; Bauer, M.A. NCBI's Conserved Domain Database and Tools for Protein Domain Analysis. *Curr. Protoc. Bioinform.* **2020**, *69*. [CrossRef]
51. Letunic, I.; Bork, P. 20 years of the SMART protein domain annotation resource. *Nucleic Acids Res.* **2017**, *4i6*, D493–D496. [CrossRef] [PubMed]
52. Goodstein, D.M.; Shu, S.; Howson, R.; Neupane, R.; Hayes, R.D.; Fazo, J.; Mitros, T.; Dirks, W.; Hellsten, U.; Putnam, N. Phytozome: A comparative platform for green plant genomics. *Nucleic Acids Res.* **2011**, *40*, D1178–D1186. [CrossRef]
53. Kelley, L.A.; Mezulis, S.; Yates, C.M.; Wass, M.N.; Sternberg, M.J. The Phyre2 web portal for protein modelling, prediction and analysis. *Nat. Protoc.* **2015**, *10*, 845–858. [CrossRef] [PubMed]
54. Gasteiger, E.; Hoogland, C.; Gattiker, A.; Duvaud, S.E.; Wilkins, M.R.; Appel, R.D.; Bairoch, A. Protein Identification and Analysis Tools in the ExPASy Server. *Methods Mol. Biol.* **1999**, *112*, 531–552.
55. Ding, S.; Cai, Z.; Du, H.; Wang, H. Genome-Wide Analysis of TCP Family Genes in *Zea mays* L. Identified a Role for ZmTCP42 in Drought Tolerance. *Int. J. Mol. Sci.* **2019**, *20*, 2762. [CrossRef]
56. Bailey, T.L.; Boden, M.; Buske, F.A.; Frith, M.; Grant, C.E.; Clementi, L.; Ren, J.; Li, W.W.; Noble, W.S. MEME Suite: Tools for motif discovery and searching. *Nucleic Acids Res.* **2009**, W202–W208. [CrossRef]
57. Wang, G.; Wang, T.; Jia, Z.; Xuan, J.; Pan, D.; Guo, Z.; Zhang, J. Genome-Wide Bioinformatics Analysis of MAPK Gene Family in Kiwifruit (*Actinidia Chinensis*). *Int. J. Mol. Sci.* **2018**, *19*, 2510. [CrossRef]
58. Jiang, W.; Yang, L.; He, Y.; Zhang, H.; Li, W.; Chen, H.; Ma, D.; Yin, J. Genome-wide identification and transcriptional expression analysis of superoxide dismutase (SOD) family in wheat (*Triticum aestivum*). *PeerJ* **2019**, *7*, 8062. [CrossRef]
59. Zhou, X.; Zhu, X.; Shao, W.; Song, J.; Jiang, W.; He, Y.; Yin, J.; Ma, D.; Qiao, Y. Genome-Wide Mining of Wheat DUF966 Gene Family Provides New Insights into Salt Stress Responses. *Front. Plant Sci.* **2020**, *11*, 9838. [CrossRef] [PubMed]
60. Gu, Z.; Gu, L.; Eils, R.; Schlesner, M.; Brors, B. Circlize implements and enhances circular visualization in R. *Bioinformatics* **2014**, *30*, 2811–2812. [CrossRef]
61. Sudhir, K.; Glen, S.; Koichiro, T. MEGA7: Molecular Evolutionary Genetics Analysis Version 7.0 for Bigger Datasets. *Mol. Biol. Evol.* **2016**, *33*, 1870–1874.
62. Letunic, I.; Bork, P. Interactive Tree of Life (iTOL) v4: Recent updates and new developments. *Nucleic Acids Res.* **2019**, *47*, W256–W259. [CrossRef]
63. Lescot, M.; Déhais, P.; Thijs, G.; Marchal, K.; Moreau, Y.; Van de Peer, Y.; Rouzé, P.; Rombauts, S. PlantCARE, a database of plant cis-acting regulatory elements and a portal to tools for in silico analysis of promoter sequences. *Nucleic Acids Res.* **2002**, *30*, 325–327. [CrossRef] [PubMed]
64. Fonseca, D.M.D.; Hand, T.W.; Han, S.; Gerner, M.Y.; Zaretsky, A.G.; Byrd, A.L.; Harrison, O.J.; Ortiz, A.M.; Quinones, M.; Trinchieri, G. Microbiota-Dependent Sequelae of Acute Infection Compromise Tissue-Specific Immunity. *Cell* **2015**, *163*, 354–366. [CrossRef]
65. Jiang, W.; Geng, Y.; Liu, Y.; Chen, S.; Cao, S.; Li, W.; Chen, H.; Ma, D.; Yin, J. Genome-wide identification and characterization of SRO gene family in wheat: Molecular evolution and expression profiles during different stresses. *Plant Physiol. Biochem.* **2020**, *154*, 590–611. [CrossRef]
66. Livak, K.J.; Schmittgen, T.D. Analysis of relative gene expression data using real-time quantitative PCR and the 2(-Delta Delta C(T)) Method. *Methods* **2001**, *25*, 402. [CrossRef] [PubMed]
67. Zhu, X.; Qi, T.; Yang, Q.; He, F.; Tan, C.; Ma, W.; Voegele, R.T.; Kang, Z.; Guo, J. Host-Induced Gene Silencing of the MAPKK Gene PsFUZ7 Confers Stable Resistance to Wheat Stripe Rust. *Plant Physiol.* **2017**, *175*, 1853–1863. [CrossRef]
68. Zhang, P.; Jia, Y.; Shi, J.; Chen, C.; Ye, W.; Wang, Y.; Ma, W.; Qiao, Y. The WY domain in the Phytophthora effector PSR1 is required for infection and RNA silencing suppression activity. *N. Phytol.* **2019**, *223*, 839–852. [CrossRef]
69. Chen, C.; He, B.; Liu, X.; Ma, X.; Liu, Y.; Yao, H.Y.; Zhang, P.; Yin, J.; Wei, X.; Koh, H.J. Pyrophosphate-fructose 6-phosphate 1-phosphotransferase (PFP1) regulates starch biosynthesis and seed development via heterotetramer formation in rice (*Oryza sativa* L). *Plant Biotechnol. J.* **2020**, *18*, 83–95. [CrossRef]

International Journal of
Molecular Sciences

Article

Heterologous Expression of the Melatonin-Related Gene *HIOMT* Improves Salt Tolerance in *Malus domestica*

Kexin Tan †, Jiangzhu Zheng †, Cheng Liu, Xianghan Liu, Xiaomin Liu, Tengteng Gao, Xinyang Song, Zhiwei Wei, Fengwang Ma * and Chao Li *

State Key Laboratory of Crop Stress Biology for Arid Areas/Shaanxi Key Laboratory of Apple, College of Horticulture, Northwest A & F University, Yangling, Xianyang 712100, China; kexin163_2020@163.com (K.T.); jiayouzjz@163.com (J.Z.); liucheng175303479@163.com (C.L.); baimoliuxianghan@163.com (X.L.); lxm0603@nwafu.edu.cn (X.L.); 15829268875@163.com (T.G.); songlina12345@163.com (X.S.); weizhiwei89@126.com (Z.W.)

* Correspondence: fwm64@nwsuaf.edu.cn (F.M.); lc453@163.com (C.L.)

† These authors contribute equally in this study (co-first author).

Citation: Tan, K.; Zheng, J.; Liu, C.; Liu, X.; Liu, X.; Gao, T.; Song, X.; Wei, Z.; Ma, F.; Li, C. Heterologous Expression of the Melatonin-Related Gene *HIOMT* Improves Salt Tolerance in *Malus domestica*. *Int. J. Mol. Sci.* **2021**, *22*, 12425. https://doi.org/10.3390/ijms222212425

Academic Editor: Setsuko Komatsu

Received: 23 October 2021
Accepted: 15 November 2021
Published: 17 November 2021

Abstract: Melatonin, a widely known indoleamine molecule that mediates various animal and plant physiological processes, is formed from N-acetyl serotonin via N-acetylserotonin methyltransferase (ASMT). ASMT is an enzyme that catalyzes melatonin synthesis in plants in the rate-determining step and is homologous to hydroxyindole-O-methyltransferase (HIOMT) melatonin synthase in animals. To date, little is known about the effect of *HIOMT* on salinity in apple plants. Here, we explored the melatonin physiological function in the salinity condition response by heterologous expressing the homologous human *HIOMT* gene in apple plants. We discovered that the expression of melatonin-related gene (*MdASMT*) in apple plants was induced by salinity. Most notably, compared with the wild type, three transgenic lines indicated higher melatonin levels, and the heterologous expression of *HIOMT* enhanced the expression of melatonin synthesis genes. The transgenic lines showed reduced salt damage symptoms, lower relative electrolyte leakage, and less total chlorophyll loss from leaves under salt stress. Meanwhile, through enhanced activity of antioxidant enzymes, transgenic lines decreased the reactive oxygen species accumulation, downregulated the expression of the abscisic acid synthesis gene (*MdNCED3*), accordingly reducing the accumulation of abscisic acid under salt stress. Both mechanisms regulated morphological changes in the stomata synergistically, thereby mitigating damage to the plants' photosynthetic ability. In addition, transgenic plants also effectively stabilized their ion balance, raised the expression of salt stress–related genes, as well as alleviated osmotic stress through changes in amino acid metabolism. In summary, heterologous expression of *HIOMT* improved the adaptation of apple leaves to salt stress, primarily by increasing melatonin concentration, maintaining a high photosynthetic capacity, reducing reactive oxygen species accumulation, and maintaining normal ion homeostasis.

Keywords: hydroxyindole-O-methyltransferase gene; melatonin; ROS; ABA; ion homeostasis; amino acids

1. Introduction

Apple (*Malus domestica* Borkh.) is one of the main fruits in the world, the cultivation in China is mainly concentrated in bohai Bay and Loess Plateau, which are two dominant producing areas. Soil salinization in these areas is one of the main obstacles to the expansion of apple eugenic cultivation areas [1]. The sustainable development of apple industry urgently needs the support of apple salt tolerance strategy and technology.

Soil salinity is a progressively serious issue worldwide, as it hinders plant development and causes yield losses [1]. The detrimental influence of NaCl stress on apple plants has two primary aspects [2]. First, the excessive sodium ions accumulation results in a significant decrease in the effective use of water [3], second, Na^+ and Cl^- ions toxic effects lead to ion imbalances [4]. The resulting disturbances in most physiological processes include growth inhibition, reduction in photosynthetic capacity, destruction of membranes, changes in enzymatic activities, and ionic imbalances [5]. These phenomena emphasize the urgency of improving apple plants salt stress tolerance.

Photosynthesis supplies the energy source for plant normal growth, while the photosynthetic apparatus are sensitive to external stress [6]. Salinity stress reduces leaf photosynthetic capacity through both stomatal and non-stomatal mechanisms [7]. Stomatal limitations result from exposure to high salt concentrations; plants become dehydrated due to a reduction of osmotic potential and turgor pressure, causing a decrease in stomatal aperture and reducing CO_2 assimilation capacity, which in turn reduce net photosynthesis [8]. Non-stomatal limitations are related to damage to photosystem II (PSII), CO_2 assimilation rate, and electron transport chain (ETC) efficiency [9].

Plants take up specialized approaches to cope with high salt concentrations. Reactive oxygen species (ROS) production is one of the initial reactions caused by salinity [10], and excessive ROS accumulation can initiate lipid peroxidation processes and increase the concentration of malondialdehyde (MDA), ultimately increasing the plasma membrane permeability, causing intracellular electrolytes leakage [11]. To mitigate ROS damage to various tissues, plants trigger salt-induced antioxidant systems, increasing the activity of superoxide dismutase (SOD), peroxidase (POD), ascorbate peroxidase (APX), and related enzymes. Together, these antioxidant enzymes interact to scavenge ROS and improve plant salt tolerance [12]. Studies have shown that ROS accumulation can act as a positive modifier to mediate abscisic acid (ABA) signal and response. As an abiotic stress phytohormone, ABA has a central function in salt stress defense [13]. Under salt stress, the CLAVATA3/ESR-RELATED 25 (CLE25) peptide may be secreted and moved to the shoots where it is identified by the receptors BARELY ANY MERISTEM 1 (BAM1) and BAM3, thus promoting the expression of the ABA biosynthesis gene *NCED3* [14]. This is one of the mechanisms of plant root from the dehydration signal under salinity condition.

The sequestration of toxic ions and the biosynthesis of osmotic substances in plants under salt stress are related to improved osmotic stress tolerance [15]. The intracellular ion balance does not allow toxic ions to accumulate in the cytoplasm and requires net Na^+ and Cl^- uptake and subsequent vacuolar separation [16]. Appropriate Na^+/K^+ homeostasis maintains the normal physiological metabolism of plants [17], and the absorption, biosynthesis, and transformation of amino acids under stress can alleviate associated damage [18]. As a crucial mechanism for plants response to salt stress, SOS transports excess Na^+ to the extracellular space using the plasma membrane proton gradient [15,19]. NHX family members act on the tonoplast membrane, and the activity of NHX1, NHX2, and NHX4 hinges on the proton gradient to achieve Na^+ compartmentalization [20]. The activity of low-affinity K channel (AKT) is also closely associated with salt tolerance. Free amino acid concentrations also respond sensitively to salinity stress. For example, it is common that physiological response of different plants accumulating Pro under salinity condition [21]. Salt stress results in upregulation of genes in the aromatic amino acid (AAA) biosynthetic pathway, increasing the AAA concentration and the production of related secondary metabolites such as auxins, alkaloids, and flavonoids, thereby improving abiotic stress tolerance [22].

Multitude studies have demonstrated that melatonin is an indole molecule with multiple regulatory functions that has a considerable role in protecting plants from abiotic stresses, including salinity–alkalinity, drought, temperature extremes, oxidative stress, and UV radiation [23–27]. The biosynthetic pathway of plant melatonin is well characterized. First, tryptophan decarboxylase (TDC) promotes the conversion of tryptophan to tryptamine. Then serotonin is produced by the hydroxylation reaction of tryptophan 5-hydroxylase (T5H), and finally, arylalkylamine-N-acetyltransferase (AANAT) and N-acetylserotonin methyltransferase (ASMT) catalyze the synthesis of plant melatonin [26,27]. At present, there are abundant researches show that exogenous melatonin can improve plant resistance under salt-alkali stress [24–26]. Among them, Li et al. [24] found that exogenous melatonin by enhancing the activity of antioxidant enzymes and stabilizing the ion balance to enhance the salt tolerance of *Malus hupehensis*. Recently, several studies have documented the physiological functions of melatonin in plants with heterologous overexpression of melatonin biosynthetic genes [28]. For example, Huang et al. [29] demonstrated that transfer ovine *AANAT*, *HIOMT* genes into switchgrass improved its growth and salinity endurance under salt stress. Droxyindole-O-methyltransferase (*HIOMT*), encodes the last enzyme catalyzes the synthesis of animal melatonin, is a homolog of *ASMT* in apple plants, plays a rate-limiting role in the melatonin synthesis pathway. Its function is to convert N-acetyl serotonin into melatonin [30,31]. In our research team, Liu et al. [31] previously found that ectopic expression *AANAT* and *HIOMT* in apple plants mainly by scavenging reactive oxygen species and increasing total phenolic content to improve resistance to UV-B stress. Although the regulation of *HIOMT* on the UV-B stress response of apples has been described, the mechanism by which *HIOMT* acts in *Malus domestica* under salinity condition has not yet been known clearly. Here, we used hydroponic methods to explore in detail the mechanisms by which *HIOMT* heterologous expression impacts the salinity response of *Malus domestica*.

2. Results

2.1. Heterologous Expression of HIOMT Improved Apple Salt Stress Tolerance and Increased Melatonin Concentration in Apple Leaves

In GL-3 (*Malus domestica* Borkh.) plants, salinity induced the heterologous expression of the *HIOMT* homolog *MdASMT*; its expression was highest at 9 h, when it was upregulated 5.39-fold relative to unstressed controls (Figure 1A). To explore the possible physiological role of *HIOMT* in the apple salinity response, we exposed heterologous expression of the *HIOMT* lines previously obtained to salt stress under hydroponic conditions [31]. No notable deviations between the transgenic lines and the wild type (WT) under normal nutritional conditions. By contrast, all genotypes exhibited damage after NaCl treatment for 15 d. Nonetheless, all leaves of the WT exhibited poorly wilting and necrosis, while only the upper leaves of the transgenic lines exhibited brown spots or chlorosis (Figure 1B). These findings demonstrated that heterologous expression of *HIOMT* conferred improved apple plants salt stress tolerance.

To study variations in melatonin metabolism under salinity condition, we surveyed the transcript levels of four core melatonin synthesis genes. Under 1/2 nutrient solution culture condition, their expression levels were low. Under salinity conditions, all melatonin-related genes were induced, and the levels in transgenic lines were higher than those in the WT significantly (Figure 1C). We also measured melatonin concentration by LC-MS and found that it was significantly lower in WT plants in than in transgenic lines under normal and salinity conditions. The melatonin concentration upregulated by salinity and was highest in the H5 lines, reaching 1.81 ng g^{-1} FW (Figure 1D). These findings showed that heterologous expression of *HIOMT* enhanced remarkably increases in the endogenous melatonin concentration of apple leaves under salinity condition.

2.2. Heterologous Expression of HIOMT Promoted Better Growth and Development of Plants under Salinity Stress

After salt treatment, the shoot height (SH) did not differ significantly among lines, but there were differences in root length (RL), total fresh weight (TFW), and total dry weight (TDW). RL increased under salt stress, and transgenic lines had greater RL than the WT. We also measured leaf number (LN), leaf fresh weight (LFW), and leaf dry weight (LDW) (Table 1). Transgenic lines continued to grow new leaves under salt stress, and their LNs were higher than that of the WT; this explains why total plant weight was lower in WT that in the transgenic lines. In comparation with control conditions, the TFW of WT plants after salinity treatment decreased to 47.9%, whereas that of H1 decreased to 65.9%, that of H2 decreased to 68.2%, and that of H5 decreased to 65.0% (Table 1). These findings illustrated that heterologous expression of *HIOMT* may effectively moderate some of the defective influences of salinity on the growth of apple lines.

Figure 1. Heterologous expression of *HIOMT* confers enhanced salinity tolerance to apple. (**A**) The expression of *MdASMT* after salt treatment in GL-3 apple; (**B**) phenotypes of WT and transgenic lines under control and 120 mM NaCl condition. Bars: 50 mm; (**C**) The expression of MT-related synthesis genes under control and salinity conditions; (**D**) the concentration of melatonin after salinity treatment. Values are noted as means of 3 replicates ± SE. Based on Tukey' s multi-range test ($p < 0.05$), not the same lowercase letters were used to represent statistically significant differences. WT, wild type. H1, heterologous expression of *HIOMT* line 1. H2, heterologous expression of *HIOMT* line 2. H5, heterologous expression of *HIOMT* line 5. CK, control group. ST, salt stress group.

Table 1. Stem height (SH), root length (RL), leaf number (LN), leaf fresh weight (LFW), leaf dry weight (LDW), total fresh weight (TFW), total dry weight (TDW), relative growth of heterologous expression of *HIOMT* apple plants after 15 d under control, and salinity conditions.

Treatment	SH (cm)	RL (cm)	LN (No. Plant^{-1})	LFW (g)	LDW (g)	TFW (g)	TDW (g)	Relative Growth
CK-WT	13.97 ± 0.91^{a}	7.50 ± 1.78^{c}	12.50 ± 1.05^{a}	1.32 ± 0.10^{a}	0.45 ± 0.05^{a}	2.11 ± 0.16^{a}	0.62 ± 0.05^{a}	47.9%
ST-WT	9.84 ± 0.26^{b}	$8.60 \pm 1.36^{b,c}$	8.17 ± 0.75^{c}	0.51 ± 0.11^{c}	0.21 ± 0.05^{c}	1.01 ± 0.22^{c}	0.31 ± 0.06^{c}	
CK-H1	13.27 ± 1.07^{a}	$9.05 \pm 0.82^{b,c}$	13.00 ± 0.89^{a}	1.34 ± 0.16^{a}	0.43 ± 0.07^{a}	2.11 ± 0.14^{a}	0.61 ± 0.08^{a}	65.9%
ST-H1	10.27 ± 0.42^{b}	10.86 ± 2.04^{a}	10.00 ± 1.41^{b}	0.82 ± 0.21^{b}	0.32 ± 0.08^{b}	1.39 ± 0.22^{b}	0.45 ± 0.08^{b}	
CK-H2	13.32 ± 1.28^{a}	$8.68 \pm 1.39^{b,c}$	12.50 ± 1.05^{a}	1.36 ± 0.13^{a}	0.48 ± 0.06^{a}	2.20 ± 0.25^{a}	0.64 ± 0.08^{a}	68.2%
ST-H2	9.97 ± 0.33^{b}	$10.30 \pm 0.97^{a,b}$	9.83 ± 1.47^{b}	0.89 ± 0.13^{b}	0.35 ± 0.07^{b}	1.50 ± 0.15^{b}	0.48 ± 0.08^{b}	
CK-H5	13.90 ± 1.45^{a}	$8.48 \pm 1.53^{b,c}$	13.67 ± 1.63^{a}	1.40 ± 0.17^{a}	0.47 ± 0.06^{a}	2.26 ± 0.24^{a}	0.64 ± 0.08^{a}	65.0%
ST-H5	10.41 ± 0.56^{b}	$9.67 \pm 0.53^{a,b}$	9.50 ± 1.05^{b}	0.85 ± 0.12^{b}	0.34 ± 0.07^{b}	1.47 ± 0.19^{b}	0.48 ± 0.09^{b}	

Values are noted as means of 6 replicates ± SE. Based on Tukey' s multi-range test ($p < 0.05$), not the same lowercase letters were used to represent statistically significant differences.

2.3. Heterologous Expression of HIOMT Changed REL, Root Vitality, and MDA Concentration in Apple

Plant abiotic stress tolerance be assessed by basic physiological indicators such as relative electrolyte leakage rate (REL), root vitality and MDA concentration. No clearly differences in the three indicators among the control lines. After salt stress, REL was 21.76–28.41% lower in transgenic lines than in WT (Figure 2A). Root vitality increased, and transgenic lines showed a greater increase (Figure 2B). Finally, the MDA concentration of the WT was higher than that of the transgenic lines significantly, which was 1.11-fold that of the unstressed control (Figure 2C). These results showed that heterologous expression of *HIOMT* could improve the endurance of apple plants to salinity.

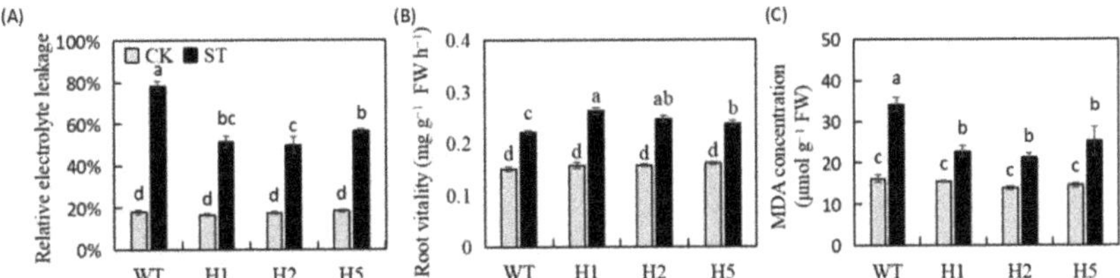

Figure 2. Heterologous expression of *HIOMT* had an impact on REL, Root vitality and MDA concentration under control and salinity conditions. (**A**) Leaf relative electrolyte leakage (REL); (**B**) root vitality and (**C**) leaf MDA concentration. Values are noted as means of 3 replicates ± SE. Based on Tukey' s multi-range test ($p < 0.05$), not the same lowercase letters were used to represent statistically significant differences.

2.4. Heterologous Expression of HIOMT Enhanced the Antioxidant Activity of Apple Plants under Salinity Stress

According to the histochemical staining results after salt stress, the WT apple leaves showed darker blue and yellow colors after staining with 3,3′-diaminobenzidine (NBT) and nitro blue tetrazolium (DAB) (Figure 3A,B). We further confirmed it by quantitative analysis. The transgenic lines accumulated significantly fewer ROS than the WT (Figure 3C,D). In addition, after salt treatment, the SOD, POD, and APX activities of all plants were significantly upregulated and were higher in the transgenic lines than in the WT obviously (Figure 3E–G). Therefore, under salinity condition, heterologous expression of *HIOMT* enhanced the activity of antioxidant enzymes, enabling apple plants to avoid excessive ROS accumulation.

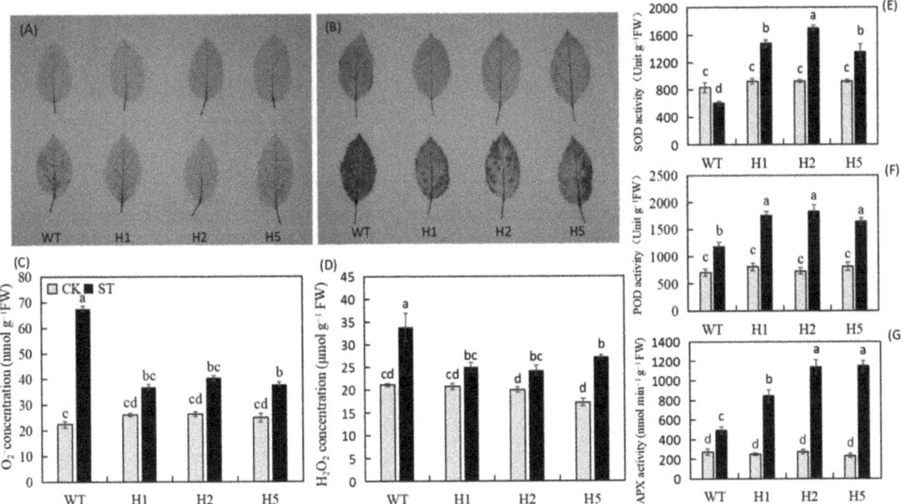

Figure 3. Accumulation of active oxygen and antioxidant enzyme in *HIOMT* transgenic and WT plants under control and salinity conditions. (**A**,**B**) Chemical staining of O_2^- and H_2O_2 in transgenic lines and WT apple leaves under control and salinity condition; (**C**,**D**) O_2^- and H_2O_2 concentrations in apple leaves with and without salinity treatment; the activity of (**E**) SOD; (**F**) POD and (**G**) APX in transgenic Lines and WT apple leaves under control and salinity condition. Values are noted as means of 3 replicates ± SE. Based on Tukey' s multi-range test ($p < 0.05$), not the same lowercase letters were used to represent statistically significant differences.

2.5. Heterologous Expression of HIOMT Enabled Apple Plants to Maintain Higher Photosynthetic Capacity under Salinity Stress

We measured the photosynthetic parameters of the plants every three days throughout the entire salinity stress process. Under control conditions, no significant differences among the genotypes in Pn (net photosynthetic rate), Ci (intercellular CO_2 concentration), gs (stomatal conductance), or Tr (transpiration rate). Under salt treatment, all four parameters had decreased rapidly in all lines by the third day, but those of the WT had decreased more (Figure 4A–D). The downward trend slowed after day 3, but the values of all parameters were significantly lower in the WT than in the transgenic plants. The Pn of the transgenic plants was about 1.7 times that of the WT. Total chlorophyll concentration also reduced in answer to salt stress, but *HIOMT* heterologous expression alleviated this decrease (Figure 4E).

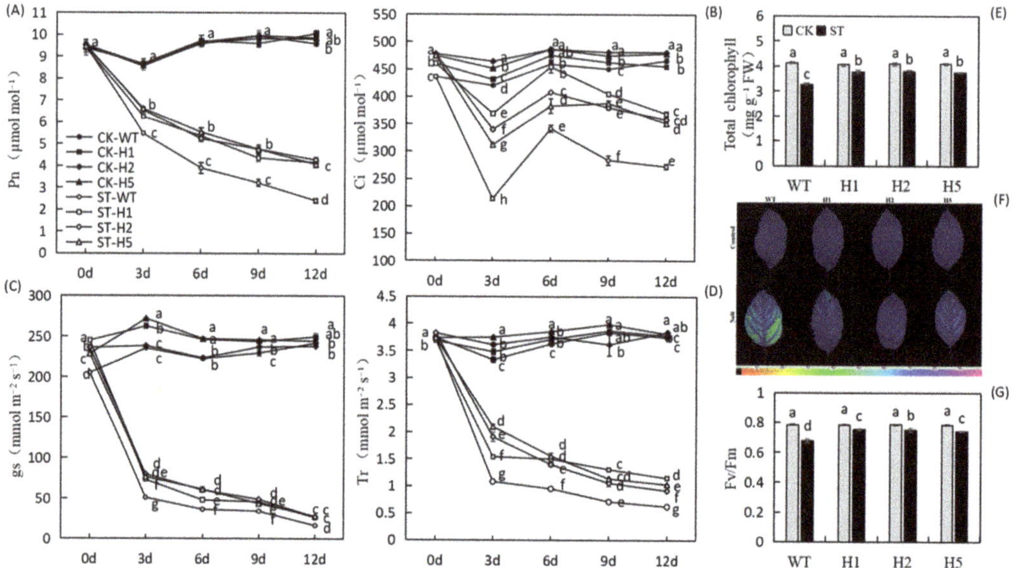

Figure 4. Influences in photosynthetic capacity of heterologous expression of *HIOMT* in apple under control and salinity conditions. (**A**) Changes in Pn; (**B**) gs; (**C**) Ci and (**D**) Tr; (**E**) the concentration of total chlorophyll; (**F**) chlorophyll fluorescence images; (**G**) Fv/Fm ratios of WT and transgenic lines treated with and without salinity. Values are noted as means of 5 replicates ± SE. Based on Tukey' s multi-range test ($p < 0.05$), not the same lowercase letters were used to represent statistically significant differences.

Photosystem II (PSII) is considered to be the primary site of photoinhibition in plants. We measured its maximum photochemical efficiency (Fv/Fm) and found that under salinity condition, the Fv/Fm values of the three transgenic lines decreased by about 3.9%, 4.5%, and 5.6%, whereas that of the WT was reduced to 0.6774, a decrease of 13.6% (Figure 4F,G). These data show that the heterologous expression of *HIOMT* could alleviate damage to plant photosynthetic capacity under salt treatment and enhance plant salt tolerance to some extent.

2.6. Heterologous Expression of HIOMT Alleviated Stomatal Closure in Apple under Salinity Stress

Plant stomata undergo a series of changes under salt stress. We found that stomatal length, width, and aperture of apple leaves decreased after salt treatment compared with normal conditions. Stomatal contraction was most obvious in the WT (Figures 5 and 6A–D). The stomatal density of all lines increased under salt stress relative to normal conditions, and there were no distinct differences among the lines. ABA mediates the response of stomatal aperture to changes in the external environment. We measured leaf ABA concentration after salt treatment and found that under normal conditions, little difference in ABA concentration among genotypes. The ABA concentration increased in all lines after salt treatment, but this increase was greatest in the leaves of WT plants; their ABA levels were 1.2 times those of the transgenic plants (Figure 6F). We measured the transcript levels of *MdNCED3*, a key gene that induces ABA biosynthesis under stress. We found that salt treatment significantly increased its expression, but its expression was significantly lower in transgenic lines than in the WT (Figure 6G). These findings indicated that heterologous expression of *HIOMT* could alleviate the ABA concentration increasement in and the decrease in stomatal aperture under salinity condition.

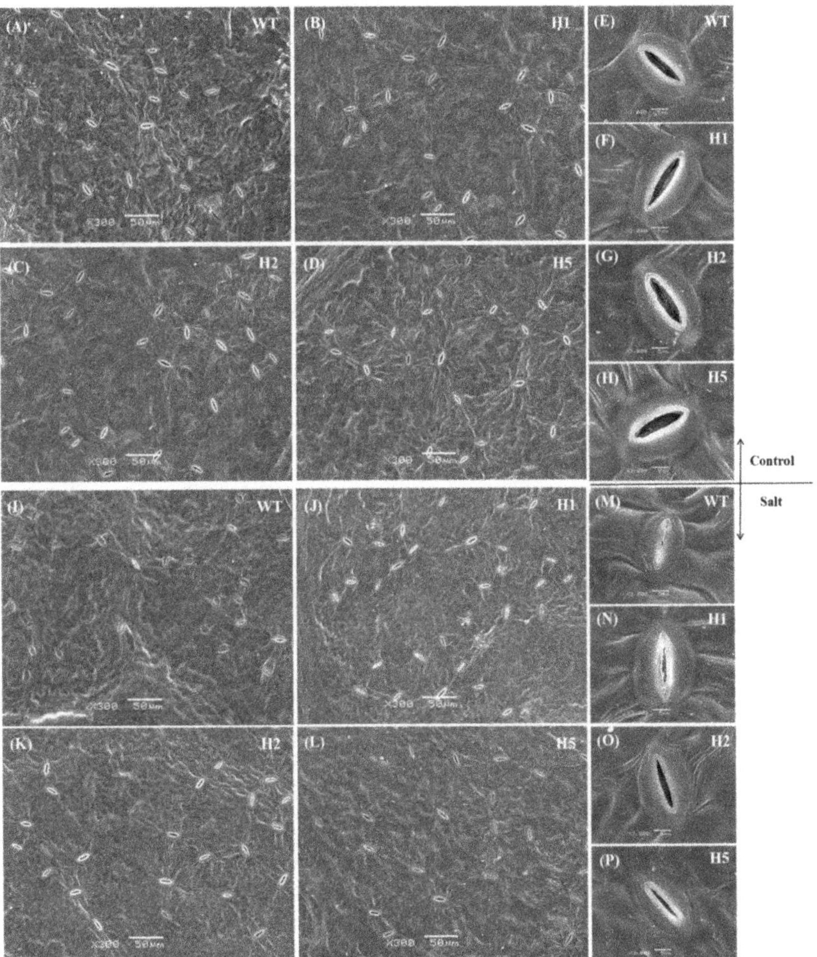

Figure 5. Changes in stomatal morphology, scanning electron microscopy (SEM) images of stomata from *HIOMT* transgenic and WT plants under control and salinity conditions. (**A–H**) Control, changes in stomatal morphology of WT, H1, H2, H5 under control condition, (**A–D**) magnification 300×, scale bars = 50 μm; (**E–H**) magnification 3000×, scale bars = 5 μm. (**I–P**) Salt, changes in stomatal morphology of WT, H1, H2, H5 under salt condition, (**I–L**) magnification 300×, scale bars = 50 μm; (**M–P**) magnification 3000×, scale bars = 5 μm.

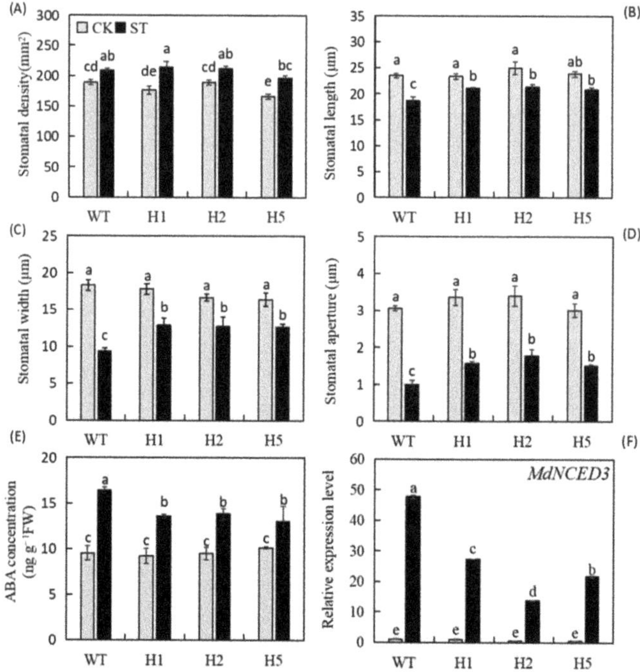

Figure 6. Changes in ABA concentration and ABA synthesis genes expression of WT and *HIOMT* plants under control and salinity conditions. Changes in (**A**) stomatal density, (**B**) stomatal length, (**C**) stomatal width and (**D**) stomatal aperture under salt stress. (**E**) the concentration of ABA; Changes in expression of (**F**) *MdNCED3*. Values are noted as means of 3 replicates ± SE. Based on Tukey' s multi-range test ($p < 0.05$), not the same lowercase letters were used to represent statistically significant differences.

2.7. Heterologous Expression of HIOMT Reduced the Na^+/K^+ Ratio in Apple Plants under Salinity Stress

Under normal conditions, levels of Na^+ ions were very low in the leaves of transgenic and WT plants, and no significant differences among the lines. The K^+ ion concentration was 23.4–26.5 mg g^{-1} DW. Plants were under NaCl stress for 15 d, the Na^+ ion concentration of all plant leaves increased significantly, but that of the three transgenic lines was significantly lower (74.8%, 93.7%, and 82.5%) than that of the WT. The K^+ ion concentration declined over time, and the final Na^+/K^+ ratio of the WT was much higher than that of the transgenic lines (Figure 7A–C). We selected key salt stress–related genes for quantitative analysis, and we found that the expression levels of *MdSOS1*, *MdSOS2*, and *MdSOS3* in the SOS pathway were all upregulated in leaves under salinity condition (Figure 7D–F). The expression levels of *MdNHX1*, *MdNHX2*, and *MdNHX4* also showed the same trend (Figure 7G–I). At the same time, the expression of a membrane protein involved in K^+ ion transport in leaves (AKT potassium transport protein) was also significantly upregulated (Figure 7J). These results showed that heterologous expression of *HIOMT* stabilized the leaf Na^+/K^+ ratio under salinity condition.

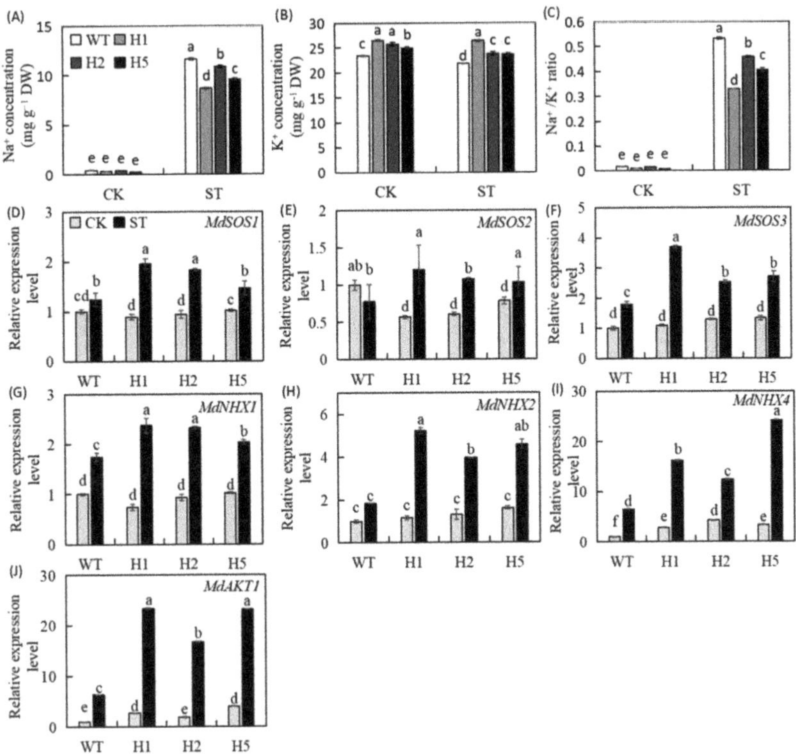

Figure 7. The changes of Na^+ and K^+ of WT and *HIOMT* plants leaves and the expression of salt-related genes of WT and *HIOMT* plants leaves under control and salinity conditions. (**A**) Changes in concentrations of Na^+; (**B**) concentrations of K^+ and (**C**) Na^+/K^+ ratios in the leaves of WT and transgenic lines. Changes in expression of (**D**) *MdSOS1*; (**E**) *MdSOS2*; (**F**) *MdSOS3*; (**G**) *MdNHX1*; (**H**) *MdNHX2*; (**I**) *MdNHX4*; (**J**) *MdAKT1*; Values are noted as means of 3 replicates ± SE. Based on Tukey' s multi-range test ($p < 0.05$), not the same lowercase letters were used to represent statistically significant differences.

2.8. Heterologous Expression of HIOMT Mediated Amino Acid Metabolism under Salinity Stress

Under control conditions, there were no clear diversities in the concentration of amino acid between the lines (Figure 8), but there were significant differences between the transgenic and WT plants under salt stress. The concentrations of alanine (Ala), aspartic acid (Asp), and threonine (Thr) were lower after stress than under control conditions. The concentration of Ala and Thr in the transgenic lines had dropped more, whereas the concentration of Asp was slightly higher in the transgenic lines than in the WT (Figure 8A,B,G). After stress, the concentration of most amino acids increased; these included lysine (Lys), phenylalanine (Phe), proline (Pro), tyrosine (Thr), and tryptophan (Tyr). Their concentration increased more in the WT and were 18.5, 7.8, 12.4, 13.1, and 10.9 times those of the unstressed control; proline concentration increased to 149.4 µg/g. The amino acid concentration of the transgenic lines were lower than those of the WT significantly (Figure 8D–F,H,I). The leucine (Leu) concentration of the WT and H5 was higher after stress than in the unstressed control; the Leu concentration of H5 increased less, the Leu concentration of H1 and H2 was reduced, and the Leu concentration of WT was the highest (Figure 8C). These results showed that heterologous expression of *HIOMT* in apples altered amino acid metabolism under salt stress.

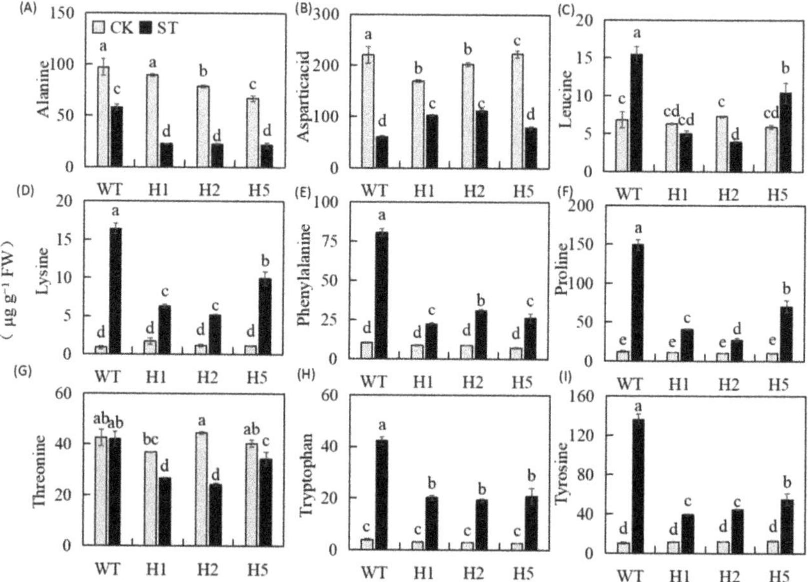

Figure 8. Changes in the concentration of amino acids in the leaves of WT and *HIOMT* plants after salt stress. The concentration of (**A**) Alanine, (**B**) Aspartic acid, (**C**) Leucine, (**D**) Lysine, (**E**) Phenylalanine, (**F**) Proline, (**G**) Threonine, (**H**) Tryptophan, (**I**) Tyrosine. Values are noted as means of 3 replicates ± SE. Based on Tukey' s multi-range test ($p < 0.05$), not the same lowercase letters were used to represent statistically significant differences.

3. Discussion

There is abundance evidence conducted studies the important roles of melatonin in plant tolerance to stressors [23,27,28]. HIOMT is an enzyme that has a rate-limiting effect in the process of human melatonin synthesis. The genes encoding HIOMT enzymes was transferred into GL-3 apple, which increased endogenous melatonin concentration in plants significantly [31]. The ability of exogenous melatonin to improve *Malus hupehensis* endurance to salinity has been reported previously [24]. However, barely known about the specific function of *HIOMT* in the apple salinity response. At this respect, we discovered that heterologous expression of *HIOMT* induced melatonin-related genes up-regulated under salt stress (Figure 1A,C), and we then used hydroponic salt treatment to understand of the mechanisms by which HIOMT functions in response to salinity stress.

We discovered that the melatonin concentration enhanced in all lines under salinity condition, but it was much lower in WT than in transgenic lines under both normal and salinity conditions (Figure 1D). This result is in accordance with antecedent research showing that changes in the external environment cause plants to activate the endogenous melatonin regulatory pathway and change their hormone levels to improve stress tolerance under various stresses, including salt stress [32–34]. A number of other studies have also confirmed that transgenic plants with foreign genes can exhibit increased melatonin production and improved stress tolerance [2,35]. Basic salt stress phenotypes include plant growth inhibition, dysplasia, metabolic disorders, and ion toxicity [36]. The clearest manifestation of salt stress is chlorosis and wilting of leaves (Figure 1B). Electrical conductivity and MDA concentration can also reflect the severity of stress damage [1]. These two indicators were lower in the transgenic lines than in the WT significantly (Figure 2A,C), and transgenic lines showed higher levels of growth, root activity, and other physiological indicators (Table 1 and Figure 2B). These findings show that the heterologous expression of *HIOMT* increased melatonin concentration, reduced the inhibitory effect of salinity on plant development, and alleviated the symptoms of salt damage.

Salinity leads to damage for the photosynthetic system, inhibits photosynthetic electron transfer, and promotes excessive ROS accumulation [11]. The antioxidant enzyme system has vital function in removing excess ROS from plant tissues [10]. Chen et al. [37] discovered that exogenous melatonin improved the activity of antioxidant enzymes in corn seeds under salinity condition, thereby increasing the salt tolerance of corn. They also discovered that after salinity stress, the concentration of H_2O_2 and O_2^- was lower in apple leaves that overexpressed *HIOMT* than in the WT, and the activities of the ROS scavenging enzymes SOD, POD, and APX were significantly higher (Figure 3A–G). These results indicated that *HIOMT* heterologous expression could increase the antioxidant enzyme activity in apple leaves, removing excess ROS and preventing plants from oxidative damage under salt stress.

Reductions in Pn under salt stress are related to changes in leaf anatomical characteristics [5]. The combination of stomatal closure, chlorophyll loss, and other metabolic changes may lead to a decline in photosynthetic capacity [2]. Here, we found that *HIOMT* heterologous expression reduced stomatal contraction (Figure 5A–P and Figure 6A–D) and total chlorophyll loss under salt stress (Figure 4E), and this may interpret the stronger photosynthetic capacity of the transgenic lines (Figure 4A–D). Some degree of stomatal closure under salt stress is a self-protective mechanism [38]. In a high-salt environment, stomatal closure prevents the water lost through transpiration from exceeding the water absorbed by the roots, thereby preventing dehydration. An increase in stomatal density enhances the capacity for CO_2 assimilation and improves the stress resistance of plants [39]. Reductions in chlorophyll concentration and chlorophyll fluorescence parameters occur simultaneously, and chlorophyll fluorescence can provide useful information on changes in photosynthetic performance under stress [40]. Here, we found that transgenic lines also had higher Fv/Fm data than the WT under salt stress (Figure 4F,G). Wei et al. [41] proved that exogenous melatonin appliance helped apple leaves to maintain a higher Fv/Fm value when subjected to photooxidative stress, a result consistent with our present findings.

ABA is closely related to plant stress response [42]. ABA accumulation can cause stomatal closure to prevent water loss while also mediating related genes to facilitate leaf osmotic regulation [43]. To explore stomatal function under salt stress, we measured leaf ABA concentration and found that it was consistent with the changes in stomatal aperture of each line (Figure 6E,F), suggesting that heterologous expression of *HIOMT* can alter the plant's endogenous ABA concentration to control changes in stomatal behavior. Sato et al. [44] reported that *NCED3* may be a key enzyme that promotes ABA biosynthesis in Arabidopsis. Our results show that *MdNCED3* was significantly upregulated under salt stress, but heterologous expression of *HIOMT* inhibited its expression, compliance with the discoveries of Li et al. [45], in which exogenous melatonin restrained the expression of the ABA synthesis gene *MdNCED3* under drought.

Under salinity condition, the absorption and regulation of Na^+ and K^+ ions is an essential ingredient of plant salt tolerance [17,46]. Here, we gauged the concentrations of Na^+ and K^+ in the leaves and found that the transgenic lines had strong salt tolerance (Figure 7A–C). This is related to their restriction of Na^+ absorption, stable K^+ concentration, and maintenance of Na^+/K^+ balance under salinity condition [24]. The SOS pathway uses the driving force of the plasma membrane proton gradient to discharge excess Na^+, thereby reducing the toxic effect of intracellular Na^+ [47]. The NHX pathway acts mainly in concert with the H^+-ATPase and H^+-PPase proton pumps on the tonoplast membrane, sequestering Na^+ from the cytoplasm into the vacuole and thereby reducing its potential toxicity [48]. Here, we discovered that salt treatment improved the expression of *SOS* genes apple plants leaves (Figure 7D–F). The expression of *NHX* genes showed a similar trend in all lines (Figure 7G–I), indicating that the Na^+ efflux ability of the transgenic lines was stronger, enabling them to effectively alleviate the accumulation of Na^+ in leaves under salinity condition. AKT1 is an internal K^+ ion current channel, and overexpression of *SsAKT1* in *Suaeda salsa* promoted the accumulation of potassium ions under salinity condition and enhanced plant salt endurance [49]. Therefore, we also measured the expression levels of *MdAKT1* genes and found that their upregulation under salt stress was higher in transgenic lines than in the WT (Figure 7J). Based on these results, we hypothesize that the heterologous expression of *HIOMT* helps to maintain Na^+/K^+ homeostasis by mediating the expression of genes connected with ion transport, thereby alleviating symptoms of leaf salt damage. The specific regulatory mechanism requires further exploration.

As the basic unit of biologically functional proteins, amino acids have vital functions in plant signal transduction and stress resistance [50]. Here, we found that the increase in Pro under salinity condition was much more in the WT than in the transgenic lines (Figure 8F). This result is consistent with the review by Mansour and Ali [51], who reported that Pro overproduction is positively correlated with the salt stress pressure experienced by plants. For example, Kanawapee et al. [52] obviously showed that the proline accumulation under salt stress showed a trend opposite to that of salt tolerance level and was consistent with changes in the Na^+/K^+ ratio. Here, we suspect that the slight increase in Pro concentration in transgenic lines under salt stress occurred because the 120 mM salt concentration caused a lighter stress, and the transgenic plants were able to adjust their osmotic potential, whereas the WT could not. In addition to proline, we also focused on three AAAs (Trp, Tyr, and Phe) whose concentration showed consistent changes with salt stress (Figure 8E,H,I). Studies have shown that diverse secondary metabolites are derived from these AAAs [22]. The synthesis of auxins and alkaloids requires Trp as a precursor material, Tyr is the precursor of betaine, and anthocyanins, flavonoids, and others are derivatives of Phe; all of these secondary metabolites play important roles in protecting plants from stress [53]. Trp, a precursor for melatonin synthesis, can be converted to tryptamine via TDC to synthesize melatonin [54]. We found that the expression level of *MdTDC1* was much higher in transgenic lines than in the WT (Figure 1B), while the Trp concentration of them was lower than WT (Figure 8H). This result may be due to Trp in the transgenic lines be converted to melatonin by *MdTDC1* and other melatonin synthases, thereby further improving the resistance of transgenic lines to salt stress. In summary, heterologous expression of *HIOMT* can directly or indirectly mediate amino acid metabolism to increase plant stress adaptability.

4. Materials and Methods

4.1. Plant Materials and Treatments

Tissue-cultured apple plants of *M. domestica* 'GL-3' lines (WT) and transgenic GL-3 plants (heterologous expression *HIOMT* lines) were acquired from the study in our laboratory previously, the heterologous expression levels of *HIOMT* (Accession#M83779) lines were increased by 52.06, 203.05 and 90.92-fold in the H1, H2 and H5 lines, respectively [31]. After 30 d of subculturing and 40 d of rooting, tissue culture seedlings were transplanted to 12×12 cm^2 nutrition bowls filled with the mixture of soil, vermiculite, perlite (v1:v2:v3 = 4:1:1). The bowls were placed in a constant temperature light incubator. After the seedlings had grown about 7–8 fully expanded true leaves, they were moved to hydroponic containers ($35 \times 28 \times 15$ cm^3) that were wrapped in black plastic and contained 6.5 L half concentration Hoagland's nutrient solution. The pH of the nutrient solution was adjusted to 6.0 ± 0.2 with H_3PO_4, and it was changed every 5 d. The culture system was designed and improved according to Li et al. previous experiments [24]. The experiments performed in a hydroponic laboratory in Northwest A & F University Yangling (34°20′ N, 108°24′ E), Shaanxi Province, China. Plants growth with condition included a 14 h photoperiod (the light intensity was 160 μmol m^{-2}·s^{-1}), 24 ± 2 °C/16 ± 2 °C day/night and $60 \pm 5\%$ relative humidity [55]. After 2 weeks of pre-adaptation, plants of consistent size and healthy were picked for treatment. The seedlings were divided into two groups, each containing 45 plants: (1) the control group (CK) was grown in half concentration Hoagland nutrient solution as the control, and (2) the salt stress group (ST) was grown in half concentration Hoagland nutrient solution supplemented with 120 mM

NaCl. Both group with a pH of 6.5 ± 0.2. The nutrient solution was updated at an interval of 3 d and treated it for 15 d.

After salt application, control and treated plant leaf tips were taken for *MdASMT* expression analysis. At the end of the treatment, the 4th to 6th leaves of the shoot tips were collected from 12 plants per line and stored at −80 °C for subsequent index determination.

4.2. Determination of Melatonin

The extraction steps of melatonin from leaves referred to the description of Pothinuch and Tongchitpakdee [56]. In brief, a 0.3-g frozen leaf tissue was weighed, each treatment was set up with 5 biological replicates. 5 mL methanol was added; the mixture was sonicated at 4 °C and 40 Hz for 40 min. The centrifuge was set to 10,000 g and 4 °C for 15 min of centrifugation. All the supernatant was collected, placed in a new centrifuge tube, and dried under nitrogen. After reconstitution in pure methanol, the melatonin concentration in the reconstituted solution was determined by high performance liquid chromatography tandem mass spectrometry (HPLC-MS/MS), the above steps refer to Zhao et al. [57]. Three analytical replicates were selected from the measurement results for analysis.

4.3. Growth Measurements

Six plants of each genotype were selected for growth measurements after 15 d of hydroponic treatment. The height from the junction of the rhizome to the top bud of the plant was measured with an iron ruler; the leaves number was accurately counted; and the fresh and dry weights of each plant were recorded [58]. The formula for calculating the relative growth rate was as follows: the average weight of the treated plants divided by the average weight of the control plants.

4.4. Measurement of Relative Electrolyte Leakage, Root Vitality, and MDA Concentration

The REL of control leaves, plant leaves under salt stress, and double distilled water were recorded as S1, S2, S0, respectively, refer to the methods of Dionisio-Sese and Tobita for relative conductivity measurement, and calculate REL = $S1-S0/S2-S0 \times 100\%$. [59]. After 15 d, white roots were harvested from 5 individual plants and stained with triphenyltetrazolium chloride as described by Gong [34]. Malondialdehyde (MDA) concentration was measured using a test kit (Suzhou Coming Biotechnology) according to the instructions of manufacturer. Three analytical replicates were selected from the measurement results for analysis.

4.5. Qualitative and Quantitative Determination of H_2O_2 and O_2^-

At the end of the treatment, 5–8 fully developed mature leaves were obtained from 5 plants per line for histochemical staining with DAB and NBT. Tissue was decolorized thoroughly to observe the accumulation of H_2O_2 and O_2^- as described previously [60]. Quantitative analysis of each concentration was performed with a test kit (Suzhou Comin Biotechnology). Three analytical replicates were selected from the measurement results for analysis.

4.6. Determination of Antioxidant Enzyme Activity

After 15 days of treatment, the frozen sample of the leaves was ground into powder, weighed 0.1 g into a 2 mL centrifuge tube, each treatment was set up with 3 biological replicates. Then used a test kit (Suzhou Comin Biotechnology) according to the instructions of manufacturer to determine the enzyme activities of SOD, POD, and APX, three replicates for each treatment [61]. Three analytical replicates were selected from the measurement results for analysis.

4.7. Quantification of Photosynthetic Parameters

During the whole process of salt treatment, every three days we measured Pn, Ci, gs, and Tr using a CIRAS-3 portable photosynthesis system (CIRAS-3, PP Systems, Amesbury, MA, USA) between 9:00 and 11:00. The instrument parameter setting referred to the description of Liu et al. [55]. For each treatment, 5–8 functional leaves at the same position are selected for the measurement of photosynthetic parameters. Five analytical replicates were selected from the measurement results for analysis.

4.8. Chlorophyll Concentration and Fv/Fm Measurements

Chlorophyll concentration was measured using the method of Arnon [62]. In brief, 8 mL 80% acetone was added to 0.1 g fresh leaves in the dark for at least 24 h to extract pigments, and the mixture was shaken 3–4 times during this period until the leaves turned white. Each treatment was set up with 3 biological replicates. The optical density values were measured at 663, 645, and 470 nm with a UV-2250 spectrophotometer (Shimadzu, Kyoto, Japan).

To measure chlorophyll fluorescence, fully expanded leaves with minimal salt damage from the same position on 5 plants were wrapped in tin foil. After 20 min of dark adaptation, the leaves were cut and placed into the chlorophyll fluorescence imaging system (IMAGING-PAM, Heinz Walz, Effeltrich, Germany) to monitor the maximum quantum efficiency of photosystem II (Fv/Fm). The parameters of the imaging system were set as follows: meas. light 3, act. light 5, ext. light 3, int 10, and FoFm 6 [63]. Five analytical replicates were selected from the measurement results for analysis.

4.9. Stomatal Observations by Scanning Electron Microscopy (SEM)

The measurement of stomatal characteristics refered to Bai et al. [64]. After treatment, three upper leaves were cut at the same leaf position for each treatment, avoiding the main and lateral veins; the leaves were cut into 5 mm × 5 mm squares, stored in pre-cooled 25% glutaraldehyde prepared in advance with 0.2 M phosphate buffered saline (PBS) (pH 7.4). The vacuum was applied with a syringe until the leaf sunk to the bottom of the liquid. The sample was wrapped with tin foil and placed in the refrigerator at 4 °C. It was then washed with PBS (pH 6.8), dehydrated with an ethanol gradient, and placed in 1 mL of isoamyl acetate. After drying and spraying with gold, leaf stomata were discerned and photographed with the scanning electron microscope at 300 and 3000 magnifications. Five pictures were taken for each treatment at different magnifications. Finally, the stomatal density, size, and degree of opening were analyzed using ImageJ software.

4.10. Determination of ABA Concentration

Determination of ABA concentration based on the description by Zhang et al. [65]. In brief, endogenous ABA was extracted using extraction liquid (methanol: isopropanol: acetic acid = 20:79:1). A 0.1-g frozen tissue was combined with 1 mL of extraction liquid, vortexed for 5 min, and then put in refrigerator at −20 °C for 12 h. The centrifuge was set to 4 °C and 12,000 rpm for 10 min of centrifugation. All the supernatant was aspirated and filtered into a sample bottle with a 0.22-μm filter, each treatment was set up with 3 biological replicates, and the ABA concentration of the supernatant was measured by HPLC-MS/MS. Three analytical replicates were selected from the measurement results for analysis.

4.11. Measurements of Sodium and Potassium Ions

At the end of the stress proceeding, leaves from 10 plants were randomly picked out for each treatment and washed thoroughly. The leaves were wiped dry with folded blotting paper. Samples were dried at 105 °C for 20 min and then transferred to a 65 °C oven for at least 72 h. After the dried tissue was ground with a mortar and a tissue lyser, it was stored in a 10-mL centrifuge tube for determination of mineral element concentration after sieving by flame spectroscopy (M410; Sherwood Scientific, Cambridge, UK). The specific steps wrote in Liang et al. [58]. Three analytical replicates were selected from the measurement results for analysis.

4.12. *Measurements of Amino Acids*

Amino acids were extracted and the concentration of them were measured as accounted by Huo et al. [61] with improvements. In brief, 0.1 g of frozen leaf sample was weighed, soaked in 1 mL 50% ethanol, shaken at 4 °C for 5 min, the centrifuge was set to 12,000 rpm, and the time was 10 min, each treatment was set up with 3 biological replicates. Impurities were filtered out, the supernatant was diluted with methanol, and stored in a sample bottle. Amino acids in the sample were measured by liquid chromatography-mass spectrometry (LC-MS, LC: AC, Exion LC; MS: QTRAP 5500, AB SCIEX), and standard curves were used to calculate amino acid concentration. Three analytical replicates were selected from the measurement results for analysis.

4.13. *qRT–PCR Analysis*

Samples were thoroughly ground in liquid nitrogen. Roughly 0.05 g of the resulting powder was weighed for RNA extraction with the Wolact Plant RNA Extraction Kit (Wickband, Hong Kong, China) according to the instructions. Then the synthesize of cDNA by the Prime Script RT reagent Kit with gDNA Eraser (Perfect Real Time). The cDNA was uniformly diluted to 200 ng/μL, and qRT-PCR quantitative analysis was performed. Each gene set three biological replicates, and each replicate used 10 μL SYBR Premix Ex Taq (TaKaRa). The ΔCt value was calculated using *EF* as the internal reference gene [66]. Primer sequences are listed in Table S1. At 0, 3, 6, 9, and 12 h and 3, 6, 9, 12 and 15 d after the initiation of salt stress, the top leaves of the plants were randomly selected for measurement. Three analytical replicates were selected from the measurement results for analysis.

4.14. *Statistical Analysis*

All values were statistically analyzed using SPSS 25.0 software, and Tukey's test was used for multiple comparisons ($p < 0.05$). Results are represented as mean ± SE.

5. Conclusions

Here, we showed that heterologous expression of *HIOMT* enhanced the tolerance of apple leaves to salt stress (Figure 9). Heterologous expression of *HIOMT* increased the expression of melatonin synthase genes under salinity condition and promoted an increase in endogenous melatonin concentration. *HIOMT*-mediated resistance to salinity was associated with reductions in ROS accumulation, maintenance of a strong photosynthetic capacity, regulation of osmotic pressure, and stabilization of the sodium–potassium balance. The heterologous expression of *HIOMT* increased the endogenous melatonin concentration, thereby improving the activity of antioxidant enzymes and inhibiting ROS accumulation. In addition, stomatal aperture could be adjusted by regulating ABA metabolism to restrain water loss and enhance carbon dioxide absorption, thereby improving plant photosynthetic capacity. *HIOMT* heterologous expression also helped to maintain the dynamic balance of intracellular ions by balancing the ratio of sodium and potassium ions and altering amino acid metabolism to alleviate osmotic stress. Our research provides evidence for melatonin-mediated salt tolerance and has crucial applications for improving the growth of horticultural crops on salinity soils.

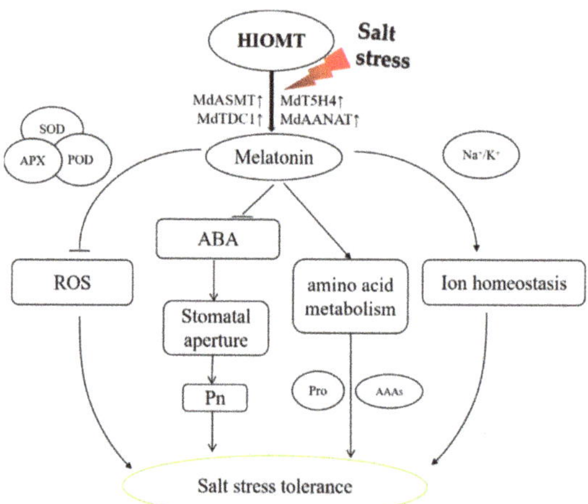

Figure 9. A scheduled model for accounting regulator function of *HIOMT* in physiological reaction of salinity in apple. Under salinity stress, heterologous expression of *HIOMT* enhanced the activity of antioxidant enzymes and inhibited ROS production; inhibited the increase of ABA, maintained stomatal aperture, and improved photosynthetic capacity; mediated amino acid metabolism; stabilized the balance of Na^+, K^+ and maintains ion homeostasis in cells. These were all through the heterologous expression of *HIOMT* resulted in an increase in endogenous melatonin. Therefore, heterologous expression of *HIOMT* improved the salt tolerance in transgenic apple plants.

Supplementary Materials: The following are available online at https://www.mdpi.com/article/10.3390/ijms222212425/s1, Table S1: Primers used in this study.

Author Contributions: C.L. (Chao Li) and K.T. designed the research. Z.W. provided the material. K.T., J.Z., C.L. (Cheng Liu), X.L. (Xianghan Liu) and X.S. performed the experiments. K.T., X.L. (Xiaomin Liu) and T.G. analyzed the data. Assisted by J.Z. and X.L. (Xiaomin Liu), K.T. drafted the manuscript. C.L. (Chao Li) revised the manuscript. F.M. and C.L. (Chao li) provided financial support for the study. All authors have read and agreed to the published version of the manuscript.

Funding: This work was supported by the National Key Research and Development Program of China (2018YFD1000303), the National Natural Science Foundation of China (31972389), and the earmarked fund for the China Agricultural Research System (CARS-27). The authors are grateful to Dr. Jason Qee for editing the language of the article.

Institutional Review Board Statement: Not applicable.

Informed Consent Statement: Not applicable.

Data Availability Statement: Not applicable.

Conflicts of Interest: The authors declare no conflict of interest.

References

1. Fu, M.; Li, C.; Ma, F. Physiological responses and tolerance to NaCl stress in different biotypes of *Malus prunifolia*. *Euphytica* **2013**, *189*, 101–109. [CrossRef]
2. Acosta-Motos, J.R.; Ortuño, M.F.; Bernal-Vicente, A.; Diaz-Vivancos, P.; Sanchez-Blanco, M.J.; Hernandez, J.A. Plant responses to salt stress: Adaptive mechanisms. *Agronomy* **2017**, *7*, 18. [CrossRef]
3. Munns, R. Genes and salt tolerance: Bringing them together. *New Phytol.* **2005**, *167*, 645–663. [CrossRef] [PubMed]
4. Dai, W.; Wang, M.; Gong, X.; Liu, J. The transcription factor FcWRKY40 of *Fortunella crassifolia* functions positively in salt tolerance through modulation of ion homeostasis and proline biosynthesis by directly regulating *SOS2* and *P5CS1* homologs. *New Phytol.* **2018**, *219*, 972–989. [CrossRef]
5. Crawford, T.; Lehotai, N.; Strand, A. The role of retrograde signals during plant stress responses. *J. Exp. Bot.* **2018**, *69*, 83–95. [CrossRef]

6. Yang, X.; Liang, Z.; Wen, X.; Lu, C. Genetic engineering of the biosynthesis of glycinebetaine leads to increased tolerance of photosynthesis to salt stress in transgenic tobacco plants. *Plant Mol. Biol.* **2007**, *66*, 73–86. [CrossRef]
7. Van Zelm, E.V.; Zhang, Y.; Testerink, C. Salt tolerance mechanisms of plants. *Annu. Rev. Plant Biol.* **2020**, *71*, 403–433. [CrossRef]
8. Barragan, V.; Leidi, E.; Andres, Z.; Rubio, L.; Luca, A.D.; Fernandez, J.; Cubero, B.; Pardo, J. Ion exchangers NHX1 and NHX2 mediate active potassium uptake into vacuoles to regulate cell turgor and stomatal function in *Arabidopsis*. *Plant Cell.* **2012**, *24*, 1127–1142. [CrossRef]
9. Sarabi, B.; Fresneau, C.; Ghaderi, N.; Bolandnazar, S.; Streb, P.; Badeck, F.; Citerne, S.; Tangama, M.; David, A.; Ghashghaie, J. Stomatal and non-stomatal limitations are responsible in down-regulation of photosynthesis in melon plants grown under the saline condition: Application of carbon isotope discrimination as a reliable proxy. *Plant Physiol. Biochem.* **2019**, *141*, 1–19. [CrossRef]
10. Choudhury, S.; Panda, P.; Sahoo, L.; Panda, S.K. Reactive oxygen species signaling in plants under abiotic stress. *Plant Signal. Behav.* **2013**, *8*, e23681. [CrossRef] [PubMed]
11. Huang, H.; Ullah, F.; Zhou, D.; Yi, M.; Zhao, Y. Mechanisms of ROS regulation of plant development and stress responses. *Front. Plant Sci.* **2019**, *10*, 800. [CrossRef] [PubMed]
12. Wang, J.; Zhang, L.; Wang, X.; Liu, L.; Lin, X.; Wang, W.; Qi, C.; Cao, Y.; Li, S.; Ren, S.; et al. *PvNAC1* increases biomass and enhances salt tolerance by decreasing Na^+ accumulation and promoting ROS scavenging in switchgrass (*Panicum virgatum* L.). *Plant Sci.* **2019**, *280*, 66–76. [CrossRef] [PubMed]
13. Niu, M.; Xie, J.; Chen, C.; Cao, H.; Sun, J.; Kong, Q.; Shabala, S.; Shabala, L.; Huang, Y.; Bie, Z. An early ABA induced stomatal closure, Na^+ sequestration in leaf vein and K^+ retention in mesophyll confer salt tissue tolerance in *Cucurbita* species. *J. Exp. Bot.* **2018**, *69*, 4945–4960. [CrossRef] [PubMed]
14. Takahashi, F.; Suzuki, T.; Osakabe, Y.; Betsuyaku, S.; Kondo, Y.; Dohmae, N.; Fukuda, H.; Yamaguchi-Shinozaki, K.; Shinozaki, K. A small peptide modulates stomatal control via abscisic acid in long-distance signalling. *Nature* **2018**, *556*, 235–238. [CrossRef]
15. Deinlein, U.; Stephan, A.B.; Horie, T.; Luo, W.; Xu, G.; Schroeder, J.I. Plant salt-tolerance mechanisms. *Trends Plant Sci.* **2014**, 371–379. [CrossRef]
16. Li, C.; Wei, Z.; Liang, D.; Zhou, S.; Li, Y.; Liu, C.; Ma, F. Enhanced salt resistance in apple plants overexpressing a malus vacuolar Na^+/H^+ antiporter gene is associated with differences in stomatal behavior and photosynthesis. *Plant Physiol. Biochem.* **2013**, *70*, 164–173. [CrossRef]
17. Mahi, H.E.; Pérez-Hormaeche, J.; Luca, A.D.; Villalta, I.; Espartero, J.; Gámez-Arjona, F.M.; Fernández, J.L.; Bundó, M.; Mendoza, I.; Mieulet, D.; et al. A Critical Role of Sodium Flux via the Plasma Membrane Na^+/H^+ Exchanger SOS1 in the Salt Tolerance of Rice. *Plant Physiol.* **2019**, *180*, 1046–1065. [CrossRef]
18. Zhu, J. Salt and drought stress signal transduction in plants. *Annu Rev. Plant Biol.* **2002**, *53*, 247–273. [CrossRef]
19. Shi, H.; Quintero, F.; Pardo, J.; Zhu, J. The putative plasma membrane Na^+/H^+ antiporter SOS1 controls long-distance Na^+ transport in plants. *Plant Cell.* **2002**, *14*, 465–477. [CrossRef]
20. Su, Q.; Zheng, X.; Tian, Y.; Wang, C. Exogenous Brassinolide Alleviates Salt Stress in *Malus hupehensis* Rehd. by Regulating the Transcription of NHX-Type $Na^+(K^+)/H^+$ Antiporters. *Front. Plant Sci.* **2020**, *11*, 00038. [CrossRef]
21. Ren, X.; Qi, G.; Feng, H.; Zhao, S.; Wu, W. Calcineurinb-like protein CBl10 directly interacts with AKT1 and modulates K^+ homeostasis in *Arabidopsis*. *Plant J.* **2013**, *74*, 258–266. [CrossRef]
22. Oliva, M.; Guy, A.; Galili, G.; Dor, E.; Schweitzer, R.; Amir, R.; Hacham, Y. Enhanced production of aromatic amino acids in tobacco plants leads to increased phenylpropanoid metabolites and tolerance to stresses. *Front. Plant Sci.* **2020**, *11*, 2110. [CrossRef]
23. Arnao, M.B.; Hernández-Ruiz, J. Functions of melatonin in plants: A review. *J. Pineal Res.* **2015**, *59*, 133–150. [CrossRef]
24. Li, C.; Ping, W.; Wei, Z.; Dong, L.; Liu, C.; Yin, L.; Jia, D.; Fu, M.; Ma, F. The mitigation effects of exogenous melatonin on salinity-induced stress in *Malus hupehensis*. *J. Pineal Res.* **2012**, *53*, 298–306. [CrossRef]
25. Wei, L.; Zhao, H.; Wang, B.; Wu, X.; Lan, R.; Huang, X.; Chen, B.; Chen, G.; Jiang, C.; Wang, J. Exogenous melatonin improves the growth of rice seedlings by regulating redox balance and ion homeostasis under salt stress. *J. Plant Growth Regul.* **2021**, 1–14. [CrossRef]
26. Back, K. Melatonin metabolism, signaling and possible roles in plants. *Plant J.* **2020**, *105*, 376–391. [CrossRef]
27. Dhole, A.; Shelat, H. Phytomelatonin: A plant hormone for management of stress. *J. Anal. Pharm. Res.* **2018**, *7*, 188–190. [CrossRef]
28. Arnao, M.B.; Hernández-Ruiz, J. Melatonin: A new plant hormone and/or a plant master regulator? *Trends Plant Sci.* **2019**, *24*, 38–48. [CrossRef]
29. Huang, Y.; Liu, S.; Yuan, S.; Guan, C.; Tian, D.; Cui, X.; Zhang, Y.; Yang, F. Overexpression of ovine *AANAT* and *HIOMT* genes in switchgrass leads to improved growth performance and salt-tolerance. *Sci. Rep.* **2017**, *7*, 1–13. [CrossRef]
30. Tan, D.; Hardeland, R.; Back, K.; Manchester, L.; Alatorre-Jiménez, M.; Reiter, R. On the significance of an alternate pathway of melatonin synthesis via 5-methoxytryptamine: Comparisons across species. *J. Pineal Res.* **2016**, *61*, 27–40. [CrossRef]
31. Liu, X.; Wei, Z.; Gao, T.; Zhang, Z.; Tan, K.; Li, C.; Feng, M. Ectopic expression of *AANAT* or *HIOMT* improves melatonin production and enhances UV-B. *Fruit Res.* **2021**, *1*, 4–13.
32. Li, C.; Liang, B.; Chang, C.; Wei, Z.; Zhou, S.; Ma, F. Exogenous melatonin improved potassium content in *Malus* under different stress conditions. *J. Pineal Res.* **2016**, *61*, 218–229. [CrossRef]

33. Li, J.; Yuan, F.; Liu, Y.; Zhang, M.; Chen, M. Exogenous melatonin enhances salt secretion from salt glands by upregulating the expression of ion transporter and vesicle transport genes in *Limonium bicolor*. *BMC Plant Biol.* **2020**, *20*, 493. [CrossRef]
34. Gong, X.; Shi, S.; Dou, F.; Song, Y.; Ma, F. Exogenous melatonin alleviates alkaline stress in *Malus hupehensis* Rehd. by Regulating the Biosynthesis of Polyamines. *Molecules* **2017**, *22*, 1542. [CrossRef]
35. Yan, Y.; Sun, S.; Zhao, N.; Yang, W.; Shi, Q.; Gong, B. *COMT1* overexpression resulting in increased melatonin biosynthesis contributes to the alleviation of carbendazim phytotoxicity and residues in tomato plants. *Environ. Pollut.* **2019**, *252*, 51–61. [CrossRef]
36. Liang, W.; Ma, X.; Wan, P.; Liu, L. Plant salt-tolerance mechanism: A review. *Biochem. Biophys. Res. Commun.* **2018**, *495*, 286–291. [CrossRef]
37. Chen, Y.; Mao, J.; Sun, L.; Huang, B.; Ding, C.; Gu, Y.; Liao, J.; Hu, C.; Zhang, Z.; Yuan, S.; et al. Exogenous melatonin enhances salt stress tolerance in maize seedlings by improving antioxidant and photosynthetic capacity. *Physiol Plant* **2018**, *164*, 349–363. [CrossRef]
38. Zhang, Y.; Kaiser, E.; Li, T.; Marcelis, L. Salt Stress Slows Down Dynamic Photosynthesis Mainly through Osmotic Effects on Dynamic Stomatal Behavior. *Chemistry* **2021**. Available online: https://europepmc.org/article/ppr/ppr351620 (accessed on 1 October 2021).
39. Harrison, E.L.; Cubas, L.A.; Gray, J.; Hepworth, C. The influence of stomatal morphology and distribution on photosynthetic gas exchange. *Plant J.* **2019**, *101*, 768–779. [CrossRef] [PubMed]
40. Acosta-Motos, J.R.; Diaz-Vivancos, P.; Álvarez, S.; Fernández-García, N.; Sanchez-Blanco, M.J.; Hernández, J.A. Physiological and biochemical mechanisms of the ornamental Eugenia *myrtifolia* L. plants for coping with NaCl stress and recovery. *Planta* **2015**, *242*, 829–846. [CrossRef]
41. Wei, Z.; Gao, T.; Liang, B.; Zhao, Q.; Ma, F.; Li, C. Effects of exogenous melatonin on methyl viologen-mediated oxidative stress in apple leaf. *Int. J. Mol. Sci.* **2018**, *19*, 316. [CrossRef] [PubMed]
42. Duan, L.; Dietrich, D.; Ng, C.H.; Chan, P.M.; Bhalerao, R.; Bennett, M.J.; Dinneny, J.R. Endodermal ABA signaling promotes lateral root quiescence during salt stress in *Arabidopsis* seedlings. *Plant Cell* **2013**, *25*, 324–341. [CrossRef] [PubMed]
43. Bharath, P.; Gahir, S.; Raghavendra, A.S. Abscisic acid-induced stomatal closure: An important component of plant defense against abiotic and biotic stress. *Front. Plant Sci.* **2021**, *12*, 324. [CrossRef] [PubMed]
44. Sato, H.; Takasaki, H.; Takahashi, F.; Suzuki, T.; Iuchi, S.; Mitsuda, N.; Ohme-Takagi, M.; Ikeda, M.; Seo, M.; Yamaguchi-Shinozaki, K.; et al. *Arabidopsis thaliana NGATHA1* transcription factor induces ABA biosynthesis by activating *NCED3* gene during dehydration stress. *Proc. Natl. Acad. Sci. USA* **2018**, *115*, E11178–E11187. [CrossRef]
45. Li, C.; Tan, D.X.; Liang, D.; Chang, C.; Jia, D.F.; Ma, F.W. Melatonin mediates the regulation of ABA metabolism, free-radical scavenging, and stomatal behaviour in two *Malus* species under drought stress. *J. Exp. Bot.* **2015**, *66*, 669–680. [CrossRef]
46. Abbas, Z.K.; Mobin, M. Comparative growth and physiological responses of two wheat (*Triticum aestivum* L.) cultivars differing in salt tolerance to salinity and cyclic drought stress. *Arch. Agron. Soil Sci.* **2016**, *62*, 745–758. [CrossRef]
47. Huang, Y.; Cui, X.; Cen, H.; Wang, K.; Zhang, Y. Transcriptomic analysis reveals vacuolar Na^+ (K^+)/H^+ antiporter gene contributing to growth, development, and defense in switchgrass (*Panicum virgatum* L.). *BMC Plant Biol.* **2018**, *18*, 1–13. [CrossRef]
48. Bassil, E.; Coku, A.; Blumwald, E. Cellular ion homeostasis: Emerging roles of intracellular NHX Na^+/H^+ antiporters in plant growth and development. *J. Exp. Bot.* **2012**, *63*, 5727–5740. [CrossRef]
49. Duan, H.; Ma, Q.; Zhang, J.; Hu, J.; Bao, A.; Wei, L.; Wang, Q.; Luan, S.; Wang, S. The inward-rectifying K^+ channel *SsAKT1* is a candidate involved in K^+ uptake in the halophyte *Suaeda salsa* under saline condition. *Plant Soil* **2015**, *395*, 173–187. [CrossRef]
50. Batista-Silva, W.; Heinemann, B.; Rugen, N.; Nunes-Nesi, A.; Araújo, W.L.; Braun, H.P.; Hildebrandt, T.M. The role of amino acid metabolism during abiotic stress release. *Plant Cell Environ.* **2019**, *42*, 1630–1644. [CrossRef]
51. Mansour, M.; Ali, E.F. Evaluation of proline functions in saline conditions. *Phytochemistry* **2017**, *140*, 52–68. [CrossRef]
52. Kanawapee, N.; Sanitchon, J.; Lontom, W.; Threerakulpisut, P. Evaluation of salt tolerance at the seedling stage in rice genotypes by growth performance, ion accumulation, proline and chlorophyll content. *Plant Soil* **2012**, *358*, 235–249. [CrossRef]
53. Lynch, J.H.; Dudareva, N. Aromatic amino acids: A complex network ripe for future exploration. *Trends Plant Sci.* **2020**, *25*, 670–681. [CrossRef]
54. Michard, E.; Simon, A. Melatonin's antioxidant properties protect plants under salt stress. *Plant Cell Environ.* **2020**, *43*, 2587–2590. [CrossRef]
55. Liu, X.; Jin, Y.; Tan, K.; Zheng, J.; Gao, T.; Zhang, Z.; Zhao, Y.; Ma, F.; Li, C. *MdTyDc* overexpression improves alkalinity tolerance in *Malus domestica*. *Front. Plant Sci.* **2021**, *12*, 213. [CrossRef]
56. Pothinuch, P.; Tongchitpakdee, S. Melatonin contents in mulberry (*morus* spp.) leaves: Effects of sample preparation, cultivar, leaf age and tea processing. *Food Chem.* **2011**, *128*, 415–419. [CrossRef]
57. Zhao, Y.; Tan, D.; Lei, Q.; Chen, H.; Wang, L.; Li, Q.; Gao, Y.; Kong, J. Melatonin and its potential biological functions in the fruits of sweet cherry. *J. Pineal Res.* **2013**, *55*, 79–88. [CrossRef]
58. Liang, B.; Gao, T.; Zhao, Q.; Ma, C.; Chen, Q.; Wei, Z.; Ma, F. Effects of exogenous dopamine on the uptake, transport, and resorption of apple ionome under moderate drought. *Front. Plant Sci.* **2018**, *9*, 755. [CrossRef]
59. Dionisio-Sese, M.L.; Tobita, S. Antioxidant responses of rice seedlings to salinity stress. *Plant Sci.* **1998**, *135*, 1–9. [CrossRef]
60. Zhou, K.; Hu, L.; Li, Y.; Chen, X.; Zhang, Z.; Liu, B.; Li, P.; Gong, X.; Ma, F. *MdUGT88F1*-mediated phloridzin biosynthesis regulates apple development and valsa canker resistance. *Plant Physiol.* **2019**, *180*, 2290–2305. [CrossRef]

61. Huo, L.; Guo, Z.; Wang, P.; Zhang, Z.; Jia, X.; Sun, Y.; Sun, X.; Gong, X.; Ma, F. *MdATG8i* functions positively in apple salt tolerance by maintaining photosynthetic ability and increasing the accumulation of arginine and polyamines. *Environ. Exp. Bot.* **2020**, *172*, 103989. [CrossRef]
62. Arnon, D. Copper enzymes in isolated chloroplasts. *Plant Physiol.* **1949**, *24*, 1–15. [CrossRef] [PubMed]
63. Deng, C.; Zhang, D.; Pan, X.; Chang, F.; Wang, S. Toxic effects of mercury on PSI and PSII activities, membrane potential and transthylakoid proton gradient in Microsorium pteropus. *J. Photochem. Photobiol. B Biol.* **2013**, *127*, 1–7. [CrossRef] [PubMed]
64. Bai, T.; Li, C.; Chao, L.; Dong, L.; Ma, F. Contrasting hypoxia tolerance and adaptation in malus species is linked to differences in stomatal behavior and photosynthesis. *Physiol. Plant.* **2013**, *147*, 514–523. [CrossRef]
65. Zhang, X.Z.; Zhao, Y.B.; Wang, G.P.; Chang, R.F.; Li, C.M.; Shu, H.R. Dynamics of endogenous cytokinins during phase change in *Malus domestica* Borkh. *Acta Hort.* **2008**, *774*, 29–33. [CrossRef]
66. Perini, P.; Pasquali, G.; Margis-Pinheiro, M.; Oliviera, P.; Revers, L. Erratum to: Reference genes for transcriptional analysis of flowering and fruit ripening stages in apple (*Malus domestica* Borkh.). *Mol. Breed.* **2014**, *34*, 829–842. [CrossRef]

International Journal of
Molecular Sciences

MDPI

Article

Functional Characterization of *PsnNAC036* under Salinity and High Temperature Stresses

Xuemei Zhang [1], Zihan Cheng [1], Wenjing Yao [1,2], Kai Zhao [1], Xueyi Wang [1] and Tingbo Jiang [1,*]

1 State Key Laboratory of Tree Genetics and Breeding, Northeast Forestry University, Harbin 150040, China; zhangxuemei199111@gmail.com (X.Z.); zxm_19910906@sina.com (Z.C.); yaowenjing@njfu.edu.cn (W.Y.); zhangdi_2002@sina.com (K.Z.); wxy20211029@163.com (X.W.)
2 Co-Innovation Center for Sustainable Forestry in Southern China/Bamboo Research Institute, Nanjing Forestry University, Nanjing 210037, China
* Correspondence: tbjiang@yahoo.com

Abstract: Plant growth and development are challenged by biotic and abiotic stresses including salinity and heat stresses. For *Populus simonii* × *P. nigra* as an important greening and economic tree species in China, increasing soil salinization and global warming have become major environmental challenges. We aim to unravel the molecular mechanisms underlying tree tolerance to salt stress and high temprerature (HT) stress conditions. Transcriptomics revealed that a *PsnNAC036* transcription factor (TF) was significantly induced by salt stress in *P. simonii* × *P. nigra*. This study focuses on addressing the biological functions of *PsnNAC036*. The gene was cloned, and its temporal and spatial expression was analyzed under different stresses. *PsnNAC036* was significantly upregulated under 150 mM NaCl and 37 °C for 12 h. The result is consistent with the presence of stress responsive *cis*-elements in the *PsnNAC036* promoter. Subcellular localization analysis showed that *PsnNAC036* was targeted to the nucleus. Additionally, *PsnNAC036* was highly expressed in the leaves and roots. To investigate the core activation region of *PsnNAC036* protein and its potential regulatory factors and targets, we conducted trans-activation analysis and the result indicates that the C-terminal region of 191–343 amino acids of the *PsnNAC036* was a potent activation domain. Furthermore, overexpression of *PsnNAC036* stimulated plant growth and enhanced salinity and HT tolerance. Moreover, 14 stress-related genes upregulated in the transgenic plants under high salt and HT conditions may be potential targets of the *PsnNAC036*. All the results demonstrate that *PsnNAC036* plays an important role in salt and HT stress tolerance.

Keywords: *Populus simonii* × *P. nigra*; *PsnNAC036*; transcription factor; salt stress; HT tolerance

Citation: Zhang, X.; Cheng, Z.; Yao, W.; Zhao, K.; Wang, X.; Jiang, T. Functional Characterization of *PsnNAC036* under Salinity and High Temperature Stresses. *Int. J. Mol. Sci.* **2021**, *22*, 2656. https://doi.org/10.3390/ijms22052656

Academic Editor: Mirza Hasanuzzaman

Received: 24 January 2021
Accepted: 1 March 2021
Published: 6 March 2021

Publisher's Note: MDPI stays neutral with regard to jurisdictional claims in published maps and institutional affiliations.

1. Introduction

Various abiotic stresses such assoil salinity, drought, extreme temperature, and heavy metals affect plant growth, development, and productivity [1]. Many important economic trees are particularly sensitive to these environmental stresses, such as *Populus. simonii* × *P. nigra*. *P. simonii* × *P. nigra* is an endemic plant found in the Yellow River basin and the northern part of mainland in China [2]. It has been used as an urban afforestation tree species in northeast and northwest China. Importantly, the wood is widely used for paper, fiber, matchsticks, and building materials [3]. In recent years, the influence of adverse environmental conditions on the growth of *P. simonii* × *P. nigra* has become more and more severe. Therefore, it is urgent to improve the stress tolerance of this important wood plant through molecular genetics and breeding.

The NAC (NAM, ATAF, and CUC) family is one of the most important transcription factor (TF) gene families in plants. Most NAC TFs are reported to participate in regulation of plant growth and developmental processes [4], including shoot apical meristem formation [5], seed and embryo development [6], lateral root development [7], cell division [8], and leaf senescence [9]. Besides development, NAC TFs also play vital roles in plant response to abiotic stresses, such as salinity, heat, cold, and drought [10]. There are 138 NAC TFs in *Arabidopsis thaliana*, 289 in *Populus trichocarpa*, and 42 in *Nicotiana tabacum* according to PlantTFDB (http://planttfdb.gao-lab.org/family.php?fam=NAC accessed on 4 March 2021). Regarding Arabidopsis, there are a large number of studies of NAC TFs in stress responses. For example, *ANAC019*, *ANAC055*, and *ANAC072* were induced by drought, salinity, and abscisic acid (ABA), and they play a vital role in ABA signaling and osmotic stress [11]. *ANAC078* was confirmed to positively regulate anthocyanin biosynthesis under high light [12]. *ANAC002* (*ATAF1*) can be induced by drought, salinity, ABA, methyl jasmonate, and wounding. Overexpression of the *ATAF1* led to increased Arabidopsis sensitivity to ABA, salt and oxidative stresses [13]. In recent years, NAC TFs have been characterized in poplar to be associated with a stress response. For example, *NAC13* was significantly induced in the roots of 84K poplar by salt stress, and *NAC13* overexpression enhanced poplar salt tolerance [14]. In addition, *PeNAC036* in *P. euphratica* was strongly induced by drought, ABA, and salt, and played a positive role in abiotic stress responses [15]. Transgenic Arabidopsis overexpressing a poplar *NAC57* gene displayed higher seed germination, superoxide dismutase, and peroxidase activities under salt stress than wild type (WT) plants [16]. Therefore, studies of poplar NAC TFs are important to understand the molecular mechanisms of stress response and how to enhance stress tolerance in woody plants.

NAC proteins contain a highly conserved NAM domain at the N-terminus, which can be divided into five subdomains known as A-E, and a highly divergent activation domain at the C-terminus [17]. Based on the conserved N-terminal NAM domain, the NAC family was divided into 18 subfamilies from NAC-a to r in populous [10]. The *PsnNAC036* in this study belongs to the NAC-d subfamily. It was cloned and its relative expression levels were analyzed under various treatments. A total of eight transgenic tobacco lines overexpressing the TF were obtained. Morphological and biochemical measurements indicate that *PsnNAC036* can enhance salinity and HT tolerance of transgenic tobacco plants. RT-qPCR results showed that overexpression of *PsnNAC036* upregulated the expression of 14 stress-related genes in tobacco. These results demonstrate that *PsnNAC036* plays an important role in improving plant salinity and HT tolerance.

2. Results

2.1. Transcript Analysis of 289 NAC TF Genes in Populus

To analyze changes of expression levels of the NAC TFs, the fragment per kilobases per million reads (FPKM) of 289 NAC transcription factor genes from *Populus simonii* × *P. nigra* leaves were retrieved from an RNA-seq dataset [18], which has been deposited in NCBI SRA (accession SRP267437). The salt-stress responsive NAC TFs were visualized using a heatmap (Figure 1A). Based on a fold change of more than 1.2 and a false discovery rate (FDR) smaller than 0.05, 37 differentially expressed NACs were identified (Figure 1B). Among these genes, *PsnNAC036* was significantly upregulated after salt stress treatment of the *P. simonii* × *P. nigra* seedlings compared to control.

2.2. Bioinformatics and Gene Expression Analysis of the PsnNAC036 Gene

To analyze the cDNA and encoded amino acids sequences of *PsnNAC036*, we isolated *PsnNAC036* gene from *P. simonii* × *P. nigra*. The results showed that the coding sequence of the *PsnNAC036* is 1029 bp in length (Supplementary Figure S1A), and encodes a 343 aa protein with 16.2% alpha helix, 15.8% extended strand, 4.0% beta turn, and 64.0% random coil. Five highly homologous genes were found in the NCBI database according to amino acid sequence blast, including *PtrNAC036* from *Populus trichocarpa*, XP_011029436 from *Populus euphratica* which has 155 NAC TFs, *PtrNAC044* from *Populus trichocarpa*, *ANAC072* (AT4G27410.2) from *A. thaliana* and *Nta009260* from *Nicotiana tabacum* (Supplementary Figure S1B). They shared 98.5, 97.1, 79.4, 70.5, and 62.0% amino acid sequence similarity with the *PsnNAC036*, respectively (Supplementary Figure S1C). According to the sequence alignment, these NACs all contain five highly conserved motifs MA, MB, MC, MD, and

ME, which constitute a NAM domain at the N-terminus. Among them, MC and MD were known to bind to DNA [19]. In addition, they all have a predicted nuclear localization signal (NLS) region between 119–150 amino acids (Supplementary Figure S1C).

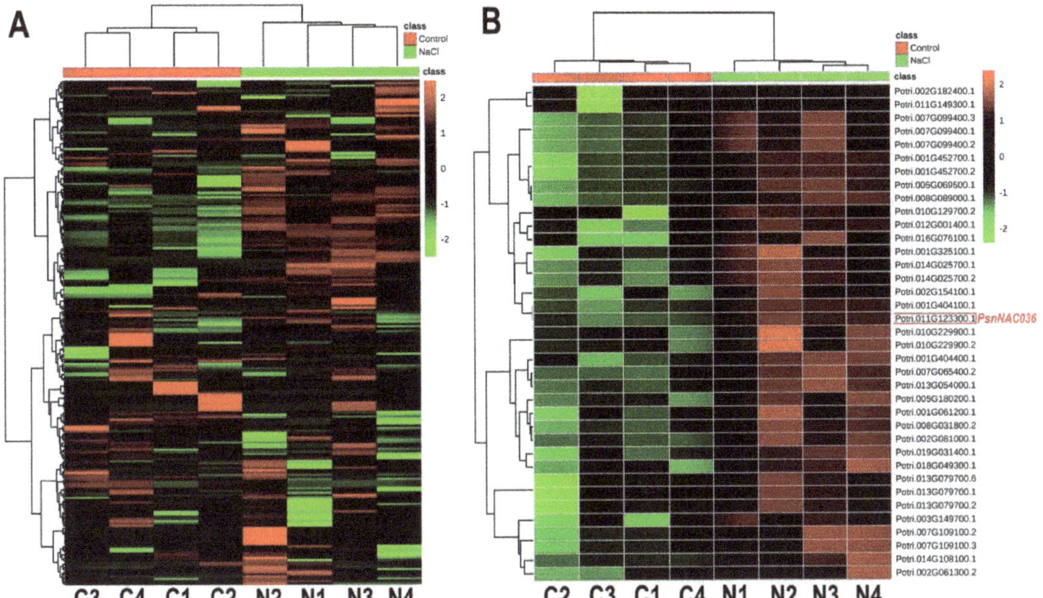

Figure 1. Heatmaps of the relative expression of *NAC* genes in control and NaCl treated *P. simonii* × *P. nigra* seedlings. (**A**) Relative expression patterns of 289 *NACs* in salt stress versus control. (**B**) Heatmap of the 37 differentially expressed *NACs*. Red and green colors indicate high and low expression, respectively. The colored scale bar represents fold changes of transcription levels betweenNaCl stress and control. C1, C2, C3, and C4 represent the four biological replicates under control conditions; N1, N2, N3, and N4 represent the four biological replicates under NaCl stress.

To investigate temporal expression patterns of *PsnNAC036* under different stress conditions, we treated the *P. simonii* × *P. nigra* seedlings with NaCl, ABA, HT, cold, or drought, and harvested the leaves and roots at 0, 3, 6, 12, 24, and 48 h, respectively. The RT-qPCR results showed that the relative expression level of *PsnNAC036* was upregulated in the leaves after different treatments, and it peaked at 12 h. In particular, the gene expression was significantly upregulated after salinity and HT treatments. The *PsnNAC036* expression showed similar expression patterns in the root tissues and the relative expression level of the gene in the root reached 15.2 and 8.8 times higher under NaCl and HT at 12 h than control, respectively (Figure 2).

2.3. Characterization of the PsnNAC036 Promoter Sequence

Promoters contain different transcriptional regulatory elements that can be recognized and bound by RNA polymerases and transcription factors. Some of the regulatory elements play important roles in response to external stimuli [20]. To explore the structure of *PsnNAC036* promoter, the promoter sequence of *PsnNAC036* (from −1 to −1726 bp) in length was isolated from the *P. simonii* × *P. nigra*. Different *cis*-elements in the promoter sequence were predicted (Supplementary Figure S2), including ABRE, Box 4, CGTCA-motif, G-box, I-box, W-box, MYC, O_2-site, etc. by PlantCARE (Supplementary Table S1). Then the promoter was cloned into the pBI121 vector to drive the *GUS* gene expression. To further investigate the promoter activity, we obtained stable transgenic tobacco lines expressing *GUS* under the control of the *PsnNAC036* promoter. *GUS* histochemical staining showed that only young leaves of transgenic tobacco showed a light blue color (Figure 3). The same pattern was observed under cold and drought stresses. After ABA treatment, the *GUS* activity was mainly expressed in the roots, while after 150 mM NaCl treatment for 12 h the *GUS* activity was found throughout the seedlings, especially in the leaves. A similar phenomenon was observed under HT stress, in spite of lower staining intensity than with NaCl treatment. The results confirmed that the *PsnNAC036* gene was responsive to NaCl and HT stresses.

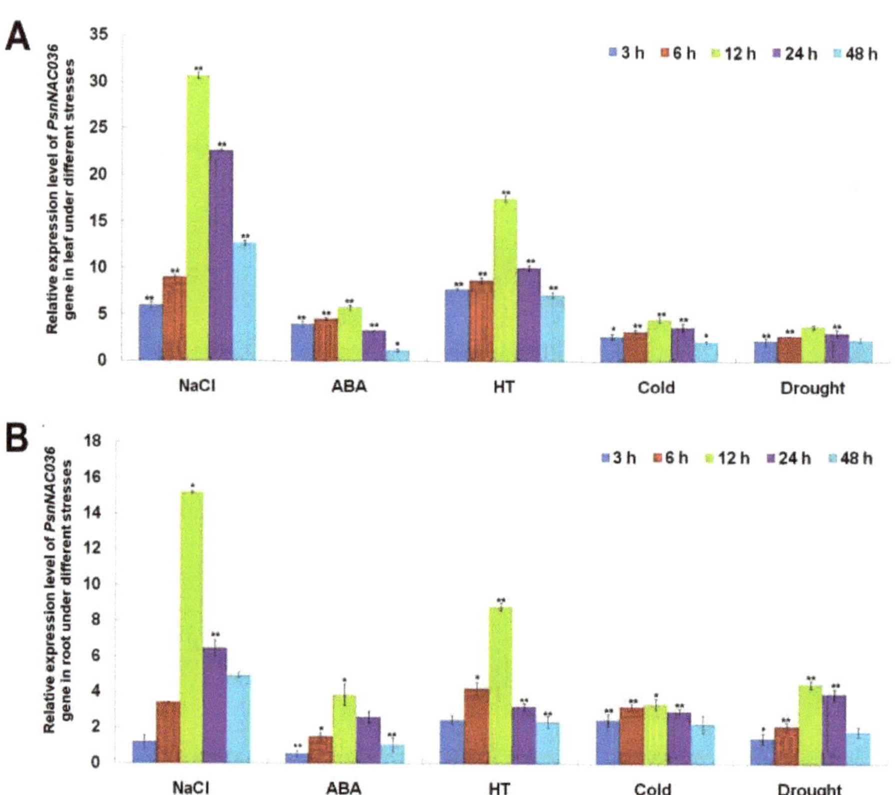

Figure 2. Expression analysis of the *PsnNAC036* under various treatments. Relative expression levels of *PsnNAC036* in leaves (**A**) and roots (**B**) after NaCl (150 mM), ABA (50 μM), high temperature (HT, 37 °C), cold (4 °C), or drought treatments. Expression levels in the control samples were normalized to 1. The data are from three independent experiments. Student's *t*-test: *t*: * $p < 0.05$; ** $p < 0.01$. Error bars indicate mean ± standard deviation.

2.4. PsnNAC036 Protein Is Localized to the Nucleus and Potent Activation Domain in C-terminal Domain

To determine the predicted nuclear localization of the *PsnNAC036* protein (http://cello.life.nctu.edu.tw/cello2go/ accessed on 4 March 2021), the *PsnNAC036* ORF without the termination codon was fused with GFP (Figure 4A). *35S::PsnNAC036-GFP* and positive control *35S::GFP* were transformed into onion epidermal cells by particle bombardment, respectively. As shown in Figure 4B, the fluorescent signal of *PsnNAC036-GFP* was only found in the nucleus; however, the fluorescent signal of the GFP control was distributed throughout the cell. The results clearly showed that the *PsnNAC036* protein was indeed targeted to the nucleus.

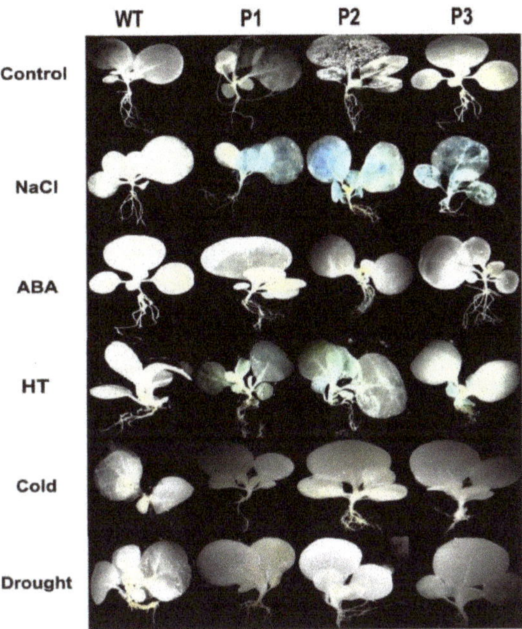

Figure 3. *GUS* activity analysis of the *PsnNAC036* promoter under different treatments.

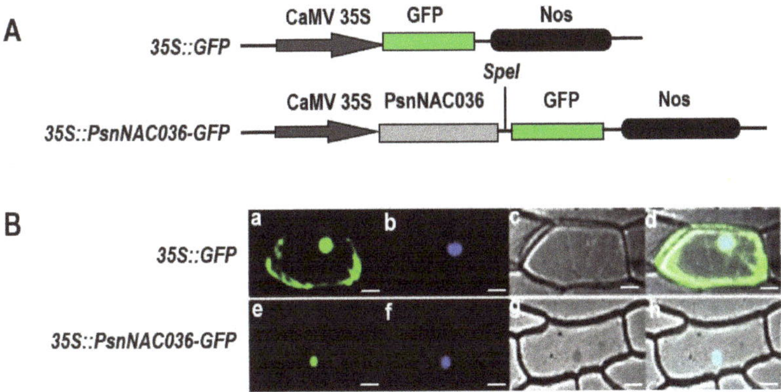

Figure 4. Subcellular localization of the *PsnNAC036* protein. (**A**) Schematic map of the T-DNA inserted in the *35S::GFP* binary vector. (**B**) The *35S::PsnNAC036-GFP* fusion construct and the positive control *35S::GFP* plasmid were introduced into onion epidermal cells by particle bombardment. GFP fluorescence was observed by confocal laser scanning microscopy. (**a**,**e**) are fluorescence images observed in a dark field (green); (**b**,**f**) are 2-(4-Amidinophenyl)-6-indolecarbamidine dihydrochloride (DAPI) staining, which is specific for the nucleus (blue); (**c**,**g**) are light images observed in bright field; (**d**,**h**) are merged images of dark field and bright field. Scale bar = 20 μm.

To test the transcriptional activity of *PsnNAC036*, the full length ORF of the *PsnNAC036* sequence was inserted into the pGBKT7 vector.As shown in the transactivation result (Figure 5A), *PsnNAC036* clearly acted as a TF. To determine the functional domains required for activating transcription, we used a yeast assay system to test the activation domains of *PsnNAC036*. Deletion analysis indicates that the constructs including fragments with amino acids 191–241, 242–292, or 293–343 resulted in transactivation, strongly suggesting that the entire region of 191–343 possesses transactivation activity (Figure 5B).

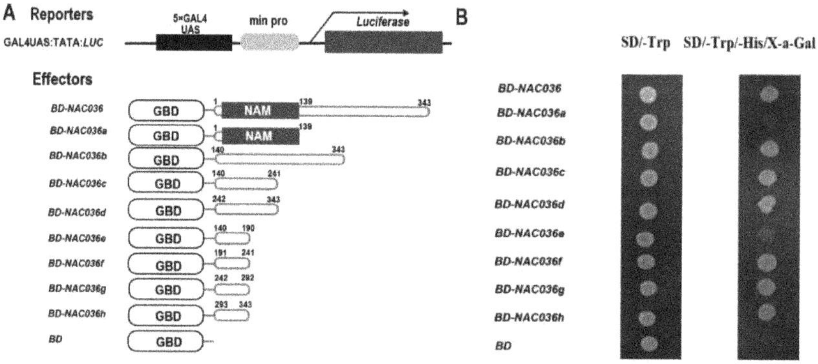

Figure 5. Transactivation analysis of *PsnNAC036* protein. Transactivation analysis of *PsnNAC036* protein. (**A**) Schematic diagrams of the effector and reporter constructs. The reporter construct includes GAL4 binding sites UAS and minimal CaMV35S promoter (min pro) upstream from the luciferase reporter gene. The effector constructs including *BD-NAC036* and GAL4-BD were fused with full-length *PsnNAC036; BD-NAC036a*, N-terminal domain (1–139 aa); *BD-NAC036b*, C-terminal domain (140–343 aa); *BD-NAC036c*, half of *BD-NAC036b* (140–241 aa); *BD-NAC036d*, 242–343 aa; *BD-NAC036e*, half of *BD-NAC036c* (140–190 aa); *BD-NAC036f*, 191–241 aa; *BD-NAC036g*, half of *BD-NAC036d* (242–292 aa); and *BD-NAC036h*, 51 aa of C-terminal domain (293–343 aa). BD was the negative control for pGBKT7. (**B**) Yeast assay.

2.5. *PsnNAC036* Enhanced Tolerance to Salt and HT Stresses in Transgenic Tobacco

Transgenic tobacco lines overexpressing the *PsnNAC036* (Figure 6A) and WT were screened for the presence of transgenes by gDNA PCR with the gene specific primers F1 (5′-ATGGGACTGCAAGAAACAGACC-3′) and R1 (5′-TCACTGCCTAAACCCATACCCA-3′) and RT-PCR with primers F2 (5′- CTTGAATCCTCTCGCAAAAGTG-3′) and R2 (5′-GAGCCGGTCATCAATCTCTGTC-3′). The housekeeping gene *actin* (aF: 5′-GCTTGCTTA CATTGCTCTCGAC-3′, aR: 5′-TGCTTCCGGCTCTGATGTTGTG-3′) for RT-PCR was designed from internal actin gene (GenBank accession number: X69885). As shown in Supplementary Figure S3, *PsnNAC036* fragments were amplified in the transgenic tobacco lines and had the same length as the positive control, but were not detected in the non-transgenic WT.

To investigate the functions of the *PsnNAC036*, transgenic lines T1, T2, T3, and WT were grown in MS, MS with 150 mM NaCl, and MS under 37 °C for two weeks. Under the control condition the plants all grew well. However, the root length of the transgenic lines was about 1.12 ± 0.07 times longer than WT. Under high NaCl or HT treatments, the growth of WT was obviously affected, and the size of WT was smaller than the transgenic lines (Figure 6A). Additionally, the root length of transgenic lines was approximately 1.22 ± 0.03 and 1.26 ± 0.01 times longer than WT under NaCl and HT treatments, respectively (Figure 6B). Besides, after salt treatment, the root length ratios of WT, T1, T2, and T3 (to the transgenic lines) were 72.29%, 77.58%, 78.61%, and 79.40%, respectively, compared to the control plants. Among them, T1 and T3 were more significant than WT. After HT treatment, the root length ratios of WT, T1, T2, and T3 (to the transgenic lines) were 40.95%, 43.95%, 47.53%, and 46.50%, respectively, compared to the control plants. These results demonstrate that the transgenic tobacco plants overexpressing *PsnNAC036* exhibited significant NaCl and HT tolerance.

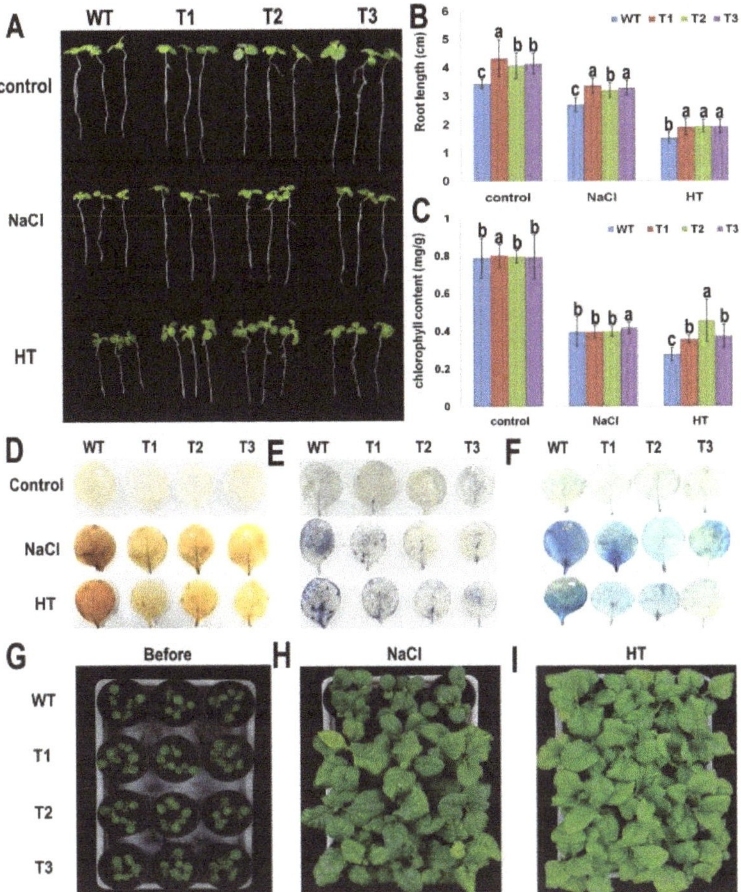

Figure 6. Morphological analysis and histochemical staining of transgenic tobacco lines under NaCl and HT stresses. (**A**) Growth of *PsnNAC036* transgenic lines T1, T2, and T3 seedlings in comparison with WT under control, 150 mM NaCl and 37 °C high temperature (HT) conditions. (**B**) Root length assay of the transgenic lines, compared to WT under control, 150 mM NaCl and HT conditions. (**C**) Chlorophyll content of transgenic tobacco lines and WT plants. The statistical analysis was done using a one-way analysis of variance (ANOVA) with a post-hoc with Tukey lines with an alpha value of 0.05. Different letters indicate significant difference between sites. (**D**) 3,3′-Diaminobenzidine (DAB) staining; (**E**) Nitroblue tertazolium (NBT) staining; (**F**) Evans blue staining; (**G**) Four-week-old tobacco plants in soil right before treatments; (**H**) 200 mM NaCl treatment for two weeks; (**I**) HT treatment for two weeks.

2.6. Changes in Leaf Chlorophyll Content and Physiological Indexes

Leaf chlorophyll is central for light capture and energy exchange between the biosphere and the atmosphere [21], and it serves as a physiological index closely related to plant metabolism and stress resistance [22]. To detect the changes of leaf chlorophyll content and physiological indexes in different tobacco lines, chlorophyll a and chlorophyll b were extracted and quantified from leaves of the transgenic lines and WT plants after salt and HT treatments.The results of chlorophyll content from T1, T2, and T3 lines compared with WT showed that there is no significant difference under normal and high salt conditions, while after HT treatment, the total chlorophyll content of transgenic tobacco lines was 1.29–1.65 fold higher than that of the WT plants (Figure 6C). The results of physiological parameters showed that under normal conditions, the activities of superoxide dismutase (SOD) and peroxidase (POD), as well as proline content were nearly the same in the transgenic lines and WT. However, under salt treatment, SOD and POD activities and proline content were all significantly higher in transgenic lines than in WT. MDA contents of WT were also approximately 1.2- and 1.6-times larger than those of the transgenic plants under normal and salt stress conditions, respectively (Supplementary Figure S4).

2.7. *Histochemical Staining and Growth Assay of the PsnNAC036 Transgenic Plants*

Histochemical staining including diaminobenzidine (DAB), nitroblue tetrazolium (NBT) and Evans blue was carried out to investigate the degree of oxidative damage and cell death. The transgenic and WT plants were treated with NaCl and HT for 12 h. Under control conditions, there was no significant difference between the transgenic and WT leaves. However, under NaCl and HT conditions, intense coloring was observed in the WT leaves after NBT, DAB, and Evans blue staining, in contrast to light staining in the transgenic lines (Figure 6D–F). The results showed that accumulation of reactive oxygen species (ROS) and cell death in the transgenic plants were greatly reduced compared to the WT NaCl and HT treatments.

Under normal growth conditions, the WT and *PsnNAC036* transgenic lines grew similarly. However, after the one-month-old seedlings were treated with 200 mM NaCl or kept at 37 °C for two weeks, the growth of WT was severely retarded, compared to transgenic lines, especially after high salt treatment (Figure 6G–I).

2.8. *PsnNAC036 Alters the Expression of Stress-Related Genes*

To elucidate molecular functions of the *PsnNAC036*, relative expression levels of 14 genes related to stress response in the transgenic plants were analyzed after NaCl and HT treatments by RT-qPCR. As shown in Figure 7, these genes were significantly upregulated in the transgenic plants compared to WT under control and stress treatments. Expression of *NtSOD* that conferred osmotic stress tolerance was found to increase by about 26.8 folds in transgenic lines under control conditions, 85.1-fold under salt stress, and 38.7-fold under HT stress, compared to the WT. The relative expression level of *NtPOD* was upregulated by 4.2-fold, 31.6-fold, and 15.4-fold compared to the WT under control, salinity, and HT, respectively. Furthermore, *NtP5CS* and *NtLEA5* involved in osmotic adjustment and membrane protection [23] were also upregulated by 1.4- and 4.3-fold in the transgenic tobacco, compared to WT, respectively (Figure 7F,H). Under high salt stress, their expression levels were upregulated by 5.4- and 10.8-fold, respectively. Under HT, their expression was significantly upregulated by 10.1- and 51.7-fold, respectively. *NtPPO* encodes a polyphenol oxidase involved in plant stress tolerance. Under normal conditions, the transcript of *NtPPO* in transgenic lines was upregulated by 5.1-fold, while under NaCl and HT, it was upregulated by 2.6- and 1.2-fold, respectively, compared to WT. Moreover, the transcript levels of *NtERD10A*, *NtERD10B*, *NtERD10C*, and *NtERD10D* [24] also showed higher expression (4.4-, 2.6-, 1.3-, and 6.5-fold, respectively) in transgenic plants compared to the WT (Figure 7H,I). The fold-change values were 3.3, 14.9, 4.0, and 9.3 under salt stress, and 12.2, 10.2, 3.9, and 10.0 under HT treatment, respectively. Also, *NtHKT555*, *NtHKT586*, and *NtSOS* encode Na^+/H^+ antiporters [25], and their expression levels were all higher in the transgenic lines than in the WT under control or treatments. Interestingly, *NtNCED1*, involved in the biosynthesis of SA, JA, and ABA [26], was significantly upregulated (41.9, 35.0, and 17.0 fold in the transgenic lines under control, NaCl and HT conditions, respectively). The upregulation of these stress responsive genes may account for the NaCl and HT tolerance of the transgenic plants.

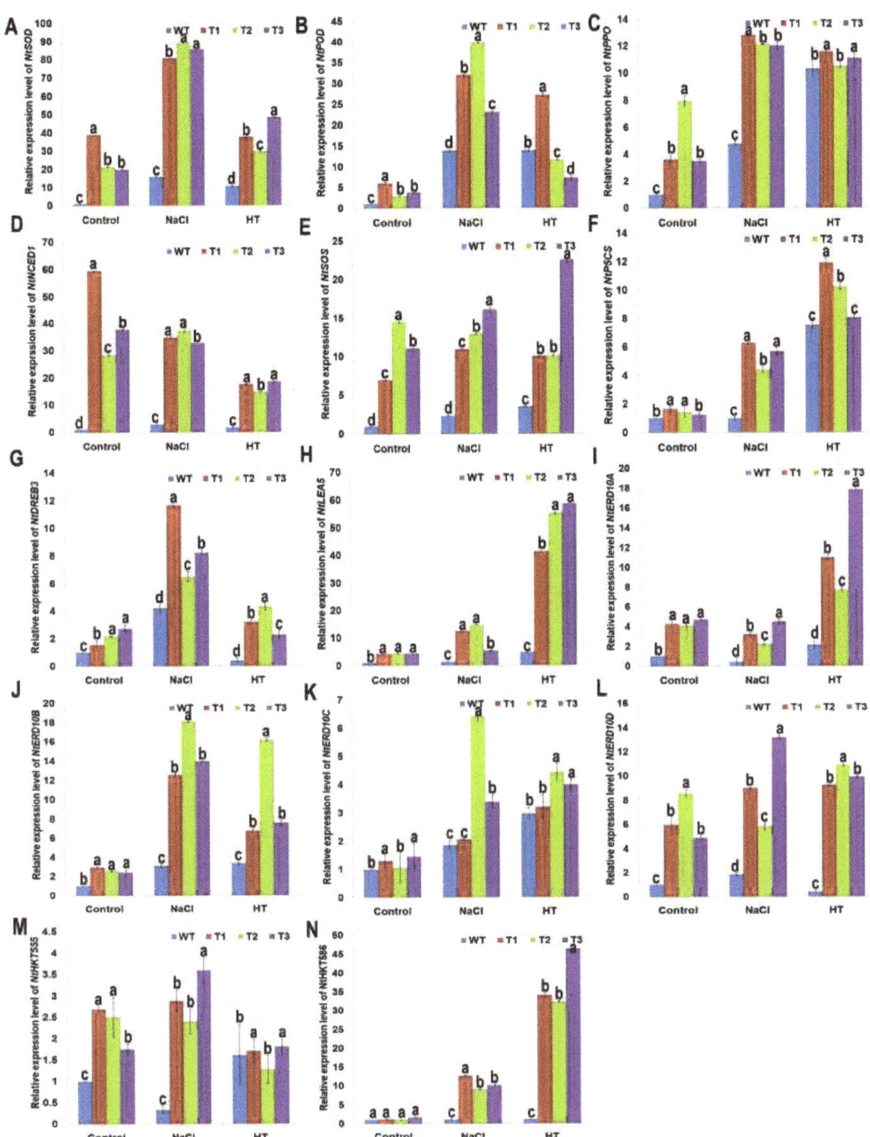

Figure 7. Genes expression profiling of stress-responsive genes in the *PsnNAC036* transgenic plants under NaCl and HT treatments. The relative fold change in expression of (**A**) *NtSOD*; (**B**) *NtPOD*; (**C**) *NtPPO*; (**D**) *NtNCED1*; (**E**) *NtSOS*; (**F**) *NtP5CS*; (**G**) *NtDREB3*; (**H**) *NtLEA5*; (**I**) *NtERD10A*; (**J**) *NtERD10B*; (**K**) *NtERD10C*; (**L**) *NtERD10D*; (**M**) *NtHKT555*; (**N**) *NtHKT586* genes in the *PsnNAC036* transgenic lines under control, high salt and heat conditions. The statistical analysis was done using a one-way ANOVA with a post-hoc with Tukey lines with an alpha value of 0.05. Different letters indicate significant difference between sites.

3. Discussions

With the intensive farming and climate change, salinity and heat have become serious environmental challenges that reduce agricultural productivity of many crops world-wide. Salinity causes osmotic stress and cellular toxicity to plants [27], and heatstress often leads to protein misfolding and degradation that will affect critical cellular reactions and processes [28]. In this study, we found that salt and heat stresses can negatively affect the growth of *P. simonii* × *P. nigra* (Supplementary Figure S5). It is therefore intriguing to investigate molecular mechanisms underlying the stress responses of this woody plant.

The NAC family first identified in Petunia is one of the largest plant-specific TF families [29]. Many NAC family members have been shown to play important roles in gene regulation under environmental stresses [30–33]. In the reference plant Arabidopsis, many studies have elucidated the functions of NAC TFs in plant response to various abiotic stresses. For example, *ANAC019* functions as an upstream regulator of several drought-induced genes such as *DREB2A*, *DREB2B*, *ARF2*, *MYB21*, *MYB24*, and thereby plays an important role in stress response and floral development [34]. NAC TFJUNG-BRUNNEN1 (*ANAC042*) was found to respond to H_2O_2 treatment and enhance plant thermo-tolerance [35]. In addition, overexpression of *ANAC069* can decrease plant ROS scavenging capability and proline biosynthesis, while knock down of *ANAC069* improves plant salt and osmotic stress tolerance [36], indicating *ANAC069* as a negative regulator of plant stress tolerance. In Arabidopsis, the homologous gene of *PsnNAC036* is *ANAC072* (AT4G27410.2, *RD26*), which has been reported to respond to drought, salinity, ABA, SA, and MeJA [37,38]. In this study, we provide several lines of evidence, showing that the *PsnNAC036* can improve plant salinity and heattolerance. Besides, the results of RT-qPCR of seedlings under ABA, cold and drought stress treatments also showed that *PsnNAC036* expression can be induced by these abiotic stresses. Further studies are needed to explore the *PsnNAC036* gene response to other abiotic stresses and determine whether it may have similar functions to the Arabidopsis *ANAC072* homolog.

Plant promoter includes a conservative basic core promoter region and upstream *cis*-acting elements that are crucial to the specificity and activity of gene transcription [39]. In this study, promoter region of *PsnNAC036* gene which is 1726 bp isolated. According to the analysis of PlantCARE software, the promoter contains a variety of *cis*-elements. Most of these elements are related to stress response, such as as-1 belongs to oxidative stress-responsive element, DRE core is involved in dehydration response, and MYC is related to chilling response. Treating *PsnNAC036* promoter transgenic tobacco with salt, cold, heat, ABA, and drought, we found that the *GUS* activity is mainly expressed under salt and HT treatment. This suggested that there may exist other elements in the *PsnNAC036* promoter that contribute to respond high salt and HT stress.

Production of ROS including the superoxide anion (O_2^-), hydrogen peroxide (H_2O_2), and hydroxyl radicals (OH^-) are associated with plant salt stress responses [40]. MDA is a cellular indicator of lipid peroxidation when plants experience oxidative stress caused by environmental challenges, such as salinity and heat [41]. To maintain the redox homeostasis, plants regulate antioxidant enzymes and molecules [42]. SOD and POD are key antioxidant enzymes in ROS-scavenging [43]. In addition, salt stress also causes osmotic stress.Proline is known to function as an important osmolyte in adjusting cellular osmotic balance [44]. In this study, we measured several physiological parameters including SOD, POD activity, proline, and MDA contents in *PsnNAC036* transgenic plants and WT under salt treatment, and found that SOD, POD, and proline content in transgenic lines were increased after stress treatment and MDA content was lower when compared to WT. These results demonstrate that *PsnNAC036* can improve the salt tolerance by regulating the redox and osmotic processes. Also, since DAB, NBT, and Evans blue are dyes commonly used to detect ROS levels and cell death in plants [45,46], the staining results clearly show that overexpression of *PsnNAC036* reduced ROS accumulation and cell death in plants (Figure 7). Our findings are consistent with previous reports that NAC proteins play a crucial role in mediating gene regulation of the anti-oxidative system under stress conditions, thereby conferring higher stress tolerance [47].

Salinity depresses plant growth through lowering water content and excessive accumulation of predominant ions including sodium (Na^+), chloride (Cl^-), calcium (Ca^{2+}), potassium (K^+), and hydrogen carbonate (HCO_3^-) [48]. It has been reported that salt stress can cause the accumulation of Na^+ and Cl^- while inhibiting the uptake of K^+ and Ca^{2+}, which would perturb osmotic homeostasis. Here we measured the H^+-antiporter-related gene *NtSOS* and an isoform of high-affinity K^+ transporter *NtHKT555/586* are critical in improving the salinity tolerance of plants [49,50]. Under stress treatment, the enhanced expression levels of these genes in transgenic tobacco lines showed that *PsnNAC036* might affect the expression of the genes participating in ion exchange to maintain the osmotic homeostasis in plants.

To further understand the regulatory functions of the *PsnNAC036* TF, relative expression levels of 14 stress-related genes (*NtSOD, NtPOD, NtPPO, NtNCED1, NtP5CS, NtDREB3, NtSOS, NtLEA5, NtERD10A/B/C/D*, and *NtHKT555/586*) were quantified in the *PsnNAC036* transgenic lines and WT under control, salinity, and heat conditions (Figure 7). *NtSOD* and *NtPOD* activities are known to be regulated at their transcription levels [51]. *NtPPO* is a polyphenol oxidase gene involved in stress tolerance in many plants [52]. *NtNECD1* plays an important role in ABA biosynthesis [53]. *NtP5CS* encodes a key enzyme in proline biosynthesis [54]. *NtLEA5* is known to protect osmotic adjustment membrane under stress conditions [55]. *NtDREB3* regulates stress-responsive genes by interacting with specific *cis*-acting elements [56]. *NtERD10A/B/C/D* are associated with dehydration response [54]. The results clearly showed that *PsnNAC036* upregulates these stress responsive genes, leading to enhanced salinity and heat stress tolerance. However, how *PsnNAC036* is activated and regulates the stress-related genes is not known. Investigating how *PsnNAC036* interacts with DNA and proteins would be an exciting future research direction.

4. Materials and Methods

4.1. Plant Materials and Stress Treatments

P. simonii × *P. nigra* seedlings were grown on 1/2 MS (Murashige and Skoog medium) plant medium (pH5.8), in a growth chamber with a temperature of 25 ± 1 °C, light of 160 μmol/m^{2}s and 16/8-h light/dark cycle, and relative humidity of approximately 65%. The one-month-old plants were grown hydroponically and those with new roots and leaves of similar sizes were selected for 150 mM NaCl treatment for 24 h. The control (water) and treated leaves from eight plants (four biological replicates each) were frozen in liquid nitrogen and provided to Beijing GENAWIZ Company (www.genewiz.com accessed on 4 March 2021) for RNA sequencing using Illumina HiSeq2500. RNA-Seq data analysiswas carried out as previously described [18] and the heatmap was generated using MetaboAnalyst (https://www.metaboanalyst.ca accessed on 4 March 2021, Version 5.0). The RNA-Seq data have been deposited in NCBI SRA with the accession number SRP267437.

4.2. Cloning and Sequence Analysis of PsnNAC036

Total RNA was extracted from the leaves, and then reversely transcribed into cDNA as a template for PCR amplification. Based on the RNA-Seq analysis, *PsnNAC036* was significantly upregulated in response to high salt treatment. From the PlantTFDB (http://planttfdb.gao-lab.org/family.php?sp=Ptr&fam=NAC accessed on 4 March 2021, Version 5.0), we identified a highly homologous gene of *PsnNAC036* with gene ID Potri.011G123300.1 and designed a pair of primers F1: 5′-ATGGGACTGCAAGAAACAGACC-3′ and R1: 5′-TCACTGCCTAAACCCATACCCA-3′ to clone the *PsnNAC036* gene from *P. simonii* × *P. nigra*. The amino acid sequence of *PsnNAC036* was used for multi-sequence alignment analysis by BioEdit (Version 7.2). Conserved domain and motifs analyses were conducted using MEME (http://meme-suite.org/tools/meme, Version 5.3.3). Nuclear localization signals (NLS) were predicted with cNLS Mapper (http://nls-mapper.iab.keio.ac.jp/cgi-bin/NLS_Mapper_form.cgi#opennewwindow accessed on 4 March 2021, Version 2012). A phylogenetic tree was constructed using MEGA 7.

4.3. *PsnNAC036* Gene Expression Analysis

To analyze the relative expression levels of *PsnNAC036* under different stresses, the seedlings were treated with NaCl, ABA, HT, cold, ordrought for 0, 3, 6, 12, 24, and 48 h, respectively. For the salt and ABA treatment, seedlings were incubated in 150 mM NaCl and 50 μM ABA solution, respectively. For drought stress, the seedlings were exposed to air on filter paper for dehydration. For high or low temperature treatment, seedlings were kept in 37 °C or 4 °C illumination incubators. There was only one variable per treatment compared to the control. Leaf and root tissues were harvested with four biological replicatesat each time point for RNA extraction and RT-qPCR. *Actin* (AF: 5′-ACCCTCCAATCCAGACACTG-3′; AR: 5′- TTGCTGACCGTATGAGCAAG-3′) was used as the reference gene in RT-qPCR. The relative expression level in different samples was calculated using $2^{-\Delta\Delta Ct}$ method.

4.4. GUS Activity of *PsnNAC036* Promoter

DNA was extracted from theleaves using a NuCleanPlantGen DNA Kit (CWBIO, Beijing, China). To clone promoter of *PsnNAC036* from *P. simonii* × *P. nigra*, the promoter sequence of the *PtrNAC036* gene (https://phytozome.jgi.doe.gov/jbrowse/index.html accessed on 4 March 2021, Populus trichocarpa v3.0)was taken as a reference for designing primers pF: 5′-GCGAAGCTTGTTAGGTGCCGAATCTCCGGTGTCC-3′ and pR: 5′-CGCGGATCCCGGGTGAAACCGAAATCCCGGTGGC-3′, which contain *Hind* and *BamHI* restriction enzyme sites, respectively (underlined). The promoter sequence was input in PlantCARE (http://bioinformatics.psb.ugent.be/webtools/plantcare/html/ accessed on 4 March 2021) for *cis*-acting element prediction. The promoter sequence was cloned into a pBI121 vector to replace its CaMV35S promoter. Then the recombinant vector was transferred into *Agrobacterium tumefaciens* EHA105. Using a leaf disc method [57,58], it was then transformed into tobacco plants. Three homozygous transgenic lines P1, P2, and P3 and WT (control) were treated with 150 mM NaCl, 50 μM ABA, HT (37 °C), cold (4 °C), or drought for 12 h, respectively. Histochemical assays of *GUS* enzyme activity driven by the *PsnNAC036* promoter were conducted as previously described [59].

4.5. Subcellular Localization of the PsnNAC036 Protein

The *PsnNAC036* open reading frame (ORF) lacking the stop codon was cloned using primers F3:5′-GGGTCGACTGACTAGTATGGGACTGCAAGAAACAGACC-3′, and R3: TGCTCACCATACTAGTCTGCCTAAACCCATACCCACTT-3′, which contain a *SpeI* restriction site (underlined).Then the sequence was ligated into the linearized pBI121 vector that contains a green fluorescent protein (GFP)using an In-Fusion HD Cloning Kit (TaKaRa, China). The fusion plasmid *35S::PsnNAC036-GFP* and the control plasmid *35S::GFP* were introduced into onion epidermis according to a published method [14]. The GFP fluorescence was observed using a confocal laser scanning microscope (LSM 700, Zeiss, Germany). The results were reproduced three times, and each one had 10 biological replicates.

4.6. Transcriptional Activation Assay of the PsnNAC036 Protein

The ORF of *PsnNAC036* (1–343amino acids (aa)) and the different truncations of the *PsnNAC036* encoding the N-terminal NAM domain 1–139 aa, non-conserved domain 140–343 aa, and the short ORF fragments 140–241 aa, 242–343 aa, 140–190 aa, 191–241 aa, 242–292 aa, and 293–343 aa were inserted into the pGBKT7 vector, respectively. These constructs were named as *BD-NAC036, BD-NAC036a, BD-NAC036b, BD-NAC036c, BD-NAC036d, BD-NAC036e, BD-NAC036f, BD-NAC036g*, and *BD-NAC036h*, respectively (Figure 3). The negative control *pGBKT7* plasmid (*BD*) and all the constructs were transformed into Y2H yeast strain. Transformants were grown for 3–5 days on a selective medium without *Trp* and *His*. *β*-Galactosidase assay was performed on filter lifts of the colonies to detect activation of the *lacZ* reporter gene.

4.7. *PsnNAC036 Overexpression Vector Construction for Plant Transformation*

On the basis of the *PsnNAC036* sequence, we designed a pair of primers F4: 5′-GCGTCTAGAATGGGACTGCAAGAAACAGACC-3′ and R4: 5′-GCGGAGCTCTGGGTAT GGGTTTAGGCAGTGA-3′, which contain *XbaI* and *SacI* restriction sites, respectively (underlined). Both the amplified fragment and the plant binary expression vector pBI121 were double digested by *XbaI* and *SacI* and then were ligated together by T4 ligase. The recombinant plasmid *PsnNAC036*-pBI121 was confirmed by PCR using the vector primers pBI121F1:5′-CCATCGTTGAAGATGCCTCTGC-3′ and pBI121R1:5′-CTCTTCGCTATTAC GCCAGCTGG-3′. Positive clones were transferred into *Agrobacterium tumefaciens* EHA105 for plant genetic transformation.

4.8. *Generation of PsnNAC036 Transgenic Tobacco Plants*

One-month-old tobacco seedlings were used for *PsnNAC036* transformation by a leaf disc method [2,19]. *A.tumefaciens* EHA105 containing the recombinant vector *PsnNAC036-pBI121* were cultured in LB liquid medium with rifampicin (50 mg/mL) and kanamycin (50 mg/mL) until $OD_{600} = 0.6$. Tobacco leaves were cut into 1 × 1 cm squares and incubated in the bacterial solution for 15 min.The leaves were then cultured in MS media with kanamycin (100 mg/mL) for selection. Total RNA of each line wasextracted for PCR with the specific primers F1 and R1. WT was regarded as a negative control and the *PsnNAC036-pBI121* plasmid as a positive control (P). Three independent homozygous transgenic lines T1, T2, and T3 were selected for subsequent experiments.

4.9. *Root Length of the PsnNAC036 Transgenic Tobacco*

To determine salinity and HT tolerance of thetransgenic tobacco overexpressing *PsnNAC036*, seeds of the transgenic lines T1, T2, T3, and WT were grown on the MS medium for seven days. Then they were transplanted into MS containing 150 mM NaCl (salt treatment) or grown in a 37 °C (HT treatment) illumination incubator. Root length was measured two weeks after the transplant.

4.10. *Chlorophyll Content and Physiological measurement*

Chlorophyll a and chlorophyll b were extracted and quantified from leaves of the transgenic lines and WT plants after salt and HT treatments following a previous method [60]. Total chlorophyll content was calculated according to the following equations: Chlorophyll a = $[(12.72 \times A_{663} - 2.69 \times A_{645})V/W]$, Chlorophyll b = $[(22.88 \times A_{645} - 4.68 \times A_{663})V/W]$, total chlorophyll content = Chlorophyll a +Chlorophyll b, where A_{663} and A_{645} are absorbance at 663 nm and 645 nm, respectively. V is the final volume of chlorophyll extract in 80% acetone, and W is fresh weight of leaves. Biochemical analyses of superoxide dismutase (SOD), peroxidase (POD), proline content, and malondialdehyde (MDA) were conducted according to previous methods [14].

4.11. *Histochemical Analyses of the PsnNAC036 Transgenic Tobacco*

Cell death, accumulation of hydrogen peroxide (H_2O_2) and superoxide (O_2^-) in the transgenic and WT plants under high saltor HT were visualized through histochemical staining with Evans blue, 3,3′-diaminobenzidine (DAB) and nitro-blue tetrazolium (NBT), respectively. Two-week-old seedlings were irrigated with water (control), 150 mM NaCl solution or grown in 37 °C HT for 12 h. Leaves were excised for histochemical staining as previously described [10].

4.12. *Salt and HT Treatments of the PsnNAC036 Transgenic Tobacco*

The seedlings of WT and transgenic tobacco lines on MS medium were transplanted in the soil oncethe fourth leaf appeared. The seedlings were grown under normal growth conditions for a total of two weeks, and then were subjected to 200 mM NaCl irrigation or 37 °C HT treatment for another two weeks.

4.13. Expression Analysis of Stress-Related Genes in PsnNAC036 Transgenic Tobacco

To further understand the functions the *PsnNAC036* TF, leaves of the transgenic lines and WT under 200 mMNaCl or 37 °C HT for 12 h were used for RT-qPCR. Relative expression levels ofstress-related genes including superoxide dismutase (*NtSOD*), peroxidase (*NtPOD*), polyphenol oxidase (*NtPPO*), 9-cis-epoxycarotenoid dioxygenase1 (*NtNCED1*), 1-pyrroline-5-carhoxylate synthetase (*NtP5CS*), regulatory proteins (*NtDREB3*), plasmalemma Na^+/H^+ antiporter (*NtSOS*), late-embryogenesis-abundant protein5 (*NtLEA5*), early responsive to dehydration (*NtERD10A/B/C/D*) and Na^+ antiporter genes (*NtHKT555/586*) were quantified. RT-qPCR was conducted according to a published method [14]. The primer sequences used for the above genes can be found in Supplementary Table S2.

5. Conclusions

In this study, we analyzed the expression changes of NAC TFs in *P. simonii* × *P. nigra* under 150 mM salt treatment and isolated a significantly upregulated NAC gene, *PsnNAC036*. RT-qPCR and promoter *GUS* analyses showed that the gene was highly responsive to stresses, especially salinity and heat. Overexpression of *PsnNAC036* improved salinity and HT tolerance of transgenic plants. RT-qPCR results revealed that overexpression of the *PsnNAC036* TF regulated the expression of several stress-related genes in the transgenic plants. Furthermore, we validated that the C-terminal domain (191–343 amino acids) was the potent activator of *PsnNAC036*. These results have demonstrated that *PsnNAC036* functions as a transcriptional activatorand plays an important role in plant salinity and HT tolerance. As such, it is a potential molecular target for crop biotechnology and marker-based breeding for enhanced resilience and yield.

Supplementary Materials: Supplementary materials can be found at https://www.mdpi.com/1422-0067/22/5/2656/s1.

Author Contributions: T.J. designed research. X.Z. conducted experiments and wrote the manuscript. Z.C., K.Z. and X.W. performed in data analysis. W.Y. revised the manuscript. All authors read and approved the manuscript.

Funding: This work was supported by the Fundamental Research Funds for the Central Universities (2572019AA06) and Natural Science Foundation of Jiangsu Province (BK20190748).

Acknowledgments: The authors thank Lisa David from University of Florida for critical reading and editing of the manuscript.

Conflicts of Interest: The authors declare no conflict of interest.

References

1. Ku, Y.; Sintaha, M.; Cheung, M.; Lam, H. Plant Hormone Signaling Crosstalks between Biotic and Abiotic Stress Responses. *Int. J. Mol. Sci.* **2018**, *19*, 3206. [CrossRef]
2. Yao, W.; Wang, S.; Zhou, B.; Jiang, T. Transgenic poplar overexpressing the endogenous transcription factor *ERF76* gene improves salinity tolerance. *Tree Physiol.* **2016**, *36*, 896–908. [CrossRef] [PubMed]
3. Wang, X.; Chen, J.; Xia, D. Research progress on elasticity modulus nondestructive examination of wood and glulam structures. *J. Cent. South Univ. For. Technol.* **2013**, *33*, 149–153.
4. Nuruzzaman, M.; Sharoni, A.M.; Kikuchi, S. Roles of NAC transcription factors in the regulation of biotic and abiotic stress responses in plants. *Front. Microbiol.* **2013**, *4*, 248. [CrossRef] [PubMed]
5. Kim, S.G.; Kim, S.Y.; Park, C.M. A membrane-associated NAC transcription factor regulates salt-responsive flowering via FLOWERING LOCUS T in *Arabidopsis*. *Planta* **2007**, *226*, 647–654. [CrossRef] [PubMed]
6. Duval, M.; Hsieh, T.-F.; Kim, S.Y.; Thomas, T.L. Molecular characterization of AtNAM: A member of the Arabidopsis NAC domain superfamily. *Plant. Mol. Biol.* **2002**, *50*, 237–248. [CrossRef]
7. Xie, Q.; Frugis, G.; Colgan, D.; Chua, N.H. Arabidopsis *NAC1* transduces auxin signal downstream of *TIR1* to promote lateral root development. *Genes Dev.* **2000**, *14*, 3024–3036. [CrossRef]
8. Kim, H.J.; Nam, H.G.; Lim, P.O. Regulatory network of NAC transcription factors in leaf senescence. *Curr. Opin. Plant. Biol.* **2016**, *33*, 48–56. [CrossRef] [PubMed]
9. Breeze, E.; Harrison, E.; McHattie, S.; Hughes, L.; Hickman, R.; Hill, C. High-resolution temporal profiling of transcripts during Arabidopsis leaf senescence reveals a distinct chronology of processes and regulation. *Plant. Cell* **2011**, *23*, 873–894. [CrossRef]

10. Hu, R.; Qi, G.; Kong, Y.; Kong, D.; Gao, Q.; Zhou, G. Comprehensive Analysis of NAC Domain Transcription Factor Gene Family in *Populustrichocarpa*. *BMC Plant Biol.* **2010**, *10*, 1–23. [CrossRef]
11. Li, X.; Li, L.; Liu, X.; Zhang, B.; Zheng, W.; Ma, W. Analysis of physiological chaaracteristics of abscisic acid sensitivity and salt resistance in Arabidopsis ANAC mutants (*ANAC019*, *ANAC072* and *ANAC055*). *Biotechnol. Biotechnol. Equip.* **2012**, *26*, 2966–2970. [CrossRef]
12. Morishita, T.; Kojima, Y.; Maruta, T.; Nishizawa-Yokoi, A.; Yabuta, Y.; Shigeoka, S. Arabidopsis NAC Transcription Factor, *ANAC078*, Regulates Flavonoid Biosynthesis under High-light. *Plant Cell Physiol.* **2009**, *50*, 2210–2222. [CrossRef]
13. Wu, Y.; Deng, Z.; Lai, J.; Zhang, Y.; Yang, C.; Yin, B. Dual function of Arabidopsis *ATAF1* in abiotic and biotic stress responses. *Cell Res.* **2009**, *19*, 1279–1290. [CrossRef] [PubMed]
14. Zhang, X.; Cheng, Z.; Zhao, K.; Yao, W.; Sun, X.; Jiang, T. Functional characterization of poplar *NAC13* gene in salt tolerance. *Plant Sci.* **2019**, *281*, 1–8. [CrossRef]
15. Lu, X.; Zhang, X.; Duan, H.; Lian, C.; Liu, C.; Yin, W. Three stress-responsive NAC transcription factors from *Populuseuphratica* differentially regulate salt and drought tolerance in transgenic plants. *Physiol. Plant.* **2018**, *162*, 73–97. [CrossRef] [PubMed]
16. Yao, W.; Zhao, K.; Cheng, Z.; Li, X.; Zhou, B.; Jiang, T. Transcriptome Analysis of Poplar Under Salt Stress and Over-Expression of Transcription Factor *NAC57* Gene Confers Salt Tolerance in Transgenic *Arabidopsis*. *Front. Plant Sci.* **2018**, *9*, 1121. [CrossRef] [PubMed]
17. Pinheiro, G.L.; Marques, C.S.; Costa, M.D.B.L.; Reis, P.A.B.; Alves, M.S.; Carvalho, C.M. Complete inventory of soybean NAC transcription factors: Sequence conservation and expression analysis uncover their distinct roles in stress response. *Gene* **2009**, *444*, 10–23. [CrossRef]
18. Yao, W.; Li, C.; Lin, S.; Wang, J.; Zhou, B.; Jiang, T. Transcriptome analysis of salt-responsive and wood-associated NACs in *Populus simonii* × *Populus nigra*. *BMC Plant Biol.* **2020**, *20*, 317. [CrossRef] [PubMed]
19. Puranik, S.; Sahu, P.P.; Srivastava, P.S.; Prasad, M. NAC proteins: Regulation and role in stress tolerance. *Trends Plant Sci.* **2012**, *17*, 369–381. [CrossRef]
20. Andersson, R.; Sandelin, A. Determinants of enhancer and promoter activities of regulatory elements. *Nat. Rev. Genet.* **2020**, *21*, 71–87. [CrossRef] [PubMed]
21. Croft, H.; Chen, J.M.; Wang, R. The global distribution of leaf chlorophyll content. *Remote Sens. Environ.* **2020**, *236*, 111479. [CrossRef]
22. Wu, X.; Jia, Q.; Ji, S. Gamma-aminobutyric acid (GABA) alleviates salt damage in tomato by modulating Na^+ uptake, the GAD gene, amino acid synthesis and reactive oxygen species metabolism. *BMC Plant Biol.* **2020**, *20*, 1–21. [CrossRef]
23. Han, D.; Zhang, Z.; Ding, H.; Chai, L.; Liu, W.; Li, H. Isolation and characterization of *MbWRKY2* gene involved in enhanced drought tolerance in transgenic tobacco. *J. Plant Interact.* **2018**, *13*, 163–172. [CrossRef]
24. Jin, C.; Li, K.; Xu, X.; Zhang, H.P.; Chen, H.X. A Novel NAC Transcription Factor, *PbeNAC1*, of *Pyrusbetulifolia* Confers Cold and Drought Tolerance via Interacting with PbeDREBs and Activating the Expression of Stress-Responsive Genes. *Front. Plant Sci.* **2017**, *8*, 1049. [CrossRef] [PubMed]
25. Zhao, Z.; Zhang, G.; Zhou, S.; Ren, Y.; Wang, W. The improvement of salt tolerance in transgenic tobacco by overexpression of wheat F-box gene *TaFBA1*. *Plant Sci.* **2017**, *259*, 71–85. [CrossRef]
26. Yang, T.; Xu, Z.; Lv, R.; Zhu, L.; Peng, Q.; Qiu, L. N gene enhances resistance to Chili veinal mottle virus and hypersensitivity to salt stress in tobacco. *J. Plant Physiol.* **2018**, *230*, 92–100. [CrossRef]
27. Liang, W.J.; Ma, X.L.; Wan, P.; Liu, L.Y. Plant salt-tolerance mechanism: A review. *Biochem. Biophys. Res. Commun.* **2018**, *495*, 286–291. [CrossRef]
28. Sedaghatmehr, M.; Thirumalaikumar, V.P.; Kamranfar, I.; Marmagne, A.; Masclaux-Daubresse, C.; Balazadeh, S. A regulatory role of autophagy for resetting the memory of heat stress in plants. *Plant Cell Environ.* **2019**, *42*, 1054–1064. [CrossRef] [PubMed]
29. Christianson, J.A.; Dennis, E.S.; Llewellyn, D.J.; Wilson, I.W. ATAF NAC transcription factors: Regulators of plant stress signaling. *Plant Signal.Behav.* **2010**, *5*, 428–432. [CrossRef]
30. Hong, Y.; Zhang, H.; Huang, L.; Li, D.; Song, F. Overexpression of a Stress-Responsive NAC Transcription Factor Gene *ONACO22* Improves Drought and Salt Tolerance in Rice. *Front. Plant Sci.* **2016**, *7*, 4. [CrossRef]
31. Shen, S.; Zhang, Q.; Shi, Y.; Sun, Z.; Zhang, Q.; Hou, S. Genome-Wide Analysis of the NAC Domain Transcription Factor Gene Family in Theobroma cacao. *Genes* **2020**, *11*, 35. [CrossRef] [PubMed]
32. Wang, L.; Li, Z.; Lu, M.; Wang, Y. *ThNAC13*, a NAC Transcription Factor from Tamarix hispida, Confers Salt and Osmotic Stress Tolerance to Transgenic Tamarix and Arabidopsis. *Front. Plant Sci.* **2017**, *8*, 635. [CrossRef] [PubMed]
33. Yang, X.; He, K.; Chi, X.; Chai, G.; Wang, Y.; Jia, C. Miscanthus NAC transcription factor *MlNAC12* positively mediates abiotic stress tolerance in transgenic Arabidopsis. *Plant Sci.* **2018**, *277*, 229–241. [CrossRef]
34. Sukiran, N.L.; Ma, J.C.; Ma, H.; Su, Z. *ANAC019* is required for recovery of reproductive development under drought stress in Arabidopsis. *Plant Mol. Biol.* **2019**, *99*, 161–174. [CrossRef]
35. Shahnejat-Bushehri, S.; Mueller-Roeber, B.; Balazadeh, S. Arabidopsis NAC transcription factor *JUNGBRUNNEN1* affects thermomemory-associated genes and enhances heat stress tolerance in primed and unprimed conditions. *Plant Signal. Behav.* **2012**, *7*, 1518–1521. [CrossRef]
36. He, L.; Shi, X.; Wang, Y.; Guo, Y.; Yang, K.; Wang, Y. Arabidopsis *ANAC069* binds to C A/G CG T/G sequences to negatively regulate salt and osmotic stress tolerance. *Plant Mol. Biol.* **2017**, *93*, 369–387. [CrossRef] [PubMed]

37. Deng, R.; Zhao, H.; Xiao, Y.; Huang, Y.; Yao, P.; Lei, Y. Cloning, Characterization, and Expression Analysis of Eight Stress-Related NAC Genes in Tartary Buckwheat. *Crop Sci.* **2019**, *59*, 266–279. [CrossRef]
38. Pang, X.; Xue, M.; Ren, M.; Nan, D.; Wu, Y.; Guo, H. Ammopiptanthus mongolicus stress-responsive NAC gene enhances the tolerance of transgenic Arabidopsis thaliana to drought and cold stresses. *Genet. Mol. Biol.* **2019**, *42*, 624–634. [CrossRef]
39. Liu, S.; Liu, C.; Wang, X. Seed-specific activity of the *Arabidopsis* β-glucosidase 19 promoter in transgenic *Arabidopsis* and tobacco. *Plant Cell Rep.* **2020**, *40*, 213–221. [CrossRef]
40. Huang, H.; Ullah, F.; Zhou, D.X.; Yi, M.; Zhao, Y. Mechanisms of ROS regulation of plant development and stress responses. *Front. Plant Sci.* **2019**, *10*, 800. [CrossRef]
41. Kong, W.; Liu, F.; Zhang, C.; Zhang, J.; Feng, H. Non-destructive determination of Malondialdehyde (MDA) distribution in oilseed rape leaves by laboratory scale NIR hyperspectral imaging. *Sci. Rep.* **2016**, *6*, 1–8. [CrossRef] [PubMed]
42. Schieber, M.; Chandel, N.S. ROS Function in Redox Signaling and Oxidative Stress. *Curr. Biol.* **2014**, *24*, R453–R462. [CrossRef] [PubMed]
43. Peng, J.; Li, Z.; Wen, X.; Li, W.; Shi, H.; Yang, L.; Zhu, H. Salt-Induced Stabilization of EIN3/EIL1 Confers Salinity Tolerance by Deterring ROS Accumulation in *Arabidopsis*. *PLoS Genet.* **2014**, *10*, e1004664. [CrossRef]
44. Ábrahám, E.; Hourton-Cabassa, C.; Erdei, L. Methods for determination of proline in plants. *Plant Stress Toler.* **2010**, 317–331. [CrossRef]
45. Szymanska, K.P.; Polkowska-Kowalczyk, L.; Lichocka, M.; Maszkowska, J.; Dobrowolska, G. SNF1-Related Protein Kinases SnRK2.4 and SnRK2.10 Modulate ROS Homeostasis in Plant Response to Salt Stress. *Int. J. Mol. Sci.* **2019**, *20*, 143. [CrossRef]
46. Zafar, S.A.; Hameed, A.; Ashraf, M.; Khan, A.S.; Ziaul, Q.; Li, X. Agronomic, physiological and molecular characterisation of rice mutants revealed the key role of reactive oxygen species and catalase in high-temperature stress tolerance. *Funct. Plant Biol.* **2020**, *47*, 440–453. [CrossRef] [PubMed]
47. Ambastha, V.; Chauhan, G.; Tiwari, B.S.; Tripathy, B.C. Execution of programmed cell death by singlet oxygen generated inside the chloroplasts of *Arabidopsisthaliana*. *Protoplasma* **2020**, *257*, 841–851. [CrossRef]
48. Dashti, A.; Khan, A.A.; Collins, J.C. Effects of Salinity on Growth, Ionic Relations and Solute Content of *SorghumBicolor* (L.) Monench. *J. Plant Nutr.* **2009**, *32*, 1219–1236. [CrossRef]
49. Arif, Y.; Singh, P.; Siddiqui, H.; Bajguz, A.; Hayat, S. Salinity induced physiological and biochemical changes in plants: An omic approach towards salt stress tolerance. *Plant Physiol. Biochem.* **2020**, *156*, 64–77. [CrossRef]
50. Qiu, Q.S.; Barkla, B.J.; Vera-Estrella, R.; Zhu, J.K.; Schumaker, K.S. Na^+/H^+ exchange activity in the plasma membrane of Arabidopsis. *Plant Physiol.* **2003**, *132*, 1041–1052. [CrossRef] [PubMed]
51. Li, W.; Dang, C.; Ye, Y.; Wang, Z.; Hu, L.; Zhang, F. Overexpression of Grapevine *VvIAA18* Gene Enhanced Salt Tolerance in Tobacco. *Int. J. Mol. Sci.* **2020**, *21*, 1323. [CrossRef]
52. Aziz, E.; Batool, R.; Akhtar, W.; Rehman, S.; Gregersen, P.L.; Mahmood, T. Expression analysis of the polyphenol oxidase gene in response to signaling molecules, herbivory and wounding in antisense transgenic tobacco plants. *3 Biotech* **2019**, *9*. [CrossRef]
53. Singh, N.K.; Shukla, P.; Kirti, P.B. A CBL-interacting protein kinase AdCIPK5 confers salt and osmotic stress tolerance in transgenic tobacco. *Sci. Rep.* **2020**, *10*. [CrossRef]
54. Li, X.; Zhuang, K.; Liu, Z.; Yang, D.; Ma, N.; Meng, Q. Overexpression of a novel NAC-type tomato transcription factor, *SlNAM1*, enhances the chilling stress tolerance of transgenic tobacco. *J. Plant Physiol.* **2016**, *204*, 54–65. [CrossRef] [PubMed]
55. Liu, Q.; Zhong, M.; Li, S.; Pan, Y.; Jiang, B.; Jia, Y. Overexpression of a chrysanthemum transcription factor gene, *DgWRKY3*, in tobacco enhances tolerance to salt stress. *Plant Physiol. Biochem.* **2013**, *69*, 27–33. [CrossRef] [PubMed]
56. Ma, X.; Zhang, B.; Liu, C.; Tong, B.; Guan, T.; Xia, D. Expression of a populus histone deacetylase gene *84KHDA903* in tobacco enhances drought tolerance. *Plant Sci.* **2017**, *265*, 1–11. [CrossRef] [PubMed]
57. Cheng, Z.; Zhang, X.; Zhao, K.; Zhou, B.; Jiang, T. Ectopic expression of a poplar gene *NAC13* confers enhanced tolerance to salinity stress in transgenic *Nicotiana tabacum*. *J. Plant Res.* **2020**, *133*, 727–737. [CrossRef] [PubMed]
58. Yao, W.; Wang, L.; Zhou, B.; Wang, S.; Li, R.; Jiang, T. Over-expression of poplar transcription factor *ERF76* gene confers salt tolerance in transgenic tobacco. *J. Plant Physiol.* **2016**, *198*, 23–31. [CrossRef]
59. Yao, W.; Wang, S.; Zhou, B.; Wang, J.; Jiang, T. Characterization of *ERF76* promoter cloned from *Populus simonii* × *P. nigra*. *Acta Physiol. Plant* **2017**, *39*. [CrossRef]
60. Shivakrishna, P.; Reddy, K.A.; Rao, D.M. Effect of PEG-6000 imposed drought stress on RNA content, relative water content (RWC), and chlorophyll content in peanut leaves and roots. *Saudi J. Biol. Sci.* **2018**, *25*, 285–289.

International Journal of
Molecular Sciences

MDPI

Article

Haplotype Analysis of *BADH1* by Next-Generation Sequencing Reveals Association with Salt Tolerance in Rice during Domestication

Myeong-Hyeon Min [1,†], Thant Zin Maung [1,†], Yuan Cao [1], Rungnapa Phitaktansakul [1], Gang-Seob Lee [2], Sang-Ho Chu [3], Kyu-Won Kim [2] and Yong-Jin Park [1,3,*]

1 Department of Plant Resources, College of Industrial Science, Kongju National University, Yesan 32439, Korea; mmh7272@gmail.com (M.-H.M.); tzmaung.yau2009@gmail.com (T.Z.M.); yuancao2017@gmail.com (Y.C.); phitaktansakul2017@gmail.com (R.P.)

2 Agricultural Biotechnology Department, National Institute of Agricultural Sciences, Rural Development Administration, Jeonju 54874, Korea; kangslee@korea.kr (G.-S.L.); kyuwonkim@kongju.ac.kr (K.-W.K.)

3 Center of Crop Breeding on Omics and Artificial Intelligence, Kongju National University, Yesan 32439, Korea; sanghochu76@gmail.com

* Correspondence: yjpark@kongju.ac.kr; Tel.: +82-41-330-1201; Fax: +82-41-330-12

† These authors contributed equally to this work.

Citation: Min, M.-H.; Maung, T.Z.; Cao, Y.; Phitaktansakul, R.; Lee, G.-S.; Chu, S.-H.; Kim, K.-W.; Park, Y.-J. Haplotype Analysis of *BADH1* by Next-Generation Sequencing Reveals Association with Salt Tolerance in Rice during Domestication. *Int. J. Mol. Sci.* **2021**, *22*, 7578. https://doi.org/ijms22147578

Academic Editors: Mirza Hasanuzzaman and Masayuki Fujita

Received: 29 April 2021
Accepted: 8 July 2021
Published: 15 July 2021

Publisher's Note: MDPI stays neutral with regard to jurisdictional claims in published maps and institutional affiliations.

Abstract: Betaine aldehyde dehydrogenase 1 (*BADH1*), a paralog of the fragrance gene *BADH2*, is known to be associated with salt stress through the accumulation of synthesized glycine betaine (GB), which is involved in the response to abiotic stresses. Despite the unclear association between *BADH1* and salt stress, we observed the responses of eight phenotypic characteristics (germination percentage (GP), germination energy (GE), germination index (GI), mean germination time (MGT), germination rate (GR), shoot length (SL), root length (RL), and total dry weight (TDW)) to salt stress during the germination stage of 475 rice accessions to investigate their association with *BADH1* haplotypes. We found a total of 116 SNPs and 77 InDels in the whole *BADH1* gene region, representing 39 haplotypes. Twenty-nine haplotypes representing 27 mutated alleles (two InDels and 25 SNPs) were highly ($p < 0.05$) associated with salt stress, including the five SNPs that have been previously reported to be associated with salt tolerance. We observed three predominant haplotypes associated with salt tolerance, Hap_2, Hap_18, and Hap_23, which were Indica specific, indicating a comparatively high number of rice accessions among the associated haplotypes. Eight plant parameters (phenotypes) also showed clear responses to salt stress, and except for MGT (mean germination time), all were positively correlated with each other. Different signatures of domestication for *BADH1* were detected in cultivated rice by identifying the highest and lowest Tajima's D values of two major cultivated ecotypes (Temperate Japonica and Indica). Our findings on these significant associations and *BADH1* evolution to plant traits can be useful for future research development related to its gene expression.

Keywords: betaine aldehyde dehydrogenase 1 (BADH1); salt stress; domestication; cultivated rice; wild rice

1. Introduction

Soil salinity has become important as one of the major constraints affecting rice production worldwide, and accordingly, breeding approaches using marker-assisted selection or genetic engineering to produce salt-tolerant varieties are also being developed [1]. Soil salinization is a serious problem in rice cultivation [2], especially at the germination stage.

A homologous gene of betaine aldehyde dehydrogenase 2, also known as *BADH1*, was reported to be a candidate gene with a close correlation with salt tolerance [3]. Glycine betaine (GB) has been reported as an osmo-protectant compound [1] that can be synthesized in many living organisms, including plants and animals, in response to abiotic stresses, such as salt, drought, and temperature [4]. Glycine betaine can protect protein structure and enzyme activity and stabilize membranes to establish osmotic and ionic stress in plants [3]. GB is also widely distributed in bacteria, algae, and higher plants such as sugar beet and cotton [3,5]. In higher plants, GB can be synthesized by the two-step oxidation of choline by ferritin-dependent choline monooxygenase (CMO) and betaine aldehyde dehydrogenase (*BADH*). The enzyme betaine aldehyde dehydrogenase (*BADH*) has been reported to be responsible for GB synthesis, and many plant species have been identified as potential GB accumulators [4].

However, although some reports on *BADH1* suggested its association with salt tolerance, the function of *BADH1* is unclear, and there are even conflicting reports from various studies on the relationship between *BADH1* and salt stress tolerance in rice. The possibility has been considered that GB cannot accumulate in rice due to the lack of a functional CMO gene [6]. Protein modified by SNP substitutions of the *BADH1* gene indicated its specific association with aroma rather than salt tolerance [7]. Despite the unclear and conflicting reports on the relationship between *BADH1* and salt stress tolerance in rice, there are some indirect findings regarding the relationship. The gene expression of *BADH1* in rice indicates that the gene encodes a key enzyme for GB biosynthesis and is closely related to salt tolerance [8]. Increased levels of *BADH1* transcripts in salt-stressed Japonica and Indica nonfragrant rice varieties [9,10] suggest the possible association of the *BADH1* gene with salt tolerance through an undetermined mechanism.

Recent developments in sequencing technology have simplified the analyses of single nucleotide polymorphisms (SNPs) and insertions and deletions (InDels), which are the basis for allele differentiation. Haplotype analysis, which includes a set of linked SNPs, is more informative than the analysis of a single SNP in determining the associations with phenotypes [7]. Extensive research on the *BADH1* gene has not yet been conducted at the genetic and molecular levels [11], and only seven SNPs in certain exons (Supplementary Figure S1) have been reported [12].

Here, to confirm the association between *BADH1* and salt tolerance, we used sophisticated sequencing technology (1) to investigate the genetic diversity of *BADH1*, (2) to examine the haplotype variation within the gene region of *BADH1*, and (3) to observe the association between the main haplotypes of *BADH1* and salt tolerance at the germination stage of tested rice accessions. We also conducted population genetic studies, such as nucleotide diversity, population structure, and Tajima's D, and phylogenetic studies.

2. Results

2.1. Discovery of Genetic Variations in BADH1

To investigate how many variations and types occurred, we conducted variant calling of 475 rice accessions and extracted all the variants within the gene region of *BADHI* by using VCFtools. The results revealed four different types of variants, single nucleotide polymorphism (SNP), insertion (Ins), deletion (Del), and structure variation (SV), and we identified variant numbers for the classified subpopulations of cultivated rice and wild rice (Table 1). According to the summarized number, we observed that SNPs represented the highest number of variants for all classified subpopulations, among which the wild showed the highest number. Among the cultivated subpopulations, Indica and Aus had the same number of SNPs (23), representing the highest value except when compared with the wild rice (105). The wild rice group showed a higher number of identified variants, 38 insertions, 36 deletions, and 2 structural variations (the detailed observed number of variants for each wild rice accession is provided in Supplementary Table S2). Overall, we noticed that the wild rice had a higher number of different variants than any of the cultivated subpopulations.

Table 1. Summary of the total variations (SNPs, InDels, and SV) detected in the *BADH1* gene region of 421 cultivated and 54 wild rice accessions. Te: Temperate; Tr: Tropical.

Ecotype	SNP	Insertion	Deletion	Structure Variation	Total Accessions
Te_Japonica	0	1	0	0	279
Tr_Japonica	1	1	0	0	26
Indica	23	2	1	0	102
Aus	23	2	0	0	9
Aromatic	0	1	0	0	2
Admixture	15	2	0	0	3
Wild	105	38	36	2	54

2.2. *Population Structure, Principal Component Analysis (PCA), and Fixation Index (F_{ST} Test) of BADH1*

To verify the subgroups (wild and cultivated subgroups) followed by genotypic variants, we conducted a Bayesian Analysis of Population Structure (BAPS) version 6.0 and PCA. The structure was implemented by increasing K values from 2 to 7 (Figure 1A). Except for K = 2, the cultivated rice was clearly separated from the wild, but their subpopulations were mixed in K values of 3 and 4. Temperate Japonica and Tropical Japonica were mixing always in all K values. At K = 3 and 4, Indica was also admixed with the Japonica group, showing its mixed structure with other minor cultivated subgroups (Aus, Aromatic, and Admixture). All the cultivated subpopulations and wild subpopulation were clearly separated from each other at K values 5, 6, and 7, but the internal subgroups of the wild variants were mixing while the others' internal subgroups were separated. Overall, the wild had internal subgroups at every K value in spite of their clear separation from cultivated subgroups.

We conducted a PCA analysis to multivariate the original variant datasets of *BADH1* into a dimensional scaling display (Figure 1B) and observed the associations between the classified ecotypes. According to the display, the wild, Indica, and Temperate Japonica should have been grouped separately but in actuality were not at all. Some Indica were mixed with Tropical Japonica, Admixture and Aus while Temperate Japonica were mixed with Aromatic, some wild, and Indica. Overall, within the cultivated subpopulations, the associations were greatly distant compared with their distance from wild rice.

To make sure the population analysis and further finding, we used a sophisticated method, F_{ST} value (fixation index), which indicates the genetic distance or differentiation between the populations or subpopulations. We calculated F_{ST} values for the classified ecotypes of cultivated rice and checked their distances from the wild within the *BADHI* region. The highest F_{ST} value (0.6786) was found between Temperate Japonica and Tropical Japonica, followed by the pair, Temperate Japonica–Indica (0.6062), while the lowest was represented between Tropical Japonica and the wild (0.0376) (Figure 1C). F_{ST} values between the cultivated subpopulations (both Japonica, Indica) are higher than those between cultivated and wild rice.

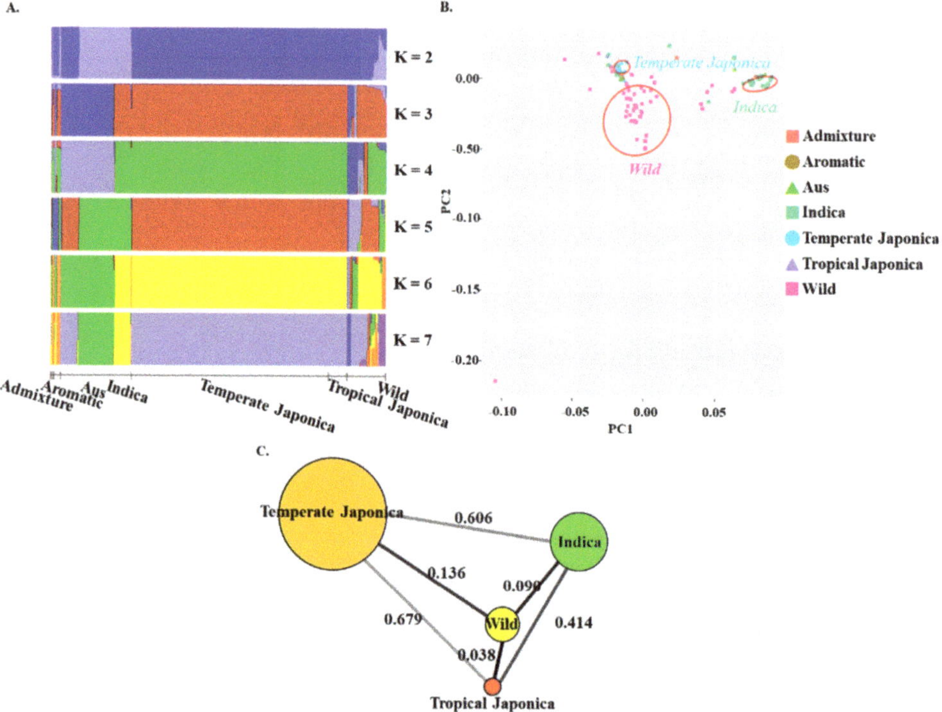

Figure 1. Population structure and diversity in *BADH1* (*Os04g0464200*). (**A**) The population structure of all subgroups based on the *BADH1* region using fastStructure to estimate individual ancestry and admixture proportions assuming K populations. With increasing K (number of populations) values from 1 to 10 with 10 iterations each, we analyzed the population structure for each K value (from K = 2 to 7). (**B**) Principal component analysis (PCA) analysis of seven classified ecotypes. Indica and Japonica clustered into clearly defined groups, and other subgroups were located around the two groups. PCA was performed using TASSEL and plotted with ggplot2. (**C**) F_{ST} values between the four classified ecotypes, indicating circle size by the number of accessions. The dark line between each pair represents their genetic distance.

2.3. Genetic Diversity of BADH1

We calculated the nucleotide diversity value of *BADHI* in 475 rice accessions to investigate the degree of polymorphisms at different segregating sites by means of comparing different genotype sequences. The diversity values were analyzed based on the classified subpopulations/groups to be compared. We also calculated diversity values for the whole rice collection and for major subgroups/ecotypes, namely, Temperate Japonica, Tropical Japonica, and Indica. We omitted other cultivated minor subgroups, aus, aroma, and admixture, due to the very low number of rice accessions. To observe the clear diversity level, we compared the resulting diversity values (π) of classified cultivated ecotypes to those of wild as well as whole accessions. We found that Indica showed its highest diversity value (Figure 2A) across the *BADHI* gene region, while both Japonicas showed the lowest nucleotide diversity. The nucleotide diversity of wild group was between that of the two higher-diversity groups and the two lower-diversity groups. As shown in (Supplementary Table S3A), Indica (0.00435) was much higher in value than the lowest Temperate Japonica (0.00005). The nucleotide diversity of Indica and wild rice were similar each other.

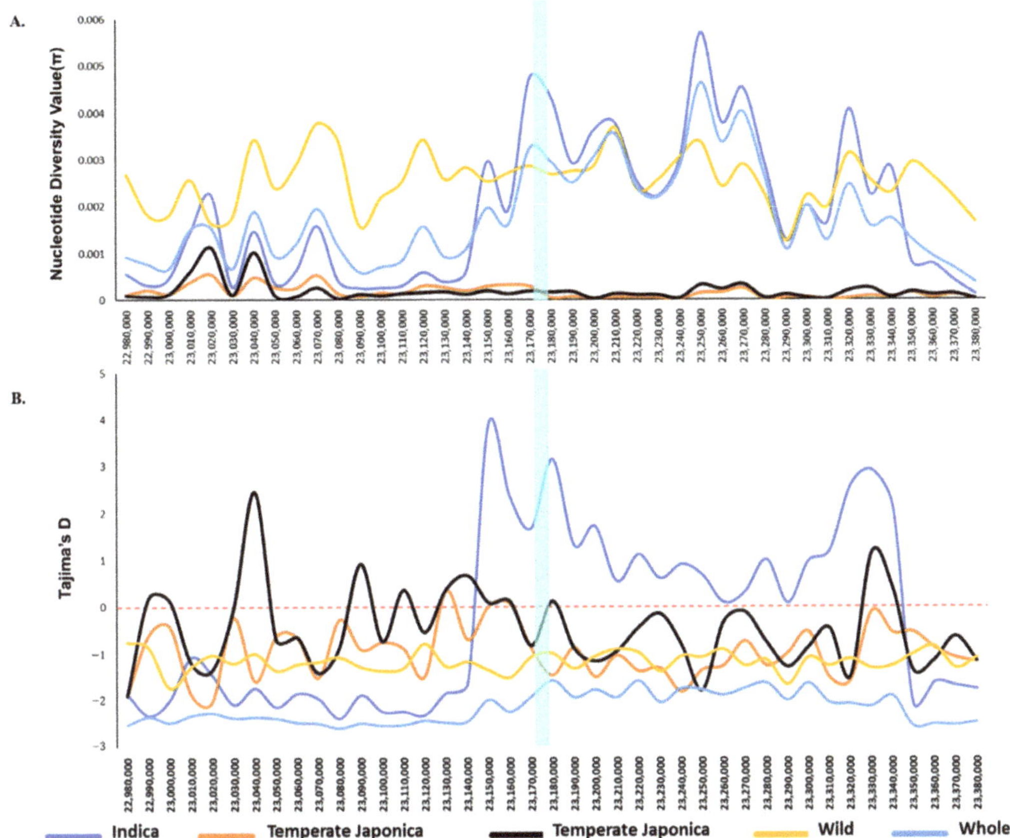

Figure 2. Nucleotide diversity (π) values and Tajima's D values of *BADH1* (*Os04g0464200*) in 475 rice accessions. (**A**) Nucleotide diversity (π) indicating different pi (π) values at segregating sites of different classified groups determined with a 10 kb sliding window within the *BADH1* region (**B**) Tajima's D values indicating different variations among the classified groups identified by a 10 kb sliding window within the *BADH1* region. Cyan color indicates the *BADH1* gene region; each colored line represents different rice groups in terms of ecotypes.

To investigate the differences between the observed nucleotide diversity and expected nucleotide diversity due to selection, we calculated Tajima's D values, which were determined by a pairwise comparison and their segregation sites number for the same classified groups. The calculated values ranged from the lowest, −1.4551 (Temperate Japonica) to the highest, 3.1529 (Indica) (Figure 2B and Supplementary Table S3B). The signatures of two directional selective sweeps were observed for two major ecotypes of cultivated subpopulations. The Tajima's D over *BADH1* for Indica, one of major cultivated subpopulations, was particularly higher than the others, indicating that *BADH1* of Indica had undergone balancing selection; whereas Temperate Japonica, another major type of cultivated rice, showed the lowest value, indicating that *BADH1* of Temperate Japonica had undergone purifying selection. Although the nucleotide diversity of Indica and wild rice is similar, a balancing selection was observed only in Indica.

2.4. Phylogenetic Study of BADH1

We constructed a phylogenetic tree to observe the evolutionary relationships of *BADHI* in 475 rice accessions by their genotypic differences or similarities in terms of ecotypes/subpopulations (Figure 3). This analysis was conducted mainly to observe the evolutionary studies of cultivated rice accessions and their relatedness to different wild accessions based on ecotype. The cultivated subgroups were dispersed across different separated tree branches of wild species. Indica (33.3%) was associated with *O. nivara* as well as with some aus in the same clade. Temperate Japonica was rooted separately and directly from common ancestors. Most of the wild rice was genetically distant from cultivated ecotypes, especially *O. meridionalis*, *O. punctata*, and *O. longistaminata*.

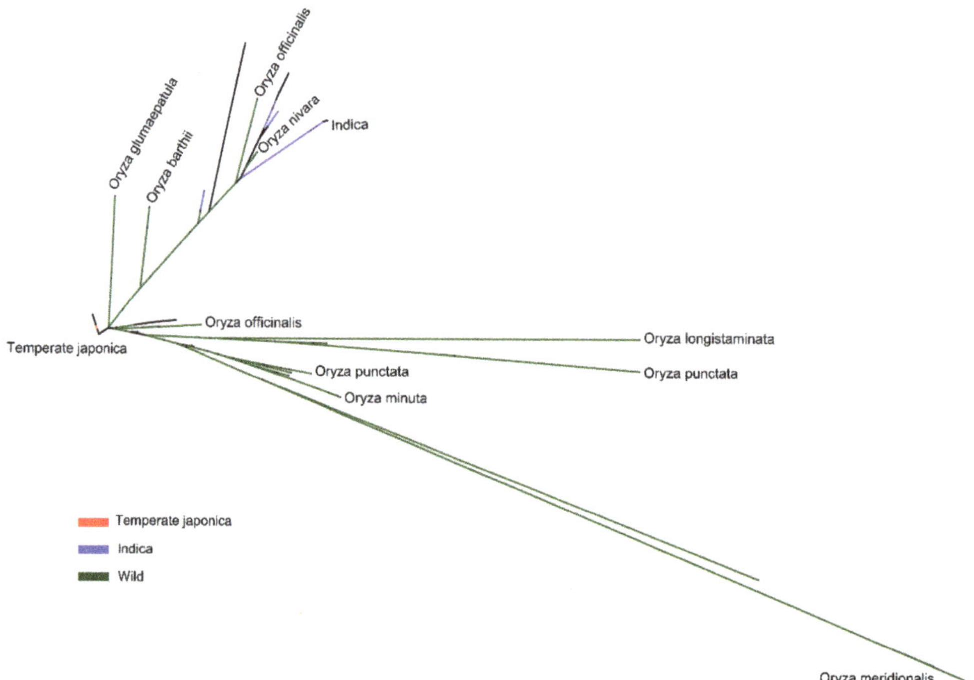

Figure 3. Phylogenetic tree of *BADH1* gene in 475 rice accessions. The classified groups of cultivated rice were considered only in terms of major ecotypes, namely, Indica, Temperate Japonica, and Tropical Japonica, and their phylogenic relationship with wild rice. The tree was constructed for genetically different sequences of the *BADH1* gene, using MEGAX software. The reliability of the neighbor-joining phylogeny output was estimated using bootstrap analysis with 1000 permutations.

2.5. Haplotype Diversity

To identify the association between *BADH1* and salt tolerance at the haplotype level, we conducted haplotype diversity analysis on the whole genomes of 475 rice accessions. We analyzed haplotype diversity only on 421 types of cultivated rice (Supplementary Table S4) first and then compared it to that of 54 types of wild rice (Supplementary Table S5). There were 116 SNPs and 77 InDels covering all the identified exons and introns within the *BADHI* gene region.

In cultivated rice, we verified 39 haplotypes (hereafter referred to as "Hap") representing genetically identified variants, but in Figure 4, we show only functional SNPs (fSNPs) observed in exons (maf < 0.03). Then, we also classified specific rice ecotype (Indica, Temperate Japonica, and Tropical Japonica) for all the rice accessions under each haplotype, and their respective accession numbers were also provided together with their haplotype number. Based on those total identified rice accession under each haplotype, we considered five major haplotypes indicated in the largest numbers of rice accessions, Hap_2, Hap_3, Hap_4, Hap_18, and Hap_23, by more than five accessions. The total number of rice accessions separately represented by each subpopulation were also listed by haplotype; then, we specifically focused on two such two haplotypes, Hap_18 and Hap_23, for their associated functional alleles (mutation sites). Hap_3 was the first major haplotype (62.7% of all cultivated rice) by which all the rice accessions showed the same sequence as the reference. Hap_18 (40 Indica) and Hap_23 (14 Indica) belonged only to Indica, showing the same fSNPs, G/T in exon 11 and A/T in exon 4. Referring to those haplotypes, we investigated wild haplotypes (in this case, haplotyping of wild rice) to examine the genetic variation in *BADHI,* and we found six wild haplotypes showing the same SNP (G/T) as Hap_18, only four of which showed the same SNP (A/T) as Hap_23 (Supplementary Table S5A). In the wild, we found many functional SNPs that addressed almost all 50 verified haplotypes, but only six haplotypes had the same SNP substitutions as those of cultivated haplotypes (Indica). When the cultivated and wild haplotypes were compared in terms of InDel variants, we found no InDel variants in exon regions, but in the intron regions, we found two InDel variants (Supplementary Table S5B).

To determine the genetic association of the *BADHI* gene among the classified subpopulations of cultivated rice and wild rice, we constructed a network using previously identified haplotypes. In this case, we investigated the association of five major cultivated haplotypes (due to their highest number of rice accessions they occupied) with wild rice accessions. We generated 50 haplotypes referring to 54 accessions of 21 different species. Using all-wild haplotypes and selected cultivated haplotypes, we constructed TCS network in the PopART program (Figure 5). Hap_3 and Hap_4 were grouped in the same clade. Two cultivated haplotypes, Hap_2 and Hap_3,4 (combined haplotype for Hap_3 and Hap_4), were closely related by the smallest number of mutational steps. Only one wild haplotype, belonging to *Oryza australiensis−1*, was related to cultivated Hap_3,4 at a closer distance than all others. Interestingly, two cultivated haplotypes, Hap_18 and Hap_23, both belonging only to Indica, were also distantly associated with wild haplotypes and even members of the same group of cultivated haplotypes, Hap_2 and Hap_3,4. All the identified wild haplotypes were genetically far from each other in the gene region of *BADH1*. This relatedness of wild haplotypes indirectly agreed with the discovery of the highest number of different variants (Table 1 and Table S2).

2.6. Screening and Evaluation of Salt Tolerance Phenotypes

BADHI is an important synthetic enzyme involved in the response mechanism to environmental stresses, especially salt tolerance. To perform a deep study of the responses of this gene *BADHI* to plant characteristics, we screened eight major plant parameters, germination percentage (GP), germination energy (GE), germination index (GI), mean germination time (MGT), germination rate (GR), shoot length (SL), root length (RL), and total dry weight (TDW), of 417 cultivated rice plants under a range of salt stress conditions (200 mM NaCl) together with the corresponding control (0 mM NaCl) during their germination stage (Supplementary Table S6). Descriptive statistics were first checked among the calculated mean values of major traits under both conditions (Table 2). Then, pairwise correlations were also checked by Pearson coefficients among the major traits (Supplementary Table S7). The statistical analysis revealed that all the tested phenotypes (parameters), except MGT and TDW, indicated their responses (in terms of lower mean values) to salt treatment compared with their respective values under the control condition. Data ranges were varied based on the trait types, and all these phenotypes (traits) during seed germination were obviously and negatively influenced by salt treatment. The response of each trait to salt treatment was consistent with our previous finding where rice seedlings were negatively affected regarding shoot length (SL), root length (RL), and total dry weight (TDW) by salinity stress [13]. In the case of pairwise comparison by their correlation coefficients under the control treatment, traits such as GP, GE, GI, and GP were positively and significantly correlated with each other, while all four parameters were negatively and significantly correlated with MGT.

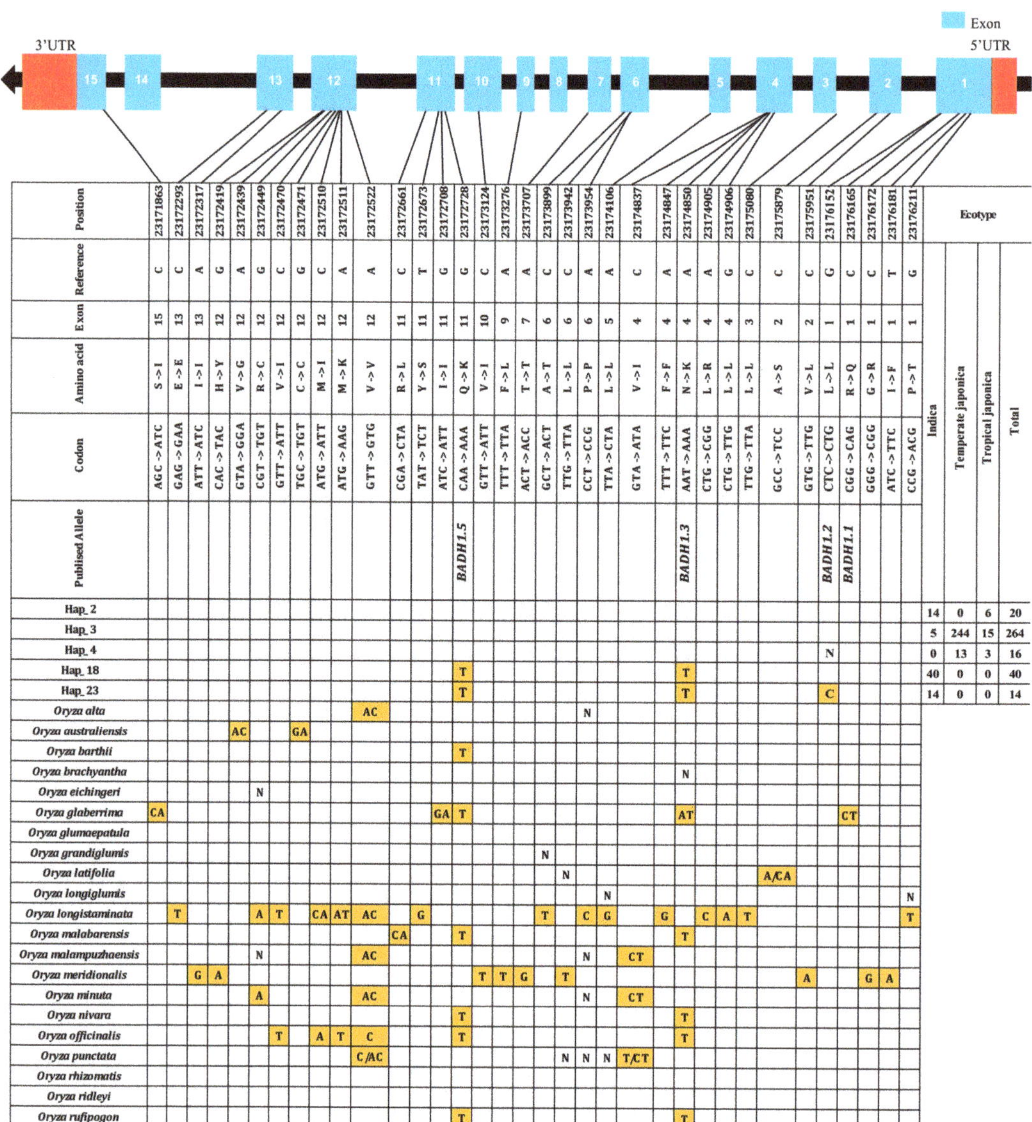

Figure 4. Gene structure and haplotype analysis of *BADH1* (*Os04g0464200*) in gene-coding regions of selected cultivated rice and all wild species. The five cultivated haplotypes representing the highest number of rice accessions were selected, and their genetic association was investigated by SNPs or InDels in wild rice. Here, we used one wild rice accession as a haplotype. Orange-colored SNPs indicate all identified SNPs observed as nonsynonymous and synonymous substitutions. Blank cells refer to the same nucleotide mentioned in reference. "N" indicates the positions of unknown nucleotides. The two-digit number indicated under ecotype (Indica, Temperate Japonica, and Tropical Japonica) refers to the number of rice accessions identified in each haplotype.

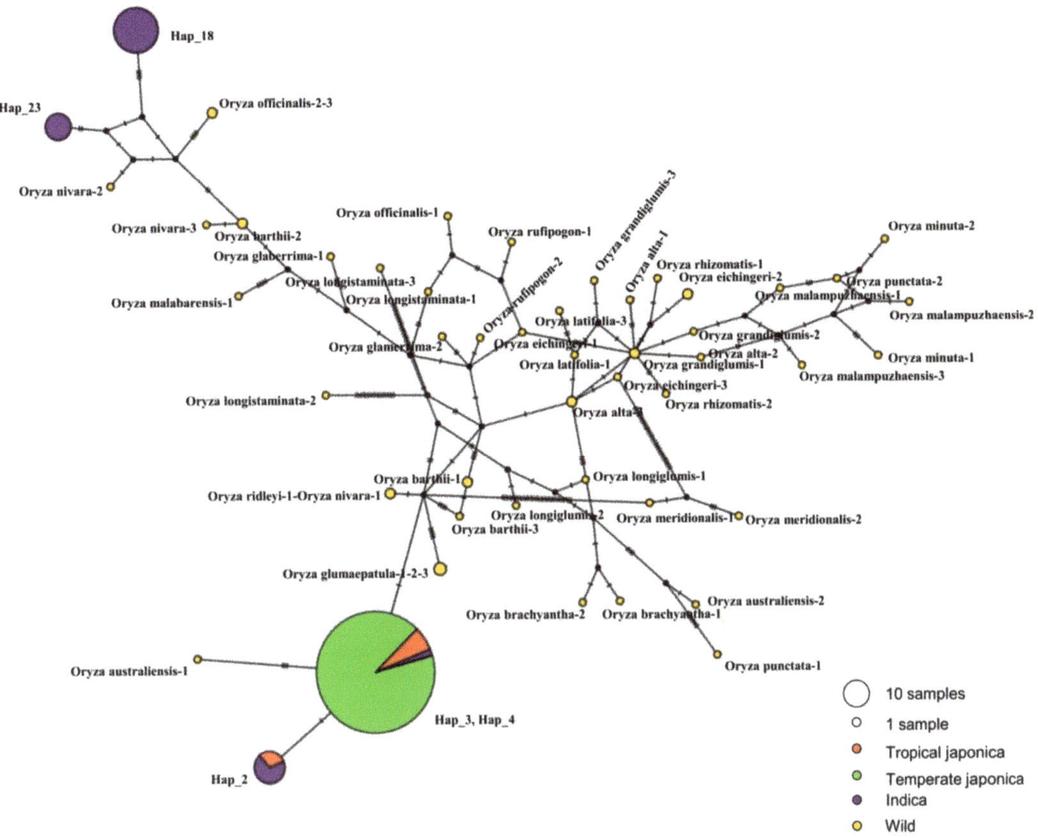

Figure 5. Haplotype network visualizing possible genotypic relationships of major cultivated haplogroups compared to wild rice haplotypes within the *BADH1* region. The size of each circle is proportional to the accession numbers encompassed, and different colors indicate its ecotype. The median vectors indicated by the black circular dot is a hypothetical sequence used to connect the existing similar sequences.

2.7. Test/Control Ratio of Eight Major Plant Parameters

The test/control ratio (the ratio of the measured value in the test condition to the measured value in the control) in the salt tolerance experiment of each trait was calculated from the ratio of recorded values under the test condition (200 mM NaCl) to those of the control (0 mM NaCl). We screened the calculated relative values of all eight phenotypic parameters and analyzed their significant differences among ecotypes (Figure 6A–H) and Supplementary Table S9A). The analyzed values revealed that almost all plant parameters had the same trend of response to salt treatment among the ecotypes. For example, there were no significant differences in the relative values of GE, GI, GR, and MGT (Figure 6B–E), but in TDW, Temperate Japonica had a significant difference from with the other ecotypes (Figure 6H). In the case of RL and SL, Japonica ecotypes had higher traits in response to salt treatment (Figure 6F,G), and again in GP, Japonica plants were significantly different from Aromatic, Aus, and Indica (Figure 6A). Although ecotype groups did not differ from those of each other in some traits, if they were observed to be different, Japonica ecotypes were mostly significantly or comparatively differentiated from others.

Table 2. Descriptive statistics for the mean value of the traits in the control and salt-treated (200 mM) rice accessions. GP: Germination Percentage; GE: Germination Energy; GI: Germination Index; GMT: Mean Germination Time; GR: Germination Rate; SL: Shoot Length; RL: Root Length; TDW: Total Dry Weight; SD: Standard Deviation; IQR: Interquartile Range.

Trait	Salinity Level (mM)	Mean ± SD	Range	Median	IQR
GP	200	21.57 ± 7.94	0–30	24.33	29.517
	0	28.47 ± 2.72	0–30	29	29.585
GE	200	3.17 ± 5.37	0–27	0.33	28.0218
	0	22.93 ± 10.19	0–30	28	48.131
GI	200	3.55 ± 1.84	0–9.24	3.59	7.5618
	0	8.49 ± 2.28	0–14.83	9.27	13.21
MGT	200	7.31 ± 1.57	3.34–10	7.24	6.7405
	0	3.87 ± 1.83	2–10	3.14	6.1583
GR	200	0.14 ± 0.03	0.1–0.3	0.14	0.2735
	0	0.29 ± 0.07	0.1–0.5	0.32	0.4516
SL	200	0.36 ± 0.27	0.19–2.34	0.25	2.1924
	0	10.13 ± 2.93	0.98–19.43	9.89	15.911
RL	200	0.87 ± 0.51	0.2–8.11	0.81	6.7159
	0	5.92 ± 1.31	1.86–11.01	5.67	7.854
TDW	200	0.68 ± 0.08	0.36–1.05	0.68	0.9325
	0	0.51 ± 0.07	0.26–0.78	0.5	0.695

2.8. Association of BADH1 Haplotypes and Plant Parameters under Salt Stress

Haplotype analysis of all cultivated rice is presented above. Here, we analyzed the association between the phenotype data and the gene region of *BADH1* in a general linear model (GLM) by the TASSEL 5 software program. The resulting marker positions were selected based on higher *p*-values. After identifying markers at the *p*-value (<5%) within the gene region of *BADH1*, we found 27 marker positions correlating to recorded plant major traits, and several traits belonged to one marker position (Supplementary Table S8A,B). Those identified marker positions were checked and merged with the previously extracted variants (SNPs/InDels) represented by 39 haplotypes.

In the case of haplotypes that met such identified marker positions for eight major phenotypes, we found that 10 of 39 haplotypes did not have any mutated variants, so only 29 haplotypes were represented by the identified marker positions for major traits (Supplementary Figure S2). Among those 29 haplotypes correlated to the positions identified in the analysis of salt-tolerant phenotypic traits, we observed that there were five major haplotypes, indicating that their highest rice accession numbers were represented by Temperate Japonica, Indica, and Tropical Japonica. We analyzed the associations of these five major haplotypes (Hap_2, Hap_3, Hap_4, Hap_18, and Hap_23) with each of eight previously identified plant parameters (Figure 7A–H and Figure S9B). We used Scheffe's test to indicate the significant difference among the comparison of those haplotypes for each parameter at *p*-values (<0.05). For the plant parameters rGP and rGI, we identified Hap_18 as a significant group, which was significantly affected by salt treatment (Figure 7A,C). Figure 7D,E, representing rMGT and rGR, showed nonsignificant differences in plant responses to salt treatment. Hap_4 (Figure 7F) was significantly different from the others in rRL analysis. For other plant parameters (Figure 7B,G,H), no highly significant responses were found among haplotypes, but Hap_4 was highly significantly different from Hap_2 and Hap_18 in rGE and was highly significantly different from Hap_2, Hap_18, and Hap_23 in rSL analysis. Hap_2 was also highly significantly different from Hap_18 in rTDW. Overall, all the tested plant parameters were relatively influenced by salt stress at the germination stage. Interestingly, among the selected haplotypes to be analyzed for association testing, two haplotypes, Hap_18 and Hap_23, belonged to only Indica rice accessions. Then, according to rGP and rGI, a significant effect of salt was observed in Hap_18.

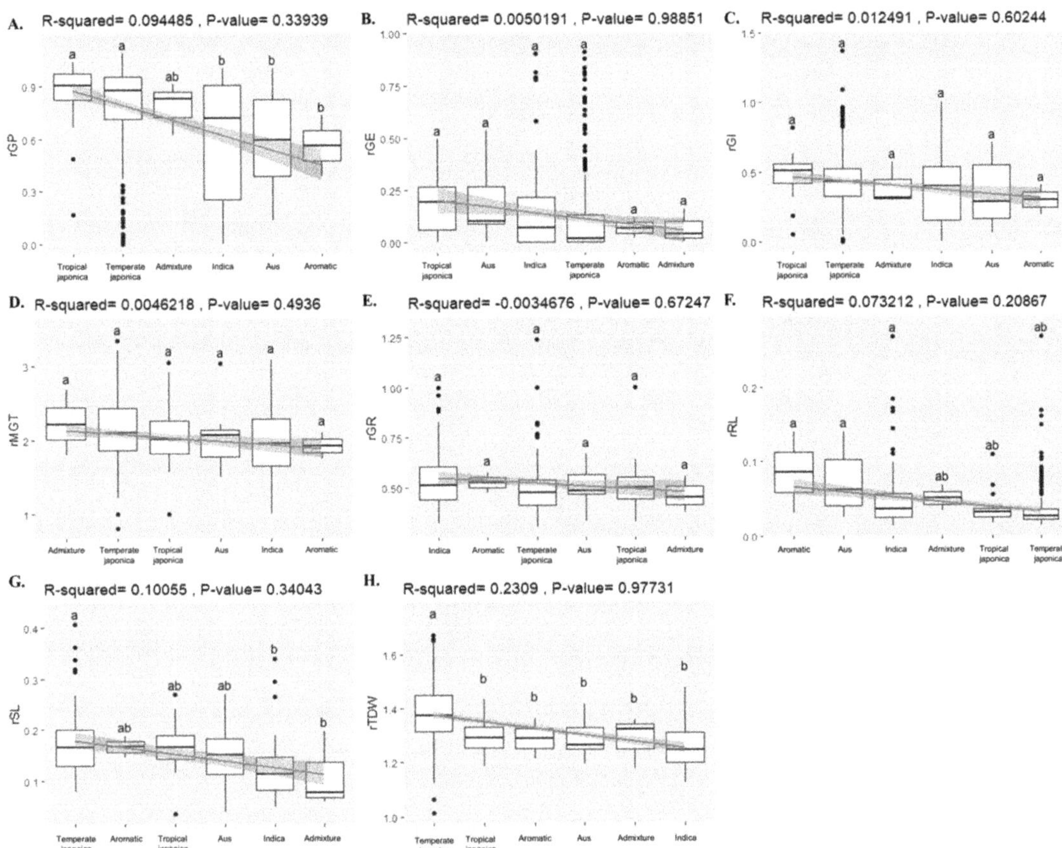

Figure 6. Test/Control ratio of eight major plant parameters distribution group by ecotypes. (**A–H**) Relative variations of salt tolerance-related traits among the classified ecotypes, indicating significant level by statistical test. The order of classified ecotypes was sorted (from left to right) by their relative/statistical mean value (from highest to lowest) of each trait. Each parameter was drawn in a boxplot using 417 rice accession for which phenotype data were surveyed. For each parameter, the analysis was performed at a significant level of p-value < 0.05, and the values were compared among the classified rice ecotypes using Scheffe's test and indicated on the boxplot of each ecotype. Abbreviations: rGP, relative germination percentage; rGE, relative germination energy; rGI, relative germination index; rGMT, relative germination mean time; rGR, relative germination rate; rRL, relative root length; rSL, relativce shoot length; rTDW, relative total dry wight.

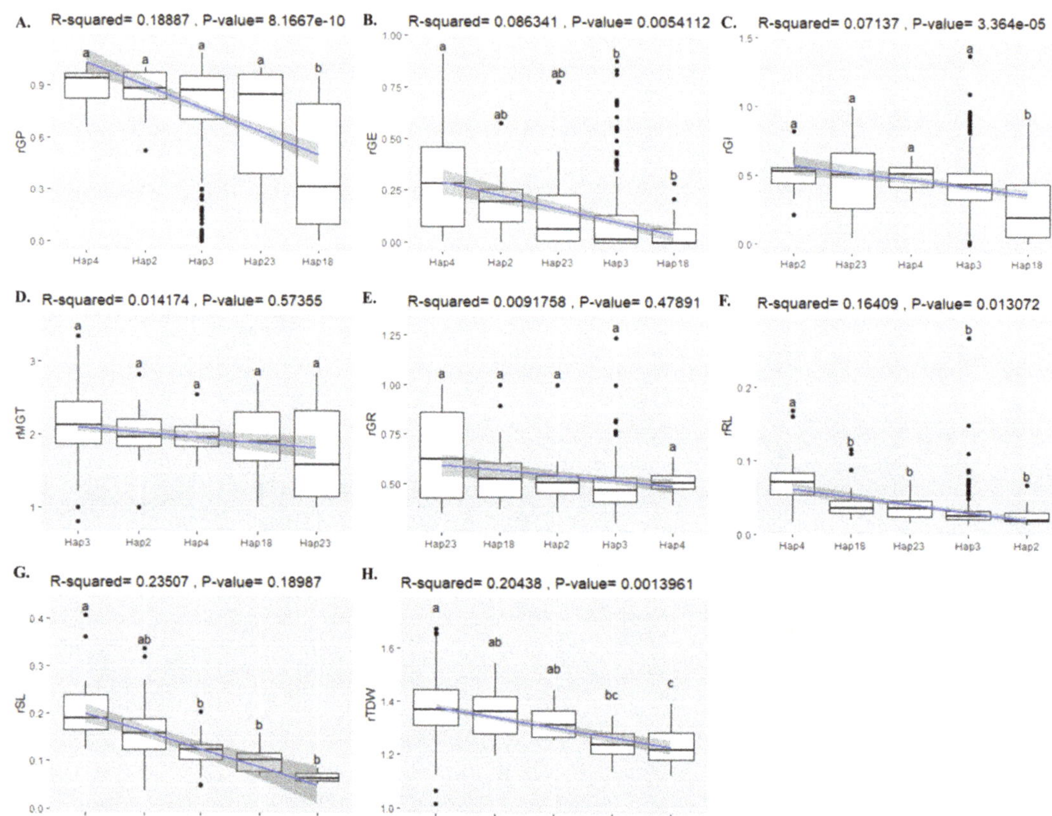

Figure 7. Association of selected *BADH1* haplotypes and eight plant parameters under salt treatment. (**A–H**) Relative variations of salt tolerance-related traits among the selective predominant haplotypes, indicating significant level by statistical test. The order of these predominant haplotypes was sorted (from left to right) by their relative/statistical mean value (from highest to lowest) of each trait. Among the previously identified 39 haplotypes of cultivated rice, only five predominant haplotypes were selected for comparison in association to each salt tolerant parameters due to the rice accession number to which they belonged. For association level to each plant parameter, the selected haplotypes were compared by using Scheffe's test at a significant level (p-value < 0.05). Abbreviations: rGP, relative germination percentage; rGE, relative germination energy; rGI, relative germination index; rGMT, relative germination mean time; rGR, relative germination rate; rRL, relative root length; rSL, relativce shoot length; rTDW, relative total dry wight.

3. Discussion

Whether *BADH1* is mainly associated with aroma or salt stress in rice is unclear, since no association was found between *BADH1* haplotypes and salt tolerance [7], but aromatic rice has a specific association with *BADH1*. A recent paper again indicated that the *BADH1* transcript level was comparatively increased during salt treatment [14] and that it, as a homolog of the *BADH2* gene, can also be induced by environmental factors, such as salt [15]. We also investigated the association of these two orthologous genes, *BADH1* and *BADH2*, by tracing back their ancestral histories in 19 rice species (Supplementary Figure S3). According to phylogenetic display, we noticed that both genes were localized in Japonica rice species but in different clades. As a result of their evolution from a common ancestor, the two *BADH* enzymes are high in sequence homology. However, their transcriptional responses to salt would still be implicated by upregulated expression of *BADH1* to salt and drought stress [16] as well as their specific association with aroma by protein modeling [7].

The inconsistent findings could be attributable to differences either in rice germplasm materials or the growth stages investigated in their studies.

We used whole genome data of 475 Korean rice accessions collected worldwide to investigate the genetic diversity and domestication information on *BADH1* and plant parameters of salt tolerance with an association study. In this case, we picked up only the gene region of *BADH1* (23171516–23176332), in another way, chromosome 4; then, possible genetic variations were discovered by using VCFtools. Variant calling on *BADH1* revealed 116 SNPs, 38 Ins, and 39 Dels, indicating that a higher number of variants were found in wild rice than in cultivated rice. This may be because compared to Asian rice, wild rice has a complex domestication history that was only recently reconstructed [12,17,18]. Haplotype analysis revealed 39 haplotypes covering 116 SNPs and 77 InDels in both exons and introns, including the untranslated region (UTR). Only 27 genetic variants (SNPs and InDels) represented the identified marker positions analyzed by a generalized linear model (GLM) for the association of salt tolerance traits and the *BADH1* gene region (Supplementary Figure S2). Only five SNPs that were localized in exons and all five SNPs have been introduced by previous reports (Supplementary Figure S1). Three SNPs by our previous study were represented by substitutions such as G/C in exon 1, A/G in exon 6, and C/A in exon 12 [19], while two other SNP substitutions, A/T (exon 4) and G/T (exon 11), were discovered by Singh et al. [7]. There remained 22 (two indels and 20 SNPs) genetic variants in introns, which represented 29 haplotypes overall. In terms of ecotypes for those haplotypes, Indica showed the highest accession number, representing three haplotypes (Hap_2, 18, and 23) with 14 new variants (one InDel and 13 SNPs) in introns. Thailand et al. have reported that the expression of the *BADH1* gene in Indica correlates with salt stress and other environmental stresses, such as plasmolysis, temperature, and light [15]. Rice and its major trait domestication have been independent of two different species, African rice (*Oryza glaberrima* Steud) and Asian rice (*Oryza sativa* L.) [20], although there could be a third domestication event by recent archeological evidence in the Amazon [21]. Genomics has produced unprecedented amounts of datasets for deeper insights into domestication studies [20], and whole-genome resequencing of rice DNA could result in tracing back the details of its domestication history by population genomic analysis [22]. Resequencing our previous whole-genome data also revealed different domestication patterns of *BADH1* and *BADH2* [19].

Here, we could find that the clue of our result of the association between genotype/haplotype and salt tolerance was due to the process of adaptation for human use rather than a random process based on two independent analytic results, population structure and selective sweep. As expected, clear separations of classified cultivated subpopulations from wild were observed at most of the K values, especially at 5, 6, and 7, indicating the existence of cultivated ecotypes. The cultivated ecotypes especially between Japonica and Indica were clearly separated from each other by PC1 and PC2 in PCA, and a following analysis of population differentiation via F_{ST} showed a considerably high F_{ST} value between Japonica and Indica, indicating genetic isolation from each other. The general phenomenon of domestication in plants or animals is the reduction of genetic diversity via genetic erosion [23], and different domestication pathways of rice genes have recently been updated through modifications of morphological traits, physiological characteristics, and ecological adaptability from the wild into modern cultivated rice [24]. In our study, we found a relatively high genetic diversity in Indica compared to Japonica rice groups. This may be due to the increase of heterozygotes by human-mediated hybridization during cultivation or breeding programs of Indica rice accessions. There was an interesting finding we noticed: most of the haplotypes were Indica rice accessions that had genetic markers (alleles) associated to salt-tolerant plant parameters. These findings also seemed to be clues of highly diverse speciation of Indica rice. Tajima's D, which signifies selective sweep by observed frequency polymorphisms relative to expectation, showed the lowest value for Temperate Japonica and the highest value for Indica, suggesting the opposite directional signification between Temperate Japonica (purifying selection) and

Indica (balancing selection). The above two results of opposite directional signification suggest that the association between *BADH1* and salt tolerance may be the result of the independent domestication of Japonica and Indica involving their genetic isolation. Similar findings but opposite domestication signatures for Japonica (Tajima's D = 4.47) and Indica (Tajima's D = −1.33) were reported by GWAS analysis [25]; one potential candidate gene *OsSTL1* (salt tolerance level) identified on chromosome 4 was higher in allele frequency in Indica than Japonica, improving the overall salt tolerance. However, our previous results reported an inconsistent finding: the lack of domestication in Asian rice might be due to the small amount of difference detected in the signature of selective sweep for Japonica (Figure 1B) [19].

Germination is considered to be one of the most critical steps in the life cycle of a crop. Due to salinity problems at the germination stage [2], the breeding of salt-tolerant varieties, especially during germination, has been increasing in agricultural countries around the globe. Many research experiments on the effects of environmental conditions have been reported by many research groups, particularly the effects of salinity, drought, cold, heat, light intensity, and CO_2 at the molecular level [15]. In this study, we found clear physical responses of salt-tolerant plant parameters to 200 mM NaCl compared to the control (0 mM NaCl). All the phenotypes showed clear responses to salt treatment, and except for MGT, all phenotypes were positively correlated with each other in salt-treated conditions (Table 3). In particular, the differences in the recorded values of phenotypes under the control and salinity conditions (0 mM and 200 mM NaCl) indicated that rice growth during the germination stage was significantly inhibited by salt stress, which resulted in a very low germination energy and index (GE and GI), as well as root length (RL) and shoot length (SL), which in turn reduced plant density and even yield. Therefore, the development of salt-tolerant rice germplasms would be indispensable to maintain good plant growth beginning in the early stages of rice cultivation.

Table 3. Plant parameters (phenotypes) and formulae for their calculation.

Parameters (Phenotypes)	**Formula**
Germination percentage (GP)	(number of germinated seeds/total number of seeds) × 100
Germination energy (GE)	(number of germinated seeds on day 4/total number of seeds) × 100
Germination index (GI)	$\Sigma (N_d/d)$
Mean germination time (MGT)	$\Sigma (d \times n)/\Sigma n_d$ n_d: the number of germinated seeds on each day d: number of days after the start of the experiment
Germination rate (GR)	$\Sigma N/\Sigma (N \times d)$ N: the number of seeds that germinated on day d d: the days during the experiment
Root length (RL) Shoot length (SL) Total dry weight TDW)	Length of root after 10 days Length of shoot after 10 days Total dry weight of shoot and root (80 °C for 24 h)
Relative GP Relative GE Relative GI Relative MGT Relative GR Relative RL Relative SL Relative TDW	GP in condition/GP in control GE in condition/GE in control GI in condition/GI in control MGT in condition/MGT in control GR in condition/GR in control RL in condition/RL in control SL in condition/SL in control TDW in condition/TDW in control

All the plant parameters showed responses to salt stress in all selected haplotypes (Figure 7). *BADH* has been playing several important functions in plants, and besides them, it is also considered as an associated gene for many types of abiotic stress tolerance, including drought, osmolarity, submergence, temperature, chilling, ultraviolet radiation, and so on [26]. Among different environmental stresses, the primary response to *BADH1* expression in Indica rice was induced by salt treatment within 24 h [7]. The effects of salt treatment on the germination percentage (GP) and germination index (GI) also highlighted the Indica haplotype (Hap_18) as significant among the selected haplotypes, as the germination percentage and index were both obviously affected by salinity. One Philippines research group identified a total of 28 SNPs (seven of which were in exons) in the gene region of *BADH1* under salt stress screening [14]. After the association test between haplotype groups and salt tolerance traits, we found 15 SNPs of the *BADH1* gene in Indica-specific Hap_18, which was one of the significant haplotypes associated with salt stress. Once all the identified and associated SNPs were verified, we observed that there were only two nonsynonymous substitutions, G/T (exon 11) resulting in amino acid change of 'Glutamine (Gln) to Lysine (Lys)' and A/T (exon 4) in the amino acid transition of 'Asparagine (Asn) to Lysine (Lys)', and all the remaining were found in different introns. We supposed that these Indica-specific SNPs would be associated with the main functional properties of the *BADH1* gene under salt treatment because of the previous findings that *BADH1* transcript levels were increased in salt-stressed Indica and Japonica nonfragrant rice varieties through its high response to osmotic stress [9].

In conclusion, different directional selections indicated the *BADH1* domestication signature among the classified populations. A greater range of genetic differentiation was observed in the *BADH1* gene region of the cultivated and wild rice group, providing useful genetic information for the upcoming breeding programs of this gene *BADH1*-related varieties development. Haplotyping revealed a list of cultivated haplotypes for a sequence of different mutated variants, of which five major haplotypes showed their associations with salt-conditioned rice seedlings (plant traits) by 27 significant marker positions (SNPs). Hap_18 and Hap_23 represented major groups for Indica rice accessions that covered significant intronic and exonic marker positions (SNPs and Indels) for salt tolerance-related plant traits, which can be future functional *BADH1* alleles in the breeding program of new varieties development.

4. Materials and Methods

4.1. Plant Materials

A heuristic set of 421 cultivated rice accessions represented by 3 original variety types (landrace, weedy, bred) (Supplementary Table S1) previously collected worldwide and generated by the National GenBank of the Rural Development Administration (RDA-GenBank, Republic of Korea) using the Power Core program [27] was selected for whole-genome resequencing [13]. An additional set of 54 wild rice accessions was also shared by the International Rice Research Institute (IRRI) in 2017.

For these 421 Asian cultivated and 54 wild rice accessions, field experiments were conducted in the Departmental Field of the Plant Resources Department, Kongju National University (Yesan Campus) in 2016 and 2017. The landrace, weedy, and bred cultivated rice set included 6 different ecotypes, 279 Temperate Japonica, 26 Tropical Japonica, 102 Indica, 9 Aus, 2 Aromatic, and 3 Admixture accessions (Supplementary Table S1). Cultural practices in field management were performed as recommended.

4.2. DNA Extraction, Resequencing, and Variant Calling

Fifteen-day-old young samples (green leaves) were taken from all tested plants for DNA extraction by the CTAB (cetyltrimethylammonium bromide) method, and then, genomic DNA was stored in a refrigerator at 4 °C until use [28]. Qualified DNA was used for whole-genome resequencing of the collected rice accessions with an average coverage of approximately 15X on the Illumina HiSeq 2500 Sequencing Systems Platform. The DNA library was generated by using the TruSeq Nano DNA kit, following a specified protocol (part no. 15,041,110 rev. D). The decoded sequences were saved in FastQ file format. VCFtools (variant call format) [29] was used to remove missing values and heterozygotes from raw data saved in FastQ. To compare the output sequences among the accessions, the high-quality reads after removing missing values and heterozygotes were aligned in the International Rice Genome Sequencing Project (IRGSP) 1.0 rice genome sequence. The alignment of the reads was saved in binary alignment map (BAM) format. Duplicate reads aligned in multiple locations were removed using PICARD version 1.88 [30]. Then, variant (SNP/InDel) calling was performed using the Genome Analysis Tool Kit (GATK) tools version 4.0.1.2 [31] to extract the variant regions from the BAM file. The extracted mutations were saved in VCF file format and filtered using VCFtools to remove false-positive SNPs/InDels. The raw sequence data were deposited in the NCBI GeneBank database (accession number: MZ544903-MZ545377).

4.3. Population Structure, Principal Component Analysis (PCA), and Phylogenetic Study

To determine the population structure and existence of subpopulations, we conducted population structure analysis and principal component analysis using 475 rice accessions. We converted annotated variants of *BADH1* into a PLINK file by using VCFtools, and using the PLINK analysis toolset, bed files were recreated, and two additional two files (.bim and .fam) were incorporated by using Python script (structure.py) in the fastStructure package tools [32] within a range of increasing K values from 2 to 7. The admixed patterns of defined populations (population structure) were implemented using average Q-values by the POPHELPER [33] analytical tool in the R program. To plot the similarity or differences among genetic variations in the identified subpopulations, principal component analysis (PCA) was performed in the R program. A list of principal components (PCs) referring to variants was generated from TASSEL 5 [34], and the relatedness among the groups was plotted in 3D scatterplots. A phylogenetic analysis was conducted in MEGAX [35] by the neighbor-joining method, and a tree was drawn in FigTree version 1.4.3 (http://tree.bio.ed.ac.uk/software/figtree, accessed on 18 January 2021).

4.4. Nucleotide Diversity, Tajima's D, and Fixation Index (F_{ST})

To determine the genetic diversity, differentiation, and variation differences, we calculated the respective values for the nucleotide diversity (π), Tajima's D, and the fixation index (F_{ST}). Using VCFtools, variant files were picked within the gene region of *BADHI* for the classified representative types/subpopulations to be compared. The sliding window sizes used for nucleotide diversity (π) and Tajima's D test were each 10 kb, and the values were compared in multiple ways. F_{ST} values were also calculated to determine the genetic differentiation between and among the identified groups or subpopulations of 475 rice accessions.

4.5. Haplotype Diversity Analysis

We conducted whole-genome haplotype diversity analysis on *BADHI* using a variant annotation file to identify the association between *BADH1* and salt tolerance at the haplotype level. We divided our sample group into two separate groups, cultivated rice and wild rice, and performed haplotyping individually. Haplotyping of cultivated rice was first conducted and then compared to the results for wild rice with a minor allele frequency (maf) filter of <0.03 maf. The sequence data from both groups were aligned in MEGAX together with the reference sequence adapted from RAP-DB (https://rapdb.dna.affrc.go.jp, accessed on 30 January 2021), and a haplotype list was generated by DnaSP version 6.0 [36] Using the filtered and aligned genome sequences, a TCS network [37] was constructed in PopART [38].

4.6. Screening of Salt Tolerance Phenotypes

First, a total of 120 seeds of each rice accession were washed in water, surface-sterilized in 1% sodium hypochlorite solution for 20 min, and then rinsed three times with deionized distilled water. Thirty seeds were taken from each rice accession and placed in 9 cm diameter Petri dishes supplemented with 200 mM NaCl solution for salt stress and two layers of filter paper underneath. Then, Petri dishes were stored in incubators at 30 °C with 40% relative humidity for 10 days. Every 2 days, the NaCl solution in Petri dishes was renewed to maintain the concentration, and the germination conditions were checked every day. Filter papers were replaced as necessary. Once the plumule emergence was 2 mm long, we started to measure it as the germination index (GI). After 10 days, we measured the root length (RL) and shoot length (SL) of the seedlings. Then, the total dry weight (TDW) of roots and shoots was also measured after drying at 80 °C for 24 h. The data of this treatment were collected for three replicates, and 0 mM NaCl was used as a control. The number of germinated seeds was counted every day after treatment for up to 10 days, and the germination percentage (GP) was calculated. Germination energy (GE) was observed and recorded daily for 4 days, and the values were calculated. The formulae we used are summarized in Table 3.

4.7. Statistical Analysis

The recorded phenotypic data were first calculated in Microsoft Excel (2010) and statistically analyzed in SPSS version 20.0 using Pearson correlation coefficients. Haplotypic and phenotypic data files were prepared and imported to TASSEL 5.0 [33] for the association test. The general linear model (GLM), containing the SNP tested as a fixed effect, was applied to test the association between phenotypic variation and haplotypes. The association between phenotype and genotype was obtained by using Scheffe's test at the significance level (p-value < 0.05).

Supplementary Materials: Supplementary Materials can be found at https://www.mdpi.com/article/10.3390/ijms22147578/s1.

Author Contributions: Conceptualization, M.-H.M. and Y.-J.P.; methodology, R.P. and S.-H.C.; investigation, M.-H.M., G.-S.L. and S.-H.C.; data curation, M.-H.M. and K.-W.K.; writing—original draft preparation, M.-H.M., Y.C. and T.Z.M.; writing—review and editing, T.Z.M. and K.-W.K.; project administration, Y.-J.P. All authors have read and agreed to the published version of the manuscript.

Funding: This work was supported by two National Research Foundation of Korea (NRF) grants funded by the Korean government (MSIT) (No. NRF-2017R1A2B3011208 and No. NRF-2017R1E1A1A01075282), the "Cooperative Research Program for Agriculture Science and Technology Development (Project No. PJ015935)" Rural Development Administration, Republic of Korea, and a research grant from Kongju National University in 2020.

Data Availability Statement: Data is contained within the article or Supplementary Materials.

Conflicts of Interest: The authors declare no conflict of interest.

References

1. Ashraf, M.; Foolad, M. Roles of glycine betaine and proline in improving plant abiotic stress resistance. *Environ. Exp. Bot.* **2007**, *59*, 206–216. [CrossRef]
2. Zhu, J.-K. Plant salt tolerance. *Trends Plant Sci.* **2001**, *6*, 66–71. [CrossRef]
3. Moghaieb, R.E.; Saneoka, H.; Fujita, K. Effect of salinity on osmotic adjustment, glycinebetaine accumulation and the betaine aldehyde dehydrogenase gene expression in two halophytic plants, Salicornia europaea and Suaeda maritima. *Plant Sci.* **2004**, *166*, 1345–1349. [CrossRef]
4. McNeil, S.D.; Rhodes, D.; Russell, B.L.; Nuccio, M.L.; Shachar-Hill, Y.; Hanson, A.D. Metabolic modeling identifies key constraints on an engineered glycine betaine synthesis pathway in tobacco. *Plant Physiol.* **2000**, *124*, 153–162. [CrossRef] [PubMed]
5. Rhodes, D.; Hanson, A. Quaternary ammonium and tertiary sulfonium compounds in higher plants. *Annu. Rev. Plant Biol.* **1993**, *44*, 357–384. [CrossRef]
6. Shirasawa, K.; Takabe, T.; Takabe, T.; Kishitani, S. Accumulation of glycinebetaine in rice plants that overexpress choline monooxygenase from spinach and evaluation of their tolerance to abiotic stress. *Ann. Bot.* **2006**, *98*, 565–571. [CrossRef] [PubMed]

7. Singh, A.; Singh, P.K.; Singh, R.; Pandit, A.; Mahato, A.K.; Gupta, D.K.; Tyagi, K.; Singh, A.K.; Singh, N.K.; Sharma, T.R. SNP haplotypes of the BADH1 gene and their association with aroma in rice (Oryza sativa L.). *Mol. Breed.* **2010**, *26*, 325–338. [CrossRef]
8. Hasthanasombut, S.; Ntui, V.; Supaibulwatana, K.; Mii, M.; Nakamura, I. Expression of Indica rice OsBADH1 gene under salinity stress in transgenic tobacco. *Plant Biotechnol. Rep.* **2010**, *4*, 75–83. [CrossRef]
9. Nakamura, T.; Yokota, S.; Muramoto, Y.; Tsutsui, K.; Oguri, Y.; Fukui, K.; Takabe, T. Expression of a betaine aldehyde dehydrogenase gene in rice, a glycinebetaine nonaccumulator, and possible localization of its protein in peroxisomes. *Plant J.* **1997**, *11*, 1115–1120. [CrossRef] [PubMed]
10. Fitzgerald, T.L.; Waters, D.L.E.; Henry, R.J. The effect of salt on betaine aldehyde dehydrogenase transcript levels and 2-acetyl-1-pyrroline concentration in fragrant and non-fragrant rice (Oryza sativa). *Plant Sci.* **2008**, *175*, 539–546. [CrossRef]
11. Wang, W.; Mauleon, R.; Hu, Z.; Chebotarov, D.; Tai, S.; Wu, Z.; Li, M.; Zheng, T.; Fuentes, R.R.; Zhang, F. Genomic variation in 3,010 diverse accessions of Asian cultivated rice. *Nature* **2018**, *557*, 43. [CrossRef]
12. Wang, M.; Yu, Y.; Haberer, G.; Marri, P.R.; Fan, C.; Goicoechea, J.L.; Zuccolo, A.; Song, X.; Kudrna, D.; Ammiraju, J.S. The genome sequence of African rice (Oryza glaberrima) and evidence for independent domestication. *Nat. Genet.* **2014**, *46*, 982. [CrossRef] [PubMed]
13. Kim, T.-S.; He, Q.; Kim, K.-W.; Yoon, M.-Y.; Ra, W.-H.; Li, F.P.; Tong, W.; Yu, J.; Oo, W.H.; Choi, B. Genome-wide resequencing of KRICE_CORE reveals their potential for future breeding, as well as functional and evolutionary studies in the post-genomic era. *BMC Genom.* **2016**, *17*, 1–13. [CrossRef] [PubMed]
14. Lapuz, R.R.; Javier, S.; Aquino, J.D.C.; Undan, J. Gene Expression and Sequence Analysis of BADH1 Gene in CLSU Aromatic Rice (Oryza sativa L.) Accessions Subjected to Drought and Saline Condition. *J. Nutr. Sci. Vitaminol.* **2019**, *65*, S196–S199. [CrossRef]
15. Hasthanasombut, S.; Paisarnwipatpong, N.; Triwitayakorn, K.; Kirdmanee, C.; Supaibulwatana, K. Expression of OsBADH1 gene in Indica rice (Oryza sativa L.) in correlation with salt, plasmolysis, temperature and light stresses. *Plant Omics* **2011**, *4*, 75–81.
16. Niu, X.; Zheng, W.; Lu, B.-R.; Ren, G.; Huang, W.; Wang, S.; Liu, J.; Tang, Z.; Luo, D.; Wang, Y. An unusual posttranscriptional processing in two betaine aldehyde dehydrogenase loci of cereal crops directed by short, direct repeats in response to stress conditions. *Plant Physiol.* **2007**, *143*, 1929–1942. [CrossRef] [PubMed]
17. Cubry, P.; Tranchant-Dubreuil, C.; Thuillet, A.-C.; Monat, C.; Ndjiondjop, M.-N.; Labadie, K.; Cruaud, C.; Engelen, S.; Scarcelli, N.; Rhoné, B. The rise and fall of African rice cultivation revealed by analysis of 246 new genomes. *Curr. Biol.* **2018**, *28*, 2274–2282.e2276. [CrossRef] [PubMed]
18. Choi, J.Y.; Zaidem, M.; Gutaker, R.; Dorph, K.; Singh, R.K.; Purugganan, M.D. The complex geography of domestication of the African rice Oryza glaberrima. *PLoS Genet.* **2019**, *15*, e1007414. [CrossRef] [PubMed]
19. He, Q.; Yu, J.; Kim, T.-S.; Cho, Y.-H.; Lee, Y.-S.; Park, Y.-J. Resequencing reveals different domestication rate for BADH1 and BADH2 in rice (Oryza sativa). *PLoS ONE* **2015**, *10*, e0134801.
20. Wambugu, P.W.; Ndjiondjop, M.-N.; Henry, R. Genetics and Genomics of African Rice (Oryza glaberrima Steud) Domestication. *Rice* **2021**, *14*, 6. [CrossRef]
21. Hilbert, L.; Neves, E.G.; Pugliese, F.; Whitney, B.S.; Shock, M.; Veasey, E.; Zimpel, C.A.; Iriarte, J. Evidence for mid-Holocene rice domestication in the Americas. *Nat. Ecol. Evol.* **2017**, *1*, 1693–1698. [CrossRef] [PubMed]
22. Cubry, P.; Vigouroux, Y. Population genomics of crop domestication: Current state and perspectives. *Popul. Genom.* **2018**, 685–707.
23. Doebley, J.F.; Gaut, B.S.; Smith, B. The molecular genetics of crop domestication. *Cell* **2006**, *127*, 1309–1321. [CrossRef] [PubMed]
24. Xu, R.; Sun, C.J.T.C.J. What happened during domestication of wild to cultivated rice. *Crop. J.* **2021**, *9*, 564–576. [CrossRef]
25. Yuan, J.; Wang, X.; Zhao, Y.; Khan, N.U.; Zhao, Z.; Zhang, Y.; Wen, X.; Tang, F.; Wang, F.; Li, Z. Genetic basis and identification of candidate genes for salt tolerance in rice by GWAS. *Sci. Rep.* **2020**, *10*, 1–9. [CrossRef] [PubMed]
26. Yang, C.; Zhou, Y.; Fan, J.; Fu, Y.; Shen, L.; Yao, Y.; Li, R.; Fu, S.; Duan, R.; Hu, X.; et al. SpBADH of the halophyte Sesuvium portulacastrum strongly confers drought tolerance through ROS scavenging in transgenic Arabidopsis. *Plant Physiol. Biochem.* **2015**, *96*, 377–387. [CrossRef]
27. Kim, K.-W.; Chung, H.-K.; Cho, G.-T.; Ma, K.-H.; Chandrabalan, D.; Gwag, J.-G.; Kim, T.-S.; Cho, E.-G.; Park, Y.-J. PowerCore: A program applying the advanced M strategy with a heuristic search for establishing core sets. *Bioinformation* **2007**, *23*, 2155–2162. [CrossRef] [PubMed]
28. Doyle, J.J. Isolation ofplant DNA from fresh tissue. *Focus* **1990**, *12*, 39–40.
29. Danecek, P.; Auton, A.; Abecasis, G.; Albers, C.A.; Banks, E.; DePristo, M.A.; Handsaker, R.E.; Lunter, G.; Marth, G.T.; Sherry, S.T. The variant call format and VCFtools. *Bioinformatics* **2011**, *27*, 2156–2158. [CrossRef]
30. Picard Toolkit. Broad Institute, GitHub Repository. 2019. Available online: http://Broadinstitute.Github.Io/Picard (accessed on 30 January 2021).
31. Van der Auwera, G.A.; O'Connor, B.D. *Genomics in the Cloud: Using Docker, GATK, and WDL in Terra*; O'Reilly Media, Inc.: Sebastopol, CA, USA, 2020.
32. Raj, A.; Stephens, M.; Pritchard, J.K. fastSTRUCTURE: Variational inference of population structure in large SNP data sets. *Genetics* **2014**, *197*, 573–589. [CrossRef]
33. Francis, R.M. Pophelper: An R package and web app to analyse and visualize population structure. *Mol. Ecol. Resour.* **2017**, *17*, 27–32. [CrossRef] [PubMed]
34. Bradbury, P.J.; Zhang, Z.; Kroon, D.E.; Casstevens, T.M.; Ramdoss, Y.; Buckler, E.S. TASSEL: Software for association mapping of complex traits in diverse samples. *Bioinformatics* **2007**, *23*, 2633–2635. [CrossRef] [PubMed]

35. Kumar, S.; Stecher, G.; Li, M.; Knyaz, C.; Tamura, K. MEGA X: Molecular evolutionary genetics analysis across computing platforms. *Mol. Biol. Evol.* **2018**, *35*, 1547–1549. [CrossRef] [PubMed]
36. Rozas, J.; Ferrer-Mata, A.; Sánchez-DelBarrio, J.C.; Guirao-Rico, S.; Librado, P.; Ramos-Onsins, S.E.; Sánchez-Gracia, A. DnaSP 6: DNA sequence polymorphism analysis of large data sets. *Mol. Biol. Evol.* **2017**, *34*, 3299–3302. [CrossRef]
37. Clement, M.; Posada, D.; Crandall, K. TCS: A computer program to estimate gene genealogies. *Mol. Ecol.* **2000**, *9*, 1657–1659. [CrossRef]
38. Leigh, J.W.; Bryant, D. Popart: Full-feature software for haplotype network construction. *Methods Ecol. Evol.* **2015**, *6*, 1110–1116. [CrossRef]

International Journal of
Molecular Sciences

Article

The Jacalin-Related Lectin HvHorcH Is Involved in the Physiological Response of Barley Roots to Salt Stress

Katja Witzel [1,2,*,†], Andrea Matros [2,†,‡], Uwe Bertsch [2,§], Tariq Aftab [2,‖], Twan Rutten [2], Eswarayya Ramireddy [3,¶], Michael Melzer [2], Gotthard Kunze [2] and Hans-Peter Mock [2]

1 Leibniz Institute of Vegetable and Ornamental Crops, 14979 Großbeeren, Germany
2 Leibniz Institute of Plant Genetics and Crop Plant Research, Stadt Seeland, 06466 Gatersleben, Germany; andrea.matros@julius-kuehn.de (A.M.); uwe.bertsch@louisenlund.de (U.B.); tarik.alig@gmail.com (T.A.); rutten@ipk-gatersleben.de (T.R.); melzer@ipk-gatersleben.de (M.M.); kunzeg@ipk-gatersleben.de (G.K.); mock@ipk-gatersleben.de (H.-P.M.)
3 Dahlem Centre of Plant Sciences, Institute of Biology/Applied Genetics, Free University of Berlin, 14195 Berlin, Germany; eswar.ramireddy@iisertirupati.ac.in
* Correspondence: witzel@igzev.de
† These authors contributed equally to this work.
‡ Present address: Institute for Resistance Research and Stress Tolerance, Julius Kühn-Institute, 06484 Quedlinburg, Germany.
§ Present address: IB World School, Stiftung Louisenlund, 24357 Güby, Germany.
‖ Present address: Department of Botany, Aligarh Muslim University, Aligarh 202002, India.
¶ Present address: Department of Biology, Indian Institute of Science Education and Research Tirupati, Tirupati 517507, India.

Citation: Witzel, K.; Matros, A.; Bertsch, U.; Aftab, T.; Rutten, T.; Ramireddy, E.; Melzer, M.; Kunze, G.; Mock, H.-P. The Jacalin-Related Lectin HvHorcH Is Involved in the Physiological Response of Barley Roots to Salt Stress. *Int. J. Mol. Sci.* **2021**, 22, 10248. https://doi.org/10.3390/ijms221910248

Academic Editor: Mirza Hasanuzzaman

Received: 25 August 2021
Accepted: 18 September 2021
Published: 23 September 2021

Abstract: Salt stress tolerance of crop plants is a trait with increasing value for future food production. In an attempt to identify proteins that participate in the salt stress response of barley, we have used a cDNA library from salt-stressed seedling roots of the relatively salt-stress-tolerant cv. Morex for the transfection of a salt-stress-sensitive yeast strain (*Saccharomyces cerevisiae* YSH818 Δ*hog1* mutant). From the retrieved cDNA sequences conferring salt tolerance to the yeast mutant, eleven contained the coding sequence of a jacalin-related lectin (JRL) that shows homology to the previously identified JRL horcolin from barley coleoptiles that we therefore named the gene *HvHorcH*. The detection of HvHorcH protein in root extracellular fluid suggests a secretion under stress conditions. Furthermore, HvHorcH exhibited specificity towards mannose. Protein abundance of HvHorcH in roots of salt-sensitive or salt-tolerant barley cultivars were not trait-specific to salinity treatment, but protein levels increased in response to the treatment, particularly in the root tip. Expression of HvHorcH in *Arabidopsis thaliana* root tips increased salt tolerance. Hence, we conclude that this protein is involved in the adaptation of plants to salinity.

Keywords: abiotic stress; apoplast; functional screening; *Hordeum vulgare*; salinity; seedling

1. Introduction

The area of farmland worldwide that can be cultivated with crop plants is declining because of increasing drought periods and the increasing salinity of the soil in several regions of the earth [1]. Global climate change, which is predicted to be accompanied by prolonged and intensified drought periods, is likely to aggravate this situation even further. Intensified irrigation attempts to combat drought ultimately increase soil salinity and thus eventually impede farmland cultivation when salinity reaches threshold levels that can no longer be tolerated by crop plants [2]. It is therefore an eminent goal for a global sustainable food supply to improve the salt stress tolerance of crop plants in order to push these thresholds of soil salinity upwards so that more farmland with high-salinity soil will still be amenable to agriculture.

Barley is regarded as one of the more salt-stress-tolerant crop species [3,4], although the level of tolerance may vary considerably between diverse barley cultivars [5]. Therefore, the comparison of proteomes and transcriptomes of contrasting barley cultivars with regard to their salt stress tolerance bears the potential to identify candidate proteins and their respective coding and regulatory genes that play a role in salt stress tolerance of barley. A collection of salt stress candidate genes discovered by a proteomics approach using different barley cultivars has been published previously [6–9]. However, the profiling of low-abundant, low-solubilizing, or small proteins in untargeted approaches is still challenging due to intrinsic technical limitations of proteomic analyses. Hence, we constructed a cDNA library made from the salt-stressed roots of the more tolerant barley cv. Morex to search for barley proteins related to stress tolerance using a functional screening procedure that involved a salt-stress-susceptible yeast mutant. The basis for the yeast complementation screening approach applied here is formed by a global commonality of the signal transduction pathways and mechanisms responsible for osmotic stress tolerance in both yeasts and plants [10–14]. A yeast screen of a cDNA library generated from salt-stressed sugar beet leaves led to the identification of a serine O-acetyltransferase conferring yeast osmotolerance [15]. In addition, new candidates for salt stress response were obtained by screening a date palm [16] and a *Thellungiella halophila* cDNA library [17]. The feasibility of this approach for *Escherichia coli* was also shown. The expression of a cDNA library derived from salt-treated *Solanum commersonii* in *E. coli* identified novel genes involved in the stress response [18,19]. A lectin receptor-like kinase was isolated from screening an *E. coli* expression library generated from salt-stressed pea seedlings [20].

Using a similar salt stress tolerance complementation assay, we identified cDNAs for a jacalin-related lectin (JRL) with homology to the JRL horcolin from barley coleoptiles. The corresponding protein was isolated from barley roots based on its mannose affinity and characterized regarding its salt-stress-induced expression in barley root tissues as well as in *Arabidopsis thaliana*. HvHorcH1 association with barley salt stress tolerance was carried out by studying the HvHorcH1 protein expression levels in salt-sensitive and salt-tolerant barley cultivars. Further, transgenic Arabidopsis with root-tip specific expression of HvHorcH1 were generated.

2. Results

2.1. Screening for Salt Stress-Related Genes from Barley cv. Morex Roots in Salt Sensitive Yeast

In order to identify gene products that participate in the salt stress tolerance of barley, we used a cDNA library from salt-stressed roots of the relatively salt-stress-tolerant cv. Morex for the transformation of the salt-stress-sensitive yeast strain YSH818. A total of 51 independent yeast transformants, containing 21 different barley cDNAs, were isolated that showed increased salt tolerance as compared to the empty vector control (Table 1). Several cDNAs represented complete open reading frames, but also cDNAs lacking 5′ or 3′ ends were found. For further validation, cDNAs were used for the retransformation of YSH818, and salt stress testing was extended up to 3% NaCl in the growth medium. Figure 1 shows exemplarily the results for two jacalin-related lectin 31 (JRL) cDNAs (09_818-20-10-10-5, 37_818-28-8-10-5, both coding for HORVU7Hr1G059330), where a vector control is contrasted to the transformed cells. Here, the JRL31-containing yeast cells show clearly increased survival at higher salt concentrations of 2% and 3% NaCl, even at higher dilutions of the initial inoculum, when compared to the vector control.

Table 1. Barley cDNAs conferring salt tolerance to a salt-sensitive yeast mutant. Given are the BLAST result (barlex CDS_HC May2016 database using blastx), the putative annotation and putative function of the gene, and the number of isolates from yeast transformants.

Annotation	Gene	Score (Bits)	E Value	Function	Number of Independent Yeast Transformants
jacalin-related lectin 31 (HvHorcH)	HORVU7Hr1G059330	296	7×10^{-103}	carbohydrate binding	10
cellulose-synthase like D2	HORVU2Hr1G042350	502	2×10^{-174}	cellulose biosynthesis	6
peptide-N(4)-(N-acetyl-β-glucosaminyl)asparagine amidase	HORVU2Hr1G048680	507	3×10^{-177}	protein deglycosylation	3
WD-40 repeat family protein	HORVU3Hr1G115170	376	4×10^{-128}	translation	3
phytanoyl-CoA dioxygenase domain-containing protein 1	HORVU4Hr1G007050	439	3×10^{-156}	fatty acid metabolism	3
unknown function	HORVU4Hr1G070280	64	8×10^{-14}	unknown	3
histone H2A 6	HORVU7Hr1G100110	171	1×10^{-52}	DNA binding	3
30S ribosomal protein S11	HORVU7Hr1G104220	256	8×10^{-88}	translation	3
exocyst complex component 6	HORVU0Hr1G006630	367	2×10^{-122}	vesicle transport	2
membrane steroid binding protein 1	HORVU1Hr1G045630	276	8×10^{-93}	vesicle transport	2
histone H2A 2	HORVU2Hr1G043860	145	1×10^{-44}	DNA binding	2
unknown function	HORVU7Hr1G032340	159	2×10^{-48}	unknown	2
cytochrome P450 superfamily protein	HORVU1Hr1G069310	124	2×10^{-43}	gibberellin catabolic process	1
GDSL esterase/lipase	HORVU2Hr1G025800	280	4×10^{-93}	fatty acid metabolism	1
adenine nucleotide alpha hydrolases-like superfamily protein	HORVU2Hr1G042480	69	4×10^{-16}	stress response	1
ribosomal protein S24e family protein	HORVU4Hr1G058010	207	1×10^{-67}	unknown	1
potassium transporter 26	HORVU4Hr1G058080	39	7×10^{-04}	ion transport	1
26S proteasome non-ATPase regulatory subunit 4 homolog	HORVU4Hr1G063820	370	3×10^{-126}	proteolysis	1
ATP-dependent zinc metalloprotease FtsH 2	HORVU5Hr1G063340	68	2×10^{-15}	proteolysis	1
peptidyl-prolyl cis-trans isomerase	HORVU6Hr1G012570	295	3×10^{-102}	protein folding	1
Rad23 UV excision repair protein family	HORVU7Hr1G042100	221	4×10^{-70}	mRNA catabolism	1

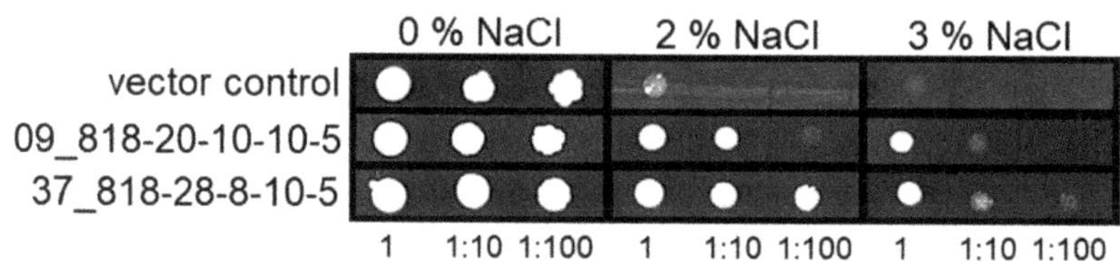

Figure 1. Salt stress tolerance screening of transfected YSH818 Δ*hog1* yeast cells plated on agar containing increasing salt concentrations. Yeast cells were either transfected with cDNA expression vector alone (upper row) or vectors containing two *HvHorcH* (HORVU7Hr1G059330) cDNAs. The respective inocula were deposited in tenfold dilution steps, from left to right for each strain and salt condition.

Currently, the putative function of several of the isolated cDNAs is unknown (Table 1). Other functional groups found in the analysis included cDNAs involved in translation, cellulose biosynthesis, DNA binding, fatty acid metabolism, vesicle transport, proteolysis, and stress response, among others. Ten barley cDNA sequences were retrieved that all were highly homologous to the coding sequence of a yet unannotated barley JRL31 HORVU7Hr1G059330, which shows a 50% sequence identity with the previously identified mannose-specific JRL horcolin (HORVU1Hr1G000160) from barley coleoptiles [21].

Figure S1 shows an amino acid sequence alignment of those sequences. Hence, the isolated JRL31 were termed as horcolin homolog HvHorcH (HORVU7Hr1G059330).

A further interesting candidate identified in the screen, besides JRL31, was cellulose-synthase like D2 (HORVU2Hr1G042350), identified in six independent yeast transformants. However, for all further experiments, we focused on JRL31 cDNAs.

2.2. *In Silico Gene Expression Pattern of HvHorcH in Barley Roots*

The expression level of *HvHorcH* gene transcripts was analysed using the Expression Atlas (Expression Atlas. Available online: https://www.ebi.ac.uk/gxa/home, accessed on 24 August 2021), a repository of RNAseq expression data, which is reproduced in Figure S2, and compared to that of horcolin. The highest transcript levels for horcolin were found for shoots and germinated embryos. The distribution pattern of HvHorcH was characterized by an accumulation mainly in roots, indicating a different functionality as compared to horcolin.

2.3. *Cellular and Subcellular Localization of HvHorcH*

Based on transcript information, HvHorcH is expressed mainly in roots. In order to investigate the localization within the root, we used the polyclonal HvHorcH antibody, in combination with a fluorescently labeled anti-rabbit-IgG antibody, in the salt-stressed roots of cv. Morex. A non-uniform tissue distribution was observed with a strong labeling of the root cap (Figure 2).

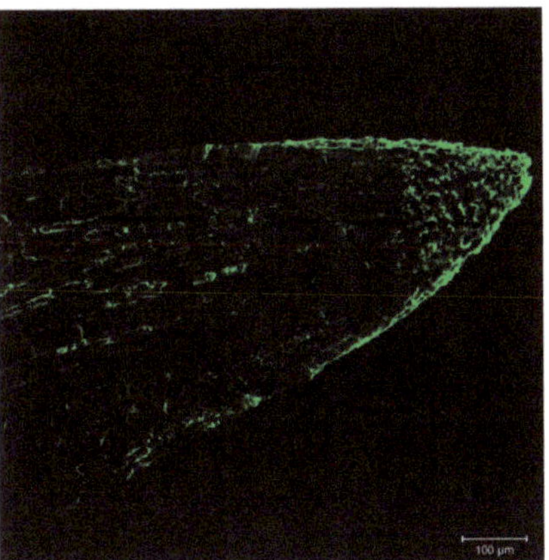

Figure 2. Immunolocalization of HvHorcH on barley cv. Morex root tips after salt stress treatment. HvHorcH was found mainly in the root cap.

Localization in the apoplast has been described for the homologous horcolin [21] as well as for the JRL helja from *Helianthus annuus* seeds [22]. In order to study the subcellular distribution of HvHorcH, we prepared an apoplast fraction of salt-stressed roots of cv. Morex. The presence of HvHorcH in apoplastic fluid of salt-stressed roots of cv. Morex was confirmed immunologically with antibodies specific for HvHorcH. The apoplastic fluid contained only minor contamination with cytosolic proteins, as exemplified by the presence of glyceraldehyde-3-phosphate dehydrogenase (GAPDH, Figure 3). Since HvHorcH lacks a signal sequence for an export via the standard secretory pathway through ER and Golgi, its secretion into the apoplast of roots may follow a non-canonical route.

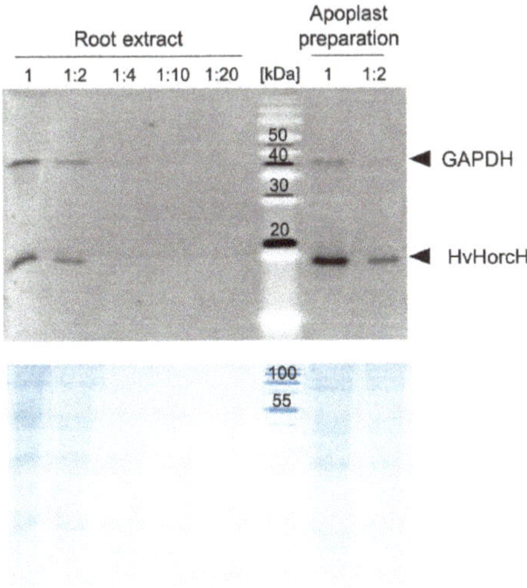

Figure 3. Detection of HvHorcH in the root apoplast. Anti-GAPDH and anti-HvHorcH stained Western blot (top) of an SDS-PAGE (bottom: Coomassie stained) with dilution series of cv. Morex native root extracts or apoplastic fluids. For the apoplastic fluid preparation, an infiltration vacuum of 20 mPa was applied with a vacuum pump under manometer control. All extracts were prepared from salt stress (150 mM NaCl)-treated roots. Two µg of protein of root extract or apoplast preparation was loaded to the SDS-PAGE as undiluted sample and then diluted as indicated in the figure.

2.4. Mannose-Binding of HvHorcH Protein

In order to test whether HvHorcH has a similar binding behaviour to mannose as horcolin, a mannose affinity chromatography was performed. HvHorcH was enriched from salt-stressed cv. Morex roots to apparent homogeneity in a two-step purification process employing anion exchange (AEX) and mannose affinity chromatography. Fractions from this purification procedure were evaluated by immunoblotting using an anti-HvHorcH polyclonal antiserum and by SDS-PAGE (Figure S3). A protein band with a molecular weight (MW) of about 15 kDa, which corresponds to the theoretical MW of HvHorcH of 15.1 kDa, was detected both in the native root extract (R) and after desalting and buffer exchange (R_d). Subsequent AEX chromatography resulted in the enrichment of HvHorcH in a number of fractions during elution with a linear NaCl gradient (Figure S3, F_{13} to F_{16}). These fractions were combined (P) and subjected to dialysis (P_d). By the following mannose affinity chromatography, a fraction highly enriched in HvHorcH could be generated. Thereby, the majority of the target protein eluted within the first fraction ($E1_1$) of the first elution with citrate buffer (elution buffer 1). For the recombinant expressed HvHorcH protein, a signal in the range of 15 to 18 kDa has been detected, which can be explained by the attached poly-histidine affinity tag.

2.5. Selection of Barley Lines with Different Salt Stress Tolerance Levels

In order to test whether HvHorcH expression levels are associated with barley salt stress tolerance, we screened doubled-haploid (DH) offspring lines of the cross between the more tolerant cv. Morex and the more sensitive cv. Steptoe. The latter has been previously screened for salt tolerance in a germination assay [23]. DH lines showing a more sensitive response such as cv. Steptoe or a more tolerant one as compared to cv. Morex were evaluated in a hydroponic assay for salt tolerance at the seedling stage. Previously, we found that the development of the third leaf was indicative of a contrasting response toward salinity [8]. Hence, this parameter was used to assess the response of tested accessions. Four accessions were isolated showing a comparable contrasting response to the parental lines cvs. Steptoe and Morex with respect to the length and fresh weight of the third leaf (Figure 4a,b). With increasing NaCl concentration, the growth was inhibited in all accessions, but the growth inhibition of DH14 and DH43 was stronger at all levels tested as compared to DH187 and DH198. The leaf fresh weight at 150 mM NaCl was reduced to 10% and 8% in DH14 and DH43, respectively, and to 27% and 46% in DH187 and DH198, respectively, as compared to the control treatment. Under the same conditions, the remaining growth rate was 4% in cv. Steptoe and 41% in cv. Morex [8], indicating that the response of selected offspring lines was comparable to that of the parent cultivars. In our previous work, we observed pronounced differences of salt-stress-mediated alterations in root architecture between the tolerant cv. Morex and the sensitive cv. Steptoe [6]. Additionally, the transcript levels of *HvHorcH* (Figure S2) and the protein levels (Figure 2) were shown to be highest in roots.

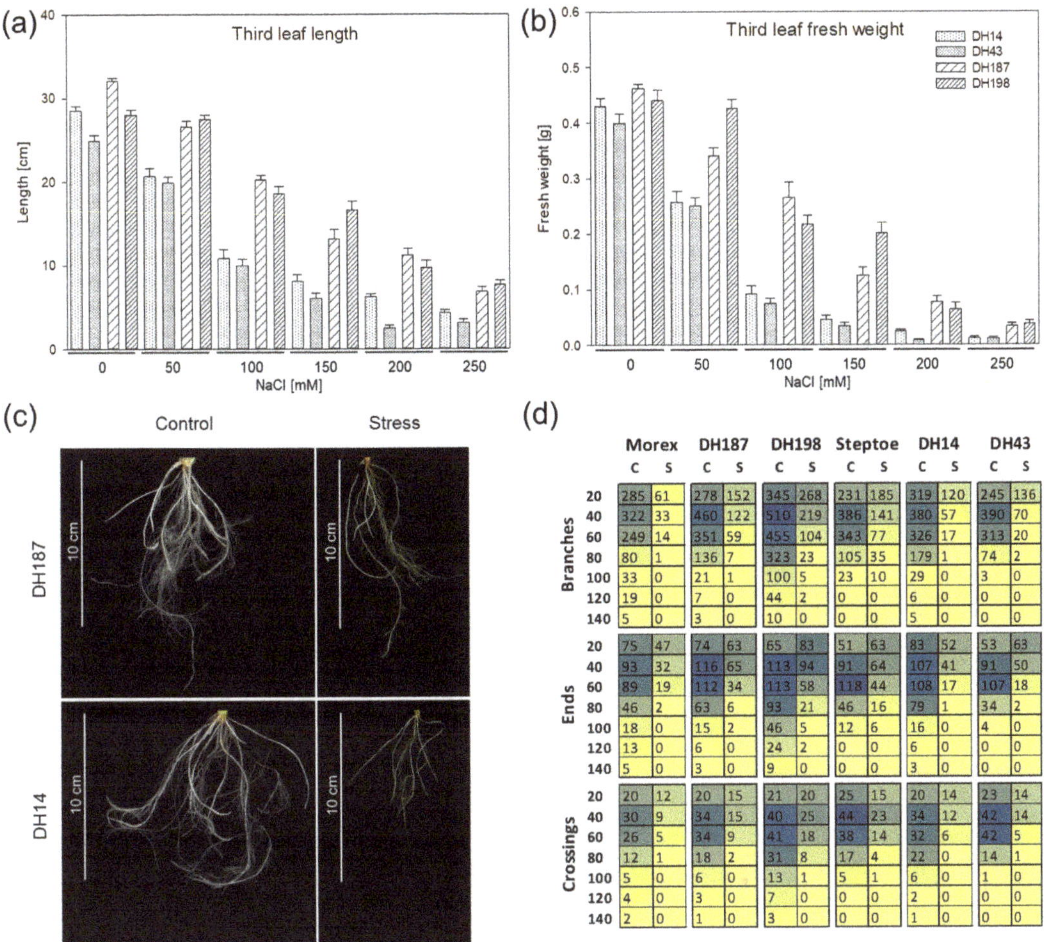

Figure 4. The effect of NaCl on the seedling growth and root growth of salt-tolerant (cv. Morex, DH 187, DH198) and salt-sensitive (cv. Steptoe, DH14, DH43) accessions of the Steptoe × Morex mapping population. (**a**,**b**) Salt tolerance was recorded as indicated by biomass production of the third leaf. The data represent the means of 17–20 plants per treatment, with the standard error shown as error bars. Bars filled with dots indicate salt-sensitive accessions, while bars filled with stripes represent tolerant ones. (**c**) Representative photographical images of control plants and plants under salt stress from DH187 and DH14. (**d**) Specific features of root architecture including branches, ends, and crossing points were measured in distinct circular segments centered on the root base. Values refer to the distance from the root base in mm (from 20 to 140). Mean values are shown in a color code representation. Blue indicates high numbers, while beige represents low values.

Thus, we additionally investigated root architecture responses in the selected DH lines under control conditions and salt stress (Figure 4c). Morphological parameters (branches, ends, and crossings) were evaluated after ten days exposure to 150 mM NaCl from digital root images and displayed as a function of the distance from the root base (Figure 4d). Under control conditions, the more tolerant cv. Morex, DH187 and DH198, displayed a more elaborate root system containing higher numbers of ends and crossings in greater distance from the root base (>100 mm, control). Additionally, the roots of those lines were more branched far from the root base under the control treatment. When grown under salt stress, all lines tested showed a strong shortening of roots and a high reduction in number of branches, ends, and crossings. Nevertheless, the highest capability for maintaining root growth and differentiation was observed for the salt-tolerant accession DH198, indicated by the longest and most branched root system under salt stress (Figure 4d).

2.6. HvHorcH Protein Abundance Is Highly Increased in Root Tips under Salt Stress

The distribution of HvHorcH across the roots of the six selected barley lines was analyzed using the polyclonal antibody against HvHorcH. Protein abundance in roots and root tips after salt stress imposition as compared to control conditions was investigated by protein gel blot analysis (Figure 5). The antibody clearly detected a protein in the molecular weight range of 15 kDa, matching the calculated molecular weight of 15.1 kDa for HvHorcH. Under control conditions, all accessions (except DH14) accumulated higher amounts of HvHorcH in total roots than in root tips. A marked increase in HvHorcH abundance was shown for all accessions when grown at 150 mM NaCl. Notably, this increase was much higher in root tips as compared to total roots, indicated by higher fold change values for all accessions in this tissue. However, no concerted up-regulation of the protein in the salt-tolerant or salt-sensitive accessions was observed for roots and root tips. Thus, we assume that HvHorcH1 is involved in the general osmotic response mechanisms during growth under salt stress in barley.

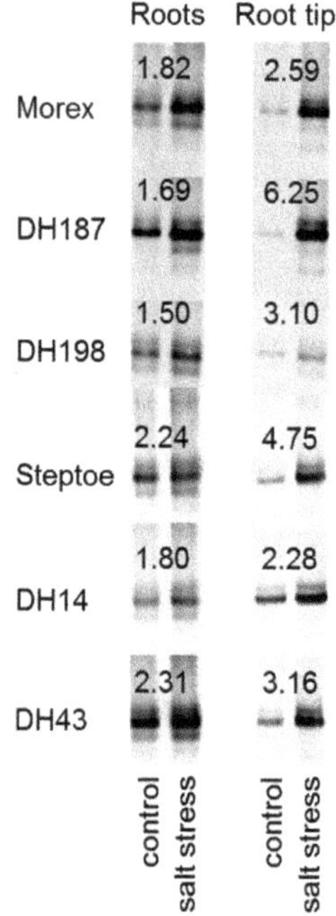

Figure 5. HvHorcH abundance in roots and root tips of tolerant (cv. Morex, DH 187, DH198) or sensitive (cv. Steptoe, DH14, DH43) barley accessions with contrasting salt stress response. Equal amounts of total protein (2 µg) were separated by SDS-PAGE on 15% gels, and HvHorcH was immunologically detected by protein gel blot analysis. Pixel intensities of detected bands were integrated. The mean fold change between control and salt stress (150 mM NaCl) conditions calculated from three independent experiments is given on top of the bands. Representative images from one experiment are shown.

2.7. HvHorcH Enhances Salt Tolerance in Transgenic Arabidopsis

To further study the function of *HvHorcH* during salt stress, a transgenic approach in *Arabidopsis thaliana* was applied. Since the protein showed a root-tip-specific expression under salinity in barley, a root-tip-specific promoter was selected to drive *HvHorcH* gene expression, coupled to eGFP as the reporter gene. Three independent lines expressing *HvHorcH* in the Col-0 background were selected for further studies based on the strength of the reporter signal (Figure S4). When grown in the presence of different levels of NaCl, the fresh mass production of the rosette in the transgenic lines 3-8, 5-7 and 9-6 was more advanced than in the wildtype Col-0 (Figure 6a). When root phenotypes of plants grown in the presence of 50 mM NaCl were analyzed, an increased primary root length of transgenic lines was observed as compared to wildtype plants (Figure 6b). These results indicate that expression of HvHorcH in root tips enhances salt tolerance in *A. thaliana*.

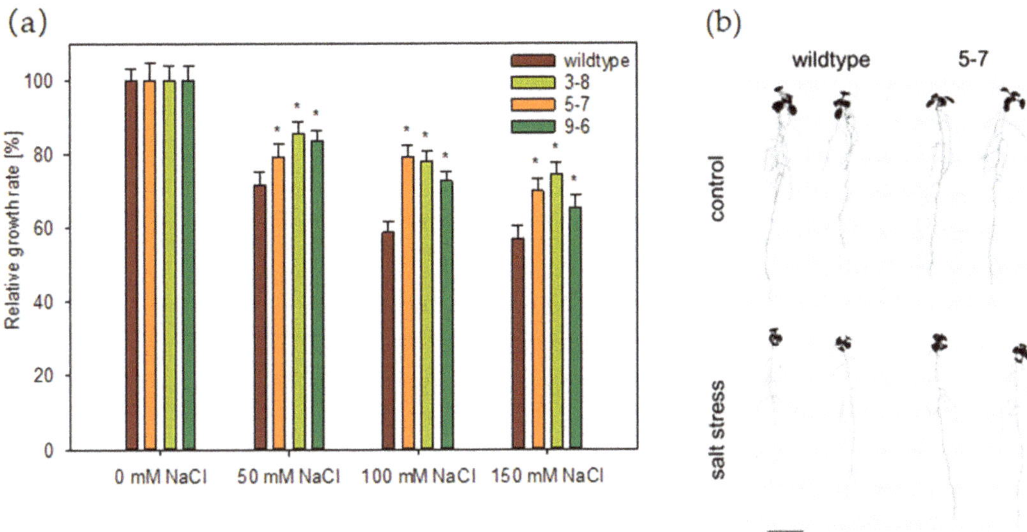

Figure 6. Salt stress response of *Arabidopsis thaliana* plants expressing *HvHorcH* in root tips. (**a**) Quantification of the relative rosette growth rate in *A. thaliana* wildtype Col-0 and *HvHorcH* transgenic plants (lines 3-8, 5-7, 9-6) under three different levels of salt stress, grown in soil culture. The data represent the means of about 80 plants per treatment and genotype, originating from three independent experiments, with the standard error shown as error bars. Asterisks (*) indicate significant differences between the wildtype Col-0 and *HvHorcH* transgenic plants (*t*-test, $p < 0.01$) (**b**) Root phenotypes of a representative transgenic line pGSTu25:HvHorcH:GFP #5-7 compared with the wildtype after 7 d of growth on an agar plate in the absence (top) or presence (bottom) of 50 mM NaCl (bar = 1 cm).

3. Discussion

The functional screening of barley salt-stress-related genes in yeast showed that, to a certain extent, plant proteins can substitute for components of the stress signaling pathway or the cellular defense mechanisms that are blocked in the Δ*hog1*-mutant yeast used in this study. One of the cDNAs conferring salt tolerance to yeast encoded the membrane steroid binding protein 1 (MSBP). In our group, this protein was recently identified in a comparative root plasma membrane proteomic analysis of barley cvs. Steptoe and Morex to be involved in the adaptation of root architecture to salinity [6]. However, numerous genes isolated in our present study are not yet described to be involved in the barley salt stress response and hence represent novel candidates for further evaluation.

Most multiple cDNA isolations were derived for a JRL 31 (named HvHorcH), which was therefore selected for further analysis. HvHorcH belongs to the group of jacalin-related lectins (JRLs), which are carbohydrate-binding proteins. In general, lectins cover a diverse group of protein families with multiple functions that bind reversibly to specific mono- or oligosaccharides without altering the substrate [24]. JRLs are a subgroup that share one or more conserved domains with jacalin, a lectin isolated from jackfruit [25]. Some JRLs are involved in the response towards biotic stresses [26–28], while others play a role in developmental processes [29]. Numerous studies have noticed the response of JRLs to abiotic stresses, such as high light [30] and salt stress [31,32]. In barley, several JRLs are described with different domain organization and expression patterns. Lem2 is a JRL with two jacalin domains, which is expressed in shoots, and its induction by salicylic acid suggests an involvement in systemic acquired resistance [33]. Chimeric JRLs JRG1.1, JRG1.2, and JRG1.3 (dirigent protein) consist of one jacalin and one dirigent domain. They are assumed to be involved in plant disease response since inducibility by jasmonic acid was demonstrated [34]. Chitin-binding ability was shown for BLc3, a lectin that accumulates in root tips and in the embryo [35].

The proteins HL2 and horcolin are characterized by a single jacalin domain. HL2 is expressed in shoots and transcripts accumulate under high light conditions and after methyl jasmonate treatment [30]. In contrast to this, horcolin transcript and protein levels are induced specifically in etiolated coleoptiles, and the protein is localized to the cell wall and can be extracted in apoplastic washing fluid [21]. HL2 and horcolin belong to the subgroup of mannose-binding JRL, a property also demonstrated for HvHorcH in our study. We have detected HvHorcH in apoplastic fluid preparations from roots, indicating a possible rhizodeposition of this protein, a hypothesis that is further underlined by the root-tip-specific accumulation of HvHorcH under salt stress conditions. This accumulation pattern was observed for all investigated genotypes, pointing toward a role of HvHorcH in the general osmotic response mechanisms during growth under salt stress in barley.

In addition, the mannose-specific JRL helja from *H. annuus* was localized to the seedling apoplast, where its release protects the roots from pathogenic fungi through cell agglutination [22,36,37]. A role in plant defense against pathogens was also proposed for chitin-binding JRL found in barley root tips, which share a high sequence similarity to wheat germ agglutinin [35,38,39]. In contrast to this, reports on wheat germ agglutinin itself underline its role in establishing ABA-controlled unspecific defense responses since they accumulate in wheat roots in response to drought stress [40] and salt stress [41], while exogenous wheat germ agglutinin is able to protect the seedling from salt-stress-induced growth arrest [41]. While the expression of JRL in the meristem of drought-stressed banana is reduced [42], the levels of rice JRL salT increase in response to drought and salt stress, although mainly in the shoot [43,44]. When testing rice cultivars of contrasting salt tolerance, salT protein accumulated nonspecifically, indicating no correlation between salt stress response and protein levels [45]. Further, the rice JRL MRL was also inducible after imposition of salt stress and even higher in roots as compared to shoots [46].

As we have found in our study that the accumulation of HvHorcH in response to salt stress is not correlated to the tolerance or sensitivity of the tested barley genotypes, we also assume that HvHorcH has a general protective role in the root tip. A non-uniform distribution was observed with strong labeling of the root cap, a similar distribution to that found for other root-specific lectins [35,38]. The rhizodeposition of HvHorcH by root tips could modify the osmotic potential in its vicinity in order to maintain meristematic activity under stress conditions, as shown for wheat germ agglutinin in wheat [47] and barley [48]. Recently, two JRL were detected in extracellular vesicles isolated from the *A. thaliana* leaf apoplast, indicating that JRL are secreted by extracellular vesicles [49]. We have shown here that HvHorcH is functional in *A. thaliana* root tips and that transgenic plants exhibit a greater tolerance towards salt stress. However, the precise functional mechanism behind the protective role of HvHorcH in roots remains to be elucidated. To examine the binding specificities of HvHorcH more profoundly, carbohydrate microarrays should be performed next [50].

A further candidate identified in our yeast screen was a cellulose-synthase-like D2 gene. This gene is 99% identical to cellulose-synthase-like *HvCslF3*, a gene involved in the biosynthesis of (1,3;1,4)-β-D-glucan, which is the major cell wall constituent in monocots and is the most abundant structural polysaccharide of the mature barley grain [51]. So far, ten *HvCslF* genes are reported [52], with *HvCslF3* not being involved in grain development, based on transcript profiling data [51,53] but being highly expressed in the coleoptile and root tips [52]. One of the most devastating effects of salt stress in plants is the induction of detrimental changes in the cell wall, e.g., disturbed cellulose synthesis [54], and reduced pectin cross-linking and cell wall integrity [55]. Often, an altered composition of cell wall components in response to salt stress is observed [56], although the precise role of (1,3;1,4)-β-D-glucan in stress adaptation remains to be elucidated. Notably, the function of C-type lectins as receptors of linear (1,3)-β-D-glucan and its role in triggering diverse immune responses in humans, invertebrates, and microorganisms has been established [57]. Similarly, the chitin elicitor receptor kinase 1 (CERK1), initially reported to directly bind chitin oligosaccharides, has been suggested to function as a co-receptor for linear (1,3)-β-D-glucan, triggering fungal immune responses in *A. thaliana*, recently [58]. Recently, the multivalent binding of horcolin to high-mannose glycans has been demonstrated [59]. Whether a similar function may be assigned to HvHorcH remains speculation at this point. However, the observed longer primary roots in *HvHorcH* expressing *A. thaliana* and the larger root systems of salt-tolerant barley genotypes hint toward effective salt stress tolerance mechanisms related to maintaining root growth under saline conditions in both species, which may involve HvHorcH and HvCslF3. In our study, we have used two groups of barley genotypes with contrasting salt stress response to distinguish specific or nonspecific stress responses. The same genotypes should now be tested for in-depth characterization of HvCslF3 and other candidates isolated by our yeast screen, such as those involved in stress response or ion transport in order to identify stress-response-related genes for plant improvement.

4. Materials and Methods

4.1. Plant Material and Salt Stress Treatment

Barley cultivars Steptoe, Morex as well as their offsprings DH14, DH43, DH187, and DH198 were grown in hydroponic culture as described by Witzel et al. [8]. Salt stress treatment began six days after germination by the NaCl concentration being increased step-wise (50 mM NaCl steps every other day) until 150 mM NaCl. Root material was harvested at ten days for cDNA library construction, at five and ten days for qPCR measurements, and at seven days for immunodetection, after this final level had been reached. Material from five individual plants was combined to form, for each of the cultivars, one batch of non-treated (control) roots, and a second one of salt-stressed roots. Three independent experiments were performed. For the screening of doubled-haploid lines of the Steptoe × Morex population, plants were grown as described above but supplied with 50, 100, 150, 200, or 250 mM NaCl for ten days. Seventeen to twenty plants per genotype and treatment were sampled, and shoot biomass was recorded. This experiment was performed once.

To assess the transgene effect on plant biomass production, *A. thaliana* wild-type Col-0 and the same three transgenic lines were grown on non-sterile standardized plant growth substrate (Fruhstorfer Erde type P, Hawita, Vechta, Germany) with a pH of 6.0 in a climate chamber under short-day conditions (8 h light/16 h dark, 22 °C, 40–60% humidity, 300 μmol m^{-2} s^{-1}). After 2 weeks, plants were transferred into single pots and cultivated for further two weeks. Salt stress treatment was performed as described above for barley, with increasing concentration in a step-wise manner up to 150 mM NaCl and maintained for one week. At harvest, the rosette fresh weight of 25–27 plants per treatment and genotype was measured. These experiments were performed in triplicate. Statistical evaluation of rosette biomass was calculated using paired t-test, implemented in Sigma Plot 14.0 (Systat Software, Frankfurt am Main, Germany).

For analyzing root morphological effects, seeds of *A. thaliana* wild-type Col-0 and three transgenic lines expressing *HvHorcH* were surface sterilized. Seeds were germinated and grown on half-strength Murashige–Skoog (MS) medium (pH 5.8) and supplemented with 1.5% *w/v* sucrose for 2 weeks. Then, seedlings were transferred to a medium supplemented with or without 50 mM NaCl. Plants were maintained at 18/20 °C under an 8 h light/16 h dark rhythm (300 μmol m^{-2} s^{-1}) for one week. These experiments were performed in triplicate.

4.2. Yeast Strain and Cultivation

The salt-sensitive yeast *Saccharomyces cerevisiae* strain YSH818 (MATa *leu2-3/112 ura3-1 trp1-1 his3-11/15 ade2-1 can1-100 GAL SUC2 hog1Δ::LEU2*) [60] was used for barley cDNA testing. Yeast cultures were grown under selective conditions on synthetic dextrose minimal medium (0.67% bacto-yeast nitrogen base without amino acids, 1.6% bacto-agar) at 30 °C with 2% glucose as carbon source and the addition of amino acids [9].

4.3. CDNA Library Construction and cDNA Isolation

A cDNA expression library was constructed in the *S. cerevisiae* strain YSH818 using the Clone Miner cDNA Library Construction Kit (Thermo Fisher Scientific, Waltham, United States), according to the manufacturer's instructions. RNA was extracted from roots of salt-stress-treated barley cv. Morex using the RNeasy Plant Mini Kit (Qiagen, Hilden, Germany), and mRNA was enriched by using the polyA Spin™ mRNA Isolation Kit (New England Biolabs, Ipswich, MA, United States). A total of 2 μg of mRNA was used for cDNA library construction, leading to the integration of cDNAs into pDONOR222 vector. The expression library (8×10^5 colony forming units, average insert size of 950 bp) was subsequently shuttled into pYES-DEST52 vector for yeast transformation, as described before [9]. Transformed yeast cells were plated onto the medium as described above, with additional 2% NaCl. Transformants with increased salt tolerance after four days of growth, as compared to the empty vector control, were isolated; cDNA was extracted, sequenced, and used for retransforming YSH818 control cells. For the re-evaluation of cDNAs, 2% and 3% NaCl were added to the medium to impose salt stress.

4.4. Database Search and Sequence Alignments

Barley cDNAs that were extracted from yeast cells were sequenced, and sequences were compared to barlex CDS_HC May2016 database [61] using blastx (BARLEX. https://apex.ipk-gatersleben.de/apex/f?p=284:10, accessed on 24 August 2021). The amino acid sequence alignment was performed using the T-Coffee program and clustalW algorithm [62]. A conserved domain database search was performed using NCBI's protein–protein blast [63].

4.5. Root Morphological Analysis

After a ten-day exposure to either control conditions or 150 mM NaCl, the plant roots of all six barley cultivars were transferred to a flat water-filled tray. After detangling the roots using a toothpick, images were captured by a conventional digital camera. Root structure was determined from the images. Branch points (branches), end points (ends), and points of intersection (crossings) were detected as described [6]. Numbers of branches, ends, and crossings were graphically displayed as a function of the distance from the root base in mm by a Sholl analysis [64]. Three biological experiments were performed, and for each cultivar and treatment, the roots of four to five plants were analyzed, respectively.

4.6. Expression of Recombinant HvHorcH in Escherichia coli

For expression in *E. coli*, the barley EST clone GCW003B01r, coding for HvHorcH, was retrieved from CR-EST database (CR-EST. https://apex.ipk-gatersleben.de/apex/f?p=CREST:1:14690193541182::NO:::, accessed on 11 June 2009) and introduced into the pCR 2.1. Topo vector (Thermo Fisher Scientific, USA). By PCR, a SalI site at the 3′-end and a BamHI site at the 5′-end were introduced. The resulting fragment was cloned into the pQE30 expression vector by using the QIAexpress Type IV Kit (Qiagen, Hilden, Germany) and transformed into *E. coli* strain XL1Blue (Stratagene, Amsterdam, The Netherlands). Recombinant HvHorcH was purified using affinity chromatography on Ni-NTA agarose

(Qiagen, Hilden, Germany) following the manufacturer's instructions. For immunodetection, polyclonal antibodies were raised against recombinant HvHorcH in rabbits.

4.7. Protein Extraction, SDS-Polyacrylamide Gel Electrophoresis (PAGE), and Gel Blot Analysis

Proteins were isolated from total roots and root tips from plant material harvested seven days after final salt stress (150 mM NaCl) application. The root material from five plants of each genotype was combined for protein extraction, and all the experiments were performed in triplicates. The frozen root material was homogenized under liquid nitrogen to a fine powder, mixed with TCA/acetone solution (10% (*w*/*v*) TCA, 0.07% (*w*/*v*) 2-mercaptoethanol in acetone) in a ratio of 100 mg to 1 mL, and incubated for 45 min at –20 C°. The precipitate was pelleted by centrifugation, washed twice with 0.07% (*w*/*v*) 2-mercaptoethanol in acetone, and dried in a vacuum centrifuge. Protein precipitates were dissolved in loading buffer (56 mM Na_2CO_3, 56 mM dithioerythritol, 0,1% (*w*/*v*) sodium dodecyl sulfate, 12% saccharose, and 0,01% bromophenol blue). Quantification was performed using a Bradford assay (Bio-Rad, Hercules, CA, United States), using bovine serum albumin as a standard.

Samples of each extract with equal amounts of protein (2 µg) were subjected to SDS-polyacrylamide gel electrophoresis [65] on 15% acrylamide gels and transferred to an Immobilon-P PVDF membrane (Merck Millipore, Darmstadt, Germany) using a semi-dry blotting apparatus (Schütt, Göttingen, Germany). Membranes were probed with anti-HvHorcH polyclonal antiserum followed by the incubation with infrared dye-coupled secondary antibody (Li-Cor, Bad Homburg, Germany). Documentation and quantification was performed on a Li-Cor scanner driven by Odyssey software v3.0 (Li-Cor, Bad Homburg, Germany). The MagicMark™ XP Western Protein Standard (Thermo Fisher Scientific, Waltham, MA, United States) served as the protein standard.

4.8. Microscopy

For immune-histochemical examination, 2 mm long root tip sections were collected. After microwave-assisted fixation and LR White resin infiltration using a PELCO e BioWave® Pro+ (TedPella, Redding, CA, United States), according to Table S1, resin was polymerized at 60 °C for 48h.

Median longitudinal root sections of 0.5 µm thickness were cut on an Ultracut UCT instrument (Leica, Nussloch, Germany) and mounted on 8-well poly-L lysine coated slides (Thermo Scientific, Waltham, CA, United States). To avoid non-specific antibody binding, sections were blocked at room temperature (RT) for 20 min in PBS buffer containing 3% (*w/v*) bovine serum albumin and 0.1% Tween, followed by washing with washing buffer (0.1% bovine serum albumin and 0.05% Tween 20 in PBS) for 2 min at RT. After primary antibody incubation with polyclonal rabbit anti-HvHorcH (see Section 4.7, 1:10 and 1:100 diluted) for 45 min at 37 °C, probes were washed five times for 5 min at RT with PBS buffer. Incubation with secondary antibody goat anti-rabbit Alexa fluor 488 (MoBiTec, Göttingen, Germany; dilution 1:200) for 30 min at 37 °C was followed by five final washes with a washing buffer for 5 min at RT. Sections were mounted in antifade (1 mL 0.1 M Tris, pH 9.0, 9 mL glycerin, 50 mg n-propyl gallate). Immunolabeling was analyzed using a LSM 510META confocal microscope (Carl Zeiss, Jena, Germany). Alexa 488 was visualized with a 488 nm laser line in combination with a 505–530 nm bandpass filter. For controls, sections were incubated with Alexa 488 only.

4.9. Collection of Apoplastic Fluid

Apoplast infiltration with 1 × phosphate buffered saline of pH 7.0 containing 300 mM mannose was carried out by submerging the roots in the buffer in a 50 mL falcon tube and infiltrating under a vacuum of exactly 20 mPa generated by a motor pump for 30 s. The centrifugation of the surface-dried roots was carried out in a 5 mL syringe placed within a 15 mL falcon tube at a centrifugation speed of 400 × *g*. The presence of HvHorcH and GAPDH (Agrisera AB, Vännäs, Sweden) in the respective fractions was validated by immunoblotting, as described before.

4.10. *Affinity Enrichment of HvHorcH by Mannose Binding*

Roots from salt-stressed cv. Morex plants were ground to a fine powder under liquid nitrogen. Proteins were isolated on ice by grinding the root material with a cold extraction buffer (20 mM tris(hydroxymethyl) aminomethane (Tris) HCl, pH 7; 300 mM mannose; 1 mM dithiothreitol, and with protease (cOmplete™) and phosphatase inhibitor (PhosSTOP™, Roche Diagnostics GmbH, Mannheim, Germany)) in a ratio of 2.5 g to 4 mL. Insoluble material was sedimented by centrifugation (10,000× *g*, 4 °C, 15 min). The supernatant was transferred to a fresh tube and the pellet re-extracted with 2 mL of cold extraction buffer. After centrifugation (10,000× *g*, 4 °C, 15 min), both supernatants were combined, resulting in about 5 mL of crude extract. The extract was desalted, small contaminants removed, and the buffer exchanged to an anion exchange buffer (20 mM 2-(N-morpholino)ethanesulfonic acid, 1 mM dithiothreitol, adjusted to pH 6.5 with 1 N NaOH) by gel filtration on PD10 columns (GE Healthcare, Chicago, IL, USA) as instructed by the manufacturer. After elution from PD10 columns, an extract volume of about 7 mL was reached. Anion exchange chromatography was performed on a 6 mL Ressource™Q column (GE Healthcare) equilibrated with five column volumes (CV) of anion exchange buffer and utilizing an automated Äkta start protein purification system (GE Healthcare, Chicago, IL, USA). The protein extract was loaded on the column in anion exchange buffer at a flow rate of 2 mL/min. Sequential elution of bound proteins was realized by a linear gradient from 0 to 500 mM of NaCl in anion exchange buffer, and fractions of 3 mL each were collected. SDS-PAGE and protein gel blot analysis of the purified fractions were performed as stated above. Fractions containing HvHorcH were combined (about 18 mL) and subjected to dialysis (4 °C, overnight) against a minimum of 40 volumes of affinity chromatography buffer (50 mM Tris HCl, pH 7.2; 100 mM NaCl; 1 mM $MgCl_2$) utilizing 2 mL PlusOneTM Mini Dialysis Kit tubes with a 1 kDa cut-off (GE Healthcare, Chicago, PD, United States) according to the manufacturer's instruction. The resulting fraction volume was stable with about 18 mL. Affinity chromatography was performed using D-mannose agarose (Sigma-Aldrich, United States) with a protein-binding capacity of about 40 mg/mL. A 20 mL empty Poly-prep® gravity-flow column (Bio-Rad, Hercules, CA, United States) was filled with 0.6 mL of D-mannose agarose matrix and equilibrated with 20 CV of affinity chromatography buffer. The dialyzed protein extract (18 mL) was added and the column sealed and incubated at 4 °C on an overhead shaker for a minimum of one hour. The column was then placed on a gravity stand and the flow-through was collected. Unbound proteins were washed off with 13 CV of affinity chromatography buffer (8 mL). Then, two times 1 mL of elution buffer 1 (50 mM MES, pH 6.5 adjusted with 1 N NaOH, 300 mM D-mannose, and with protease (cOmplete™, Roche Diagnostics GmbH, Mannheim, Germany) and phosphatase inhibitor (PhosSTOP™, Roche Diagnostics GmbH, Mannheim, Germany)) was applied, and fractions were collected (2 × 1 mL). Next, 1 mL of elution buffer 2 (50 mM citric acid, pH 2.4 adjusted with 1 N NaOH, 100 mM NaCl, 1 mM $MgCl_2$, 300 mM D-mannose) was applied, and the fraction was collected (1 mL). SDS-PAGE and protein gel blot analysis of the purified fractions were performed as stated above. HvHorcH was mainly found in the first fraction of elution 1, which was stored at −20 °C until further usage.

4.11. *Generation of Transgenic Arabidopsis Plants*

For the generation of transgenic *A. thaliana* plants, the Col-0 ecotype was used as the genetic background. Plants were grown in a greenhouse under long-day (16 h of light/8 h of darkness) conditions at 22 °C. For the generation of pGSTu:HvHorcH;eGFP transgenic plants, 1.36 kb pGSTu25 promoter (pGSTu25) and eGFP were subcloned together with HvHorcH ORF (see Section 4.6) into Gateway® Donor vectors. All three components were introduced into pB7m34GW expression vector using LR Clonase™ II Plus (Thermo

Scientific, Waltham, MA, United States). Constructs were subsequently introduced into *Agrobacterium tumefaciens* strain GV3101, and wild-type plants were transformed using the floral dip method [66]. Transgenic lines were selected on medium containing phosphinothricin (PPT).

Supplementary Materials: The following are available online at https://www.mdpi.com/article/10.3390/ijms221910248/s1.

Author Contributions: K.W., A.M., G.K. and H.-P.M. contributed to the conception of the research; K.W., A.M., U.B., T.A., T.R. and E.R. conducted the research; K.W., A.M., U.B., T.R., E.R., M.M. and H.-P.M. analyzed the data. K.W., A.M. and H.-P.M. wrote the manuscript with contributions from all authors. All authors have read and agreed to the published version of the manuscript.

Funding: This research received no external funding.

Institutional Review Board Statement: Not applicable.

Informed Consent Statement: Not applicable.

Data Availability Statement: The data presented in this study are available in the article and Supplementary Materials.

Acknowledgments: We thank Heike Ernst, Kirsten Hoffie, Christa Kallas, Petra Linow, Monika Wiesner, and Annegret Wolf for their excellent technical assistance. Andreas Börner (Genebank, Leibniz Institute of Plant Genetics and Crop Plant Research) is acknowledged for providing grains of barley accessions.

Conflicts of Interest: The authors declare that the research was conducted in the absence of any commercial or financial relationships that could be construed as a potential conflict of interest.

References

1. Butcher, K.; Wick, A.F.; DeSutter, T.; Chatterjee, A.; Harmon, J. Soil salinity: A threat to global food security. *Agron. J.* **2016**, *108*, 2189–2200. [CrossRef]
2. Roy, S.J.; Negrao, S.; Tester, M. Salt resistant crop plants. *Curr. Opin. Biotechnol.* **2014**, *26*, 115–124. [CrossRef] [PubMed]
3. Francois, L.E.; Maas, E.V. Crop Response and Management of Salt-Affected Soils. In *Handbook of Plant and Crop Stress*; Marcel Dekker Press Inc.: New York, NY, USA, 1999; pp. 169–201.
4. Munns, R.; Tester, M. Mechanisms of salinity tolerance. *Annu. Rev. Plant Biol.* **2008**, *59*, 651–681. [CrossRef] [PubMed]
5. Dwivedi, S.L.; Ceccarelli, S.; Blair, M.W.; Upadhyaya, H.D.; Are, A.K.; Ortiz, R. Landrace germplasm for improving yield and abiotic stress adaptation. *Trends Plant Sci.* **2016**, *21*, 31–42. [CrossRef] [PubMed]
6. Witzel, K.; Matros, A.; Møller, A.L.B.; Ramireddy, E.; Finnie, C.; Peukert, M.; Rutten, T.; Herzog, A.; Kunze, G.; Melzer, M.; et al. Plasma membrane proteome analysis identifies a role of barley membrane steroid binding protein in root architecture response to salinity. *Plant Cell Environ.* **2018**, *41*, 1311–1330. [CrossRef]
7. Witzel, K.; Matros, A.; Strickert, M.; Kaspar, S.; Peukert, M.; Mühling, K.H.; Börner, A.; Mock, H.-P. Salinity stress in roots of contrasting barley genotypes reveals time-distinct and genotype-specific patterns for defined proteins. *Mol. Plant* **2014**, *7*, 336–355. [CrossRef]
8. Witzel, K.; Weidner, A.; Surabhi, G.-K.; Börner, A.; Mock, H.-P. Salt stress-induced alterations in the root proteome of barley genotypes with contrasting response towards salinity. *J. Exp. Bot.* **2009**, *60*, 3545–3557. [CrossRef]
9. Witzel, K.; Weidner, A.; Surabhi, G.-K.; Varshney, R.K.; Kunze, G.; Buck-Sorlin, G.H.; Börner, A.; Mock, H.-P. Comparative analysis of the grain proteome fraction in barley genotypes with contrasting salinity tolerance during germination. *Plant Cell Environ.* **2010**, *33*, 211–222. [CrossRef]
10. Garciadeblas, B.; Haro, R.; Benito, B. Cloning of two SOS1 transporters from the seagrass *Cymodocea nodosa*. SOS1 transporters from *Cymodocea* and *Arabidopsis* mediate potassium uptake in bacteria. *Plant Mol. Biol.* **2007**, *63*, 479–490. [CrossRef]
11. Hohmann, S. Osmotic stress signaling and osmoadaptation in yeasts. *Microbiol. Mol. Biol. Rev.* **2002**, *66*, 300–372. [CrossRef]
12. Nozawa, A.; Miwa, K.; Kobayashi, M.; Fujiwara, T. Isolation of *Arabidopsis thaliana* cDNAs that confer yeast boric acid tolerance. *Biosci. Biotechnol. Biochem.* **2006**, *70*, 1724–1730. [CrossRef]
13. Quintero, F.J.; Ohta, M.; Shi, H.Z.; Zhu, J.K.; Pardo, J.M. Reconstitution in yeast of the Arabidopsis SOS signaling pathway for Na^+ homeostasis. *Proc. Natl. Acad. Sci. USA* **2002**, *99*, 9061–9066. [CrossRef] [PubMed]
14. Serrano, R.; Culianz-Macia, F.A.; Moreno, V. Genetic engineering of salt and drought tolerance with yeast regulatory genes. *Sci. Hortic.* **1999**, *78*, 261–269. [CrossRef]
15. Mulet, J.M.; Alemany, B.; Ros, R.; Calvete, J.J.; Serrano, R. Expression of a plant serine O-acetyltransferase in *Saccharomyces cerevisiae* confers osmotic tolerance and creates an alternative pathway for cysteine biosynthesis. *Yeast* **2004**, *21*, 303–312. [CrossRef] [PubMed]

16. Patankar, H.V.; Al-Harrasi, I.; Al-Yahyai, R.; Yaish, M.W. Identification of candidate genes involved in the salt tolerance of date palm (*Phoenix dactylifera* L.) based on a yeast functional bioassay. *DNA Cell Biol.* **2018**, *37*, 524–534. [CrossRef] [PubMed]
17. Liu, N.; Chen, A.-P.; Zhong, N.-Q.; Wang, F.; Wang, H.-Y.; Xia, G.-X. Functional screening of salt stress-related genes from *Thellungiella halophila* using fission yeast system. *Physiol. Plant.* **2007**, *129*, 671–678. [CrossRef]
18. Batelli, G.; Massarelli, I.; van Oosten, M.; Nurcato, R.; Vannini, C.; Raimondi, G.; Leone, A.; Zhu, J.K.; Maggio, A.; Grillo, S. Asg1 is a stress-inducible gene which increases stomatal resistance in salt stressed potato. *J. Plant Physiol.* **2012**, *169*, 1849–1857. [CrossRef]
19. Massarelli, I.; Cioffi, R.; Batelli, G.; de Palma, M.; Costa, A.; Grillo, S.; Leone, A. Functional screening of plant stress-related cDNAs by random over-expression in *Escherichia coli*. *Plant Sci.* **2006**, *170*, 880–888. [CrossRef]
20. Joshi, A.; Dang, H.Q.; Vaid, N.; Tuteja, N. Pea lectin receptor-like kinase promotes high salinity stress tolerance in bacteria and expresses in response to stress in planta. *Glycoconj. J.* **2010**, *27*, 133–150. [CrossRef]
21. Grunwald, I.; Heinig, I.; Thole, H.H.; Neumann, D.; Kahmann, U.; Kloppstech, K.; Gau, A.E. Purification and characterisation of a jacalin-related, coleoptile specific lectin from Hordeum vulgare. *Planta* **2007**, *226*, 225–234. [CrossRef]
22. Pinedo, M.; Orts, F.; de Oliveira Carvalho, A.; Regente, M.; Soares, J.R.; Gomes, V.M.; de la Canal, L. Molecular characterization of Helja, an extracellular jacalin-related protein from *Helianthus annuus*: Insights into the relationship of this protein with unconventionally secreted lectins. *J. Plant Physiol.* **2015**, *183*, 144–153. [CrossRef] [PubMed]
23. Weidner, A.; Varshney, R.K.; Buck-Sorlin, G.H.; Stein, N.; Graner, A.; Börner, A. QTLs for salt tolerance in three different barley mapping populations. In Proceedings of the 13th International EWAC Conference, Prague, Czech Republic, 27 June–1 July 2005; Börner, A., Pánková, K., Snape, J.W., Eds.
24. Bellande, K.; Bono, J.-J.; Savelli, B.; Jamet, E.; Canut, H. Plant lectins and lectin receptor-like kinases: How do they sense the outside? *Int. J. Mol. Sci.* **2017**, *18*, 1164. [CrossRef] [PubMed]
25. Bunn-Moreno, M.M.; Campos-Neto, A. Lectin(s) extracted from seeds of *Artocarpus integrifolia* (jackfruit): Potent and selective stimulator(s) of distinct human T and B cell functions. *J. Immunol.* **1981**, *127*, 427–429.
26. Esch, L.; Schaffrath, U. An update on jacalin-like lectins and their role in plant defense. *Int. J. Mol. Sci.* **2017**, *18*, 11. [CrossRef]
27. Xiang, Y.; Song, M.; Wei, Z.; Tong, J.; Zhang, L.; Xiao, L.; Ma, Z.; Wang, Y. A jacalin-related lectin-like gene in wheat is a component of the plant defence system. *J. Exp. Bot.* **2011**, *62*, 5471–5483. [CrossRef]
28. Al Atalah, B.; Smagghe, G.; van Damme, E.J.M. Orysata, a jacalin-related lectin from rice, could protect plants against biting-chewing and piercing-sucking insects. *Plant Sci.* **2014**, *221–222*, 21–28. [CrossRef]
29. Xiao, J.; Li, C.H.; Xu, S.J.; Xing, L.J.; Xu, Y.Y.; Chong, K. JACALIN-LECTIN LIKE1 Regulates the Nuclear Accumulation of GLYCINE-RICH RNA-BINDING PROTEIN7, Influencing the RNA Processing of *FLOWERING LOCUS C* Antisense Transcripts and Flowering Time in Arabidopsis. *Plant Physiol.* **2015**, *169*, 2102–2117.
30. Menhaj, A.R.; Mishra, S.K.; Bezhani, S.; Kloppstech, K. Posttranscriptional control in the expression of the genes coding for high-light-regulated HL#2 proteins. *Planta* **1999**, *209*, 406–413.
31. Zhang, W.; Peumans, W.J.; Barre, A.; Astoul, C.H.; Rovira, P.; Rougé, P.; Proost, P.; Truffa-Bachi, P.; Jalali, A.A.; Van Damme, E.J. Isolation and characterization of a jacalin-related mannose-binding lectin from salt-stressed rice (*Oryza sativa*) plants. *Planta* **2000**, *210*, 970–978. [PubMed]
32. Song, M.; Xu, W.Q.; Xiang, Y.; Jia, H.Y.; Zhang, L.X.; Ma, Z.Q. Association of jacalin-related lectins with wheat responses to stresses revealed by transcriptional profiling. *Plant Mol. Biol.* **2014**, *84*, 95–110. [CrossRef]
33. Abebe, T.; Skadsen, R.W.; Kaeppler, H.F. A proximal upstream sequence controls tissue-specific expression of Lem2, a salicylate-inducible barley lectin-like gene. *Planta* **2005**, *221*, 170–183. [CrossRef] [PubMed]
34. Lee, J.; Parthier, B.; Löbler, M. Jasmonate signalling can be uncoupled from abscisic acid signalling in barley: Identification of jasmonate-regulated transcripts which are not induced by abscisic acid. *Planta* **1996**, *199*, 625–632. [CrossRef]
35. Lerner, D.R.; Raikhel, N.V. Cloning and characterization of root-specific barley lectin. *Plant Physiol.* **1989**, *91*, 124–129. [CrossRef] [PubMed]
36. Regente, M.; Taveira, G.B.; Pinedo, M.; Elizalde, M.M.; Ticchi, A.J.; Diz, M.S.S.; Carvalho, A.O.; de la Canal, L.; Gomes, V.M. A sunflower lectin with antifungal properties and putative medical mycology applications. *Curr. Microbiol.* **2014**, *69*, 88–95. [CrossRef] [PubMed]
37. Pinedo, M.; Regente, M.; Elizalde, M.; Quiroga, I.Y.; Pagnussat, L.A.; Jorrin-Novo, J.; Maldonado, A.; de la Canal, L. Extracellular sunflower proteins: Evidence on non-classical secretion of a jacalin-related lectin. *Protein Pept. Lett.* **2012**, *19*, 270–276. [CrossRef]
38. Olbrich, A.; Hillmer, S.; Hinz, G.; Oliviusson, P.; Robinson, D.G. Newly formed vacuoles in root meristems of barley and pea seedlings have characteristics of both protein storage and lytic vacuoles. *Plant Physiol.* **2007**, *145*, 1383–1394. [CrossRef] [PubMed]
39. Cammue, B.; Stinissen, H.M.; Peumans, W.J. Lectin in vegetative tissues of adult barley plants grown under field conditions. *Plant Physiol.* **1985**, *78*, 384–387. [CrossRef]
40. Cammue, B.P.A.; Broekaert, W.F.; Kellens, J.T.C.; Raikhel, N.V.; Peumans, W.J. Stress-induced accumulation of wheat germ agglutinin and abscisic acid in roots of wheat seedlings. *Plant Physiol.* **1989**, *91*, 1432–1435. [CrossRef]
41. Shakirova, F.M.; Bezrukova, M.V.; Aval'baev, A.M.; Fatkhutdinova, R.A. Control mechanisms of lectin accumulation in wheat seedlings under salinity. *Russ. J. Plant Physiol.* **2003**, *50*, 301–304. [CrossRef]

42. Mattos-Moreira, L.A.; Ferreira, C.F.; Amorim, E.P.; Pirovani, C.P.; de Andrade, E.M.; Filho, M.A.C.; da Silva Ledo, C.A. Differentially expressed proteins associated with drought tolerance in bananas (*Musa* spp.). *Acta Physiol. Plant.* **2018**, *40*, 60. [CrossRef]
43. Claes, B.; Dekeyser, R.; Villarroel, R.; Vandenbulcke, M.; Bauw, G.; Vanmontagu, M.; Caplan, A. Characterization of a rice gene showing organ-specific expression in response to salt stress and drought. *Plant Cell* **1990**, *2*, 19–27. [PubMed]
44. Garcia, A.B.; Engler, J.A.; Claes, B.; Villarroel, R.; van Montagu, M.; Gerats, T.; Caplan, A. The expression of the salt-responsive gene sal T from rice is regulated by hormonal and developmental cues. *Planta* **1998**, *207*, 172–180. [CrossRef] [PubMed]
45. de Souza, G.A.; Ferreira, B.S.; Dias, J.M.; Queiroz, K.S.; Branco, A.T.; Bressan-Smith, R.E.; Oliveira, J.G.; Garcia, A.B. Accumulation of SALT protein in rice plants as a response to environmental stresses. *Plant Sci.* **2003**, *164*, 623–628.
46. Hirano, K.; Teraoka, T.; Yamanaka, H.; Harashima, A.; Kunisaki, A.; Takahashi, H.; Hosokawa, D. Novel mannose-binding rice lectin composed of some isolectins and its relation to a stress-inducible salT gene. *Plant Cell Physiol.* **2000**, *41*, 258–267. [CrossRef] [PubMed]
47. Bezrukova, M.V.; Fatkhutdinova, R.A.; Shakirova, F.M. Protective effect of wheat germ agglutinin on the course of mitosis in the roots of *Triticum aestivum* seedlings exposed to cadmium. *Russ. J. Plant Physiol.* **2016**, *63*, 358–364. [CrossRef]
48. Bezrukova, M.V.; Lubyanova, A.R.; Fatkhutdinova, R.A. The involvement of wheat and common bean lectins in the control of cell division in the root apical meristems of various plant species. *Russ. J. Plant Physiol.* **2011**, *58*, 174–180. [CrossRef]
49. Rutter, B.D.; Innes, R.W. Extracellular vesicles isolated from the leaf apoplast carry stress-response proteins. *Plant Physiol.* **2017**, *173*, 728–741. [CrossRef]
50. Chang, C.-F. Carbohydrate Microarray. In *Carbohydrates*; Chang, C.-F., Ed.; IntechOpen: London, UK, 2012.
51. Wilson, S.M.; Burton, R.A.; Collins, H.M.; Doblin, M.S.; Pettolino, F.A.; Shirley, N.; Fincher, G.B.; Bacic, A. Pattern of deposition of cell wall polysaccharides and transcript abundance of related cell wall synthesis genes during differentiation in barley endosperm. *Plant Physiol.* **2012**, *159*, 655–670. [CrossRef]
52. Little, A.; Lahnstein, J.; Jeffery, D.W.; Khor, S.F.; Schwerdt, J.G.; Shirley, N.J.; Hooi, M.; Xing, X.; Burton, R.A.; Bulone, V. A novel (1,4)-β-linked glucoxylan is synthesized by members of the cellulose synthase-like F gene family in land plants. *ACS Cent. Sci* **2019**, *5*, 73–84. [CrossRef]
53. Burton, R.A.; Jobling, S.A.; Harvey, A.J.; Shirley, N.J.; Mather, D.E.; Bacic, A.; Fincher, G.B. The genetics and transcriptional profiles of the cellulose synthase-like HvCslF gene family in barley. *Plant Physiol.* **2008**, *146*, 1821–1833. [CrossRef]
54. Endler, A.; Kesten, C.; Schneider, R.; Zhang, Y.; Ivakov, A.; Froehlich, A.; Funke, N.; Persson, S. A mechanism for sustained cellulose synthesis during salt stress. *Cell* **2015**, *162*, 1353–1364. [CrossRef]
55. Byrt, C.S.; Munns, R.; Burton, R.A.; Gilliham, M.; Wege, S. Root cell wall solutions for crop plants in saline soils. *Plant Sci.* **2018**, *269*, 47–55. [CrossRef]
56. Tenhaken, R. Cell wall remodeling under abiotic stress. *Front. Plant Sci.* **2015**, *5*, 771. [CrossRef] [PubMed]
57. Legentil, L.; Paris, F.; Ballet, C.; Trouvelot, S.; Daire, X.; Vetvicka, V.; Ferrieres, V. Molecular interactions of β-(1->3)-glucans with their receptors. *Molecules* **2015**, *20*, 9745–9766. [CrossRef]
58. Melida, H.; Sopena-Torres, S.; Bacete, L.; Garrido-Arandia, M.; Jorda, L.; Lopez, G.; Munoz-Barrios, A.; Pacios, L.F.; Molina, A. Non-branched β-1,3-glucan oligosaccharides trigger immune responses in *Arabidopsis*. *Plant J.* **2018**, *93*, 34–49. [CrossRef] [PubMed]
59. Jayaprakash, N.G.; Singh, A.; Vivek, R.; Yadav, S.; Pathak, S.; Trivedi, J.; Jayaraman, N.; Nandi, D.; Mitra, D.; Surolia, A. The barley lectin, horcolin, binds high-mannose glycans in a multivalent fashion, enabling high-affinity, specific inhibition of cellular HIV infection. *J. Biol. Chem.* **2020**, *295*, 12111–12129. [CrossRef] [PubMed]
60. Albertyn, J.; Hohmann, S.; Thevelein, J.M.; Prior, B.A. Gpd1, which encodes glycerol-3-phosphate dehydrogenase, is essential for growth under osmotic-stress in *Saccharomyces cerevisiae*, and its expression is regulated by the high-osmolarity glycerol response pathway. *Mol. Cell. Biol.* **1994**, *14*, 4135–4144. [CrossRef] [PubMed]
61. Colmsee, C.; Beier, S.; Himmelbach, A.; Schmutzer, T.; Stein, N.; Scholz, U.; Mascher, M. BARLEX–The Barley Draft Genome Explorer. *Mol. Plant* **2015**, *8*, 964–966. [CrossRef] [PubMed]
62. Di Tommaso, P.; Moretti, S.; Xenarios, I.; Orobitg, M.; Montanyola, A.; Chang, J.M.; Taly, J.F.; Notredame, C. T-Coffee: A web server for the multiple sequence alignment of protein and RNA sequences using structural information and homology extension. *Nucleic Acids Res.* **2011**, *39*, W13–W17. [CrossRef]
63. Marchler-Bauer, A.; Derbyshire, M.K.; Gonzales, N.R.; Lu, S.; Chitsaz, F.; Geer, L.Y.; Geer, R.C.; He, J.; Gwadz, M.; Hurwitz, D.I.; et al. CDD: NCBI's conserved domain database. *Nucleic Acids Res.* **2015**, *43*, D222–D226. [CrossRef]
64. Sholl, D.A. Dendritic organization in the neurons of the visual and motor cortices of the cat. *J. Anat.* **1953**, *87*, 387–406. [PubMed]
65. Laemmli, U.K. Cleavage of structural proteins during assembly of head of bacteriophage T4. *Nature* **1970**, *227*, 680–685. [CrossRef] [PubMed]
66. Clough, S.J.; Bent, A.F. Floral dip: A simplified method for *Agrobacterium*-mediated transformation of *Arabidopsis thaliana*. *Plant J.* **1998**, *16*, 735–743. [CrossRef] [PubMed]

International Journal of
Molecular Sciences

Article

A C2H2-Type Zinc-Finger Protein from *Millettia pinnata*, MpZFP1, Enhances Salt Tolerance in Transgenic Arabidopsis

Zhonghua Yu [1,2,†], Hao Yan [3,†], Ling Liang [1,4], Yi Zhang [1], Heng Yang [1,4], Wei Li [2,5], Jaehyuck Choi [2], Jianzi Huang [3,*] and Shulin Deng [1,6,7,*]

1 Guangdong Provincial Key Laboratory of Applied Botany, South China Botanical Garden, Chinese Academy of Sciences, Guangzhou 510650, China; yuee.zhonghua2000@gmail.com (Z.Y.); liangling20@mails.ucas.ac.cn (L.L.); yizhang@scbg.ac.cn (Y.Z.); hengy@scbg.ac.cn (H.Y.)
2 Department of Landscape Architecture, PaiChai University, Deajeon 35345, Korea; liwei7137@gmail.com (W.L.); jhchoi@pcu.ac.kr (J.C.)
3 Guangdong Provincial Key Laboratory for Plant Epigenetics, College of Life Sciences and Oceanography, Shenzhen University, Shenzhen 518060, China; yanhaonihao@sina.com
4 College of Life Sciences, University of Chinese Academy of Sciences, Beijing 100049, China
5 School of Biological and Food Processing Engineering, Huanghuai University, Zhumadian 463000, China
6 Xiaoliang Research Station for Tropical Coastal Ecosystems, South China Botanical Garden, Chinese Academy of Sciences, Guangzhou 510650, China
7 National Engineering Research Center of Navel Orange, Gannan Normal University, Ganzhou 341000, China
* Correspondence: biohjz@szu.edu.cn (J.H.); sldeng@scbg.ac.cn (S.D.)
† These authors contributed equally to this work.

Citation: Yu, Z.; Yan, H.; Liang, L.; Zhang, Y.; Yang, H.; Li, W.; Choi, J.; Huang, J.; Deng, S. A C2H2-Type Zinc-Finger Protein from *Millettia pinnata*, MpZFP1, Enhances Salt Tolerance in Transgenic Arabidopsis. *Int. J. Mol. Sci.* **2021**, *22*, 10832. https://doi.org/10.3390/ijms221910832

Academic Editors: Mirza Hasanuzzaman and Masayuki Fujita

Received: 31 August 2021
Accepted: 3 October 2021
Published: 7 October 2021

Abstract: C2H2 zinc finger proteins (ZFPs) play important roles in plant development and response to abiotic stresses, and have been studied extensively. However, there are few studies on ZFPs in mangroves and mangrove associates, which represent a unique plant community with robust stress tolerance. *MpZFP1*, which is highly induced by salt stress in the mangrove associate *Millettia pinnata*, was cloned and functionally characterized in this study. MpZFP1 protein contains two zinc finger domains with conserved QALGGH motifs and targets to the nucleus. The heterologous expression of *MpZFP1* in Arabidopsis increased the seeds' germination rate, seedling survival rate, and biomass accumulation under salt stress. The transgenic plants also increased the expression of stress-responsive genes, including *RD22* and *RD29A*, and reduced the accumulation of reactive oxygen species (ROS). These results indicate that MpZFP1 is a positive regulator of plant responses to salt stress due to its activation of gene expression and efficient scavenging of ROS.

Keywords: C2H2 zinc finger protein; heterologous expression; *Millettia pinnata*; salt tolerance

1. Introduction

Arable lands are suffering from continuous salinization at an annual rate of ~10% due to environmental changes and poor cultural practices. Approximately 50% of the total cultivated land area was predicted to be salinized by the year 2050 worldwide [1]. Salt stress has multiple deleterious effects on plant growth and is a major environmental factor reducing crop productivity. Thus, improving the salt tolerance of crops through genetic modification is a potential approach for optimum economic sustainability [1,2]. The generation of salt-resistant crops relies on the discovery of plant stress-responsive mechanisms and the availability of genetic resources.

Pongamia (*Millettia pinnata* syn. *Pongamia pinnata*) belongs to the semi-mangrove (or mangrove associate) growing within intertidal zones in tropical and subtropical regions, and possesses a high degree of salt tolerance [3,4]. Unlike true mangroves, Pongamia does not feature the salty glands or other specialized morphological traits required to endure salinity stress. It is suggested that the mechanisms that Pongamia uses to cope with the high saline environment are tightly linked to gene regulation and protein function [3–5].

Therefore, investigating the molecular mechanisms of Pongamia's adaptation to saline environments offers promising insight into stress-tolerant crop breeding, and the salt tolerance genes derived from Pongamia may be highly efficient in the genetic modification of crops. The physiological mechanisms through which Pongamia responds to salt stress have been extensively studied [3–6]. These reports indicated that the hydrophobic cell-wall barriers and vacuolar sequestration of Na^+ were among the key mechanisms conferring salt tolerance in Pongamia, and that osmolytes (myo-inositol and mannitol) and phytohormone (zeatin and jasmonic acid) were increased in salt-treated Pongamia [5,6]. Nevertheless, these analyses of Pongamia's transcriptional regulation mechanisms are still insufficient for understanding its molecular responses to salinity stress. Transcriptome profiles of leaf and/or root tissues of Pongamia have been conducted to address the issue [7,8], while the detailed functional study of salt stress responsive gene is scarce. To date, only two salt-tolerant Pongamia genes, *MpCHI* and *MpCML40*, have been cloned and characterized [9,10]. The heterologous expression of *MpCHI* in yeast (*Saccharomyces cerevisiae*), which encodes the chalcone isomerase involved in flavonoid biosynthesis, could enhance the salt tolerance capacity of recipient cells [9]. Recently, we characterized a salt-induced Calmodulin-like gene, *MpCML40*, that enabled transgenic yeast and Arabidopsis to become salt-tolerant. The transgenic *MpCML40* Arabidopsis did not show any growth retardation, which differed from most genetically modified stress-tolerant plants [10].

Through transcriptional regulation, the complex network of plant development and abiotic stress responses is orchestrated by transcription factors (TFs), such as Zinc finger proteins (ZFPs) [11–13]. TFs bind to specific sequences in the promoters of their target genes, thereby regulating gene expression and affecting biological phenotypes. ZFPs form one of the largest TF families in plants and have been sub-classified into nine major types, C2H2, C2HC, C3H, C4, C6, C2HC5, C3HC4, C4HC3, and C8, according to the number and order of the cysteine and histidine residues, which bind tetrahedrally to zinc ions [11,14]. C2H2 ZFPs, also called classical or TFIIIA-type fingers, are one of the best-characterized DNA-binding motifs found in plant transcription factors, typically containing one to four conserved QALGGH motifs in zinc finger helices [11,14,15]. The Arabidopsis genome contains 176 C2H2-type ZFP genes and can be divided into three clades: A, B, and C. Clade C is further split into three sub-clades, C1, C2, and C3, based on sequence divergence [15,16]. The first C2H2-type ZFP gene discovered in plants was *EPF1*, from Petunia [12]. EPF1-related proteins have been demonstrated to enhance tolerance to abiotic stresses in several plant species [17–20].

In the past two decades, many C2H2 ZFP genes have been identified and studied in model plants, such as Arabidopsis, rice (*Oryza sativa*), soybean (*Glycine max*), and tomato (*Solanum lycopersicum*) [17,18,20–22]. The overexpression of Arabidopsis C2H2 zinc finger protein STZ/ZAT10 enhanced tolerance to salinity, heat and osmotic stress in transgenic Arabidopsis [18,23]. Another Arabidopsis C2H2 zinc finger protein, ZAT12, has been reported to be involved in oxidative, osmotic, salinity, high light, heat, and cold stress response [24,25]. The overexpression of rice C2H2 zinc finger protein genes, *ZFP179*, *ZFP182*, *ZFP245*, and *ZFP252*, has been shown to participate in salt, drought, and cold stress response [26–29]. The heterologous expression of the soybean C2H2 zinc finger gene *SCOF-1* has been shown to enhance low temperature stress tolerance in Arabidopsis and sweet potato [30,31]. Another soybean C2H2, GmZAT4, enhanced PEG and NaCl stress tolerance in transgenic Arabidopsis [32]. *GsZFP1*, a C2H2 zinc finger protein gene from wild soybean (*Glycine soja*), was induced by ABA and abiotic stress treatments. Transgenic *GsZFP1* Arabidopsis plants showed increased tolerance to cold and drought stresses by activating cold stress-responsive genes and ABA biosynthesis-related genes [33,34]. Transgenic alfalfa (*Medicago sativa* L.) expressing *GsZFP1* showed enhanced tolerance to cold and salt stresses [35]. In tomatoes, C2H2 zinc finger proteins have been reported to regulate trichome formation [36], ascorbic acid synthesis and salt tolerance [37], and mating system transition [38]. The heterologous expression of two C2H2 zinc finger protein genes, *CgZFP1* from chrysanthemum [39] and *IbZFP1* from sweet potato [19], improved salinity and drought tolerance in Arabidopsis. Some C2H2 ZFPs function in both biotic stress response and abiotic stress response. *CaZFP1*, a pathogen-induced pepper C2H2 zinc finger gene, endowed the transgenic Arabidopsis with drought tolerance and resistance against infection by *Pseudomonas syringae* [40]. Most of the available information suggests that ZFP genes from crops or other plant species are highly important for response to abiotic stresses, while few reports are available for mangroves or mangrove associates.

In the present study, we identified a ZFP gene, *MpZFP1*, from Pongamia. MpZFP1 belongs to the C1-2i subclass of C2H2 ZFPs containing two highly conserved QALGGH motifs. Under salt treatment, the *MpZFP1* gene was highly induced in roots. The heterologous expression of *MpZFP1* in Arabidopsis strongly enhanced the salt tolerance of transgenic plants, probably through efficient ROS scavenging.

2. Results

2.1. Cloning and Sequence Analysis of MpZFP1

The full-length cDNA of the *MpZFP1* gene was cloned by 5′ and 3′ rapid amplification of cDNA ends (RACE) assays with four gene specific primers, based on the sequence of an EST from our previous study [7]. The full-length sequence of cDNA, which was deposited in GenBank under accession number MZ934391, comprised a 543 bp open reading frame (ORF), a 111 bp 5′ untranslated region (UTR), and a 177 bp 3′ UTR (Figure 1A). The corresponding protein contained 180 amino acids, and conserve domain analysis showed that *MpZFP1* contained two C2H2 zinc finger motifs (Figure 1A). A BLASTP homolog search of GenBank indicated that the deduced *MpZFP1* sequences showed high similarity with the predicted ZFPs of *Glycine max* (XP_003552680.1 67%), *Solanum lycopersicum* (XP_004239776.1, 57%), *Petunia X hybirda* (BAA21923.1, 55%), *Vitis vinifera* (XP_002284111.1, 55%), *Ricinus communis* (XP_002528469.1, 53%), and Arabidopsis (AT2G37430.1/ZAT11, 43%) (Figure 1B). All of these proteins contained two C2H2 zinc finger domains with QALGGH-conserved motifs and a short hydrophobic region consisting of core DLNL sequence, DLN-box. Apart from MpZFP1, in which the first Leu mutated to Pro, other proteins contained a consensus LXLXLX EAR motif (Figure 1B). The Arabidopsis ZAT11 belongs to the C1-2i subclade, which includes 20 members [15]. To classify the possible Arabidopsis ortholog of MpZFP1, a phylogenetic tree based on the amino acid sequences of MpZFP1 and the Arabidopsis C1-2i subclass proteins was constructed. The architecture of the phylogenetic trees suggested that the Arabidopsis C1-2i could be divided into three subsets, ZAT7-8/ZAT11-12/ZAT16-18, together with MpZFP1 merged into subset I (Figure 1C).

2.2. MpZFP1 Localizes in the Nucleus

Classic transcription factors locate in the nucleus and bind to specific DNA sequences to modulate gene expression. To address the subcellular localization of MpZFP1 protein, the ORF of *MpZFP1* was cloned and then inserted into the pCAMBIA-1302 expressing vector containing the green fluorescent protein gene (GFP) under the control of the CaMV35S promoter. The MpZFP1-GFP fusion protein was expressed in tobacco (*Nicotiana benthamiana*) leaves by the Agrobacterium tumefaciens-mediated transient expression system. The leaves were stained with DAPI ((4′,6-diamidino-2-phenylindole), which binds to double-stranded DNA as a nucleus marker, two hours before fluorescence microscopy detection. The results showed that the MpZFP1-GFP was co-localized with the DAPI signals (Figure 2). These results suggested that *MpZFP1* localized in the nucleus.

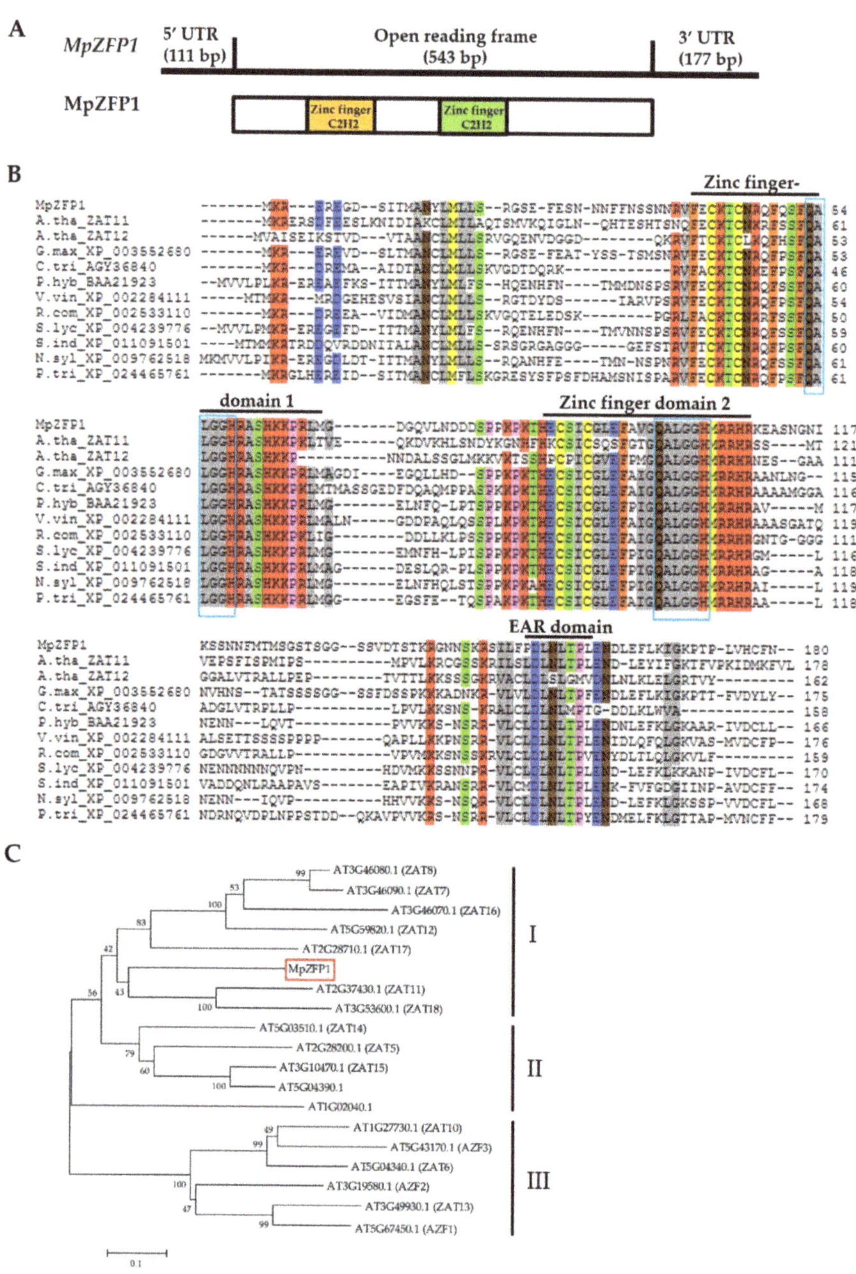

Figure 1. Analysis of the *MpZFP1* gene from Pongamia. (**A**) Structure of *MpZFP1* gene (upper panel) and MpZFP1 protein (lower panel). (**B**) Sequence alignment of MpZFP1 protein with its homologs from other species. The conserved zinc finger domains and EAR motif were marked and the two QALGGH consensus sequences are highlighted with a blue frame. Sequences of Glycine max (XP_003552680.1), Citrus trifoliata (AGY36840), Solanum lycopersicum (XP_004239776.1), Petunia X hybirda (BAA21923.1), Vitis vinifera (XP_002284111.1), Ricinus communis (XP_002528469.1), Nicotiana sylvestris (XP_009762518.1), Populus trichocarpa (XP_024465761.1), Sesamum indicum (XP_011091501.1), and Arabidopsis (AT2G37430.1/ZAT11, AT5G59820.1/ZAT12) were downloaded from GenBank. (**C**) Phylogenetic tree of MpZFP1 with Arabidopsis C1-2i clade C2H2 proteins. MpZFP1 is marked by a red box.

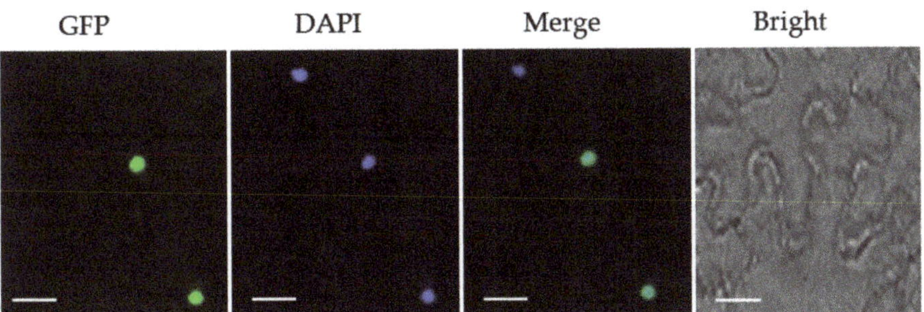

Figure 2. Subcellular localization of *MpZFP1-GFP*. Subcellular localization of *MpZFP1*-GFP was assayed with DAPI in tobacco leaf epidermal cells. The fluorescence signals were detected 48 h after infiltration. Bar = 50 μM.

2.3. Expression of MpZFP Is Induced by NaCl

The previous transcriptome study indicated that *MpZFP1* transcripts accumulated in the salt-treated Pongamia cDNA library [7]. To investigate the dynamic changes of the gene expression during the early stage of salt stress in detail, we treated the Pongamia seedlings with a 500 mM salt solution for 0 h, 2 h, 4 h, 8 h, and 12 h. All the samples were collected at the same time, and the RNAs from the leaves and roots were extracted separately. Quantitative RT-PCR analysis showed that the expression of *MpZFP1* was significantly induced by NaCl. The expression of *MpZFP1* was higher in the roots compared to the leaves, and it reached its highest level (~56 folds compared to no stress) at 4 h of NaCl stress, followed by a decrease in the roots (Figure 3). Comparatively, the expression peaked at 8 h of NaCl stress in the leaves (~6 folds compared to no stress), and the induction levels were much lower in the leaves than in the roots (Figure 3). These results indicated that the MpZFP1 mainly functioned in the roots, especially under salt stress.

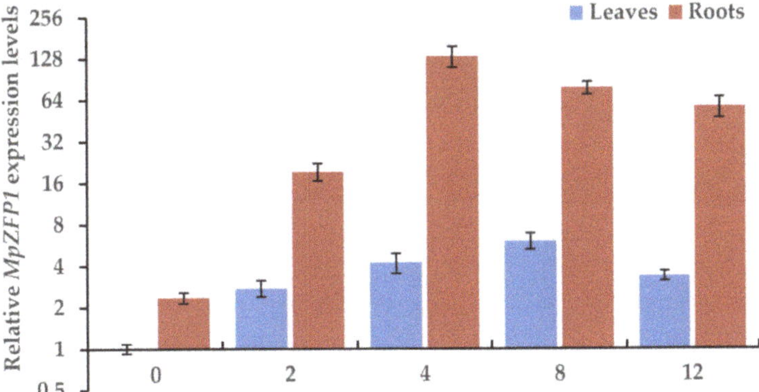

Figure 3. Relative expression levels of *MpZFP1* in Pongamia seedlings under salt stress. Relative expression levels of *MpZFP1* gene in leaves and roots after 500 mM NaCl treatment for 0, 2, 4, 8, 12 h was analyzed by quantitative RT-PCR. *MpActin* and *Mp18S* genes were used as internal references. All data were normalized to 0 h of leaves. Error bars show mean values (±SD) of three independent samples.

2.4. Heterologous Expression of MpZFP1 Strongly Enhances Salt Tolerance in Arabidopsis

First, we generated the transgenic Arabidopsis plants carrying *MpZFP1* expressed from a CaMV (Cauliflower Mosaic Virus)-35S promoter to investigate possible functions of *MpZFP1* in salt stress response. The transgenic plants were generated by the method of Agrobacterium-mediated transformation [41]. Three genetically stable transgenic lines were selected for further experiments, and the significantly high expression of *MpZFP1* was confirmed by RT-PCR (Figure 4A). No obvious difference in development was observed in transgenic plants compared to wild-type plants under normal conditions. We checked the germination rate of the seeds for the *35S::MpZFP1* transgenic and wild-type Arabidopsis under salt stress. High concentrations of NaCl strongly inhibited seed germination in the wild-type plants. The germination rate of the wild-type Arabidopsis reduced to around 60%, 16%, and 4% under 150 mM, 200 mM, and 250 mM salt stress, respectively (Figure 4B,C). The 150 mM of NaCl had no obvious effects on seed germination in the transgenic plants; more than 90% of seeds were able to germinate under this condition (Figure 4B,C). The germination rate of the three transgenic plants varied from 70% to 87% under 200 mM of NaCl (Figure 4B,C). Moreover, the seeds of the transgenic line 2 and 3 plants showed a germination rate of more than 20% on the medium containing 250 mM of NaCl (Figure 4B,C).

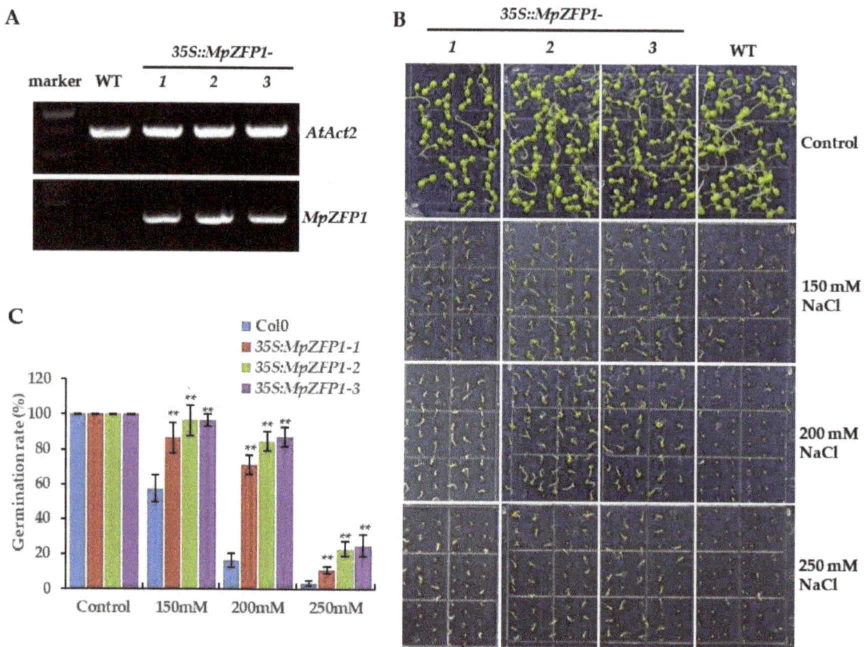

Figure 4. Ectopic expression of *MpZFP1* in Arabidopsis increases the seed germination of transgenic plants under salt stress. (**A**) Expression levels of *MpZFP1* in three independent *MpZFP1* heterologous expression lines. *AtAct2* was used as a control. (**B**) Typical phenotype of wild type and transgenic Arabidopsis seeds germinated on 1/2 MS medium contain 0 (Control), 150, 200, or 250 mM NaCl. (**C**) Quantification of (**B**), data shown are mean values of at least 50 individuals ± SD. ** $p < 0.01$ (Student's *t*-test).

Secondly, we measured the root lengths of the wild-type and transgenic seedlings under salt stress. The seedlings were cultured for three days on a normal 1/2 MS agar medium and then transferred to the medium supplemented with 0 or 125 mM NaCl. After one week of treatments, the root lengths did not show differences between the wild-type and transgenic seedlings in the medium without additional NaCl (Figure 5A,C). Under treatment with 125 mM NaCl, the root growth of all the plants was inhibited. However, the roots of the transgenic seedlings were significantly longer than those of the wild-type seedlings (Figure 5B,C).

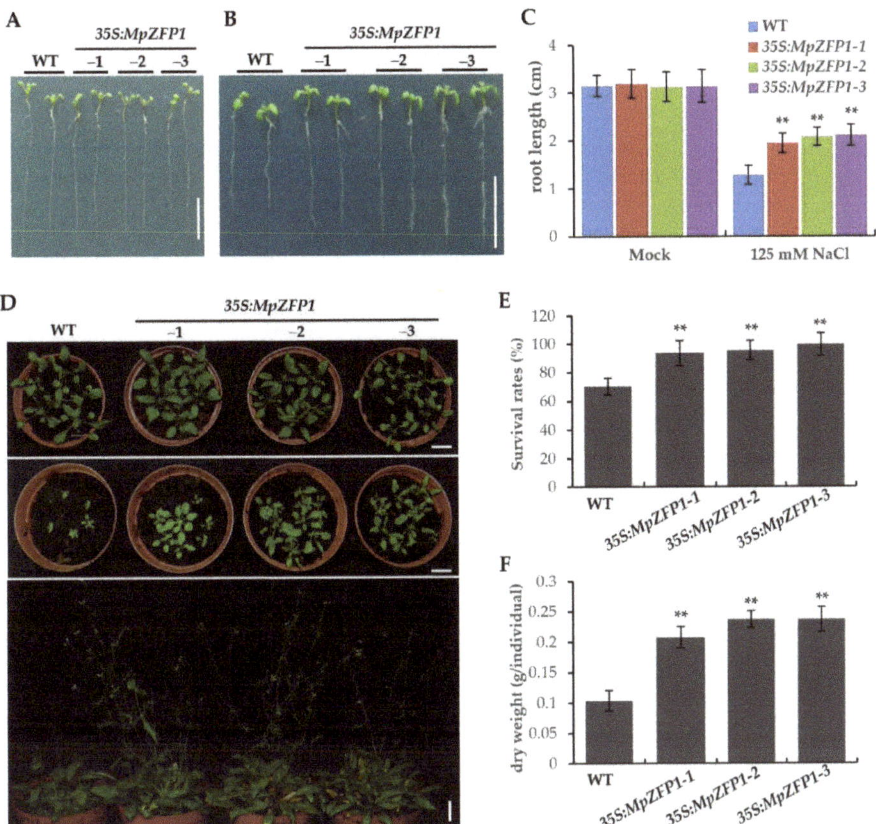

Figure 5. Ectopic expression of *MpZFP1* in Arabidopsis increases the salt tolerance of transgenic plants. Typical root length phenotype of 10-day wild-type and *MpZFP1* heterologous-expression Arabidopsis seedling grown on 1/2 MS medium (**A**) or 1/2 MS medium with 125 mM NaCl (**B**), bar = 1 cm. (**C**) Quantification of (**A**,**B**). Root length for the seedlings grown on normal (Mock) or 125 mM NaCl medium were measured. (**D**) Growth phenotype of 10-day recovery from non-treatment (upper panel) or 125 mM NaCl treated (middle panel) and growth phenotype of 30-day recovery from salt treatment (lower panel). (**E**) Survival rates of wild-type (WT) and *MpZFP1* transgenic plants after 10-day recovery from salt stress. (**F**) Dry weight of survived plants that recovered after 30 days. Data shown are mean values of at least 50 individuals ± SD (**C**,**E**,**F**). ** $p < 0.01$ (Student's *t*-test).

Lastly, we monitored the growth phenotype of the *MpZFP1* transgenic plants under salt stress. We germinated and cultured the seedlings on a normal 1/2 MS agar medium for five days and then transferred them to the medium supplemented with 0 or 125 mM NaCl. The seedlings were then transferred to pots with soil after ten days of treatment. For the plants without salt stress, the transgenic lines grew as well as the wild type after ten days growth in soil (Figure 5D, upper panel). However, the transgenic plants performed much better in both survival rate and growth vigor after ten days of recovery (Figure 5D (middle panel) and Figure 5E). Almost all the transgenic plants survived the stress, but ~30% of the wild-type plants were killed by ten days of salt treatment (Figure 5E). All the surviving transgenic plants flowered and had seed sets after 30 days of recovery, but the wild-type plants were still in vegetative growth or had just started to flower (Figure 5D, lower panel). The transgenic plants had significantly greater biomass compared to the wild-type plants after 30 days of recovery from salt stress (Figure 5F).

Collectively, the results indicated that the *MpZFP1* heterologous expression endowed the transgenic plants with salt tolerance.

2.5. Heterologous Expression of MpZFP1 in Arabidopsis Enhances the Expression of Stress-Responsive Genes

Transcriptional reprogramming is a major mechanism through which plants respond to stress, and the induction of stress-responsive genes is a hallmark of stress acclimation in plants. To investigate whether the ectopic expression of *MpZFP1* activated stress marker genes in the transgenic plants, we quantified the expression levels of three well-studied abiotic stress responsive genes, *RD22*, *RD29A*, and *RD29B*, in the transgenic plants with or without salt stress by quantitative RT-PCR. The expression levels of *RD22* and *RD29A* were significantly higher in the transgenic plants in both normal growth conditions and stress conditions (Figure 6A,B). The expression levels of RD29B were slightly lower in the transgenic plants under normal growth conditions but were higher under salt stress than in the wild-type plants (Figure 6C). Generally, the expression levels of the three stress-responsive genes were activated to higher levels by the expression of *MpZFP1*, especially under salt stress.

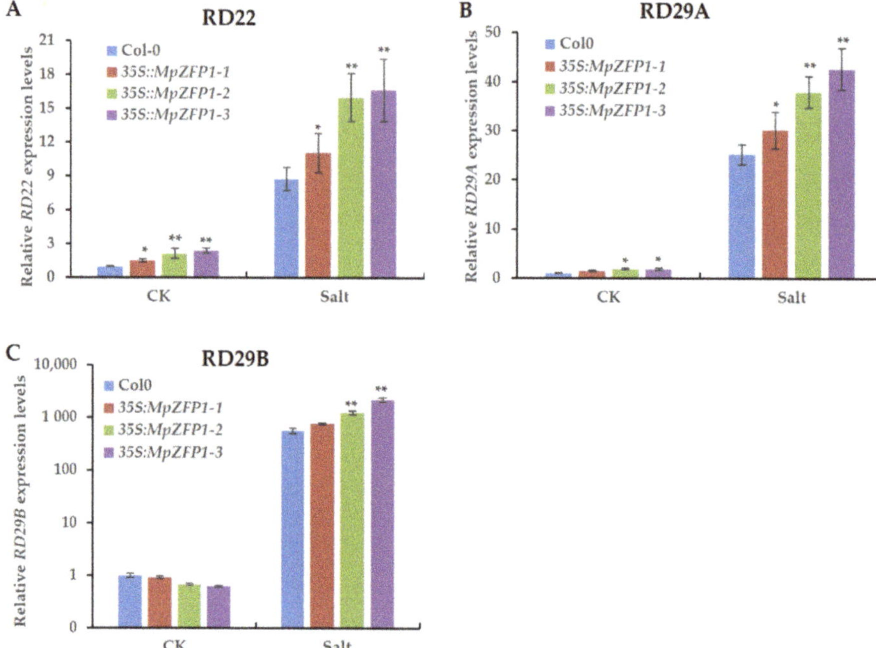

Figure 6. Expression levels of stress-responsive genes in wild type and 35S:*MpZFP1* transgenic plants. The expression levels of *RD22* (**A**), *RD29A* (**B**), and *RD29B* (**C**) of two-week-old Arabidopsis plants with (Salt) or without (CK) 200 mM NaCl treatment for three hours. *AtACT2* (*ACTIN2*) was used as a control. Mean values and standard deviations of three biological replicates are shown. * $p < 0.05$, ** $p < 0.01$ (Student's *t*-test).

2.6. Heterologous Expression of MpZFP1 in Arabidopsis Improves ROS Scavenging

Abiotic stresses such as salt, drought, and high temperature usually lead to secondary oxidative stress resulting in reactive oxygen species (ROS) overaccumulation. Excess ROS cause damage to DNA, protein, and lipid membranes, which may ultimately lead to cell death. To analyze the ROS induced by salt stress, NBT (nitroblue tetrazolium) and DAB (3, 3′-diaminobenzidine) staining assays were performed to detect the contents of H_2O_2 and O_2^- under mock or 200 mM-NaCl treatment. The leaves of the transgenic plants showed lighter staining colors than those of the wild-type plants did after salt treatment (Figure 7A,B), indicating lower contents of H_2O_2 and O_2^- in the transgenic plants.

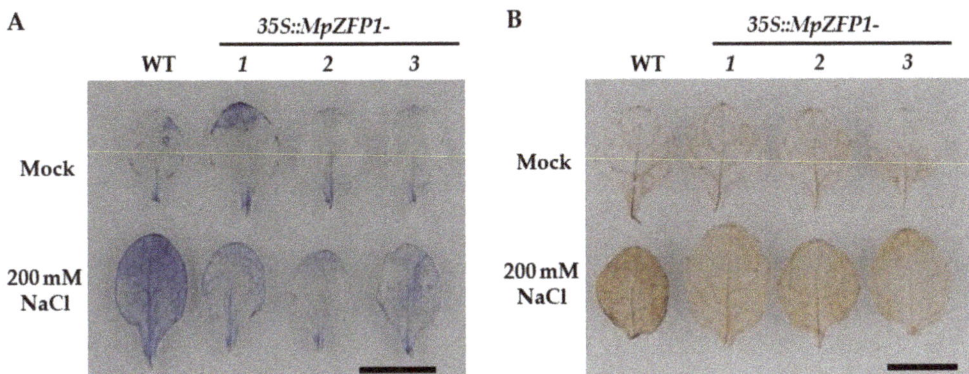

Figure 7. NBT and DAB staining of wild-type and 35S:*MpZFP1* transgenic Arabidopsis. The rosette leaves of three-week-old Arabidopsis were used for staining. The leaves were soaked in 1/2 MS liquid medium containing 0 mM (mock) or 200 mM of NaCl for 3 h and then transferred into NBT (**A**) or DAB (**B**) solution for staining. The experiments were repeated twice and at least 10 leaves for each genotype were assayed. The data obtained are shown here. Bar = 1 cm.

3. Discussion

Zinc finger proteins constitute one of the largest families that have been claimed to play critical roles in plant development and response to environmental stresses. The zinc finger domain can bind to DNA or RNA, and serve as a mediator of protein–protein interaction, which contributes to the functional diversity of this protein family [13–15,20]. Although C2H2 ZFPs have been extensively studied and found to be involved in responses to abiotic stresses, such as drought, salt, high temperatures, cold and high light [11,20,42], there have been few reports on this type of protein in mangroves or mangrove associates, which represent a unique plant community with high and robust stress tolerance.

We cloned and functionally characterized the first stress-responsive ZFPs, MpZFP1, from Pongamia, a typical mangrove associate. Phylogenetic analysis showed that the MpZFP1 belonged to the C1-2i type of C2H2, with the highest similarity with ZAT11/12 in Arabidopsis (Figure 1). As the other homolog in C1-2i, the MpZFP1 has two dispersed zinc finger domains. The other motif that embodies this family is a short hydrophobic region consisting of the core LXLXL sequence, the EAR motif, at the C-terminus [15,18]. The EAR domain, the smallest known repressive domain, was initially found in APETALA2 (AP2)/ETHYLENE RESPONSE FACTOR (ERF) proteins [43]. Many plant C1-2i C2H2 ZFPs containing the EAR motif have also been shown to function as repressors, such as AZF1/2/3, ZAT7, and ZAT10/11/12 in Arabidopsis [15,18,43]. The first Leu mutated to Pro residue in the core LXLXL motif in MpZFP1 (Figure 1B), which brought uncertainty to the issue of whether MpZFP1 functioned as a transcription repressor in the same way as its close homolog ZAT12. Through microarray study, many cold positive regulators were found to be suppressed by ZAT12, which was consistent with its function as a repressor [25]. However, ZAT12 played positive roles in cold tolerance, with its full function yet to be uncovered.

Since the transcripts of *MpZFP1* were highly induced by salt stress (Figure 2), we focused on its function in salt stress response. Arabidopsis transgenic plants harboring 35S:*MpZFP1* were generated and three independent lines varied in expression levels, marked as line 1 to 3 from, low to high (Figure 4A), were chosen for stress tolerance assays. We first checked the germination rate of *35S:MpZFP1* heterologous-expression plants under different salt concentrations. There was no difference between the wild type and the three lines of transgenic plants when germinated on the control medium without additional NaCl

(Figure 4B,C). The germination rates of the transgenic lines were much higher than those of the wild type under salt stress, with line 3 showing the best performance, which was consistent with the expression levels of *MpZFP1* (Figure 4B,C). The inhibition of root growth under salt stress was alleviated by the expression of *MpZFP1* (Figure 5A–C). The transgenic plants also showed a better survival rate and biomass accumulation after salt stress (Figure 5D–F). These results indicate that the MpZFP1 is a positive regulator of salt stress tolerance. To gain insight into the molecular function of the MpZFP1, we examined the expression level of several stress-responsive genes in transgenic plants. The expression levels of *RD22* and *RD29A* were activated by MpZFP1 even under normal growth conditions (Figure 6A,B). A similar result was reported for the GsZFP1, a C2H2 protein from the wild soybean. The expression of *RD22* and *RD29A* was also activated by the heterologous expression of GsZFP1 in Arabidopsis [33,34]. Unlike *RD29A*, which is activated by the AP2/ERF transcription factor DREB, *RD29B* is a direct target of bZIP transcription factors, such as ABI5 and ABFs [44]. A slightly decreased expression level was detected for *RD29B* in *MpZFP1* transgenic plants (Figure 6C), probably through the inhibition of *ABFs* and *ABI5*, whose expression was suppressed by GsZFP1 in Arabidopsis [34].

Abiotic stresses always lead to the over-accumulation of ROS, which causes significant damage to the cells. The MpZFP1 close Arabidopsis homolog ZAT12 plays critical roles in reactive oxygen and abiotic stress signaling [24]. To investigate the possible roles of MpZFP1 in the regulation of ROS levels under salt stress, we used NBT and DAB staining to monitor the contents of H_2O_2 and O_2^-. The leaves of the *MpZFP1* transgenic plants showed much lighter staining colors under 200 mM-NaCl treatment, which indicated that MpZFP1 might enhance salt tolerance through efficient ROS scavenging. Taken together, these results indicated that the closely related C2H2 ZFPs from diverse plants functioned conservatively.

4. Materials and Methods

4.1. Plant Materials and Growth Conditions

The Arabidopsis thaliana ecotype Columbia (Col-0) was used as a wild type to generate *35S:MpZFP1* transgenic plants. The transgenic plants were generated by agrobacterium-mediated flora dipping [41]. The Arabidopsis seeds were germinated and cultivated on 1/2 MS agar plates (half MS basal salts supply with 1% sucrose and 0.8% agar) in a tissue culture room, with a long-day (16 h light and 8 h dark) photoperiod, at 22 °C. Ten-day-old seedlings were transferred into pots with soil and grown in a growth chamber, with a long-day photoperiod, at 22 °C. The Pongamia seeds were soaked in tap water at 28 °C in a growth cabinet until radicles appeared. These germinated seeds were then planted in soil for further growth.

4.2. Full-Length cDNA Cloning, Motif Prediction and Phylogenetic Analysis

SMARTer^{TM} RACE cDNA Amplification Kit (Takara Bio, Otsu, Japan) was used for 5′ and 3′ RACE. The putative *MpZFP1* fragment from the Pongamia transcriptome dataset was used as a template for designing gene-specific internal primers. A nested PCR protocol was carried out with the primers listed in Table 1. The 5′ and 3′ ends of cDNA were sequenced and assembled into full-length cDNA. The conserved motifs of MpZFP1 and homologous protein were analyzed using SMART (Simple Modular Architecture Research Tool) and SMS online tools [45]. MpZFP1 and 18 Arabidopsis C1-2i proteins (download from TAIR) were used for phylogenetic analysis. The phylogenetic tree was constructed using the Neighbor-Joint method implemented in the MEGA X program [46].

Table 1. Primers used in this study.

Name	Sequences (5′–3′)	Description
MpActin_F	AGAGCAGTTCTTCAGTTGAG	RT-PCR
MpActin_R	TCCTCCAATCCAGACACTAT	RT-PCR
Mp18s_RtF	GCTCGTAGTTGGACCTTG	RT-PCR
Mp18s_RtR	TTCGCAGTTGTTCGTCTT	RT-PCR
MpZFP1_RtF	TTTGCTGTAGGACAAGCTTTGGGA	RT-PCR
MpZFP1_RtR	CGGGAAACAAAATTGATCTCTTGCT	RT-PCR
RD22_RtF	ACGTCAGGGCTGTTTCCAC	RT-PCR
RD22_RtR	TACTTCTGTTTGTGACACACC	RT-PCR
RD29A_RtF	TTCCGTTGAAGAGTCTCCAC	RT-PCR
RD29A_RtR	AACAAAACACACATAAACATCC	RT-PCR
RD29B_RtF	CCACGGTCCGTTGAAGAGTC	RT-PCR
RD29B_RtR	CAAAAACACAAACATTCAAAAGC	RT-PCR
AtAct2_RtF	GACCTTTAACTCTCCCGCTATG	RT-PCR
AtAct2_RtR	GAGACACACCATCACCAGAAT	RT-PCR
Long-UPM	CTAATACGACTCACTATAGGGCAAGCAGTGGTATCAACGCA	RACE
Short-UPM	CTAATACGACTCACTATAGGGC	RACE
NUP	AAGCAGTGGTATCAACGCAGAGT	RACE
MpZFP1_5′GSP	TGGCTTCTTATGGCTTGCACGGT	RACE
MpZFP1_5′NGSP	TGGCGGTTACATGTCTTGCACTCGAAG	RACE
MpZFP1_3′GSP	GGACAAGCTTTGGGAGGCCACATGA	RACE
MpZFP1_BamHIF	TATGGATCCATGAAGAGAGAAAGGGAAGGT	clone
MpZFP1_SacIR	CACGAGCTCTCAATTGAAACAATGAACCAAAG	clone

4.3. Subcellular Localization Analysis

The *MpZFP1-GFP* fusion expression vector was transferred into agrobacteria strain GV3101. The four-week-old *N. benthamiana* leaves were used for agrobacteria-mediated transient expression. GFP fluorescence was taken after 48 h of infiltration using a LEICA SP8 STED 3X fluorescence microscope confocal system. Two hours before microscope detection, the infiltrated leaves were stained with DAPI to mark the nucleus.

4.4. RNA Extraction and Quantitative Real-Time PCR

One-month-old Pongamia seedlings were transferred from soil into 1/2 MS liquid medium. After overnight culture, the normal 1/2 MS liquid medium was replaced by a 1/2 MS liquid medium containing 500 mM NaCl. The seedling samples were collected 0, 2, 4, 8, 12 h after treatments. Two-week-old Arabidopsis were used for stress-responsive gene analysis. The seedlings were transferred to a 1/2 MS liquid medium from a 1/2 MS agar plate one day before treatment. The samples were frozen by liquid nitrogen after 3 h treatment with or without 200 mM NaCl in a fresh 1/2 MS liquid medium. The total RNAs were isolated using TRIzol™ Plus (Takara Bio, Otsu, Japan) following the manufacturer's protocol. About 1000 ng of total RNA were digested by DNase I for 30 min at 37 °C before reverse transcription. The DNase digestion was terminated by the addition of 25 mM EDTA and followed by incubation at 70 °C for 10 min. A first-strand cDNA synthesis was performed using an oligo(dT) 18 primer and GoScript™ Reverse Transcriptase (Promega, Madison, WI, USA). Subsequently, qRT-PCR was performed on a Roche LightCycler 480 with gene-specific primers and an SYBR Green mix (Takara Bio, Otsu, Japan). All primers used in the qRT-PCR are listed in Table 1.

4.5. Phenotype Analysis of Wild-Type and 35S:MpZFP1 Transgenic Arabidopsis Plants

For the germination rate assays, at least 100 seeds of wild-type and transgenic plants were sowed on a 1/2 MS medium containing different concentrations of NaCl (0, 150, 200, and 250 mM). After two days of stratification at 4 °C in the dark, the seeds were transferred to light for the assessment of their germination rates. A seed was considered as germinated when the radical protruded through its envelope.

For the root length assay, the seedlings were germinated and grown for three days on a normal 1/2 MS agar medium, and then transferred to the medium containing 0 or 125 mM of NaCl. The root length was measured by ImageJ software after 7 days of treatment. At least 30 plants for each genotype in each biological repetition were checked.

For the growth assay, the seedlings were germinated and grown for five days on a normal 1/2 MS agar medium, and then transferred to the medium containing 0 or 125 mM of NaCl. The seedlings were transferred to pots with soil after 10 days of treatment. The survival rate was scored after 10 days of recovery. To address the influence of salt stress on biomass, the 30-day-recovery seedlings were dried in a drying oven for three days at 85 °C and their weights were measured. At least 30 plants for each genotype in each biological repetition were checked.

4.6. DAB and NBT Staining

Three-week-old Arabidopsis leaves grown on a 1/2 MS agar plate were used for DAB staining and NBT staining. The leaves were vacuumed in a 1/2 MS liquid medium containing 200 mM of NaCl for 5 min and then soaked for another 4 h. Staining was performed by vacuuming the leaves in a 1 mg/mL DAB solution or a 0.2% NBT solution for 5 min and then staining for another 4 h.

5. Conclusions

C2H2 ZFPs have been reported to regulate responses to abiotic stress in a number of plants. We functionally characterized a nucleus-localized C2H2 ZFP, *MpZFP1*, which was involved in salt tolerance in Pongamia in the study. The ectopic expression of *MpZFP1* in Arabidopsis greatly enhanced the salt tolerance of transgenic plants through the activation of stress-responsive gene expression and ROS scavenging. Abiotic stresses, including salt stress, frequently impose constraints on plant distribution and growth performance, which threatens food security. The transgenic lines grew as well as the wild-type plants under normal growth conditions, which makes them ideal candidates for the breeding of stress-tolerant crops by using genetic modification.

Author Contributions: Conceptualization, S.D. and J.H.; data curation, H.Y. (Hao Yan), L.L. and H.Y. (Heng Yang); formal analysis, Y.Z., J.H. and S.D.; funding acquisition, S.D. and J.H.; investigation, L.L., H.Y. (Heng Yang), W.L. and H.Y. (Hao Yan); resources, J.H., S.D. and J.C.; supervision, S.D. and J.H.; writing—original draft, Z.Y. and S.D.; writing—review & editing, Y.Z., J.H. and S.D. All authors have read and agreed to the published version of the manuscript.

Funding: This research was supported by grants from the Natural Science Foundation of Guangdong Province (2021A1515012434 and 2021A1515010622) and Shenzhen Stable Support Project for Colleges and Universities (No. 20200813135146001).

Institutional Review Board Statement: Not applicable.

Informed Consent Statement: Not applicable.

Conflicts of Interest: The authors declare no conflict of interest.

References

1. Jamil, A.; Riaz, S.; Ashraf, M.; Foolad, M.R. Gene Expression Profiling of Plants under Salt Stress. *Crit. Rev. Plant Sci.* **2011**, *30*, 435–458. [CrossRef]
2. Munns, R.; Tester, M. Mechanisms of Salinity Tolerance. *Annu. Rev. Plant Biol.* **2008**, *59*, 651–681. [CrossRef]

3. Wang, L.; Mu, M.; Li, X.; Lin, P.; Wang, W. Differentiation between true mangroves and mangrove associates based on leaf traits and salt contents. *J. Plant Ecol.* **2011**, *4*, 292–301. [CrossRef]
4. Marriboina, S.; Sengupta, D.; Kumar, S.; Reddy, A.R. Physiological and molecular insights into the high salinity tolerance of *Pongamia pinnata* (L.) pierre, a potential biofuel tree species. *Plant Sci.* **2017**, *258*, 102–111. [CrossRef] [PubMed]
5. Marriboina, S.; Reddy, A.R. Hydrophobic cell-wall barriers and vacuolar sequestration of Na+ ions are among the key mechanisms conferring high salinity tolerance in a biofuel tree species, *Pongamia pinnata* L. pierre. *Environ. Exp. Bot.* **2020**, *171*, 103949. [CrossRef]
6. Marriboina, S.; Sharma, K.; Sengupta, D.; Yadavalli, A.D.; Sharma, R.P.; Attipalli, R.R. Evaluation of high salinity tolerance in *Pongamia pinnata* (L.) Pierre by a systematic analysis of hormone-metabolic network. *Physiol. Plant.* **2021**. [CrossRef] [PubMed]
7. Huang, J.; Lu, X.; Yan, H.; Chen, S.; Zhang, W.; Huang, R.; Zheng, Y. Transcriptome characterization and sequencing-based identification of salt-responsive genes in *Millettia pinnata*, a semi-mangrove plant. *DNA Res.* **2012**, *19*, 195–207. [CrossRef]
8. Wegrzyn, J.L.; Whalen, J.; Kinlaw, C.S.; Harry, D.E.; Puryear, J.; Loopstra, C.A.; Gonzalez-Ibeas, D.; Vasquez-Gross, H.A.; Famula, R.A.; Neale, D.B. Transcriptomic profile of leaf tissue from the leguminous tree, *Millettia pinnata. Tree Genet. Genomes* **2016**, *12*, 44. [CrossRef]
9. Wang, H.; Hu, T.; Huang, J.; Lu, X.; Huang, B.; Zheng, Y. The expression of *Millettia pinnata* chalcone isomerase in Saccharomyces cerevisiae salt-sensitive mutants enhances salt-tolerance. *Int. J. Mol. Sci.* **2013**, *14*, 8775–8786. [CrossRef]
10. Zhang, Y.; Huang, J.; Hou, Q.; Liu, Y.; Wang, J.; Deng, S. Isolation and Functional Characterization of a Salt-Responsive Calmodulin-Like Gene MpCML40 from Semi-Mangrove *Millettia pinnata*. *Int. J. Mol. Sci.* **2021**, *22*, 3475. [CrossRef]
11. Han, G.; Lu, C.; Guo, J.; Qiao, Z.; Sui, N.; Qiu, N.; Wang, B. C2H2 Zinc Finger Proteins: Master Regulators of Abiotic Stress Responses in Plants. *Front. Plant Sci.* **2020**, *11*, 115. [CrossRef]
12. Takatsuji, H.; Mori, M.; Benfey, P.N.; Ren, L.; Chua, N.H. Characterization of a zinc finger DNA-binding protein expressed specifically in Petunia petals and seedlings. *EMBO J.* **1992**, *11*, 241–249. [CrossRef]
13. Riechmann, J.L.; Heard, J.; Martin, G.; Reuber, L.; Jiang, C.; Keddie, J.; Adam, L.; Pineda, O.; Ratcliffe, O.J.; Samaha, R.R.; et al. Arabidopsis transcription factors: Genome-wide comparative analysis among eukaryotes. *Science* **2000**, *290*, 2105–2110. [CrossRef]
14. Miller, J.; McLachlan, A.D.; Klug, A. Repetitive zinc-binding domains in the protein transcription factor IIIA from Xenopus oocytes. *EMBO J.* **1985**, *4*, 1609–1614. [CrossRef]
15. Xie, M.; Sun, J.; Gong, D.; Kong, Y. The Roles of Arabidopsis C1-2i Subclass of C2H2-type Zinc-Finger Transcription Factors. *Genes* **2019**, *10*, 653. [CrossRef]
16. Englbrecht, C.C.; Schoof, H.; Böhm, S. Conservation, diversification and expansion of C2H2 zinc finger proteins in the Arabidopsis thaliana genome. *BMC Genom.* **2004**, *5*, 39. [CrossRef]
17. Hu, X.; Zhu, L.; Zhang, Y.; Xu, L.; Li, N.; Zhang, X.; Pan, Y. Genome-wide identification of C2H2 zinc-finger genes and their expression patterns under heat stress in tomato (*Solanum lycopersicum* L.). *PeerJ* **2019**, *7*, e7929. [CrossRef]
18. Sakamoto, H.; Maruyama, K.; Sakuma, Y.; Meshi, T.; Iwabuchi, M.; Shinozaki, K.; Yamaguchi-Shinozaki, K. Arabidopsis Cys2/His2-type zinc-finger proteins function as transcription repressors under drought, cold, and high-salinity stress conditions. *Plant Physiol.* **2004**, *136*, 2734–2746. [CrossRef] [PubMed]
19. Wang, F.; Tong, W.; Zhu, H.; Kong, W.; Peng, R.; Liu, Q.; Yao, Q. A novel Cys2/His2 zinc finger protein gene from sweetpotato, IbZFP1, is involved in salt and drought tolerance in transgenic Arabidopsis. *Planta* **2016**, *243*, 783–797. [CrossRef] [PubMed]
20. Wang, K.; Ding, Y.; Cai, C.; Chen, Z.; Zhu, C. The role of C2H2 zinc finger proteins in plant responses to abiotic stresses. *Physiol Plant* **2019**, *165*, 690–700. [CrossRef] [PubMed]
21. Agarwal, P.; Arora, R.; Ray, S.; Singh, A.K.; Singh, V.P.; Takatsuji, H.; Kapoor, S.; Tyagi, A.K. Genome-wide identification of C2H2 zinc-finger gene family in rice and their phylogeny and expression analysis. *Plant Mol. Biol.* **2007**, *65*, 467–485. [CrossRef]
22. Yuan, S.; Li, X.; Li, R.; Wang, L.; Zhang, C.; Chen, L.; Hao, Q.; Zhang, X.; Chen, H.; Shan, Z.; et al. Genome-Wide Identification and Classification of Soybean C2H2 Zinc Finger Proteins and Their Expression Analysis in Legume-Rhizobium Symbiosis. *Front. Microbiol.* **2018**, *9*, 126. [CrossRef]
23. Mittler, R.; Kim, Y.; Song, L.; Coutu, J.; Coutu, A.; Ciftci-Yilmaz, S.; Lee, H.; Stevenson, B.; Zhu, J.K. Gain- and loss-of-function mutations in Zat10 enhance the tolerance of plants to abiotic stress. *FEBS Lett.* **2006**, *580*, 6537–6542. [CrossRef]
24. Davletova, S.; Schlauch, K.; Coutu, J.; Mittler, R. The zinc-finger protein Zat12 plays a central role in reactive oxygen and abiotic stress signaling in Arabidopsis. *Plant Physiol.* **2005**, *139*, 847–856. [CrossRef]
25. Vogel, J.T.; Zarka, D.G.; Van Buskirk, H.A.; Fowler, S.G.; Thomashow, M.F. Roles of the CBF2 and ZAT12 transcription factors in configuring the low temperature transcriptome of Arabidopsis. *Plant J. Cell Mol. Biol.* **2005**, *41*, 195–211. [CrossRef]
26. Huang, J.; Sun, S.-J.; Xu, D.-Q.; Yang, X.; Bao, Y.-M.; Wang, Z.-F.; Tang, H.-J.; Zhang, H. Increased tolerance of rice to cold, drought and oxidative stresses mediated by the overexpression of a gene that encodes the zinc finger protein ZFP245. *Biochem. Biophys. Res. Commun.* **2009**, *389*, 556–561. [CrossRef] [PubMed]
27. Huang, J.; Zhang, H.-S. The plant TFIIIA-type zinc finger proteins and their roles in abiotic stress tolerance. *Yi Chuan* **2007**, *29*, 915–922. [CrossRef] [PubMed]
28. Sun, S.J.; Guo, S.Q.; Yang, X.; Bao, Y.M.; Tang, H.J.; Sun, H.; Huang, J.; Zhang, H.S. Functional analysis of a novel Cys2/His2-type zinc finger protein involved in salt tolerance in rice. *J. Exp. Bot.* **2010**, *61*, 2807–2818. [CrossRef] [PubMed]

29. Xu, D.-Q.; Huang, J.; Guo, S.-Q.; Yang, X.; Bao, Y.-M.; Tang, H.-J.; Zhang, H.-S. Overexpression of a TFIIIA-type zinc finger protein gene ZFP252 enhances drought and salt tolerance in rice (*Oryza sativa* L.). *FEBS Lett.* **2008**, *582*, 1037–1043. [CrossRef]
30. Kim, J.C.; Lee, S.H.; Cheong, Y.H.; Yoo, C.M.; Lee, S.I.; Chun, H.J.; Yun, D.J.; Hong, J.C.; Lee, S.Y.; Lim, C.O.; et al. A novel cold-inducible zinc finger protein from soybean, SCOF-1, enhances cold tolerance in transgenic plants. *Plant J. Cell Mol. Biol.* **2001**, *25*, 247–259. [CrossRef]
31. Kim, Y.H.; Kim, M.D.; Park, S.C.; Yang, K.S.; Jeong, J.C.; Lee, H.S.; Kwak, S.S. SCOF-1-expressing transgenic sweetpotato plants show enhanced tolerance to low-temperature stress. *Plant Physiol. Biochem. PPB* **2011**, *49*, 1436–1441. [CrossRef]
32. Sun, Z.; Liu, R.; Guo, B.; Huang, K.; Wang, L.; Han, Y.; Li, H.; Hou, S. Ectopic expression of GmZAT4, a putative C2H2-type zinc finger protein, enhances PEG and NaCl stress tolerances in Arabidopsis thaliana. *3 Biotech* **2019**, *9*, 166. [CrossRef]
33. Luo, X.; Bai, X.; Zhu, D.; Li, Y.; Ji, W.; Cai, H.; Wu, J.; Liu, B.; Zhu, Y. GsZFP1, a new Cys2/His2-type zinc-finger protein, is a positive regulator of plant tolerance to cold and drought stress. *Planta* **2012**, *235*, 1141–1155. [CrossRef]
34. Luo, X.; Cui, N.; Zhu, Y.; Cao, L.; Zhai, H.; Cai, H.; Ji, W.; Wang, X.; Zhu, D.; Li, Y.; et al. Over-expression of GsZFP1, an ABA-responsive C2H2-type zinc finger protein lacking a QALGGH motif, reduces ABA sensitivity and decreases stomata size. *J. Plant Physiol.* **2012**, *169*, 1192–1202. [CrossRef] [PubMed]
35. Tang, L.; Cai, H.; Ji, W.; Luo, X.; Wang, Z.; Wu, J.; Wang, X.; Cui, L.; Wang, Y.; Zhu, Y.; et al. Overexpression of GsZFP1 enhances salt and drought tolerance in transgenic alfalfa (*Medicago sativa* L.). *Plant Physiol. Biochem. PPB* **2013**, *71*, 22–30. [CrossRef]
36. Chang, J.; Yu, T.; Yang, Q.; Li, C.; Xiong, C.; Gao, S.; Xie, Q.; Zheng, F.; Li, H.; Tian, Z.; et al. Hair, encoding a single C2H2 zinc-finger protein, regulates multicellular trichome formation in tomato. *Plant J. Cell Mol. Biol.* **2018**, *96*, 90–102. [CrossRef] [PubMed]
37. Li, Y.; Chu, Z.; Luo, J.; Zhou, Y.; Cai, Y.; Lu, Y.; Xia, J.; Kuang, H.; Ye, Z.; Ouyang, B. The C2H2 zinc-finger protein SlZF3 regulates AsA synthesis and salt tolerance by interacting with CSN5B. *Plant Biotechnol. J.* **2018**, *16*, 1201–1213. [CrossRef] [PubMed]
38. Shang, L.; Song, J.; Yu, H.; Wang, X.; Yu, C.; Wang, Y.; Li, F.; Lu, Y.; Wang, T.; Ouyang, B.; et al. A mutation in a C2H2-type zinc finger transcription factor contributed to the transition towards self-pollination in cultivated tomato. *Plant Cell* **2021**, koab201. [CrossRef]
39. Gao, H.; Song, A.; Zhu, X.; Chen, F.; Jiang, J.; Chen, Y.; Sun, Y.; Shan, H.; Gu, C.; Li, P.; et al. The heterologous expression in Arabidopsis of a chrysanthemum Cys2/His2 zinc finger protein gene confers salinity and drought tolerance. *Planta* **2012**, *235*, 979–993. [CrossRef]
40. Kim, S.H.; Hong, J.K.; Lee, S.C.; Sohn, K.H.; Jung, H.W.; Hwang, B.K. CAZFP1, Cys2/His2-type zinc-finger transcription factor gene functions as a pathogen-induced early-defense gene in Capsicum annuum. *Plant Mol. Biol.* **2004**, *55*, 883–904. [CrossRef]
41. Clough, S.J.; Bent, A.F. Floral dip: A simplified method for Agrobacterium -mediated transformation of Arabidopsis thaliana. *Plant J.* **1998**, *16*, 735–743. [CrossRef]
42. Kiełbowicz-Matuk, A. Involvement of plant C2H2-type zinc finger transcription factors in stress responses. *Plant Sci.* **2012**, *185–186*, 78–85. [CrossRef] [PubMed]
43. Kazan, K. Negative regulation of defence and stress genes by EAR-motif-containing repressors. *Trends Plant Sci.* **2006**, *11*, 109–112. [CrossRef] [PubMed]
44. Ma, N.L.; Rahmat, Z.; Lam, S.S. A review of the "Omics" approach to biomarkers of oxidative stress in *Oryza sativa*. *Int. J. Mol. Sci.* **2013**, *14*, 7515–7541. [CrossRef] [PubMed]
45. Stothard, P. The sequence manipulation suite: JavaScript programs for analyzing and formatting protein and DNA sequences. *BioTechniques* **2000**, *28*, 1102–1104. [CrossRef]
46. Kumar, S.; Stecher, G.; Li, M.; Knyaz, C.; Tamura, K. MEGA X: Molecular Evolutionary Genetics Analysis across Computing Platforms. *Mol. Biol. Evol.* **2018**, *35*, 1547–1549. [CrossRef] [PubMed]

International Journal of
Molecular Sciences

Article

Microtubule Dynamics Plays a Vital Role in Plant Adaptation and Tolerance to Salt Stress

Hyun Jin Chun [1,†], Dongwon Baek [2,†], Byung Jun Jin [3,†], Hyun Min Cho [3], Mi Suk Park [2], Su Hyeon Lee [3], Lack Hyeon Lim [3], Ye Jin Cha [3], Dong-Won Bae [4], Sun Tae Kim [5], Dae-Jin Yun [6] and Min Chul Kim [1,2,3,*]

1 Institute of Agriculture & Life Science, Gyeongsang National University, Jinju 52828, Korea; hj_chun@hanmail.net
2 Plant Molecular Biology and Biotechnology Research Center, Gyeongsang National University, Jinju 52828, Korea; dw100@hanmail.net (D.B.); misugip@hanmail.net (M.S.P.)
3 Division of Applied Life Science (BK21 Four), Gyeongsang National University, Jinju 52828, Korea; scv5789@naver.com (B.J.J.); hmcho86@gnu.ac.kr (H.M.C.); leesuhyeon86@gmail.com (S.H.L.); dlafkrgus@gnu.ac.kr (L.H.L.); cdw3280@naver.com (Y.J.C.)
4 Central Instrument Facility, Gyeongsang National University, Jinju 52828, Korea; bdwon@gnu.ac.kr
5 Department of Plant Bioscience, Life and Industry Convergence Research Institute, Pusan National University, Miryang 50463, Korea; stkim71@pusan.ac.kr
6 Department of Biomedical Science & Engineering, Konkuk University, Seoul 05029, Korea; djyun@konkuk.ac.kr
* Correspondence: mckim@gnu.ac.kr; Tel.: +82-55-772-1874
† These authors contributed equally to this study.

Citation: Chun, H.J.; Baek, D.; Jin, B.J.; Cho, H.M.; Park, M.S.; Lee, S.H.; Lim, L.H.; Cha, Y.J.; Bae, D.-W.; Kim, S.T.; et al. Microtubule Dynamics Plays a Vital Role in Plant Adaptation and Tolerance to Salt Stress. *Int. J. Mol. Sci.* **2021**, *22*, 5957. https://doi.org/10.3390/ijms22115957

Academic Editors: Mirza Hasanuzzaman and Masayuki Fujita

Received: 6 May 2021
Accepted: 27 May 2021
Published: 31 May 2021

Abstract: Although recent studies suggest that the plant cytoskeleton is associated with plant stress responses, such as salt, cold, and drought, the molecular mechanism underlying microtubule function in plant salt stress response remains unclear. We performed a comparative proteomic analysis between control suspension-cultured cells (A0) and salt-adapted cells (A120) established from *Arabidopsis* root callus to investigate plant adaptation mechanisms to long-term salt stress. We identified 50 differentially expressed proteins (45 up- and 5 down-regulated proteins) in A120 cells compared with A0 cells. Gene ontology enrichment and protein network analyses indicated that differentially expressed proteins in A120 cells were strongly associated with cell structure-associated clusters, including cytoskeleton and cell wall biogenesis. Gene expression analysis revealed that expressions of cytoskeleton-related genes, such as *FBA8*, *TUB3*, *TUB4*, *TUB7*, *TUB9*, and *ACT7*, and a cell wall biogenesis-related gene, *CCoAOMT1*, were induced in salt-adapted A120 cells. Moreover, the loss-of-function mutant of *Arabidopsis TUB9* gene, *tub9*, showed a hypersensitive phenotype to salt stress. Consistent overexpression of *Arabidopsis TUB9* gene in rice transgenic plants enhanced tolerance to salt stress. Our results suggest that microtubules play crucial roles in plant adaptation and tolerance to salt stress. The modulation of microtubule-related gene expression can be an effective strategy for developing salt-tolerant crops.

Keywords: salt stress; salt adaptation; proteomics; microtubules; tubulin

1. Introduction

Plant adaptation to environmental stress is regulated by cascades of molecular networks, including stress perception, signal transduction, metabolic adjustment, and the regulation of stress-responsive gene expressions, to reestablish cellular homeostasis, such as osmotic and ionic homeostasis, and protect proteins and cell membranes by using heat shock proteins (Hsps), chaperones, late embryogenesis abundant (LEA) proteins, osmoprotectants, and free-radical scavengers [1]. Plant cells have adapted to salt stress by changing cell wall composition [2,3]. Extensin, a significant cell wall glycoprotein, is cross-linked with phenolics by reactive oxygen species (ROS) accumulation to stiffen the cell wall when plant cells are exposed to salt stress [2]. The RhEXP4, expansin A4 of rose, overexpressing *Arabidopsis* plants show increased seed germination, root growth, and several lateral roots under salt stress conditions [4]. A higher pectin content in root tips enhances plant tolerance to salt stress by increasing root growth compared with the cell wall composition in two soybean cultivars [5]. The dysfunction of *Arabidopsis* AtCSLD5, a pectin biosynthesis enzyme in *sos6* (salt overly sensitive 6) mutant, enhances plant sensitivity to salt stress [6]. The *Arabidopsis* CC1 and CC2 proteins, along with cellulose synthases, interact with microtubules (MTs) and are essential for seedling growth under salt stress conditions [7]. However, the molecular mechanisms of changing cell wall dynamics by salt stress, including signal transduction and cell wall integrity pathways, remain unclear.

The plant cytoskeleton comprises the systemic polymers between actin filaments and MTs [8,9]. MTs form heterodimers by polymerization between α-tubulin and β-tubulin [10]; also, actin filaments polymerize to form filamentous structures by G-actin [11]. The adaptive mechanisms of the plant cytoskeleton to salt stress are varied by organization, dynamics, and cellular processes [12,13]. MTs play essential roles in the cell cycle, cell growth, and stress response by forming highly dynamic polymers [14,15]. Additionally, MT depolymerization and reorganization are essential for enhancing plant tolerance to salt stress [16]. In plants, response to salt stress, calcium ions, abscisic acid (ABA), and ROS as signaling molecules are associated with cortical MT array organization [17–19]. Cytosolic-increased calcium induces MT depolymerization by regulating calcium channels in the salt stress response [18]. The plant hormone, ABA, influences the organization and stability of cortical MTs [18]. ABA promotes the ectopic derivative of root cells by depolymerizing and reorganizing cortical MTs and activating MT depolymerization in guard cells during stomatal closure [20,21]. ROS induces the MTs' reorganization through MT disassembly and the formation of irregular MT polymers [19]. When ROS homeostasis is collapsed by salt stress, tubulin forms a modified structural state by assembling non-typical tubulin structures [22]. Propyzamide-hypersensitive 1 (PHS1), a mitogen-activated protein kinase phosphatase, phosphorylates α-tubulin and elevates MT depolymerization to salt stress [23]. When plants are exposed for a long period to salt stress, cortical MT reorganization is induced by the depolymerization and reassembly of MT networks [17]. The MT-associated proteins, 65-1 and MAP65-1, facilitate MT polymerization and bundling, enhance MT-stabilizing activity, and expedite cortical MT recovery by binding phosphatidic acid under salt stress conditions [24]. Although the role of MTs during plants' responses to salt stress has been much studied, the mechanism of the actin cytoskeleton is less understood. The actin cytoskeleton leads to assembly and bundle formation in response to a short period of salt stress; however, long-term exposure to salt stress or exposure to high salt stress induces the disassembly of the actin cytoskeleton [25]. Salt stress regulates the cellular process of actin dynamics via the salt overly sensitive (SOS) pathway and calcium signaling [26]. The *arp2* (actin-related protein 2) mutant showed a hypersensitive phenotype to salt stress by increasing mitochondria-dependent $[Ca^{2+}]_{cyt}$ levels [27].

Salt stress affects the expression levels of MT-associated genes and proteins. The loss-of-function mutants of prefoldin subunits 3 (PFD3) and PFD5 showed a hypersensitive phenotype to salt stress by decreasing expression levels of α-tubulin and β-tubulin [28]. The 26S proteasome degrades MT-associated protein SPIRAL1 (SPR1)-stabilizing MT in response to salt stress [16]. Proteomics analyses reveal that the plant adaptation to salt stress is associated with complex networks of protein expression and post-translational modifications [29–31]. Functional profiling of various proteins by a comparative proteomic approach has made it possible to characterize essential proteins involved in salt tolerance in various plant species, including *Thellungiella halophila* [32], *Halogetong lomeratus* [33], *Tangut Nitraria* [34], canola [35], sesame [36], and rice [37]. Adaptation to salt stress is a congested process in the whole plant and cellular levels and needs to adjust the transcription of various genes that trigger protein profile change [38,39]. Thus, quantitative analysis of expressed proteins by proteomics is valuable for understanding the molecular mechanisms underlying plant adaptation and tolerance to salt stress.

Our previous metabolite profiling study using salt-adapted *Arabidopsis* callus suspension-cultured cells reveals that various cellular processes, including cell wall thickening, play essential roles in plant salt adaptation [40]. In this proteomics study, we revealed that major differentially expressed proteins (DEPs) identified from salt-adapted cells were functionally associated with cytoskeleton and cell wall biogenesis. Structural and morphological changes of plant cells mediated by cytoskeleton and cell wall biogenesis functions are vital for adaptation and tolerance to salt stress.

2. Results

2.1. Morphological Features of Salt-Adapted Callus Suspension-Cultured Cells

Plants exhibit growth inhibition and impediment of tissue development in response to salt stress because of a deficit of cell wall extensibility [41]. When we compared morphologies between control cells (A0) and salt-adapted cells (A120; adapted to 120 mM NaCl), we observed that the A120 cells showed distinct morphological changes compared with A0 cells, including spherical or ellipsoidal and isodiametric shapes (Figure 1a). Additionally, newly divided A120 cells stuck together in small clumps. Vacuole size and the cytoplasmic volume in A120 cells were significantly reduced compared with those in A0 cells (Figure 1a). These data suggested that plant suspension cells have changed their morphology to adapt to long periods of salt stress. To understand the molecular mechanism underlying cell morphology changes during salt adaptation, we identified DEPs in salt-adapted A 120 cells by proteomics analysis. Additionally, we characterized their biological functions by molecular genetic analysis using *Arabidopsis* mutants and transgenic rice plants (Figure 1b).

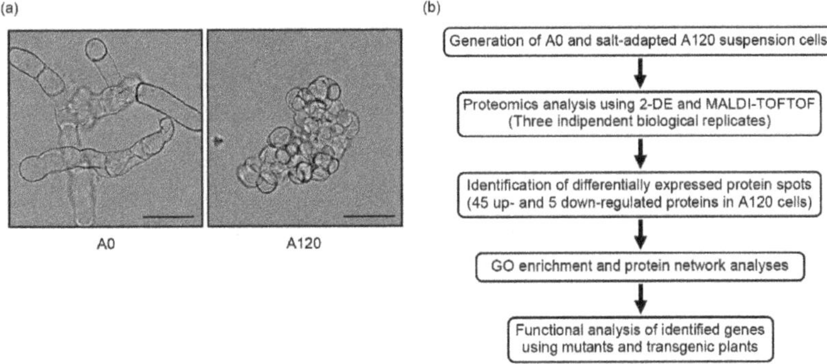

Figure 1. Characterization of *Arabidopsis* salt-adapted cells. (**a**) The morphological phenotype of control suspension cells (A0) grown in normal MS medium and salt-adapted cells (A120) grown in high salt MS medium with 120 mM NaCl. The photograph was taken using a microscope after 2 weeks of subculture. Scale bars indicate 50 μm. (**b**) Experimental scheme of proteomic and functional analyses.

2.2. Overview of Proteomic Profiles in Salt-Adapted Cells

Crude proteins were extracted from A0 and A120 cells grown in normal media (for A0 cells) and saline media with 120 mM NaCl (for A120 cells) for 8 days after subculture using the trichloroacetic acid/acetone/phenol extraction protocol [42] and quantified using a 2D-Quant Kit (GE Healthcare, Waukesha, WI, USA). Representative two-dimensional gel electrophoresis (2-DE) images from three biological replicates of A0 and A120 cells are displayed in Figure 2. With a cut-off point as a p-value of <0.05 for the differential expression between A0 and A120 cells, 50 DEP spots were identified by matrix-assisted laser desorption/ionization time-of-flight mass spectrometry (MALDI-TOF/TOF MS) (Table 1). When comparing expression levels in A120 cells with those in A0 cells, we identified 45 induced spots and 5 reduced spots in A120 cells (Table 1). Fifty DEPs identified in A120 cells were classified into functional categories based on gene ontology (GO) analysis using the PANTHER program (http://pantherdb.org) (Figure 3). The DEPs were included in "binding" (42.6%), "catalytic activity" (42.6%), "structural molecule activity" (8.5%), and "translation regulator activity" (6.4%) categories in the molecular function (Figure 3a, red color). In the biological process, DEPs were included in five categories, which are "biological regulation" (1.9%), "cellular process" (55.6%), "localization" (1.9%), "metabolic process" (33.3%), and "response to stimulus" (7.4%) (Figure 3a, green color). The DEPs were included in "cellular, anatomical entity" (46.6%), "intracellular" (48.3%), and "protein-containing complex" (5.2%) categories in the cellular component (Figure 3a, black bar). In the analysis of protein class, the two largest proportions of DEPs belonged to the "metabolite interconversion enzyme" (41.5%) and "cytoskeletal protein" (14.6%) classes (Figure 3b). Above these, DEPs were included in "calcium binding protein" (4.9%), "chaperone" (12.2%), "gene-specific transcriptional regulator" (2.4%), "nucleic acid metabolism protein" (2.4%), "protein modifying enzyme" (7.3%), "protein-binding activity modulator" (2.4%), and "translational protein" (9.8%) classes (Figure 3b). In the analysis of pathway class, the largest proportion of DEPs belonged to the "cytoskeletal regulation by Rho GTPase" (30.8%) class (Figure 3c). Above these, DEPs were associated with "apoptosis signaling pathway" (7.7%), "cell cycle" (7.7%), "de novo purine biosynthesis" (7.7%), "fructose galactose metabolism" (7.7%), "glycolysis" (7.7%), "S-adenosylmethionine biosynthesis" (15.4%), "TCA cycle" (7.7%), and "ubiquitin-proteasome pathway" (7.7%) (Figure 3c). Thus, our results suggested that critical cellular changes during plant adaptation to salt stress were related to cytoskeletal regulation and metabolite processes.

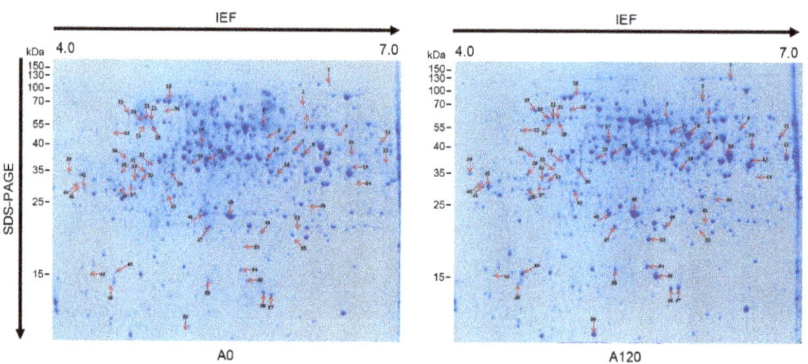

Figure 2. Proteome analysis of control A0 and salt-adapted A120 cell lines. The 50 differentially expressed protein (DEP) spots (45 up- and 5 down-regulated in A120 cells) were identified by 2-DE and MALDI-TOF/TOF analyses.

Table 1. Identification of DEPs between control A0 and salt-adapted A120 *Arabidopsis* callus suspension-cultured cells.

Spot No. [a]	Locus No.		Protein Name	Theo. Mr/pI [b]	Queries Matched [c]	Scores [d]	Expect	Fold (A0 vs. A120)
1, 28	AT1G56070	LOS1	Low expression of osmotically responsive genes 1	95.10/ 5.89	31	566	3.70×10^{-50}	1.554
2	AT1G62740	HOP2	Stress-inducible protein, putative	67.63/ 6.24	22	349	1.90×10^{-28}	2.538
3	AT4G13940	SAHH1	S-adenosyl-L-homocysteine hydrolase 1	53.97/ 5.66	27	522	9.40×10^{-46}	−1.133
4	AT1G51710	UBP6	Ubiquitin carboxyl-terminal hydrolase 6	54.00/ 5.82	27	467	3.00×10^{-40}	2.473
5	AT4G01850	SAM-2	S-adenosylmethionine synthase 2	43.63/ 5.67	28	616	3.70×10^{-55}	1.932
6	AT2G36880	MAT3	Methionine adenosyltransferase 3	42.93/ 5.76	33	699	1.90×10^{-63}	1.249
7	AT1G77120	ADH1	Alcohol dehydrogenase class-P	41.84/ 5.83	27	798	2.40×10^{-73}	−1.614
8	AT4G02930		GTP binding Elongation factor Tu family protein	49.61/ 6.25	33	1100	1.50×10^{-103}	1.129
9	AT3G51800	CPR	Metallopeptidase M24 family protein	43.28/ 6.36	17	280	1.50×10^{-21}	4.172

Table 1. *Cont.*

Spot No. [a]	Locus No.		Protein Name	Theo. Mr/pI [b]	Queries Matched [c]	Scores [d]	Expect	Fold (A0 vs. A120)
10	AT5G14780	FDH1	Formate dehydrogenase, chloroplastic/mitochondrial	42.67/ 7.12	25	560	1.50×10^{-49}	2.048
11	AT4G26910		Dihydrolipoamide succinyltransferase	50.03/ 9.21	13	185	4.70×10^{-12}	2.026
12, 50	AT3G04120	GAPC1	Glyceraldehyde-3-phosphate dehydrogenase 1, cytosolic	37.01/ 6.62	30	1040	1.50×10^{-97}	4.510
13	AT5G43330	MDH2	Lactate/malate dehydrogenase family protein	35.98/ 7.00	24	675	$4.70 * 10^{-61}$	1.407
14	AT5G23540	RPN11	26S proteasome non-ATPase regulatory subunit 14 homolog	34.39/ 6.31	18	241	1.20×10^{-17}	5.121
15	AT3G52930	FBA8	Fructose-bisphosphate aldolase 8, cytosolic	38.86/ 6.05	31	1090	1.50×10^{-102}	1.031
16	AT5G65020	ANNAT2	Annexin D2, calcium binding proteins	36.36/ 5.76	27	339	1.9×10^{-27}	1.426
17	AT2G47470	PDIL2-1	Disulfide isomerase-like (PDIL) protein	39.81/ 5.80	20	576	3.70×10^{-51}	1.308
18	AT5G02500	HSP70-1	*Arabidopsis thaliana* heat shock cognate protein 70-1	57.54/ 5.01	34	844	5.90×10^{-78}	−4.195
19	AT1G21750	PDIL1-1	Disulfide isomerase-like (PDIL) protein	55.85/ 4.81	23	457	3.00×10^{-39}	1.466
20, 21	AT5G62700	TUB3	Tubulin beta chain 3	51.27/ 4.73	38	778	2.40×10^{-71}	3.032
22	AT2G29550	TUB7	Tubulin beta-7 chain	51.34/ 4.74	36	674	5.90×10^{-61}	3.857
23	AT5G38470	RAD23D	Rad23 UV excision repair protein family	40.10/ 4.58	17	382	$9.40 * 10^{-32}$	2.453
24	AT4G20890	TUB9	Tubulin beta-9 chain	50.31/ 4.69	38	715	4.70×10^{-65}	1.649
25	AT5G44340	TUB4	Tubulin beta chain 4	50.36/ 4.76	31	490	$1.50 \times \times 10^{-42}$	2.158
26	AT4G37910	HSP70-9	Heat shock 70 kDa protein 9, mitochondrial	73.32/ 5.51	29	625	4.70×10^{-56}	2.615
27	AT5G09810	ACT7	Actin 7	41.94/ 5.31	32	1100	1.50×10^{-103}	1.110
29	AT1G35720	ANNAT1	Annexin D1, calcium binding proteins	36.30/ 5.21	27	754	5.90×10^{-69}	2.469
30	AT1G79230	STR1	Thiosulfate/3-mercaptopyruvate sulfurtransferase 1, mitochondrial	42.15/ 5.95	21	529	1.90×10^{-46}	1.487
31	AT3G53970		Probable proteasome inhibitor	32.15/ 4.94	15	329	1.90×10^{-26}	3.353
32	AT1G62380	ACO2	1-aminocyclopropane-1-carboxylate oxidase 2	36.39/ 4.98	21	722	9.40×10^{-66}	−2.900
33, 34	AT4G20260	PCAP1	Plasma membrane associated cation-binding protein 1	18.98/ 9.88	9	168	2.40×10^{-10}	−1.309
37	AT5G38480	GRF3	14-3-3-like protein GF14 psi	32.00/ 4.91	8	149	1.90×10^{-08}	1.052
38	AT4G04020	PAP1	Probable plastid-lipid-associated protein 1, chloroplastic	34.99/ 5.45	23	786	3.70×10^{-72}	5.711
40	AT4G10480	NACα4	Nascent polypeptide-associated complex (NAC), alpha subunit family protein	23.10/ 4.25	8	202	9.40×10^{-14}	2.031
41, 42	AT4G02450	P23-1	HSP20-like chaperones superfamily protein	25.38/ 4.46	10	248	2.40×10^{-18}	3.315
43	AT4G34050	CCoAOMT1	S-adenosyl-L-methionine-dependent methyltransferases superfamily protein	29.25/ 5.13	20	659	1.90×10^{-59}	7.222
45	AT1G47420	SDH5	Succinate dehydrogenase subunit 5, mitochondrial	28.15/ 6.19	15	410	1.50×10^{-34}	5.433
46, 48	AT3G55440	CTIMC	Triosephosphate isomerase, cytosolic	27.38/ 5.39	26	904	5.90×10^{-84}	1.155
47	AT5G20720	CPN20	20 kDa chaperonin, chloroplastic	26.79/ 8.86	8	87	0.033	1.638
49	AT5G26667	UMK3	P-loop containing nucleoside triphosphate hydrolases superfamily protein	22.58/ 5.79	13	298	2.40×10^{-23}	2.088
51	AT1G02930	GSTF6	Glutathione S-transferase F6	23.47/ 5.80	18	467	$3.00 * 10^{-40}$	1.795
52	AT3G22630	PBD1	Proteasome subunit beta type-2-A	22.64/ 5.95	15	138	$2.40 * 10^{-07}$	1.952
53	AT4G38680	CSP2	Glycine-rich protein 2, *Arabidopsis thaliana* cold shock protein 2	19.49/ 5.62	5	97	0.0031	4.117
54	AT3G62030	CYP20-3 /ROC4	Peptidyl-prolyl cis-trans isomerase CYP20-3, chloroplastic	26.73/ 8.63	16	418	2.40×10^{-35}	1.402

Table 1. *Cont.*

Spot No. [a]	Locus No.		Protein Name	Theo. Mr/pI [b]	Queries Matched [c]	Scores [d]	Expect	Fold (A0 vs. A120)
55	AT1G26630	ELF5A-2	Eukaryotic translation initiation factor 5A-1 (eIF-5A 1) protein	17.36/ 5.55	7	232	9.40×10^{-17}	2.141
56	AT5G59880	ADF3	Actin-depolymerizing factor 3	16.03/ 5.93	10	545	4.70×10^{-48}	1.290
57	AT3G53990	F5K20_290	Adenine nucleotide alpha hydrolases-like superfamily protein	17.90/ 5.66	21	575	4.70×10^{-51}	1.502
58	AT3G23490	CYN	Cyanate hydratase	18.64/ 5.49	15	574	5.90×10^{-51}	1.316
59	AT4G13850	RBG2	Glycine-rich RNA-binding protein 2, mitochondrial	14.74/ 6.73	7	268	2.40×10^{-20}	13.161
60	AT5G18060	SAUR23	SAUR-like auxin-responsive protein family	72.78/ 5.87	21	69	1.9	2.814

[a] The number of identification spots. [b] Theoretical mass (Mr, kDa) and pI of identified proteins. Theoretical values were retrieved from the protein database. [c] Number of matched peptides. [d] The mascot scores.

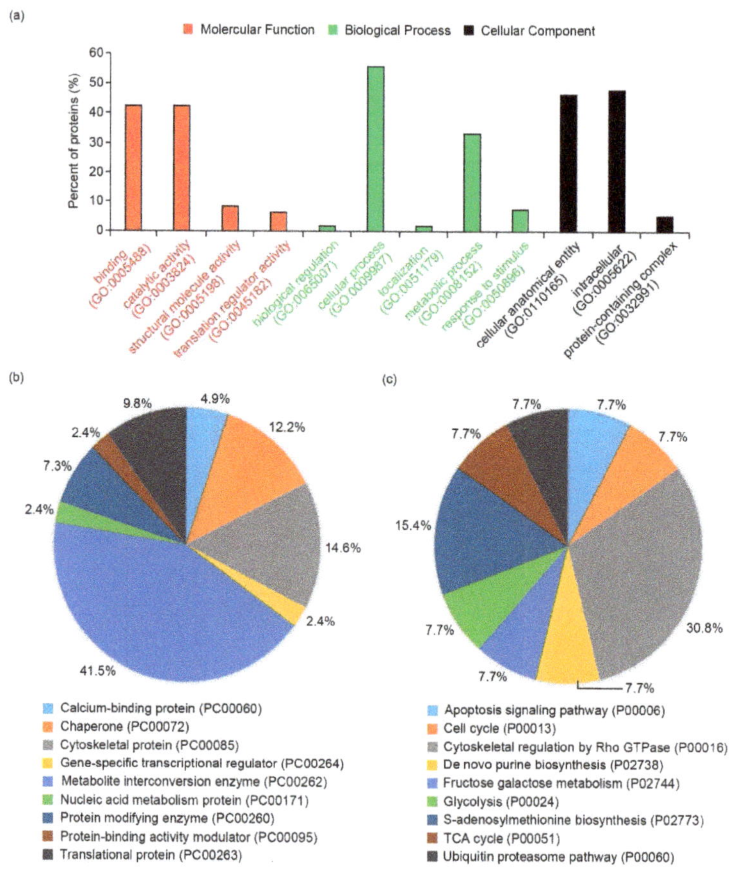

Figure 3. Functional classification of DEPs using the PANTHER database. (**a**) The bar chart of molecular function (red), biological process (green), and cellular components (black) represented is based on the PANTHER GO analysis. (**b**) Bar chart of the PANTHER protein classes in DEPs. (**c**) Bar chart of the PANTHER pathway classes in DEPs. The input percentage was calculated on the basis of the number of proteins mapped to the GO term divided by all protein numbers in the lists of DEPs (*Arabidopsis thaliana* IDs from the NCBI database).

2.3. Functional Network Analysis of Differentially Expressed Proteins

To understand the biological functions and modes of action of 50 DEPs in plant salt adaptation, we analyzed putative physical interactions of DEPs using the Cytoscape software platform (https://cytoscape.org/) (accessed on 1 April 2021) and the IntAct database (https://www.ebi.ac.uk/intact/) (accessed on 1 April 2021) (Figure 4). The Cytoscape with large databases of protein–protein, protein–DNA, and genetic interactions is a powerful software for studying the prediction of a physical interaction network in model organisms [43]. Out of 50 DEPs, the physical interactions of 34 DEPs were identified from this analysis. The largest cluster was the "cell structure-associated cluster," including 12 DEPs (red ellipse) in the functional network. The proteins in this cluster were mainly involved in the regulation of cell structures, including both cytoskeleton functions, such as actin filaments (ACT7, ADF3, and FBA8) and MTs (TUB3, TUB4, and TUB9), and secondary cell wall biogenesis (CCoAOMT1) (Figure 4). Even though TUB7 and PCAP1 proteins were highly induced in A120 cells (Table 1), they were not identified in functional network analysis. This is probably due to the lack of physical interaction information identified so far. PCAP1, also known as MT-destabilizing protein 25 (MDP25), functions as a negative regulator in hypocotyl cell elongation [44]. Additionally, other DEPs physically interacted with various functional proteins clustered in the ROS-associated cluster (CTIMC and ANNAT1; green ellipse), drought- and ABA-associated cluster (GRF3 and RBG2; purple ellipse), temperature-associated cluster (HSP70-1, HSP70-9, HOP2, and CSP2; orange ellipse), and transcriptional/translational system-associated cluster (PBD1, RPN11, PAP1, and CPN20; gray box) (Figure 4). The connectivity of protein interaction networks suggested that significant cellular and molecular changes in plant adaptation to salt stress might be associated with the plant cytoskeleton and cell wall biogenesis, affecting cell structure changes.

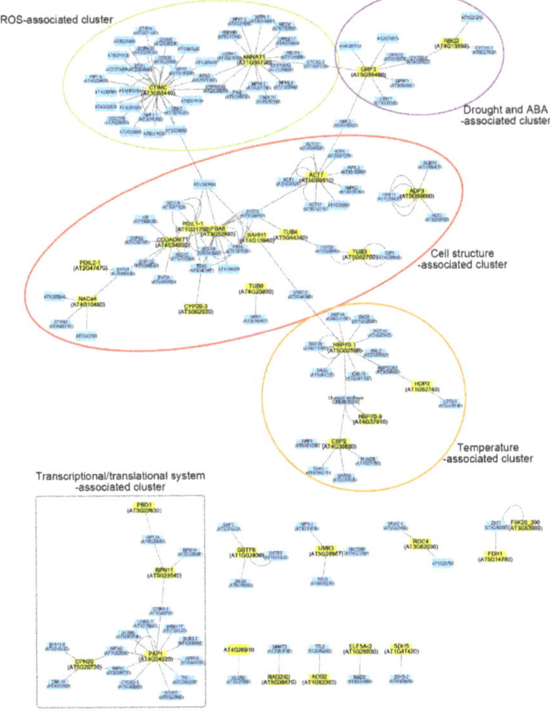

Figure 4. Interaction network analysis with DEPs by the Cytoscape software platform using IntAct molecular interaction database. Yellow squares indicate proteins of the input lists, whereas blue squares indicate proteins of physical interaction.

2.4. Expression Patterns of Cytoskeleton-Related Genes in Salt-Adapted Cells

To confirm the results of proteomics and bioinformatics analyses suggesting the crucial roles of cell structure-related proteins in plant salt adaptation (Table 1 and Figure 4), we tested the expressions of 12 genes encoding DEPs. This belonged to the cytoskeleton and cell wall biogenesis functions between A0 and A120 cells using quantitative real-time PCR (qRT-PCR). We also included two cytoskeleton-related genes, TUB7 and PCAP1, in the gene expression analysis.

The expression of cytoskeleton-related genes, including AT3G52930 (FBA8), AT5G62770 (TUB3), AT5G44340 (TUB4), AT2G29550 (TUB7), AT4G20890 (TUB9), and AT5G09810 (ACT7), and cell wall-related gene, AT4G34050 (CCoAOMT1), was significantly induced in A120 cells (Figure 5). However, the expression of other genes, such as AT4G13940 (SAHH1), AT4G20260 (PCAP1), and AT4G10480 (NACα4), in this cluster was decreased in A120 cells. Those of AT2G47470 (PDIL2-1), AT1G21750 (PDIL1-1), AT3G62030 (CYP20-3), and AT5G59880 (ADF3) were similar in A0 and A120 cells (Figure 5). Among 14 genes tested, FBA8, TUB3, TUB4, TUB7, TUB9, ACT7, and CCoAOMT1 were induced in mRNA (Figure 5) and protein levels (Table 1) in salt-adapted A120 cells. Although TUB7 was not identified in functional network analysis (Figure 4), its mRNA was more abundant in A120 cells (Figure 5). Since four TUB genes, TUB3, TUB4, TUB7, and TUB9, were induced in the protein and mRNA levels in A120 cells, MT-related proteins play essential roles in plant adaptation to salt stress.

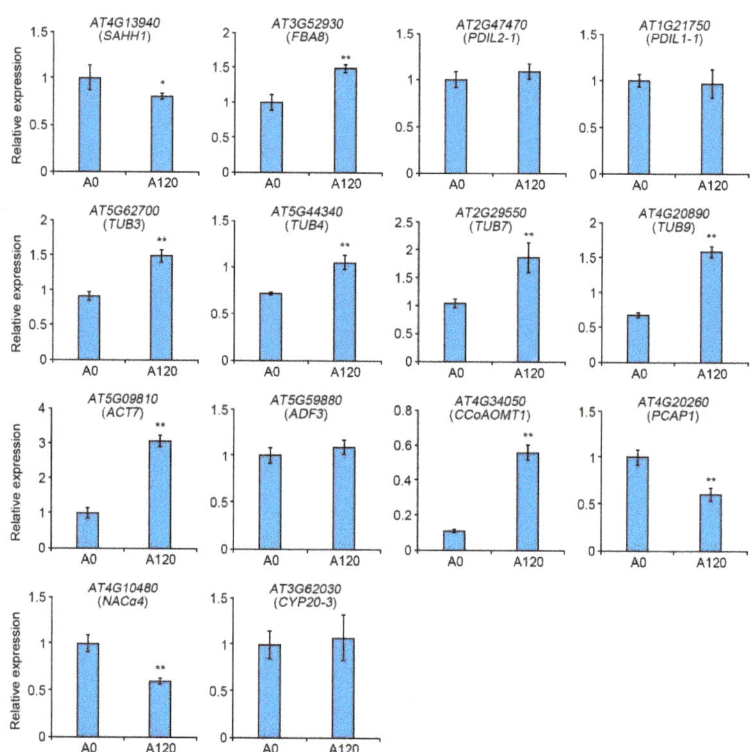

Figure 5. Transcript levels of cell structure-related genes in control (A0) and salt-adapted (A120) cells. Total RNAs were extracted from the A0 and A120 cells. Transcript levels were determined by quantitative real-time PCR (qRT-PCR). UBQ10 was used as a quantitative control for qRT-PCR. Error bars represent the standard deviation (SD) of three independent replicates. Asterisks indicate significant differences in the A0 cells (* p-value < 0.5; **, p-value ≤ 0.01, Student's t-test).

2.5. The Effect of the Loss-of-Function β-Tubulin Genes in Salt Stress Response

Recent evidence indicates that the regulation of MTs' destabilization and reorganization is essential for plant adaptation to salt stress [17,45,46]. Furthermore, the largest protein family among the cell structure-associated cluster is related to the β-tubulin family proteins, including TUB3, TUB4, TUB7, and TUB9 (Table 1 and Figure 4). To characterize the physiological functions of β-tubulin in salt stress responses, we isolated T-DNA insertion mutants of *Arabidopsis* β-tubulin genes (*tub3*, SALK_073132; *tub4*, SALK_204506; *tub7*, SALK_026797; *tub9*, SALK_015876). Wild-type (WT, ecotype Col-0) and four *tub* mutant plants were grown on MS medium for 5 days and then transferred to the soil to test mutant phenotypes under salt stress conditions. After 9 days, we supplied water containing 130 mM NaCl to the soil once a week for 4 weeks. The *tub4* mutants displayed strongly tolerant phenotypes, such as enhanced plant height and late wilting of leaves, to salt stress compared to WT plants (Figure 6b). In contrast, *tub9* mutants were hypersensitive to salt stress with a quickly wilting phenotype compared with WT plants (Figure 6d). Furthermore, *tub3* and *tub7* mutants showed similar phenotypes, including plant height and wilting of leaves, to WT plants under salt stress conditions (Figure 6a,c). These results suggested that both TUB4 and TUB9 play significant roles in plant adaptation and tolerance to salt stress, but their mode of function differs.

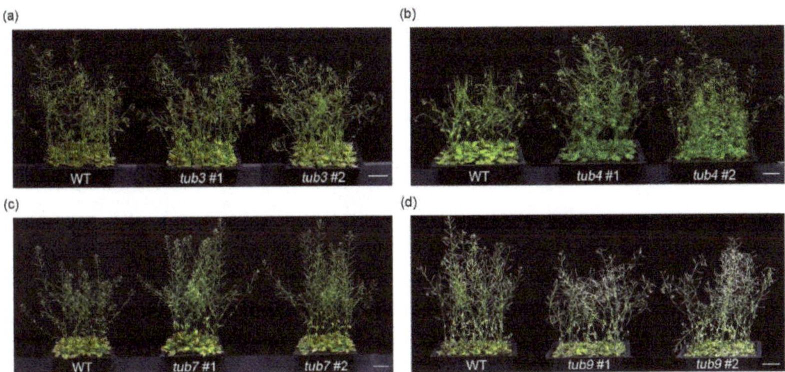

Figure 6. Characterization of loss-of-function mutants of *Arabidopsis* β-tubulin genes in salt stress. The 2-week-old wild-type (WT), tub3 (**a**), tub4 (**b**), tub7 (**c**), and tub9 (**d**) mutants grown on soil were treated with 130 mM NaCl for four weeks. Photos show a representative of 12 to 16 individual test plants. Scale bars indicate 5 cm.

2.6. The Effect of TUB9 Overexpression in Rice during Salt Stress

The hypersensitive phenotype of *Arabidopsis tub9* mutants to salt stress suggests that the overexpression of *Arabidopsis TUB9* gene can enhance crop tolerance to salt stress. To confirm this, we generated transgenic rice plants overexpressing *Arabidopsis TUB9* gene under the control of the *CaMV 35S* promoter (*TUB9*-OX). The *Arabidopsis TUB9*-OX construct was transformed into rice ("Ilmi" cultivar) embryogenic callus, and three independent *TUB9*-OX T_1 lines were selected by hygromycin B resistance and RT-PCR analysis. Under normal conditions, *TUB9*-OX transgenic plants were shorter than WT plants (Figure 7a). Besides plant height, other morphological phenotypes of TUB9-OX transgenic plants were comparable with WT plants. Ten-day-old WT and *TUB9*-OX transgenic plant seedlings were transferred into MS liquid media containing 120 mM NaCl. After 7 days of salt treatment, salt-treated WT and *TUB9*-OX transgenic plants were recovered in liquid MS medium without NaCl for 10 days. The *TUB9*-OX transgenic plants had greener leaves and higher heights than WT plants (Figure 7b). The number of rice transgenic plants with green leaves in WT and *TUB9*-OX transgenic lines was counted in the recovery stage after salt treatment to calculate the survival rate. The survival rate of *TUB9*-OX transgenic plants was approximately 40%; however, most WT plants had no green leaves (Figure 7b,c). These results suggested that *Arabidopsis TUB9* gene functions as a positive regulator in plant adaptation to salt stress and can enhance plant tolerance to salt stress.

Figure 7. Characterization of *TUB9*-overexpressing transgenic rice plants. (**a**) Comparison of plant growth in wild-type (WT, "Ilmi" cultivar) and *TUB9*-overexpressing transgenic rice plants (*TUB9*-OX). Photographs were taken at maturity (milky ripening) stages. (**b**) Comparison of plant tolerance in WT and *TUB9*-OX plants to salt stress. Ten-day-old seedlings were treated with 120 mM NaCl for 7 days. After NaCl treatment, plants were recovered in MS medium without NaCl. Scale bars indicate 5 cm. (**c**) Quantitative analysis of survival rate of WT and *TUB9*-OX plants after salt stress treatment. Error bars represent the standard deviation (SD) of three independent replicates of the same experiment. Asterisks represent significant differences in the WT (** $p \leq 0.01$, Student's *t*-test).

3. Discussion

Salt stress disrupts cell division in leaves and roots through various cellular mechanisms, such as calcium ion, ROS, and ABA-dependent responses [47]. The changes in cellular morphology, such as cell proliferation and cell expansion, are essential for plant adaptation and tolerance to salt stress [47,48]. However, cellular and molecular mechanisms of morphological changes during salt adaptation have not been well elucidated. This study demonstrated that salt-adapted A120 cells showed morphological changes, such as spherical or ellipsoidal and isodiametric shapes, compared with control A0 cells (Figure 1a). Results of GO and network analysis using proteomics data showed that many DEPs identified from salt-adapted cells were associated with regulating cell structures, including cytoskeleton and cell wall biogenesis (Figures 3 and 4). Moreover, our gene expression and molecular genetic analyses revealed that β-tubulin family proteins play positive and negative roles in plant adaptation and tolerance to salt stress (Figures 6 and 7). Our results suggest that β-tubulin MTs are vital components in modulating plant adaptation and tolerance to salt stress.

3.1. Molecular Functions of Differentially Expressed Proteins in Salt-Adapted Cells

This study elucidated the molecular mechanisms underlying plant adaptation to prolonged salt stress by comparative proteomics between control and salt-adapted cells. The previous proteomics studies conducted using suspension cells demonstrated that molecular mechanisms of suspension cells in salt stress response are complicated but similar to those studied at the whole plant level [49,50]. Using proteomics, we identified 50 DEPs, including 45 up-regulated and 5 down-regulated proteins, in salt-adapted cells compared with control cells (Table 1). Functional network analysis revealed that the identified DEPs were included in various functional clusters, but many of them in cell structure-associated clusters, including cytoskeleton and cell wall biogenesis functions (Figure 4).

3.1.1. Cell Structure-Associated Cluster

The plant cell surface comprises the cell wall, plasma membrane, and cytoskeleton [41]. Plant cytoskeletons play essential functions in plant tolerance and survival to salt stress [17,18]. Many up-regulated proteins in A120 cells were MTs and actin filament-related proteins (Table 1 and Figure 4). The ACT7 (AT5G09810), ADF3 (AT5G59880), and FBA8 (AT3G52930) proteins were involved in the actin cytoskeleton. Actin cytoskeletons are composed of two classes, which are vegetative (ACT2, ACT7, and ACT8) and reproductive (ACT1, ACT3, ACT4, ACT11, and ACT12). ACT7 transcription is high in vegetative organs and induced by auxin [51]. The act11 mutant decreases pollen germination and increases pollen tube growth by increasing the actin turnover rate [52]. However, the loss-of-function ACT2 mutant vegetative class affects root hair growth but is not complemented by overexpressing ACT7, even if they are of the same classes [53]. Additionally, ACT7 physically interacts with ACT1, ACT11, ACT12, actin-depolymerizing factor 6 (ADF6), and actin-interacting protein 1-2 (AIP1-2) (Figure 4). ADF3 (actin-depolymerizing factor 3) depolymerizes F-actin and acts as a crucial regulator in plant defense response to biotic stress [54]. Abiotic stresses also regulate the protein and gene expression of ADFs. OsADF proteins in rice leaves are highly accumulated because of drought stress [55]. OsADF3 protein is induced by salt stress in two rice cultivars (Oryza sativa L. cv. Nipponbare and Oryza sativa L. cv. Tainung 67) [56,57]. FBA8, which encodes fructose-bisphosphate aldolase 8, is involved in actin polymerization and various abiotic stress responses, such as salt, drought, ABA, and temperature stresses [50,58]. MT dynamics, polymerization and depolymerization, are necessary for cellular processes of plant tolerance and adaptation to salt stress [17]. Our results revealed that TUB3, TUB4, TUB7, and TUB9 proteins, involved in MT depolymerization and reorganization, play vital roles in plant adaptation and tolerance to salt stress (Table 1 and Figure 4). It was also reported that TUA6 (α-chain tubulin 6) and TUB2 (β-chain tubulin 2) proteins are highly expressed in Arabidopsis roots in response to salt stress [59].

CCoAOMT1 (AT4G34050), encoding caffeoyl-coA o-methyltransferase 1, plays an essential role in lignin biosynthesis and salt stress response [60]. *ccoaomt1* mutants showed a hypersensitive phenotype to salt and drought stresses [60,61]. SAHH1 (AT4G13940), encoding S-adenosyl-L-homocysteine hydrolase 1, is involved in the interaction between cytokinin and DNA methylation. Protein disulfide isomerase-like (PDIL) protein 1-1 (PDIL1-1, AT1G21750), which plays essential roles in ER trafficking, is associated with the response to salt stress. The loss-of-function mutant of PDIL2-1 (AT2G47470) disrupts pollen tube growth by delaying embryo development [62]. In addition, protein disulfide isomerase genes in maize are induced by abiotic stress, such as salt, drought, ABA, and H_2O_2 [63]. Cyclophilin CYP20-3 (ROC4, AT3G62030), which plays an important role in redox regulation, is involved in salt stress response [64]. CYP20-3 protein composes the sophisticated and reticular connective networks with FBA8, SAHH1, CCoAOMT1, PDIL1-1, and PDIL2-1 (Figure 4). However, nascent polypeptide-associated complex (NAC) alpha subunit family protein (NACα4 and AT4G10480) is involved in the cell structure-associated cluster (Figure 4). NACα4 function is not well known; however, the NAC complex plays a vital role in abiotic stress responses, such as drought and salt in barley [65].

3.1.2. ROS-Associated Cluster

Plants depend on cellular signaling and pathways via the reestablishment of ROS homeostasis in salt stress adaptation [66]. CTIMC (AT3G55440), which encodes triosephosphate isomerase (TPI), plays an essential role in redox regulation and is induced in response to salt stress by reactive carbonyl species (RCS). CTIMC protein forms a complex network in abiotic stress signaling through redox regulation by ROS-related proteins, including protein detoxification (DTX) proteins, L-type lectin receptor kinases 32 (LECRK32), plasma membrane intrinsic protein 1-5 (PIP1-5), nitrate transporter 1.1 (NPF1.1), cysteine-rich RLK 2 (CRK2), and acyl-lipid desaturase 2 (ADS2) (Figure 4). ANNAT1 (AT1G35720), annexin protein, has peroxidase activity and is involved in various abiotic stresses, such as salt, drought, and ABA. The *annat1* mutant showed tolerance to salt and drought stress by regulating ABA and proline biosynthesis [67]. ANNAT1 protein forms a complex network in various signaling pathways by controlling ROS-related proteins, including NPF6.1,

7.2, 8.2, 8.3, ADS2, respiratory burst oxidase homolog D (RBOHD), LysM Receptor-Like Kinase1 (CERK1), PIN5, cyclic nucleotide-gated channels 17 (CNGC17), proline transporter 1 (PROT1), and sugar transporter 7 (SWEET7) (Figure 4).

3.1.3. Drought- and ABA-Associated Cluster

Growth regulating factor 3 (GRF3) influences plant tolerance to drought stress and organ growth by increasing leaf size. Glycine-rich RNA-binding protein 2 (RGB2/GRP2) affects seed germination in an ABA-independent manner under salt stress. GRF2 and RGB2 proteins are related to various plant stress-associated proteins, such as CBL-interacting protein kinases (CIPK12), cyclin-B 2-2 (CYCB2-2), ABA-responsive element-binding protein 3 (AREB3/DPBF3), cytosolic invertase 1 (CINV1), and cyclin-H 1-1 (CYCH1-1) (Figure 4).

3.1.4. Temperature-Associated Cluster

Heat shock 70 kDa protein 1 (HSP70-1), a key component in protein folding, plays a vital role in stomatal closure and seed germination and response to ABA stress. Mitochondrial HSP70-9 protein is involved in iron–sulfur protein biogenesis. Cold shock protein 2 (CRP2) protein plays different roles as a negative regulator in response to cold stress and as a positive regulator in salt stress response. Hsp70-Hsp90 organizing protein 2 (HOP2) influences plant adaptation to prolonged heat stress. These four proteins compose functional networks via physical interaction with chaperon regulators, including heat stress transcription factor A-1 (HSF1A), suppressor of G2 allele of skp1 (SGT1) homolog B (SGT1B), Bcl-2-associated athanogene 3 (BAG3) and 5 (BAG5), and cytokinin response factor 3 (CRF3) (Figure 4).

3.1.5. Transcriptional/Translational System-Associated Cluster

Proteasome subunit beta type 2-A (PBD1) and regulatory particle non-ATPase 11 (RPN11) play an important role in the plant ubiquitin–proteasome system via protein degradation and stabilization. Plastid-lipid-associated protein 1 (PAP1) contributes to the protection of photosystem II (PSII) and in response to ABA stress. Chloroplast co-chaperonin 20 (CPN20) acts as a negative regulator in ABA signaling. These four proteins compose functional networks via physical interaction with protein stability- or gene transcription-related proteins, including 26S proteasome regulatory subunit 4 homolog A (RPT2A), RPN1A, trithorax-related protein 5 (ATXR5), importin alpha isoform 6 (IMPA6), calmodulin-like protein 18 (CML18), sensitive to proton rhizotoxicity 2 (STOP2), authentic response regulator 14 (ARR14), budding uninhibited by benzymidazol 3.3 (BUB3.3), and WRKY17 (Figure 4).

3.2. *The Role of Microtubules in Plant Adaptation and Tolerance to Salt Stress*

MTs are fixed in the plasma membrane and composed of a greater part of plant interphase arrays [9,15]. The cortical MT arrays are involved in plant response to various abiotic stresses, especially salt stress [9,18]. Plants increase salt tolerance by regulating depolymerization and reorganization of the cortical MTs [48]. MAP65-1 acts as a positive regulator in plant salt tolerance by promoting cortical MT reorganization [68]. Calcium ions reorganize the damage of MT arrays in the salt stress response of plant cells [17]. The loss-of-function sos3, a calcium sensor in the salt stress response, mutant shows hypersensitivity to salt stress due to the irregular organization of MTs [26]. Plants with salt-susceptible phenotypes have a lower concentration of calcium ions than that of salt-tolerant plants [69]. Our proteomic analysis showed that the four β-tubulin family proteins, including TUB3, TUB4, TUB7, and TUB9, were induced in salt-adapted A120 cells compared with control A0 cells (Table 1 and Figure 4). Additionally, the mRNA levels of TUB3, TUB4, TUB7, and TUB9 genes were higher in A120 cells than in A0 cells (Figure 5). Our results suggest that the elevation of β-tubulin mRNAs and protein levels can affect MT functions and enhance plant adaptation to salt stress. In our molecular genetic analysis, the loss-of-function tub4 mutant showed enhanced tolerance to salt stress. In contrast, the tub9 mutant was more hypersensitive than WT plants (Figure 6). The overexpression of TUB9 in rice plants enhanced the plant's tolerance to salt stress (Figure 7). Interestingly, tub4 and tub9 mutant plants showed opposite phenotypes in response to transiently applied salt stress, even though TUB4 and TUB9 protein levels were higher in cells that have adapted to salt stress for a long time. These results suggest that TUB4 and TUB9 proteins play different roles in plant responses to short-term and long-term salt stresses. It was also reported that short-term and long-term salt stress have different effects on the actin filament assembly and disassembly [25]. It would be worthwhile to dissect the biological functions of TUB4 and TUB9 in plant adaptation and tolerance to salt stress in further studies.

Altogether, our results suggest that β-tubulin proteins play different roles in plant adaptation and tolerance to salt stress by regulating MT depolymerization and reorganization. Therefore, changes in MT dynamics in plant cells would be essential for cellular processes to enhance the adaptation and tolerance to salt stress. Furthermore, morphological changes in salt-adapted suspension cells are at least partly due to the changes in MT dynamics.

4. Materials and Methods

4.1. Growth Conditions of Callus Suspension Cells

Salt-adapted callus suspension cells were generated from *Arabidopsis thaliana* (Col-0 ecotype) roots as described in detail in a previous study [40]. Callus suspension cells were maintained at 23 °C in the dark with gentle shaking (140 rpm).

4.2. Proteomic Profiling Using Two-Dimensional Gel Electrophoresis

Total protein was isolated from 5 g of A0 and salt-adapted cells (A120) using trichloroacetic acid/acetone/phenol extraction protocol described in detail in a previous study [42]. Total soluble proteins were quantified using the 2D-Quant Kit (Amersham Biosciences Europe GmbH, Freiburg, Germany). Two-dimensional gel electrophoresis was performed with Protean IEF cell (Bio-Rad, Hercules, CA, USA) for the first-dimensional isoelectric focusing using immobilized pH gradient strips (24 cm, pH 4–7; Bio-Rad Laboratories, Hercules, CA, USA), and with the Protean Xi-II Cell system (Bio-Rad Laboratories, Hercules, CA, USA) for the second-dimensional sodium dodecyl sulfate–polyacrylamide gel electrophoresis. After Coomassie brilliant blue staining, gel images were taken using a GS-800 Imaging Densitometer Scanner (Bio-Rad Laboratories, Hercules, CA, USA) and analyzed using PDQuest v.7.2.0 (Bio-Rad Laboratories, Hercules, CA, USA). All experiments were performed in three independent biological replicates, and the volume of each spot was detected and normalized to a relative density. Proteins showing a statistically significant difference ($p < 0.05$) between A0 and A120 cells were identified. For protein identification, differential protein spots visualized in the gel were excised and subjected to in-gel digestion as described previously [42]. Protein identification was performed by MALDI-TOF/TOF MS using the ABI 4800 Plus TOF-TOF Mass Spectrometer (Applied Biosystems, Framingham, MA, USA). Fifty proteins were identified, of which peptide and fragment mass tolerance was fixed at 100 ppm. The high confidence interval displayed statistically reliable search scores (more than 95% confidence) corresponding to protein's experimental isoelectric point (pI) and molecular weight.

4.3. Bioinformatics Analysis

The functional classification of DEPs identified in proteomics was performed using the PANTHER classification system (http://www.pantherdb.org/) (accessed on 3 April 2021). We used network-based enrichment by Cytoscape software platform to forecast physical interactions of DEPs (https://cytoscape.org/) (accessed on 1 April 2021) using the IntAct database (https://www.ebi.ac.uk/intact/) (accessed on 1 April 2021).

4.4. Analysis of Quantitative Real Time PCR (qRT-PCR)

Total RNA was extracted from A0 and A120 cells using the RNeasy Plant Kit (Qiagen, Valencia, CA, USA) following the manufacturer's protocol. To remove genomic DNA contaminants, extracted RNA was treated with DNaseI (Thermo Fisher Scientific, Waltham, MA, USA). One µg of total RNA was used the first strand of cDNA synthesis using a cDNA synthesis kit (Invitrogen, Carlsbad, CA, USA), according to the manufacturer's protocol.

The qRT-PCR analysis was performed using the QuantiMix SYBR (PhileKorea, Seoul, Korea), and the relative values of indicated gene expression were automatically calculated using the CFX96 real-time PCR detection system (Bio-Rad Laboratories, Hercules, CA, USA) by applying normalization of the expression of *UBQ10*. The qRT-PCR was performed using the following conditions: 50 °C for 10 min, 95 °C for 10 min; followed by 50 cycles at 95 °C for 15 s, 60 °C for 15 s, and 72 °C for 15 s. The gene specific primers in qRT-PCR analysis are listed in Supplementary Table S1.

4.5. Plant Materials and Growth Conditions

Oryza sativa L. ("Ilmi" cultivar) and *Arabidopsis thaliana* (Col-0 ecotype) plants were used in all experiments. Rice plants were grown under natural light conditions in a greenhouse at 25–30 °C. The *tub3* (SALK_073132), *tub4* (SALK_204506), *tub7* (SALK_026797), and *tub9* (SALK_015876) mutants were obtained from Arabidopsis Biological Resource Center (https://www.arabidopsis.org/) (accessed on 24 February 2014). *Arabidopsis* plants were grown in a growth chamber in long-day conditions (16 h light/8 h dark) at 23 °C.

4.6. Generation of Transgenic Rice Plants

To generate the transgenic rice plants overexpressing the *Arabidopsis TUB9* gene, we cloned the full-length cDNA (1335 bp) of the *Arabidopsis TUB9* gene into *pH2GW7* vector under the control of *CaMV 35S* promoter. The *TUB9*-OX construct was introduced into *Agrobacterium tumefaciens* (LBA4404) by electroporation. We used a modified version of the general rice-transformation protocol [70]. Transgenic *TUB9*-OX (T1) plants were selected on MS medium containing hygromycin B and then transferred to soil and allowed to self-pollinate.

4.7. Salt Stress Treatment

In *Arabidopsis*, 5-day-old WT seedlings, *tub3*, *tub4*, *tub7*, and *tub9* plants grown on MS media were transferred to soil. After 9 days, we supplied water containing 130 mM NaCl to the soil once a week for 4 weeks. Photographs of each representative of 12–16 individual plants were taken to analyze plant phenotypes. In rice, 10-day-old WT seedlings and *TUB9*-OX plants germinated in MS media containing hygromycin B were transferred into MS liquid medium with 120 mM NaCl. After 7 days, plants were recovered in MS solution without NaCl for 10 days. Photographs were taken to represent 8–10 individual plants to analyze plant phenotypes.

4.8. Statistical Analyses

Statistical analyses, including Student's *t*-test, were performed using Excel 2010. qRT-PCR analysis was performed in three independent biological replicates, and the average values of $2^{\Delta\Delta}$CT were used to determine expression differences. Data were indicated as means ± standard deviation (SD). Error bars indicate SD.

5. Conclusions

This study suggests that the morphological changes of plant cells are an essential cellular process for adaptation to prolonged salt stress. We revealed that various protein families involved in various cellular processes play a role in salt adaptation response using proteomic analysis. Furthermore, gene expression and molecular genetic analyses demonstrated that β-tubulin proteins play an important role in plant adaptation and tolerance to salt stress. Altogether, our results suggest that the dynamics of depolymerization and reorganization of tubulin MTs play critical roles in plant adaptation to salt stress.

Supplementary Materials: The following are available online at https://www.mdpi.com/article/10.3390/ijms22115957/s1, Table S1: Primers used for qRT-PCR.

Author Contributions: H.J.C., D.B., B.J.J. and M.C.K. designed and performed the experiments, analyzed data, and wrote the manuscript. H.M.C., S.H.L., M.S.P., L.H.L., Y.J.C., D.-W.B. and S.T.K. performed experiments. D.-J.Y. and M.C.K. discussed and commented on results and revised the manuscript. All authors have read and agreed to the published version of the manuscript.

Funding: This research was supported by the Basic Science Research Program through the National Research Foundation of Korea (NRF) funded by the Ministry of Education (2015R1A6A1A03031413 (M.C.K.), 2020R1I1A1A01067256 (H.J.C.), and 2016R1D1A1B01011803 (D.B.)).

Institutional Review Board Statement: Not Applicable.

Informed Consent Statement: Not Applicable.

Data Availability Statement: The data presented in this study are available on request from the corresponding author.

Conflicts of Interest: The authors declare no conflict of interest.

References

1. Vinocur, B.; Altman, A. Recent advances in engineering plant tolerance to abiotic stress: Achievements and limitations. *Curr. Opin. Biotechnol.* **2005**, *16*, 123–132. [CrossRef] [PubMed]
2. Tenhaken, R. Cell wall remodeling under abiotic stress. *Front. Plant Sci.* **2015**, *5*, 771. [CrossRef]
3. Zhu, J.-K. Abiotic Stress Signaling and Responses in Plants. *Cell* **2016**, *167*, 313–324. [CrossRef]
4. Lü, P.; Kang, M.; Jiang, X.; Dai, F.; Gao, J.; Zhang, C. *RhEXPA*$_4$, a rose expansin gene, modulates leaf growth and confers drought and salt tolerance to *Arabidopsis*. *Planta* **2013**, *237*, 1547–1559. [CrossRef] [PubMed]
5. An, P.; Li, X.; Zheng, Y.; Matsuura, A.; Abe, J.; Eneji, A.E.; Tanimoto, E.; Inanaga, S. Effects of NaCl on Root Growth and Cell Wall Composition of Two Soya bean Cultivars with Contrasting Salt Tolerance. *J. Agron. Crop. Sci.* **2014**, *200*, 212–218. [CrossRef]
6. Zhu, J.; Lee, B.-H.; Dellinger, M.; Cui, X.; Zhang, C.; Wu, S.; Nothnagel, E.A.; Zhu, J.-K. A cellulose synthase-like protein is required for osmotic stress tolerance in *Arabidopsis*. *Plant J.* **2010**, *63*, 128–140. [CrossRef]
7. Endler, A.; Kesten, C.; Schneider, R.; Zhang, Y.; Ivakov, A.; Froehlich, A.; Funke, N.; Persson, S. A Mechanism for Sustained Cellulose Synthesis during Salt Stress. *Cell* **2015**, *162*, 1353–1364. [CrossRef]
8. Henty-Ridilla, J.L.; Li, J.; Blanchoin, L.; Staiger, C.J. Actin dynamics in the cortical array of plant cells. *Curr. Opin. Plant Biol.* **2013**, *16*, 678–687. [CrossRef] [PubMed]
9. Ehrhardt, D.W.; Shaw, S. Microtubule Dynamics and Organization in The Plant Cortical Array. *Annu. Rev. Plant Biol.* **2006**, *57*, 859–875. [CrossRef]
10. Dixit, R.; Cyr, R. The Cortical Microtubule Array: From Dynamics to Organization. *Plant Cell* **2004**, *16*, 2546–2552. [CrossRef]
11. Blanchoin, L.; Boujemaa-Paterski, R.; Henty, J.L.; Khurana, P.; Staiger, C.J. Actin dynamics in plant cells: A team effort from multiple proteins orchestrates this very fast-paced game. *Curr. Opin. Plant Biol.* **2010**, *13*, 714–723. [CrossRef]
12. Li, J.; Staiger, C.J. Understanding Cytoskeletal Dynamics During the Plant Immune Response. *Annu. Rev. Phytopathol.* **2018**, *56*, 513–533. [CrossRef]
13. Wang, X.; Mao, T. Understanding the functions and mechanisms of plant cytoskeleton in response to environmental signals. *Curr. Opin. Plant Biol.* **2019**, *52*, 86–96. [CrossRef]
14. Smith, L.G.; Oppenheimer, D.G. Spatial Control of Cell Expansion by The Plant Cytoskeleton. *Annu. Rev. Cell Dev. Biol.* **2005**, *21*, 271–295. [CrossRef]
15. Hashimoto, T.; Kato, T. Cortical control of plant microtubules. *Curr. Opin. Plant Biol.* **2006**, *9*, 5–11. [CrossRef] [PubMed]
16. Wang, S.; Kurepa, J.; Hashimoto, T.; Smalle, J.A. Salt Stress–Induced Disassembly of *Arabidopsis* Cortical Microtubule Arrays Involves 26S Proteasome–Dependent Degradation of SPIRAL1. *Plant Cell* **2011**, *23*, 3412–3427. [CrossRef] [PubMed]
17. Wang, C.; Li, J.; Yuan, M. Salt Tolerance Requires Cortical Microtubule Reorganization in *Arabidopsis*. *Plant Cell Physiol.* **2007**, *48*, 1534–1547. [CrossRef]
18. Wang, C.; Zhang, L.-J.; Huang, R.-D. Cytoskeleton and plant salt stress tolerance. *Plant Signal. Behav.* **2011**, *6*, 29–31. [CrossRef] [PubMed]
19. Livanos, P.; Galatis, B.; Quader, H.; Apostolakos, P. Disturbance of reactive oxygen species homeostasis induces atypical tubulin polymer formation and affects mitosis in root-tip cells of *Triticum turgidum* and *Arabidopsis thaliana*. *Cytoskeleton* **2012**, *69*, 1–21. [CrossRef] [PubMed]
20. Jiang, Y.; Wu, K.; Lin, F.; Qu, Y.; Liu, X.; Zhang, Q. Phosphatidic acid integrates calcium signaling and microtubule dynamics into regulating ABA-induced stomatal closure in *Arabidopsis*. *Planta* **2013**, *239*, 565–575. [CrossRef]

21. Takatani, S.; Hirayama, T.; Hashimoto, T.; Takahashi, T.; Motose, H. Abscisic acid induces ectopic outgrowth in epidermal cells through cortical microtubule reorganization in *Arabidopsis thaliana*. *Sci. Rep.* **2015**, *5*, 11364. [CrossRef]
22. Bogoutdinova, L.R.; Lazareva, E.M.; Chaban, I.A.; Kononenko, N.V.; Dilovarova, T.; Khaliluev, M.R.; Kurenina, L.V.; Gulevich, A.A.; Smirnova, E.A.; Baranova, E.N. Salt Stress-Induced Structural Changes Are Mitigated in Transgenic Tomato Plants Over-Expressing Superoxide Dismutase. *Biology* **2020**, *9*, 297. [CrossRef]
23. Fujita, S.; Pytela, J.; Hotta, T.; Kato, T.; Hamada, T.; Akamatsu, R.; Ishida, Y.; Kutsuna, N.; Hasezawa, S.; Nomura, Y.; et al. An Atypical Tubulin Kinase Mediates Stress-Induced Microtubule Depolymerization in *Arabidopsis*. *Curr. Biol.* **2013**, *23*, 1969–1978. [CrossRef]
24. Zhang, Q.; Lin, F.; Mao, T.; Nie, J.; Yan, M.; Yuan, M.; Zhang, W. Phosphatidic Acid Regulates Microtubule Organization by Interacting with MAP65-1 in Response to Salt Stress in *Arabidopsis*. *Plant Cell* **2012**, *24*, 4555–4576. [CrossRef] [PubMed]
25. Wang, C.; Zhang, L.; Yuan, M.; Ge, Y.; Liu, Y.; Fan, J.; Ruan, Y.; Cui, Z.; Tong, S.; Zhang, S. The microfilament cytoskeleton plays a vital role in salt and osmotic stress tolerance in *Arabidopsis*. *Plant Biol.* **2009**, *12*, 70–78. [CrossRef] [PubMed]
26. Ye, J.; Zhang, W.; Guo, Y. *Arabidopsis* SOS3 plays an important role in salt tolerance by mediating calcium-dependent microfilament reorganization. *Plant Cell Rep.* **2013**, *32*, 139–148. [CrossRef] [PubMed]
27. Zhao, Y.; Pan, Z.; Zhang, Y.; Qu, X.; Zhang, Y.; Yang, Y.; Jiang, X.; Huang, S.; Yuan, M.; Schumaker, K.S.; et al. The Actin-Related Protein2/3 Complex Regulates Mitochondrial-Associated Calcium Signaling during Salt Stress in *Arabidopsis*. *Plant Cell* **2013**, *25*, 4544–4559. [CrossRef] [PubMed]
28. Rodríguez-Milla, M.A.; Salinas, J. Prefoldins 3 and 5 Play an Essential Role in *Arabidopsis* Tolerance to Salt Stress. *Mol. Plant* **2009**, *2*, 526–534. [CrossRef]
29. Barkla, B.J.; Vera-Estrella, R.; Pantoja, O. Progress and challenges for abiotic stress proteomics of crop plants. *Proteomics* **2013**, *13*, 1801–1815. [CrossRef]
30. Silveira, J.A.; Carvalho, F.E. Proteomics, photosynthesis and salt resistance in crops: An integrative view. *J. Proteom.* **2016**, *143*, 24–35. [CrossRef]
31. Lv, X.; Chen, S.; Wang, Y. Advances in Understanding the Physiological and Molecular Responses of Sugar Beet to Salt Stress. *Front. Plant Sci.* **2019**, *10*, 1431. [CrossRef]
32. Wang, X.; Chang, L.; Wang, B.; Wang, D.; Li, P.; Wang, L.; Yi, X.; Huang, Q.; Peng, M.; Guo, A. Comparative Proteomics of Thellungiella halophila Leaves from Plants Subjected to Salinity Reveals the Importance of Chloroplastic Starch and Soluble Sugars in Halophyte Salt Tolerance. *Mol. Cell. Proteom.* **2013**, *12*, 2174–2195. [CrossRef] [PubMed]
33. Wang, J.; Meng, Y.; Li, B.; Ma, X.; Lai, Y.; Si, E.; Yang, K.; Xu, X.; Shang, X.; Wang, H.; et al. Physiological and proteomic analyses of salt stress response in the halophyte H alogeton glomeratus. *Plant Cell Environ.* **2015**, *38*, 655–669. [CrossRef] [PubMed]
34. Cheng, T.; Chen, J.; Zhang, J.; Shi, S.; Zhou, Y.; Lu, L.; Wang, P.; Jiang, Z.; Yang, J.; Zhang, S.; et al. Physiological and proteomic analyses of leaves from the halophyte Tangut Nitraria reveals diverse response pathways critical for high salinity tolerance. *Front. Plant Sci.* **2015**, *6*, 30. [CrossRef] [PubMed]
35. Shokri-Gharelo, R.; Noparvar, P.M. Molecular response of canola to salt stress: Insights on tolerance mechanisms. *Peer J* **2018**, *6*, e4822. [CrossRef]
36. Zhang, Y.; Wei, M.; Liu, A.; Zhou, R.; Li, D.; Dossa, K.; Wang, L.; Zhang, Y.; Gong, H.; Zhang, X.; et al. Comparative proteomic analysis of two sesame genotypes with contrasting salinity tolerance in response to salt stress. *J. Proteom.* **2019**, *201*, 73–83. [CrossRef]
37. Frukh, A.; Siddiqi, T.O.; Khan, M.I.R.; Ahmad, A. Modulation in growth, biochemical attributes and proteome profile of rice cultivars under salt stress. *Plant Physiol. Biochem.* **2020**, *146*, 55–70. [CrossRef]
38. Parker, R.; Flowers, T.J.; Moore, A.L.; Harpham, N.V.J. An accurate and reproducible method for proteome profiling of the effects of salt stress in the rice leaf lamina. *J. Exp. Bot.* **2006**, *57*, 1109–1118. [CrossRef]
39. Sobhanian, H.; Aghaei, K.; Komatsu, S. Changes in the plant proteome resulting from salt stress: Toward the creation of salt-tolerant crops? *J. Proteom.* **2011**, *74*, 1323–1337. [CrossRef]
40. Chun, H.J.; Baek, D.; Cho, H.M.; Jung, H.S.; Jeong, M.S.; Jung, W.-H.; Choi, C.W.; Lee, S.H.; Jin, B.J.; Park, M.S.; et al. Metabolic Adjustment of *Arabidopsis* Root Suspension Cells During Adaptation to Salt Stress and Mitotic Stress Memory. *Plant Cell Physiol.* **2018**, *60*, 612–625. [CrossRef]
41. Le Gall, H.; Philippe, F.; Domon, J.-M.; Gillet, F.; Pelloux, J.; Rayon, C. Cell Wall Metabolism in Response to Abiotic Stress. *Plants* **2015**, *4*, 112–166. [CrossRef]
42. Kwon, Y.S.; Kim, S.G.; Chung, W.S.; Bae, H.; Jeong, S.W.; Shin, S.C.; Jeong, M.-J.; Park, S.-C.; Kwak, Y.-S.; Bae, D.-W.; et al. Proteomic analysis of Rhizoctonia solani AG-1 sclerotia maturation. *Fungal Biol.* **2014**, *118*, 433–443. [CrossRef]
43. Shannon, P.; Markiel, A.; Ozier, O.; Baliga, N.S.; Wang, J.T.; Ramage, D.; Amin, N.; Schwikowski, B.; Ideker, T. Cytoscape: A Software Environment for Integrated Models of Biomolecular Interaction Networks. *Genome Res.* **2003**, *13*, 2498–2504. [CrossRef] [PubMed]
44. Li, J.; Wang, X.; Qin, T.; Zhang, Y.; Liu, X.; Sun, J.; Zhou, Y.; Zhu, L.; Zhang, Z.; Yuan, M.; et al. MDP25, A Novel Calcium Regulatory Protein, Mediates Hypocotyl Cell Elongation by Destabilizing Cortical Microtubules in *Arabidopsis*. *Plant Cell* **2011**, *23*, 4411–4427. [CrossRef] [PubMed]
45. Shoji, T.; Suzuki, K.; Abe, T.; Kaneko, Y.; Shi, H.; Zhu, J.-K.; Rus, A.; Hasegawa, P.M.; Hashimoto, T. Salt Stress Affects Cortical Microtubule Organization and Helical Growth in *Arabidopsis*. *Plant Cell Physiol.* **2006**, *47*, 1158–1168. [CrossRef] [PubMed]

46. Yang, Y.; Guo, Y. Elucidating the molecular mechanisms mediating plant salt-stress responses. *New Phytol.* **2018**, *217*, 523–539. [CrossRef]
47. Qi, F.; Zhang, F. Cell Cycle Regulation in the Plant Response to Stress. *Front. Plant Sci.* **2020**, *10*, 1765. [CrossRef] [PubMed]
48. Ma, H.; Liu, M. The microtubule cytoskeleton acts as a sensor for stress response signaling in plants. *Mol. Biol. Rep.* **2019**, *46*, 5603–5608. [CrossRef]
49. Liu, D.; Ford, K.L.; Roessner, U.; Natera, S.; Cassin, A.M.; Patterson, J.H.; Bacic, A. Rice suspension cultured cells are evaluated as a model system to study salt responsive networks in plants using a combined proteomic and metabolomic profiling approach. *Proteomics* **2013**, *13*, 2046–2062. [CrossRef]
50. Wang, J.; Yao, L.; Li, B.; Meng, Y.; Ma, X.; Lai, Y.; Si, E.; Ren, P.; Yang, K.; Shang, X.; et al. Comparative Proteomic Analysis of Cultured Suspension Cells of the Halophyte Halogeton glomeratus by iTRAQ Provides Insights into Response Mechanisms to Salt Stress. *Front. Plant Sci.* **2016**, *7*, 110. [CrossRef] [PubMed]
51. McDowell, J.M.; Huang, S.; McKinney, E.C.; An, Y.Q.; Meagher, R.B. Structure and evolution of the actin gene family in Ar-abidopsis thaliana. *Genetics* **1996**, *142*, 587–602. [CrossRef] [PubMed]
52. Chang, M.; Huang, S. *Arabidopsis*ACT11 modifies actin turnover to promote pollen germination and maintain the normal rate of tube growth. *Plant J.* **2015**, *83*, 515–527. [CrossRef]
53. Kandasamy, M.K.; McKinney, E.C.; Meagher, R.B. A Single Vegetative Actin Isovariant Overexpressed under the Control of Multiple Regulatory Sequences Is Sufficient for Normal *Arabidopsis* Development. *Plant Cell* **2009**, *21*, 701–718. [CrossRef]
54. Mondal, H.A.; Louis, J.; Archer, L.; Patel, M.; Nalam, V.J.; Sarowar, S.; Sivapalan, V.; Root, D.D.; Shah, J. *Arabidopsis* Actin-Depolymerizing Factor3 Is Required for Controlling Aphid Feeding from the Phloem. *Plant Physiol.* **2018**, *176*, 879–890. [CrossRef] [PubMed]
55. Salekdeh, G.H.; Siopongco, J.; Wade, L.J.; Ghareyazie, B.; Bennett, J. Proteomic analysis of rice leaves during drought stress and recovery. *Proteomics* **2002**, *2*, 1131–1145. [CrossRef]
56. Yan, S.; Tang, Z.; Su, W.; Sun, W. Proteomic analysis of salt stress-responsive proteins in rice root. *Proteomics* **2005**, *5*, 235–244. [CrossRef] [PubMed]
57. Huang, Y.-C.; Huang, W.-L.; Hong, C.-Y.; Lur, H.-S.; Chang, M.-C. Comprehensive analysis of differentially expressed rice actin depolymerizing factor gene family and heterologous overexpression of *OsADF3* confers *Arabidopsis Thaliana* drought tolerance. *Rice* **2012**, *5*, 33. [CrossRef] [PubMed]
58. Lu, W.; Tang, X.; Huo, Y.; Xu, R.; Qi, S.; Huang, J.; Zheng, C.; Wu, C. Identification and characterization of fructose 1,6-bisphosphate aldolase genes in *Arabidopsis* reveal a gene family with diverse responses to abiotic stresses. *Gene* **2012**, *503*, 65–74. [CrossRef] [PubMed]
59. Jiang, Y.-Q.; Yang, B.; Harris, N.S.; Deyholos, M.K. Comparative proteomic analysis of NaCl stress-responsive proteins in *Arabidopsis* roots. *J. Exp. Bot.* **2007**, *58*, 3591–3607. [CrossRef] [PubMed]
60. Chun, H.J.; Baek, D.; Cho, H.M.; Lee, S.H.; Jin, B.J.; Yun, D.-J.; Hong, Y.-S.; Kim, M.C. Lignin biosynthesis genes play critical roles in the adaptation of *Arabidopsis* plants to high-salt stress. *Plant Signal. Behav.* **2019**, *14*, 1625697. [CrossRef]
61. Chun, H.; Lim, L.; Cheong, M.; Baek, D.; Park, M.; Cho, H.; Lee, S.; Jin, B.; No, D.; Cha, Y.; et al. *Arabidopsis* CCoAOMT1 Plays a Role in Drought Stress Response via ROS- and ABA-Dependent Manners. *Plants* **2021**, *10*, 831. [CrossRef] [PubMed]
62. Wang, H.; Boavida, L.C.; Ron, M.; McCormick, S. Truncation of a Protein Disulfide Isomerase, PDIL2-1, Delays Embryo Sac Maturation and Disrupts Pollen Tube Guidance in *Arabidopsis thaliana*. *Plant Cell* **2009**, *20*, 3300–3311. [CrossRef] [PubMed]
63. Zhu, C.; Luo, N.; He, M.; Chen, G.; Zhu, J.; Yin, G.; Li, X.; Hu, Y.; Li, J.; Yan, Y. Molecular Characterization and Expression Profiling of the Protein Disulfide Isomerase Gene Family in *Brachypodium distachyon* L. *PLoS ONE* **2014**, *9*, e94704. [CrossRef]
64. Dominguez-Solis, J.R.; He, Z.; Lima, A.; Ting, J.; Buchanan, B.B.; Luan, S. A cyclophilin links redox and light signals to cysteine biosynthesis and stress responses in chloroplasts. *Proc. Natl. Acad. Sci. USA* **2008**, *105*, 16386–16391. [CrossRef]
65. Maršálová, L.; Vítámvás, P.; Hynek, R.; Prášil, I.T.; Kosová, K. Proteomic Response of Hordeum vulgare cv. Tadmor and Hordeum marinum to Salinity Stress: Similarities and Differences between a Glycophyte and a Halophyte. *Front. Plant Sci.* **2016**, *7*, 1154. [CrossRef]
66. Yang, Y.; Guo, Y. Unraveling salt stress signaling in plants. *J. Integr. Plant Biol.* **2018**, *60*, 796–804. [CrossRef]
67. Huh, S.M.; Noh, E.K.; Kim, H.G.; Jeon, B.W.; Bae, K.; Hu, H.-C.; Kwak, J.M.; Park, O.K. *Arabidopsis* Annexins AnnAt1 and AnnAt4 Interact with Each Other and Regulate Drought and Salt Stress Responses. *Plant Cell Physiol.* **2010**, *51*, 1499–1514. [CrossRef] [PubMed]
68. Zhou, S.; Chen, Q.; Li, X.; Li, Y. MAP65-1 is required for the depolymerization and reorganization of cortical microtubules in the response to salt stress in *Arabidopsis*. *Plant Sci.* **2017**, *264*, 112–121. [CrossRef]
69. Errabii, T.; Gandonou, C.B.; Essalmani, H.; Abrini, J.; Idaomar, M.; Senhaji, N.S. Effects of NaCl and mannitol induced stress on sugarcane (Saccharum sp.) callus cultures. *Acta Physiol. Plant.* **2007**, *29*, 95–102. [CrossRef]
70. Hiei, Y.; Ohta, S.; Komari, T.; Kumashiro, T. Efficient transformation of rice (*Oryza sativa* L.) mediated by Agrobacterium and sequence analysis of the boundaries of the T-DNA. *Plant J.* **1994**, *6*, 271–282. [CrossRef]

MDPI
St. Alban-Anlage 66
4052 Basel
Switzerland
Tel. +41 61 683 77 34
Fax +41 61 302 89 18
www.mdpi.com

International Journal of Molecular Sciences Editorial Office
E-mail: ijms@mdpi.com
www.mdpi.com/journal/ijms

www.ingramcontent.com/pod-product-compliance
Lightning Source LLC
LaVergne TN
LVHW071554170726
843515LV00008B/2290

* 9 7 8 3 0 3 6 5 4 1 3 8 9 *